T0181592

Optical WDM Networks

OPTICAL NETWORKS SERIES

Series Editor
Biswanath Mukherjee, *University of California, Davis*

Other books in the series:

TRAFFIC GROOMING IN OPTICAL WDM MESH NETWORKS
Keyao Zhu, Hongyue Zhu, Biswanath Mukherjee, ISBN 0-387-25432-3
SURVIVABLE OPTICAL WDM NETWORKS
Canhui (Sam) Ou and Biswanath Mukherjee, ISBN 0-387-24498-0
OPTICAL BURST SWITCHED NETWORKS
Jason P. Jue and Vinod M. Vokkarane, ISBN 0-387-23756-9

Optical WDM Networks

Biswanath Mukherjee
Department of Computer Science
University of California
Davis, CA 95616 U.S.A.

Email: mukherjee@cs.ucdavis.edu
Tel: +1-530-752-4826
Fax: +1-530-752-4767

Biswanath Mukherjee
Department of Computer Science
University of California
Davis, CA 95616 U.S.A.

Optical WDM Networks

Series Editor:

Biswanath Mukherjee
Department of Computer Science
University of California
Davis, CA 95616 U.S.A.

ISBN 978-1-4899-7883-7 ISBN 978-0-387-29188-8 (eBook)
DOI: 10.1007/978-0-387-29188-8

Printed on acid-free paper.

Printed in the United States of America.

9 8 7 6 5 4 3 2 1 SPIN 11051329

springeronline.com

To my family:

my parents Dhirendranath and Gita Mukherjee,

my wife Supriya Mukherjee,

and my daughters Bipasha and Suchitra Mukherjee.

Preface

New Materials

Before discussing the book's topic, intended audience, etc., we remark that another book, entitled *Optical Communication Networks*, was published by the same author in 1997. Eight years is a long time in our fast-moving field. Relative to the old book, a brief outline of the new materials in the new book is provided below, followed by brief descriptions of what has been revamped and what has been deleted.

New Chapters:

- Optical Access Networks (Chapter 5)
- Optical Metro Networks (Chapter 6)
- Survivable Optical Networks (Chapter 11)
- Light-Trees: Optical Multicasting (Chapter 12)
- Traffic Grooming in Optical Mesh Networks (Chapters 13 and 14)
- Impairment-Aware Routing (Chapter 15)
- Optical Packet Switching (Chapter 17)
- Optical Burst Switching (Chapter 18)

Revamped Materials:

- Chapter 1 (Introduction) has been enhanced with six new sections to lead the discussion on telecom network overview, business models, traffic models, role of software (and computer science) in optical networks, etc.
- Chapter 2 (Enabling Technologies) has been updated, especially with a significant amount of new switching material.
- The materials on single-hop and multihop networks (Chapters 3 and 4) have been condensed, but still retained. Because these topics are quite mature, we see less activity on them in academic research and industry today. But we all know that what seems practical today might not be so tomorrow, and vice versa. Thus, a student of optical networking must be educated on this subject; and a practitioner may find this material to be of easy reference.
- Network Control and Management (Chapter 16) has undergone very significant revision.

- Routing and wavelength assignment (RWA) (Chapter 7) has been enhanced with discussions on the routing subproblem and wavelength-assignment heuristics.
- Chapter 9 has also been enhanced with material on virtual-topology adaptation under dynamic traffic.
- The materials in Chapters 8 (Elements of Virtual-Topology Design) and 10 (Wavelength Conversion) have been retained.

Deleted Materials:

- The material on optical time-division multiplexing (OTDM) and code-division multiplexing (OCDM) has been deleted; thus, the current book's focus is solely on WDM networks, as reflected in its title.
- The following chapters have also been deleted: (i) wavelength distributed data interface (WDDI), a multi-wavelength version of FDDI; (ii) all-optical cycle elimination in all-optical networks; and (iii) optimal amplifier placement in all-optical networks.

Interested readers should consult the old book [Mukh97] for these materials.

The Topic

The basic premise of our subject – optical communication networks – is that, as more users start to use our data networks, and as their usage patterns evolve to include more bandwidth-intensive networking applications such as data browsing on the world wide web, Java applications, video conferencing, etc., there emerges an acute need for very high-bandwidth transport network facilities, whose capabilities are much beyond those that current networks (e.g., today's Internet) can provide. Given that a single-mode fiber's potential bandwidth is nearly 50 terabits per second (Tbps), which is nearly four orders of magnitude higher than electronic data rate (of a few tens of Gigabits per second (Gbps) today), every effort should be made to tap into this huge opto-electronic bandwidth mismatch. Wavelength-division multiplexing (WDM) is an approach that can exploit this bandwidth mismatch by requiring that each end-user's equipment operates only at electronic rate, but multiple WDM channels from different end-users may be multiplexed on the same fiber.

Research and development on optical WDM networks have matured considerably over the past decade, with commercial deployment of WDM systems being common. Most such deployments are for WDM optical transmission systems with wavelength channel counts of 32 to 64. Channel counts as high as 160 are practical today, and expected to increase to 320 per fiber strand. While WDM transmission systems are quite mature, the technologies for optical WDM switching, e.g., optical crossconnects (OXCs) are significantly behind. While a lot of efforts have been spent on building *all-optical* (or *transparent*) switches, the switching that is mature today is based on *optical-electronic-optical (OEO)* (or *opaque*) technology. However, besides these hardware platforms, note that *intelligence* in any network (including an optical network) comes from "software," namely for network control, management, signaling, traffic engineering (or routing), network planning (or optimization), etc. Unfortunately, the role of software – or more generally, the role of computer science and engineering – has been paid little attention in optical network research and development. As a result, the progress in the field of optical networking has not been commensurate with its promise from a decade back.

The above problem needs to be rectified. And everyone with a stake in the field of optical networking (researchers, government funding agencies, telecom network operators, equipment vendors, etc.) needs to realize the important roles which *optics*, *electronics*, and *software* play in building a successful optical network (see Fig. 1.2). These players also need to be mindful of the importance of "cross-layer design" issues involving the physical layer (optics and electronics), the network layer (architecture), and the applications (software) (see Fig. 1.3).

Interest on this topic has been growing to better understand the issues and challenges in designing such networks. It is anticipated that the next generation of the Internet will employ WDM-based optical backbones.

Intended Audience

The intended audience of this book includes researchers, industry practitioners, and graduate students (both as a graduate textbook and for doctoral students' research).

Many electrical engineering, computer engineering, and computer science programs around the world have been offering a graduate course on this topic. That is, research and development on optical communication networks have matured significantly to the extent that some of these principles have moved from the research laboratories to the formal (graduate) classroom setting. While there are several good books that deal mainly with the physical-layer issues of optical communications, e.g., [RaSi01, DuDF02, Agra04], few books exist on the "networking" side of optical networking. However, these books are also getting outdated by the fast pace of research and development in the field. These observations led me to write this updated textbook on this topic.

I expect that instructors will find the book useful for teaching a graduate course on optical networking. At my home institution, University of California, Davis, I regularly teach a one-quarter graduate course, ECS 259, "Optical Communication Networks," using materials from the book. Actually, I find that the book's contents are longer than what one can cover in a typical quarter-long course (30 hours of lecture) or a semester-long course (45 hours of lecture). Thus, I wish to provide the following helpful guidelines for those who may wish to skip some materials.

- Each chapter is typically organized in a stand-alone and modular fashion, so any of them can be skipped, if so desired.

- Chapters 3 and 4 on single-hop and multihop networks can be skipped.
- Courses which are more computer-science oriented may wish to treat enabling technologies (Chapter 2) at an overview level and skip Chapter 15 on impairment-aware routing.
- Some instructor may wish to skip the longer-term research problems on optical packet switching (OPS) and optical burst switching (OBS) (Chapters 17 and 18).
- Chapter 12 (Optical Multicasting) may also be skipped by some instructors.
- Some instructors may have less interest in metro networks (Chapter 6) and/or access networks (Chapter 5), but I don't recommend skipping Chapter 5.
- For teaching short courses (of duration 15 hours +/- 5 hours), I have typically used the following materials: Chapter 1 (Introduction), Chapter 5 (Access), Chapters 7-9 (RWA and virtual-topology design), Chapter 11 (survivability), Chapters 13-14 (Traffic Grooming), and some aspects of Chapter 16 (Network Control and Management).

Since the major developments in optical communication networks have started to capture the imagination of the computing, telecommunications, and opto-electronic industries, I expect that industry professionals will find this book useful as a well-rounded reference. Through my own industry relationships, I find that there exists a large group of people who are experts in physical-layer optics, and who wish to learn more about network architectures, protocols, and the corresponding engineering problems in order to design new state-of-the-art optical networking products. This group of people is also who I had in mind while developing this book.

Organization Principles and Unique Features

Writing a book on optical networking is not easy since such a book has to cover materials that span several disciplines ranging from physics to electrical engineering to computer science to operations research to telecom business models, and also since the field itself is evolving very rapidly. This has been a very challenging exercise.

This book is *not* intended to cover in any detail the physical-layer aspects of optical communications; readers should consult an appropriate book, e.g., [RaSi01, DuDF02, Agra04], for such material. We summarize these "enabling technologies" in a single chapter (Chapter 2). Our treatment of the material in this chapter should allow us to uncover the unique strengths and limitations of the appropriate technologies, and then determine how the characteristics of the physical devices may be exploited in pragmatic network architectures, while compensating for the device limitations or "mismatches".

An important organizing principle that I have attempted to keep in mind while developing this book is that research, development, and education on optical communication networks should allow tight coupling between network architectures and device capabilities. To date, research on optical network architectures has taught us that, without a sound knowledge of device capabilities and limitations, one can produce architectures which may be unrealizable; similarly, research on new optical devices, conducted without the concept of a useful system, can lead to sophisticated technology with limited or no usefulness.

The book is organized into three parts. Part I introduces WDM and its enabling technologies. Part II is devoted to WDM local, access, and metro optical network architectures. Part III discusses a wide variety of problems in wavelength-routed WDM optical networks for wide-area coverage. The appendices on "where to learn more" and a "Glossary of Important Terms" should also be beneficial to many readers. More details on the book's organization can be found in Section 1.15.

The most unique feature of this book is its timeliness in capturing the state of the art in the fast-moving field of optical WDM networking. Other major salient aspects of the book are its breadth of coverage, depth of analytical treatment, clear identification of recent developments and open problems, an extensive bibliography (535 references), an extensive number of end-of-chapter exercises (a total of 301), an extensive set of illustrations (353 figures, 56 tables), etc.

A solutions manual is currently being developed. Instructors may obtain a copy from the publisher. Please see additional web-based features below.

Web Enhancements

This book is "web-enhanced," i.e., we have (or plan to have) material such as the following available through the web (see author's web address: *http://networks.cs.ucdavis.edu/~mukherje*):

1. table of contents,
2. list of figures,
3. color versions of some figures – some of the figures were drawn originally in color, and the color versions may be more informative than the black-and-white versions appearing in the book,
4. Chapter 1 in its entirety,
5. introductory material for the other chapters,
6. posting of corrections, revisions, updates, etc.,
7. easy mechanism for readers to provide feedback to the author and the publisher, and
8. if and when possible, additional material as deemed useful, e.g., simulations for some single-hop protocols, multicasting, virtual-topology embedding, etc.; allowing animation and remote operation of simulations via parameter selection by the user using Java.

Acknowledgements

Although my name is the only one to appear on the cover, the combination of efforts from a large number of individuals is required to produce a quality book.

First and foremost, I wish to thank Dr. Anpeng Huang for the countless hours of time and effort he spent in getting the book to its current form. Without Dr. Huang's efforts, this book would not have been produced.

Much of the book's material is based on research that I have conducted over the years with my graduate students and research scientists visiting my laboratory, and I would like to acknowledge them as follows: Amitabha Banerjee for optical access networks (Chapter 5); Dhritiman Banerjee for the wavelength-routing material (Chapters 7-8); Mike Borella for the enabling-technologies review (Chapter 2); Wonhong Cho for optical metro networks (Chapter 6); Aysegul Gencata for virtual-topology adaptation (Chapter 9); Anpeng Huang for optical burst switching (Chapter 18); Yurong (Grace) Huang for impairment-aware routing (Chapter 15); Jason Iness for multihop networks and wavelength conversion (Chapters 4 and 10); Jason Jue for enabling technologies and single-hop networks (Chapters 2 and 3); Glen Kramer for optical access networks (Chapter 5); Canhui (Sam) Ou for survivability and traffic grooming (Chapters 11 and 14); Byrav Ramamurthy for wavelength conversion (Chapter 10); S. Ramamurthy for wavelength routing and survivability (Chapters 7 and 11); Laxman Sahasrabuddhe for survivability and optical multicasting (Chapters 11 and 12); Narendra Singhal for optical multicasting (Chapter 12); Srini Tridandapani for channel sharing in multihop networks (Chapter 4); Jian Wang for optical metro networks (Chapter 6); Wushao Wen for survivability (Chapter 11); Shun Yao for optical packet switching (Chapter 17); Hui Zang for wavelength routing and network control & management (Chapters 7 and 16); Jing Zhang for survivability (Chapter 11); Hongyue Zhu for traffic grooming and metro networks (Chapters 6, 13, and 14); and Keyao Zhu for traffic grooming (Chapters 13 and 14). I would also like to acknowledge Dr. Nasir Ghani for some of the material on optical metro networks (Chapter 6).

A number of additional individuals who I have collaborated with over the years, who have enabled me to better understand the subject matter, and who I would like to acknowledge are the following: Subrata Banerjee, Debasish Datta, Sudhir Dixit, Christoph Gauger, Jon Heritage, Feiling Jia, Sungchang Kim, Behzad Moslehi, Massimo Tornatore, Yinghua Ye, and Ben Yoo.

I would like to thank the following students in my ECS 259 classes (formerly ECS 289I) for developing many of the end-of-chapter problems: Dragos Andrei, Amitabha Banerjee, Steven Cheung, Fred Clarke, Bill Coffman, Anpeng Huang, Yurong (Grace) Huang, Jason Jue, Arijit Mukherjee, Raja Mukhopadhyay, Canhui (Sam) Ou, Vijoy Pandey, Byrav Ramamurthy, S. Ramamurthy, Smita Rai, Laxman Sahasrabuddhe, Suman Sarkar, Huan Song, Lei Song, Feng Wang, Raymond Yip, Hongyue Zhu, and Keyao Zhu. Murat Azizoglu and Debasish Datta also contributed several problems, while Anpeng Huang, Jason Jue, and Laxman Sahasrabuddhe "organized" these exercises; their efforts are highly appreciated.

Special thanks are due to Anpeng Huang and Byrav Ramamurthy for the book's "electronic organization," several figure drawings, and answers to my LaTeX2e questions; and to Ken Victa for developing the graphical user interface (GUI) to visualize and draw the results of our optimization studies.

This book wouldn't have been possible without the support of my research on optical communication networks from several funding agencies as follows: National Science Foundation (NSF) (Grant Nos. NCR-9205755, NCR-9508239, ECS-9521249, ANI-9805285, ANI-9986665, ANI-0207864, INT-0323384, ANI-0435525, and ANI-0520190); Defense Advanced Research Projects Agency (DARPA) (Contract Nos. DABT63-92-C-0031 and DAAH04-95-1-0487); Air Force Office of Scientific Research (AFOSR) (Grant No. 89-0292); Los Alamos National Laboratory; Agilent; Alcatel; Bellsouth; Cisco; ETRI; Fujitsu; Nokia; Optivision, Inc.; ONI Systems; Pacific Telesis; SBC; Sprint; and University of California MICRO Program.

Quality control for a book can be ensured through independent technical reviews. In this regard, the following reviewers are deeply appreciated for their time, effort, suggestions for improvement, and feedback on the book's manuscript: Nasir Ghani, Jason Jue, George Rouskas, Suresh Subramanium, Ioannis Tomkos, and Lena Wosinska.

I wish to acknowledge the people at Springer who I interfaced with – Alex Greene and Melissa Guasch – for their assistance during the book's production. The book's camera-ready version was produced by the author with assistance from Dr. Anpeng Huang using the LaTeX2e document-processing system.

Finally, I wish to thank my family members for their constant encouragement and support: my father Dhirendranath Mukherjee, my mother Gita Mukherjee, my wife Supriya, and my daughters Bipasha and Suchitra. I couldn't have done it without them!

I welcome email from readers who wish to provide any sort of feedback: errors, comments, criticisms, and suggestions for improvements.

Biswanath Mukherjee

Davis, California, U.S.A.

mukherje@cs.ucdavis.edu

http://networks.cs.ucdavis.edu/~mukherje

Contents

List of Figures

List of Tables

Part I

Introduction

Chapter

1

Optical Networking: Principles and Challenges

1.1 Introduction

After experiencing rapid growth during the late 90s, the telecom industry in general, and optical networking in particular, has been experiencing some challenging times over the past several years. (An analysis of the underlying reasons will be offered later in this chapter.) Nevertheless, even though the telecom business market is unsettled today (but showing signs of improvement), we need to be ready with the appropriate technologies and engineering solutions to meet the growing bandwidth needs of our information society.

Optical networking using wavelength-division multiplexing (WDM) – the term WDM will be explained shortly in this chapter – is the technology of choice for meeting these growing demands [Mukh97, Mukh00]. While there may be an abundance of dark fiber and WDM transmission capacity today, we believe that there is – and there will continue to be – a tremendous need for optical switching equipment, namely high-capacity and high-density optical crossconnects (OXCs), for managing high-capacity optical signals. These technologies can be exploited by various categories of telecom businesses as outlined later in this chapter.

The rest of this chapter is organized as follows. Section 1.2 provides an overview of telecom networks. Section 1.3 discusses various categories of telecom business models. Section 1.4 makes the case for the important role software plays in bringing cost-effective and intelligent optical networking to the marketplace. Section 1.5 emphasizes the role of cross-layer design, analysis, and thinking for successful deployment of optical networks. Section 1.6 discusses the role of traffic engineering vs. network engineering vs. network planning in our networking investigations. Section 1.7 tries to clarify the question: "What is an Optical Network?" Section 1.8 starts with the basic characteristics of optics which can be exploited for optical networks. Section 1.9 clarifies the terms xDM vs xDMA. Section 1.10 introduces wavelength-division multiplexing (WDM). Section 1.11 outlines WDM networking evolution. Section 1.12 illustrates some WDM network constructions. Section 1.13 discusses some economic studies indicating the benefits of WDM. Section 1.14 outlines some important research problems and challenges in the WDM networking field today. Section 1.15 concludes this chapter with a "road map" of the rest of the book.

1.2 Telecom Network Overview

Figure 1.1 provides an overview of telecommunication networks. They consist of the access network, the metropolitan-area (or regional) network, and the backbone network.

The access network enables end-users (businesses and residential customers) to get connected to the rest of the network infrastructure. The access network spans a distance of a few kilometers (perhaps up to 20 km as some local exchange carriers (LECs) seem to prefer). Our current solutions for access are dial-up modems, higher-speed lines (such as T1/E1), digital subscriber line (DSL), and cable modem. However, the access network continues to be a bottleneck, and users require (and are demanding) higher bandwidth to be delivered to their machines. How to provide this high bandwidth in an inexpensive manner is a key R&D priority. Passive optical networks (PONs) based on inexpensive, proven, and ubiquitous Ethernet technology (and referred to as EPON) seem an attractive proposition for this market segment. PON technology in general, and EPON in particular, will be studied in Chapter 5.

The metro-area network typically spans a metropolitan region, cover-

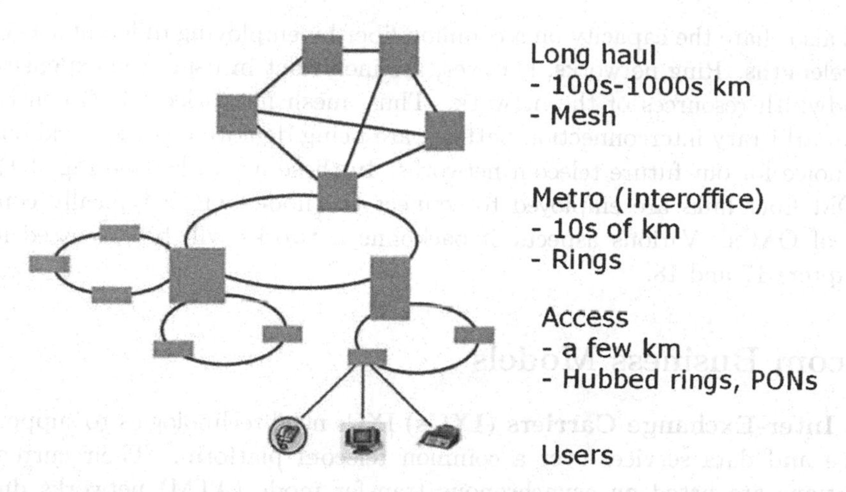

Figure 1.1 Telecom network overview [Gers00].

ing distances of a few tens to a few hundreds of kilometers. Given the deep-rooted legacy of SONET/SDH ring networks[1] in our Incumbent LEC's (ILEC's) networks, multi-wavelength versions of these rings are being deployed for our metro networks. Even in the access network, some people are inclined to deploy hubbed rings, where a "head-end" serves as a central office (CO) node to coordinate the networking activities of the end-users located at the other nodes of the ring. Important characteristics of optical metro networks will be discussed in Chapter 6.

The backbone network spans long distances, e.g., each link could be a few hundreds to a few thousands of kilometers in length. The backbone network can be set up to provide nationwide or global coverage. (For national networks covering smaller geographical areas, naturally, the link lengths will be shorter.) Most telecom backbone networks are deployed today as an interconnection of "stacked" SONET/SDH rings, in which the fibers support multiple wavelengths using WDM transmission equipment; however, by "tying" together several wavelengths on different fiber segments, one can create logical rings, and these rings can "meet" one another at some junction nodes,

[1]SONET = Synchronous Optical Network. SDH = Synchronous Digital Hierarchy. SONET and SDH are very similar technologies employed in our telecom networks, with SONET being the prevalent standard in North America, while SDH is the dominant technology in most of the rest of the world. Often times, when we state SONET, we mean both SONET and SDH.

and also share the capacity on a common fiber by employing different sets of wavelengths. Ring networks, however, are inefficient in using the expensive bandwidth resources of the network. Thus, mesh networks, which consist of an arbitrary interconnection pattern, are being deployed as the backbone of choice for our future telecom networks. In these networks (see Fig. 1.1), WDM fiber links are employed to connect the nodes which typically consist of OXCs. Various aspects of backbone networks will be discussed in Chapters 17 and 18.

1.3 Telecom Business Models

(a) Inter-Exchange Carriers (IXCs) IXCs need technologies to support voice and data services over a common telecom platform. Their current solutions are based on asynchronous transfer mode (ATM) networks due to ATM's ability to support voice and data over an integrated platform, excellent traffic-engineering capabilities, and quality-of-service (QoS) features. However, an estimated 60-70% of an IXC's traffic is "pass-through" at each node, and expensive SONET add-drop multiplexers (ADMs) and digital crossconnects (DXCs) are used to send OC-192 (approx. 10 Gbps) signals through the node[2]. The longer-term outlook for IXCs is that ATM is less attractive due to its cost and due to the advancements being made in Internet Protocol (IP) and the Generalized Multi-Protocol Label Switching (GMPLS) approach for routing. IXCs are expected to increasingly deploy WDM and OXC technologies and move to GMPLS-based networks. The potential market size of this business is very large.

(b) Next-Generation National/Global Carriers or "Greenfield Networks" (GNs)
The objective of the next-generation national/global carriers is to provision high-bandwidth circuits throughout their greenfield networks (GNs) to serve as the carriers' carrier (wherever needed) and to offer wholesale services to ISPs. The GNs' backbone networks are based on efficient mesh architectures instead of inefficient SONET-ring-based architectures. Their initial solutions are based on employing terabit routers with ATM switches for traffic aggregation and management, and SONET ADMs and DXCs for cir-

[2]OC = Optical Carrier, a SONET terminology. A channel rate of OC-n equals n x 51.84 Mbps. Transmission systems operating at OC-192 (approx. 10 Gbps) are quite common today.

cuit provisioning. Their follow-on solutions are expected to replace SONET ADMs and DXCs with OXCs to switch and provision wavelengths and possibly connections of sub-wavelength granularity (called traffic grooming, which is discussed in Chapters 13 and 14). Unfortunately, recent market conditions have been unkind to this telecom business segment because the cost of deploying GNs is very high, OXC technologies are still maturing, competition among GNs has been quite fierce (due to abundance of capital during the late 90s), telecom entities have been finding it hard to generate revenue from data services, and the delay in return on investment (ROI) has been making investors nervous about continued support for GNs.

(c) National/Global Internet Service Providers (ISPs) National/global ISPs require that they have technologies which can provide for quick and easy deployment of flexible and scalable bandwidth throughout their networks. They need transparency, high level of reliability, security, and potential reconfigurability. Their current solutions employ the "routed-edge, switched-core" architecture based on IP routers and ATM switches, where ATM is used for its traffic management and statistical-multiplexing capabilities. The ISPs' longer-term objective is to use GMPLS for simplifying the operation and management of complex backbones, and to employ WDM and OXC solutions for their ability to lease wavelengths and high-capacity sub-wavelength-granularity channels.

(d) Incumbent Local-Exchange Carriers (ILECs), Competitive Local Exchange Carriers (CLECs), Carrier Hotels, etc.

The LECs need to integrate disparate geographic and service-oriented networks into one common platform. The carrier hotels need to provide neutral co-location space. The LECs' current solutions are mainly to employ SONET and ATM-based networks. We expect that, in the near future, OXC and core optical network equipment will allow the management of the LECs' core infrastructure and establish a platform for delivering high-bandwidth circuits (using WDM). The WDM and OXC technologies will also enable IXCs, GNs, LECs, ISPs, etc. to lease, trade, buy, and sell bandwidth on demand through their co-location at carrier-neutral internet exchange points (namely carrier hotels). Thus, it seems very logical that the "carrier hotel" model should become a growing business.

Based on these discussions, we can identify the following four business models, which map roughly, but not exactly, to the above four categories, as follows [Varm01].

- *Business Model #1:* An ISP that owns the network from the "ground up" (i.e., to the duct) and only delivers IP-based services.

- *Business Model #2:* The business owns the layer-one infrastructure and sells services to customers who may themselves resell to others (e.g., Greenfield Network (GN) operators).

- *Business Model #3:* An ISP that leases fiber or transport capacity from a third part, and only delivers IP-based services (e.g., National/Global ISPs).

  Note that BM #2 and BM #3 are complimentary, and both are integrated in case of BM #1.

  Thus, BM #1 applies to network-planning (NP) and traffic-engineering (TE) problems where the network operator has, at its disposal, all of the network resources; and the input traffic it has to handle consists of data packets. (NP is also referred to as Network Design (ND) in the literature.)

  BM #2's problems are similar except that its input traffic consists of circuit requests (which are generated by BM #3).

  In case of BM #3, the input traffic is packet data, but BM #3 may not have full visibility into the physical network infrastructure. In fact, BM #3 may even lease capacities (for its different "logical links", say between its Internet Protocol (IP) routers) from multiple BM #2's for operating different parts of its networks.

  As examples, we shall study the applications of (1) BM #1 to virtual-topology design problems (in Chapters 8 and 9); (2) BM #3 to virtual-topology adaptation problems for ISPs (in Chapter 9); and (3) BM #2 to numerous other problems such as routing and wavelength assignment (RWA), survivability, etc. (in Chapters 7 and 10-16).

- *Business Model #4:* The business is a *bandwidth broker.* It provides "match-making" by enabling a variety of ISPs (BM #3) to lease bandwidth from a variety of network operators (BM #2). Thus, an ISP's customer may have its traffic being actually carried over third-party networks.

1.4 Optical Networking: Role of Optics, Electronics, and Software

"Disruptive technology" is a common phrase used by pundits (and non-pundits) to state what it will take to build the next generation of our optical networks. The focus of these pundits (and, more frequently, the non-pundits) is on making quantum leaps in optical device technologies, without paying attention to all three "pillars" - optics, electronics, and software - which are all very important to build a successful optical network. For example, optical networks and optical computers (do you remember optical computers and the amount of R&D dollars poured into them?) are examples of network systems, as opposed to optical devices (e.g., lasers) or optical subsystems (e.g., transponder arrays).

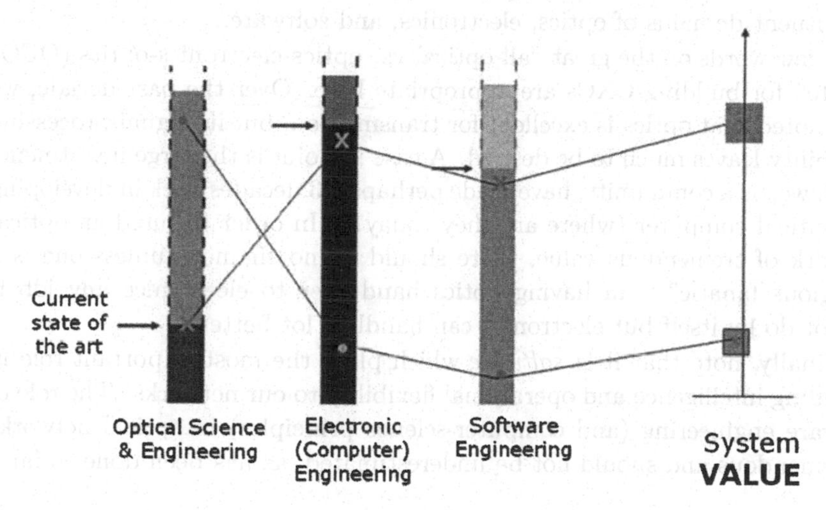

Figure 1.2 System: value proposition.

In Fig. 1.2, we show that the state of the art in *optical science and engineering* is in its infancy and has a lot of room for growth. This is in sharp contrast with the states of the art in *computer engineering (electronics)* and *software engineering*, both of which are quite mature (and will continue to mature further). It is our belief that Optical Networking System Proposition "A" (marked by • in Fig. 1.2), which is based on disruptive optical technology (and hence is riskier) and which does not pay attention to the industry-standard electronics and software, would lead to a system with limited value.

Market risk aside, we note that there is also a tremendous *technology risk* associated with Proposition A. With numerous optical scientists leading the technology charge in optical networking companies and paying attention mainly to the "beauty" of optics, Proposition A systems have had difficulty to create a large system value.

Optical Networking System Proposition "B" (marked by × in Fig. 1.2), on the other hand, is based on the philosophy that a technologist can create enormous system value by building a network which exploits the state-of-the-art optical devices, and supporting them with state-of-the-art electronics and software. The technologist's intellectual property is created by determining how much of the art to employ in each from the three "dimensions" – optics, electronics, and software – so that the solution is cost-effective, flexible, and scalable. Thus, instead of disruptive technology in the optical domain, one should think in terms of creating a disruptive integrated technology from all constituent domains of optics, electronics, and software.

A few words on the great "all-optical vs. optics-electronics-optics (OEO) debate" for building OXCs are appropriate here. Over the past decade, we have noted that optics is excellent for transmission, but its signal-processing capability leaves much to be desired. A case in point is the large investments which we, as a community, have made perhaps 2-3 decades back in developing the optical computer (where are they today?). In order to build an optical network of tremendous value, there should be no shame – unless one is a "religious fanatic" – in having optics hand over to electronics any job it cannot do by itself but electronics can handle a lot better.

Finally, note that it is *software* which plays the most important role in providing intelligence and operational flexibility to our networks. The role of software engineering (and computer-science principles) in optical networks is tremendous and should not be underestimated, as has been done so far.

1.5 Cross-Layer Design

Continuing our discussions along the above lines, note the need for tight coupling between network architectures and device capabilities. To date, research on optical network architectures have taught us that, without a sound knowledge of device capabilities and limitations, one can produce architectures which may be unrealizable; conversely, research on new optical devices, conducted without the concept of a useful system, can lead to sophisticated technology with limited or no usefulness. Thus, the bottom half of Fig. 1.3

indicates that the relationship between the network layer[3] and the physical layer is neither top-down, nor bottom-up; instead, we characterize the relationship as "push-pull" where both layers need to influence each other, thereby leading to "cross-layer" design (and analysis) of such systems and networks.

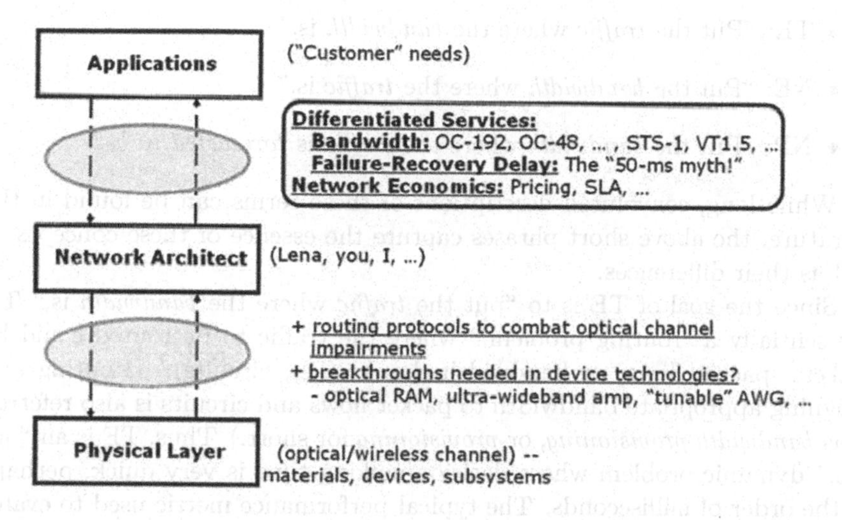

Figure 1.3 Optical network architecture: extending our "boundaries".

The same "cross-layer" principles apply to the interworkings between the network layer and the application layer which sits on top. Again, the relationship here is of a "push-pull" nature, e.g., the network layer should be able to offer new networking primitives to the applications based on emerging device technologies and which the (new) applications may be able to exploit. Conversely, the application layer should also be able to drive the network layer, demanding from it new networking paradigms, which, in turn, may determine prioritization of research problems for the physical layer.

Some example (and important) research problems at these cross-layer boundaries are indicated in Fig. 1.3. Thus, the network architect, who is the main audience of this book, needs to expand its boundaries to produce more-effective research.

[3]The term "layer" is used as an abstraction here, and the layers in our current discussion do not necessarily map to the standard layers used in various network architectures.

1.6 Traffic Engineering (TE) vs. Network Engineering (NE) vs. Network Planning (NP)

Consider the following brief descriptions of the above terms (original source unknown).

- TE: "Put the *traffic* where the *bandwidth* is."
- NE: "Put the *bandwidth* where the *traffic* is."
- NP: "Put the *bandwidth* where the *traffic* is *forecasted to be.*"

While long, convoluted descriptions of these terms can be found in the literature, the above short phrases capture the essence of these concepts as well as their differences.

Since the goal of TE is to "put the *traffic* where the *bandwidth* is," TE is essentially a "routing problem," where the traffic to be routed could be packets, packet flows, or bandwidth chunks (i.e., circuits). (Routing and assigning appropriate bandwidth to packet flows and circuits is also referred to as *bandwidth provisioning*, or *provisioning* for short.) Thus, TE is an "on-line," dynamic problem whose decision-making time is very quick, perhaps on the order of milliseconds. The typical performance metric used to evaluate a TE algorithm is "blocking probability" (by (implicitly) assuming that the network is operating at "steady-state") although other metrics are also important such as volume of control overhead, convergence time (to reduce routing instability), etc.

By transposing the words "traffic" and "bandwidth" in the TE description, an accurate description of NE is obtained. As a network continues its operation, and as traffic on it builds up (perhaps asymmetrically), certain parts of the network may become more congested due to increasing traffic, and these parts may need "help" in the form of additional capacity to relieve the congestion.[4] The NE decision-making time is perhaps on the order of

[4] In this regard, note that blocking-probability calculations apply to a network that is operating in "steady state," where, for example, traffic demands arrive, hold for a certain time, and then leave, but the overall traffic intensity over time remains constant. In an operational network, on the other hand, even though we expect that traffic demands will arrive, hold for a certain time, and then leave, we also expect that the average intensity of traffic offered to the network will steadily increase with time (due to more users getting on their network and also due to evolution in their networking needs towards higher-bandwidth applications, longer-holding-time applications, etc.) Thus, blocking-probability calculations do not apply since traffic is not in steady state any more. Instead, we need

weeks or months. Because some part of a network may experience capacity exhaust sooner than other parts of the network, the capacities of the various links in an operational network may be quite asymmetric. Thus, a typical performance metric for a NE problem could be "exhaustion probability" which determines when a current network, given a traffic-growth pattern, will run into capacity exhaust. Thus, the network operator will need to incur more Capital Expenditure (CapEx) to handle this capacity-exhaust problem. This (NE) is a very realistic problem in our operational networks; and, unfortunately, it has been underestimated in the academic research literature.

The NP description is almost the same as that of NE, except that an additional phrase "*forecasted to be*" is appended. NP corresponds to the planning (i.e., design) of a network from scratch, with a decision-making timescale of perhaps a few years. A sample NP problem is the following: Given a set of traffic demands between various pairs of nodes (which is also called a "traffic matrix"), design the network for minimum cost, i.e., determine how much capacity to put on each link of the network, as well as routing of traffic through the network links. (Note that the typical performance/optimization metric for a NP problem is "cost.") As an example, the cost could be the sum of the (bandwidth) cost of all the links. In the brief NP problem description (above), no statement was made about the connectivity between network nodes (i.e., "network topology"). By default, the network topology (or graph) may be given. But an additional dimension to the NP problem would be to also determine the topology (while achieving minimum cost). In addition, there exists a "dual" version of the NP problem (as is the case with most optimization problems). This "dual" problem can be stated as follows: Given the traffic demands, and the maximum cost (including perhaps the topology and capacity of each link), determine how to establish the demands so that the network throughput (in terms of carried demands) is maximized. (Thus, not all demands may be successfully established in this "dual version of the NP problem.)

Finally, let us state the "real" Network Planning (NP) problem. Generally, for NP purposes, the traffic forecasts are not a "one-snapshot demand matrix." It is expected that the network will experience traffic growth, so additional capacity must be added to the network with time (recall the characteristics of NE). As an example, a network planner may be given annual

to consider "exhaustion probability," which is a time duration at the end of which the network is expected to run into capacity exhaust [NaSi03].

traffic forecasts over a N-year period (say N=5). The NP problem would be to design the network (at low cost) for Year 1, and to plan how/where to add capacity (and perhaps reprovision some of the demands) in subsequent years in order to ensure that (1) the initial equipment chosen can "grow" in capacity over the N years, and (2) the growth can be performed for minimum capital expenditure (CapEx)[5] over the network's lifetime. (Actually, the CapEx calculation is not as straightforward because, for each year, the Network Planner needs to incur some additional CapEx, while previous years' CapEx's need to be depreciated appropriately.)

Unfortunately, we have not seen any real Network Planning (NP) problems attacked in the research literature.

A summary of TE vs. NE vs. NP is provided in Fig. 1.4.

- **Traffic Engineering (TE)**
  - "Put the traffic where the bandwidth is"
- **Network Engineering (NE)**
  - "Put the bandwidth where the traffic is"
- **Network Planning (NP)**
  - "Put the bandwidth where the traffic is *forecasted to be*"

- **TE** - online, dynamic, provisioning problem, ms time scale
- **NE** - intermediate problem, months time scale
- **NP** - offline, static, dimensioning problem, 5-yr time scale

Figure 1.4 TE vs. NE vs. NP.

[5]Capital Expenditure (CapEx) and Operational Expenditure (OpEx) are common terms used by network operators (and by providers of any service, in general). As the names indicate, CapEx refers to the equipment expenses, while OpEx refers to expenses incurred in maintaining the equipment, including human resources. Typically, OpEx is several times larger than CapEx, say 8-10 times, so a major goal is to reduce OpEx with new and intelligent equipment (recall role of software). Nevertheless, a major network operator's CapEx can be quite high, perhaps several tens of billions of dollars annually, so reducing CapEx is also a significant concern.

1.7 What is an Optical Network?

Many general networking professionals ask this question. They wonder what role networking (or computer science and computer engineering, in general) has to play in the field of optical networks. Some of them are unclear about whether or not optical networking is just a mere extension of optics.

It is **NOT NECESSARILY** all optical
" " " " packet switched

Characteristics of an optical network

- **Transmission: optical**
- **Switching:** could be optical, could be electronic, could be hybrid
  could be circuit, could be packet, could be burst

Most Promising Approach Today

- Electronic circuit switching with sub-lambda granularity (STS-1, STS-3, ...)

Example Utility for IP Networking

- Connect any two IP routers (geographically far apart) with a direct ("virtual") bandwidth pipe... of whatever capacity (STS-1, ... , STS-192)
- Increase (or decrease or delete) the capacity on demand
- Dynamically control the "topology" connecting the IP routers
- Create a "separated control network" (of whatever bandwidth)
- ...

Figure 1.5 Definition of an optical network.

To answer this question, we first refer the reader to Fig. 1.5. We all know that a network consists of a collection of nodes interconnected by links (perhaps in an arbitrary topology). The links require "transmission equipment," while the nodes require "switching equipment." Technology developments to date have taught us that optics is fantastic for transmission, e.g., an optical amplifier can simultaneously amplify all of the signals on multiple wavelength channels (perhaps as high as 160) on a single fiber link, independent of how many of these wavelengths are currently carrying live traffic. However, many attempts at developing all-optical switches have indicated that optical switching is still in its infancy.

Thus, an optical network is not necessarily all-optical: the transmission is certainly optical, but the switching could be optical, or electrical, or hybrid (see Chapters 13 and 14).

Also, an optical is not necessarily packet-switched (as in Chapter 17). It could switch circuits (Chapters 7-12, 15-16), or sub-wavelength-granularity bandwidth pipes (Chapters 13-14), or "bursts," where a burst is a collection of packets (Chapter 18).

Based on the maturity of various optical technologies, the most-prevalent deployment of optical networks today consists of optical-electrical-optical (OEO) switches (also called *opaque* switches), with each input operating at OC-192 (approx. 10 Gbps) rate. However, inside the OEO switch, each input channel can be demultiplexed into STS-1 timeslots[6] (each of approx. capacity 51.84 Mbps), and the switch can perform switching at STS-1 granularity. Thus, a network operator can support a variety of connection requests ranging in bandwidth from STS-1 to OC-192.

So, how do such technologies benefit the operator of an Internet Protocol (IP) network – or any data network, for that matter – which needs to carry data packets? Herein, we note that the "links" in IP networks are "leased lines" of fixed bandwidth and, typically, they are leased for a long duration, say a year. Now, would it not be a tremendous benefit if the IP network operator had the capability to *"dial" for bandwidth on demand*[7]? The IP network operator could increase the bandwidth, or decrease it, or even delete it, based on its current needs; thus, the IP network will be operating with only the bandwidth it needs at any point in time. And the IP network "topology" (also called the "grid") can be reconfigured on demand, as necessary.

As an example, consider that two users (say atmospheric scientists), located at the two coasts of the USA, need to exchange some large files. Under present mode of operation (PMO) of today's data networks, a simple "traceroute" from Davis to Boston (see Fig. 1.6) indicates that the file transfers may encounter 20+ router hops, at each of which there exists the possibility of buffering (and hence delay), as well as loss (due to buffer overflow). In future mode of operation (FMO) of data networks, by exploiting the underlying support from optical-networking technologies, one should be able to "dial up" a fat bandwidth pipe (of appropriate capacity and duration) to complete the task. It is not necessary that all such applications need to accomplished in "one hop." However, by establishing such fat pipes using optical-networking principles, the data network operator can reduce

[6]Synchronous Transport Signal (STS) is the electrical-level transmission frame structure in SONET. A STS-N frame is usually carried by an OC-N transmission link.

[7]Just like we pick up the phone, dial any destination around the world, and expect to get the connection established in a few minutes... and we are not tied to a long-term (annual) lease, as seems to be the case with most high-capacity "leased bandwidth lines" today.

```
2:33pm pepper (~) 27 : traceroute aland.bbn.com
traceroute to aland.bbn.com (68.22.232.249), 30 hops max, 40 byte packets
 1  169.237.4.254 (169.237.4.254) 6.961 ms 0.699 ms 0.448 ms
 2  169.237.246.238 (169.237.246.238) 0.623 ms 0.661 ms 0.691 ms
 3  area2-13-area2.ucdavis.edu (128.120.2.49) 0.926 ms 0.685 ms 0.849 ms
 4  area2-area0.ucdavis.edu (128.120.0.133) 0.846 ms 0.642 ms 0.591 ms
 5  area0-ucd.ucdavis.edu (128.120.0.114) 67.471 ms 67.633 ms 70.661 ms
 6  inet-oak-isp--ucd-ge.cenic.net (137.164.24.233) 7.581 ms 8.460 ms 11.830 ms
 7  f5.ba01.b003070-1.sfo01.atlas.cogentco.com (38.112.6.225) 73.018 ms 65.948 ms 55.990 ms
 8  p10-0.core01.sjc03.atlas.cogentco.com (66.28.4.133) 69.680 ms 53.346 ms 74.431 ms
 9  eq2-g4-2-1.eqsjca.sbcglobal.net (151.164.249.185) 70.071 ms 83.845 ms 82.674 ms
10  ex1-p9-0.eqsjca.sbcglobal.net (151.164.191.201) 71.123 ms 65.604 ms 64.093 ms
11  bb1-p6-0.crsfca.sbcglobal.net (151.164.41.9) 71.285 ms 68.963 ms 68.596 ms
12  core1-p4-0.crsfca.sbcglobal.net (151.164.240.133) 136.819 ms 182.120 ms 217.642 ms
13  core1-p5-0.crskut.sbcglobal.net (151.164.42.11) 95.375 ms 63.709 ms 71.437 ms
14  core1-p2-0.crdnco.sbcglobal.net (151.164.243.242) 102.403 ms 71.730 ms 91.980 ms
15  core1-p3-0.crkcmo.sbcglobal.net (151.164.188.34) 84.839 ms 67.509 ms 62.889 ms
16  core2-p11-0.crchil.sbcglobal.net (151.164.240.118) 80.109 ms 101.998 ms 82.066 ms
17  bb2-p5-0.sfldmi.ameritech.net (151.164.242.130) 76.808 ms 65.863 ms 74.787 ms
18  bb1-p15-0.sfldmi.ameritech.net (151.164.40.181) 74.925 ms 108.616 ms 103.503 ms
19  bb1-p3-0.lgtpmi.sbcglobal.net (151.164.40.101) 112.321 ms 96.089 ms 80.460 ms
20  dist1-vlan30.lgtpmi.ameritech.net (65.42.245.97) 92.278 ms 72.988 ms 78.367 ms
21  rback3-g1-0.lgtpmi.sbcglobal.net (65.42.245.230) 73.354 ms 83.651 ms 77.640 ms
22  adsl-68-22-232-254.dsl.lgtpmi.ameritech.net (68.22.232.254) 86.725 ms 100.490 ms 90.429 ms
23  adsl-68-22-232-249.dsl.lgtpmi.ameritech.net (68.22.232.249) 85.333 ms 94.014 ms 90.203 ms
2:41pm pepper (~) 28 :
```

Figure 1.6 **Traceroute** example.

the number of router hops significantly – not necessarily from 20+ down to 1, but to some small number, say 3, possibly depending on administrative domains, peering relationships between operators, service-level agreements (SLAs), etc. When traffic encounters fewer router hops, it should experience better performance (in terms of lower delay, lower delay jitter, lower packet-loss probability, etc.

Yet another application of optical-networking principles to facilitate IP data networks is to create a "separated control network plane" of whatever bandwidth the network operator desires. Note that signaling in IP networks is "in-band," so (short) control packets may have to contend with (long) data packets for transmission bandwidth. As data entities that need to get transferred over our networks get longer, control packets are expected to encounter more contention for bandwidth. Thus, one can create a separated control network by setting aside a wavelength (or a sub-wavelength-granularity bandwidth chunk using a traffic-grooming principle) on each link for this purpose, so that control packets have their own dedicated network.

The optical-networking paradigms are maturing sufficiently to facilitate such operations indicated above.

1.8 Optical Networking: Need + Promise = Challenge!

Life in our increasingly information-dependent society requires that we have access to information at our finger tips *when we need it, where we need it, and in whatever format we need it.* The information is provided to us through our global mesh of communication networks, whose current implementations – such as the present-day *Internet* and ATM networks – unfortunately, don't have the capacity to support the foreseeable bandwidth demands.

Optical fiber technology can meet our above need because of its potentially limitless capabilities, as follows [ShVe04]:

1. huge bandwidth (over 50 terabits per second (Tbps)),
2. low signal attenuation (as low as 0.2 dB/km), so repeater-less transmission over long distances is possible,
3. immunity to electromagnetic interference,
4. high security of signal because of no electromagnetic radiation, so difficult to eavesdrop,
5. no crosstalk and interferences between fibers in the same cable,
6. low signal *distortion*, suitable for carrying digital information,
7. low power requirement,
8. low material usage, small space requirement, light weight, non-flammable, cost-effective, and
9. high electrical resistance, so safe to use near high-voltage equipment or between areas with different earth potentials.

Our challenge now is to turn the promise of optical fiber technology to reality to meet our information networking demands of the foreseeable future.

Thus, the basic premise of this book's topic – viz., optical communication networks – is that, as more and more users start to use our data networks, and as their usage patterns evolve to include more and more bandwidth-intensive networking applications such as data browsing on the *world wide web (WWW)*, java applications, video conferencing, etc., there emerges an acute need for very high-bandwidth transport network facilities, whose capabilities are much beyond those that current networks can provide.

Given that a single-mode fiber's potential bandwidth is nearly 50 Tbps, which is nearly 3-4 orders of magnitude higher than electronic data rates of a few tens of gigabits per second (Gbps), every effort should be made to tap into this huge opto-electronic bandwidth mismatch.

Realizing that the maximum rate at which an end-user – which can be a workstation or a gateway that interfaces with lower-speed subnetworks – can access the network is limited by electronic speed (to a few Gbps), the key in designing optical communication networks in order to exploit the fiber's huge bandwidth is to introduce concurrency among multiple user transmissions into the network architectures and protocols. In an optical communication network, this concurrency may be provided according to either wavelength or frequency [*wavelength-division multiplexing (WDM)*], time slots [*time-division multiplexing (TDM)*], or wave shape [spread spectrum, *code-division multiplexing (CDM)*].

Optical TDM and CDM are somewhat futuristic technologies today. Under (optical) TDM, each end-user should be able to synchronize to within one time slot. The optical TDM bit rate is the aggregate rate over all TDM channels in the system, while the optical CDM chip rate may be much higher than each user's data rate. As a result, both the TDM bit rate and the CDM chip rate may be much higher than electronic processing speed, i.e., some part of an end user's network interface must operate at a rate higher than electronic speed. Thus, TDM and CDM are relatively less attractive than WDM, since WDM – unlike TDM or CDM – has no such requirement.

Specifically, WDM is the favorite multiplexing technology for practical optical communication networks since all of the end-user equipment needs to operate only at the bit rate of a WDM channel, which can be chosen arbitrarily, e.g., peak electronic processing speed. Hence, this book will concentrate on WDM networks. In an earlier book by this author [Mukh97], Chapter 16 was devoted to optical TDM and CDM.

1.9 xDM vs. xDMA

We have introduced the term xDM where x = {W, T, C} for wavelength, time, and code, respectively. Sometimes, any one of these techniques may be employed for multiuser communication in a *multiple access* environment, e.g., for broadcast communication in a local-area network (LAN) (to be examined in Section 1.12.1)[8].

Thus, a *local optical network* that employs wavelength-division multiplexing is referred to as a *wavelength-division multiple access (WDMA) network*; and TDMA and CDMA networks are defined similarly.

In this book, we will rarely use the term xDMA; instead, we will use the term *xDM* to refer to the multiplexing strategy, and employ the term *xDM network* to refer to the corresponding network, either for local-area or for wide-area (switched) applications. (Corresponding example WDM networks will be introduced in Sections 1.12.1 and 1.12.2, respectively.)

1.10 Wavelength-Division Multiplexing (WDM)

Wavelength-division multiplexing (WDM) is an approach that can exploit the huge opto-electronic bandwidth mismatch by requiring that each end-user's equipment operate only at electronic rate, but multiple WDM channels from different end-users may be multiplexed on the same fiber. Under WDM, the optical transmission spectrum (see Fig. 1.7, more discussion on which will be provided in Chapter 2) is carved up into a number of non-overlapping *wavelength* (or *frequency*) bands, with each wavelength supporting a single communication channel operating at whatever rate one desires, e.g., peak electronic speed. Thus, by allowing multiple WDM channels to coexist on a single fiber, one can tap into the huge fiber bandwidth, with the corresponding challenges being the design and development of appropriate network architectures, protocols, and algorithms. Also, WDM devices are easier to implement since, generally, all components in a WDM device need to operate only at electronic speed; as a result, several WDM devices are available in the marketplace today, and more are emerging.

Research and development on optical WDM networks have matured considerably over the past decade. They are being increasingly deployed by telecom network operators all over the world. The next generation of the Internet is employing WDM-based optical backbones, leading to IP-over-WDM networks. In such a network, end-users – to whom the architecture and operation of the backbone will be transparent except for significantly improved response times – will attach to the network through a wavelength-sensitive switching/routing node (details of which will become clearer later in the book). An end-user in this context need not necessarily be a terminal equipment, but the aggregate activity from a collection of terminals –

[8]For example, Ethernet is a *multiple access* protocol employed on a broadcast bus.

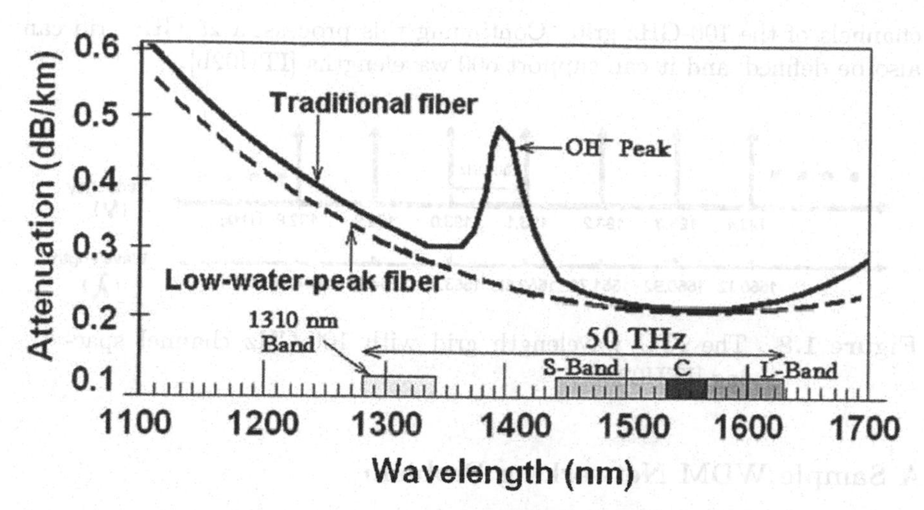

Figure 1.7 The low-attenuation regions of an optical fiber.

including those that may possibly be feeding in from other regional and/or local subnetworks – so that the end-user's aggregate activity on any of its transmitters is close to the peak electronic transmission rate.

Let us examine below wavelength standards and a sample networking problem on such a WDM network.

1.10.1 ITU Wavelength Grid

There is a strong need for the standardization of WDM systems so that WDM components and equipments from difference vendors can inter-operate with one another. Thus, industry standards for wavelengths have been developed under the leadership of the International Telecommunications Union (ITU) [ITU02b]. A standard set of wavelengths, called the ITU grid, has been defined to coincide with the 1550-nm low-loss region of the fiber. Specifically, this grid is anchored at a frequency of 193.1 THz (which corresponds to a wavelength of 1552.52 nm). There is a 100-GHz grid, which means that spacing between adjacent channels is 100 GHz, which corresponds approximately to 0.8-nm wavelength channel spacing around the anchor frequency. A few channels of this grid around the anchor channel are shown in Fig. 1.8.

For denser packing of channels, a 50-GHz grid has also been defined around the same reference frequency of 193.1 THz [ITU02b]. The 50-GHz grid is obtained by adding a channel exactly half way between two adjacent channels of the 100-GHz grid. Continuing this process, a 25-GHz grid can also be defined, and it can support 600 wavelengths [ITU02b].

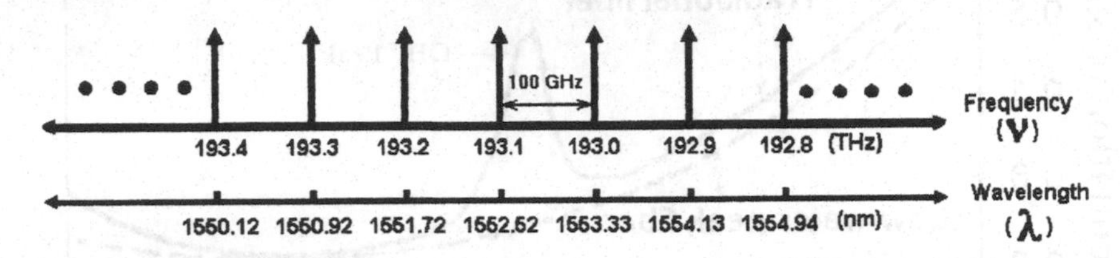

Figure 1.8 The ITU wavelength grid with 100-GHz channel spacing [ITU02b].

1.10.2 A Sample WDM Networking Problem

End-users in a fiber-based WDM backbone network may communicate with one another via *optical (WDM) channels*, which are referred to as *lightpaths.* A *lightpath* may span multiple fiber links, e.g., to provide a "circuit-switched" interconnection between two nodes which may have a heavy traffic flow between them and which may be located "far" from each other in the physical fiber network topology. Each intermediate node in the lightpath essentially provides an optical bypass facility to support the lightpath[9].

In an N-node optical WDM network, if each node is equipped with $N-1$ transceivers [transmitters (lasers) and receivers (filters)] and if there are enough wavelengths on all fiber links, then every node pair could be connected by a lightpath, and there is no networking problem to solve. However, note that the network size (N) should be scalable, transceivers are relatively expensive so that each node may be equipped with only a few of them, technological constraints dictate that the number of WDM channels that can be supported in a fiber be limited to W (whose value is a few tens today, perhaps as high as 160 in some systems, and is expected to improve with time and technological breakthroughs), and switches with large port count (i.e., large W) are very costly. Thus, only a limited number of lightpaths may be set up on the network.

[9]Note that, in the past, we have referred to a lightpath as an all-optical entity because that is how these networks were expected to evolve about a decade back. However, as has been explained in Section 1.4, producing all-optical networks consisting of all-optical switches has been very hard, so current optical networks are based on *opaque* (OEO) switches. Thus, a lightpath is still an optical circuit-switched entity, but it is not necessarily all-optical.

Under such a network setting, a challenging networking problem is that, given a set of lightpaths that need to be established on the network, and given a constraint on the number of wavelengths (or else the switch cost will become very high), determine the routes over which these lightpaths should be set up and also determine the wavelengths that should be assigned to these lightpaths so that the maximum number of lightpaths may be established. While shortest-path routes may be most preferable, note that this choice may have to be sometimes sacrificed, in order to allow more lightpaths to be set up. Thus, one may allow several alternate routes for lightpaths to be established. Lightpaths that cannot be set up due to constraints on routes and wavelengths are said to be blocked, so the corresponding network optimization problem is to minimize this blocking probability.

In this regard, note that, normally, a lightpath operates on the same wavelength across all fiber links that it traverses, in which case the lightpath is said to satisfy the *wavelength-continuity constraint.* Thus, two lightpaths that share a common fiber link should not be assigned the same wavelength. However, if a switching/routing node is also equipped with a *wavelength converter facility* (which is a "built-in property of opaque OEO switches), then the *wavelength-continuity constraints* disappear, and a lightpath may switch between different wavelengths on its route from its origin to its termination.

This particular problem, referred to as the Routing and Wavelength Assignment (RWA) problem, will be examined in detail in Chapter 7, while the general topic of *wavelength-routed networks* will be studied in Part III (Chapters 7 through 18).

Returning to our sample networking problem, note that designers of next-generation optical networks must be aware of the properties and limitations of optical fibers and devices in order for their corresponding protocols and algorithms to take advantage of the full potential of WDM. Often, a network designer may approach the WDM architectures and protocols from an overly simplified, ideal, or traditional-networking point of view. Unfortunately, this may lead an individual to make unrealistic assumptions about the properties of fiber and optical components, and hence may result in an unrealizable or impractical design. The goal of this book is to clarify the properties of WDM optical components and present the WDM networking architectures and challenges.

1.11 WDM Networking Evolution

1.11.1 Point-to-Point WDM Systems

WDM technology has been deployed by telecomm network operators worldwide for point-to-point communications. They are being increasingly deployed for ring and mesh networks as well. These deployments are driven by increasing demands for communication bandwidth. When the demand exceeds the capacity in existing fibers, WDM is turning out to be a more cost-effective alternative compared to laying more fibers. An example study [MePD95] compared the relative costs of upgrading the transmission capacity of a point-to-point transmission link from OC-48 (2.5 Gbps)[10] to OC-192 (10 Gbps) via three possible solutions:

1. installation/burial of additional fibers and terminating equipment (the "multifiber" solution);
2. a four-channel "WDM solution" (see Fig. 1.9) where a WDM multiplexer (mux) combines four independent data streams, each on a unique wavelength, and sends them on a fiber; and a demultiplexer (demux) at the fiber's receiving end separates out these data streams; and
3. OC-192, a "higher-electronic-speed" solution.

The analysis in [MePD95] showed that, for distances lower than 50 km for the transmission link, the "multi-fiber" solution is the least expensive; but for distances longer than 50 km, the "WDM" solution's cost is the least with the cost of the "higher-electronic-speed" solution not that far behind.

WDM transmission systems (i.e., mux/demux in point-to-point links) have been available as products from many vendors for nearly a decade now. Channel counts of 64 are quite common in such products, with the maximum channel count being 160 today, and expected to increase to 320 soon.

[10] The terminology OC-n is a widely used telecommunications jargon for SONET systems. "OC" stands for "optical channel" which is unfortunate since it has almost nothing to do with our type of *optics* research and development; it simply specifies electronic data rates. "OC-n" stands for a data rate of $n \times 51.84$ megabits per second (Mbps) approximately; so, for example, OC-48, OC-192, and OC-768 correspond to approximate data rates of 2.5 Gbps, 10 Gbps, and 40 Gbps, respectively. OC-768 is the next milestone in highest achievable electronic communication speed for practical deployment.

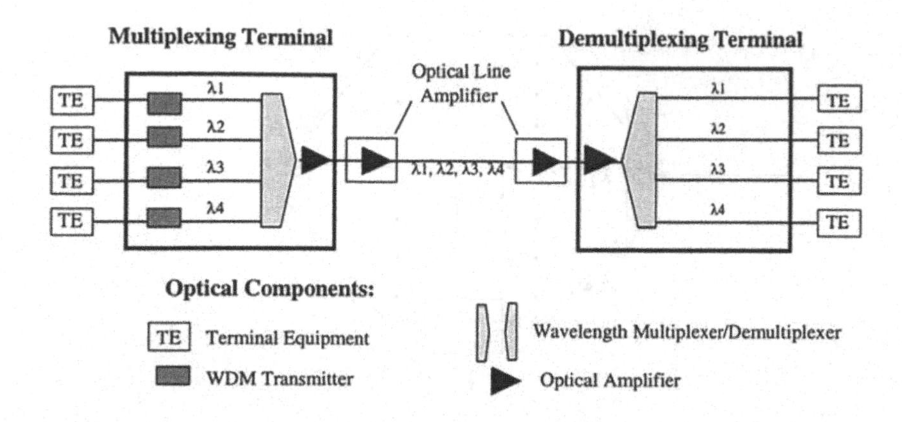

Figure 1.9 A four-channel point-to-point WDM transmission system with amplifiers.

1.11.2 Wavelength Add/Drop Multiplexer (WADM)

A Wavelength Add/Drop Multiplexer (WADM) [also referred to as on Optical Add/Drop Multiplexer (OADM)] (see Section 2.6.1) is shown in Fig. 1.10. It consists of a demux, followed by a set of 2 × 2 switches – one switch per wavelength – followed by a mux. The WADM can be essentially "inserted" on a physical fiber link. If all of the 2 × 2 switches are in the "bar" state, then all of the wavelengths flow through the WADM "undisturbed." However, if one of the 2 × 2 switches is configured into the "cross" state (as is the case for the λ_i switch in Fig. 1.10) via electronic control (not shown in Fig. 1.10), then the signal on the corresponding wavelength is "dropped" locally, and a new data stream can be "added" on to the same wavelength at this WADM location. More than one wavelength can be "dropped and added" if the WADM interface has the necessary hardware and processing capability.

1.11.3 Fiber and Wavelength Crossconnects – Passive Star, Passive Router, and Active Switch

In order to have a "network" of multiwavelength fiber links, we need appropriate fiber interconnection devices. These devices fall under three broad categories:

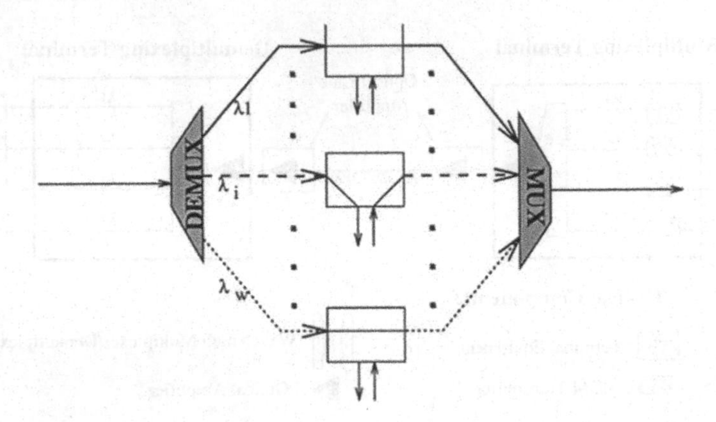

Figure 1.10 A Wavelength Add/Drop Multiplexer (WADM).

- passive star (see Fig. 1.11),
- passive router (see Fig. 1.12), and
- active switch (see Fig. 1.13).

The *passive star* is a "*broadcast*" device, so a signal that is inserted on a given wavelength from an input fiber port will have its power equally divided among (and appear on the same wavelength on) all output ports. As an example, in Fig. 1.11, a signal on wavelength λ_1 from Input Fiber 1 and another on wavelength λ_4 from Input Fiber 4 are broadcast to all output ports. A "collision" will occur when two or more signals from the input fibers are simultaneously launched into the star on the same wavelength. Assuming as many wavelengths as there are fiber ports, an $N \times N$ passive star can route N simultaneous connections through itself.

A *passive router* can separately route each of several wavelengths incident on an input fiber to the same wavelength on separate output fibers, e.g., wavelengths λ_1, λ_2, λ_3, and λ_4 incident on Input Fiber 1 are routed to the same corresponding wavelengths to Output Fibers 1, 2, 3, and 4, respectively, in Fig. 1.12. Observe that this device allows *wavelength reuse*, i.e., the same wavelength may be spatially reused to carry multiple connections through the router. The wavelength on which an input port gets routed to an output port depends on a "*routing matrix*" characterizing the router; this matrix is determined by the internal "connections" between the demux and mux

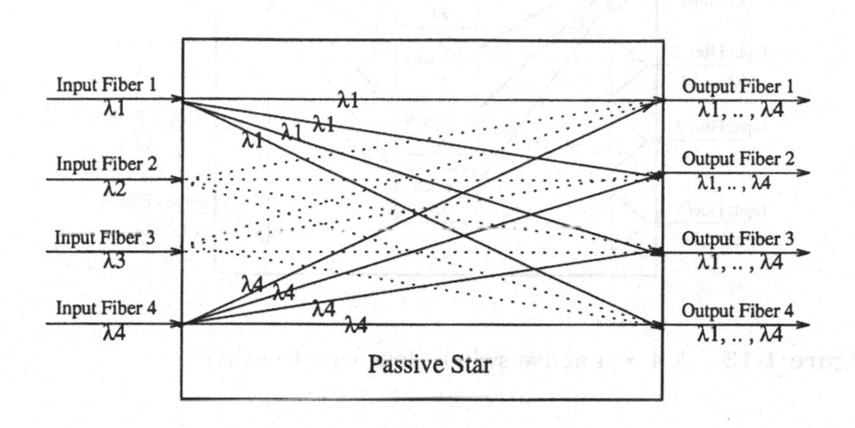

Figure 1.11 A 4 × 4 passive star.

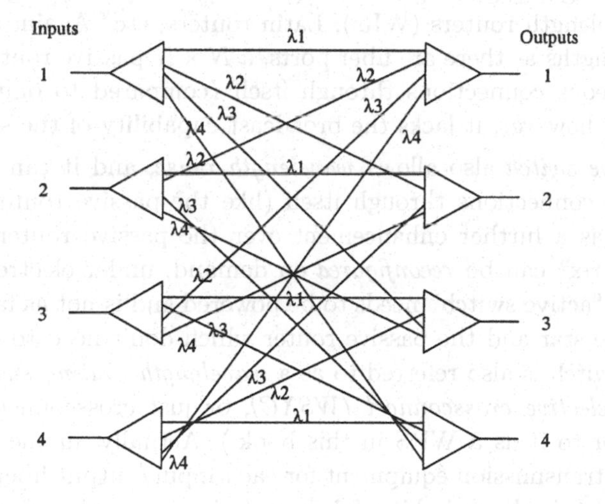

Figure 1.12 A 4 × 4 passive router (four wavelengths).

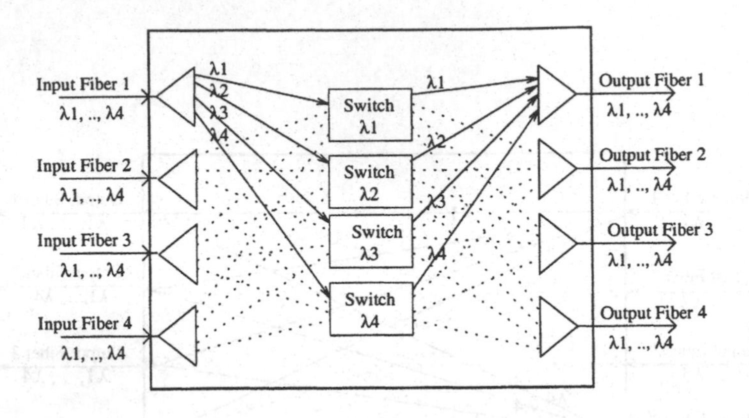

Figure 1.13 A 4×4 active switch (four wavelengths).

stages inside the router (see Fig. 1.12). The routing matrix is "fixed" and cannot be changed. Such routers are commercially available, and are also known as waveguide grating routers (WGRs), arrayed waveguide grating (AWG), wavelength routers (WRs), Latin routers, etc. Again, assuming as many wavelengths as there are fiber ports, a $N \times N$ passive router can route N^2 simultaneous connections through itself (compared to only N for the passive star); however, it lacks the broadcast capability of the star.

The *active switch* also allows *wavelength reuse*, and it can support N^2 simultaneous connections through itself (like the passive router). But the active star has a further enhancement over the passive router in that its "routing matrix" can be *reconfigured* on demand, under electronic control. However the "active switch" needs to be powered and is not as fault-tolerant as the passive star and the passive router which don't need to be powered. The active switch is also referred to as a *wavelength-routing switch (WRS)*, *wavelength-selective crossconnect (WSXC)*, or just crossconnect for short. (We will refer to it as a WRS in this book.) Actually, in the WRS, if we set aside the transmission equipment for each input/output fiber, then what we are left with is the switching fabric (and its supporting circuitry). The latter – i.e., the switching fabric (and its supporting circuitry) – is called the optical crossconnect (OXC). Depending on the context, we shall also use the term OXC in the book, wherever appropriate.

The active switch can be enhanced with an additional capability, viz., a wavelength may be converted to another wavelength just before it enters the mux stage before the output fiber (see Fig. 1.13). A switch equipped with such a wavelength-conversion facility is more capable than a WRS, and it is referred to as a *wavelength-convertible switch*, *wavelength-interchanging crossconnect (WIXC)*, etc.

The passive star is used to build local WDM networks, while the active switch is used for constructing wide-area wavelength-routed networks. The passive router has mainly found application as a mux/demux device, and it is also used in optical access networks employing WDM.

1.11.4 Development of WDM Networks

The first generation of WDM networks provides only the point-to-point physical links, which are either static or manually configured (see Fig. 1.9. The technical issues of the first-generation WDM include design and development of WDM lasers and optical amplifiers (OAMP) [Liu02].

The second generation of WDM is capable of establishing connection-oriented end-to-end lightpaths in the optical layer by introducing optical add/drop elements (WADM or OADM) and optical crossconnects (OXC). The ring and mesh topologies can be implemented using these OADMs and OXCs. The lightpaths are operated and managed based on a virtual topology over the physical fiber topology, and the virtual topology can be reconfigured dynamically in response to traffic changes. The technical issues of second-generation WDM include the development of OADM and OXC, wavelength conversion, routing and wavelength assignment (RWA), interoperability among WDM networks, network control and management (recall the role of software) and so on.

Both first-generation and second-generation WDM networks have been deployed in various carriers' operational networks [Liu02].

The third generation of WDM is expected to support a connectionless optical network. The key issues include the development of optical access network (such as passive optical network (PON)), and optical switching technologies, generically referred to as Optical "X" Switching (OXS), where X = P (for packet), B (for burst), L (for label), F (for flow), C (for cluster or circuit), etc. Some of these techniques, namely, OPS and OBS will be discussed in Chapters 17 and 18.

Figure 1.14 shows the WDM network evolution. Traffic granularity refers to both the volume of the traffic and the size of each traffic unit. Traffic in access networks is aggregated/multiplexed before it rides over backbone networks [Liu02].

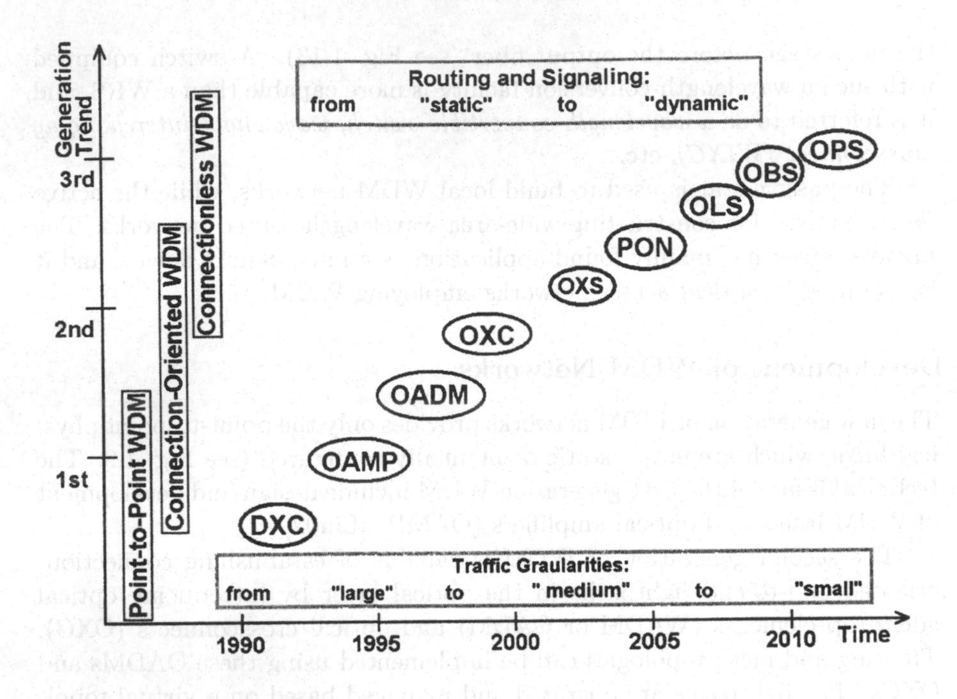

Figure 1.14 WDM network evolution. Note: DXC = Digital Cross-connect, a TDM product.

1.12 WDM Network Constructions

1.12.1 Broadcast-and-Select (Local) Optical WDM Network

A local WDM optical network may be constructed by connecting network nodes via two-way fibers to a passive star, as shown in Fig. 1.15. A node sends its transmission to the star on one available wavelength, using a laser which produces an optical information stream. The information streams from multiple sources are optically combined by the star and the signal power of each stream is equally split and forwarded to all of the nodes on their receive fibers. A node's receiver, using an optical filter, is tuned to only one of the wavelengths; hence, it can receive the information stream. Communication between sources and receivers may follow one of two methods: (1) *single-hop*, or (2) *multihop* (which will be studied in Chapters 3 and 4). Also, note that, when a source transmits on a particular wavelength

λ_1, more than one receiver can be tuned to wavelength λ_1, and all such receivers may pick up the information stream. Thus, the passive-star can support "*multicast*" services.

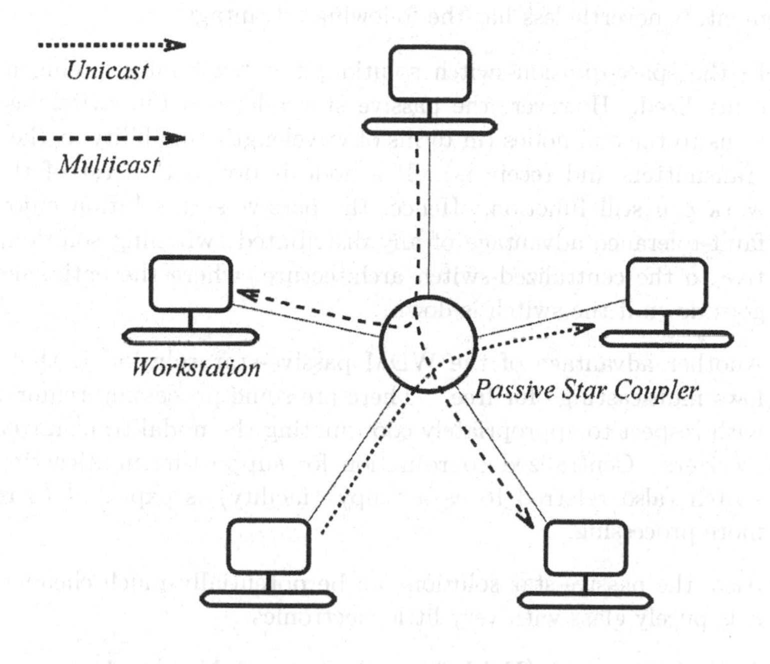

Figure 1.15 A passive-star-based local optical WDM network.

Passive-Star-Based Optical WDM LAN vs. Centralized Non-blocking-Switch-Based LAN

Consider the passive-star-based optical WDM LAN in Fig. 1.15. If there are N nodes in the system and as many wavelengths as nodes, and also if the bit rate of each WDM channel (and hence of each electronic interface) is B bps, then the aggregate information-carrying capacity of the LAN is upper-bounded by $N \times B$ bps.

Now consider the same network topology as in Fig. 1.15, but the passive star is replaced by a centralized, nonblocking, space-division switch, where the notion of WDM does not exist. Each of the N nodal interfaces still operate at B bps, all on the "same wavelength," so that the aggregate system capacity is still $N \times B$ bps, due to "space-division multiplexing" at the nonblocking switch. So, how does this architecture compare with the passive-star-based WDM LAN solution?

While the passive-star WDM solution cannot boast any capacity enhancement, it nevertheless has the following advantages:

1. In the space-division-switch solution, the "switching intelligence" is centralized. However, the passive star relegates the switching functions to the end nodes (in terms of wavelength tunability at the nodal transmitters and receivers). If a node is down, the rest of the network can still function. Hence, the passive-star solution enjoys the fault-tolerance advantage of any distributed switching solution, relative to the centralized-switch architecture, where the entire network goes down if the switch is down.

2. Another advantage of the WDM passive-star solution is that it allows multicasting "for free." There are some processing requirements with respect to appropriately coordinating the nodal transmitters and receivers. Centralized coordination for supporting multicasting in a switch (also referred to as a "copy" facility) is expected to require more processing.

3. Also, the passive-star solution can be potentially much cheaper since it is purely glass with very little electronics.

1.12.2 Wavelength-Routed (Wide-Area) Optical Network

A wavelength-routed (wide-area) optical WDM network is shown in Fig. 1.16. The network consists of a *optical switching fabric*, comprising "active switches" connected by fiber links to form an arbitrary *physical topology*. Each end-user is connected to an active switch via a fiber link. The combination of an end-user and its corresponding switch is referred to as a network node. Actually, each switching node may support multiple end-users (or multiple edge devices).

Each node (at its access station) is equipped with a set of transmitters and receivers, both of which may be wavelength tunable. A transmitter at a node sends data into the network and a receiver receives data from the network.

The basic mechanism of communication in a wavelength-routed network is a *lightpath*. A lightpath is an optical communication channel between two nodes in the network, and it may span more than one fiber link. The intermediate nodes in the fiber path route the lightpath in the optical domain

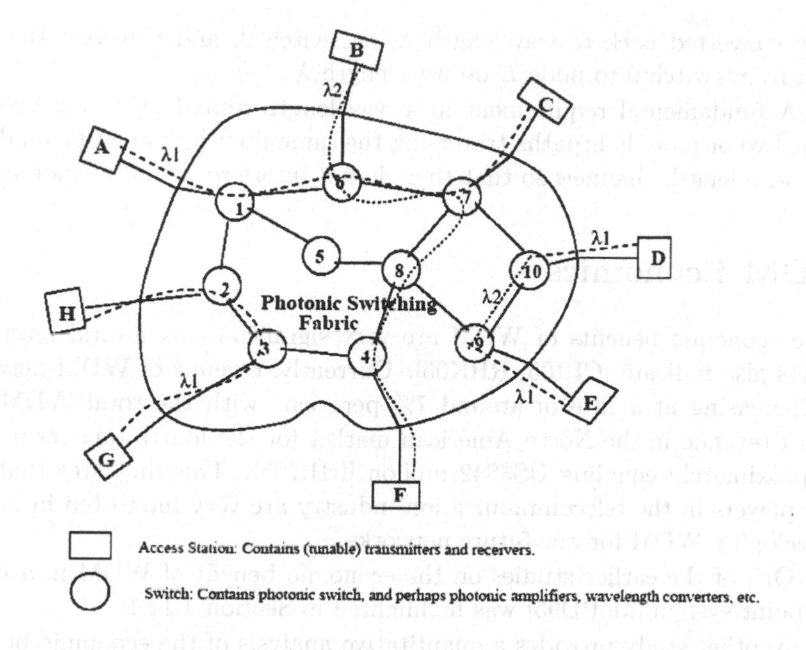

Figure 1.16 A wavelength-routed (wide-area) optical WDM network.

using their active switches (which could possibly be opaque OEO switches, if necessary). The end-nodes of the lightpath access the lightpath with transmitters and receivers that are tuned to the wavelength on which the lightpath operates. For example, in Fig. 1.16, lightpaths are established between nodes A and C on wavelength channel λ_1, between B and F on wavelength channel λ_2, and between H and G on wavelength channel λ_1. The lightpath between nodes A and C is routed via active switches 1, 6, and 7. (Note the wavelength reuse for λ_1.)

In the absence of any wavelength-conversion device, a lightpath is required to be on the same wavelength channel throughout its path in the network; this requirement is referred to as the *wavelength-continuity property* of the lightpath. This requirement may not be necessary if we also have wavelength converters in the network. For example, in Fig. 1.16, the lightpath between nodes D and E traverses the fiber link from node D to switch 10 on wavelength λ_1, gets converted to wavelength λ_2 at switch 10, traverses the fiber link between switch 10 and switch 9 on wavelength λ_2, gets converted back to wavelength λ_1 at switch 9, and traverses the fiber link from switch 9 to node E on wavelength λ_1.

A fundamental requirement in a wavelength-routed optical network is that two or more lightpaths traversing the same fiber link must be on different wavelength channels so that they do not interfere with one another.

1.13 WDM Economics

The economic benefits of WDM are very significant, as several recent reports also indicate [CIR05, RHK05]. Currently, revenue of WDM networks is increasing at a rate of around 7% per year, with the total WDM network revenue in the North American market for the fourth quarter in 2004 approximately equaling US$842 million [RHK05]. This indicates that major players in the telecommunication industry are very interested in further developing WDM for our future networks.

One of the earlier studies on the economic benefit of WDM in a point-to-point system [MePD95] was highlighted in Section 1.11.1.

Another study provides a quantitative analysis of the economic benefits of using WDM in a telephone company's Local Exchange Carrier (LEC) network under traffic demands projected over a multi-year period [Bala96b]. An aggressive traffic demand model was used which multiplied the normal traffic projection by 10, assuming that heavy deployment of high-speed networking technology will occur by the end of the projected duration. Synchronous Optical Network (SONET) terminal equipment was assumed to operate between OC-3 rate (155 Mbps) and OC-48 rate (2.5 Gbps).

The cost of introducing WDM in the terminal electronics – i.e., the cost of WDM transmitters and receivers – was assumed to be negligible. Costs for electronic equipment was averaged over several vendor-supplied data. The following costs were modeled: optical amplifier cost = 35% of a OC-48 terminal's cost, WADM cost = 40% of a OC-48 terminal's cost. Only a moderate cost decrease was modeled for WDM devices over the next few years. Two fiber cost models were employed: one considered cost for fiber material and cabling only; the other assumed structure exhaust and computed costs for new conduits and trenching, whenever required.

The above model, further details of which can be found in [Bala96b], was applied to three metropolitan-area networks belonging to various telephone companies. Using fiber cost model 1 (no conduit exhaust), *the cost savings of using WDM (vs. no WDM) ranged from 16% to 36%, with the actual "dollar* *values" of the savings ranging from \$86,000,000 to \$151,000,000.* Fiber cost model 2 (which models extra cost for conduit exhaust) was applied to only one of the three networks, and it yielded a *cost savings of 33% for using WDM (relative to no WDM), with the actual "dollar values" of the savings equalling \$224,000,000.*

It is therefore safe to say that WDM is here to stay! WDM standardization efforts, e.g., to set up a standard set of wavelengths to facilitate WDM equipment interoperability, are currently in progress under the watch of the International Telecommunications Union (ITU-T).

1.14 Sample Research Problems

A sampling of research topics which are of immediate interest to both academe and industry today are indicated below.

(a) Access: Ethernet Passive Optical Network (EPON) (Chapter 5). Ethernet in the First Mile (EFM) is an attractive idea for access networks because of Ethernet's ubiquity and low cost. By employing passive optical components, an inexpensive and high-capacity transport mechanism can be created. These are the basis for Ethernet Passive Optical Networks (EPONs) for next-generation access networks.

(b) Metro: Multi-wavelength Rings, ROADMs (Chapter 6). An optical (wavelength) add-drop multiplexer (OADM) is an important device for optical metro networks. Most multiwavelength-ring networks being deployed today employ fixed (or static) OADMs (FOADMs) because FOADM technology is quite mature today. However, reconfigurable OADMs (ROADMs) are more powerful because a network based on tunable ROADMs can easily adapt to fluctuating traffic demands. It is, therefore, essential that appropriate network architectures and operational algorithms be designed that can fully exploit the power of ROADMs (which are beginning to emerge).

(c) Backbone: Provisioning Connections of Different Bandwidth Granularities (Chapters 13 and 14). The bandwidth of a wavelength channel is quite high [10 Gbps (OC-192) today, and expected to grow to 40 Gbps (OC-768) soon]. However, only a fraction of customers are expected to need such high bandwidth although the aggregate bandwidth demand over all customers is expected to fill up the capacity of a wavelength channel. Many customers will be content with a lower bandwidth: STS-1 (51.84 Mbps), OC-3, OC-12, OC-48, etc. Thus, efficiently provisioning customer connections with diverse bandwidth needs is an important problem and is known as the traffic-grooming problem. More R&D efforts need to be devoted to this topic.

(d) Backbone: Fault Management (Chapters 11, 14, and 16). In a WDM mesh network, fault-management techniques are: (a) protection and (b) restoration. In protection, spare capacity is reserved during call setup. In restoration, the spare capacity that is available after the fault occurrence is utilized for rerouting the disrupted connections. By implementing appropriate fault-management techniques, service providers can broaden their service portfolio to support varying degrees of service guarantees, based on customer requirements.

(e) Global ISP: Dynamic Network Planning, Topology Engineering (Chapters 8 and 9). Bandwidth is becoming a commodity, so that bandwidth can be sold, purchased, or leased on a universal scale. Bandwidth markets allow ISPs to drastically reduce their capital and operational expenditure, by employing on-demand bandwidth provisioning. A control plane architecture that can support on-demand bandwidth provisioning is being standardized, and we expect to see signaling solutions for dynamic traffic engineering soon.

For longer-term research, we consider the following topics to be important.

(f) Network Architectures and Algorithms to Combat Optical Signal-Quality Impairments (Chapter 15). In a long-haul network, instead of regenerating an optical signal on a wavelength at every intermediate node, where should the signal be regenerated optimally? Mechanisms to monitor and model signal quality as well as network algorithms to compute appropriate routes and regenerator points are important research problems.

(g) Light-Trees: Optical Multicasting (Chapter 12). How can we map multicast applications and their needs to the underlying optical hardware, namely multicast-capable OXCs using appropriate network architectures and algorithms?

(h) Optical Packet Switching (OPS) and Optical Burst Switching (OBS) (Chapters 17 and 18). How can we build optical packet and burst switches (and the corresponding network architectures) that can perform fast and efficient packet/burst processing in the optical domain?

1.15 Road Map – Organization of the Book

This book is organized into three parts – Introduction; Local, Access, and Metro Networks; and Wavelength-Routed (Wide-Area) Optical Networks – and each part contains several chapters. Each part, as well as each chapter, has been organized as a "stand-alone" entity so that the interested and/or advanced reader can directly go to the part or chapter that is of interest.

Part I – Introduction – consists of this chapter and the next which deals with enabling technologies, viz., the building blocks for constructing WDM networks. Chapter 2 is essentially a summary of the physical aspects of WDM systems, and is written from the point of view of a computer scientist/engineer. For further details on WDM device technologies, the reader should consult a book that exclusively deals with these topics, e.g., [RaSi01, DuDF02, Agra04]. Readers familiar with device technologies and wishing to study WDM network architectures could skip Chapter 2.

Part II – Local, Access, and Metro Networks – contains four chapters. Chapters 3 examines single-hop network architectures and protocols. The corresponding treatment of multihop networks is provided in Chapter 4. Chapter 5 is devoted to the study of optical access networks with emphasis on the Ethernet passive optical network (EPON). Chapter 6 examines the problems associated with optical metro network.

Part III – Wavelength-Routed (Wide-Area) Optical Networks – consists of 12 chapters.

Chapter 7 deals with the solution to the Routing and Wavelength Assignment (RWA) problem outlined earlier in this chapter (Section 1.10.2). This section also introduced the concept of a *lightpath.* The set of lightpaths in a wavelength-routed network forms a "virtual topology," (1) which may be operated as a "virtual Internet," viz., a *packet-switched electronic overlay* on top of the *WDM optical layer*; (2) which may be optimized based on prevailing traffic conditions; and (3) which may be reconfigured on demand when the pattern of offered traffic changes. Such topics related to optimal virtual topology design are studied in Chapters 8 and 9. Chapter 10 examines the issues related to wavelength conversion, reviews several approaches, and reports on a simulation-based study on "sparse wavelength conversion" under which conversion capabilities (which are costly) are sparsely sprinkled through the network.

Chapter 11 provides a reviews of the large body of research on the design of survivable optical WDM networks and related issues. Chapter 12 investigates "light-trees" for supporting optical multicasting in such networks.

Chapters 13 and 14 are devoted to traffic grooming where low-speed connections are efficiently multiplexed on high-capacity wavelength channels; static and dynamic traffic grooming, grooming-node architectures, hierarchical traffic grooming, survivable traffic grooming, data-over-SONET/SDH, etc. are the topics covered by these chapters.

Chapter 15 attempts to address the cross-layer design topic at the "device-network interface"; specifically, this chapter examines the not-so-desirable properties of optics, namely optical channel impairments, and tries to correct for these "mismatches" using intelligent networking algorithms. Chapter 16 studies various network control, management, and signaling protocols that are needed to make optical WDM networks a reality. Chapters 17 and 18 deal with optical packet switching (OPS) and optical burst switching (OBS), respectively.

Problem sets are included at the end of each chapter. There are two appendices containing (1) information on further reading, and (2) a glossary of important terms. Finally, all citations are included in a common bibliography at the end of the book.

Exercises

1.1. What are the advantages of fiber optic technology in communication systems?

1.2. In order to take full advantage of the huge bandwidth available on fiber, various multiplexing techniques such as WDM, TDM, and CDM can be used which allow multiple users to share the bandwidth on a single fiber. Compare and contrast these multiplexing techniques. Why is WDM the most promising choice for optical communication networks?

1.3. Consider two regions, 1200-1400 nm and 1450-1650 nm, in a fiber low-loss spectrum. Calculate the actual bandwidth provided by each region. (Hint: Use the identity $f = v/\lambda$ where $v = 2.0 \times 10^8$ m/s. Note that velocity of light in vacuum is approx. 3.0×10^8 m/s, and the velocity of signals in fiber is approx. two-thirds of this value.)

1.4. What is the bandwidth of a 1-nm signal at 1500 nm? At 1350 nm? Give an approximate relation for the bandwidth of a $\Delta\lambda$ nm signal at λ nm.

1.5. Consider the three solutions for upgrading the transmission capacity of a link from OC-48 to OC-192. Suppose the cost of installing additional fiber is \$100 per meter, the cost of each transciever is \$1000, and the cost of a WDM multiplexer/demultiplexer is \$10,000. Determine the maximum length for which you would want to use the multi-fiber solution.

1.6. Give the advantages and disadvantages of the following wavelength crossconnects:
(a) passive star,
(b) passive router, and
(c) active switch.

1.7. Consider the passive star of Fig. 1.11, the passive router of Fig. 1.12, and the active switch of Fig. 1.13. Which of these devices can support the following simultaneous connections? (Assume that TDM is not used, but multicasting is allowed.)

(a) Wavelength λ_1 from input fiber 1 to output fiber 1,
Wavelength λ_1 from input fiber 1 to output fiber 2,
Wavelength λ_2 from input fiber 2 to output fiber 1.

(b) Wavelength λ_2 from input fiber 1 to output fiber 2,
Wavelength λ_2 from input fiber 2 to output fiber 1,
Wavelength λ_3 from input fiber 3 to output fiber 1.

(c) Wavelength λ_1 from input fiber 1 to output fiber 1,
Wavelength λ_2 from input fiber 2 to output fiber 1,
Wavelength λ_3 from input fiber 1 to output fiber 3.

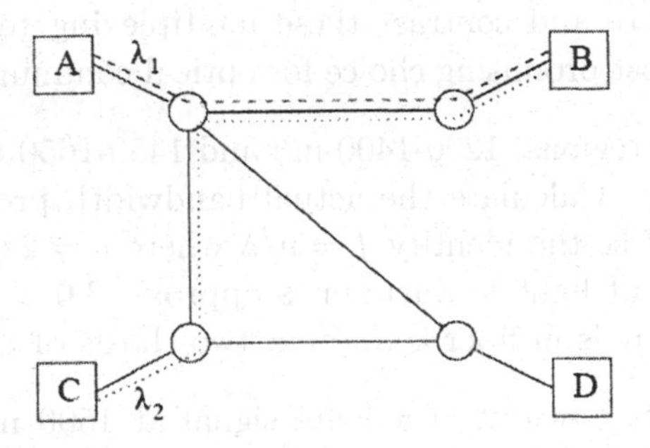

Figure 1.17 A wavelength-routed WDM network.

1.8. The routing matrix for an $N \times N$ passive router is called an $N \times N$ Latin Square.
(a) Identify which of the following 2 × 2 matrices below are Latin Squares.

$$\Lambda_1 = \begin{bmatrix} \lambda_1 & \lambda_2 & \lambda_3 & \lambda_4 \\ \lambda_2 & \lambda_3 & \lambda_4 & \lambda_1 \\ \lambda_4 & \lambda_1 & \lambda_3 & \lambda_2 \\ \lambda_3 & \lambda_4 & \lambda_2 & \lambda_1 \end{bmatrix}$$

$$\Lambda_2 = \begin{bmatrix} \lambda_1 & \lambda_2 & \lambda_3 & \lambda_4 \\ \lambda_2 & \lambda_3 & \lambda_4 & \lambda_1 \\ \lambda_3 & \lambda_4 & \lambda_1 & \lambda_2 \\ \lambda_4 & \lambda_1 & \lambda_2 & \lambda_3 \end{bmatrix}$$

(b) How many distinct 3×3 Latin Squares are there?

1.9. Consider the active switch of Fig. 1.13.
(a) What is the size of each switching element in the center?
(b) Is it possible to construct a 4×4 switch out of 2×2 switches?

1.10. Consider the network of Fig. 1.15. Suppose we replace the passive star by a TDM switch. Compare this new architecture with the previous architecture.

1.11. Consider the simple wavelength-routed optical WDM network shown in Fig. 1.17. Two connections have been established: A-B on wavelength λ_1, and C-B on wavelength λ_2. Establish the connections D-B and C-D while using the minimum number of wavelengths. How would your solution change if wavelength conversion is available at each node?

1.12. Suppose the optical switching fabric in Fig. 1.16 is replaced by a passive-star coupler. What is the minimum number of wavelengths required to maintain the connections shown in the figure?

Chapter 2

Enabling Technologies: Building Blocks

2.1 Introduction

This chapter is an introduction to WDM device issues. The reader needs no background in optics or advanced physics. For a more advanced and/or detailed discussion of WDM devices, we suggest that the reader refer to the references cited in this section. We highly recommend [KaBW96, Hech99, SaLu99, Keis00, RaSi01, DuDF02, Alwa04, Agra04, Pala04, Hech04].

This chapter presents an overview of optical fiber and devices such as couplers, optical receivers and filters, optical transmitters, optical amplifiers, and optical switches. The chapter attempts to condense the physics behind the principles of optical transmission in fiber in order to provide some background for the nonexpert. In addition, WDM network design issues are discussed in relation to the advantages and limits of optical devices. Finally, we demonstrate how these optical components can be used to create various WDM network architectures.

2.2 Optical Fiber

Fiber possesses many characteristics that make it an excellent physical medium for high-speed networking. Figure 2.1 shows the attenuation (and dispersion) characteristics of optical fiber.

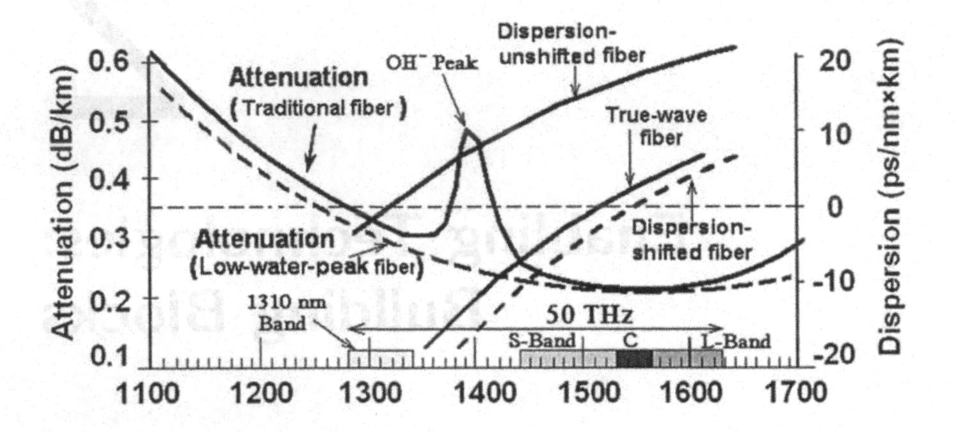

Figure 2.1 The low-attenuation regions of an optical fiber.

For traditional fiber, centered at approximately 1310 nm is a window of 200 nm in which attenuation is less than 0.5 dB/km. The total bandwidth in this region is about 25 THz. Centered at 1550 nm is a window of similar size, with attenuation as low as 0.2 dB/km, which consists of three bands, i.e., *S-band* (1460-1530 nm), *C-band* (1530-1560 nm), and *L-band* (1560-1630 nm). Combined, these two windows provide a theoretical upper bound of 50 THz of bandwidth[1]. The dominant loss mechanism in good fibers is Rayleigh scattering, while the peak in loss in the 1400 nm neighborhood is due to hydroxyl ion (OH^-) impurities in the fiber. Other sources of loss include material absorption and radiative loss.

Besides traditional fibers, *full-spectrum fiber* has also attracted a lot of attention in the industry, because of its permanently reduced water peak, as well as additional enhanced specifications in the L-band. Full-spectrum ap-

[1]However, usable bandwidth is limited by fiber nonlinearities (Section 2.2.5), spectrum of optical amplifiers (see Section 2.5), etc.

plications involve simultaneous (WDM) transmission in multiple operating windows (1270 to 1610 nm) over a single fiber. Full-spectrum fibers provide more useable wavelengths than standard single-mode fiber and therefore more bandwidth per fiber. Specifically, *low-water-peak fibers* have attenuation specifications in line with the attenuation values in other transmission windows. Fibers with low-water-peak attenuation may use the 1360 to 1480 nm range without the severe loss previously experienced in traditional standard single-mode fibers. Industry standards organizations have established new classes of standard single-mode fibers that require the average attenuation at 1383 nm after hydrogen aging to be less than or equal to the specified attenuation at 1310 nm. The most widely recognized examples of such a standard are ITU-T G.652.C and D.

By using these large low-attenuation windows for data transmission, the signal loss for a set of one or more wavelengths can be made very small, thus reducing the number of amplifiers and repeaters needed. Considering factors such as *low attenuation, wide window, and availability of optical amplifiers*, the 1550-nm window is preferred for long-haul wide-area applications. In single-channel long-distance experiments, optical signals typically have been sent over 80 km without amplification. Besides its enormous bandwidth and low attenuation, fiber also offers low error rates. Fiber optic systems typically operate at bit-error rates (BERs) of less than 10^{-15}.

The small size and thickness of fiber allows more fiber to occupy the same physical space as copper, a property which is desirable when installing local networks in buildings. Fiber is flexible, light, reliable in corrosive environments, and deployable at short notice (which makes it particularly favorable for military communication systems). Also, fiber transmission is immune to electromagnetic interference, and does not cause signal interference between fibers. Finally, fiber is made from one of the cheapest and most readily available substances on earth, namely glass. This makes fiber environmentally sound, unlike copper.

2.2.1 Optical Transmission in Fiber

Before discussing optical components, it is essential to understand the characteristics of the optical fiber itself. Fiber is essentially a thin filament of glass which acts as a waveguide. A waveguide is a physical medium or a path which allows the propagation of electromagnetic waves, such as light. Due to the physical phenomenon of *total internal reflection*, light can propagate the length of a fiber with little loss, which is illuminated as following.

Light travels through vacuum at a speed of $c = 3 \times 10^8$ m/s. Light can also travel through any transparent material, but the speed of light will be slower in the material than in a vacuum. Let c_{mat} be the speed of light for a given material. The ratio of the speed of light in vacuum to that in a material is known as the material's *refractive index* (n), and is given by: $n_{\text{mat}} = c/c_{\text{mat}}$. Given that $n_{mat} = 1.5$ approximately for glass, the velocity of signal propagation in a fiber approximately equals 2×10^8 m/s, which corresponds to a signal propagation delay of 5 μs/km.

When light travels from material of a given refractive index to material of a different refractive index (i.e., when refraction occurs), the angle at which the light is transmitted in the second material depends on the refractive indices of the two materials as well as the angle at which light strikes the interface between the two materials. Due to Snell's Law, $n_a sin\theta_a = n_b sin\theta_b$, where n_a and n_b are the refractive indices of the first substance and the second substance, respectively; θ_a is the angle of incidence, or the angle with respect to normal that light hits the surface between the two materials; and θ_b is the angle of light in the second material. However, if $n_a > n_b$ and θ_a is greater than some critical value, the rays are reflected back into substance a from its boundary with substance b.

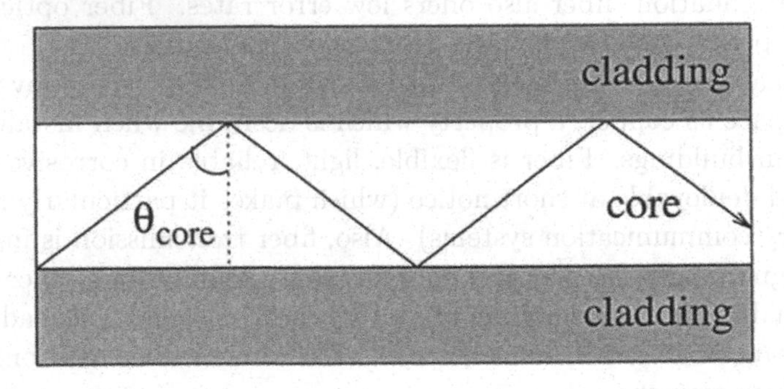

Figure 2.2 Light traveling via total internal reflection within a fiber.

Looking at Figs. 2.2 and 2.5, we see that the fiber consists of a core completely surrounded by a cladding (both the core and the cladding consist of glass of different refractive indices). Let us first consider a *step-index fiber*, in which the change of refractive index at the core-cladding boundary is a step function. If the refractive index of the cladding is less than that of the core, then *total internal reflection* can occur in the core, and light can propagate through the fiber (as shown in Fig. 2.2). The angle above which total internal reflection will take place is known as the *critical angle*, and is given by θ_{core} which corresponds to $\theta_{\text{clad}} = 90°$. From Snell's Law, we have:

$$\sin\theta_{\text{clad}} = \frac{n_{\text{core}}}{n_{\text{clad}}}\sin\theta_{\text{core}}$$

The critical angle is then:

$$\theta_{\text{crit}} = \sin^{-1}\left(\frac{n_{\text{clad}}}{n_{\text{core}}}\right). \tag{2.1}$$

So, for total internal reflection, we require:

$$\theta_{\text{crit}} > \sin^{-1}\left(\frac{n_{\text{clad}}}{n_{\text{core}}}\right)$$

In other words, for light to travel down a fiber, the light must be incident on the core-cladding surface at an angle greater than θ_{crit}.

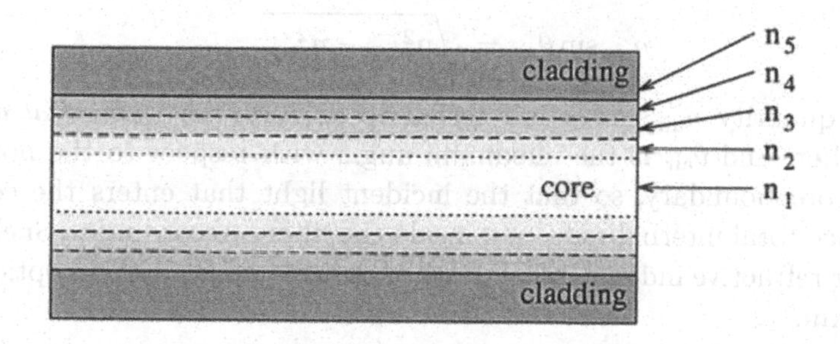

Figure 2.3 Graded-index fiber.

In some cases, the fiber may have a *graded index* in which the interface between the core and the cladding undergoes a gradual change in refractive index with $n_i > n_{i+1}$ (Fig. 2.3). A *graded-index* fiber reduces the minimum θ_{crit} required for total internal reflection, and also helps to reduce the *intermodal dispersion* in the fiber. Intermodal dispersion will be discussed in the following sections.

In order for light to enter a fiber, the incoming light should be at an angle such that the refraction at the air-core boundary results in the transmitted light being at an angle for which total internal reflection can take place at

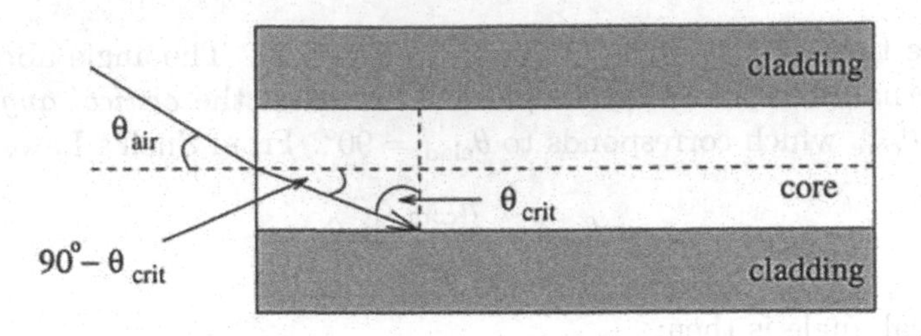

Figure 2.4 Numerical aperture of a fiber.

the core-cladding boundary. As shown in Fig. 2.4, the maximum value of θ_{air} can be derived from:

$$\begin{aligned} n_{\text{air}} \sin\theta_{\text{air}} &= n_{\text{core}} \sin(90^\circ - \theta_{\text{crit}}) \\ &= n_{\text{core}}\sqrt{1 - \sin^2\theta_{\text{crit}}} \end{aligned} \tag{2.2}$$

From Eqn. (2.1), since $\sin\theta_{\text{crit}} = n_{\text{clad}}/n_{\text{core}}$, we can rewrite Eqn. (2.2) as:

$$n_{\text{air}} \sin\theta_{\text{air}} = \sqrt{n_{\text{core}}^2 - n_{\text{clad}}^2} \tag{2.3}$$

The quantity $n_{\text{air}} \sin\theta_{\text{air}}$ is referred to as NA, the *numerical aperture* of the fiber, and θ_{air} is the maximum angle with respect to the normal at the air-core boundary, so that the incident light that enters the core will experience total internal reflection inside the fiber. According to Snell's Law and fiber refractive index, typical delay of optical propagation in optical fiber is 5 μs/km.

2.2.2 Single-Mode vs. Multimode Fiber

A mode in an optical fiber corresponds to one of possibly many ways in which a wave may propagate through the fiber. It can also be viewed as a standing wave in the transverse plane of the fiber. More formally, a mode corresponds to a solution of the wave equation which is derived from Maxwell's equations and is subject to boundary conditions imposed by the optical fiber waveguide.

An electromagnetic wave propagating along an optical fiber consists of an electric field vector, **E**, and a magnetic field vector, **H**. Each field can be broken down into three components. In the cylindrical coordinate system, these components are E_ρ, E_ϕ, E_z, H_ρ, H_ϕ, and H_z, where ρ is the component of the field which is normal to the wall (core-cladding boundary) of the fiber, ϕ is the component of the field which is tangential to the wall of the fiber, and z is the component of the field which is in the direction of propagation. Fiber modes are typically referred to using the notation HE_{xy} (if $H_z > E_z$), or EH_{xy} (if $E_z > H_z$), where x and y are both integers. For the case $x = 0$, the modes are also referred to as transverse-electric (TE) in which case $E_z = 0$, or transverse-magnetic (TM) in which case $H_z = 0$.

Although total internal reflection may occur for any angle θ_{core} which is greater than θ_{crit}, light will not necessarily propagate for all of these angles.

For some of these angles, light will not propagate due to destructive interference between the incident light and the reflected light at the core-cladding interface within the fiber. For other angles of incidence, the incident wave and the reflected wave at the core-cladding interface constructively interfere in order to maintain the propagation of the wave. The angles for which waves do propagate correspond to *modes* in a fiber. If more than one mode may propagate through a fiber, the fiber is called multimode. In general, a larger core diameter or higher operating frequencies allow a greater number of modes to propagate.

The number of modes supported by a multimode optical fiber is related to the normalized frequency V which is defined as:

$$V = k_0 a \sqrt{n_{\text{core}}^2 - n_{\text{clad}}^2} \tag{2.4}$$

where $k_0 = 2\pi/\lambda$, a is the radius of the core, and λ is the wavelength of the propagating light in vacuum. In multimode fiber, the number of modes, m, is given approximately by:

$$m \approx \frac{1}{2} V^2. \tag{2.5}$$

The advantage of multimode fiber is that its core diameter is relatively large; as a result, injection of light into the fiber with low coupling loss[2] can be accomplished by using inexpensive, large-area light sources, such as light-emitting diodes (LEDs).

The disadvantage of multimode fiber is that it introduces the phenomenon of *intermodal dispersion*. In multimode fiber, each mode propagates at a different velocity due to different angles of incidence at the core-cladding boundary. This effect causes different rays of light from the same source to

[2]Coupling loss measures the power loss experienced when attempting to direct light into a fiber.

arrive at the other end of the fiber at different times, resulting in a pulse which is spread out in the time domain. Intermodal dispersion increases with the propagation distance. The effect of intermodal dispersion may be reduced through the use of *graded-index* fiber, in which the region between the cladding and the core of the fiber consists of a series of gradual changes in the index of refraction (see Fig. 2.3). However, even with graded-index multimode fiber, intermodal dispersion may still limit the bit rate of the transmitted signal and the distance that the signal can travel.

One way to limit intermodal dispersion is to reduce the number of modes. From Eqns. (2.4) and (2.5), we observe that this reduction in the number of modes can be accomplished by reducing the core diameter, by reducing the numerical aperture, or by increasing the wavelength of the light.

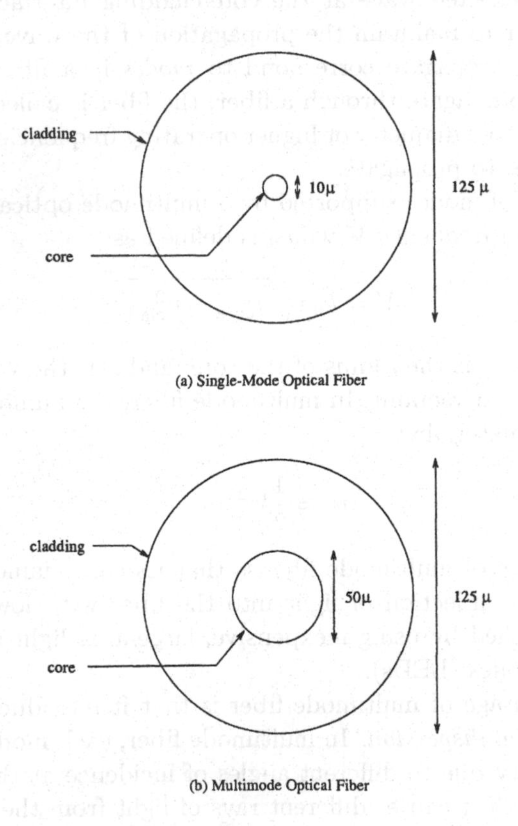

Figure 2.5 Single-mode and multimode optical fibers.

By reducing the fiber core to a sufficiently small diameter and by reducing the numerical aperture, it is possible to capture only a single mode in the fiber. This single mode is the HE_{11} mode, also known as the *fundamental mode.* Single-mode fiber usually has a core size of about 10 μm, while multimode fiber typically has a core size of 50 to 100 μm (see Fig. 2.5). A step-index fiber will support a single mode if V in Eqn. (2.4) is less than 2.4048 [Ishi91].

Thus, single-mode fiber eliminates intermodal dispersion, and can, hence, support transmission over much longer distances. However, it introduces the problem of concentrating enough power into a very small core. LEDs cannot couple enough light into a single-mode fiber to facilitate long distance communications. Such a high concentration of light energy may be provided by a semiconductor laser, which can generate a narrow beam of light.

2.2.3 Attenuation in Fiber

Attenuation in optical fiber leads to a reduction of the signal power as the signal propagates over some distance. When determining the maximum distance that a signal can propagate for a given transmitter power and receiver sensitivity, one must consider attenuation. Receiver sensitivity is the minimum power required by a receiver to detect the signal. Let $P(L)$ be the power of the optical pulse at distance L km from the transmitter and A be the attenuation constant of the fiber (in dB/km). Attenuation is characterized by [Henr85]:

$$P(L) = 10^{-AL/10}P(0) \tag{2.6}$$

where $P(0)$ is the optical power at the transmitter. For a link length of L km, $P(L)$ must be greater than or equal to P_r, the receiver sensitivity. From Eqn. (2.6), we get:

$$L_{\max} = \frac{10}{A}\log_{10}\frac{P(0)}{P_r} \tag{2.7}$$

The maximum distance between the transmitter and the receiver (or the distance between amplifiers[3]) depends more heavily on the constant A than on the optical power launched by the transmitter. Referring back to Fig. 2.1, we note that the lowest attenuation ($\approx$ 0.2 dB/km) occurs at approximately

[3]The amplifier sensitivity is usually equal to the receiver sensitivity, while the amplifier output is usually equal to the optical power at a transmitter.

1550 nm. In optical communication system, the traditional value of optical propagation is 80km without amplification[RaSi01]. With developments of new fibers and optical communication systems, the traditional propagation distance is surpassed in both physical system and research field. In March 2005, a vendor announced a new generation of optical amplifiers by increasing amplifier spacing from the traditional 80 km to 160 km.

2.2.4 Dispersion in Fiber

Dispersion is the widening of a pulse duration as it travels through a fiber. As a pulse widens, it can broaden enough to interfere with neighboring pulses (bits) on the fiber, leading to intersymbol interference. Dispersion thus limits the bit rate and the maximum transmission rate on a fiber-optic channel.

As mentioned earlier, one form of dispersion is *intermodal dispersion*. This is caused when multiple modes of the same signal propagate at different velocities along the fiber. Intermodal dispersion does not occur in a single-mode fiber.

Another form of dispersion is *chromatic dispersion* (see Fig. 2.1). Chromatic dispersion represents the fact that different colors or wavelengths travel at different speeds, even within the same mode. In a dispersive medium, the index of refraction is a function of the wavelength. Thus, if the transmitted signal consists of more than one wavelength, certain wavelengths will propagate faster than other wavelengths. Chromatic dispersion is the result of *material dispersion*, *waveguide dispersion*, and/or *profile dispersion*. *Material dispersion* results from the different velocities of each wavelength in a material. Since no laser can create a signal consisting of an exact single wavelength, or more precisely, since any information carrying signal will have a nonzero spectral width (range of wavelengths/frequencies in the signal), material dispersion will occur in most systems[4]. *Waveguide dispersion* is caused because the propagation of different wavelengths depends on waveguide characteristics, such as the indices and shape of the fiber core and cladding. *Profile dispersion* is caused by the variation of refractive index with respect to wavelength. These first two forms of dispersion are universal in optical fibers. Although the single-mode fiber (SMF) can perfectly eliminate several types of dispersion (which the multimode fiber cannot), *chromatic dispersion* and *polarization mode dispersion* (PMD) still need to be dealt with.

[4]Even if an unmodulated source consisted of a single wavelength, the process of modulation would cause a spread of wavelengths.

Polarization mode dispersion (PMD) is another complex optical effect that can occur in single-mode optical fibers. Single-mode fibers support two perpendicular polarizations of the original transmitted signal. If a fiber were perfectly round and free from all stresses, both polarization modes would propagate at exactly the same speed, resulting in zero PMD. However, practical fibers are not perfect. The two perpendicular polarizations may travel at different speeds and, consequently, arrive at the end of the fiber at different times (one perpendicular polarization direction is the fast axis, and the other one is the slow axis). The difference in arrival times between the axes is known as PMD. Like chromatic dispersion, PMD causes digitally-transmitted pulses to spread out as the polarization modes arrive at their destination at different times. For digital high-bit-rate transmissions, this effect of PMD can lead to bit errors at the receiver or limit the receiver sensitivity. The maximum acceptable dispersion penalty is usually 2 dB, though it is possible for a system to tolerate a larger dispersion penalty if the optical attenuation is low.

The above discussion mainly referred to the most prevalent fiber type, namely, dispersion-unshifted SMF. Currently, there are a number of special designs of optical fibers available, which offer lower dispersion than the dispersion-unshifted SMF. For example, in a dispersion-shifted fiber (e.g., non-zero dispersion-shifted fiber (NZDSF), dispersion-compensated fiber (DCF), etc.), the core and cladding are designed such that the waveguide dispersion is negative with respect to the material dispersion, thus the total *chromatic dispersion* is set to zero (see Fig. 2.1).

2.2.5 Nonlinearities in Fiber

Nonlinear effects in fiber may potentially have a significant impact on the performance of WDM optical communication systems. Nonlinearities in fiber may lead to attenuation, distortion, and cross-channel interference. In a WDM system, these effects place constraints on the spacing between adjacent wavelength channels, and they limit the maximum power per channel, the maximum bit rate, and the system reach.

Nonlinear Refraction

In an optical fiber, the index of refraction depends on the optical intensity of signals propagating through the fiber [Chra90]. Thus, the phase of the light at the receiver will depend on the phase of the light sent by the transmitter, the length of the fiber, and the optical intensity. Two types of nonlinear effects caused by this phenomenon are self-phase modulation (SPM) and cross-phase modulation (XPM).

SPM is caused by variations in the power of an optical signal and results in variations in the phase of the signal. The amount of phase shift introduced by SPM is given in the equation:

$$\phi_{NL} = n_2 k_0 L |E|^2 \tag{2.8}$$

where n_2 is the nonlinear coefficient for the index of refraction, $k_0 = 2\pi/\lambda$, L is the length of the fiber, and $|E|^2$ is the optical intensity. In phase-shift-keying (PSK) systems, SPM may lead to a degradation of the system performance, since the receiver relies on the phase information. SPM also leads to the spectral broadening of pulses, as explained below. Instantaneous variations in a signal's phase caused by changes in the signal's intensity will result in instantaneous variations of frequency around the signal's central frequency. For very short pulses, the additional frequency components generated by SPM combined with the effects of material dispersion will also lead to the spreading or compression of the pulse in the time domain and affect the maximum bit rate and the bit error rate.

Cross-phase modulation (XPM) is a shift in the phase of a signal caused by the change in intensity of a signal propagating at a different wavelength. XPM can lead to asymmetric spectral broadening, and combined with dispersion, may also affect the pulse shape in the time domain.

Although XPM may limit the performance of fiber-optic systems, it may have advantageous applications as well. XPM can be used to modulate a pump signal at one wavelength from a modulated signal on a different wavelength. Such techniques can be used in wavelength conversion devices and are discussed in Section 2.7.

Stimulated Raman Scattering

Stimulated Raman Scattering (SRS) is caused by the interaction of light with molecular vibrations. Light incident on the molecules creates scattered light at a longer wavelength than that of the incident light. A portion of the light traveling at each frequency in a Raman-active fiber is downshifted across a region of lower frequencies. The light generated at the lower frequencies is called the Stokes wave. The range of frequencies occupied by the Stokes wave is determined by the *Raman gain spectrum*[5] which covers a range of around 40 THz below the frequency of the input light. In silica fiber, the Stokes wave has a maximum gain at a frequency of around 13.2 THz less than the input signal.

The fraction of power transferred to the Stokes wave grows rapidly as the power of the input signal is increased. Under very high input power, SRS will cause almost all of the power in the input signal to be transferred to the Stokes wave.

In multiwavelength systems, the shorter-wavelength channels will lose some power to each of the higher-wavelength channels within the Raman gain spectrum. To reduce the amount of loss, the power on each channel needs to be below a certain level. In [Chra84], it is shown that, in a 10-channel system with 10-nm channel spacing, the power on each channel should be kept below 3 mW to minimize the effects of SRS. In Section 2.5.4, SRS will be utilized as a basis of Raman amplifier.

Stimulated Brillouin Scattering

Stimulated Brillouin scattering (SBS) is similar to SRS, except that the frequency shift is caused by sound waves rather than molecular vibrations [Chra90]. Other characteristics of SBS are that the Stokes wave propagates in the opposite direction of the input light, and SBS occurs at relatively low input powers for wide pulses (greater than 1 μs), but has negligible effect for short pulses (less than 10 ns) [Agra01]. The intensity of the scattered light is much greater in SBS than in SRS, but the frequency range of SBS, on the order of 10 GHz, is much lower than that of SRS. Also, the gain bandwidth of SBS is only on the order of 100 MHz.

To counter the effects of SBS, one must ensure that the input power is below a certain threshold. Also, in multiwavelength systems, SBS may induce crosstalk between channels. Crosstalk will occur when two counter-propagating channels differ in frequency by the Brillouin shift, which is around 11 GHz for wavelengths at 1550 nm. However, the narrow gain bandwidth of SBS makes SBS crosstalk fairly easy to avoid.

[5]The Raman gain spectrum typically describes the measured Raman-gain coefficient for silica fibers as a function of the frequency shift at a pump wavelength of 1.0 μm.

Four-Wave Mixing

As the bit rate of optical data streams in fibers increases, four-wave mixing (FWM) is one principal among nonlinear effects in pulse propagation.

FWM causes inter-channel crosstalk and is worst-case for equally-spaced WDM channels. FWM penalty can be mitigated by using fiber with high local dispersion (SMF, NZDSF) or unequally spaced channels [FTCM94]. Even if using NZDSF can mitigate the FWM penalty, the minimum channel spacing has a lower limit.

However, there are some significant motivations that are taking advantages of FWM in WDM networks. For example, FWM can be used to provide wavelength conversion. More information will be shown in Section 2.7.

Summary

Nonlinear effects in optical fibers may potentially limit the performance of WDM optical networks. Such nonlinearities may limit the optical power on each channel, the maximum number of channels, and the maximum transmission rate, and constrain the spacing between different channels.

It is shown that in a WDM system using channels spaced 10 GHz apart and a transmitter power of 0.1 mW per channel, a maximum of about 100 channels can be obtained in the 1550-nm low-attenuation region [Chra90].

However, there are some applications, in which nonlinear effects are exploited, such as Raman amplifier, and wavelength conversion by using FWM.

There are some efforts to combat nonlinearities by using dispersion. An example is an optical soliton[6] system which is suitable for a compensation between dispersion and nonlinear effects. The main limitation of soliton systems is fiber loss. With decreasing power, compensation between dispersion and nonlinearities is no longer achieved [CCEL04].

The details of optical nonlinearities are very complex, and beyond the scope of this book. However, they are a major limiting factor in the available number channels in a WDM system, especially those operating over distances greater than 30 km [Chra90]. The existence of these nonlinearities suggests that WDM protocols which limit the number of nodes to the number of channels do not scale well. For further details on fiber nonlinearities, the reader is referred to [Agra01].

[6] An optical soliton is a pulse with specific shape, power, and duration.

2.2.6 Optical Fiber Couplers

Coupler is a general term that covers all devices that combine light into or split light out of a fiber. Optic fiber couplers can be either active or passive devices. The difference between active and passive couplers is that a passive coupler redistributes the optical signal without optical-to-electrical conversion. Active couplers, however, are electronic devices that split or combine the signal electrically and use fiber optic detectors and sources for input and output. The passive coupler is the most popular because it requires no external power to operate. Thus, passive couplers are referred in next sections. A splitter is a coupler that divides the optical signal on one fiber to two or more fibers. The most common splitter is a 1×2 splitter, as shown in Fig. 2.6(a).

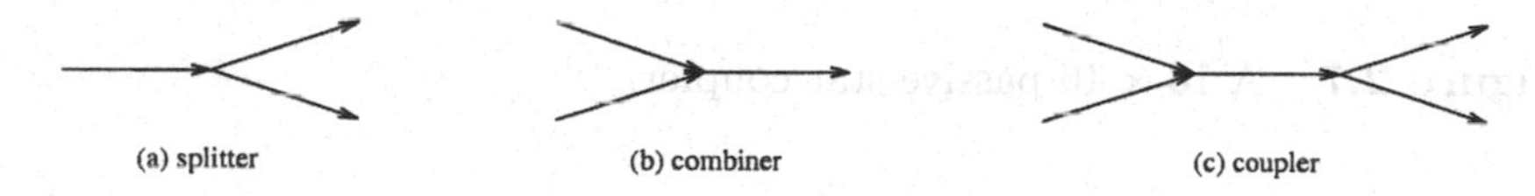

Figure 2.6 Splitter, combiner, and coupler.

The *splitting ratio*, α, is the amount of power that goes to each output. For a two-port splitter, the most common splitting ratio is 50:50, though splitters with any ratio can be manufactured [Powe93]. Combiners (see Fig. 2.6(b)) are the reverse of splitters, and when turned around, a combiner can be used as a splitter. An input signal to the combiner suffers a power loss of about 3 dB. A 2×2 coupler (see Fig. 2.6(c)), in general, is a 2×1 combiner followed immediately by a 1×2 splitter, which has the effect of broadcasting the signals from two input fibers onto two output fibers. One implementation of a 2×2 coupler is the *fused biconical tapered coupler* which basically consists of two fibers fused together.

In addition to the 50:50 power split incurred in a coupler, a signal also experiences *return loss.* If the signal enters an input of the coupler, roughly half of the signal's power goes to each output of the coupler. However, a small amount of power is reflected in the opposite direction and is directed back to the inputs of the coupler, in which the amount of power returned by a coupler is typically 40-50 dB below the input power.

Another type of loss is *insertion loss.* One source of insertion loss is the loss incurred when directing the light from a fiber into the coupler device; ideally, the axes of the fiber core and the coupler input port must be perfectly aligned, but full perfection may not be achievable due to the very small dimensions.

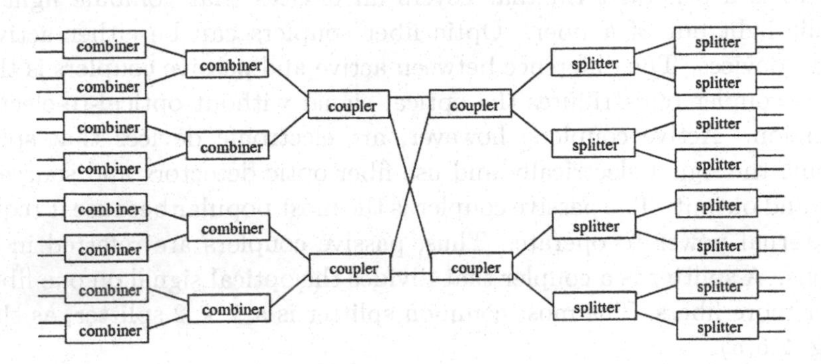

Figure 2.7 A 16×16 passive-star coupler..

The passive-star coupler (PSC) is a multiport device in which light coming into any input port is broadcast to every output port. The PSC is attractive because the optical power that each output receives P_{out} equals:

$$P_{\text{out}} = \frac{P_{\text{in}}}{N} \tag{2.9}$$

where P_{in} is the optical power introduced into the star by a single node and N is the number of output ports of the star. Note that this expression ignores the *excess loss*, caused by flaws introduced in the manufacturing process, that the signal experiences when passing through each coupling element.

One way to implement the PSC is to use a combination of splitters, combiners, and couplers as shown in Fig. 2.7. (However, this implementation of a PSC and the implementation of the 2×2 coupler in Fig. 2.6(c) are not the most power efficient. As an exercise, an alternative, power-efficient design of the PSC can be explored.) Another implementation of the star coupler is the integrated-optics planar star coupler in which the star coupler and waveguides are fabricated on a semiconductor, glass (silica), or polymer substrate. A 19×19 star coupler on silicon has been demonstrated with excess loss of around 3.5 dB at a wavelength of 1300 nm [DHKK89], and an 8×8 star coupler with an excess loss of 1.6 dB at a wavelength of 1550 nm has been demonstrated in [OkTa91].

2.3 Optical Transmitters

In order to understand how a tunable optical transmitter works, we must first understand some of the fundamental principles of lasers and how they work. Then, we will discuss various implementations of tunable lasers and their properties. Good references on tunable laser technology include [Gree93, Brac90, LeZa89, RaSi01].

2.3.1 How a Laser Works

The word *laser* is an acronym for *Light Amplification by Stimulated Emission of Radiation.* The key words are stimulated emission, which is what allows a laser to produce intense high-powered beams of coherent light (light which contains one or more distinct frequencies).

In order to understand stimulated emission, we must first acquaint ourselves with the energy levels of atoms. Atoms that are stable (in the ground state) have electrons that are in the lowest possible energy levels. In each atom, there are a number of discrete levels of energy that an electron can have; thus, we refer to them as states. In order to change the level of an electron in the ground state, the atom must absorb energy. When an atom absorbs energy, it becomes excited, and moves to a higher energy level. At this point, the electron is unstable, and usually moves quickly back to the ground state by releasing a *photon*, a particle of light.

However, there are certain materials whose states are *quasi-stable*, which means that the substances are likely to stay in the excited state for longer periods of time. By applying enough energy (either in the form of an optical pump or in the form of an electrical current) to a substance with quasi-stable states for a long enough period of time, *population inversion* occurs, which means that there are more electrons in the excited state than in the ground state. As we shall see, this inversion allows the substance to emit more light than it absorbs.

Figure 2.8 shows a general representation of the structure of a laser. The laser consists of two mirrors which form a cavity (the space between the mirrors), a lasing medium which occupies the cavity, and an excitation device. The excitation device applies current to the lasing medium, which is made of a quasi-stable substance. The applied current excites electrons in the lasing medium, and when an electron in the lasing medium drops back to the ground state, it emits a photon of light. The photon will reflect off the mirrors at each end of the cavity, and will pass through the medium again.

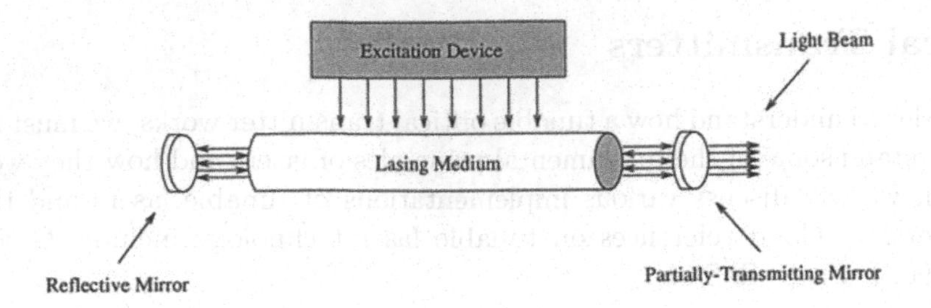

Figure 2.8 The general structure of a laser.

Stimulated emission occurs when a photon passes very closely to an excited electron. The photon may cause the electron to release its energy and return to the ground state. In the process of doing so, the electron releases another photon which will have the same direction and coherence as the stimulating photon. Photons for which the frequency is an integral fraction of the cavity length will coherently combine to build up light at the given frequency within the cavity. Between "normal" and stimulated emission, the light at the selected frequency builds in intensity until energy is being removed from the medium as fast as it is being inserted. The mirrors feed the photons back and forth, so further stimulated emission can occur and higher intensities of light can be produced. One of the mirrors is partially transmitting, so that some photons will escape the cavity in the form of a narrowly focused beam of light. By changing the length of the cavity, the frequency of the emitted light can be adjusted.

The frequency of the photon emitted depends on its change in energy levels. The frequency is determined by the equation:

$$f = \frac{E_i - E_f}{h} \tag{2.10}$$

where f is the frequency of the photon, E_i is the initial (quasi-stable) state of the electron, E_f is the final (ground) state of the electron, and h is Planck's constant. In a gas laser, the distribution for $E_i - E_f$ is given by an exponential probability distribution, known as the *Boltzmann distribution*, which changes depending on the temperature of the gas. Although many frequencies are possible, only a single frequency, which is determined by the cavity length, is emitted from the laser.

Semiconductor Diode Lasers

The most useful type of laser for optical communications is the semiconductor diode laser. The simplest implementation of a semiconductor laser is the bulk laser diode, which is a p-n junction with mirrored edges perpendicular to the junction (see Fig. 2.9). To understand the operation of the semiconductor diode requires a brief diversion into semiconductor physics.

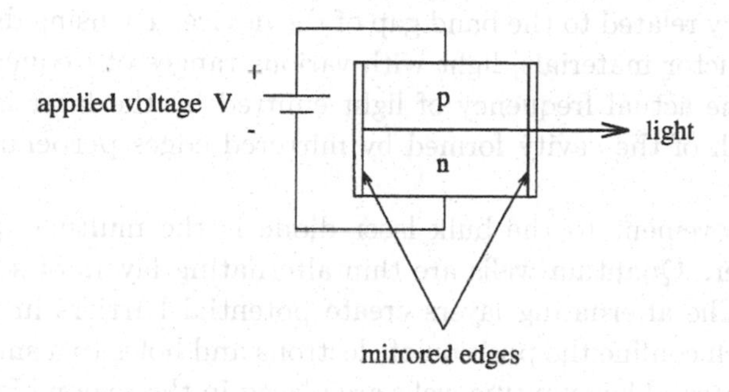

Figure 2.9 Structure of a semiconductor diode laser.

In semiconductor materials, electrons may occupy either the valence band or the conduction band. The valence band and conduction band are analogous to the ground state and excited state of an electron mentioned in the previous section. The valence band corresponds to an energy level at which an electron is not free from an atom. The conduction band corresponds to a level of energy at which an electron has become a free electron and may move freely to create current flow. The region of energy between the valence band and the conduction band is known as the *band gap*. An electron may not occupy any energy levels in the band-gap region. When an electron moves from the valence band to the conduction band, it leaves a vacancy, or *hole*, in the valence band. When the electron moves from the conduction band to the valence band, it recombines with the hole and may produce the spontaneous emission of a photon. The frequency of the photon is given by Eqn. (2.10), where $E_i - E_f$ is the band-gap energy. The distribution of the energy levels which electrons may occupy is given by the *Fermi-Dirac distribution* in [Fuku98].

A semiconductor may be *doped* with impurities to increase either the number of electrons or the number of holes. An n-type semiconductor is doped with impurities which provide extra electrons. These electrons will remain in the conduction band. A p-type semiconductor is doped with impurities which increase the number of holes in the valence band. A p-n junction is formed by layering p-type semiconductor material over n-type semiconductor material.

In order to produce stimulated emission, a voltage is applied across the p-n junction to forward bias the device and cause electrons in the "n" region to combine with holes in the "p" region, resulting in light energy being released at a frequency related to the band gap of the device. By using different types of semiconductor materials, light with various ranges of frequencies may be released. The actual frequency of light emitted by the laser is determined by the length of the cavity formed by mirrored edges perpendicular to the p-n junction.

An improvement to the bulk laser diode is the multiple-quantum-well (MQW) laser. Quantum wells are thin alternating layers of semiconductor materials. The alternating layers create potential barriers in the semiconductors which confine the position of electrons and holes to a smaller number of energy states. The quantum wells are placed in the region of the p-n junction. By confining the possible states of the electrons and holes, it is possible to achieve higher-resolution, low-linewidth lasers (lasers which generate light with a very narrow frequency range).

2.3.2 Tunable and Fixed Lasers

The previous section provided an overview of a generic model of a laser, but the transmitters used in WDM networks often require the capability to tune to different wavelengths. This section briefly describes some of the more-popular, tunable and fixed, single-frequency laser designs.

Laser Characteristics

Some of the physical characteristics of lasers which may affect system performance are *laser linewidth*, *frequency stability*, and the *number of longitudinal modes*.

The *laser linewidth* is the spectral width of the light generated by the laser. The linewidth affects the spacing of channels and also affects the amount of dispersion that occurs when the light is propagating along a fiber. As was mentioned in Section 2.2.4, the spreading of a pulse due to dispersion will limit the maximum bit rate.

Frequency instabilities in lasers are variations in the laser frequency, three examples of which are *mode hopping*, *mode shifts*, and *wavelength chirp* [MoTo93]. *Mode hopping* occurs primarily in injection-current lasers and is a sudden jump in the laser frequency caused by a change in the injection current above a given threshold. *Mode shifts* are changes in frequency due to temperature changes. *Wavelength chirp* is a variation in the frequency due to variations in injection current. In WDM systems, frequency instabilities may limit the placement and spacing of channels. In order to avoid large shifts in frequency, methods must be utilized to compensate for variations in temperature or injection current. One approach for temperature compensation is to package with the laser a thermoelectric cooler element which produces cooling as a function of applied current. The current for the thermoelectric cooler may be provided through a thermistor, which is a temperature-dependent resistor.

The *number of longitudinal modes* in a laser is the number of wavelengths that it can amplify. In lasers consisting of a simple cavity, wavelengths for which an integer multiple of the wavelength is equal to twice the cavity length will be amplified (i.e., wavelengths λ for which $n\lambda = 2L$, where L is the length of the cavity, and n is an integer). The unwanted longitudinal modes produced by a laser may result in significant dispersion; therefore, it is desirable to implement lasers which produce only a single longitudinal mode.

Some primary characteristics of interest for tunable lasers are the *tuning range*, the *tuning time*, and whether the laser is *continuously tunable* (over its tuning range) or *discretely tunable* (only to selected wavelengths). The *tuning range* refers to the range of wavelengths over which the laser may be operated. The *tuning time* specifies the time required for the laser to tune from one wavelength to another.

Mechanically-Tuned Lasers

Most mechanically-tuned lasers use a Fabry-Perot cavity that is an optical resonator in which feedback is accomplished by two parallel mirrors and adjacent to the lasing medium (i.e., an *external cavity*) to filter out unwanted wavelengths. Tuning is accomplished by physically adjusting the distance between two mirrors on either end of the cavity such that only the desired wavelength constructively interferes with its multiple reflections in the cavity. This approach to tuning results in a tuning range that encompasses the entire useful gain spectrum of the semiconductor laser [Brac90], but tuning time is limited to the order of milliseconds due to the mechanical nature of the tuning and the length of the cavity. The length of the cavity may also limit transmission rates unless an external modulator is used. External cavity lasers tend to have very good frequency stability. Fabry-Perot lasers further break down into Buried Hetero (BH) and Multi-Quantum Well (MQW) types. BH and related styles have ruled for many years, but currently MQW types are becoming very widespread. MQW lasers offer significant advantages over all former types of Fabry-Perot lasers. MQW lasers offer lower threshold current, higher slope efficiency, lower noise, better linearity, and much greater stability over temperature. As a bonus, MQW lasers have great performance margins, give laser manufacturers better yields, and reduce laser cost. One disadvantage of MQW lasers, however, is their tendency to be more susceptible to back reflections.

Acoustooptically- and Electrooptically-Tuned lasers

Other types of tunable lasers that use external tunable filters include acoustooptically and electrooptically tuned lasers. In an acoustooptic- or electrooptic-laser, the index of refraction in the external cavity is changed by using either sound waves or electrical current, respectively. The change in the index results in the transmission of light at different frequencies. In these types of tunable lasers, the tuning time is limited by the time required for light to build up in the cavity at the new frequency.

An acoustooptic laser combines a moderate tuning range with a moderate tuning time. While not quite fast enough for packet switching with multigigabit-per-second channels, the approximately 10-μs tuning time is a vast improvement over that of mechanically-tuned lasers (which have millisecond tuning times). Electrooptically-tuned lasers are expected to tune on the order of some tens of nanoseconds. Neither of these approaches allow continuous tuning over a range of wavelengths.

Injection-Current-Tuned Lasers

Injection-current-tuned lasers form a family of transmitters which allow wavelength selection via a diffraction grating. The Distributed Feedback (DFB) laser uses a diffraction grating placed in the lasing medium. In general, the grating consists of a waveguide in which the index of refraction alternates periodically between two values. Only wavelengths which match the period and indices of the grating will be constructively reinforced. All other wavelengths will destructively interfere, and will not propagate through the waveguide. The condition for propagation is given by:

$$D = \frac{\lambda}{2n}$$

where D is the period of the grating [MoTo93]. The laser is tuned by injecting a current which changes the index of the grating region.

If the grating is moved to the outside of the lasing medium, the laser is called a Distributed Bragg Reflector (DBR) laser. The DBR-based laser offers operating advantages such as a broad range of biasing conditions and the tunability of the emitted wavelength. There are various kinds of Modified-DBR lasers, such as Sampled-Grating DBR (SG-DBR), Digital-Supermode (DS-DBR) and Super-Structure Grating DBR (DBR) [GoNT04, JHBC05, WRBB05]. A SG-DBR laser with tuning range exceeding 40 nm is demonstrated in [JHBC05]. A tunable DS-DBR laser has been presented in [WRBB05], in which the full continuous tuning range of the laser can reach 45 nm. However, DBR devices suffer from the major drawback of mode hopping that occurs due to the changes in the lasing-region index of refraction with increasing driven current.

Laser Arrays

An alternative to tunable lasers is the laser array, which contains a set of fixed-tuned lasers and whose advantage/application is explained below. A laser array consists of a number of lasers which are integrated into a single component, with each laser operating at a different wavelength. The advantage of using a laser array is that, if each of the wavelengths in the array is modulated independently, then multiple transmissions may take place simultaneously.

One current laser array technique is MEMS-based vertical-cavity surface-emitting laser (VCSEL). VCSEL is a new laser structure that emits laser light vertically from its surface and has vertical laser cavity. The VCSEL's principles of operation closely resembles those of conventional edge-emitting semiconductor lasers. The core of the VCSEL is an electrically-pumped gain region, called the active region, which emits light. Layers of varying semiconductor materials above and below the gain region create mirrors. Each mirror reflects a narrow range of wavelengths back into the cavity causing light emission at a single wavelength. VCSELs are typically multi-quantum-well (MQW) devices with lasing occurring in layers only 20-30 atoms thick. In VCSELs, the Bragg reflectors with as many as 120 mirror layers form the laser reflectors.

There are many advantages of VCSELs. The small-size and high-efficiency mirrors produce a low threshold current, which is less than 1 mA. The transfer function allows stability over a wide temperature range, which is unique to this type of laser diode. The MEMS-based VCSEL realizes a fast tuning time of 1 to 10 μs. These features make the VCSEL ideal for applications that require an array of devices.

2.3.3 Optical Modulation

In order to transmit data across an optical fiber, the information must first be encoded, or modulated, onto the laser signal. Analog techniques include amplitude modulation (AM), frequency modulation (FM), and phase modulation (PM). Digital techniques include amplitude-shift keying (ASK), frequency-shift keying (FSK), and phase-shift keying (PSK).

Of these techniques, binary ASK is the preferred method of digital modulation because of its simplicity. In binary ASK, also known as on-off keying (OOK), the signal is switched between two power levels. The lower power level represents a "0" bit, while the higher power level represents a "1" bit.

In systems employing OOK, modulation of the signal can be achieved by simply turning the laser on and off (direct modulation). In general, however, this can lead to *chirp*, or variations in the laser's amplitude and frequency, when the laser is turned on. A preferred approach for high bit rates (≥ 2 Gbps) is to have an external modulator which modulates the light coming out of the laser. The external modulator blocks or passes light depending on the current applied to it.

In long-haul, high-capacity WDM systems, the use of advanced modulation formats has been an effective scheme to manage signal impairments arising from amplified spontaneous emission (ASE) noise, fiber nonlinear effects, and polarization-mode dispersion (PMD). The ideal modulation format for long-haul high-speed WDM transmission links is one that has a narrow spectral width, low susceptibility to fiber nonlinear effects, large dispersion tolerance, and a simple and cost-effective configuration for signal generation. There are a number of advanced formats that meet these criteria to varying degrees, including nonreturn-to-zero (NRZ), return-to-zero (RZ), and duobinary coding. There are also a number of variations of the RZ format, including simple RZ, carrier-suppressed RZ (CS-RZ), chirped RZ (CRZ), vestigial sideband RZ, and dispersion-managed soliton-based RZ [WeNL04].

Recently, the differential phase-shift-keyed (DPSK) format has attracted renewed interest due to its 3-dB lower requirement on optical signal-to-noise ratio and better resilience to cross-phase modulation compared with other on-off-keyed (OOK) formats mentioned above. A typical DPSK transmitter consists of a continuous-wave laser followed by two cascaded modulators. The first modulator is used to encode data via binary phase modulation. It can be a phase modulator (PM) or a dual-drive Mach-Zehnder modulator (MZM) biased at the null in its transmission curve with twice the drive voltage swing. The difference between these two methods is that a PM leaves the optical intensity constant and modulates the phase subject to its bandwidth limitations, while a MZM produces instantaneous phase jumps at the expense of some residual intensity modulation, which are of minor importance when using an RZ format. The second modulator was used to carve pulses out of the NRZ-phase modulated signal. Depending on the drive conditions of this modulator, RZ-DPSK with 33% and 50% duty cycle and CSRZ-DPSK with 67% duty cycle can be obtained correspondingly [WeNL04].

The Mach-Zehnder interferometer, described later in Section 2.4.2, can be used as a modulation device. A drive voltage is applied to one of two waveguides creating an electric field which causes the signals in the two waveguides to either be in phase or 180° out of phase, resulting in the light from the laser being either passed through the device or blocked. Mach-Zehnder amplitude modulators which offer bandwidths of up to 45 GHz are currently available [SMOI05]. One of the advantages of using integrated-optics devices such as the Mach-Zehnder interferometer is that the laser and modulator can be integrated on a single structure, which may be cost effective. Also, integrating the laser with the modulator eliminates the need for polarization control and results in low chirp.

2.3.4 Summary

Table 2.1 summarizes the characteristics of the different types of tunable transmitters. Observe that there is a trade-off between the tuning range of a transmitter and its tuning time.

2.4 Optical Receivers and Filters

Tunable optical filter technology is a key in making WDM networks realizable. Good sources of information on these devices include [Gree93, Brac90, KoCh89, RaSi01].

Table 2.1 Tunable optical transmitters and their associated tuning ranges and times.

Tunable Transmitter	*Approx. Tuning Range (nm)*	*Tuning Time*
Mechanical (external cavity)	550	1–10 ms
Acoustooptic	750	~9 μs
Electrooptic	7	1–10 ns
Injection-Current (DFB and DBR)	45	1–10 ns

2.4.1 Photodetection

In receivers employing *direct detection*, a photodetector converts the incoming optical stream into a stream of electrons. The electron stream (i.e., electrical current) is then amplified and passed through a threshold device. Whether a bit is a logical 0 or 1 depends on whether the stream is above or below a certain threshold for a bit duration. In other words, the decision is made based on whether or not light is present during the bit duration.

The basic detection devices for direct-detection optical networks are the PN photodiode (a p-n junction) and the PIN photodiode (an intrinsic material[7] is placed between "p" and "n" type material). In its simplest form, the photodiode is basically a reverse-biased p-n junction. Through the photoelectric effect, light incident on the junction will create electron-hole pairs in both the "n" and the "p" regions of the photodiode. The electrons released in the "p" region will cross over to the "n" region, and the holes created in the "n" region will cross over to the "p" region, thereby resulting in a current flow.

The alternative to direct detection is *coherent detection* in which phase information is used in the encoding and detection of signals. Coherent-detection-based receivers use a monochromatic laser as a local oscillator. The incoming optical stream, which is at a slightly different frequency from the oscillator, is combined with the signal from the oscillator, resulting in a signal at the difference frequency. This difference signal, which is in the

[7]An intrinsic material is a semiconductor material with electrical properties, which essentially has characteristics of the pure crystal, e.g., essentially silicon or germanium crystal with no measurable impurities.

microwave range which frequency range is from 1000 MHz and upward, is amplified, and then photodetected. While coherent detection is more elaborate than direct detection, the former allows the reception of weak signals from a noisy background. However, in optical systems, it is difficult to maintain the phase information required for coherent detection (see [Aziz91]). Since semiconductor lasers have nonzero linewidths, the transmitted signal consists of a number of frequencies with varying phases and amplitudes. The effect is that the phase of the transmitted signal experiences random but significant fluctuations around the desired phase. These phase fluctuations make it difficult to recover the original phase information from the transmitted signal, thus limiting the performance of coherent detection systems.

2.4.2 Tunable Optical Filters

This section discusses several types of tunable optical filters and the properties of each type, while Section 2.4.3 examines fixed-tuned optical filters. The feasibility of many local WDM networks is dependent upon the speed and range of tunable filters. Overviews of tunable filter technology can be found in [Gree93] and [RaSi01].

Filter Characteristics

Tunable optical filters are characterized primarily by their *tuning range* and *tuning time.* The *tuning range* specifies the range of wavelengths which can be accessed by a filter. A wide tuning range allows systems to utilize a greater number of channels. The *tuning time* of a filter specifies the time required to tune from one wavelength to another. Fast tunable filters are required for many WDM network architectures.

Some filters, such as the *etalon* (described in the following section), are further characterized by two parameters: *free spectral range* and *finesse.* In some filters, the transfer function, or the shape of the filter passband, repeats itself after a certain period. The period of such devices is referred to as the *free spectral range* (FSR). In other words, the filter passes every frequency which is a distance of $n \times$ FSR from the selected frequency, where n is a positive integer. For example, in Fig. 2.10, if the filter is tuned to frequency f_1, then all frequencies labeled with a 1 will be passed by the filter; tuning the filter to the next frequency, f_2, will allow all frequencies labeled with a 2 to be passed by the filter; etc. The FSR usually depends on various physical parameters in the device, such as cavity lengths or waveguide lengths.

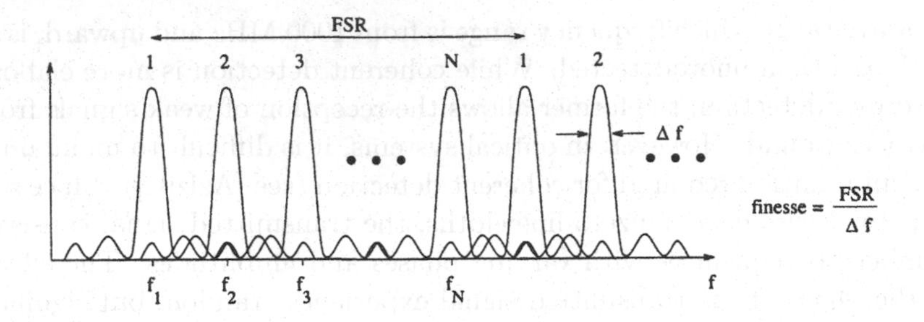

Figure 2.10 Free spectral range and finesse of a tunable filter capable of tuning to N different channels.

The *finesse* of a filter is a measure of the width of the transfer function. It is the ratio of FSR to channel bandwidth, where the channel bandwidth is defined to be the 3-dB bandwidth of a channel.

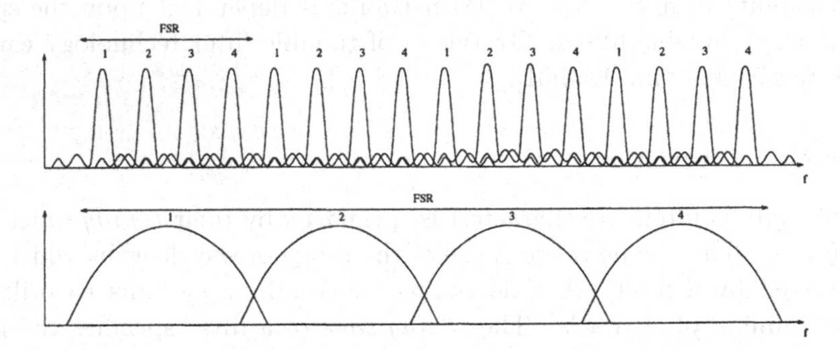

Figure 2.11 Cascading filters with different FSRs.

The number of channels in an optical filter is limited by the FSR and finesse. All of the channels must fit within one FSR. If the finesse is high, the transfer functions (passband peaks) are narrower, resulting in more channels being able to fit into one FSR. With a low finesse, the channels would need to be spaced further apart to avoid crosstalk, resulting in fewer channels. One approach to increasing the number of channels is to cascade filters with different FSRs [Gree93]. Figure 2.11 shows the filter passbands for a high-resolution filter and a low-resolution filter, each with four channels within a FSR. By cascading these filters, up to 16 unique channels may be resolved.

The Etalon

An etalon can be one of the following two types:

- First, a single of cavity is formed by two parallel flat mirrors (planar etalon). Almost all etalons are planar etalons.
- Second, a single of cavity consists of two identical spherical mirrors with their concave sides facing each other and with the distance between the mirrors equal to each mirror's radius of curvature (confocal or spherical etalon). This kind of etalons is much less common and will not be elaborated in this section.

In planar etalon filters, a typical type is the Fabry-Perot etalon, which is the simplest form of a Fabry-Perot interferometer. Its primary optical property is that if a monochromatic light ray travels back and forth between two mirrors and distance between mirrors equals an integral number of wavelengths, then the light passes through the etalon. Many modifications (e.g., multicavity and multipass) to the etalon can be made to improve the number of resolvable channels. In a multipass filter, the light passes through the same cavity multiple times, while in a multicavity filter, multiple etalons of different FSRs are cascaded to effectively increase the finesse.

The Fabry-Perot etalon can be made to virtually access the entire low-attenuation region of the fiber and can resolve very narrow passbands. But, it has a tuning time on the order of tens of milliseconds due to its mechanical tuning. This makes it unsuitable for many packet-switched applications in which the packet duration is much smaller than the tuning time.

For applications to optical WDM communication systems, the FSR of the Fabry-Perot Etalon corresponds to the typical channel spacing of 100 GHz or 50 GHz, thus requiring cavity lengths of 1–2 mm. Furthermore, the Fabry-Perot filter ideally covers a whole communication band, which is typically tens of nanometers large. All-fiber Fabry-Perot devices with finesse reaching 240 and covering a spectral band of 26 nm are presented in [SlDL03].

The Mach-Zehnder Chain

In a Mach-Zehnder (MZ) interferometer, a splitter splits the incoming wave into two waveguides, and a combiner recombines the signals at the outputs of the waveguides (see Fig. 2.12). An adjustable delay element controls the optical path length in one of the waveguides, resulting in a phase difference between the two signals when they are recombined. Wavelengths for which the phase difference is 180° are filtered out. By constructing a chain of these elements, a single desired optical wavelength can be selected.

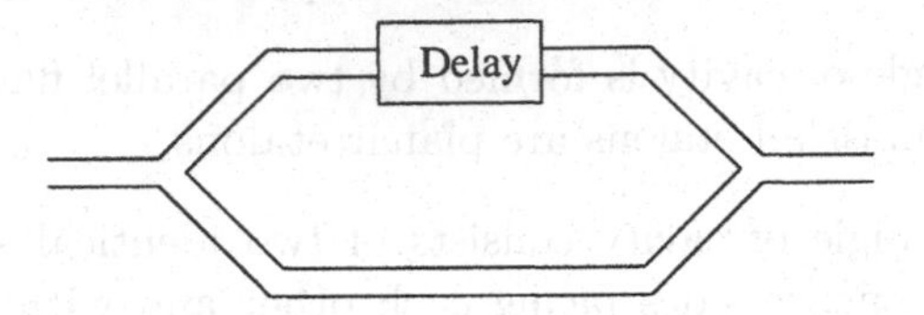

Figure 2.12 Structure of a Mach-Zehnder interferometer.

While the MZ chain may be a low-cost device because it can be fabricated on semiconductor material, its tuning time is still on the order of milliseconds, and its tuning control is complex, requiring that the setting of the delay element in each stage of the MZ chain be based on the settings in previous stages of the chain [Gree93]. The high tuning time is due to thermal elements used to implement the delay elements. Recent advances have produced a fast tuning MZ filter, which exhibits the total free-spectral range of 16 nm, an extinction ratio of 20 dB, and a 3-dB transmission bandwidth of 32 GHz [SMOI05].

Acoustooptic Filters

Using acoustooptic filters leads to a fast tuning time. Radio frequency (RF) waves are passed through a transducer. The transducer is a piezoelectric crystal that converts sound waves to mechanical movement. The sound waves change the crystal's index of refraction, which enables the crystal to act as a grating. Light incident upon the transducer will diffract at an angle that depends on the angle of incidence and the wavelength of the light [Gree93]. By changing the RF waves, a single optical wavelength can be chosen to pass through the material while the rest of the wavelengths destructively interfere.

The tuning time of the acoustooptic filter is limited by the flight time of the *surface acoustic wave (SAW)* to about 10 μs [NewF94]. However, the tuning range for acoustooptic filters covers the entire 1300 nm to 1560 nm spectrum. This tuning range potentially allows about 100 channels.

If more than one RF wave is passed through the grating simultaneously, more than one wavelength can be filtered out [CSBH89]. This allows the filter to be effectively tuned to several channels at the same time. However, the received signal is the superposition of all of the received wavelengths; therefore, if more than one of those channels is active, crosstalk will occur. The selection of up to five wavelengths was reported in [CSBH89].

An acousto-optic tunable filter (AOTF) is one of the most suitable tunable filters for wavelength tunable lasers because it has a wide tuning range (100 nm) and fast switching speed (several microseconds). Additionally, an AOTF provides stable operation against shock and vibration owing to its non-mechanical structure in [TTHD04]. A vendor recently announced its Acousto-Optic Filter whose tuning range is up from 380 to 750 nm with rapid spectrum scanning from 6 to 9 μs in March 2005.

One drawback of acoustooptic filters is that, because of their wide transfer function, they are unable to filter out crosstalk from adjacent channels if the channels are closely spaced. Therefore, the use of acoustooptic filters in a multiwavelength system places a constraint on the channel spacing, thus limiting the allowable number of channels.

Electrooptic Filters

Since the tuning time of the acoustooptic filter is limited by the speed of sound, crystals whose indices of refraction can be changed by electrical currents can be used. Electrodes, which rest in the crystal, are used to supply current to the crystal. The current changes the crystal's index of refraction, which allows some wavelengths to pass through while others destructively interfere [Gree93]. Since the tuning time is limited only by the speed of electronics, tuning time can be on the order of several nanoseconds.

Liquid-Crystal Fabry-Perot Filters

The design of a liquid-crystal filter is similar to the design of a Fabry-Perot filter, but the cavity consists of a liquid crystal (LC). The refractive index of the LC is modulated by an electrical current to filter out a desired wavelength, as in an electrooptic filter. These filters have low power requirements and are inexpensive to fabricate. The filter speed of LC filter technology promises to be high enough to handle high-speed packet switching in WDM networks.

In [Hira05], a new filter is presented that uses electrooptic material instead of LC in Fabry-Perot tuning filter. The free spectral range (FSR) of the filters is about 10 nm, tunable range is about 10 nm, loss is 2.2 dB, finesse is 150, and tuning speed takes only 1 μs.

2.4.3 Fixed Filters

An alternative to tunable filters are fixed filters or grating devices. Grating devices typically filter out one or more different wavelength signals from a single fiber. Such devices may be used to implement optical multiplexers and demultiplexers or receiver arrays.

Grating Filters

One implementation of a fixed filter is the diffraction grating. The diffraction grating is essentially a flat layer of transparent material (e.g., glass or plastic) with a row of parallel grooves cut into it [Hech04]. The grating separates light into its component wavelengths by reflecting light incident with the grooves at all angles. At certain angles, only one wavelength adds constructively; all others destructively interfere. This allows us to select the wavelength(s) we want by placing a filter tuned to the proper wavelength at the proper angle. Alternatively, some gratings are transmissive rather than reflective and are used in tunable lasers (see DFB lasers in Section 2.3.2).

An alternative implementation of a demultiplexer is the *arranged waveguide grating* (AWG) (also know as *waveguide grating router* (WGR)), in which only one input is utilized. AWGs will be discussed in Section 2.6.5.

Fiber Bragg Gratings (FBG)

In a fiber Bragg grating, a periodical variation of the index of refraction is directly photo-induced in the core of an optical fiber. A Bragg grating will reflect a given wavelength of light back to the source while passing the other wavelengths. Two primary characteristics of a Bragg grating are the reflectivity and the spectral bandwidth. Typical spectral bandwidths are on the order of 0.1 nm, while a reflectivity in excess of 99% is achievable [ISII95]. While inducing a grating directly into the core of a fiber leads to low insertion loss, a drawback of Bragg gratings is that the refractive index in the grating varies with temperature, with increases in temperature resulting in longer wavelengths being reflected. An approach for compensating for temperature variations is presented in [ASWC96]. Fiber Bragg gratings may be used in the implementation of multiplexers, demultiplexers, and tunable filters.

Thin-Film Interference Filters

Thin-film interference filters offer another approach for filtering out one or more wavelengths from a number of wavelengths. These filters are similar to fiber Bragg grating devices with the exception that they are fabricated by depositing alternating layers of low-index and high-index materials onto a substrate layer. Thin-film filter technology suffers from poor thermal stability, high insertion loss, and poor spectral profile. However, advances have been made which address some of these issues [DMNM04].

2.4.4 Summary of Optical Filtering Technologies

Table 2.2 Tunable optical filters and their associated tuning ranges and times.

Tunable Receiver	*Approx. Tuning Range (nm)*	*Tuning Time*
Fabry-Perot	500	1–10 ms
Acoustooptic	250	~10 μs
Electrooptic	16	1–10 ns
LC Fabry-Perot	50	0.5–10 μs

Table 2.3 Summary of optical filtering technologies [TzZT04].

Technology	*Loss* (dB)	*Channel Spacing* (GHz)	*Crosstalk* (dB)	*Tunability*	*Maturity*
FBG	0.5	50	30	Yes	High
TFF	1	100	12	Yes	High
AWG	6	50	30	No	High
MZI	1	50	30	No	High
AOF	3	100	15-20	Yes	Low
DCE	5	50	30	Yes	High

FBG = Fiber Bragg Gratings; TFF = Thin-Film Filters;
AWG = Arrayed Waveguide Gratings; MZI = Mach-Zehnder Interferometers;
AOF = Acousto-Optic Filters; DCE = Dynamic Channel Equalizers.

Table 2.2 summarizes the state of the art in tunable receivers. As has been stated earlier, tuning range and tuning time seem inversely proportional, except in LC Fabry-Perot filters.

Table 2.3 is a summary of several technologies used in filters with the corresponding features and specifications.

2.4.5 Channel Equalizers

The WDM systems currently under development incorporate longer spans and/or higher bit rates and/or reconfigurable add/drop, all of which lead to requirements for dynamic compensation using optical amplifiers (see Section 2.5. Two different types of compensation are generally required. To compensate for the residual non-flatness of amplifier gain profiles, resulting from imperfections in gain-flattening filters, changes in the amplifier operating conditions, and changes in channel loading, one needs a dynamic gain equalizer (DGE), which can achieve a smooth, low-ripple spectral attenuation profile that is the negative of the deviations of the amplifier gain profile from the desired profile. To compensate for unequal channel powers, resulting from dropping and adding channels, one needs a dynamic channel-power equalizer (DCE), which can achieve a flat attenuation profile across the full bandwidth of each channel, but can individually adjust the attenuation for each channel.

A DCE can not only provide a variable attenuation for each channel, but it can also achieve a sufficiently high attenuation (perhaps 40 to 50 dB) to effectively eliminate or block selected channels. Thus, it can be used as a key active element in a reconfigurable add/drop node. These devices are commonly referred to as wavelength blockers (WB) [Toml03]. In some cases, the same technology can be used to implement either the DGE functionality or the DCE functionality, but the basic design parameters for these two cases are sufficiently different that different device designs are required to accomplish the two different functionalities. In particular, the DGE plays an important role in WDM networks, because of its ability to control the power profile of the wavelength channels, hence maintaining a high quality of service (QoS) and providing more flexibility in transmission management.

The key requirements of future dynamic WDM equalizers include low insertion loss, wide bandwidth, fast equalization speed, small size, and low cost. Dynamic WDM equalizer structures include micro-opto-mechanical-system (MEMS) filters, Mach-Zehnder interferometer filters, acousto-optic filters, digital holographic filters, and liquid-crystal modulators. These structures utilize few cascaded (or parallel) optical filters, whose weights are dynamically optimized to realize smooth spectral equalization. Usually, it is hard to achieve channel-by-channel equalization unless the number of optical filters used in an equalizer subsystem is made equal to the number of WDM channels.

In [RAAE04], a dynamic WDM equalizer structure that can achieve channel-by-channel spectral equalization is proposed, while maintaining a constant insertion loss, independent of the number of WDM channels. It is based on a reflective, free-space opto-VLSI processor, which generates phase holograms to independently steer/reshape each incident WDM beam, thus realizing channel-by-channel optical attenuation. For more detailed information, we refer the reader to [Toml03, RAAE04].

2.5 Optical Amplifiers

Although an optical signal can propagate a long distance typically 80 km in current deployment before it needs amplification, optical networks, particularly for long-distance links, can benefit from optical amplifiers.

All-optical amplification may differ from optoelectronic amplification in that it may act only to boost the power of a signal, but not to restore the shape or timing of the signal. This type of amplification is known as *1R (re-amplification)*, and it provides total data transparency (the amplification process is independent of the signal's modulation format).

However, in today's digital networks (e.g., Synchronous Optical Network (SONET) and Synchronous Digital Hierarchy (SDH)), which use the optical fiber only as a transmission medium, the optical signals are amplified by first converting the information stream into an electronic data signal, and then retransmitting the signal optically. Such a process is referred to as *3R* (*re-amplification*, *re-shaping*, and *re-timing*).

The *re-shaping* of the signal reproduces the original pulse shape, eliminating much of the noise. Reshaping applies primarily to digitally-modulated signals, but in some cases may also be applied to analog signals. The *re-timing* of the signal synchronizes the signal to its original bit timing pattern and bit rate. Re-timing applies only to digitally-modulated signals.

Another approach to amplification is *2R* (*re-amplification* and *re-shaping*), in which the optical signal is converted to an electronic signal which is then used to directly modulate a laser. Research and development on all-optical *3R* regeneration is a very important topic today.

Also, in a WDM system with optoelectronic regeneration, each wavelength would need to be separated before being amplified electronically, and then recombined before being retransmitted. Thus, in order to eliminate the need for optical multiplexers and demultiplexers in amplifiers, optical amplifiers must boost the strength of optical signals without first converting them to electrical signals. A drawback is that optical noise, as well as the signal, will be amplified. Also, the amplifier introduces spontaneous emission noise.

Optical amplification uses the principle of stimulated emission, similar to the approach used in a laser. Optical amplifiers can be divided into two basic classes: optical fiber amplifiers (OFAs) and semiconductor optical amplifiers (SOAs), which will be discussed in detail in the following section. In Table 2.4, comparison between OFAs and SOAs are presented in general. Besides, there is a new kind of optical amplifier that is *Raman amplifier.* They will be discussed in detail in the following sections.

Table 2.4 Characteristics of OFAs and SOAs.

Features	*OFAs*	*SOAs*
Maximum Internal Gain (dB)	25-30	20-25
Insertion Loss (dB)	0.1-2	6-10
Polarization Sensitivity	Negligible	< 2 dB
Nonlinear Effects	Negligible	Yes
Saturation Output Power (dBm)	13-23	5-20
Noise Figure (dBm)	4-6	7-12
Integrated Circuit Compatible	No	Yes
Functional Device Possibility	No	Yes

2.5.1 Optical Amplifier Characteristics

Some basic parameters of interest in an optical amplifier are *gain*, *gain bandwidth*, *gain saturation*, *polarization sensitivity*, and *amplifier noise.*

Gain measures the ratio of the output power of a signal to its input power. Amplifiers are sometimes also characterized by *gain efficiency*, which measures the gain as a function of pump power in dB/mW, where pump is a local source of energy.

The *gain bandwidth* of an amplifier refers to the range of frequencies or wavelengths over which the amplifier is effective. In a network, the gain bandwidth limits the number of wavelengths available for a given channel spacing.

The *gain saturation* point of an amplifier is the value of output power at which the output power no longer increases with an increase in the input power. When the input power is increased beyond a certain value, the carriers (electrons) in the amplifier are unable to output any additional light energy. The saturation power is typically defined as the output power at which there is a 3-dB reduction in the ratio of output power to input power (the small-signal gain).

Polarization sensitivity refers to the dependence of the gain on the polarization of the signal. The sensitivity is measured in dB and refers to the gain difference between the TE and TM polarizations.

In optical amplifiers, the dominant source of *noise* is *amplified spontaneous emission* (ASE), which arises from the spontaneous emission of photons in the active region of the amplifier (see Fig. 2.13). The amount of noise generated by the amplifier depends on factors such as the amplifier gain spectrum, the noise bandwidth, and the *population inversion parameter* which specifies the degree of population inversion that has been achieved between two energy levels. Amplifier noise is especially a problem when multiple amplifiers are *cascaded*, e.g., in long-distance links. Each subsequent amplifier in the cascade amplifies the noise generated by previous amplifiers.

2.5.2 Semiconductor Laser Amplifier

A semiconductor laser amplifier (see Fig. 2.13) is a modified semiconductor laser, which typically has different facet reflectivity and different device length. A weak signal is sent through the active region of the semiconductor, which, via stimulated emission, results in a stronger signal emitted from the semiconductor.

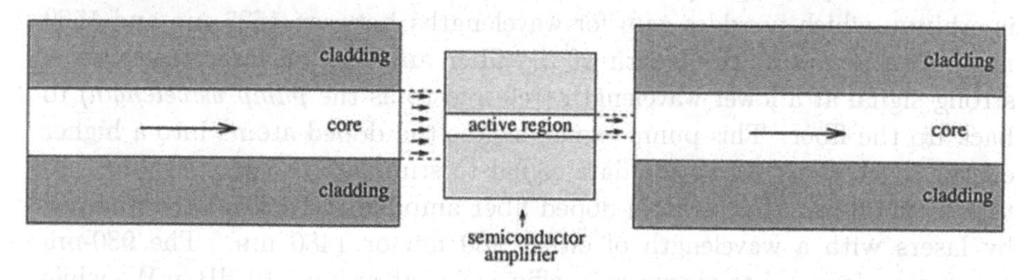

Figure 2.13 A semiconductor optical amplifier.

The two basic types of semiconductor laser amplifiers are the Fabry-

Perot amplifier, which is basically a semiconductor laser, and the traveling-wave amplifier (TWA). The primary difference between the two is in the reflectivity of the end mirrors. Fabry-Perot amplifiers have a reflectivity of around 30%, while TWAs have a reflectivity of around 0.01% [Maho93]. In order to prevent lasing in the Fabry-Perot amplifier, the bias current is operated below the lasing threshold current. The higher reflections in the Fabry-Perot amplifier cause Fabry-Perot resonances in the amplifier, resulting in narrow passbands of around 5 GHz. This phenomenon is not very desirable for WDM systems; therefore, by reducing the reflectivity, the amplification is performed in a single pass and no resonances occur. Thus, TWAs are more appropriate than Fabry-Perot amplifiers for WDM networks.

Semiconductor amplifiers based on multiple quantum wells (MQW) have been studied in [LKSK05, MTTK05, LKSK05]. These amplifiers have higher bandwidth and higher gain saturation than bulk devices. They also provide faster on-off switching times. The disadvantage is a higher polarization sensitivity.

Currently, SOAs attract more interest in both research and industry fields, and their advances are illustrated in [KaIw05, LOLH05, OYTB05]. One advantage of semiconductor amplifiers is the ability to integrate them with other components. For example, they can be used as gate elements in switches. By turning a drive current on and off, the amplifier basically acts like a gate, either blocking or amplifying the signal.

2.5.3 Doped-Fiber Amplifier

Doped-fiber amplifiers are lengths of fiber doped with an element (rare earth) which can amplify light (see Fig. 2.14). The most common doping element is erbium, which provides gain for wavelengths between 1525 nm and 1560 nm. At the end of the length of the fiber amplifier, a laser transmits a strong signal at a lower wavelength (referred to as the *pump wavelength*) to back up the fiber. This pump signal excites the doped atoms into a higher energy level. This allows the data signal to stimulate the excited atoms to release photons. Most erbium-doped fiber amplifiers (EDFAs) are pumped by lasers with a wavelength of either 980 nm or 1480 nm. The 980-nm pump wavelength has shown gain efficiencies of around 10 dB/mW, while the 1480-nm pump wavelength provides efficiencies of around 5 dB/mW. Typical gains are on the order of 25 dB. Experimentally, EDFAs have been shown to achieve gains of up to 51 dB with the maximum gain limited by internal *Rayleigh backscattering (RBS)*, which occurs when a fraction of scattered light is recaptured and back-reflected towards the launch end within the optical waveguide [HaDL92]. The 3-dB gain bandwidth for the EDFA is around 35 nm (see Fig. 2.15), and the saturation power is around 20 dBm[8] [Maho93]. A serial structured wide-band EDFA with 25 dB of flat gain over 77 nm (1528−1605 nm) and dynamic gain clamping over 13 dB of input range (25−483 W) is demonstrated experimentally in [JLWF04].

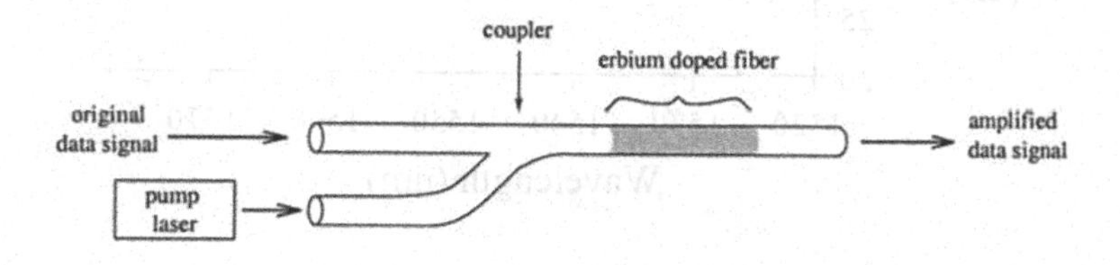

Figure 2.14 Erbium-doped fiber amplifier.

Additionally, the praseodymium-doped fluoride fiber amplifier (PDFFA) is presented in [Whit95], and Yb-Doped Fiber Amplifier is shown in [WrVa05].

A limitation to optical amplification is the unequal gain spectrum of optical amplifiers. The EDFA gain spectrum is shown in Fig. 2.15 (from [Sams97]). While an optical amplifier may provide gain across a range of wavelengths, it will not necessarily amplify all wavelengths equally. This characteristic, accompanied by the facts that optical amplifiers amplify noise as well as signal and that the active region of the amplifier can spontaneously emit photons which also cause noise, limit the performance of optical amplifiers. Thus, a multiwavelength optical signal passing through a series of amplifiers will eventually result in the power of the wavelengths being uneven.

A number of approaches to equalizing the gain of an EDFA have been studied. In [TaLa91], a notch filter (a filter which attenuates the signal at a selected frequency) centered at around 1530 nm is used to suppress the peak in the EDFA gain (see Fig. 2.15). However, when multiple EDFAs are cascaded, another peak appears around the 1560-nm wavelength. In

[8]Power is commonly expressed in decibel values regarding to losses and gains. The conversion equation between dBm and mW units is:

$$P(dBm) = 10\log(P(mW))$$

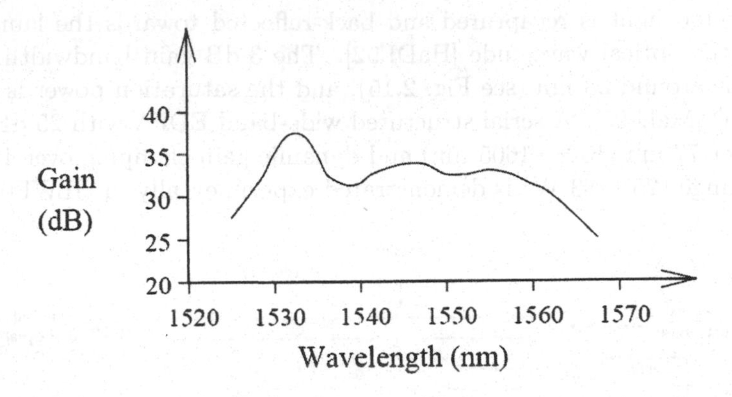

Figure 2.15 The gain spectrum of an erbium-doped fiber amplifier with input power = −40 dBm.

Table 2.5 Amplifier characteristics.

Amplifier Type	*Gain Region*	*Gain Bandwidth*	*Maximum Gain*
Semiconductor	Any	40 nm	25 dB
EDFA	1525–1560 nm	35 nm	30 dB
PDFFA	1280–1330 nm	50 nm	30 dB

[WiHw93], a notch filter centered at 1560 nm is used to equalize the gain for a cascade of EDFAs. Another approach to flattening the gain is to adjust the input transmitter power such that the power on all received wavelengths at the destination is equal [ChNT92]. A third approach to gain equalization is to demultiplex the individual wavelengths and then attenuate selected wavelengths such that all wavelengths have equal power. In [EGZJ93], this approach is applied to a WDM interoffice ring network.

2.5.4 Raman Amplifier

Besides the classical amplifiers above, Raman amplifiers have been deployed for new long-haul and ultra-long-haul fiber-optic transmission systems, making them one of the first widely commercialized nonlinear optical devices in telecommunications. Raman amplifiers have some fundamental advantages. First, Raman gain exists in every fiber, which provides a cost-effective means of upgrading from the terminal ends. Second, the gain is nonresonant, which is available over the entire transparency region of the fiber ranging from approximately 0.3 to 2 μm. The third advantage of Raman amplifiers is that the gain spectrum can be tailored by adjusting the pump wavelengths. For instance, multiple pump lines can be used to increase the optical bandwidth, and the pump distribution determines the gain flatness. Another advantage of Raman amplification is that it is a relatively broad-band amplifier with a bandwidth > 5 THz, and the gain is reasonably flat over a wide wavelength range [Moha02].

However, Raman gain requires more pump power, roughly tens of milliwatts per dB of gain, whereas the tenths of a milliwatt per dB is required for EDFAs for small signal powers. This disadvantage, combined with the scarcity of high-power pumps at appropriate wavelengths, slowed down the development of the Raman amplifier during the commercialization of EDFAs in the early 1990s. Then, in the mid-1990s, the development of suitable high-power pumps sparked renewed interest. Researchers were quick to demonstrate some of the advantages that Raman amplifiers have over EDFAs, particularly when the transmission fiber itself is turned into a Raman amplifier. This, in turn, fueled further advances in Raman pump technology. Now, Raman amplification is an accepted technique for enhancing system performance.

The schematic diagram of a Raman amplifier is shown in Fig. 2.16. When an optical field is incident on a molecule, the bound electrons oscillate at the optical frequency. This induced oscillating dipole moment produces optical radiation at the same frequency, with a phase shift that leads to the medium's refractive index. Simultaneously, the molecular structure itself is oscillating at the frequencies of various molecular vibrations. Therefore, the induced oscillating dipole moment also contains the sum and difference frequency terms between the optical and vibrational frequencies. These terms give rise to Raman scattered light in the re-radiated field. In a solid-state quantum-mechanical description, optical photons are inelastically scattered by quantized molecular vibrations called optical phonons. Photon energy is lost (the molecular lattice is heated) or gained (the lattice is cooled), shifting the frequency of the light. The components of scattered light that are shifted to lower frequencies are called Stokes lines, while those shifted to higher fre-

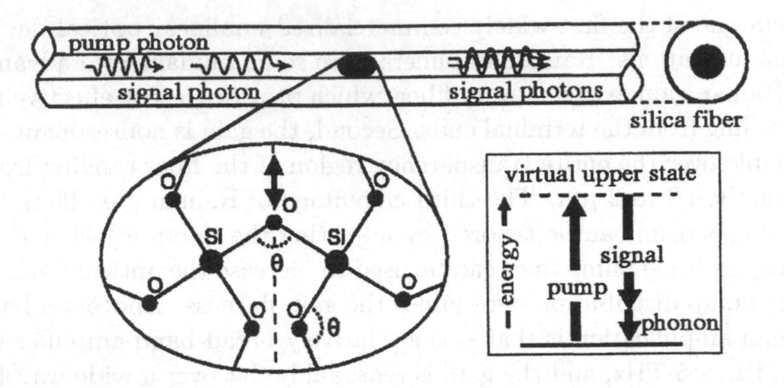

Figure 2.16 Schematic depicting amplification by stimulated Raman scattering in an optical silica fiber. The Raman Stokes interaction between a pump and signal photon and the silica molecules converts the pump into a replica of the signal photon, producing an optical phonon.

quencies are called anti-Stokes lines. The frequency shift is equal to the oscillation frequency of the lattice phonon that is created or annihilated. (The anti-Stokes process is not mentioned further in this book as it is typically orders of magnitude weaker than the Stokes process in the context of optical communications, making it irrelevant.) Raman scattering can occur in all materials, but in silica glass the dominant Raman lines are due to the bending motion of the Si-O-Si bond (see bond angle in Fig. 2.16). Raman scattering can also be stimulated by signal light at an appropriate frequency shift from a pump, leading to stimulated Raman scattering (SRS). In this process, pump and signal light are coherently coupled by the Raman process. In a quantum-mechanical description, shown in the energy-level diagram in Fig. 2.16, a pump photon is converted into a second signal photon that is an exact replica of the first one, and the remaining energy produces an optical phonon. The initial signal photon, therefore, has been amplified. This process is considered nonresonant because the upper state is a short-lived virtual state.

Raman amplification has enabled dramatic increases in the reach and capacity of lightwave systems. Novel Raman pumping schemes that have recently been developed are highlighted in [Jake04].

2.6 Switching Elements

Obviously, *switching elements* are essential component of any network. The concept of switching originated from electronics field. Thus, according to the signal carriers, there are *optical switching* and *electronic switching.* In the switching granularity point of view, there are two basic classes: *circuit switching* and *cell switching.* In optical field, *circuit switching* is corresponding to *wavelength routing,* and *cell switching* is *optical packet switching* and *optical burst switching.* As far as the transparency of signals is considered, there are opaque switching and transparent switching. In the section, switching devices are classified into two basic classes: logic switching and relational switching.

Logic switching is performed by a device in which the data (or the information-carrying signal) incident on the device controls the state of the device in such a way that some Boolean function, or combination of Boolean functions, is performed on the inputs. In a logic device, format and rate of data would be changed or converted in intermediate nodes, thus, *logic switching* is also sometimes referred to *opaque switching.* Furthermore, some of its components must be able to change states or *switch* as fast as or faster than the signal bit rate [Hint90]. This ability gives the device some added flexibility but limits the maximum bit rate that can be accommodated. Based on the logic device available and ideal performance in electronic field, logic switching is primarily employed in electronic field. But, traditional optical-electronic-optical (o-e-o) conversion in today's optical networks is still widely applied due to the lack of counterpart logic devices in the optical field. It means that most current optical networks employ electronic processing and use the optical fiber only as a transmission medium. Switching and processing of data are performed by converting an optical signal back to its "native" electronic form. Such a network relies on electronic switches, i.e., logic devices. It provides a high degree of flexibility in terms of switching and routing functions for optical networks; however, the speed of electronics is unable to match the high bandwidth of an optical fiber. Also, an electrooptic conversion at an intermediate node in the network introduces extra delay and cost. These factors above have motivated a push toward the development of all-optical networks in which optical switching components are able to switch high-bandwidth optical data streams without electrooptic conversion. The emergence of *relational switching* is derived from this push.

Relational switching is to establish a relation between the inputs and the outputs. The relation is a function of the control signals applied to the device and is independent of the contents of the signal or data inputs. A property of this device is that the information entering and flowing through it cannot change or influence the current relation between the inputs and the outputs. In the class of switching devices currently being developed, the control of the switching function is performed electronically with the optical stream being transparently routed from a given input of the switch to a given output. Such transparent switching allows the switch to be independent of the data rate and format of the optical signals. Thus, the strength of a relational device, which allows signals at high bit rates to pass through it, is that it cannot sense the presence of individual bits that are flowing through itself. This characteristic is also known as *transparent switching.* Due to the limits of optical hardware, various kinds of optical switching devices basically employ *relational switching. Relational switching* provides more advantages for optical networks in terms of the optical hardware limits. In the following sections, we review a number of different optical switching elements and architectures which are typically developed in today's optical networks.

2.6.1 Optical Add-Drop Multiplexers (OADM)

Optical Add/Drop Multiplexers (OADMs) are elements that provide capability to add and drop traffic in the network (similar to SONET ADMs). They are located at sites supporting one or two (bidirectional) fiber pairs and enable a number of wavelength channels to be dropped and added, thereby reducing the number of unnecessary optoelectronic conversions, without affecting the traffic that is transmitted transparently through the node (see Fig. 2.17).

An OADM can be used in both linear and ring network architectures and in practice operates in either fixed or reconfigurable mode. In fixed OADMs, the add/drop and through channels are predetermined and can only be manually rearranged after installation. In reconfigurable OADMs, the channels that are added/dropped or pass transparently through the node can be dynamically reconfigured as required by the network. These are more complex structures but more flexible as they provide on-demand provisioning without manual intervention; therefore, they can be set up on the fly. The reduction of unnecessary optoelectronic conversions through the use of OADMs introduces significant cost savings in the network.

Reconfigurable OADMs can be divided into two categories: one is partly reconfigurable architecture and the other is fully reconfigurable architecture. In partly-reconfigurable architectures, there is capability to select the chan-

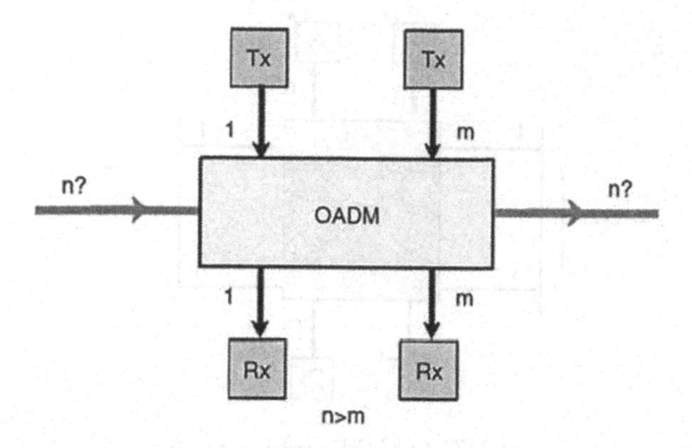

Figure 2.17 Generic OADM architecture.

nels to be added/dropped, but there is also a predetermined connectivity matrix between add/drop and through ports, restricting the wavelength-assignment function. Fully-reconfigurable OADMs provide the ability to select the channels to be added/dropped, but they also offer connectivity between add/drop and through ports, which enables flexible wavelength assignment with the use of tunable transmitters and receivers. Reconfigurable OADMs can be divided into two main generations. The first is mainly applied in linear network configurations and does not support optical path protection, while the second is applied in ring configurations and provides optical layer protection.

The two most common examples of fully-reconfigurable OADMs, i.e., *wavelength selective* (WS) and *broadcast selective* (BS) architectures, are illustrated in Figs. 2.18(a) and 2.18(b). The WS architecture utilizes wavelength demultiplexing/multiplexing and a switch fabric interconnecting express and add/drop ports, while the BS is based on passive splitters/couplers and tunable filters. The overall loss introduced by the through path of the BS solution is noticeably lower than the loss of the WS approach, significantly improving the optical signal-to-noise ratio (OSNR) of the node and therefore its concatenation performance in a practical transmission link or ring. In addition, the BS design offers superior filter concatenation performance, advanced features such as drop and continue, and good scalability in terms of add/drop percentage.

For more detailed information, we refer the reader to [TzZT04].

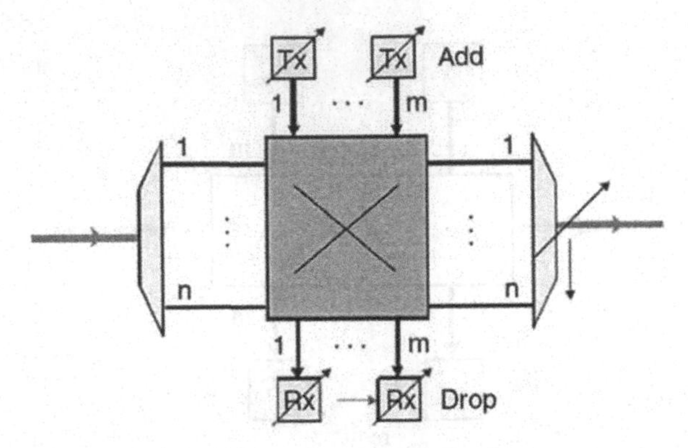

(a) WS OADM architecture

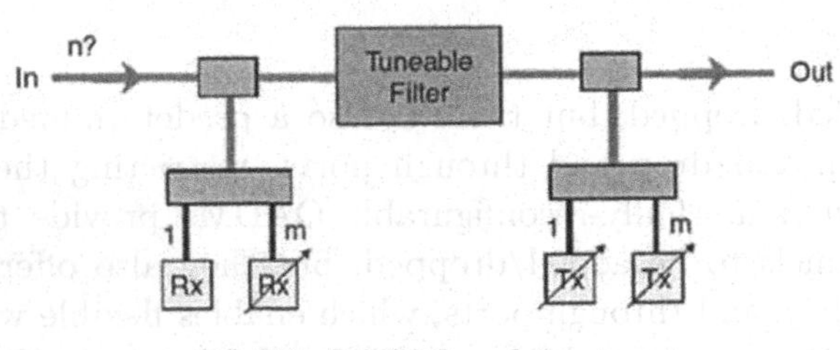

(b) BS OADM architecture

Figure 2.18 Fully-reconfigurable WS and BS OADM architectures [TzZT04].

2.6.2 Optical Cross-Connect (OXC)

An *optical crossconnect (OXC)* switches optical signals from input ports to output ports. These type of elements are usually considered to be wavelength insensitive, i.e., incapable of demultiplexing different wavelength signals on a given input fiber.

A basic crossconnect element is the 2×2 crosspoint element. A 2×2 crosspoint element routes optical signals from two input ports to two output ports and has two states: cross state and bar state (see Fig. 2.19). In the cross state, the signal from the upper input port is routed to the lower output port, and the signal from the lower input port is routed to the upper output port. In the bar state, the signal from the upper input port is routed to the

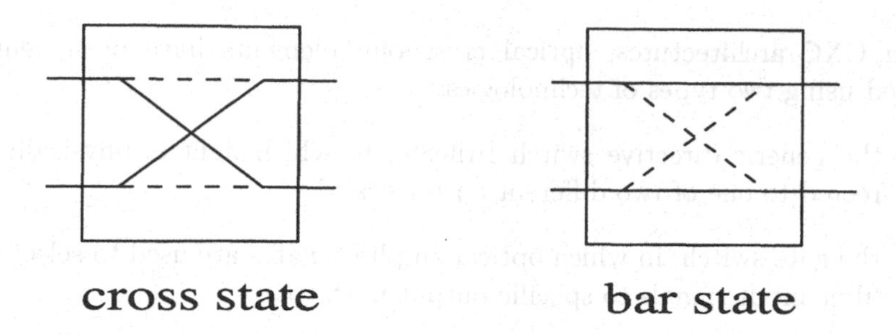

Figure 2.19 2 × 2 crossconnect elements in the cross state and bar state.

upper output port, and the signal from the lower input port is routed to the lower output port.

To support flexible path provisioning and network resilience, OXCs normally utilize a switch fabric to enable routing of any incoming channels to the appropriate output port and access to the local client traffic. The features that an OXC should ideally support are similar to these of an OADM, but OXCs need to additionally provide:

(1) strictly nonblocking connectivity between input and output ports,

(2) span and ring protection as well as mesh restoration capabilities.

A number of OXC solutions based on different technologies have been proposed to date and, depending on the switching technology and the architecture used, they are commonly divided into two main categories: opaque and transparent [TzZT04].

Opaque OXCs are either based on electrical switching technology or on optical switch fabrics surrounded by optical-electrical-optical (OEO) conversions, imposing the requirement of expensive optoelectronic interfaces. In OXCs using electrical switching, subwavelength switching granularities can be supported by providing grooming capabilities for more efficient bandwidth utilization. Opaque OXCs also offer inherent regeneration, wavelength conversion, and bit-level monitoring.

In transparent OXCs, the incoming signals are routed through an optical switch fabric without the requirement of optoelectronic conversions, thereby offering transparency to a variety of bit rates and protocols. The switching granularity may vary and support switching at the fiber level, the wavelength band level, or the wavelength channel level.

In OXC architectures, optical crosspoint elements have been demonstrated using two types of technologies:

(1) the generic directive switch [Alfe88], in which light is physically directed to one of two different outputs, and

(2) the gate switch, in which optical amplifier gates are used to select and filter input signals to specific output ports.

Directive Switches

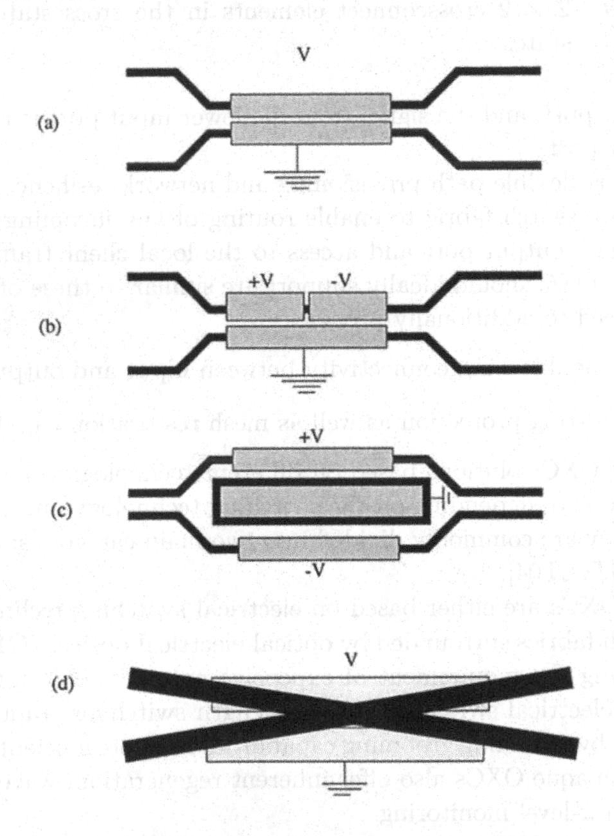

Figure 2.20 Schematic of optical crosspoint elements.

The directional coupler (Fig. 2.20(a) [Alfe88]) consists of a pair of optical channel waveguides that are parallel and in close proximity over a finite interaction length. Light input to one of the waveguides couples to the second waveguide via evanescent[9] coupling. The coupling strength corresponds to the interwaveguide separation and the waveguide mode size which depends on the optical wavelength and confinement factor[10] of the waveguide. If the two waveguides are identical, complete coupling between the two waveguides occurs over a characteristic length which depends on the coupling strength. However, by placing electrodes over the waveguides, the difference in the propagation constants in the waveguides can be sufficiently increased so that no light couples between the two waveguides. Therefore, the cross state corresponds to zero applied voltage, and bar state corresponds to a nonzero switching voltage. Unfortunately, the interaction length needs to be very accurate for good isolation, and these couplers are wavelength specific.

Switch fabrication tolerances, as well as the ability to achieve good switching for a relatively wide range of wavelengths, can be overcome by using the so-called reversed delta-beta coupler (see Fig. 2.20(b)). In this device, the electrode is split into at least two sections. The cross state is achieved by applying equal and opposite voltages to the two electrodes. This approach has been shown to be very successful [Alfe88].

The balanced bridge interferometric switch (see Fig. 2.20(c)) consists of an input 3-dB coupler, two waveguides sufficiently separated so that they do not couple, electrodes to allow changing the effective path length over the two arms, and a final 3-dB coupler. Light incident on the upper waveguide is split in half by the first coupler. With no voltage applied to the electrodes, the optical path length of the two arms enters the second coupler in phase. The second coupler acts like the continuation of the first, and all the light is crossed over to the second waveguide to provide the cross state. To achieve the bar state, voltage is applied to an electrode, placed over one of the interferometer arms to electrooptically produce a 180° phase difference between the two arms. In this case, the two inputs from the arms of the interferometer combine at the second 3-dB coupler out of phase, with the result that light remains in the upper waveguide.

The intersecting waveguide switch is shown in Fig. 2.20(d). This device can be viewed as a directional coupler (see Fig. 2.20(a)) with no gap between the waveguides in the interaction region. When properly fabricated, both cross and bar states can be electrooptically achieved with good crosstalk

[9]An evanescent wave is the part of a propagating wave which travels along or outside of the waveguide boundary.

[10]The confinement factor determines the fraction of optical power that travels within the core of the waveguide.

performance.

Other types of switches include the mechanical fiber-optic switch and the thermo-optic switch. These devices offer slow switching (about milliseconds) and may be employed in circuit-switched networks. One mechanical switch, for example, consists of two ferrules, each with polished end faces that can rotate to switch the light appropriately [Ande95]. Thermo-optic waveguide switches, on the other hand, are fabricated on a glass substrate and are operated by the use of the thermo-optic effect. One such device uses a zero-gap directional-coupler configuration with a heater electrode to increase the waveguide index of refraction [LeSu94].

Gate Switches

In the $N \times N$ gate switch, each input signal first passes through a $1 \times N$ splitter. The signals then pass through an array of N^2 gate elements, and are then recombined in $N\times$ combiners and sent to the N outputs. The gate elements can be implemented using optical amplifiers which can either be turned on or off to pass only selected signals to the outputs. The amplifier gains can compensate for coupling losses and losses incurred at the splitters and combiners. A 2×2 amplifier gate switch is illustrated in Fig. 2.21. A disadvantage of the gate switch is that the splitting and combining losses limit the size of the switch.

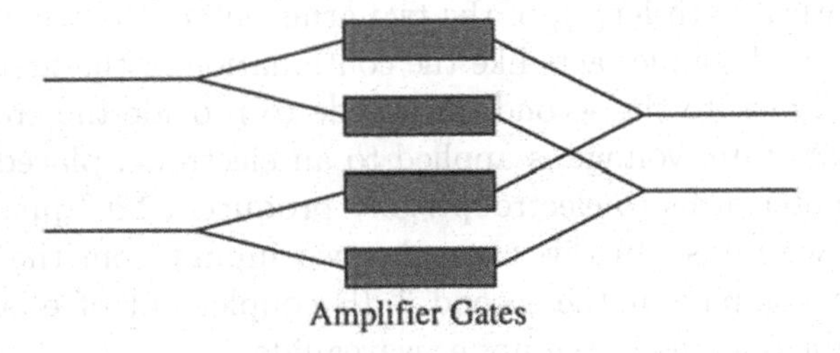

Figure 2.21 A 2×2 amplifier gate switch.

2.6.3 Clos Architecture

For more than 50 years, the Clos architecture [Clos53] has been the hands-down favorite for building multi-stage TDM switching systems. It has maintained this status, independent of the advances in switching fabrics over that same time period. It can realize the fewest switching crosspoints for the largest range of scalability while providing strict or rearrangably non-blocking traffic paths. In Fig. 2.22, 3-stage Clos architecture is presented, which is a typical application of an OXC.

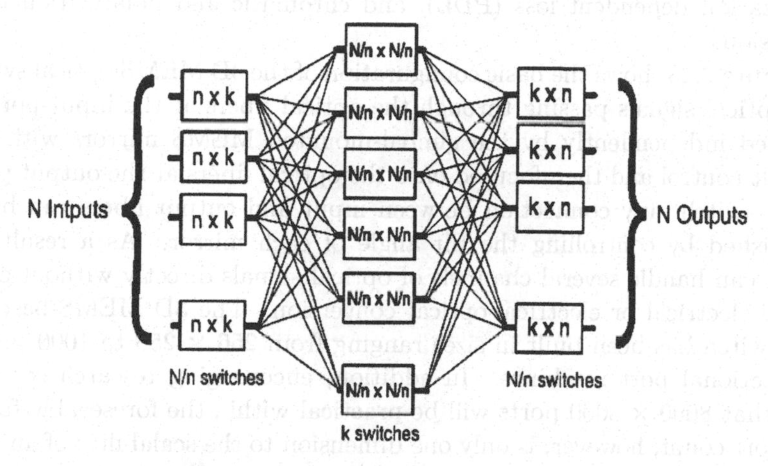

Figure 2.22 A 3-stage Clos architecture.

In Fig. 2.22, the number of 2nd stage switches depends on blocking: in fully non-blocking, $k \geq 2n - 1$; in rearrangeably non-blocking, $k \geq n$.

Advanced development of *3-stage Clos Cross-connect Switch Architecture* with up to 2048×2048 ports and 10 Gbps per port is presented by some vendors in 2005.

2.6.4 Micro-Electro Mechanical Systems (MEMS)

Currently, *micro-electro mechanical systems (MEMS)* is widely believed to be the most promising method for large-scale optical cross-connects. Optical MEMS-based switches are distinguished in being based on mirrors, membranes, and planar moving waveguides. The former two are free-space switches; the latter are waveguide switches. Furthermore, MEMS-based switches are classified into the two major approaches, i.e., 2-Dimensional and 3-Dimensional approaches. Among these classifications, the 3D optical MEMS based on mirrors is popular because it is suitable for compact, large-scale switching fabrics. The ability of this architecture to achieve input- and output-port counts of over one thousand is the primary driver of the large-scale OXCs, in which spatial parallelism is utilized. In particular, the type of switch provides high application flexibility in network design because of low and uniform insertion loss with low wavelength dependency under various operating conditions. Furthermore, this switch exhibits minimal degradation of the optical signal-to-noise ratio, which is mainly caused by crosstalk, polarization dependent loss (PDL), and chromatic and polarization mode dispersions.

Figure 2.23 shows the basic configuration of the 3D MEMS optical switch. The optical signals passing through the optical fibers at the input port are switched independently by the gimbal-mounted MEMS mirrors with two-axis tilt control and then focused onto the optical fibers at the output ports. In the switch, any connection between input and output fibers can be accomplished by controlling the tilt angle of each mirror. As a result, the switch can handle several channels of optical signals directly without costly optical-electrical or electrical-optical conversion. The 3D MEMS-based O-O-O switch has been built in sizes ranging from 256 × 256 to 1000 × 1000 bi-directional port machines. In addition, encouraging research seems to show that 8000 × 8000 ports will be practical within the foreseeable future. The port count, however, is only one dimension to the scalability of an O-O-O switch. An O-O-O switch is also scalable in terms of throughput. A truly all-optical switch is bit-rate and protocol independent. The combination of thousands of ports and bit-rate independence results in a theoretically future-proof switch with unlimited scalability.

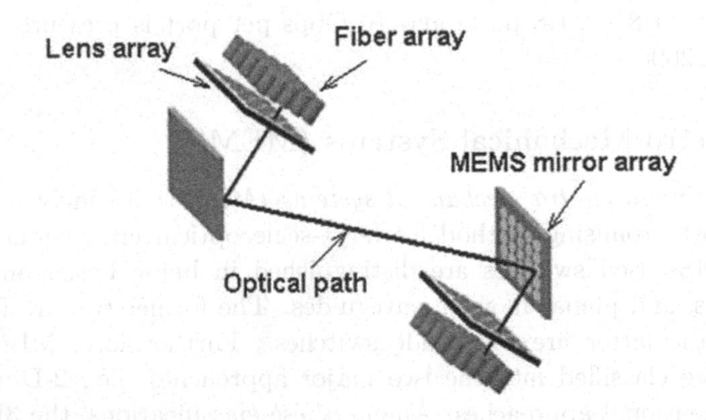

Figure 2.23 Schematic diagram of a 3D MEMS optical switch.

Recent developments of 3D MEMS for optical switching are shown in [YaYT05]. They can lead to compact and stable optical crossconnect switches for simple, fast, and flexible wavelength applications in today's optical networks.

Optical MEMS are miniature devices with optical, electrical, and mechanical functionalities at the same time, fabricated using batch process techniques derived from microelectronic fabrication. Optical MEMS provide intrinsic characteristics for very low crosstalk, wavelength insensitivity, polarization insensitivity, and scalability. Comparison of optical MEMS vs. other OXC switching elements are presented in [MaKu03].

2.6.5 Nonreconfigurable Wavelength Router

In this and next sections, two classes of wavelength routers are introduced, which are in fact circuit switches in the optical field. A wavelength-routing device can route signals arriving at different input fibers (ports) of the device to different output fibers (ports) based on the wavelengths of the signals. Wavelength routing is accomplished by demultiplexing the different wavelengths from each input port, optionally switching each wavelength separately, and then multiplexing signals at each output port. The device can be either *nonreconfigurable*, in which case there is no switching stage between the demultiplexers and the multiplexers, and the routes for different signals arriving at any input port are fixed (these devices are referred to as routers rather than switches), or *reconfigurable*, in which case the routing function of the switch can be controlled electronically. In this section, we will discuss wavelength routers, while Section 2.6.6 will cover reconfigurable wavelength switches.

A nonreconfigurable wavelength router can be constructed with a stage of demultiplexers which separate each of the wavelengths on an incoming fiber, followed by a stage of multiplexers which recombine wavelengths from various inputs to a single output. The outputs of the demultiplexers are hardwired to the inputs of the multiplexers. Let this router have P incoming fibers, and P outgoing fibers. On each incoming fiber, there are M wavelength channels. A 4×4 nonreconfigurable wavelength router with $M = 4$ is illustrated in Fig. 2.24. The router is nonreconfigurable because the path of a given wavelength channel, after it enters the router on a particular input fiber, is fixed. The wavelengths on each incoming fiber are separated using a grating demultiplexer. And finally, information from multiple WDM channels are multiplexed before launching them back onto an output fiber. In between the demultiplexers and multiplexers, there are direct connections from each demultiplexer output to each multiplexer input. Which wavelength on which input port gets routed to which output port depends on a "routing matrix" characterizing the router; this matrix is determined by the internal "connections" between the demultiplexers and multiplexers.

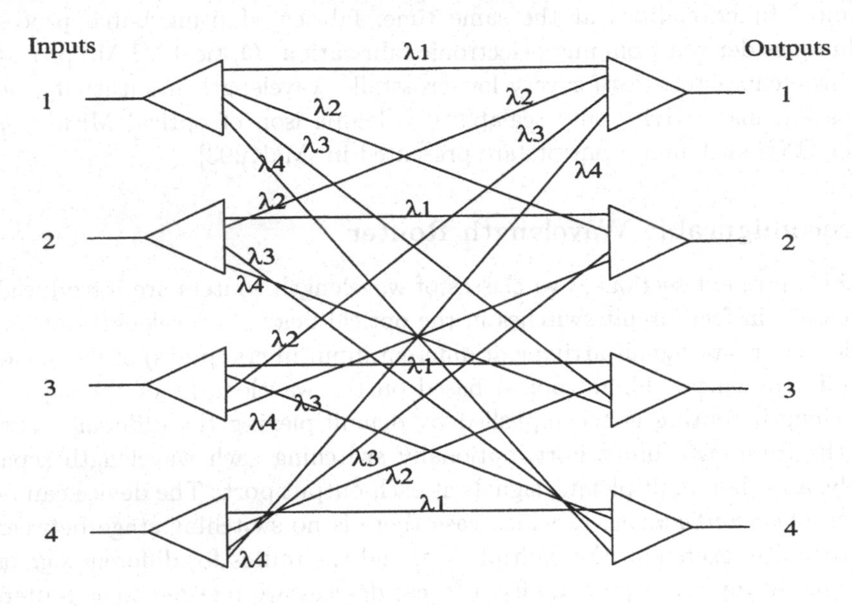

Figure 2.24 A 4 × 4 nonreconfigurable wavelength router.

Arrayed Waveguide Grating

One implementation of a wavelength router is the arrayed waveguide grating (AWG) multiplexer. An AWG provides a fixed routing of an optical signal from a given input port to a given output port based on the wavelength of the signal. Signals of different wavelengths coming into an input port will each be routed to a different output port. Also, different signals using the same wavelength can be input simultaneously to different input ports, and still not interfere with each other at the output ports. Compared to a passive-star coupler in which a given wavelength may only be used on a single input port, the AWG with N input and N output ports is capable of routing a maximum of N^2 connections, as opposed to a maximum of N connections in the passive-star coupler. Also, because the AWG is an integrated device, it can easily be fabricated at low cost. The disadvantage of the AWG is that it is a device with a fixed routing matrix which cannot be reconfigured.

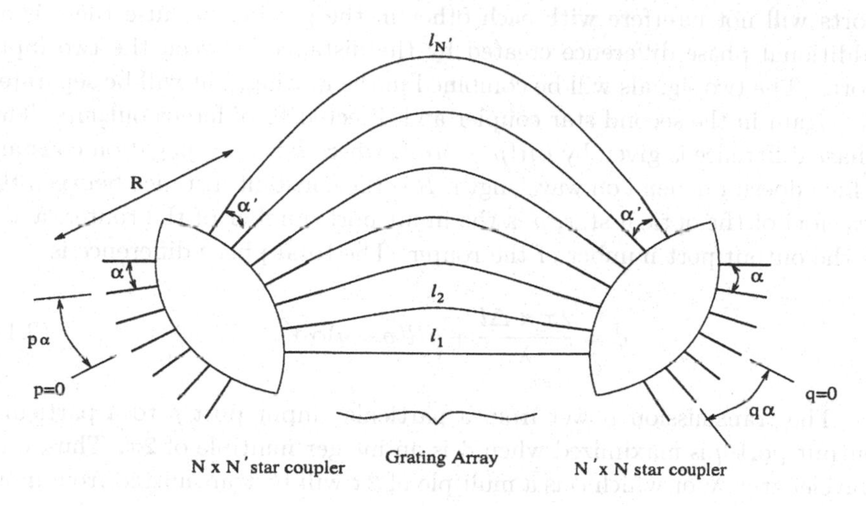

Figure 2.25 The arrayed waveguide grating (AWG).

The AWG, shown in Fig. 2.25, can be used as a nonreconfigurable wavelength-router, or it can be used to build a tunable optical transmitter or a tunable optical receiver. It consists of two passive-star couplers connected by a grating array. The first star coupler has N inputs and N' outputs (where $N \ll N'$), while the second one has N' inputs and N outputs. The inputs to the first star are separated by an angular distance of α, and their outputs are separated by angular distance α'. The grating array consists of N' waveguides, with lengths $l_1, l_2, \ldots, l_{N'}$ where $l_1 < l_2 < \cdots < l_{N'}$. The difference in length between any two adjacent waveguides is a constant Δl.

In the first star coupler, a signal on a given wavelength entering from any of the input ports is split and transmitted to its N' outputs which are also the N' inputs of the grating array. The signal travels through the grating array, experiencing a different phase shift in each waveguide depending on the length of the waveguides and the wavelength of the signal. The constant difference in the lengths of the waveguides creates a phase difference of $\beta \times \Delta l$ in adjacent waveguides, where $\beta = 2\pi n_{\text{eff}}/\lambda$ is the propagation constant in the waveguide, n_{eff} is the effective refractive index of the waveguide, and λ is the wavelength of the light. At the input of the second star coupler, the phase difference in the signal will be such that the signal will constructively recombine only at a single output port.

Two signals of the same wavelength coming from two different input ports will not interfere with each other in the grating because there is an additional phase difference created by the distance between the two input ports. The two signals will be combined in the grating, but will be separated out again in the second star coupler and directed to different outputs. This phase difference is given by $kR(p-q)\alpha\alpha'$, where k is a propagation constant which doesn't depend on wavelength, R is the constant distance between the two foci of the optical star, p is the input port number of the router, and q is the output port number of the router. The total phase difference is:

$$\phi = \frac{2\pi \times \Delta l}{\lambda} + kR(p-q)\alpha\alpha' \tag{2.11}$$

The transmission power from a particular input port p to a particular output port q is maximized when ϕ is an integer multiple of 2π. Thus, only wavelengths λ for which ϕ is a multiple of 2π will be transmitted from input port p to output port q. Alternately, for a given input port and a given wavelength, the signal will only be transmitted to the output port which causes ϕ to be a multiple of 2π.

Prototype devices have been built on silicon with $N = 11$ and $N' = 11$, giving a channel spacing of 16.5 nm; and $N = 7$ and $N' = 15$, giving a channel spacing of 23.1 nm [DrEK91]. In [ZiDJ92], a 15×15 waveguide grating multiplexer on InP is demonstrated to have a free spectral range of 10.5 nm and channel spacing of 0.7 nm in the 1550-nm region. In [OkMS95], a 64×64 arrayed-waveguide multiplexer on silicon is demonstrated with a channel spacing of 0.4 nm. WGRs with flat passbands have also been developed [OkSu96, TrBe97]. Other applications of the AWG, such as tunable transmitters and tunable receivers, are presented in [GlKW94]. These tunable components are implemented by integrating the AWG with switched amplifier elements. An amplifier element may either be activated, in which case it amplifies the signal passing through it, or it may be turned off, in which case it prevents any signals from passing through it. By using only a single input port of the AWG, each wavelength on that input port will be routed to a different output port. By placing an amplifier element at each output port of the AWG, we may filter out selected wavelengths by activating or deactivating the appropriate amplifiers. The outputs of the amplifier elements may then be multiplexed into a signal containing only the desired wavelengths.

2.6.6 Reconfigurable Wavelength-Routing Switch

A reconfigurable wavelength-routing switch (WRS), also referred to as a wavelength selective crossconnect (WSXC), uses photonic switches inside the routing element. The functionality of the reconfigurable WRS, illustrated in Fig. 2.26, is as follows. The WRS has P incoming fibers and P outgoing fibers. On each incoming fiber, there are M wavelength channels. Similar to the nonreconfigurable router, the wavelengths on each incoming fiber are separated using a grating demultiplexer.

The outputs of the demultiplexers are directed to an array of M $P \times P$ optical switches between the demultiplexer and the multiplexer stages. All signals on a given wavelength are directed to the same switch. The switched signals are then directed to multiplexers which are associated with the output ports. Finally, information streams from multiple WDM channels are multiplexed before launching them back onto an output fiber.

Space-division optical-routing switches may be built from 2×2 optical crosspoint elements [ScAl90] arranged in a banyan-based fabric. The space-division switches (which may be one per wavelength [ShCS93]) can route a signal from any input to any output on a given wavelength. Such switches based on relational devices [Hint90] are capable of switching very high-capacity signals. The 2×2 crosspoint elements that are used to build the space-division switches may be slowly tunable and they may be reconfigured to adapt to changing traffic requirements. Switches of this type can be constructed from off-the-shelf components available today.

Networks built from such switches are more flexible than passive, nonreconfigurable, wavelength-routed networks, because they provide additional control in setting up connections. The routing is a function of both the wavelength chosen at the source node, as well as the configuration of the switches in the network nodes.

Most of the switches discussed above are *relational devices*, i.e., they are useful in a circuit-switched environment where a connection may be set up over long periods of time. Besides *wavelength routing*, *Optical Packet Switching* (OPS) and *Optical Burst Switching* (OBS) have attracted interest in both research and industry. In OPS and OBS, the data path is fully optical, but the control of the switching operation is performed electronically. By employing the electronic control, relationship of data switching is established in optical domain. Essentially, OBS and OPS are based on *relational switching*, which will be studied in Chapters 17 and 18.

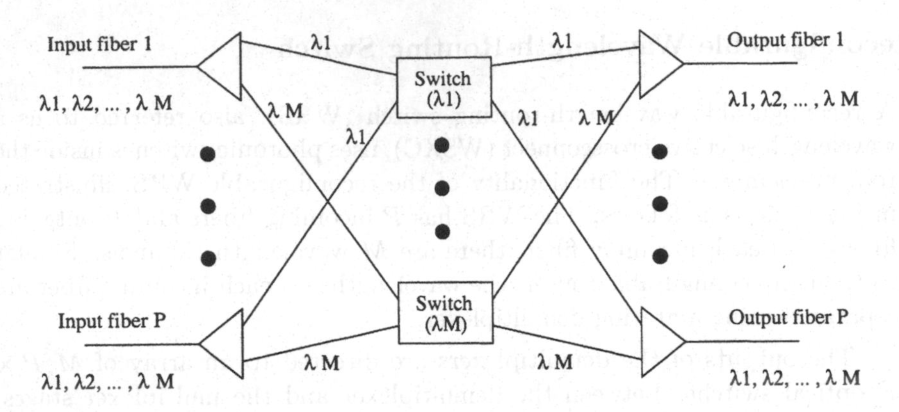

Figure 2.26 A P × P reconfigurable wavelength-routing switch with M wavelengths.

2.7 Wavelength Conversion

Consider the network in Fig. 2.27. It shows a wavelength-routed network containing two WDM crossconnects (S1 and S2) and five access stations (A through E). Three lightpaths have been set up (C to A on wavelength λ_1, C to B on λ_2, and D to E on λ_1). To establish a lightpath, we require that the *same* wavelength be allocated on all the links in the path. This requirement is known as the *wavelength-continuity constraint* (e.g., see [BaMu96]). This constraint distinguishes the wavelength-routed network from a circuit-switched network which blocks calls only when there is no capacity along any of the links in the path assigned to the call. Consider the example in Fig. 2.28(a). Two lightpaths have been established in the network: (i) between Node 1 and Node 2 on wavelength λ_1, and (ii) between Node 2 and Node 3 on wavelength λ_2. Now suppose a lightpath between Node 1 and Node 3 needs to be set up. Establishing such a lightpath is impossible even though there is a free wavelength on each of the links along the path from Node 1 to Node 3. This is because the available wavelengths on the two links are *different.* Thus, a wavelength-continuity network may suffer from higher blocking as compared to a circuit-switched network.

It is easy to eliminate the wavelength-continuity constraint, if we were able to *convert* the data arriving on one wavelength along a link into another wavelength at an intermediate node and forward it along the next link. Such a technique is referred to as *wavelength conversion.* In Fig. 2.28(b), a wave-

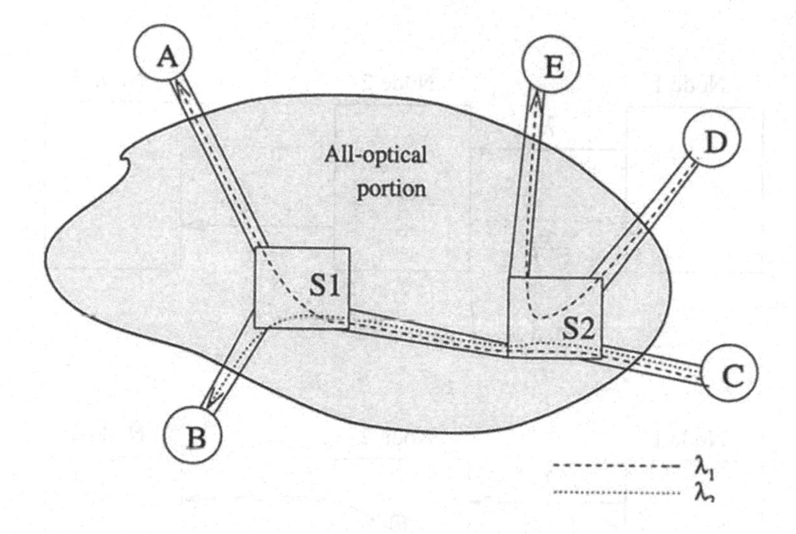

Figure 2.27 An all-optical wavelength-routed network.

length converter at Node 2 is employed to convert data from wavelength λ_2 to λ_1. The new lightpath between Node 1 and Node 3 can now be established by using the wavelength λ_2 on the link from Node 1 to Node 2, and then by using the wavelength λ_1 to reach Node 3 from Node 2. Notice that a single lightpath in such a *wavelength-convertible* network can use a different wavelength along each of the links in its path. Thus, wavelength conversion may improve the efficiency in the network by resolving the wavelength conflicts of the lightpaths. The impact of wavelength conversion on WDM wide-area network (WAN) design is further elaborated in Section 2.8.6.

The function of a wavelength converter is to convert data on an input wavelength onto a possibly different output wavelength among the N wavelengths in the system (see Fig. 2.29). In this figure and throughout this section, λ_s denotes the input signal wavelength; λ_c, the converted wavelength; λ_p, the pump wavelength; f_s, the input frequency; f_c, the converted frequency; f_p, the pump frequency; and CW, the continuous wave (unmodulated) generated as the pump signal.

An ideal wavelength converter should possess the following characteristics [DMJD96]:

- transparency to bit rates and signal formats,

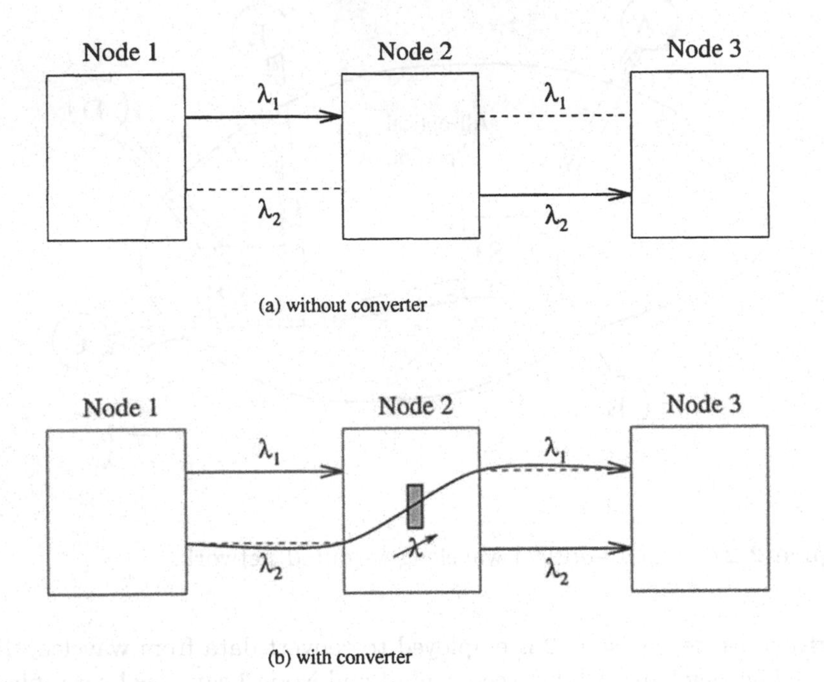

Figure 2.28 Wavelength-continuity constraint in a wavelength-routed network.

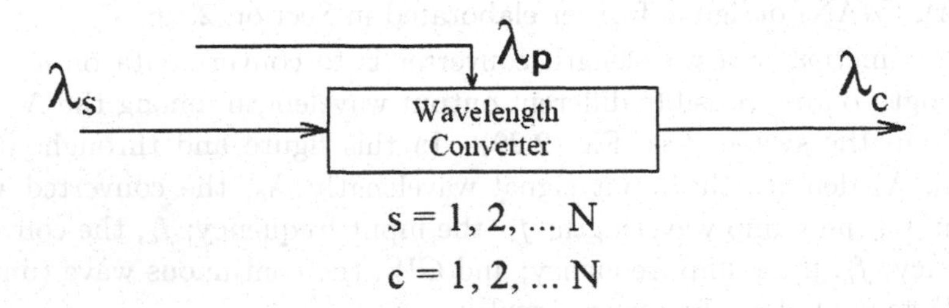

Figure 2.29 Functionality of a wavelength converter.

- fast setup time of output wavelength,
- conversion to both shorter and longer wavelengths,
- moderate input power levels,
- possibility for same input and output wavelengths (i.e., no conversion),
- insensitivity to input signal polarization,
- low-chirp output signal with high extinction ratio[11] and large signal-to-noise ratio, and
- simple implementation.

2.7.1 Wavelength Conversion Technologies

Several researchers have attempted to classify and compare the several techniques available for wavelength conversion [DMJD96, MDJD96, SaIa96, Wies96, Yoo96]. The classification of these techniques presented in this section follows that in [Wies96]. Wavelength conversion techniques can be broadly classified into two types: *opto-electronic wavelength conversion*, in which the optical signal must first be converted into an electronic signal; and *all-optical wavelength conversion*, in which the signal remains in the optical domain. All-optical conversion techniques may be subdivided into techniques which employ *coherent effects* and techniques which use *cross modulation.*

Opto-Electronic Wavelength Conversion

In opto-electronic wavelength conversion [Fuji88], the optical signal to be converted is first translated into the electronic domain using a photodetector (labeled R in Fig. 2.30, from [Mest95]). The electronic bit stream is stored in the buffer (labeled FIFO for the First-In-First-Out queue mechanism). The electronic signal is then used to drive the input of a tunable laser (labeled T) tuned to the desired wavelength of the output (see Fig. 2.30). This method has been demonstrated for bit rates up to 10 Gbps [Yoo96]. However, this method is much more complex and consumes a lot more power than the

[11] The *extinction ratio* is defined as the ratio of the optical power transmitted for a bit "1" to the power transmitted for a bit "0."

other methods described below [DMJD96]. Moreover, the process of opto-electronic (O/E) conversion adversely affects the transparency of the signal, requiring the optical data to be in a specified modulation format and at a specific bit rate. All information in the form of phase, frequency, and analog amplitude of the optical signal is lost during the conversion process.

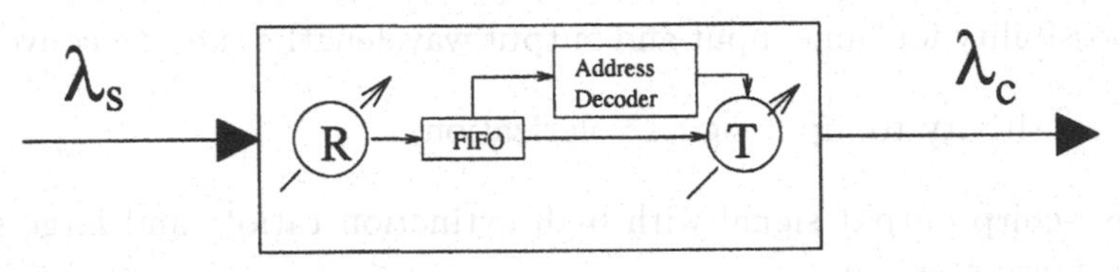

Figure 2.30 An opto-electronic wavelength converter.

Wavelength Conversion Using Coherent Effects

Wavelength conversion methods using coherent effects are typically based on wave-mixing properties(see Fig. 2.31). Wave-mixing arises from a nonlinear optical response of a medium when more than one wave is present. It results in the generation of another wave whose intensity is proportional to the product of the interacting wave intensities. Wave-mixing preserves both phase and amplitude information, offering strict transparency. It is also the only approach that allows simultaneous conversion of a set of multiple input wavelengths to another set of multiple output wavelengths and could potentially accommodate signal with high-bit rates. In Fig. 2.31, the value $n = 3$ corresponds to Four-Wave Mixing (FWM) and $n = 2$ corresponds to Difference Frequency Generation (DFG). These techniques are described below.

- *Four-Wave Mixing (FWM)*: FWM (also referred to as four-photon mixing) is a third-order nonlinearity in silica fibers, which causes three optical waves of frequencies f_i, f_j, and f_k $(k \neq i, j)$ to interact in a multichannel WDM system [TCFG95] to generate a fourth wave of frequency given by:

  $$f_{ijk} = f_i \pm f_j \pm f_k$$

  FWM is also achievable in other passive waveguides such as semiconductor waveguides and in an active medium such as a semiconductor

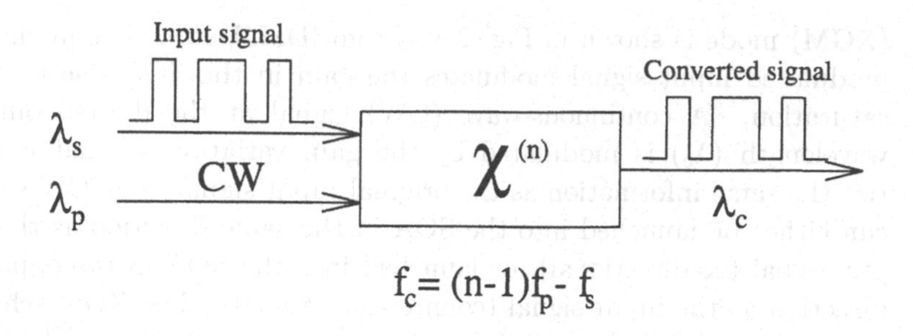

Figure 2.31 A wavelength converter based on nonlinear wave-mixing effects.

optical amplifier (SOA). Four-wave mixing (FWM) is a promising technique for wavelength conversion in optical networks owing to its ultrafast response and high transparency to bit rate and modulation format. In [CSLB05], a 3-dB conversion range over 40 nm (1535C1575 nm) is obtained with a flat conversion efficiency of -16 dB and a polarization sensitivity of less than 0.3 dB.

- *Difference Frequency Generation (DFG)*: DFG is a consequence of a second-order nonlinear interaction of a medium with two optical waves: a pump wave and a signal wave [Yoo96]. DFG is free from satellite signals which appear in FWM-based techniques. This technique offers a full range of transparency without adding excess noise to the signal [YCBK95]. It is also bidirectional and fast, but it suffers from low efficiency and high polarization sensitivity. The main difficulties in implementing this technique lie in the phase-matching of interacting waves [YCBK96] and in fabricating a low-loss waveguide for high conversion efficiency [Yoo96].

Wavelength Conversion Using Cross Modulation

Cross-modulation wavelength conversion techniques utilize active semiconductor optical devices such as semiconductor optical amplifiers (SOAs) and lasers. These techniques belong to a class known as optical-gating wavelength conversion [Yoo96].

- *Semiconductor Optical Amplifiers (SOAs) in XGM and XPM mode*: The principle behind using an SOA in the cross-gain modulation

(XGM) mode is shown in Fig. 2.32 (from [DMJD96]). The intensity-modulated input signal modulates the gain in the SOA due to gain saturation. A continuous-wave (CW) signal at the desired output wavelength (λ_c) is modulated by the gain variation so that it carries the same information as the original input signal. The CW signal can either be launched into the SOA in the same direction as the input signal (co-directional), or launched into the SOA in the opposite direction as the input signal (counter-directional). The XGM scheme gives a wavelength-converted signal that is inverted compared to the input signal. While the XGM scheme is simple to realize and offers penalty-free conversion at 10 Gbps [DMJD96], it suffers from the drawbacks due to inversion of the converted bit stream and extinction ratio degradation for the converted signal. But, new advanced developments demonstrate that no signal inversion occurs in [KLKH05].

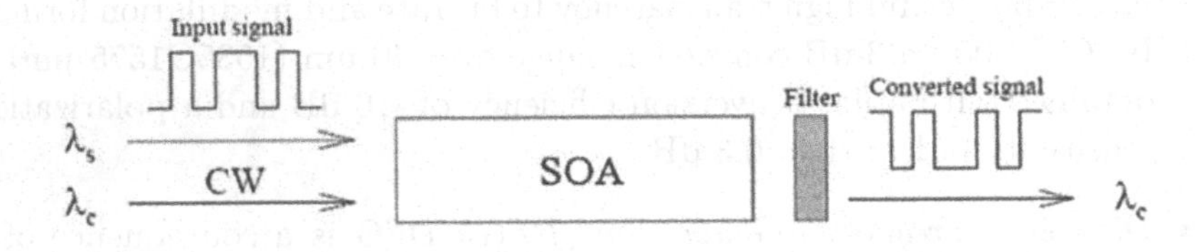

Figure 2.32 A wavelength converter using co-propagation based on XGM in an SOA.

The operation of a wavelength converter using SOA in cross-phase modulation (XPM) mode is based on the fact that the refractive index of the SOA is dependent on the carrier density in its active region. An incoming signal that depletes the carrier density will modulate the refractive index and thereby results in phase modulation of a CW signal (wavelength λ_c) coupled into the converter [DMJD96, LaPT96]. The SOA can be integrated into an interferometer so that an intensity-modulated signal format results at the output of the converter. Techniques involving SOAs in XPM mode have been proposed using nonlinear optical loop mirrors (NOLMs) [EiPW93], Mach-Zender interferometers (MZI) [DJMP94] and Michelson interferometers (MI) [MDJP94]. Figure 2.33 shows an asymmetric MZI wavelength converter based on SOA in XPM mode (from [DMJD96]). With the XPM scheme, the converted output signal can be either inverted or noninverted, unlike in the XGM scheme where the output is always inverted. The XPM scheme is also very power efficient compared to the XGM scheme [DMJD96]. A signal up-conversion utilizing an XPM effect in an all-optical SOA-MZI wavelength converter is demonstrated in [SoLS04], in which this scheme not only shows high conversion efficiency, polarization immunity, and no increase in phase noise, but also linear signal up-conversion with a low optical power requirement.

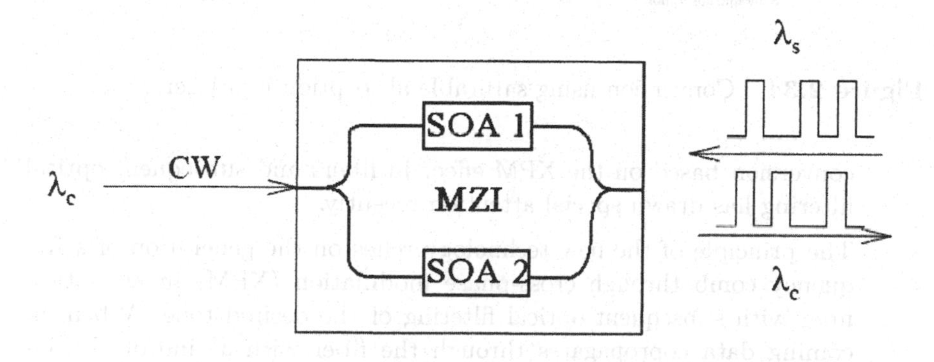

Figure 2.33 An interferometric wavelength converter based on XPM in SOAs.

- *Semiconductor Lasers*: Using single-mode semiconductor lasers, lasing-mode intensity is modulated by input signal light through lasing-mode gain saturation. The obtained output (converted) signal is inverted compared to the input signal. This gain suppression mechanism has been employed in a Distributed Bragg Reflector (DBR) laser to convert signals at 10 Gbps [YSIY96]. In the method using saturable absorption in lasers (e.g., [KOYM87]), the input signal saturates the absorption of carrier transitions near the band gap and allows the probe beam to transmit (see Fig. 2.34). This technique shows a bandwidth limit of 1 GHz due to carrier recombinations [Yoo96].

- *All-Optical Wavelength Conversion Based on XPM in Optical Fiber*: In particular, besides the classical methods described above, wavelength

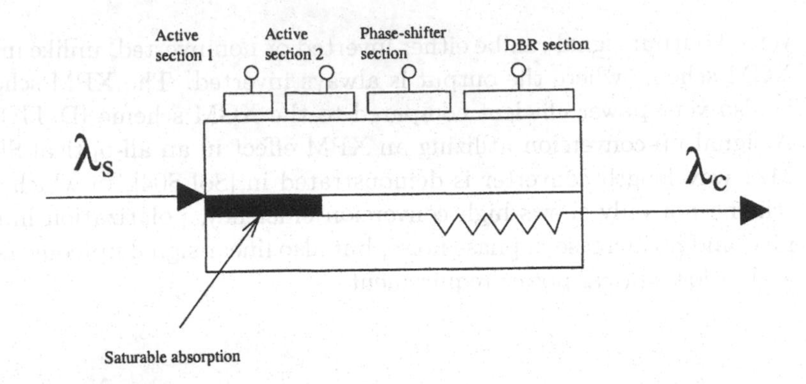

Figure 2.34 Conversion using saturable absorption in a laser.

conversion based on the XPM effect in fibers and subsequent optical filtering has drawn special attention recently.

The principle of the new technology relies on the generation of a frequency comb through cross-phase modulation (XPM) in an optical fiber with subsequent optical filtering of the desired tone. When incoming data copropagates through the fiber with a sinusoidally intensity modulated high-power optical pump signal, the data acquires a sinusoidal phase modulation from the pump through XPM, which generates multiple wavelengths spaced at the modulation frequency on both sides of the incoming signal wavelength, i.e., generating sidebands around the incoming signal wavelength. By filtering out a portion of the sidebands, a wavelength-converted signal at the output is obtained. These new components, the strengths of which depend on the phase-modulation index (i.e., the optical pump power), are imprinted with the information carried by the original wavelength without any distortion, assuming the signal bandwidth is much smaller than the pump modulation frequency. Compared to other techniques, this method has several important advantages: it is environmentally stable, operates in a wide signal wavelength range, and provides a short switching window with relatively broad signal pulses. Moreover, the scheme offers additional attractive functions such as waveform reshaping and phase-reconstruction.

A simple wavelength converter based on sinusoidal cross-phase modulation in optical fibers, which offers format- and bit-rate independence is demonstrated in [MaAT05]. Conversion to more than 20 channels is demonstrated and error-free conversion of data with bit rates of up to 10 Gbps is also reported. In [TSKK05], polarization-independent all-optical wavelength conversion is demonstrated by using cross-phase modulation in a twisted fiber and subsequent optical filtering. By twisting a 1-km-long fiber at a rate of 15 turns/m and aligning the probe light to circular polarization, the polarization sensitivity is successfully reduced from 3.5 to 0.3 dB. Error-free operation with only 1-dB penalty is realized at 40 Gbps with the input signal polarization scrambled. The demonstrated scheme also offers a novel function of restoring the degree-of-polarization of the signal light. The experimental setup is shown in Fig. 2.35.

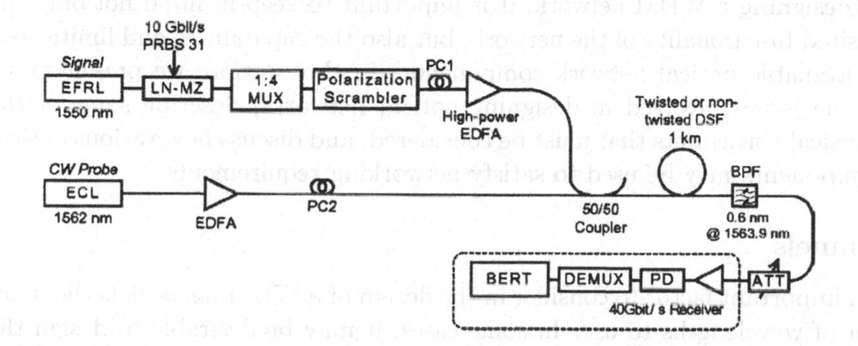

Figure 2.35 Experimental setup for demonstrating the polarization-insensitive 40-Gbps WC based on fiber XPM and optical filtering. ATT: Variable attenuator, PD: Photodetector, DEMUX: Electrical demultiplexer, BERT: BER tester.

Summary

In this subsection, we reviewed the various techniques and technologies used in the design of a wavelength converter. The actual choice of the technology to be employed for wavelength conversion in a network depends on the requirements of the particular system. However, it is clear that opto-electronic converters offer only limited digital transparency. Moreover, deploying multiple opto-electronic converters in a WDM intermediate node requires sophisticated packaging to avoid crosstalk among channels. This leads to increased cost per converter, further making this technology less attractive than all-optical converters [Yoo96]. Other disadvantages of opto-electronic converters include complexity and large power consumption [DMJD96]. Regarding all-optical wavelength conversion, there are a lot of various schemes developed in which each has own strength and week in application. Besides all-optical converters mentioned above, another kind of wavelength conversion is based on crystal material, which has also attracted more attention currently. Specially, *periodically-poled $LiNbO_3$ (PPLN)* waveguide is preferential in this kind of all-optical wavelength conversion scheme in [YuGu04].

2.8 Designing WDM Networks: Systems Considerations

In designing a WDM network, it is important to keep in mind not only the desired functionality of the network, but also the capabilities and limitations of available optical network components. In this section, we present some of the issues involved in designing optical networks, describe some of the physical constraints that must be considered, and discuss how various optical components may be used to satisfy networking requirements.

2.8.1 Channels

An important factor to consider in the design of a WDM network is the number of wavelengths to use. In some cases, it may be desirable to design the network with the maximum number of channels attainable with the current device technology, subject to tuning time requirements and cost constraints. Another approach is to assign a different wavelength to each node, although this type of network doesn't scale very well. In wide-area networks (WANs), the objective is often to minimize the number of wavelengths for a desired network topology or traffic pattern. In any case, the maximum number of wavelengths is limited by the optical device technology. The number of channels is affected primarily by the total available bandwidth or spectral range of the components and the channel spacing. In addition, channel spacing is standardized by the ITU grid.

The bandwidth of the fiber medium, as mentioned in Section 2.2, is limited to the low-attenuation regions around 1310 nm and 1550 nm. These regions have bandwidths of approximately 200 nm (25 THz) each. However, optical networks may not necessarily be able to take advantage of this entire range due to the bandwidth limitations of optical components. Amplifiers have an optical bandwidth of around 35 to 40 nm, injection-current tunable lasers have a tuning range of around 10 nm, while the tuning range of tunable receivers varies from the entire low-attenuation region of fiber for Fabry-Perot filters to around 16 nm for electrooptic filters.

Some factors which affect the channel spacing are the channel bit rates, the optical power budget, nonlinearities in the fiber, and the resolution of transmitters and receivers. We now illustrate how some of these parameters may relate to the maximum number of channels in a WDM system. We will assume that tunable transmitters and receivers are being used, and we would like to design a WDM passive-star-based network for N nodes.

Let ΔT be the tuning range of the transmitters and let ΔR be the tuning range of the receivers (both are measured in nm). The available transceiver bandwidth, BW_T is given by the frequency range in which the transmitter tuning range intersects with the receiver tuning range.

Using the identity

$$\Delta f = \frac{c\Delta\lambda}{\lambda^2}$$

the frequency needed for BW_T is

$$\Delta f = \frac{cBW_T}{\lambda^2}$$

where λ is the "center" wavelength.

If we want each channel to have a bit rate of B Gbps, $2B$ GHz of bandwidth will be needed for encoding, assuming a modulation efficiency of 2 Hz/bps. According to [Brac90], a channel spacing of at least 6 times the channel bit rate is needed to minimize crosstalk on a WDM system. Thus, if we want W channels, we need

$$2B \cdot W + 6B(W-1) \text{ GHz}$$

Thus, the maximum number of resolvable channels for this network is

$$W = \frac{\Delta f + 6B}{8B}$$

Although a maximum of W channels may be accommodated, in some cases, it may be desirable to use fewer than W channels, e.g., in a shared-channel WDM optical LAN [TrMu96, TrMu97]. A higher number of channels may provide more network capacity, but it also results in higher network costs, and in some cases, may require more complex protocols.

2.8.2 Power Considerations

In any network, it is important to maintain adequate signal-to-noise ratio (SNR) in order to ensure reliable detection at the receiver. In a WDM network, signal power can degrade due to losses such as attenuation in the fiber, splitting losses, and coupling losses. Some of the losses may be countered through the use of optical amplifiers, and an important consideration in designing a WDM network is the design and appropriate placement of amplifiers.

There are three main applications for optical amplifiers in a lightwave network [Gree93]. The first application is as a *transmitter power booster*, which is placed immediately after the transmitter in order to provide a high power signal to the network. This allows the signal to undergo splitting at couplers or to travel longer distances. The second application is as a *receiver preamplifier*, which boosts the power of a signal before detection at a receiver photodetector. The third application is as an *in-line amplifier*, which is used within the network to boost degraded signals for further propagation. Each of these situations requires the amplifier to have different characteristics. A discussion of the requirements and design of multistage EDFAs for various applications is given in [DeNa95]. Table 2.6 summarizes some of these requirements.

Table 2.6 Requirements for EDFA applications.

EDFA Application	*Gain*	*Noise Figure*	*Saturated Output*
Transmitter power booster	Moderate	Moderate	High
Preamplifier	High	Low	Moderate
In-line amplifier	High	Low	High

For in-line amplifier applications, there is the additional issue of amplifier placement. Amplifiers need to be placed strategically throughout the network in a way which guarantees that all signals are adequately amplified while minimizing the total number of amplifiers being used.

When utilizing cascades of in-line amplifiers, one must also consider issues such as ASE noise introduced by the amplifiers, and the unequal gain spectrum of the amplifiers. The accumulation of ASE noise in a cascade of amplifiers may seriously degrade the SNR. If the input signal power is too low, ASE noise may cause the SNR to fall below detectable levels; however, if the signal power is too high, the signal combined with ASE noise may saturate the amplifiers. The unequal gain spectrum of the EDFA places limitations on the usable bandwidth in WDM systems. When multiple EDFAs are cascaded, the resulting gain bandwidth may be significantly reduced from the gain bandwidth of a single EDFA. An initial bandwidth of 30 nm can potentially be reduced to less than 10 nm after a cascade of 50 EDFAs [Maho95]. That is just the reason that we proposed the elimination of all-optical cycle in [Mukh97], which will be further discussed in next section.

Although recent developments in amplifier technology have solved many of the power-loss and noise problems in optical networks, network designers should not rely solely on amplifiers for resolving power issues, but should also consider other options. For example, to avoid splitting losses in network interconnections, it might be worthwhile to consider using wavelength-routing devices, such as the wavelength-routing switch (WRS) or the arranged waveguide grating (AWG), instead of wavelength-independent devices, such as the amplifier gate switch or the passive-star coupler.

2.8.3 All-optical Cycle of Elimination

Obviously, all-optical cycle of elimination is a key issue in transparent optical networks. *All-optical cycle* is referred to a situation in which there exists the possibility of setting up unintended all-optical cycles in the optical network (i.e., a loop with no terminating electronics in it).

A transparent (wide-area) wavelength-routed optical network may be constructed by using wavelength cross-connect switches connected together by fiber to form an arbitrary mesh structure. The network is accessed through electronic stations that are attached to some of these cross-connects. These wavelength cross-connect switches have the property that they may configure themselves into unspecified states. Each input port of a switch is always connected to some output port of the switch whether or not such a connection is required for the purpose of information transfer.

The presence of these unspecified states result possibly in such an all-optical cycle. If such a cycle contains amplifiers [e.g., Erbium-Doped Fiber Amplifiers (EDFA's)], there exists the possibility that the net loop gain is greater than the net loop loss. The amplified spontaneous emission (ASE) noise from amplifiers can build up in such a feedback loop to saturate the amplifiers and result in oscillations of the ASE noise in the loop.

Such all-optical cycles as defined above must be eliminated from an optical network in order for the network to perform any useful operation. Furthermore, for the realistic case in which the wavelength cross-connects result in signal crosstalk, there is a possibility of having closed cycles with oscillating crosstalk signals.

Some algorithms are proposed in [Mukh97], which attempt to find a route for a connection and then (in a postprocessing fashion) configure switches such that all-optical cycles that might get created would automatically get eliminated. In addition, call-set-up algorithms are proposed, which avoid the possibility of crosstalk cycles in [Mukh97].

2.8.4 Additional Considerations

Other device considerations in the design of WDM networks include *crosstalk*, *dispersion*, and so on.

Crosstalk may either be caused by signals on different wavelengths (interband crosstalk or hetero-wavelength), or by signals on the same wavelength on another fiber (intraband crosstalk or home-wavelength) [Maho95]. Interband crosstalk must be considered when determining channel spacing. In some cases, intraband crosstalk may be removed through the use of appropriate narrowband filters. Intraband crosstalk usually occurs in switching nodes where multiple signals on the same wavelength are being switched from different inputs to different outputs. This form of crosstalk is more of a concern than interband crosstalk because intraband crosstalk cannot be removed through filtering, and may accumulate over a number of nodes. The degree of intraband crosstalk depends in part on the switch architecture.

As was mentioned in Section 2.2.4, dispersion in an optical communication system causes a pulse to broaden as it propagates along the fiber. The pulse broadening limits the spacing between bits, and thus limits the maximum transmission rate for a given propagation distance. Alternatively, it limits the maximum fiber distance for a given bit rate.

Apart from the device considerations mentioned above, there are architectural considerations in designing a WDM network. The topology of the physical optical fiber buried in the ground may influence the choice of which transmitter-receiver pairs to operate on which wavelengths. The need for fault-tolerance and reliability affects the choice of the network architectures. Moreover, the standards on optical wavelengths and channel spacing (e.g., ITU-T) will influence the design of the network components.

2.8.5 Elements of Local-Area WDM Network Design

A local area WDM network will typically consist of a number of nodes which are connected via two-way optical fibers either to some physical network medium or directly to other nodes. In this section, we will present some of the issues involved in selecting the hardware for both the network medium and the nodes.

The Network Medium

The simplest and most popular interconnection device for a local-area WDM network is the passive-star coupler which provides a broadcast medium (see Fig. 2.36). The broadcast capability of the star coupler combined with multiple WDM channels allows for a wide range of possible media access protocols [Mukh92a, Mukh92b]. Also, since the star coupler is a passive device, it is fairly reliable. The drawback of having a passive network medium is that the network nodes may be required to handle additional processing and may require additional hardware in order to route and schedule transmissions. The broadcast capability of the star coupler also prevents the reuse of wavelengths to create more simultaneous connections.

Network Nodes

Another important consideration in the design of a WDM network is the hardware at each node. Each node in a network typically consists of a workstation connected to the network medium via optical fiber, and the node may potentially access any of the available wavelength channels on each fiber. In designing the network interface for each node, one must select the number of transmitters and receivers, as well as the type of transmitters and receivers – fixed-tuned or tunable – to place at each node. These decisions usually depend on the protocol, degree of access, and connectivity desired in the network, as well as on practicality and cost considerations.

A WDM network protocol may either be a single-hop protocol [Mukh92a], in which communication takes place directly between two nodes without being routed through intermediate nodes; or a multihop protocol [Mukh92b], in which information from a source node to a destination node may be routed through the electronics at intermediate nodes in the network. In general, multihop networks require less tuning than single-hop networks.

At a minimum, each node must be equipped with at least one transmitter and one receiver. When both the transmitters and the receivers are

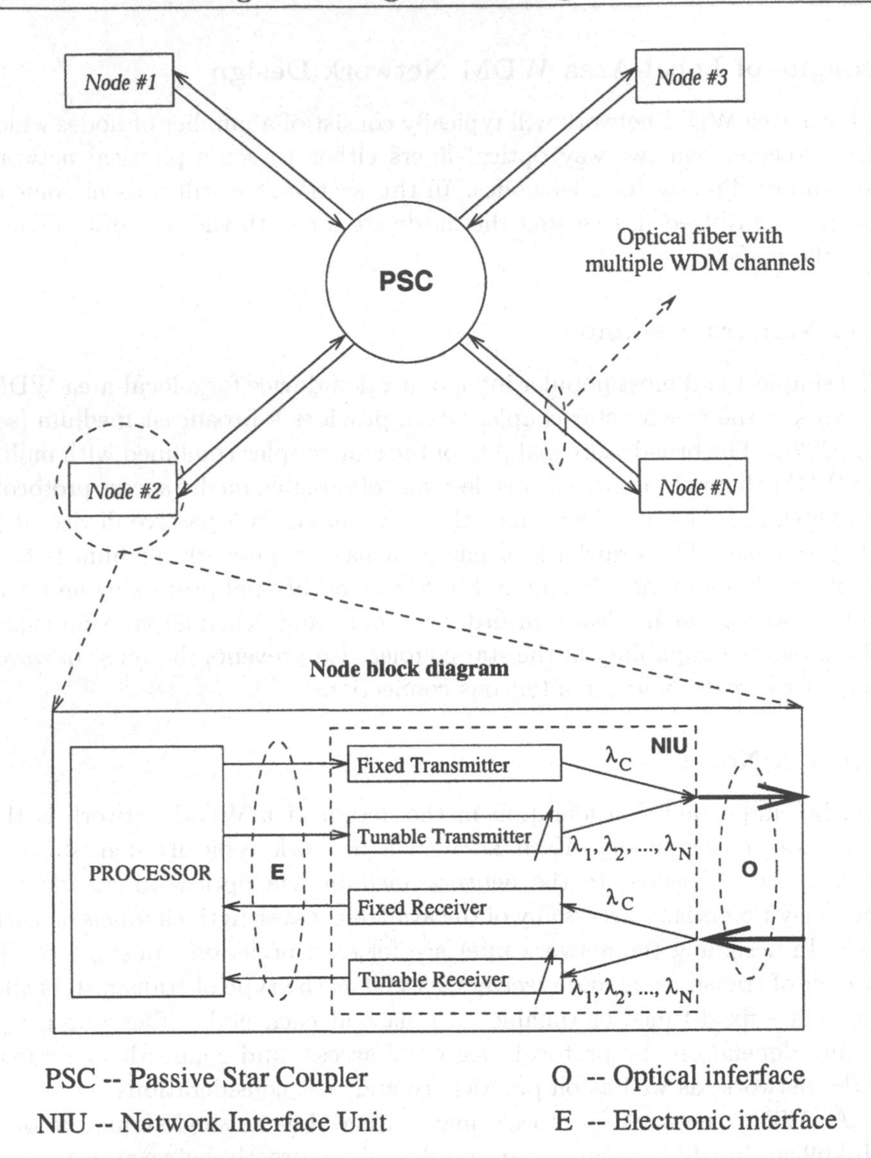

Figure 2.36 Broadcast-and-select WDM local optical network with a passive-star coupler network medium.

fixed tuned to certain wavelength channels, and there is more than one channel, then a static multihop topology must be established over the passive-star coupler. The topology is created by establishing connections between pairs of nodes on given wavelengths. An overview of multihop protocols and topologies is provided in [Mukh92b].

A more flexible approach would be to use either a tunable transmitter and/or a tunable receiver. The tunability allows the network to be dynamically reconfigured based on traffic patterns, and it also allows the implementation of single-hop protocols. A number of single-hop WDM protocols based on nodes with tunable transmitters and/or tunable receivers are presented in [Mukh92a]. Additional transmitters and receivers at each node may help to increase the connectivity of the network and may also be used to help coordinate transmissions. In some cases, the network may have a control channel which may be used for pretransmission coordination (pretransmission coordination allows a node to preannounce its transmission so that the receiving node may get ready for reception, e.g., by appropriately tuning its receiver). Each node may then be equipped with an additional fixed transmitter and an additional fixed receiver, each permanently tuned to the control channel.

The tuning latency of tunable transmitters and receivers may be an important factor in choosing components, depending on the type of network being implemented. A single-hop network generally requires tunable components to create connections on demand and requires some amount of coordination in order to have the source node's transmitter and the destination node's receiver tuned to the same channel for the duration of an information transfer. In this case, the tuning time of transmitters and receivers may have a significant impact on the performance of the network. On the other hand, most multihop networks require tunability only for infrequent reconfigurations of the network based on changing traffic patterns; thus, the tuning time of components in a multihop network is not as critical as in the case of a single-hop network.

Chronologically, the multihop concept came first precisely because fast device tuning was difficult to achieve in the early stages of developments in this field.

Node Separation in Passive-Star Coupler WDM LAN

Given the output power of the transmitters and the receiver sensitivity, we can compute the maximum distance allowable between network nodes. Assume that all nodes are D meters from the passive-star coupler (PSC), and that the input-to-output power splitting ratio of the PSC is given by Eqn. (2.9). Then, the maximum value of D such that the optical signals reaching each receiver are strong enough to be received (D_{max}) can be computed by combining Eqns. (2.9) and (2.7), so that

$$D_{max} = \frac{10}{A} \log_{10} \frac{P_t}{NP_r}$$

where P_t is the optical power of the transmitter and P_r is the minimum amount of power that the receiver needs to resolve the optical signal.

2.8.6 WDM Wide-Area Network Design Issues

Due to the limitations of optical hardware development currently, today's optical networks may not be able to take full advantage of the bandwidth provided by optical fiber. It is anticipated that the next generation of optical networks will make use of optical switching elements to allow all-optical lightpaths to be set up from a source node to a destination node, thus bypassing electronic bottlenecks at intermediate switching nodes. Also, WDM will allow multiple lightpaths to share each fiber link. The concept of WDM lightpaths is analogous to a multilane highway which can be used to bypass stoplights on city roads. Another concept in WAN design is *wavelength reuse.* Since each wavelength may be used on each fiber link in the network, multiple lightpaths which do not share any links may use the same wavelength. For example, in Fig. 2.37, wavelength λ_1 is used to set up one lightpath from node A to node C, and another lightpath from node G to node H. (Such wavelength reuse is not possible in a passive-star-based WDM network.)

The issue of setting up lightpaths and routing the lightpaths over the physical fibers and switches in a wide-area WDM network is an optimization problem in which the overall network performance must be balanced against the consumption of network resources. The degree of freedom in designing the lightpaths depends in part on the type of switching elements or crossconnects used in the access nodes or switching nodes. If wavelength-insensitive crossconnect devices are used, then each signal on a given input fiber must be routed to the same output fiber. Wavelength-sensitive switching devices offer more flexibility, allowing different wavelengths arriving on a single input fiber to be directed independently to different output fibers. However, this approach may still result in conflicts at the nodes if two signals

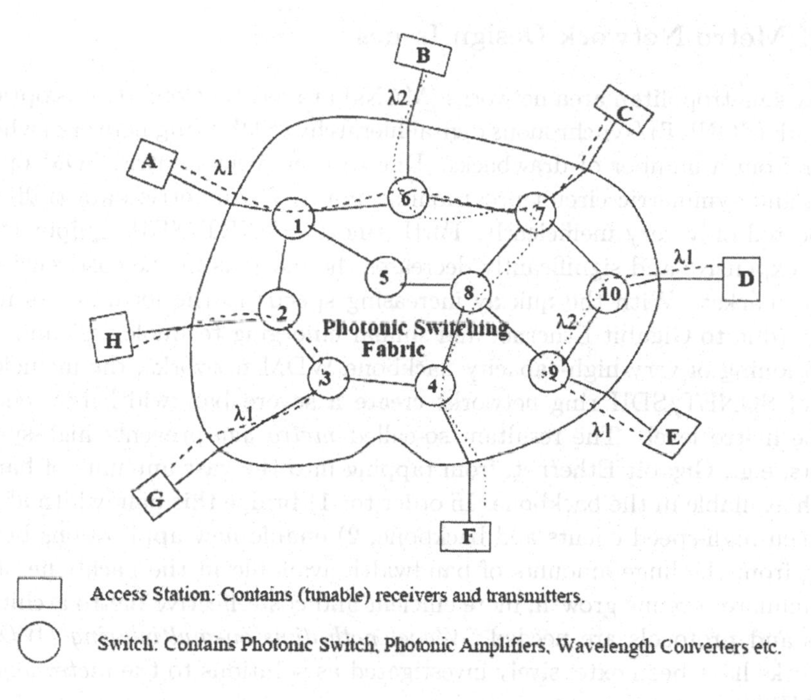

Figure 2.37 Lightpath routing in a WDM WAN.

on the same wavelength arriving on different input ports need to be routed to the same output port. The conflict can be resolved by incorporating wavelength converters at each node, and converting one of the incoming signals to a different wavelength (see Section 2.7). If wavelength-conversion facilities are not available at switching nodes, then a lightpath must have the same wavelength on all of the fiber links through which it traverses; this is referred to as the *wavelength-continuity constraint* (see Fig. 2.28). Another approach for resolving conflicts is to find an alternate route in the network for one of the two conflicting lightpaths, and in some cases an alternate wavelength.

In designing an optical network, it is important to recognize what can and cannot be accomplished by optical switching devices.

2.8.7 WDM Metro Network Design Issues

Today's metropolitan area networks (MANs) are mostly synchronous optical network (SONET)/synchronous digital hierarchy (SDH) ring networks which suffer from a number of drawbacks. Due to their voice-centric TDM operation and symmetric circuit provisioning, bursty asymmetric data traffic is supported only very inefficiently. Furthermore, SONET/SDH equipment is quite expensive and significantly decreases the margins in the cost-sensitive metro market. With the quickly increasing speeds in the local access networks (due to Gigabit Ethernet and similar emerging technologies) and the provisioning of very-high-capacity backbone WDM networks, the inefficiencies of SONET/SDH ring networks create a severe bandwidth bottleneck at the metro level. The resultant so-called *metro gap* prevents high-speed clients, e.g., Gigabit Ethernet, from tapping into the vast amounts of bandwidth available in the backbone. In order to: 1) bridge this bandwidth abyss between high-speed clients and backbone, 2) enable new applications benefiting from the huge amounts of bandwidth available in the backbone, and 3) stimulate revenue growth, more efficient and cost-effective metro architectures and protocols are needed. *Wavelength-division-multiplexing (WDM)* networks have been extensively investigated as solutions to the *metro gap* in [YHMR04].

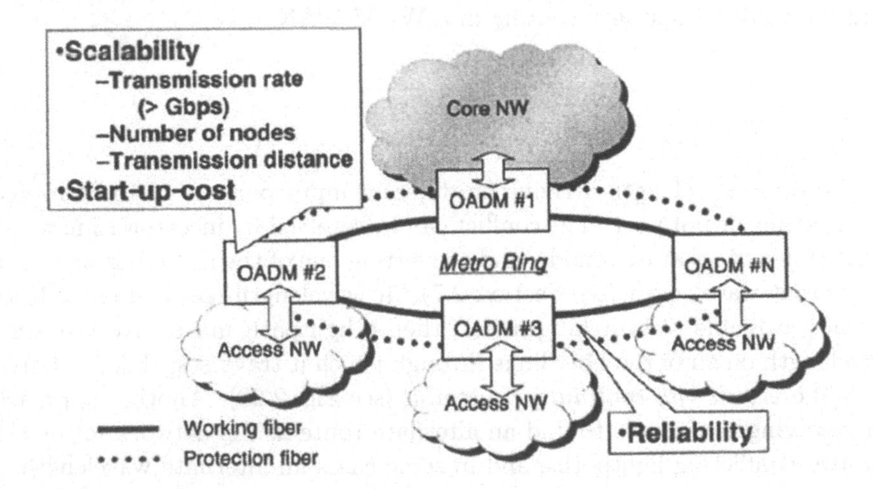

Figure 2.38 A typical WDM metro network.

As for the metro network, there are some technical issues. One is how to make best use of the scalability offered by WDM. We should design networks scalable in terms of transmission rate, number of nodes accommodated, and transmission distance (i.e., span length and total ring length). Another issue is how to reduce the start-up-cost. Some service providers are already starting to deploy scalable, cost-effective value-added metro services. But it is difficult to accurately estimate the future traffic, especially in metro networks. Therefore, an optical add/drop multiplexer (OADM) architecture with both superior scalability and low start-up-cost is necessary to install ring networks in metro areas. A typical configuration of the WDM metro network with OADMs is shown in Fig. 2.38 in [IKSF04]. The other major challenge in metro networks is to guarantee the quality of service (QoS) using intelligent traffic engineering (TE) and management schemes. By using working and protection fibers, automatic protection and supervision functions based on simple OADM are also needed to ensure service reliability. Other challenges include *operations, administration, maintenance (OAM)*, and so on.

2.8.8 Optical Access Network Design Issues

The recent explosive growth of the Internet has triggered the introduction of a broadband access network based on fiber-to-the-office (FTTO) and fiber-to-the-home (FTTH). This trend will dramatically accelerate from now on due to further progress in e-businesses such as contents delivery and TV services. To deal with the various demands, access networks requires scalability in terms of capacity and accommodation, and the flexibility with regard to physical topology. The introduction of technically mature WDM technologies that are also cost-effective in core networks to support access networks is expected to yield scalable and flexible networks.

Figure 2.39 shows a typical WDM access network in [IKSF04]. A wavelength multi/demultiplexer MUX is used to connect plural *optical network units (ONUs)* placed in homes/buildings (end users) to an *optical line terminal (OLT)*; splitting loss of the MUX can be much lower (i.e., 5 dB for 16 splits) than that of the power splitter used in TDM access (i.e., 10 dB for 8 splits). The main technical issue is to realize colorless ONUs; the ONUs should be colorless (in other words, no ONU is wavelength specific) to decrease the costs of operation, administration, and maintenance (OADM) functions, as well as the production cost since mass production becomes possible with just one specification.

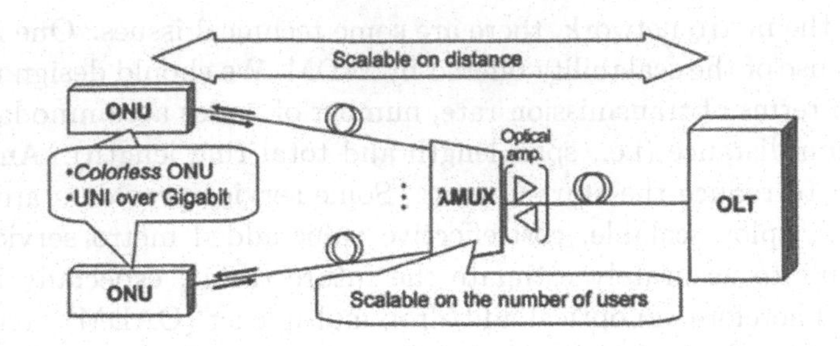

Figure 2.39 A typical WDM-access network.

Enabling technologies include: 1) employing a modulator only in the ONU; 2) employing a light source with broadband optical spectrum at each ONU; and 3) employing a tunable light source at each ONU. Since approach 3) requires wavelength setting and control in each ONU and increases the burden of OADM, approaches 1) and 2) are more attractive. Note that approach 1) has several variants; i.e., the supplied optical carriers can be generated by spectrum-slicing broadband ASE light sources, and/or a *Fabry-Perot* laser can be employed instead of the modulator in the ONU so that its wavelength is locked into that of the supplied optical carrier.

Another issue is how to make the best use of the large scalability offered by WDM. We should design scalable networks in terms of transmission distance, the number of users accommodated, and bandwidth for each user.

The another major challenge is load balancing which is highly desirable in WDM access networks. Balanced traffic distribution can be obtained using an efficient and fast access provisioning tool.

2.9 Summary

Recent advances in the field of optics have opened the way for the practical implementation of WDM networks. In this chapter, we have provided a brief overview of some of the optical WDM devices currently available or under development, as well as some insight into the underlying technology. As optical device technology continues to improve, network designers need to be ready to take advantage of new device capabilities, while keeping in mind the limitations of such devices.

Exercises

2.1. An IP-based network application will be built on top of a fiber-optic communication network. The application programmer considers two options for error correction. One option is to simply use the TCP/IP protocol. The other option is to use the UDP/IP protocol, which the programmer calculates will have a 10% higher bandwidth when transmissions are error free. The programmer has included a cyclic redundancy check in the UDP packets, but does not have a good scheme for retransmitting individual packets. Thus, when an error is detected, an average of 125 megabytes of data will have to be retransmitted. Which scheme will yield the higher average bandwidth if the fiber bit error rates is: (a) 10^{-9}? (b) 10^{-15}?

2.2. Consider a step-index fiber which has a core refractive index of 1.495. What is the maximum refractive index of the cladding in order for light entering the fiber at an angle of 60 degrees to propagate through the fiber? Air has refractive index of 1.0.

2.3. Find the numerical aperture in a graded-index fiber with two layers shown in Fig. 2.41. Compare the answer with the numerical aperture of the step-index fiber shown in Fig. 2.40. Can we use geometric optics to deal with situations where the wavelength and core diameter are of the same order of magnitude (e.g., single-mode fiber)?

2.4. Consider a step-index multimode fiber in which the refractive indices of the cladding and core are 1.35 and 1.4, respectively. The diameter of the core is 50 μm. Approximately how many modes are supported by the fiber for a signal at a wavelength of 1550 nm?

2.5. Find the approximate number of modes in a 100 μm core step-index multimode fiber at 850 nm. Assume the refractive index of glass to be 1.5 and that of the cladding to be 1.47.

2.6. Consider an optical link in which power at the transmitter is 0.1 mW and the minimum power required at the receiver is 0.08 mW. The attenuation constant for the fiber material is 0.033 dB/km. What is the maximum length of the optical link, assuming that there are no amplifiers?

2.7. Describe the various types of dispersion and explain how the effects of each type of dispersion can be reduced.

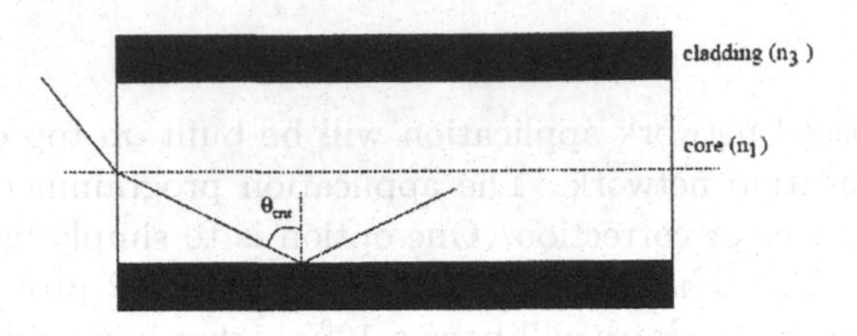

Figure 2.40 Critical angle in a step index fiber.

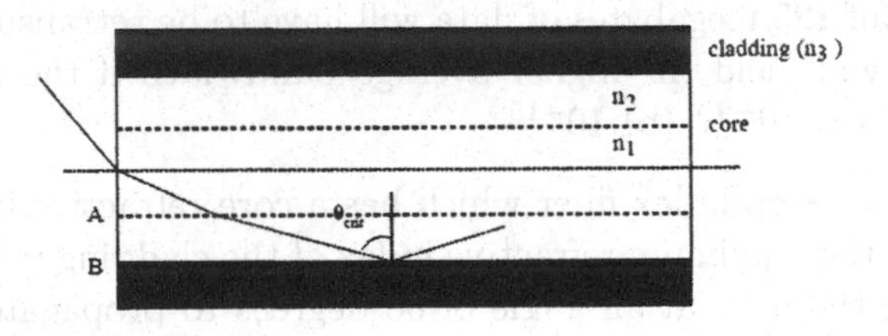

Figure 2.41 Critical angle in a graded index fiber.

2.8. Consider a single mode optical fiber with an attenuation of 0.2 dB/km and a dispersion limit of 200 Gbps·km. The transmitter power is 1 mW and the receiver sensitivity is 10^{-5} mW. The link operates at a rate of 2.4 Gbps. Assume a 10 dB power margin for losses in connectors. Calculate the maximum length of the optical link.

2.9. Suppose we have a system which has 3 channels operating at 1549.32 nm, 1554.13 nm, and 1558.98 nm. At which frequencies will we have sidebands as a result of four-wave mixing? (Use $c = 2.998 \times 10^8$.)

2.10. Consider a broadcast star network with $N = 2^k$ nodes where k is an integer. The network is built out of 2×2 couplers with excess loss β and coupling coefficient α. Each transmitter has a laser with power P_t.

(a) Find the power levels received by the receivers when a single transmitter (say transmitter T_1) transmits. That is, determine how many different power levels are received by the N receivers, and how many receivers receive each of these levels.

Hint: Construct a tree with the transmitter at the root.

(b) Suppose your goal as the network designer is to maximize the minimum power received by any of the receivers from a transmitter. Find the optimal value of α for this design criterion. Explain your answer.

(c) Now suppose you have a different design criterion. Your new goal is to maximize the expected value of the power between a random transmitter/receiver pair. Assume that each such pair is equally likely. How would you select your couplers? Explain your answer. *Hint:* You can exploit the symmetry to fix the transmitter.

2.11. Suppose a 1 mW, 1550 nm signal is transmitted across a 5 km fiber, through an 8×8 passive star coupler, and through another 15 km of fiber before reaching its destination. No amplifiers are used. What is the power of the signal at the destination?

2.12. A 16×16 passive-star hub has been constructed from combiners, couplers, and splitters as in Fig. 2.7. Each combiner, coupler, and splitter results in a 3 dB power loss. Each host is up to 10 km away from the star, with a signal attenuation of 0.2 dB/km. If each host must receive signal power of at least 0.01 mW to clearly recognize signals, how strong must each host's transmission signal be?

2.13. Consider a unidirectional fiber bus with N nodes. (Assume that N is even.) All the couplers have an excess loss β ($\ll 1$). (Let $\gamma = 1 - \beta$ and use γ in your solution instead of β.) The ith coupler has a coupling coefficient α_i for $1 \leq i \leq N$. The coupling coefficients can be independently selected. Optimize the coupling coefficients to maximize the worst-case power transfer between a transmitter and a receiver. Compare the resulting worst-case power with the case where all couplers are identical and optimized. Hint: First assume some transmitter k is the worst-off. Argue why, in the optimal solution, all the receivers to the right of k will see the same attenuation from transmitter k. Similarly all the transmitters to the left of k should be equally badly off. Obtain a recursion for α_i assuming k is known and find the value of k.

2.14. Consider the following simplified model of a direct detection binary FSK system. By using a pair of optical filters and a pair of photodetectors in the receiver, we observe two Poisson distributed photon counts: X_0 and X_1. When the data bit is 0, the parameter of X_0 is $\lambda + \lambda_d$ while that of X_1 is λ_d. (Here λ_d models the dark current.) Conversely, when the data bit is 1, X_0 has parameter λ_d and X_1 has $\lambda + \lambda_d$. X_0 and X_1 are statistically independent when conditioned on the value of the data bit.

(a) Obtain the Maximum Likelihood processing of X_0 and X_1 explicitly.

(b) Find the probability of a bit error as a function of λ and λ_d. You may leave your answer as a double series.

(c) Repeat part (b) when there is no dark current ($\lambda_d = 0$). Now your answer must have a simple form.

2.15. In this problem, you will investigate the relationship between the finesse F and the reflection coefficient R of a Fabry-Perot filter.

(a) Show that for $R \simeq 1$, the finesse can be approximated as

$$F \simeq -\frac{\pi}{\ln R}$$

(b) Find the exact and the approximate expression for R in terms of F. Evaluate the accuracy of the approximation for $F = 10$ and $F = 100$.

2.16. In this problem, you will consider the worst-case crosstalk in a WDM environment with Fabry-Perot (FP) filters. Assume that the filter is lossless ($A = 0$).

(a) Show that the power transfer function $T(f)$ of the FP can be written as

$$T(f) = \frac{1-R}{1+R} \sum_{m=-\infty}^{\infty} R^{|m|} e^{j2\pi m f/P}$$

where P is the free spectral range.

(b) Using the result in (a), show that the worst-case interference

$$C_{\max} = \sum_{i=1}^{M-1} T\left(\frac{iP}{M}\right)$$

is given by

$$C_{\max} = M \frac{1-R}{1+R} \frac{1+R^M}{1-R^M} - 1$$

(c) Using the approximation in Problem 2.15, show that when $F \gg 1$

$$C_{\max} \simeq \frac{\pi M}{2F} \coth\left(\frac{\pi M}{2F}\right) - 1$$

where $\coth(x) = (e^x + e^{-x})/(e^x - e^{-x})$. Find, numerically, the maximum value of M/F such that $C_{\max} \leq 1$. Comment on your result.

2.17. Consider a Mach-Zehnder filter chain of K cascaded filters with

$$\Delta L_i = 2^{i-1}\Delta L \quad i = 1, 2, \ldots, K$$

Show that the power transfer function of this chain is given by

$$T(f) = \frac{\sin^2(\pi M f/P)}{M^2 \sin^2(\pi f/P)}$$

where $M = 2^K$ and $P = c/\Delta L$.

2.18. Optical amplifiers saturate at high levels of output power. Suppose the saturation power of an erbium-doped fiber amplifier is 20 mW, and the amplifier gain is 5 dB/mW of pump power. The pump power is set to 5 mW. What is the largest amount of input power that can be amplified without driving the amplifier into saturation.

2.19. An optical amplifier delivers an output power P_{out} in response to an input power P_{in} as described by the following equation

$$P_{out} = a\,(1 - \exp(-bP_{in}))$$

where a and b are constants.

(a) What is the saturation power P_{sat} of this amplifier?

(b) Find the power gain of the amplifier in the linear operating region (i.e., small input power).

(c) Suppose this amplifier is to be placed in a transmission link of length L. The fiber has an attenuation factor of α per unit length, i.e., after a distance l, a factor $e^{-\alpha l}$ of the original power remains. The transmitter has a laser with power P_t. The goal is to maximize the received power P_r. Find the optimal position x (measured from the transmitter) of the amplifier. Comment on your result.
Hint: The only root of the equation $u = e^u - 1$ is at $u = 0$.

2.20. (a) Use four 2×2 optical crosspoint elements to construct a 4×4 Banyan interconnect.

(b) How many rows and how many columns of 2×2 crossbars would be required for an $N \times N$ Banyan interconnect.

(c) Suppose we have an 8×8 space-division Banyan optical routing switch. Label the inputs from 0 to 7 and label the outputs from 0 to 7. Suppose we need to simultaneously route Input 5 to Output 2 and Input 7 to Output 0. Can this routing be accomplished? If yes, give the routes through the switch, otherwise explain why the routing isn't possible.

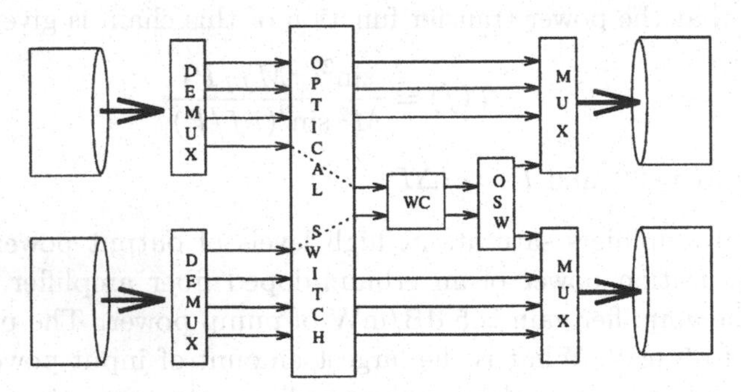

(a) Share-per-node wavelength-convertible switch architecture

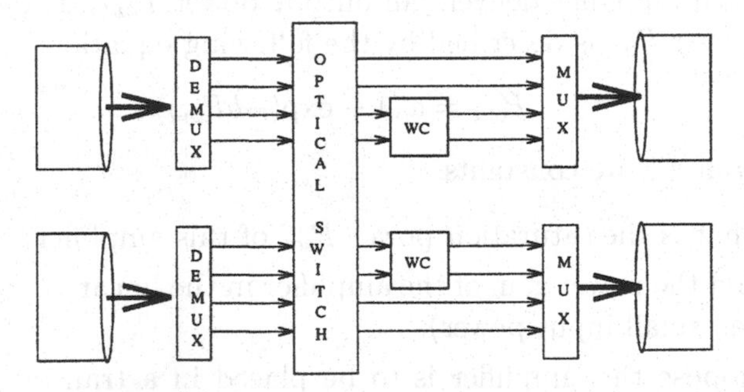

(b) Share-per-link wavelength-convertible switch architecture

Figure 2.42 Two architectures for wavelength convertible routers: (a) share-per-node, (b) share-per-link.

2.21. What are the uses of wavelength conversion in a WDM network?

Consider the two wavelength-convertible switch (WCS) architectures shown in Fig. 2.42. Construct a set of connection requests that can be routed by the share-per-node WCS and cannot be routed by the share-per-link WCS, and vice-versa.

2.22. Suppose we want to design a system with 16 channels, each channel with a rate of 1 Gbps. How much bandwidth is required for the system?

2.23. Suppose we have a fiber medium with a bandwidth of about 20 nm. The center wavelength is 0.82 μm. How many 10 GHz channels can be accommodated by the fiber? Calculate the maximum number of channels for a center wavelength of 1.5 μm.

2.24. Consider an optical communication system in which the transmitter tuning range is from 1450 nm to 1600 nm, and the receiver tuning range is from 1500 nm to 1650 nm. How many 1 Gbps channels can be supported in the system?

2.25. Consider a WDM passive-star-based network for N nodes. Let the tuning range of the transmitters be 1550 nm to 1560 nm, and let the tuning range of the receivers be 1555 nm to 1570 nm. Assume that the desired bit rate per channel is 1 Gbps. Also assume that a channel spacing of at least 10 times the channel bit rate is needed to minimize crosstalk on a WDM system. Find the maximum number of resolvable channels for this system.

2.26. In which type of network, single-hop or multihop, is a smaller tuning latency more critical? Why?

2.27. In a WDM network node, if two signals on the same wavelength arriving from different input ports need to go to the same output port, then a conflict may occur. Describe two or more methods for resolving this conflict. Discuss the advantages and disadvantages of each solution.

2.28. Figure 2.43 shows a WDM WAN constructed using AWGs at each node. Assume that there are sufficient number of fibers between the node pairs (not shown in Fig. 2.43). Show how the following connection requests can be satisfied (you may have to write a program which tries out various possibilities).

- *node* 3 $\rightarrow$ *node* 1

- *node* 3 → *node* 1
- *node* 1 → *node* 2
- *node* 1 → *node* 2
- *node* 2 → *node* 3
- *node* 2 → *node* 3

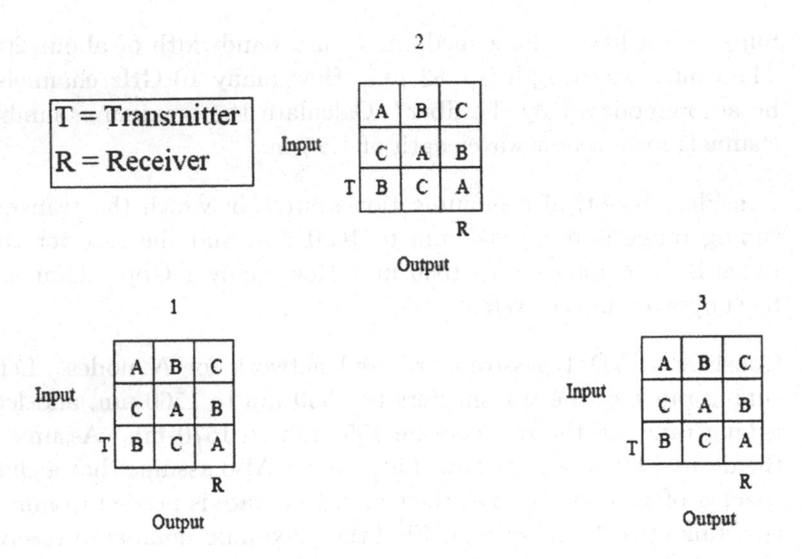

Figure 2.43 A WDM WAN constructed using AWGs at each node. (All connections begin at transmitters and end at receivers.)

2.29. Suppose we are given the network in Fig. 2.37 and have two wavelengths available. We wish to set up the following connections:

i. H-2-3-4-8-9-E
ii. C-7-8-4-F
iii. B-6-7-8-9-E
iv. D-10-7-C

At which nodes are wavelength converters required, and how many conversions are required at these nodes? Explain.

Part II

Local, Access, and Metro Networks

Chapter 3

Single-Hop Networks

3.1 Introduction

This chapter will start with a discussion on how a local lightwave network employing WDM may be constructed based on a passive-star coupler. Such networks can be broadly categorized into one of two types – single-hop and multihop. Various approaches for single-hop network design will be discussed in the remainder of this chapter. The corresponding treatment of multihop networks will be provided in Chapter 4.

3.2 Constructions of Single-Hop Networks

A *local area network (LAN)* can be constructed by exploiting the capabilities of optical technology, i.e., WDM and tunable optical transceivers (transmitters or receivers), as follows. The vast optical bandwidth of a fiber is carved up into smaller-capacity channels, each of which can operate at peak electronic processing speeds (i.e., over a small wavelength range) of, say, 10 or 40 Gbps. By tuning its transmitter(s) to one or more wavelength channels, a node can transmit into those channel(s); similarly, a node can tune its receiver(s) to receive from the appropriate channels. The system can be configured as a broadcast-and-select network in which all of the inputs from various nodes are combined in a WDM *passive-star coupler (PSC)* and the mixed optical information is broadcast to all outputs (see Fig. 3.1).

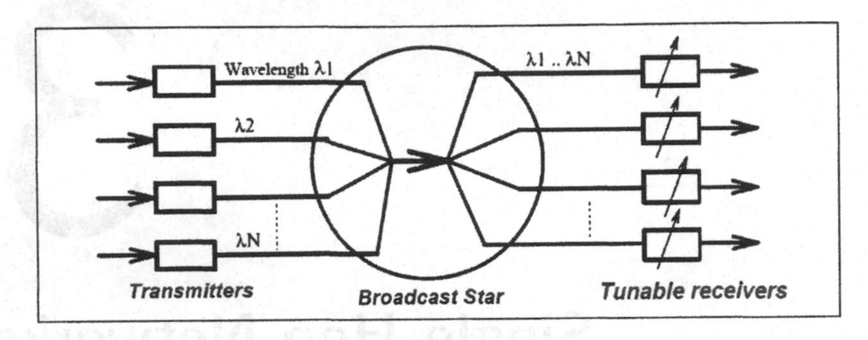

Figure 3.1 A broadcast-and-select WDM network.

An $N \times N$ star coupler can be considered to consist of an $N \times 1$ combiner followed by a $1 \times N$ splitter[1]; thus, the signal strength incident from any input can be (approximately) equally divided among all of the N outputs. The passive-star topology is attractive (1) because of its logarithmic splitting loss in the coupler (since the splitter portion of the coupler circuit is essentially a binary tree type structure) and (2) because there is no tapping or insertion loss (as in a linear bus). In addition, the passive property of the optical star coupler is important for network reliability since no power is needed to operate the coupler; also it allows information relaying without the bottleneck of electrooptic conversion.

But the key bottleneck in the PSC-based network is the channel resource limitation due to the lack of spatial wavelength reuse. Each wavelength on the PSC provides a broadcast channel from a given PSC input port to all output ports. Thus, the number of simultaneous transmissions in a PSC network is limited by the number of available wavelengths. Furthermore, since each wavelength channel is shared among multiple nodes, PSC-based networks require *media-access control (MAC)* to avoid a large number of data-packet collisions at high traffic loads. The issue will be further explained in the following sections regarding protocols.

[1]If the splitters and combiners are made out of couplers, then this arrangement may not be very power efficient. An alternate arrangement would be to have a butterfly arrangement (also called a multistage interconnection network) which has $\log_2 N$ stages of 2×2 couplers with $N/2$ couplers per stage. See Problem 3.2.

Furthermore, single-hop WDM networks have also attracted a great deal of attention to solve the *"metro gap"*. In [FaAR05], an *arranged-waveguide grating (AWG)*-based single-hop metro network architecture is proposed (see Fig. 3.2). The AWG is a passive wavelength-routing device that allows for spatial wavelength reuse, i.e., the entire set of wavelengths can be simultaneously applied at each AWG input port without resulting in collisions at the AWG output ports (see Chapter 2). This spatial wavelength reuse has been demonstrated to significantly improve the network performance for a fixed set of wavelengths compared with PSC-based single-hop networks. Unfortunately, scalability is a significant problem in an AWG-based single-hop WDM network because the number of transceivers required at each node is equal to the total number of nodes. The problem can be solved by installing optical couplers or splitters between the AWG and the nodes, and by aggregating multiple nodes before connecting them to the AWG [Bore95]. In Fig. 3.2 (from [FaAR05]), a cyclic AWG with D input ports and D output ports is considered. At each AWG input port, an $S \times 1$ combiner collects transmissions from the transmitters of S attached nodes. At each AWG output port, a $1 \times S$ splitter equally distributes the signal to S individual fibers that are attached to the receivers of the nodes. Thus, this architecture can support $N = S \times D$ transceivers.

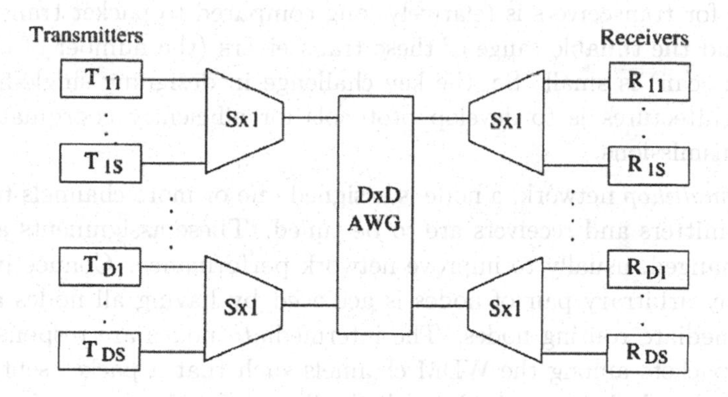

Figure 3.2 An AWG-based single-hop metro network [FaAR05].

In general, the physical topology – instead of being a star – can be a linear bus or a tree (see [Mukh92a, Mukh97]). From a network architecture perspective, all three structures – star, bus, and tree – can be considered *equivalent* since, in all of them, information from a sender to a recipient must flow through a *central funneling point*. However, the bus has an additional *attempt-and-defer capability* under which a node, before or during its transmission, can "sense" activity on the bus from *upstream transmissions* (see [Mukh92a, Mukh97]).

The optimal physical topology design problem may be referred to as the cable plant design problem to determine that the necessary power budget is satisfied [BaFG90]).

From an architectural perspective, given that the input lasers (transmitters) or the output filters (receivers) or both can be made tunable opens up a multitude of networking possibilities. The tunable transceivers are used differently depending on the type of network architecture chosen – single-hop or multihop.

In a *single-hop* network, a significant amount of dynamic coordination between nodes is required. For a packet transmission to occur, one of the transmitters of the sending node and one of the receivers of the destination node must be tuned to the same wavelength for the duration of the packet's transmission. In the single-hop environment, it is also important that transmitters and receivers be able to tune to different channels quickly so that packets may be sent or received in quick succession. Currently, the tuning time for transceivers is relatively long compared to packet transmission times, and the tunable range of these transceivers (the number of channels they can scan) is small. So, the key challenge in designing single-hop network architectures is to develop protocols for efficiently coordinating the data transmissions.

In a *multihop* network, a node is assigned one or more channels to which its transmitters and receivers are to be tuned. These assignments are only rarely changed, usually to improve network performance. Connectivity between any arbitrary pair of nodes is achieved by having all nodes also act as intermediate routing nodes. The intermediate nodes are responsible for routing packets among the WDM channels such that a packet sent out on one of the sender's transmit channels finally gets to the destination on one of the destination's receive channels, possibly after multihopping through a number of intermediate nodes. A number of different multihop architectures are possible, with a range of operational properties (e.g., ease of routing) and performance characteristics (e.g., average packet delay, number of hops that must be traversed, and efficient use of links). As can be inferred, in a single-hop network, there are no intermediate nodes in the optical path.

Both single-hop and multihop networks have their own strengths (and weaknesses) based on, for example, optical transceiver tuning capabilities. Accordingly, we study both approaches. For both single-hop and multihop networks, it is important to keep in mind that any design must be not only simple to implement (i.e., based on realistic assumptions about the properties of optical components), but also scalable to accommodate large user populations. Our attention will focus on single-hop approaches in this chapter, and will shift to multihop approaches in Chapter 4.

3.3 Characteristics of a Single-Hop System

Single-hop WDM networks, in which all nodes are connected to a central hub node, have been investigated as an all-optical architecture for local area network (LAN) applications [Mukh92a, Mukh97]. In a single-hop WDM network, due to its minimum hop distance of one (i.e., no bandwidth devoted to multihop packet forwarding) and inherent transparency, routing and signaling are not necessary and operational expenditure is minimized. Just because of their simple operation and reduced network cost, single-hop WDM networks have attracted a great deal of attention as an architecture for LAN and MAN applications.

For a single-hop system to be efficient, the bandwidth allocation among the contending nodes must be managed dynamically. Such systems can be classified into two categories – those employing pretransmission coordination vs. those not requiring any pretransmission coordination.

Pretransmission coordination systems employ a *single*[2], *shared control channel* through which nodes arbitrate their transmission requirements, and the actual data transfers take place through a number of data channels. Idle nodes may be required to monitor the control channel. Before data packet transmission or data packet reception, a node must tune its transmitter or its receiver, respectively, to the proper data channel.

Generally, no such control channel exists in systems that do not require any pretransmission coordination, and arbitration of transmission rights is performed either in a preassigned fashion or through contention-based data transmissions on the regular data channels (e.g., requiring nodes to either transmit on or receive from predetermined channels). As a result, *for a large user population whose size may be time-varying,* deterministic scheduling

[2]While at least one channel is required for control signalling, some approaches employ as many as N.

approaches fall out of favor so that pretransmission coordination may be the preferred choice.

An alternative classification of WDM systems can be developed based on whether the nodal transceivers are tunable or not. A node's network interface unit (NIU) can employ one of the following four structures:

1. Fixed Transmitter(s) and Fixed Receiver(s) — $(FT - FR)$
2. Tunable Transmitter(s) and Fixed Receiver(s) — $(TT - FR)$
3. Fixed Transmitter(s) and Tunable Receiver(s) — $(FT - TR)$
4. Tunable Transmitter(s) and Tunable Receiver(s) — $(TT - TR)$

Fixed transceivers, which can only access some predetermined channels, are readily available, but cost considerations may restrict the installation of a large number of such transceivers at each node. The $FT - FR$ structure is generally suitable for constructing multihop systems in which no dynamic system reconfiguration may be necessary, although a single-hop $FT-FR$ system (LAMBDANET) with a small number of nodes has been demonstrated [GGKV90] (see Section 3.4). $FT-FR$ and $TT-FR$ systems, because they employ fixed receivers, may not require any coordination in control channel selection between two communicating parties, while such coordination is usually necessary in systems employing $FT - TR$ and $TT - TR$ structures. If each node (actually each nodal transmitter) is assigned a different channel under the $FT - FR$ or $FT - TR$ structures, then no channel collisions will occur and simple medium access protocols can be employed, but the maximum number of nodes will be limited by the number of available channels. Systems based on the $TT - TR$ structure are the most flexible in accommodating a scalable user population, but they also have to deal with the channel-switching overhead of the transceivers.

In addition, some systems require that a node be equipped with multiple transmitters or receivers. Accordingly, the following general classification for single-hop systems can be developed:

$$\begin{cases} FT^iTT^j - FR^mTR^n & \text{no pretransmission coordination} \\ CC - FT^iTT^j - FR^mTR^n & \text{control-channel (CC) based system} \end{cases}$$

where a node has i fixed transmitters, j tunable transmitters, m fixed receivers, and n tunable receivers. In this classification, the default values of

i, j, m, and n, if not specified, will be unity. Also, wherever possible, the number of network nodes, if finite, will be denoted by M. Thus, LAMBDANET [GGKV90] is a $FT - FR^M$ system since each of the M nodes in the system requires one fixed transmitter and an array of M fixed receivers. The TT and TR portions of the classification are suppressed since the system requires no tunable transmitter or tunable receiver.

Most experimental WDM network prototypes belong to the single-hop category, and they do not employ any control channel for pretransmission coordination.

From the central hub node point of view, single-hop WDM networks are typically either based on a central passive star coupler (PSC) or a central arrayed-waveguide grating (AWG) [Kami05]. In the following section, some experimental WDM systems will be reviewed.

3.4 Experimental WDM Systems

While [Brac90] documented the state of experimental systems in 1990, several systems such as ACTS's SONATA [BLMN01], Stanford's HORNET [SWWG00] and STARNET [CAMS96], and IBM's Rainbow [JaRS93] have been developed. Initial work in this field was done by the British Telecom Research Lab (BTRL), whose experiment introduced the concept of a multiwavelength network, operating in the broadcast mode and using mechanically-tunable filters at each receiver [PaSt86]. The AT&T Bell Labs experiment was the first to demonstrate channel spacings on the order of 1 nm [OHLJ85]. The Heinrich Hertz Institute (HHI) reported the first broadcast star demonstration of video distribution using coherent lightwave technology [BBCG86]. Other demonstrations include those by AT&T [Kami90, GlSc90] and by NTT [TONT90]. Results from a number of demonstrations are reported in [Kami90] – initially with two 45-Mbps channels and employing tunable receivers, later with two 600-Mbps channels, and followed by two 1.2-Gbps channels and employing tunable transmitters as well. An experimental system employing six 200-Mbps channels, spaced by 2.2 GHz, is reported in [GlSc90]. The work in [TONT90] demonstrates a system operating with a 5-GHz-spaced (equivalent wavelength spacing $\approx$ 0.04 nm), 16-channel system based on tunable receivers, where each channel can carry 622 Mbps (enough to accommodate a high-definition television channel).

Experimental demonstrations of subcarrier-multiplexing-based systems have also been recently reported [BESO92, MWWP93]. In [MWWP93], for example, a system with one control node and two data nodes is demonstrated. Fixed-wavelength transmitters are used at each node, and a 1-Mbps control information is subcarrier-multiplexed at frequencies $f_c = 710$ MHz, $f_1 = 720$ MHz, and $f_2 = 730$ MHz at the central controller, node 1, and node 2 (which are operated at three distinct wavelengths, viz., 1526 nm, 1543 nm, and 1553 nm, respectively). A receiving node employs a tunable filter. It has a passband in the 1526-nm wavelength region (to continuously receive from the control channel) and another tunable passband to receive from either of the data nodes (depending on the control information in the f_c subcarrier).

The following subsections elaborate on some experimental demonstrations and prototypes of single-hop WDM networks.

3.4.1 SONATA

Switchless Optical Network for Advanced Transport Architecture (SONATA) was a project supported by ACTS in Europe [BLMN01]. The main objective of SONATA was to define and demonstrate a single-layer network platform for end-to-end optical connections between a large number of terminals for business and residential customers. Time and wavelength agility at terminals were heavily exploited by simplifying the network structure to only a single node providing passive routing functions and, optionally, actively controlled wavelength conversion (see Fig. 3.3). A first demonstrator to test the optical physical layer of a "switchless" network was realized by integrating wavelength-agile transmitters, wavelength-agile burst-mode receivers, a passive wavelength-routing node, a wavelength converter, and optical fiber amplifiers in gated loops. Burst-mode transmission at 622 Mbps was used. The aim of this demonstrator was to assess the capability of a "switchless" network architecture to transport data packets when end-to-end multipoint-to-multipoint connections are established. By means of *bit-error rate (BER)* measurements, the effects of optical amplifier noise, crosstalk, and wavelength conversion were evaluated, and the limits of the optical layer were analyzed.

A second demonstrator for the network aspects was realized by integrating wavelength-agile transmitters, wavelength-agile burst-mode receivers, a passive wavelength routing node, a network controller, and external services. Burst-mode transmission at 155 Mbps was used. The terminals were placed in different rooms and were able to accept data from two external sources: PC application (client service) and cell generator/analyser instrument. The aim of this phase was to show the feasibility of connection set-up and medium-access procedures using the same optical network used for the data and to show an example of interconnection of the "switchless" network with an external service [BLMN01].

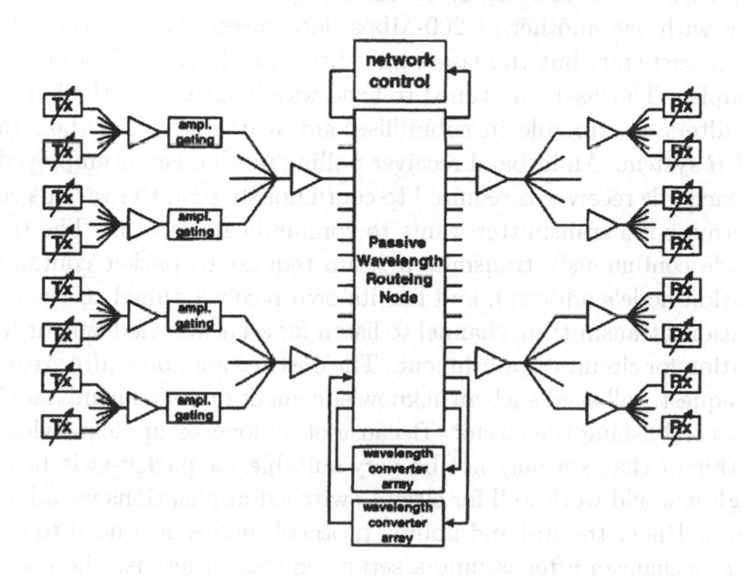

Figure 3.3 SONATA network architecture [BLMN01].

3.4.2 LAMBDANET

In LAMBDANET [GGKV90], a $FT - FR^M$ system with M nodes, each transmitter was equipped with a laser transmitting at a fixed wavelength. Via a broadcast star at the center of the network, each of the wavelengths in the network was broadcast to every receiving node. The experiment demonstrated the use of an array of M receivers at each node in the network, employing a grating demultiplexer to separate the different optical channels. Experiments report that 18 wavelengths were successfully transmitted at 2 Gbps over 57.5 km. Although each node requires M receivers, advances in *opto-electronic integrated circuits* (OEICs) impact this design [GGKV90].

3.4.3 Rainbow

In IBM's Rainbow project [DGLR90], the experimental prototype takes the form of a direct-detection, circuit-switched metropolitan-area network (MAN) backbone consisting of 32 IBM PS/2s as network nodes, communicating with one another at 200-Mbps data rates. The network structure is a broadcast-star, but the lasers and filters are housed centrally near the star coupler. The lasers are tuned to fixed wavelengths, but the Fabry-Perot etalon filters are tunable in submillisecond switching times, i.e., this is a $FT-TR$ system. An in-band receiver polling mechanism is employed under which each idle receiver is required to continuously scan the various channels to determine if a transmitter wants to communicate with it. The transmitting node continuously transmits a setup request (a packet containing the destination node's address), and has its own receiver tuned to the intended destination's transmitting channel to listen for an acknowledgement from the destination for circuit establishment. The destination node, after reading the setup request, will send such an acknowledgement on its transmitter channel, thereby establishing the circuit. Because of its long setup-acknowledgement delay, this mechanism may not be very suitable for packet-switched traffic, although it would work well for circuit-switched applications with long holding times. Under the in-band polling protocol, nodes also need to employ a *timeout* mechanism after issuing a setup request; otherwise there exists the possibility of a *deadlock*.

The Rainbow-I prototype was demonstrated at Telecom '91 in Geneva. The lessons learned from the corresponding prototype development and demonstration are provided in [Gree92].

Rainbow-II, a follow-on to Rainbow-I, is an optical MAN that supports 32 nodes, each operating at 1 Gbps, over a distance of 10 km to 20 km [HaKR96]. It uses the same optical hardware and medium access control protocol as Rainbow-I, viz., a broadcast-star architecture with each node having a fixed transmitter and a tunable receiver that follows the in-band polling protocol. The goals of Rainbow-II are: (1) to provide connectivity to host computers using standard interfaces, e.g., to interconnect supercomputers via the standard high-performance parallel interface (HIPPI) while overcoming distance limitations, (2) to deliver a throughput of 1 Gbps to the application layer, and (3) to demonstrate real computing applications requiring Gbps bandwidth. The Rainbow-II prototype was deployed as an experimental testbed at the Los Alamos National Laboratory (LANL), where performance measurements and experimentation with gigabit applications are currently being conducted [HaKR96].

3.4.4 Fiber-Optic Crossconnect

The goal of the Fiber-Optic Crossconnect (FOX) demonstration [ACGK86] was to investigate the potential of using fast tunable lasers in a parallel processing environment (with fixed receivers), i.e., this is a $TT - FR$ system. The architecture employed two star networks, one for signals traveling from the processors to the memory banks, and the other for information flowing in the reverse direction. Since the utilization of the memory accesses is relatively slow, a binary exponential backoff algorithm was used for resolving contentions, and it was shown to achieve sufficiently good performance. Since the transmitters are tunable, for applications in which data packet transmission times are in the range 100 ns to 1 μs, transmitter tuning times less than a few tens of nanoseconds will ensure reasonable efficiency.

3.4.5 STARNET

STARNET is a WDM LAN, based on the passive-star topology [CAMS96]. It supports – and allows all of its nodes to be on – two virtual subnetworks: a high-speed reconfigurable packet-switched data subnetwork, and a moderate-speed fixed-tuned packet-switched control subnetwork. Each STARNET node is equipped with a single fixed-wavelength transmitter, which employs a combined modulation technique to simultaneously send data on both subnetworks on the same transmitter carrier wavelength. Each node also has two receivers, referred to as a main receiver (which operates at high speed, viz., 2.5 Gbps) and an auxiliary receiver (which operates at a moderate speed of 125 Mbps, viz., at the rate of a fiber distributed data interface (FDDI) network). The auxiliary receiver at a node is tuned to the "previous node's transmitting wavelength" so that the moderate-speed subnetwork is a logical ring that carries control packet and is also FDDI-compatible. The main (2.5 Gbps) receiver can be tuned to any node, based on prevailing traffic conditions. The corresponding high-speed subnetwork may be operated as a multihop network (see Chapter 4) that allows electronic multihopping whenever required.

3.4.6 Other Experimental WDM Systems

Thunder and Lightning is a 30-Gbps ATM network using optical transmission and electronic switching [MeBo96]. Electronic switching, using 7.5-GHz

Galium Arsenide (GaAs) circuits, was chosen to simplify clock recovery, synchronization, routing, and packet buffering; and to facilitate the transition to manufacture.

In HYPASS [AGKV88], an extension of FOX, the receivers were made tunable as well (i.e., a $TT - TR$ system), resulting in vastly improved throughputs. Other reported experiments include BHYPASS, STAR-TRACK, passive optical loop (POL), and broadcast video distribution systems. Characteristics of these systems are discussed in [Brac90], and are not repeated here to conserve space. See also [GiTe91] for a report on Columbia's TeraNet prototype.

3.5 Other Non-Pretransmission Coordination Protocols

Among protocols that do not require any pretransmission coordination, some are based on fixed assignment of the channel bandwidth, whereas others are based on demand assignment. These protocols are categorized accordingly in the following subsections.

3.5.1 Fixed Assignment

A simple technique that allows one-hop communication is based on a fixed assignment technique, viz., time-division multiplexing (TDM) extended over a multichannel environment [ChGa88]. Each node is equipped with one tunable transmitter and one tunable receiver; hence, these systems are classified as $TT - TR$ systems. The tuning times are assumed zero and the transceiver tuning ranges are the entire set of N available channels. Time is divided into cycles, and it is predetermined at what point in a cycle and over what channel a pair of nodes is allowed to communicate.

For example, for the case of three nodes (numbered 1, 2, 3) and two channels (numbered 0 and 1), one can formulate the following channel allocation matrix which indicates a periodic assignment of the channel bandwidth, and, in which, $t = 3n$ where $n = 0, 1, 2, 3, \ldots$.

Channel No.	t	$t+1$	$t+2$
0	(1,2)	(1,3)	(2,1)
1	(2,3)	(3,1)	(3,2)

An entry (i, j) for channel k in slot l means that node i has exclusive permission to transmit a packet to node j over channel k during slot l. The allocation matrix can be generalized for an arbitrary number of nodes M and an arbitrary number of channels N [ChGa88]. The allocation matrix can be further generalized to accommodate tuning times, (in integral number of slots), via a "staggerred" approach [BoSD93].

This scheme has the usual limitations of any fixed assignment technique, i.e., it is insensitive to the dynamic bandwidth requirements of the network nodes and it is not easily scalable in terms of the number of nodes. Also, the packet delay at light loads can be high.

The above approach can be extended to a versatile time-wavelength assignment algorithm in which node i is equipped with t_i transmitters and r_i receivers, all of which are tunable over all available channels [GaGa92a]. The scheduling algorithm is mindful of the fact the transceiver tuning times are nonnegligible. Specifically, the algorithm is designed such that, given a traffic demand matrix, it will minimize the tuning times in the schedule, while also attempting to reduce the packet delay.

Arbitrary switching times and nonuniform traffic loads can also be accommodated. One approach requires the establishment of a periodic TDM frame structure consisting of a transmission subframe followed by a switching subframe, so that all of the necessary switching functions of nodal transmitters/receivers are performed during the switching subframe [GaGa92a]. Another approach distributes the nodal switching requirements all over the frame, and is hence less restrictive and more efficient [PiSa94, AzBM96, BoMu96].

A brief examination of the characteristics of these "optimal" scheduling approaches is appropriate, and we will do so by reviewing the findings in one of these studies [AzBM96]. This study considers the scheduling of an arbitrary traffic matrix with a tunable transmitter and a fixed-tuned receiver at each node. Specifically, it quantifies the degradation due to nonzero tuning time on several performance criteria such as schedule-completion time and average packet delay. For off-line scheduling, where the traffic matrix is known a priori, the effect of tuning delay is found to be small even if tuning time is as large as the packet transmission duration, and a lower bound on the expected schedule completion time is obtained. (Similar results have been reported in [PiSa94, BoMu96], with [BoMu96] allowing tuning time larger than a packet duration also.) The average packet delay is found to be insensitive to the tuning time under a near-optimal schedule where an idle transmitter tunes *just-in-time* to the appropriate channel just before its packet transmission. This study examines extensions of the approach to accommodate real-time and connection-oriented traffic.

The approach in [GaGa92a] is further extended in [GaGa92b] so that users, based on their traffic flow patterns with other users, can be grouped into separate communities. Users within a community are connected by their own local WDM star, but users can communicate with users in other communities (also in a single-hop fashion) via a remote WDM star. Again, given a traffic demand matrix, the algorithm determines the proper time-wavelength schedule.

3.5.2 Partial Fixed Assignment Protocols

The above fixed assignment protocol is too pessimistic because its main goal is to avoid both channel collision and *receiver collisions.* (A *receiver collision* occurs when a collision-free data packet transmission cannot be picked up by the intended destination since the destination's receiver may be tuned to some other channel for receiving data from some other source.) However, alternative protocols can be defined in which the channel allocation procedures are less restrictive [ChGa88].

In the Destination Allocation (DA) protocol, the number of node pairs which can communicate over a slot is increased from the earlier value of N (the number of channels) to M (the number of nodes). During a slot, a destination is still required to receive from a fixed channel, but more than one source can transmit to it in this slot. Thus, even though receiver collisions are avoided, the possibility of channel collision is introduced. For the 3-node, 2-channel case, an example slot allocation may be the following.

Channel No.	t	$t+1$
0	(1,2); (3,2)	(1,3)
1	(2,3)	(2,1); (3,1)

A Source Allocation (SA) protocol can also be defined in which the control of access to the channels is further reduced. Now, over a slot duration, N ($N \leq M$) source nodes are allowed to transmit, each over a different channel. Since a node can transmit to each of the remaining $(M-1)$ nodes, the possibility of receiver collisions is introduced. An example periodic slot allocation policy for the 3-node, 2-channel case may now be the following.

Channel No.	t	$t+1$	$t+2$
0	(1,2)	(1,2)	(2,1)
	(1,3)	(1,3)	(2,3)
1	(2,1)	(3,1)	(3,1)
	(2,3)	(3,2)	(3,2)

Finally, an Allocation Free (AF) protocol can be defined in which all source-destination pairs have full rights to transmit on any channel over any slot duration. Due to the possibility of receiver collisions, the latter two protocols (SA and AF) may not have much practical significance.

3.5.3 Random Access Protocols I

One can design random access protocols that require each node to be equipped with one tunable transmitter and one fixed receiver (i.e., it is a $TT - FR$ system). The channel on which a node will receive is directly determined by the node's address, e.g., based on the low-order bits of the node's address. The channel a receiver receives from is referred to as that node's *home channel*.

Two slotted-ALOHA protocols were proposed in [Dowd91], and both were shown to out-perform the control-channel-based slotted-ALOHA/ALOHA protocol [HaKS87] and its improved version [Mehr90] (the latter two protocols will be discussed later in Section 3.6). Under one of the protocols in [Dowd91], time is slotted on all the channels, and these slots are synchronized across all channels. A slot length equals a packet's transmission time. In the second protocol, each packet is considered to be of L minislots, and time across all channels is synchronized over minislots. Throughput calculations were performed for these two schemes, and slotting across the entire packet length was found to perform better than minislotting since the latter scheme increases the vulnerability period of a data packet transmission (just like pure ALOHA has poorer performance than slotted ALOHA). Also, not surprisingly, the maximum throughput on each data channel is found to be $1/e$, the value for the single-channel case.

3.5.4 Random Access Protocols II

A slotted-ALOHA and a random TDM protocol have been investigated in [GaKo91]. Unlike previous work, these protocols assume limited tuning range, but zero tuning time. Both of these protocols are based on slotted architectures. Any node, say node i, is equipped with a *single tunable transmitter* and *a number of fixed receivers* (i.e., this is a $TT - FR^x$ system where x is a system parameter). Let $T(i)$ and $R(i)$ be the set of wavelengths over which node i can transmit and receive, respectively. The assignment of transmitters and receivers to various nodes is performed such that the intersection of $T(i)$ and $R(j)$ is always non-null for all i and j, i.e., any two nodes can communicate with one another via one hop. The optimal node/transceiver assignment task is a challenging but open problem.

Under the slotted-ALOHA scheme, if node i wants to transmit to node j, it arbitrarily selects a channel from the set $T(i) \cap R(j)$, and transmits its packet on the selected channel with probability $p(i)$.

The random time-division multiple access (TDMA) scheme operates under the presumption that all network nodes, even though they are distributed, are capable of generating the same random number to perform the arbitration decision in a slot. It is indicated in [GaKo91] that this can be done by equipping all nodes with the same random number generator starting with the same seed. Thus, for every slot, and for each channel at a time, the distributed nodes generate the same random number, which indicates the identity of the node with the corresponding transmission right. Analytical Markov chain models for the slotted ALOHA and random TDMA schemes are formulated to determine the systems' delay and throughput performances.

3.5.5 The PAC Optical Network

In a $TT - FR$ system, packet collisions can be avoided by employing Protection-Against-Collision (PAC) switches at each node's interface with the network's star coupler. These collisions are avoided by allowing a node's transmitter access to a channel (through the PAC switch) only if the channel is available. Also, packets simultaneously accessing the same channel are denied access. The concept is similar to that in collision-avoidance stars [Alba83, SuMo89], except that collision-avoidance is now extended to a multichannel environment.

The PAC circuit probes the state of the selected channel (i.e., it performs carrier sensing) by using a n-bit burst which precedes the packet. The carrier burst is switched through a second $N \times N$ "control" star coupler, where it is combined with a fraction of all the packets coming out of the "main" star plus all carrier bursts trying to gain access to the "control" star (see

Fig. 3.4). The resulting electrical signal controls the optical switch which connects the input to the network. The switch is closed only if no energy is detected on the selected channel from other nodes. When two or more nodes try to access the channel simultaneously, all of them detect the "carrier" and their access to the network is blocked. Blocked packets are reflected back to the sender. Because of its "carrier-sensing" mechanism, this approach is sensitive to propagation delays.

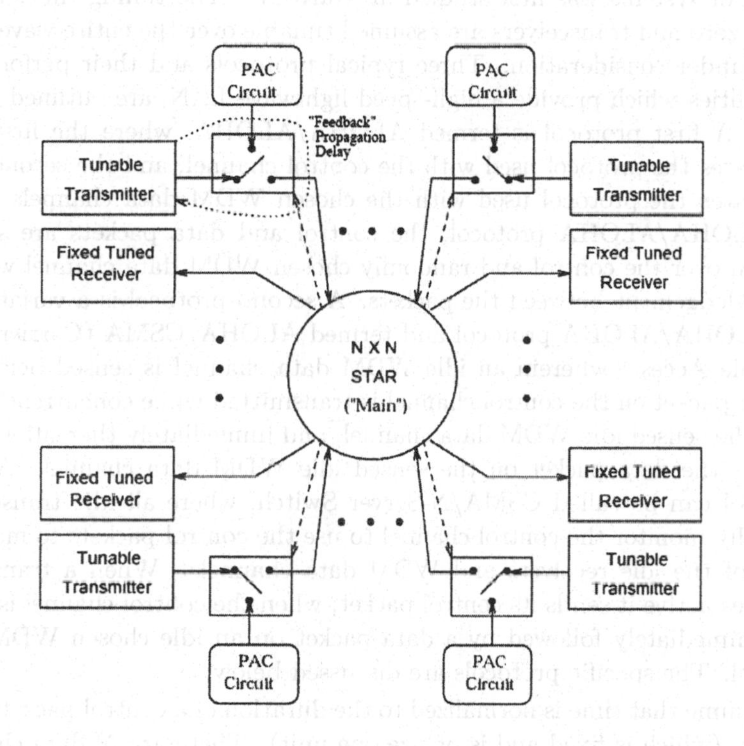

Figure 3.4 Architecture of the PAC optical packet network (the dashed lines are used to detect energy on the various channels from the "main" star).

Note that the length n of the carrier burst will influence the characteristics of this mechanism. Also, it would be preferable to co-locate the individual PAC circuits as close to the passive star (also referred to as the *hub*) as possible.

3.6 Pretransmission Coordination Protocols

3.6.1 Partial Random Access Protocols

The simplest requirement for single-hop communication is that each node be equipped with a single tunable transmitter and a single tunable receiver, and the system employ a control channel, i.e., a $CC-TT-TR$ system. Such a class of systems was first studied in [HaKS87]. The tuning times are assumed zero and transceivers are assumed tunable over the entire wavelength range under consideration. Three typical protocols and their performance capabilities which provide a high-speed lightwave LAN, are outlined in this work. A first protocol is termed ALOHA/ALOHA, where the first term designates the protocol used with the control channel, and the second term designates the protocol used with the chosen WDM data channels. With the ALOHA/ALOHA protocol, the control and data packets are sent at random over the control and randomly chosen WDM data channel without acknowledgement between the packets. A second protocol is a variation on the ALOHA/ALOHA protocol and termed ALOHA/CSMA (Carrier Sense Multiple Access) wherein an idle WDM data channel is sensed before the control packet on the control channel is transmitted while concurrently jamming the sensed idle WDM data channel, and immediately thereafter transmitting the data packet on the sensed idle WDM data channel. A third protocol can be called CSMA/N-Server Switch, where all idle transceivers normally monitor the control channel to use the control packets to maintain a list of the idle receivers and WDM data channels. When a transceiver becomes active it sends its control packet, when the control channel is found idle, immediately followed by a data packet on an idle chosen WDM data channel. The specific protocols are discussed below.

Assume that time is normalized to the duration of a control packet transmission (which is fixed and is of size one unit). There are N data channels and data packets are of fixed length, L units [HaKS87]. A control packet contains three pieces of information – the source address, the destination address, and a data channel wavelength number (which may be chosen at random by the source and on which the corresponding data packet is to be transmitted).

Under the ALOHA/ALOHA protocol, a node transmits a control packet over the control channel at a randomly selected time, after which it immediately transmits the data packet on data channel i, $1 < i < N$, which was specified in its control packet (see Fig. 3.5). Note that the "vulnerable period" of the control packet equals two time units, extending from $t_0 - 1$ to $t_0 + 1$ where t_0 is the instant at which the control packet's transmission is started. That is, any other control packet transmitted during the tagged packet's "vulnerable period" would "collide" with (and destroy) the tagged packet. (Since different nodes can be at different distances from the hub, these times are specified relative to the activity seen at the hub.) However, even if the control packet transmission is successful, the corresponding data packet may still encounter a collision. This may happen if there is another successful control packet transmission over the period $t_0 - L$ to $t_0 + L$, and the data channel chosen by that control packet is also i. Using such arguments, the throughput performance of this protocol can be analytically obtained. However, what this and the other protocols in [HaKS87] ignore is the possibility of "receiver collisions." Even if the control and data packet transmissions occur without collision, the intended receiver of the destination node might not always be able to read either the control packet or the data packet if it is tuned to some other data channel for receiving data from some other source. For a large or infinite population system, the effect of receiver collisions on the system's performance is negligible [HaKS87, JiMu92b].

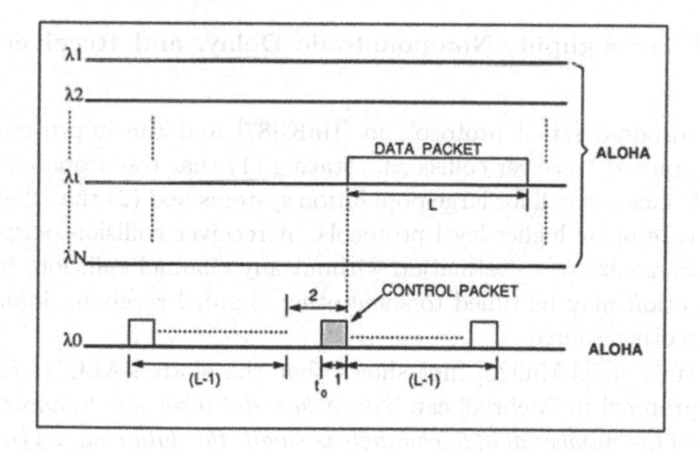

Figure 3.5 The ALOHA/ALOHA protocol.

The slotted-ALOHA/ALOHA protocol is similar, except that access to the control channel is via the slotted-ALOHA protocol. Other schemes outlined in [HaKS87] include ALOHA/CSMA, CSMA/ALOHA, and CSMA/N-server protocols. However, the main limitation of the CSMA-based schemes is that carrier sensing is based on near-immediate feedback, which may not be a practical feature of high-speed systems even for short distances in the range of a kilometer or so.

3.6.2 Improved Random Access Protocols

The approaches in Section 3.6.1 have been extended to obtain improved protocols and performance predictions [Mehr90]. The focus in is on realistic protocols which do not require any carrier sensing since the channel propagation delay in a high-speed environment may exceed the packet transmission time. Hence, slotted-ALOHA for the control channel and ALOHA and the N-server mechanism for the data channels are examined. Another improvement in the protocols is also studied. Specifically, it is required that a node delay its access to a data channel until after it learns that its transmission on the control channel has been successful. As a result, better throughput performance can be achieved, and such results for the improved protocols has been analytically demonstrated [Mehr90].

Bimodal Throughput, Nonmonotonic Delay, and Receiver Collisions

Both the original set of protocols in [HaKS87] and the improvements in [Mehr90] ignored "receiver collisions," stating (1) that the probability of receiver collisions is small for large population systems and (2) that they would be taken care of by higher-level protocols. A receiver collision occurs when a source transmits to a destination without any channel collision; however, the destination may be tuned to some other channel receiving information from some other source.

The study in [JiMu92b] first shows that the slotted-ALOHA/delayed-ALOHA protocol in [Mehr90] can have a *bimodal throughput characteristic.* Basically, *if the number of data channels is small, the data channel bandwidth is underdimensioned* and the data channels are the bottleneck. *If there is a large number of data channels, then the control channel's bandwidth is underdimensioned* and it is the bottleneck. See Fig. 3.6 for some representative throughput results.

The study in [JiMu92b] also finds a useful relationship for *optimally dimensioning the available bandwidth* (viz., properly selecting the number of

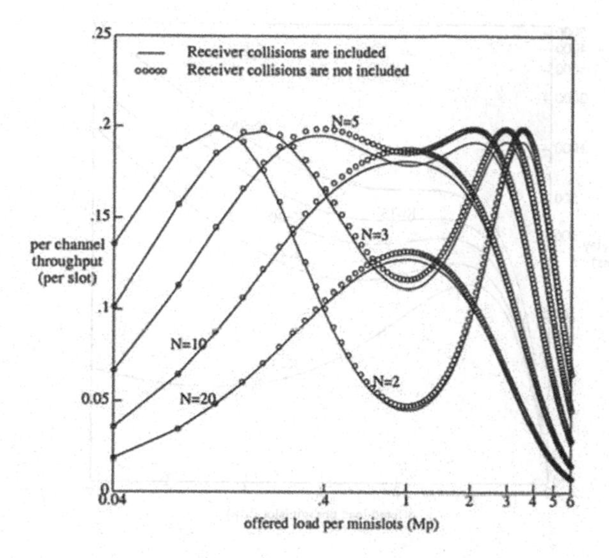

Figure 3.6 Bimodal throughput characteristics of the slotted-ALOHA/delayed-ALOHA protocol for $L = 10$ slots per data packet and N = number of data channels.

data channels) so that neither is the bottleneck. Specifically, it is required that, under the slotted-ALOHA/delayed-ALOHA protocol with L-slot data packets, the number of data channels should be given by

$$N = \left\lfloor \frac{2L-1}{e} \right\rfloor$$

Additional investigations in [JiMu92b] reveal that the system has an interesting delay characteristic, viz., that the mean packet delay is not necessarily monotonically increasing with increase in offered load or throughput. For example, for some sets of system parameters, such as short backoff period, the mean packet delay can actually reduce even though the offered load is increased. Of course, this can only happen when the throughput also decreases due to the data channel bandwidth being underdimensioned. Figure 3.7 shows some representative delay results from [JiMu92b].

The work in [JiMu92b] also studies the slotted-ALOHA/delayed-ALOHA protocol's performance degradation due to receiver collisions for finite population systems, and the corresponding throughput results are also shown in

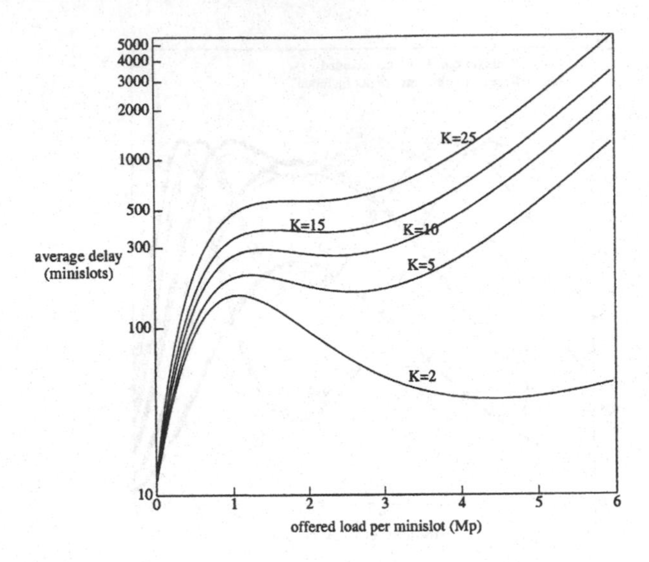

Figure 3.7 Nonmonotonic delay characteristics of the slotted-ALOHA/delayed-ALOHA protocol for L=10 slots per data packet, N=3 data channels, zero propagation delay, and different values of the backoff parameter K.

Fig. 3.6. As expected, the throughput reduction due to receiver collisions is more prominent when the system population is smaller [JiMu92b].

3.6.3 Receiver Collision Avoidance (RCA) Protocol

In the previous protocols, the main difficulty in detecting receiver collisions arose due to the simplicity of the systems, viz., the availability of only one tunable receiver per node to track both the control channel and the data channel activities. However, even for such a simple system, it turns out that, by adding some intelligence to the receivers, receiver collisions can be avoided and resolved at the data link (medium access control) layer. Thus, the Receiver Collision Avoidance (RCA) protocol [JiMu93a] operates under the same basic system parameters as before, viz., one tunable transmitter and one tunable receiver per node and a contention-based control channel. In addition, the protocol accommodates the fact that transceiver tuning times can be nonzero (of duration T slots, say). For simplicity of presentation, all nodes are assumed to be D slots away from the hub and $N = L$, but these conditions can be generalized [JiMu92a]. The protocol is briefly outlined below.

Channel Selection

Before a control packet is sent, the sender should decide which channel will be used to transmit the corresponding data packet. In order to avoid data channel collision, the RCA protocol proposes a simple and fixed data channel assignment policy. For the case $N = L$, each control slot is numbered 1 through N, periodically, as in a TDM system. Specifically, each control slot is assigned a fixed wavelength which will be the channel number on which a data packet will be transmitted if the corresponding control packet is successfully sent in that slot. Not only is this assignment scheme simple, but also it guarantees that the corresponding data channel transmission will be collision-free. For the cases $N \neq L$, please see [JiMu93a].

Node Activity List (NAL)

Each node maintains a Node Activity List (NAL) which contains information on the control channel history during the most recent $2T + L$ slots. Each entry contains the slot number and a status (Active or Quiet). If the status is Active (which means that a successful control packet is received), the corresponding NAL entry will also contain the source address, the destination address, and the wavelength selected, which are copied from the corresponding control packet. NAL may not be available (or its information is outdated) if the local receiver has been receiving on some data channel.

Packet Transmission

Consider a packet generated at transmitter i and destined for receiver j. Transmitter i will send out a control packet only if the following condition holds: node i's NAL does not contain any entry with either node i or node j as a packet destination. The control packet thus transmitted will be received back at node i after $2D$ slots, during which time node i's receiver must also be on the control channel. Based on the NAL updated by node i's receiver, if a successful control packet to node i (without receiver collision) is received during the $2T + L$ slots prior to the return of the control packet, then a receiver collision is detected and the current transmission procedure has to be aborted and restarted. Otherwise, transmitter i starts to tune its transmitter to the selected channel at time $t + 2D + 1$, and the tuning takes

T slots, after which L slots are used for data packet transmission, which is followed by another T-slot duration during which the transmitter tunes back to the control channel.

Packet Reception

The packet reception procedure is quite straightforward and is left as an exercise for the reader.

Performance

The maximum throughput achievable (over all data channels) under the RCA protocol is $1/e \approx 0.368$. This maximum is affected very slightly even when the transceiver tuning T is increased to several times the packet transmission time. For additional related work, see [JiMu93b, JiMu92a, Jia93].

3.6.4 Reservation Protocols

A single transmitter and a single receiver per node are the minimal requirements for a single-hop system, but the protocol and the system's performance can be improved by equipping nodes with multiple transceivers. The dynamic time-wavelength division multiple access (DT-WDMA) protocol [ChDR90] requires that each node be equipped with two transmitters and two receivers – one transmitter and one receiver at each node are always tuned to the control channel, each node has exclusive transmission rights on a data channel on which its other transmitter is always tuned to, and the second receiver at each node is tunable over the entire wavelength range, i.e., this is a $CC - FT^2 - FRTR$ system.

If there are N nodes, the system requires $N + 1$ channels – N for data transmission and the $(N + 1)$-th for control. Access to the control channel is TDM-based. The system is slotted with slots synchronized over all channels at the passive star (hub). A slot on the control channel consists of N minislots, one for each of the N nodes. Each minislot contains a source address field, a destination field, and an additional field by which the source node can signal the priority of the packet it has queued up for transmission, e.g., the priority information could be the delay the packet would experience from its arrival instant until the time it would reach the hub when it is transmitted. Note that control information is transmitted collision-free, and after transmitting in a control minislot, the node transmits the data packet in the following slot over its own dedicated data channel. By monitoring the control channel over a slot, a node determines if it is to receive any data over the following data slot. If a receiver finds that more than one node is transmitting data to it over the next data slot, it checks the priority fields of the corresponding minislots, and selects the one with highest priority. To receive the data packet, the node simply tunes its receiver to the source node's dedicated transmission channel. Figure 3.8 elaborates on the protocol's operation.

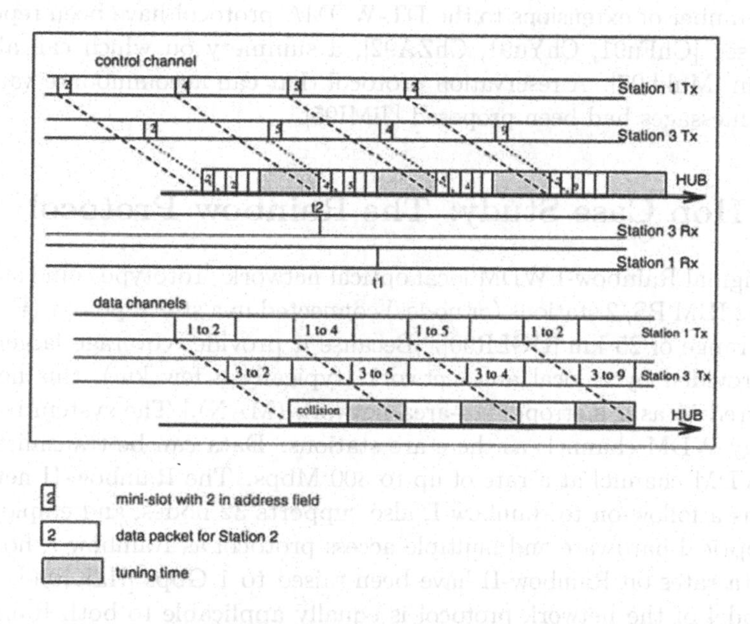

Figure 3.8 The dynamic time-wavelength division multiple access (DT-WDMA) protocol.

A novelty of this mechanism is that, even though there may be a "collision" in the sense that two or more nodes might have transmitted data packets to the same destination over a data slot duration, *exactly one of these transmissions will always be successfully received.* Also, this mechanism has an *embedded acknowledgement feature* since all other nodes can learn about successful data packet transmissions by following the same distributed arbitration protocol. In addition, the mechanism supports arbitrary propagation delays between the various nodes and the passive hub. The main limitation of the system is its scalability property since it requires that each node's transmitter have its own dedicated data channel. An additional issue is that this mechanism requires infinitely fast receivers or requires that the receiver tuning time be part of the slot duration (which may lead to a reduction of the protocol's efficiency). Without this limitation, for a large user population, the peak throughput of the system is $1 - 1/e \approx 0.632$ [ChDR90].

A number of extensions to the DT-WDMA protocol have been reported, please see [ChFu91, ChYu91, ChZA92], a summery on which can also be found in [Mukh97]. A reservation protocol that can accommodate variable-length messages had been proposed [JiMI95].

3.7 Single-Hop Case Study: The Rainbow Protocol

The original Rainbow-I WDM local optical network prototype could support up to 32 IBM PS/2 stations (or nodes) connected in a star topology (Fig. 3.9) over a range of 25 km [DGLR90]. [Because it provides coverage larger than that provided by a local area network (typically a few km), this network is referred to as a metropolitan-area network (MAN).] The system requires as many WDM channels as there are stations. Data can be transmitted on each WDM channel at a rate of up to 300 Mbps. The Rainbow-II network, which is a follow-on to Rainbow-I, also supports 32 nodes, and employs the same optical hardware and multiple access protocol as Rainbow-I; however, the data rates on Rainbow-II have been raised to 1 Gbps [HaKR96]. Thus, the model of the network protocol is equally applicable to both Rainbow-I and Rainbow-II.

In the Rainbow architecture, each node (or station) is equipped with a single fixed transmitter, which is tuned to its own unique wavelength channel, and a single tunable *Fabry-Perot* filter, which can be tuned to any wavelength ($FT - TR$). Tuning to any particular channel may take up to 25 ms. The tunable receiver scans across all the channels, looking for connection requests or acknowledgements from other stations. Rainbow's protocol is also referred to as an *in-band polling protocol.*

The Rainbow system is intended primarily as a circuit-switched network. The large filter tuning time results in a high connection setup time. This makes the system impractical for packet switching.

This section provides an analytical model for the Rainbow medium-access

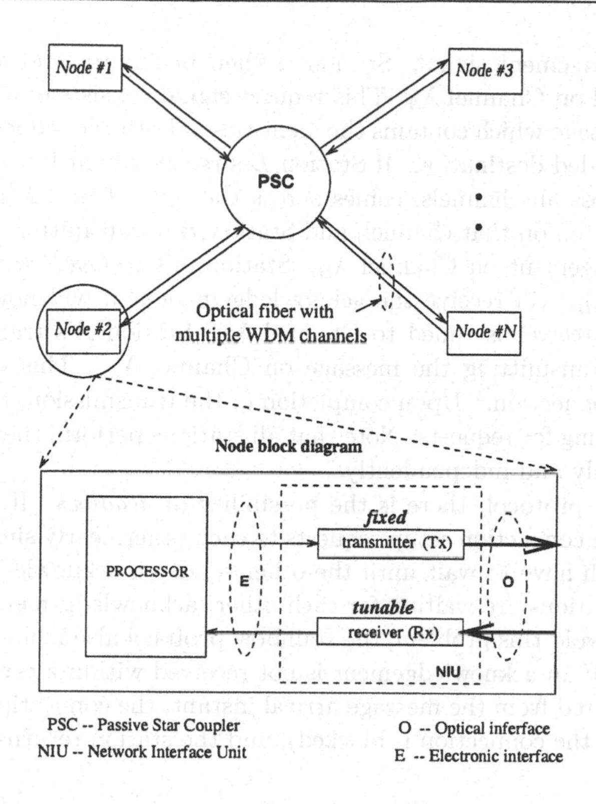

Figure 3.9 Passive-star topology for Rainbow.

protocol and presents an analysis of the system using the equilibrium point analysis (EPA) technique. The EPA technique is a means of analyzing complex systems by assuming that the system is always at an equilibrium point [FuTa83]. This technique has been successfully used to analyze a number of communication systems, e.g., satellite systems, and has been found to provide accurate results.

3.7.1 Rainbow Protocol

The signaling protocol for Rainbow is as follows. Each station is assigned its own unique channel on which its transmitter is fixed tuned. Upon the arrival of a message at Station A and destined for Station B, Station A first tunes its receiver to Channel λ_B so that it will be able to pick up Station

B's acknowledgement signal. Station A then begins to send a continuous request signal on Channel λ_A. This request signal consists of a periodically repeated message which contains the identities of both the requesting station and the intended destination. If Station B's receiver, which is continuously scanning across all channels, comes across the request on Channel λ_A, the receiver will stop on that channel, and Station B's transmitter will send out an acknowledgement on Channel λ_B. Station A's receiver, which is tuned to Channel λ_B, will receive the acknowledgement and will now know that Station B's receiver is tuned to Channel λ_A. Station A's transmitter will then begin transmitting the message on Channel λ_A. This establishes a full duplex connection.[3] Upon completion of the transmission, both stations resume scanning for requests. Note that all stations perform their operations asynchronously and independently.

With this protocol, there is the possibility of *deadlock*. If two stations begin sending connection setup requests to each other nearly simultaneously, they will both have to wait until the other sends an acknowledgement, but since both stations are waiting for each other, acknowledgements will never be sent. To avoid this problem, the Rainbow protocol also includes a *timeout mechanism*. If an acknowledgement is not received within a certain timeout period measured from the message arrival instant, the connection attempt is aborted (i.e., the connection is blocked), and the station returns to scanning mode.

3.7.2 Model of Rainbow

Round-robin (or polling) systems have been modeled and analyzed extensively in the literature (see, for example, [Taka86]). In our current setting of the Rainbow protocol, although a station's receiver is performing a round-robin operation, that operation may be interrupted by the station's transmitter. "Vacation models" do not appear to be applicable here because the system's performance characteristics are now determined by the

[3] In the actual prototype system [JaRS93], the filters used do not allow round-robin-type scanning. Instead, the filter must scan back and forth across all wavelengths (the so-called elevator-type scanning). Also, the filters are not able to tune directly to a particular wavelength. They must perform "elevator scanning" to find the appropriate channel before stopping. Here, we examine the original Rainbow system presented in [DGLR90]. We do not model the system in [JaRS93] because it is difficult to model elevator-type scanning, and also because we expect that newer filters, which do not have these limitations, may be employed in the future. In the remainder of this chapter, the term "scan" will be used to mean round-robin scanning, instead of elevator-type scanning.

transmission ("vacation") process. The modeling challenge is to relate the transmitter and receiver operations at a station in a simple manner.

Our model makes the following assumptions:

- There are N stations.
- There is no queueing. Each station has a single buffer to store a message, and any arrival to a nonempty buffer is blocked. A message departs from the buffer after it is completely transmitted.[4]
- The sending station, upon a message arrival, tunes its receiver to the channel of the target station prior to sending the connection setup request.
- Stations monitor the channels in a round robin fashion, in the sequence: $1, 2, \ldots, N, 1, 2, \ldots$.
- Time is slotted with slot length equal to 1 μs. This was chosen to provide a fine level of granularity in the system's model.
- It takes a fixed amount of time, τ slots, to tune a receiver to any particular channel.
- Messages arrive at each station according to a Bernoulli process with parameter σ, i.e., in any slot, a station with an empty slot can have a message arrival with probability σ.
- Message lengths are geometrically distributed with the average message length being $1/\rho$ slots. Message lengths are used to model connection holding times of circuits in the Rainbow prototype.
- The propagation delay between each station and the passive-star coupler is R slots. Given that signal propagation delay in fiber is approximately 5 μs/km, the value of R can be quite large, e.g., $R = 50$ slots for a station-to-star distance of 10 km.
- The timeout duration is denoted by ϕ (in time slots).
- Transmission times for request and acknowledgement messages are negligible.

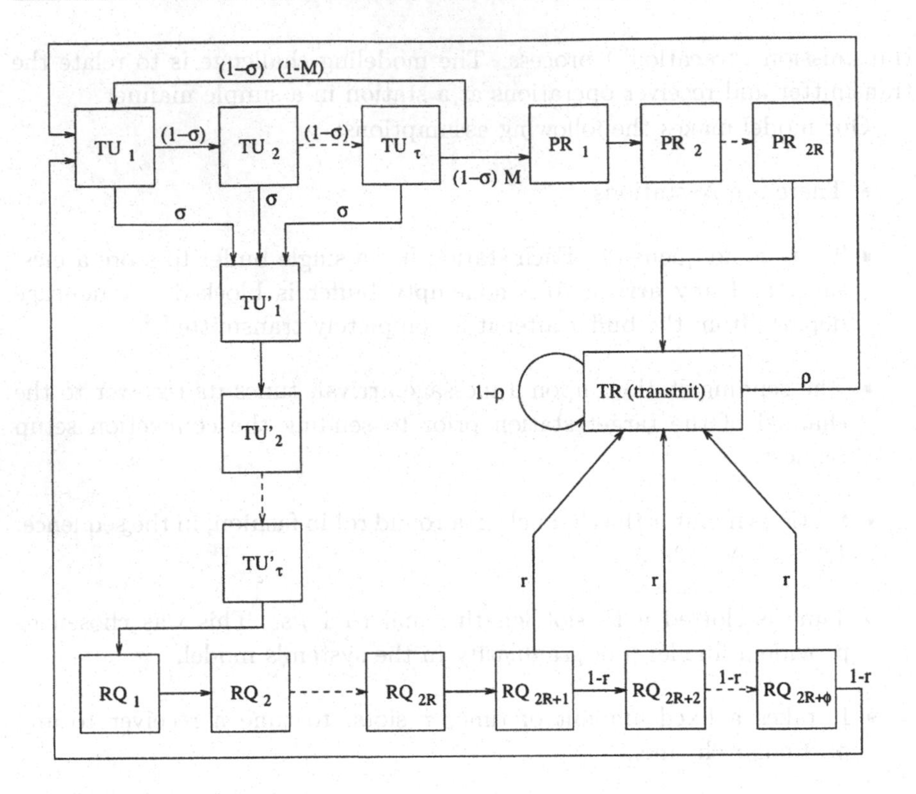

Figure 3.10 State diagram for the Rainbow model.

The state diagram for the model is given in Fig. 3.10. A station can be in any state and remains in that state for a geometrically distributed amount of time if it is in the transmission state, TR, or a fixed amount of time (one slot) if it is in any other state. A station departs from state TR with probability ρ at the end of a time slot and remains in state TR with probability $1 - \rho$. The states are defined as follows:

- $TU_1, TU_2, \ldots, TU_\tau$: These are states during which a station's receiver is scanning across the channels for requests. It takes τ time slots to

[4]The delay performance of this protocol with buffers has also been analyzed in [MoAz95a, MoAz95b] in the more general context of hybrid multiaccess, e.g., WDMA/CDMA. Our current setting on pure WDMA is a special case of the approach in [MoAz95a, MoAz95b].

tune to a particular station. From each of these states, an arrival can occur with probability σ. From the state TU_τ, if there is no arrival, the station either finds a request on the channel to which it has just completed tuning with some probability M and proceeds to send an acknowledgement, or it doesn't find a request with probability $1 - M$ and proceeds to tune its receiver to the next channel.

- $TU'_1, TU'_2, \ldots, TU'_\tau$: After an arrival occurs, the station immediately begins tuning its receiver to the channel of the destination. This process takes τ time slots. After tuning, the station begins to transmit a request.

- $RQ_1, RQ_2, \ldots, RQ_{2R+\phi}$: Upon sending a request, it takes R time slots of propagation delay for the request to reach the destination (see Fig. 3.11). At the earliest, an acknowledgement will be received after a propagation delay of $2R$ time slots. The station continues to send the request signal for a duration of ϕ time slots or until it receives an acknowledgement, whichever occurs first, where ϕ is the timeout duration. After sending a request for a duration of ϕ slots, the station must wait an additional $2R$ time slots of propagation delay for an acknowledgement. This ensures that all acknowledgements will result in a connection. If no acknowledgement is received, the current message is "timed out" and considered "lost," and the station returns to scanning mode. The probability of getting an acknowledgement is denoted by r, which is the same for each of the states RQ_{2R+1} to $RQ_{2R+\phi}$ since the system is memoryless, and an acknowledgement can be sent at any time by the acknowledging station. (The parameter r will be related to the probability M later in this analysis.) When an acknowledgement is received, the station immediately begins transmission of its message and goes into the transmission state TR.

- $PR_1, PR_2, \ldots, PR_{2R}$: The station enters these states if it finds a request while scanning. Upon identifying the request, the station sends an acknowledgement to the requesting station. The acknowledgement takes R time slots of propagation delay to reach the station requesting the connection, after which the requesting station begins its transmission. It takes another R slots of propagation delay for the message to arrive at the destination station, after which the station goes into the transmission state, TR, to receive the message. A connection will always be established if an acknowledgement has been sent.

- TR (transmission): In this state, a station is either transmitting or receiving a message. A station may stay in this state for a duration of more than one time unit and depart with probability ρ at the end of a slot. Upon completion of message transmission or reception, the station returns to the scanning operation.

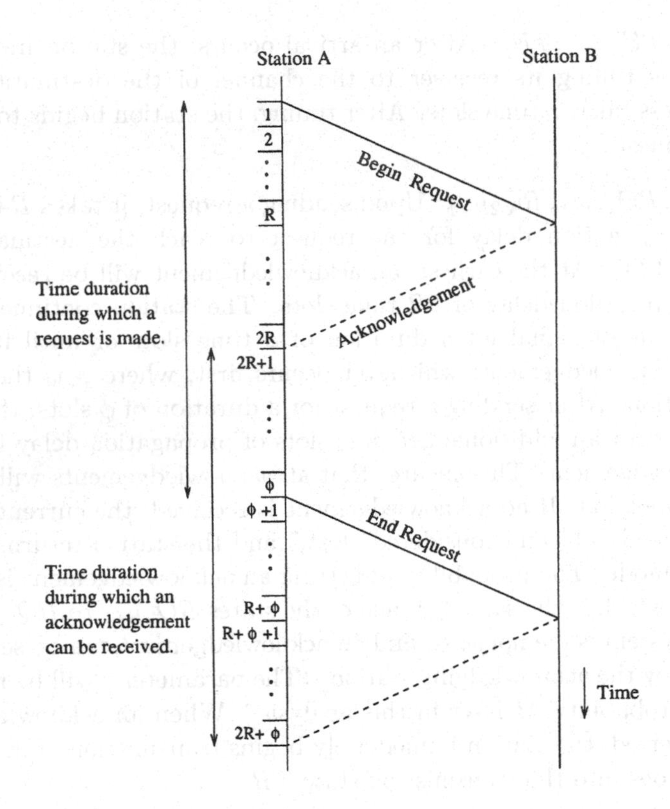

Figure 3.11 Timing for connection setup.

We define N_{TU_i} to be the expected number of stations in state TU_i, $N_{TU_i'}$ is the expected number of stations in state TU_i', N_{RQ_i} is the expected number of stations in state RQ_i, N_{PR_i} is the expected number of stations in state PR_i, and N_{TR} is the expected number of stations in the transmission state.

The system can be modeled as a Markov chain with state space vector:

$$\mathbf{N} = \{N_{TU_1}, N_{TU_2}, \ldots, N_{TU_\tau}, N_{TU'_1}, N_{TU'_2}, \ldots, N_{TU'_\tau}, N_{RQ_1}, N_{RQ_2}, \ldots, N_{RQ_{2R+\phi}}, N_{PR_1}, N_{PR_2}, \ldots, N_{PR_{2R}}, N_{TR}\}$$

It is difficult to analyze this system using Markov analysis techniques because of the very large state space. Therefore, we analyze the system at an equilibrium point using equilibrium point analysis (EPA).

3.7.3 Analysis of Rainbow

In EPA, the system is assumed to be always operating at an equilibrium point. This is an approximation since the system actually moves around the equilibrium point. At an equilibrium point of the system, the expected increase in the number of stations in any state is zero. Thus, the expected number of stations entering each state must be equal to the number of stations departing from each state in each time slot.

By writing the flow equation for each state, we obtain a set of K equations with K unknowns, where K is the number of states. Also, the flow equations can be written such that the expected number of stations in each state is expressed in terms of the expected number of stations in state TU_1. Thus, if we are able to solve for N_{TU_1}, we will have the solution for the entire system.

The flow equations can be written as follows. (The N's, which previously represented the random variables, now take on the corresponding average values.)

$$N_{TU_i} = (1-\sigma)^{i-1} N_{TU_1} \text{ for } i = 2, 3, \ldots, \tau \tag{3.1}$$

$$N_{PR_1} = N_{PR_2} = \cdots = N_{PR_{2R}} = (1-\sigma)^\tau M \times N_{TU_1} \tag{3.2}$$

$$\begin{aligned} N_{TU'_1} &= N_{TU'_2} = \cdots = N_{TU'_\tau} \\ &= N_{RQ_1} = N_{RQ_2} = \cdots \\ &= N_{RQ_{2R}} = [1-(1-\sigma)^\tau]\, N_{TU_1} \end{aligned} \tag{3.3}$$

$$N_{RQ_{2R+j}} = (1-r)^{j-1} (1-(1-\sigma)^\tau)\, N_{TU_1} \quad \text{for } j = 1, 2, \ldots, \phi \tag{3.4}$$

$$\begin{aligned}\rho \times N_{TR} &= N_{PR_{2R}} + \sum_{j=1}^{\phi} r \times N_{RQ_{2R+j}} \\ &= \left[(1-\sigma)^{\tau} M + \left\{1-(1-r)^{\phi}\right\} \times \{1-(1-\sigma)^{\tau}\}\right] \times N_{TU_1} \end{aligned} \quad (3.5)$$

We now need to solve for the unknown variables N_{TU_1}, r, and M. M is the probability that a request will be found by a scanning station. This is equal to the probability that another station is in states RQ_{R+1} to $RQ_{R+\phi}$, and that the request is intended for the scanning station. That is,

$$M = \frac{1}{N-1} \times \frac{1}{N}\left(\sum_{i=R+1}^{R+\phi} N_{RQ_i}\right) \quad (3.6)$$

Substituting Eqns. (3.3) and (3.4) into Eqn. (3.6), and simplifying yields:

$$\begin{aligned} M &= \frac{1}{N-1} \times \frac{1}{N}\left[1-(1-\sigma)^{\tau}\right] \\ &\quad \times \left\{R + \frac{1}{r}\left(1-(1-r)^{\phi-R}\right)\right\} N_{TU_1} \end{aligned} \quad (3.7)$$

The rate of flow into the active state from the request states must equal the rate of flow into the active state from the state PR_{2R}. This is because a station can only begin transmission if there is another station that will be receiving the transmission. This leads to the equation:

$$N_{PR_{2R}} = \sum_{i=1}^{\phi} r \times N_{RQ_{2R+i}} \quad (3.8)$$

which, upon substitutions from Eqns. (3.2) and (3.4), yields:

$$(1-\sigma)^{\tau} M = \left[1-(1-\sigma)^{\tau}\right]\left[1-(1-r)^{\phi}\right] \quad (3.9)$$

In steady state, the sum of the stations in each state is equal to the total number of stations in the system, i.e.,

$$N = \sum_{i=1}^{\tau} N_{TU_i} + \sum_{i=1}^{\tau} N_{TU_i'} + \sum_{i=1}^{2R+\phi} N_{RQ_i} + \sum_{i=1}^{2R} N_{PR_i} + N_{TR} \quad (3.10)$$

or

$$N = [1-(1-\sigma)^{\tau}] \times \left\{\frac{1}{\sigma} + \tau + 2R + \left(\frac{1}{r} + \frac{1}{\rho}\right)\left[1-(1-r)^{\phi}\right]\right\} N_{TU_1} + \left[\left(2R + \frac{1}{\rho}\right)(1-\sigma)^{\tau} M\right] N_{TU_1} \quad (3.11)$$

Equations (3.8), (3.9), and (3.11) can be solved simultaneously for the variables r, M, and TU_1, which can then be used to provide the steady state solution to the entire system.

The primary measures of interest are throughput, delay, and timeout probability. The normalized throughput is defined as the expected fraction of stations in the active state (which is also the fraction of a channel's bandwidth that is utilized):

$$S = \frac{N_{TR}}{N} \quad (3.12)$$

Delay is defined to be the time from a message's arrival to the system until the time that the message completes its transmission. This consists of the time required to tune to the destination station's channel, the propagation delay for the request and acknowledgement signals, the time until an acknowledgement is received, and the message transmission time. Delay is measured in slots, and is given by:

$$D = \tau + 2R + \sum_{k=1}^{\phi} k \times (1-r)^{k-1} \times r + \frac{1}{\rho} \quad (3.13)$$

The timeout probability is defined as the probability that a station will time out after entering the request mode, i.e.,

$$p_{TO} = (1-r)^{\phi} \quad (3.14)$$

Also of interest is the blocking probability, which is the probability of an arrival being blocked. This is equal to the probability that a station is not in scanning mode. The equation for blocking probability is:

$$p_{BL} = 1 - \frac{\sum_{i=1}^{\tau} N_{TU_i}}{N} \quad (3.15)$$

3.7.4 Illustrative Examples of Rainbow

For illustration purposes, a system with the following default parameters is considered:

- $N = 32$ stations,
- slot length $= 1\ \mu s$,
- $R = 50$ slots (corresponding to a 10 km distance between each station and the star coupler),
- $\tau = 1000$ slots (corresponding to a 1 ms receiver tuning time),
- $\rho = 10^{-5}$ (corresponding to a mean message length of 100 ms),
- $\sigma = 10^{-4}$ (corresponding to a message arrival rate of 100 msg/s at each station), and
- $\phi = 10^4$ slots (corresponding to a timeout of 10 ms),

In numerical examples presented below, we shall study the effect of some of these parameters on the system performance by varying them around the default values.

Figure 3.12 shows the normalized throughput versus message arrival rate for different values of timeout duration. As the arrival rate increases, the throughput will first increase as more messages become available for transmission, and then will eventually begin to decrease as the number of stations in request mode begin to outnumber the stations that are available to acknowledge requests. Note that, for a given arrival rate, there is an optimum timeout duration at which the system achieves its maximum throughput. Also, with timeout duration equaling 10 ms or higher, and other parameters as chosen, the peak system throughput is approximately 0.45, which means that, if each channel is operating at 300 Mbps (as in the Rainbow-I prototype), the effective information rate on each channel equals 135 Mbps. To verify the accuracy of the analytical model, results from a simulation of the Rainbow protocol are also included in Fig. 3.12. We find that, for higher values of message arrival rate, there is excellent agreement between the analysis and simulation, but the agreement is not so good for low message arrival rates.

In the analytical model, a request times out when the target of the request is either in request mode or engaged in a connection (transmitting

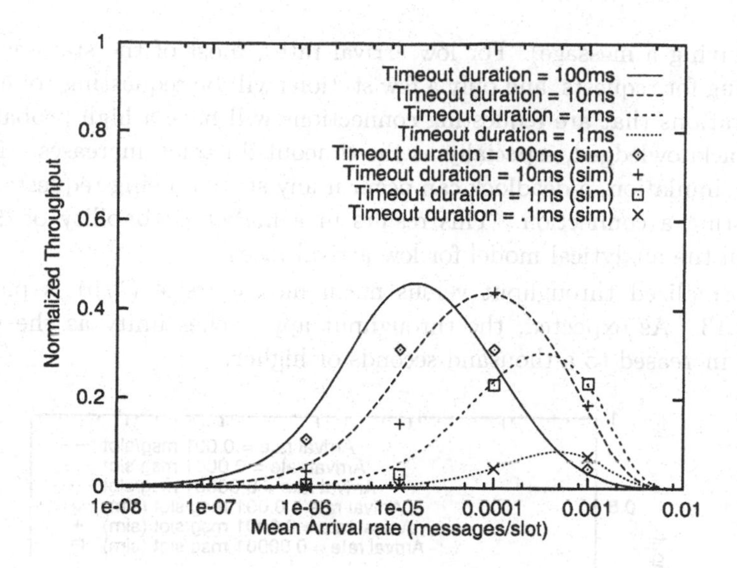

Figure 3.12 Throughput vs. arrival rate. Slot = 1 μs, $N = 32$, $R = 50$ μs, $\tau = 1$ ms, $1/\rho = 100$ ms.

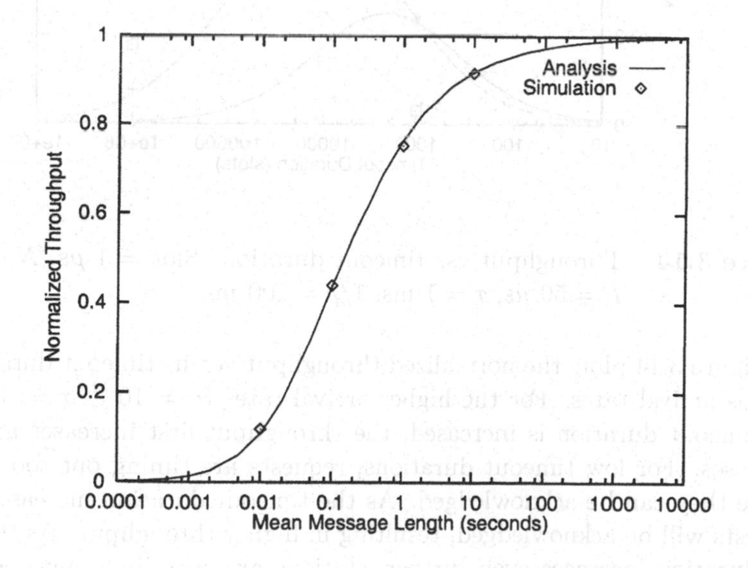

Figure 3.13 Throughput vs. message size. Slot = 1 μs, $N = 32$, $R = 50$ μs, $\tau = 1$ ms, $\sigma = 10^{-4}$ msg/slot, $\phi = 10$ ms.

or receiving a message). For low arrival rates, most of the stations will be scanning for requests, and only a few stations will be requesting connections. The stations that are requesting connections will have a high probability of being acknowledged, especially as the timeout duration increases. However, in the simulation, a deadlock can occur if any station being requested is also requesting a connection. This results in a higher probability of deadlock than in the analytical model for low arrival rates.

Normalized throughput versus mean message size ($1/\rho$) is plotted in Fig. 3.13. As expected, the throughput approaches unity as the message size is increased to a thousand seconds or higher.

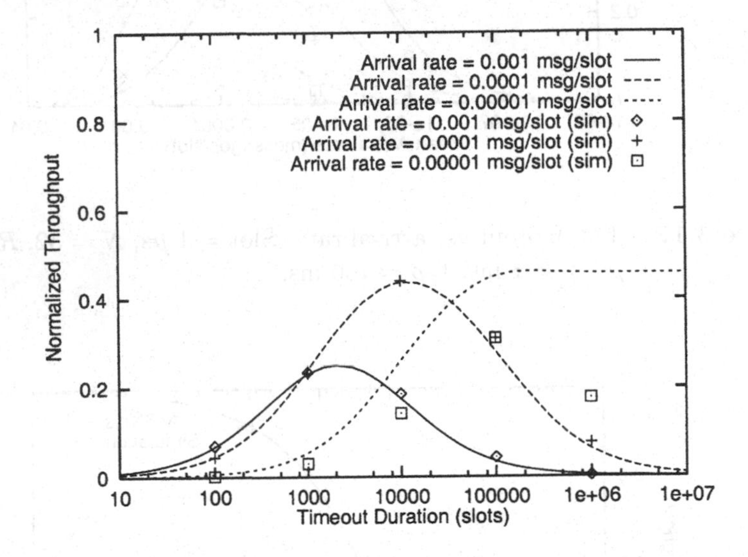

Figure 3.14 Throughput vs. timeout duration. Slot $= 1\ \mu$s, $N = 32$, $R = 50\ \mu$s, $\tau = 1$ ms, $1/\rho = 100$ ms.

Figure 3.14 plots the normalized throughput versus timeout duration for various arrival rates. For the higher arrival rates ($\sigma = 10^{-4}, \sigma = 10^{-3}$), as the timeout duration is increased, the throughput first increases and then decreases. For low timeout durations, requests are timing out too quickly, before they can be acknowledged. As the timeout duration increases, more requests will be acknowledged, resulting in higher throughput. As the timeout duration increases even further, stations are spending longer amounts of time in the request mode, resulting in fewer stations being available to acknowledge requests.

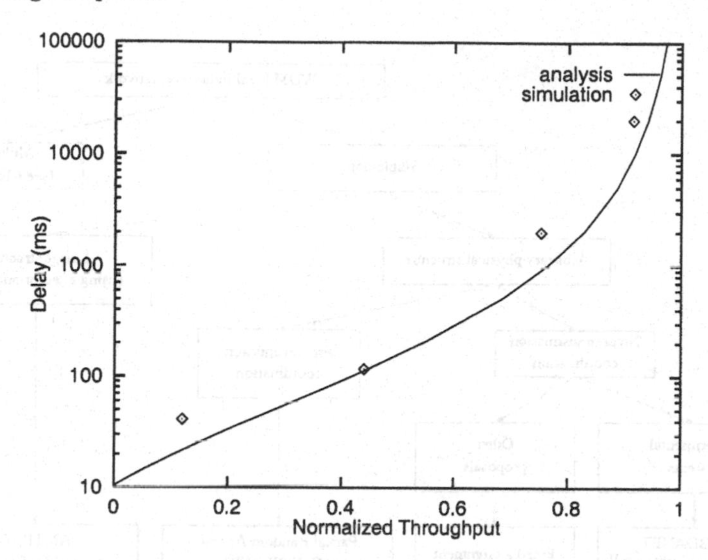

Figure 3.15 Delay vs. throughput for varying message size. Slot = 1 μs, N = 32, R = 50 μs, τ = 1 ms, $\sigma = 10^{-4}$ msg/slot, ϕ = 10 ms.

Figure 3.15 shows the average delay versus normalized throughput for increasing message size. As expected, the delay increases for larger message sizes with the $1/\rho$ term dominating Eqn. (3.13).

For more results on Rainbow performance, please see [JuBM96, Mukh97].

3.8 Summary

A large sampling of contributions made to date on the research and development of wavelength-division multiplexing (WDM) based local lightwave networks employing the single-hop approach was reviewed in this chapter. Various architectures, protocols, and experimental prototypes belonging to single-hop networks were examined. Especially, a quantitative analysis of the performance characteristics of Rainbow's *in-band polling protocol* is specifically implemented. Figure 3.16 provides a summarized classification of single-hop systems. (Material in Fig. 3.16 not discussed in this chapter can be found in [Mukh92a, Mukh97].)

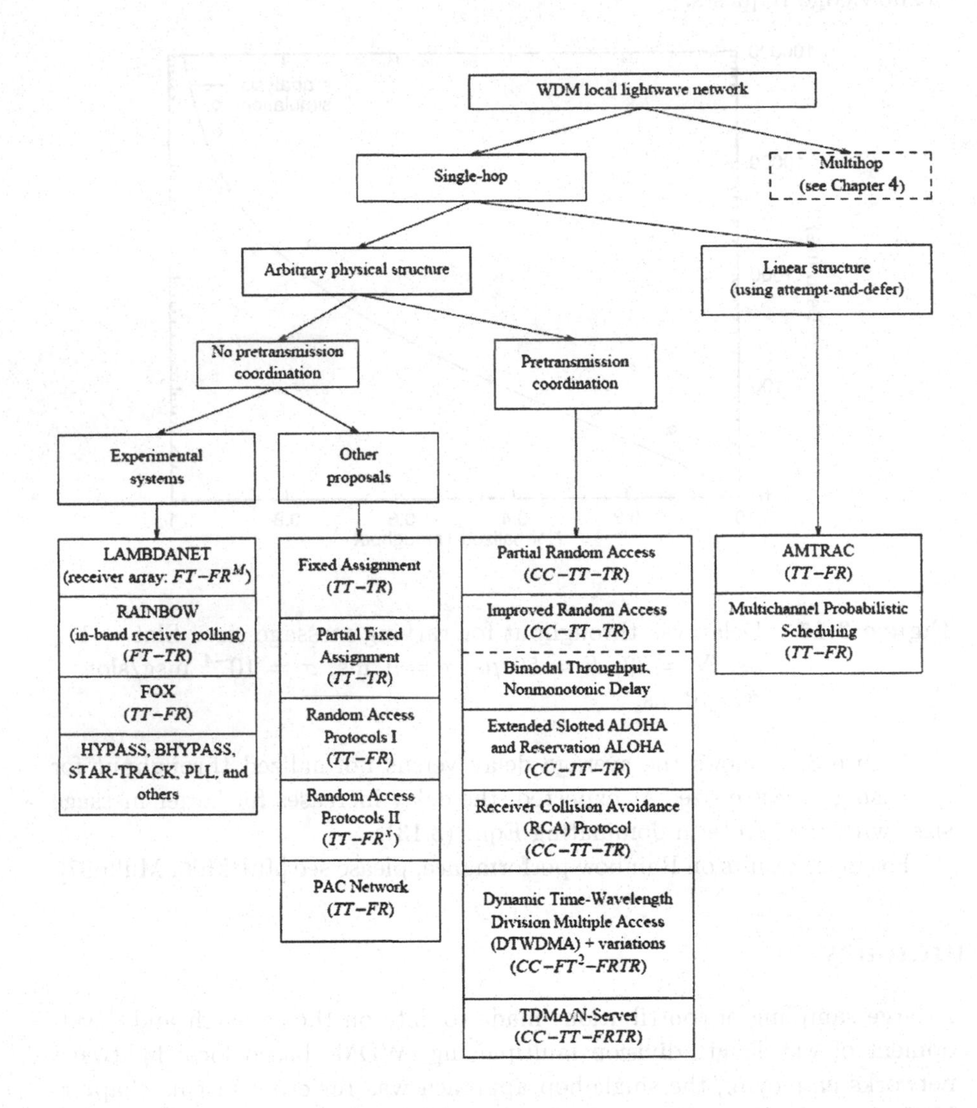

Figure 3.16 Classification of single-hop network architectures [Mukh92a].

Much more exciting research and development on single-hop systems is expected. Single-hop systems which can accommodate a large and time-varying user population, and which properly utilize the available channel capacity based on wavelength-agile transceivers (with nonzero tuning times and limited tuning ranges) are candidates for further study. It is expected that reservation-based protocols as well as protocols that can simultaneously accommodate circuit-switched traffic, narrowband packet traffic (short packets), and bulk data transfers (wideband packet traffic) will be developed. Protocols that can accommodate multicast traffic are expected to be very useful as well.

Although most of the single-hop architectures and protocols discussed in this chapter are meant for LANs/MANs, some of them can also be applied to situations where the propagation between the end users is negligible, e.g., the end-users may be co-located, e.g., as optical switching fabrics or as multiprocessor interconnects.

Exercises

3.1. Compare the physical topologies: star, bus, and tree with respect to: (a) number of simultaneous connections, (b) scalability, and (c) delay.

3.2. Consider the following two designs of a passive-star coupler using 2×2 couplers.

(a) One approach is shown in Fig. 3.1. Determine the number of 2×2 couplers needed for this design, and prove that each output gets $1/N^2$ of the input power.

(b) The second approach is the butterfly arrangement (also called a multistage interconnection network) mentioned in the text. Determine the number of 2×2 couplers needed for this design, and prove that each output gets $1/N$ of the input power.

(c) What are the trade-offs between the two designs.

3.3. Compare the maximum power loss in an N node passive-star coupler to the maximum power loss in an N node linear bus. Assume that the transmitted power is P. Consider only losses from splitting and coupling.

3.4. Compare protocols with pretransmission coordination to those protocols without pretransmission coordination.

3.5. Compare single-hop WDM systems which employ fixed transceivers to single-hop WDM systems which employ tunable transceivers.

3.6. Consider a single-hop network with four nodes and three channels.
(a) Give a schedule for the fixed channel assignment technique.
(b) Give a schedule for the Destination Allocation protocol.

3.7. Consider an optical LAN consisting of three nodes connected to a PSC. Assume two channels. Consider the following three scheduling protocols:

(a) fixed assignment,

(b) destination allocation, and

(c) source allocation.

Show the schedules for three time slots. Assume uniform traffic conditions. Calculate the maximum throughput per channel that can be achieved by employing the protocols. Now assume that a packet corresponding to every (*source*, *destination*) pair arrives with probability p every three time slots. Calculate the throughput by taking account of channel and receiver collisions. Calculate the maximum expected throughput for the protocols.

When can source/destination assignment protocols perform better than fixed assignment?

3.8. A single-hop $TT - TR$ network with 10 nodes connected via a passive star coupler is to use a fixed channel assignment. Each transmitter and receiver is capable of tuning to three noninterfering channels. How many TDM slots are required? Show that there is no interference. Assume negligible tuning times.

3.9. A partial fixed-assignment protocol is used in a passive star network. There are 10 nodes in the network and three channels available. What is the shortest time slot scheme (i.e., the scheme with the shortest period)? Avoid receiver collision, but not necessarily channel collision. When might such a scheme be desirable, and when might it not?

3.10. A fixed-assignment protocol is designed for a system with N nodes and W channels. The time slot duration is T. Find a lower bound on the expected packet delay in the system. Compare this with the lower bound on a partial fixed-assignment protocol. By what factor do these values differ?

3.11. Consider a two-channel, four-node broadcast-and-select WDM network. Each node is equipped with a single fixed transmitter and a single tunable receiver. Design a fixed assignment schedule for the following traffic matrix. Assume that the tuning time is equal to one time slot. Be sure to indicate which node is tuning its receiver during which slot.

$$A = \begin{bmatrix} 0 & 1 & 4 & 3 \\ 2 & 0 & 1 & 2 \\ 3 & 2 & 0 & 2 \\ 1 & 1 & 2 & 0 \end{bmatrix}$$

3.12. Consider a system using a fixed assignment protocol. Each node is equipped with a fixed transmitter and a tunable receiver. For arbitrary values of N (number of nodes), W (number of channels), L, (tuning time for a receiver), and traffic matrix a_{ij} where $i, j = 1, 2, \ldots, N$, find a lower bound on the optimal schedule length.

3.13. Suppose we have a system with four nodes. Each node is equipped with a single tunable transmitter and two fixed receivers. Assign channels to each of the transmitters and receivers ($T(i)$ and $R(i)$) for the random TDM protocol. What must we consider when assigning channels?

3.14. Consider a packet network in which the packet arrival process is a Poisson process with rate λ. In this network, users can start packet transmissions at times $T, 2T, 3T, \ldots$, where T is the length of a packet. All the packets that arrive during the time interval $[(n-1)T, nT)$ are transmitted during the time "slot" $[nT, (n+1)T)$. When a single packet is transmitted in a time slot, that packet is successfully received by the receiver. When two or more packets are transmitted in the same slot, a collision occurs and all the packets are lost.

(a) Find the probability P_k that k packets are transmitted in a time slot.

(b) The *throughput* S of this network is defined as the fraction of time slots which result in successful packet transfer. Find S as a function of λ and T.

(c) Find the value of λ that maximizes the throughput and the resulting throughput.

3.15. Consider the following WDM protocol. The network has a large number of nodes and W channels. Each node has W fixed-tuned receivers, one per channel, and a tunable transmitter. The time is slotted, and packets arrive according to a Poisson process with rate G packets/slot. When a node has a packet to transmit, it selects a wavelength randomly and transmits the packet in the first time slot on that wavelength.

(a) Find the probability that a wavelength is successfully utilized in a given time slot.

(b) Find the traffic rate that maximizes the throughput per channel and the resulting throughput.

(c) Find the average number of packets successfully carried by this network per time slot.

3.16. Consider a random-access $TT-FR$ system in which the home channel for node j is specified by $\lambda_{j \bmod 8}$. There are 64 nodes in this slotted ALOHA network, and the capacity of each channel is 2.5 Gbps. What is the maximum bandwidth of the network assuming uniform traffic, i.e., if every node has packets to send to every other node, what is the total of all the links bandwidth. Also, under these circumstances, what is the maximum average bandwidth of a single communication link?

3.17. In the ALOHA/ALOHA protocol, calculate the throughput as a function of the normalized offered load and calculate the maximum throughput. Assume that there are N stations, and packets arrive to each station according to a Poisson process with rate λ. Traffic is uniformly distributed across all stations. (Normalized offered load is defined as the rate of arrivals per packet transmission time.)

3.18. In slotted-ALOHA/ALOHA, what is the length of the vulnerable period for the control packet? For the data packet? Assume control packets to be one time unit in length, and data packets to be L time units in length.

3.19. In the ALOHA/ALOHA partial random-access protocol, a node transmits its data packet immediately after it transmits its control packet, regardless of whether the control packet succeeds or not. Suppose a node can transmit its data packet only when the corresponding control packet is successful. Compute the throughput per channel for the improved protocol. Assume N data channels in the system.

3.20. Explain the reason behind bimodal throughput seen in a slotted-ALOHA/delayed-ALOHA protocol. What is the optimal number of channels for a system with data packet length equal to 15 slots.

3.21. Why are CSMA protocols usually not very attractive for a single-hop WDM optical network?

3.22. Describe the procedure for receiving a packet in the Receiver Collision Avoidance (RCA) protocol.

3.23. In the Receiver Collision Avoidance (RCA) protocol, we set the size of the Node Activity List (NAL) to $2T + L$ slots. Explain why.

3.24. Describe possible scenarios in which receiver collisions may occur in the RCA protocol. Assume all nodes are D slots away from the hub, T is the tuning time, and L is the data packet length.

3.25. Consider the DT-WDMA protocol with N stations. Suppose that a packet arrives to a station in a given time slot with probability p, and no packets arrive to the station in the time slot with probability $(1-p)$. Calculate the probability that a transmitted packet will encounter a receiver collision. Assume tuning time is included in the time slot.

3.26. Show that under uniform traffic conditions and a large user population, the peak throughput in a DT-WDMA system is given by $1 - 1/e$.

3.27. In DT-WDMA, nodes can be made more intelligent by storing an $N \times N$ matrix B, called the "backlog" matrix at each node. Element b_{ij} denotes the number of packets at node i destined for node j. We can find an optimal algorithm which constructs a transmission schedule T, where $t_{ij} = 1$ denotes that node i should transmit a data packet to node j in the next slot. An optimal algorithm maximizes the number of transmission in a slot. What is the maximum throughput that can be achieved using this algorithm? (A qualitative justification is sufficient.)

3.28. Given a STARNET network of 100 nodes, and control packets of length 100 bytes, with an average distance between nodes of 100 m and negligible processing time, how long will a simple broadcast take, given that the information in the broadcast signal is embedded in the control packets?

3.29. For LAMBDANET, approximately how much of the low-loss region of bandwidth was utilized?

3.30. Professor W. D. Myer receives a small grant to set up an experimental WDM testbed. He decides to have four nodes in his network, each node with a fixed transmitter and tunable receiver.

(a) Should Professor Myer go for pretransmission control? Why or why not?

(b) Suddenly Professor Myer gets a lot of money and decides to construct a large network with pretransmission coordination, with each node having two tunable transceivers. How would this network be represented using the notation introduced in this chapter?

3.31. Consider the linear bus with attempt and defer nodes. Suppose the bus has five nodes over 5 km, and we require the received power at each node to be at least 30% of the transmitted power. How many amplifiers are required? Assume that amplifiers provide a gain of 25 dB.

3.32. An engineer is asked to come up with a transmission protocol for a single-hop network in which each node has a tunable transmitter and a tunable receiver. There are N nodes in the network and W channels. His solution employs a control channel. A frame consists of $N+L$ slots, where the length of a data packet is L. The first N slots in the frame correspond to control slots. Host i puts the number of the destination host to which it wants to transmit in slot i, $1 \leq i \leq N$. When more than one host wants to transmit to the same destination, the station with the lowest index wins. Further, the choice of channel is implicit – the first winner transmits on Channel 1, the second winner transmits on Channel 2, etc. Note that some winners may not be able to transmit because of the constraint on the number of channels. Comment on the characteristics of the protocol and suggest ways of improving it.

3.33. The Rainbow testbed utilized Fabry-Perot filters for the tunable receivers. In regards to the Rainbow medium-access control (MAC) protocol, how would you justify this choice of receiver?

3.34. Which kind of traffic is more well-suited to the Rainbow protocol – packet-switched traffic or circuit-switched traffic? Why?

3.35. Suppose station A is trying to set up a connection with station B. Draw simple state diagrams for stations A and B that illustrate this process.

3.36. Why is the timeout mechanism necessary in the Rainbow protocol? Show by example how, in the absence of a timeout mechanism, the system can become deadlocked. Give an alternate method of avoiding deadlocks in the Rainbow protocol.

3.37. Consider a Rainbow network with 32 stations. Each station is 10 m from the star coupler (i.e., propagation time is negligible), and the receiver tuning time is 1 ms. Two stations, A and B, wish to send a message to each other at the same time. What will happen assuming the protocol is as indicated in the text? Now suppose that station A

transmits its message first. Station B will then transmit its message at some random time chosen from an uniform random distribution of duration over the range 0 to 100 ms. What is the probability that station B's message will be successfully transmitted? Assume that the timeout is sufficiently large to allow the scanning of all channels.

3.38. In this exercise, we will simulate the Rainbow protocol. Consider a Rainbow system with four nodes, four channels numbered 1 through 4, and a tuning time of 10 slots. Propagation delay is 5 slots. At time = 0, all receivers are parked on channel 1. Connection hold times are long (assume infinity for this exercise), and the timeout duration is finite. The connection requests are:

- Connections ($1 \to 2$) and ($2 \to 3$) at time = 0.
- Connections ($1 \to 3$) and ($4 \to 3$) at time = 10.

Which connection(s) will succeed?

3.39. Consider a Rainbow network with 100 nodes, an average scan time of 100 μs for each channel (including tuning time as well as signal detection time), and a round-robin channel scanning algorithm on each receiver. What is the expected time required to broadcast a packet of 1000 bytes from one node to all of the other nodes? Propagation time, as well as acknowledgement transmission time, is negligible. Also assume that the network has no traffic when the broadcast occurs. Bandwidth per channel is 250 Mbps.

3.40. If we allow synchronization among the receivers and senders in the Rainbow network with 100 nodes, can you find a scheme to significantly reduce the time of a broadcast from one node to all other nodes, given the same network and assumptions of the previous problem? What is the total broadcast time required by your scheme?

3.41. What is EPA? How does the EPA technique simplify the analysis? What is lost in the simplification?

3.42. Model the following system. There are N jobs and one server. A job can be in two states only: either it is being serviced (or queued), or it is idle. The service time is exponentially distributed with parameter μ, while the "idle" time (of the jobs) is exponentially distributed with parameter λ. Develop and solve the Markov chain for the system. Using EPA, find the average number of idle nodes.

Let $N = 10, \mu = 1$ and vary λ from 0.05 to 1 in steps of 0.05. Compare the average number of idle jobs calculated from EPA analysis with those computed from the Markov chain. Explain your results.

3.43. Why is the analysis of the Rainbow protocol inaccurate for large time-out values?

3.44. Prove that if N connections are attempted in an N node Rainbow system, *all* nodes will be "locked up", i.e., either a node will be involved in a deadlock or will "wait" for a node involved in a deadlock.

3.45. Through simulation, find out the average number of nodes "locked up" due to deadlock when E simultaneous connections (chosen randomly) are attempted in a system with N nodes.

3.46. Let $\Lambda_N(e)$ equal the average number of nodes that are "locked up" when E random connections are attempted simultaneously. Show how we can use $\Lambda_N(e)$ to improve the analytical model for the Rainbow protocol.

3.47. Using the analytical model of Rainbow, find the ratio of stations which are scanning to the stations which are requesting connections.

3.48. Explain the relationship between the timeout duration and the normalized throughput in the Rainbow system. What happens to the system throughput when the timeout duration is made too large? Too small? Why?

3.49. In the Rainbow state diagram, no state information is maintained regarding specific node identities. The state diagram applies to the case in which round-robin scanning is performed (i.e., scanning order is $1, 2, \ldots, N, 1, 2, \ldots, N$, etc.). Why can't this state diagram also apply to the protocol in which elevator-type scanning ($1, 2, \ldots, N-1, N, N-1, \ldots, 2, 1, 2$, etc.) is performed?

3.50. Consider the Rainbow protocol. In order to avoid deadlock, suppose the source node, upon tuning its receiver to the destination node's channel, doesn't proceed with its connection request if it finds that the destination node is busy transmitting its own connection request.

It instead resumes scanning. How does this affect the performance of the protocol? How can this change be modeled in the state diagram shown in Fig. 3.10?

3.51. Suppose we modify the Rainbow protocol such that if a source node, upon tuning its receiver to the destination node's channel (state TU'_τ in the state diagram in Fig. 3.10), finds that the destination node is requesting a connection with the source node; and then, instead of sending a request, it sends an acknowledgement to the destination node's request. In Fig. 3.10, this transition could be modeled as a link from state TU'_τ to state PR_1. What would be the probability associated with this link? How would the flow equations change?

Chapter 4

Multihop Networks

4.1 Introduction

This chapter discusses multihop networks, covering various multihop topologies, their properties, and some related optimization problems.

4.2 Characteristics of a Multihop System

In a multihop system, unlike the case in a single-hop system, the channel to which a node's transmitter or receiver is tuned (or is to be tuned) is relatively static, and this assignment is normally not expected to change except when a new global reassignment of all transceivers is deemed to be beneficial. It is unlikely that there will be a direct path between every node pair (in which case each node in a N-node network must be equipped with $N-1$ fixed-tuned transceivers) so that, in general, traffic (e.g., a packet) from a source to a destination may have to hop through some intermediate nodes, possibly zero. Different virtual structures will have different operational features (e.g., ease of routing) and different performance characteristics (e.g., minimal average packet delay, minimal number of hops, balanced link flows, etc.)

An example multihop architecture is shown in Fig. 4.1. The physical topology is a star (Fig. 4.1(a)) while the embedded virtual topology is a

2×2 torus (Fig. 4.1(b)). Note that, in this example, node 1 can communicate with nodes 2 and 3 directly via wavelength channels w_1 and w_2, but in order to reach node 4, information from node 1 should "multihop" either through node 2 or node 3.

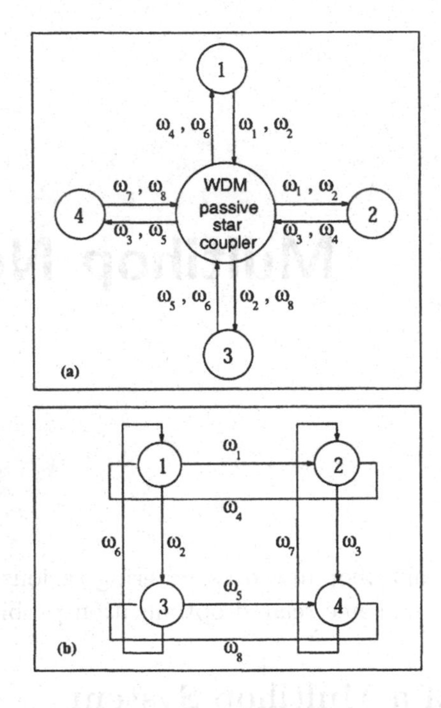

Figure 4.1 An example four-node multihop network: (a) physical topology, (b) logical topology.

While the transceiver tuning times play a vital role in determining the performance and characteristics of single-hop systems, they have little impact on multihop systems since the multihop virtual topology is essentially a static one. However, in designing a "good" multihop system, there are two other important issues which the system architect must address.

First, the virtual structure chosen must be close to "optimal" in some sense, e.g., the structure's *average (hop) distance* between nodes must be small, the average packet delay must be minimal, or the maximum flow on any link in the virtual structure must be minimal. Two nodes are at a *hop distance* of h if the *shortest* path between them requires h hops. In a multihop structure, each such hop means "travel to the star and back." The *maximum hop distance* between any two nodes is referred to as the structure's *diameter*. Multihop networks with small $\overline{h}$ and small diameter are desirable.

Second, the nodal processing complexity must also be small because the high-speed environment allows very little processing time; consequently, simple routing mechanisms must be employed. A routing-related subproblem is the buffering strategies at the intermediate nodes. Some approaches propose the use of "deflection routing" under which a packet, instead of being buffered at an intermediate node, may be intentionally misrouted but still reach its destination over a slightly longer path.

Multihop structures can be either irregular or regular. Irregular multihop structures generally address the optimality criterion directly, but the routing complexity can be large since they lack any structural connectivity pattern. Topological optimization of multihop architectures can be performed. Regular structures, because of their structured node-connectivity pattern, have simplified routing schemes; however, their regularity also constrains the set of solutions in addressing the optimality problem, and the number of nodes in a complete regular structure usually forms a special set of integers, rather than an arbitrary integer. Regular structures which have been studied include perfect shuffle (called ShuffleNet), de Bruijn graph, toroid (Manhattan Street Network, MSN), hypercube, linear dual bus, and a virtual tree. Characteristics of alternative routing strategies, including deflection routing, in ShuffleNet have also been studied quite extensively.

Attention must be paid to another piece of input to these designs, viz., the fact that the offered loads by the various nodes may not necessarily be symmetric, which is more pronounced with the proliferation of special-purpose networking equipment such as servers and gateways. Regular structures are generally amenable to uniform loading patterns, while irregular structures can generally be optimized for arbitrary workloads. The performance effect of nonuniform traffic and corresponding adaptive routing schemes to control congestion are important topics.

Finally, another pertinent issue which a multihop network architect must be mindful of is whether to employ "dedicated channels" or "shared channels." Under the case of dedicated channels, each virtual link employs a dedicated wavelength channel. However, since internode traffic may be bursty, the traffic on an arbitrary link is expected to be bursty as well, as a result of which some of the links' utilizations may be low. Consequently, the shared-channel mechanism advocates the use of two or more virtual links to share the same channel in order to improve the channel utilization. However, this also introduces the need for a multiple access protocol on the channel, viz., an arbitration mechanism that governs access rights to the channel. Issues related to shared-channel strategies are also discussed in this chapter.

4.3 Topological Optimization Studies

This section will first review the construction of optimal structures based on minimizing the maximum link flow [LaAc91], followed by optimizations based on minimization of the mean network-wide packet delay [BaFG90].

4.3.1 Flow-Based Optimization

Consider a network containing an arbitrary number of nodes N, which are indexed 1, 2, ... , N. Each node has T transmitters and T receivers. The number of wavelength channels is not a constraint in our initial discussions, or, in other words, there are $N \times T$ channels since each transmitter/receiver transmits/receivers on its own unique channel. The capacity of each WDM channel is C units (say bps). The traffic matrix is given by λ_{sd}, where λ_{sd} is the traffic flow from source node s to destination node d for $s, d = 1, 2, \ldots, N$. The flow in link ij is denoted by f_{ij}, while the fraction of the λ_{sd} traffic flowing through link ij is denoted by $f_{ij}^{s}d$. Let Z_{ij} be the number of directed channels from node i to node j. Then, the capacity of link ij equals $C_{ij} = Z_{ij} \times C$. The fraction of the (i,j)-link capacity which is utilized equals f_{ij}/C_{ij}. An arbitrary topology will have a link with maximum utilization given by:

$$max_{(i,j)} \left\{ \frac{f_{ij}}{C_{ij}} \right\}$$

Among various alternative topologies that are possible, the one that minimizes the above quantity is chosen to be the optimal interconnection pattern.

Formally, the above flow and wavelength assignment (FWA) problem can be set up as a mixed integer optimization problem with a min-max objective function subject to a set of linear constraints [LaAc91]. The main characteristic of this problem formulation is that it allows the traffic matrix to scale up by the maximum amount before its most heavily loaded link saturates. Another important characteristic is that only the node-to-node traffic intensities need to be known, and the solution is independent of the traffic type, which could be either circuit-switched or packet-switched (which, in turn, could be datagram-based or virtual-circuit-based).

Unfortunately, the search space for the connectivity diagram grows rapidly with increasing N. Hence, there exists a suboptimal and iterative algorithm which first determines a heuristic initial solution and then applies branch-exchange operations iteratively to improve the solution [LaAc91]. The initial solution, in turn, consists of a connectivity problem, which heuristically tries to maximize the one-hop path traffic (i.e., it attempts to connect nodes with more traffic between them in one hop), and this can be solved by a special version of the simplex algorithm. The second part of the initial solution is the routing problem, which can be formulated as a multicommodity flow problem with a nonlinear, nondifferentiable, convex objective function, and it can be solved by using the flow-deviation method [FrGK73]. Iterative improvement is performed by considering a number of least-utilized branches (say K) two at a time. A branch-exchange operation is performed by (1) swapping the transmitters (or receivers) of the two least-utilized branches, (2) re-solving the routing problem on the new connectivity diagram, and (3) accepting the swap if the new topology leads to a lower network-wide maximum link utilization. This procedure is repeated until no improvement is obtained.

Results obtained via the above algorithm for $N = 8$ and $T = 2$ are quite encouraging [LaAc91]. The connectivity diagrams and the corresponding link flows for uniform traffic, ring-type traffic, disconnected-type traffic, and centralized traffic do match with intuition. An improvement of the problem formulation can also accommodate the finite tuning range of the transceivers [LaAc90].

4.3.2 Delay-Based Optimization

In designing an optimal virtual topology, an alternative objective may be to minimize the mean network-wide packet delay. The packet delay has two components – the first is due to the propagation delays encountered by the packet as it hops from the source through intermediate nodes to the final destination, and the second is due to queueing at the intermediate nodes. In a high-speed environment where the channel capacity C is quite large and the link utilizations are expected to be in the light-to-moderate range, the queueing delay component can be ignorable compared to the propagation delay component which is directly dependent on the "glass distance" between the nodes [BaFG90]. Thus, this optimization also requires knowledge of the distance matrix d_{ij}, where d_{ij} is the glass distance from node i to node j per the underlying physical topology. The mean network-wide packet delay can therefore be written as:

$$\overline{D} = \sum_{i=1}^{N} \sum_{j=1}^{N} \frac{f_{ij} d_{ij}}{\nu \gamma} + \Delta$$

where ν = velocity of light in fiber, f_{ij} is the flow through link ij, $\gamma = \sum_{s=1}^{N} \sum_{d=1}^{N} \lambda_{sd}$ = total offered load to the network, and Δ is the nodal queueing delay component (see [BaFG90]).

Formally, the optimization can be stated as follows:

Given:	Traffic Matrix
	Distance Matrix
Objective:	Minimize Mean Packet Delay
Design Variables:	Virtual Topology
	Link Flows
Constraints:	Flow Conservation
	Nodal Connectivity (including the number of transmitters and receivers per node)

In [BaFG90], algorithms based on simulated annealing have been employed to solve both the dedicated-channel and the shared-channel cases, where time-division multiple access (TDMA) is employed for channel-sharing. A faster solution to the shared-channel topology has also been obtained in [BaFG90] by using genetic algorithms.

4.4 Regular Structures

Regular topologies which have been studied as candidates for multihop lightwave networks include perfect shuffle, de Bruijn graph, toroid, and hypercube. Their characteristics are outlined in the following subsections. Some general results and bound on regular multihop structures can be found in [HlKa91].

4.4.1 ShuffleNet

A (p, k) ShuffleNet can be constructed out of $N = kp^k$ nodes which are arranged in k columns of p^k nodes each (where p, k = 1, 2, 3, ...), and the

k^{th} column is wrapped around to the first in a cylindrical fashion. The nodal connectivity between adjacent columns is a p-shuffle, which is analogous to the shuffling of p decks of cards. This interconnection pattern can be defined more precisely as follows: (1) number the nodes in a column from top to bottom as 0 through $p^k - 1$, and (2) direct p arcs from node i to nodes $j, j+1, \ldots, j+p-1$ in the next column where $j = (i \bmod p^{k-1}) \cdot p$. A $(2,2)$ ShuffleNet is shown in Fig. 4.2.

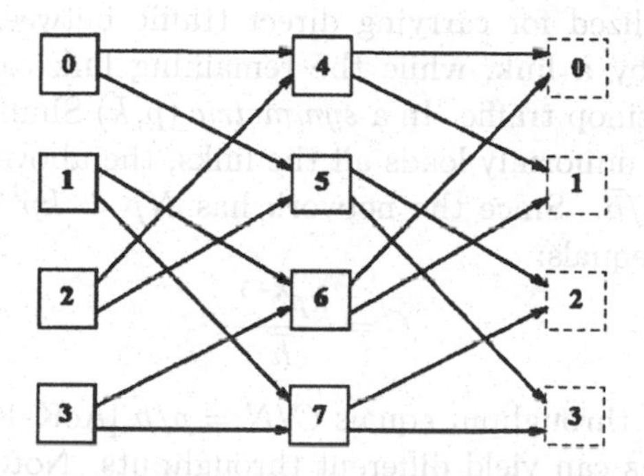

Figure 4.2 A $(2,2)$ ShuffleNet.

An important performance metric of this structure is the mean hop distance between any two randomly chosen nodes. From any "tagged" node in any column (say the first column), note that p nodes can be reached in one hop, another p^2 nodes in two hops, and so on, until all remaining $p^k - 1$ nodes in the first column are visited. Call this the first pass. In the second pass, all nodes which were not visited in the first pass can now be visited [although there can be multiple (shortest-path) routes for doing so]. For example, in the $(2,2)$ ShuffleNet in Fig. 4.2, node 6 can be reached from node 0 either via the path 0-5-3-6 or the path 0-4-1-6, both of which are "shortest paths." A preferred routing algorithm will be outlined later.

Thus, the number of nodes which are h hops away from a "tagged" node can be written as:

$$n_h = \begin{cases} p^h & h = 1, 2, \ldots, k-1 \\ p^k - p^{h-k} & h = k, k+1, k+2, \ldots, 2k-1 \end{cases}$$

Then, the average number of hops between any two randomly selected nodes, given by $\sum_{h=1}^{2k-1} h n_h / (N-1)$, can be obtained as:

$$\overline{h} = \frac{kp^k(p-1)(3k-1) - 2k(p^k-1)}{2(p-1)(kp^k-1)} \tag{4.1}$$

Note that the ShuffleNet structure's *diameter*, viz., the maximum hop distance between any two nodes, equals $2k-1$.

Due to multihopping, note that only a fraction of a link's capacity is being actually utilized for carrying direct traffic between the two specific nodes connected by a link, while the remaining link capacity is used for forwarding of multihop traffic. In a *symmetric* (p, k) ShuffleNet in which the routing algorithm uniformly loads all the links, the above utilization of any link is given by $1/\overline{h}$. Since the network has $Np = kp^{k+1}$ links, the total network capacity equals:

$$C = \frac{kp^{k+1}}{\overline{h}}$$

while the per-user throughput equals $C/N = p/\overline{h}$ [AcKa89]. Thus, different (p, k) combinations can yield different throughputs. Note that the per-user throughput may be increased by choosing a small k and a large p, so that the mean hop distance between nodes is reduced.

The following paragraphs examine alternative routing strategies that can be employed in ShuffleNet.

Simple Routing in ShuffleNet

There exist a number of approaches that deal with the routing problem in ShuffleNet. A simple addressing and fixed routing scheme is outlined first. A node in a (p, k) ShuffleNet is assigned the address (c, r) where $c \in 0$, $1, \cdots$, k-1 is the node's column coordinate [labeled 0 through $k-1$ from left to right (see Fig. 4.2)], and $r \in 0, 1, 2, \ldots, p^k - 1$ is the node's row coordinate (labeled 0 through $p^k - 1$ from top to bottom, using base-p digits). Thus, one may write $r = r_{k-1}r_{k-2}\ldots r_2r_1r_0$. This addressing scheme, along with the p-shuffle interconnection pattern, has the property that, from any node (c, r) where $r = r_{k-1}r_{k-2}\cdots r_2r_1r_0$, the row addresses of all the nodes reachable in the next column have the same first $k-1$ p-ary digits (given by $r_{k-2}r_{k-3}\cdots r_2r_1r_0$), and they differ in only the least-significant digit. For routing purposes, it is required that the destination address (c^d, r^d) be included in every packet. When such a packet arrives at an arbitrary node $(\hat{c}, \hat{r})$, then, it is removed from the network if $(c^d, r^d) = (\hat{c}, \hat{r})$ (i.e., the packet has reached its destination). Otherwise, node $(\hat{c}, \hat{r})$ determines the column distance X between itself and the packet's destination (c^d, r^d) to be:

$$X = \begin{cases} k + c^d - \hat{c} & c^d \neq \hat{c} \\ k & c^d = \hat{c} \end{cases}$$

Out of the p nodes in the next column to which node $(\hat{c}, \hat{r})$ may forward the current packet, it chooses the one whose least-significant digit is given by r^d_{X-1} (which is part of the destination node's address obtainable from the packet header). In particular, the packet is routed to the node with the identity $[(\hat{c}+1) \bmod k, \hat{r}_{k-2}\hat{r}_{k-3}\cdots\hat{r}_2\hat{r}_1\hat{r}_0 r^d_{X-1}]$. Note that this routing scheme follows the single shortest path between nodes $(\hat{c}, \hat{r})$ and (c^d, r^d) if the number of hops between them equals k or less; otherwise, it chooses one among several possible shortest paths. Also, note that the routing decision made at node $(\hat{c}, \hat{r})$ is independent of the packet's original source.

For additional details, such as (1) effect of nonuniform traffic on ShuffleNet performance [EiMe88], and (2) adaptive and deflection routing strategies in ShuffleNet [KaSh91], please see [Mukh92b, Mukh97].

4.4.2 de Bruijn Graph

While ShuffleNet has been actively studied as a candidate for multihop lightwave networks, some other structures have been investigated as well. One such structure is the de Bruijn graph [SiRa94]. A (Δ, D) de Bruijn graph ($\Delta \geq 2$, $D \geq 2$) is a directed graph with the set of nodes $\{0, 1, 2, \ldots, \Delta-1\}^D$ with an edge from node $a_1a_2\cdots a_D$ to node $b_1b_2\cdots b_D$ if and only if the following condition is satisfied:

$$b_i = a_{i+1}$$

where $a_i, b_i \in \{0, 1, 2, \ldots, \Delta-1\}$ and $1 \leq i \leq D-1$. Each node has in-degree and out-degree Δ, some of the nodes may have "self-loops," and the total number of nodes in the graph equals $N = \Delta^D$. An example $(2, 3)$ de Bruijn graph is shown in Fig. 4.3.

Simple Routing in de Bruijn Graph

A link from node A to node B can be represented by $(D+1)$ Δ-ary digits, the first D of which represent node A, and the last D digits represent node B. In a similar fashion, any path of length k can be expressed by $D+k$

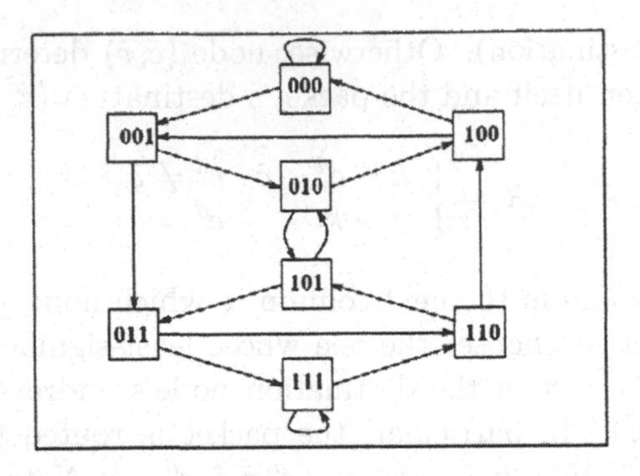

Figure 4.3 A $(2, 3)$ de Bruijn graph.

digits. In determining the shortest path from node $A = (a_1a_2 \cdots a_D)$ to node $B = (b_1b_2 \cdots b_D)$, one needs to consider the last several digits of A and the first several digits of B to obtain a perfect match over the largest possible number of digits. If this match is of size k digits, i.e., $(b_1b_2 \cdots b_{D-k}) = (a_{k+1}a_{k+2} \cdots a_D)$, then the k-hop shortest path from node A to node B is given by $(a_1a_2 \cdots a_Db_{D-k+1}b_{D-k+2} \cdots b_D)$.

An upper bound on the average hop distance between two arbitrary nodes in a de Bruijn graph follows [SiRa94]:

$$\overline{h}_{deBr} \leq D\frac{N}{N-1} - \frac{1}{\Delta - 1}$$

For a large Δ, this bound is close to the theoretical lower bound on the mean hop distance in an arbitrary directed graph with N nodes and maximum out-degree $\Delta \geq 2$ [SiRa94], thereby implying that the de Bruijn graph can be considered to be asymptotically optimal. The mean hop distances in (Δ,D) de Bruijn graphs and (p, k) ShuffleNets have been compared in [SiRa94], and it is found that, in general, for the same average number of hops, topologies based on de Bruijn graphs can support a larger number of nodes than can ShuffleNets. This is mainly due to the fact that the diameter (the maximum hop distance) in a ShuffleNet can be very large (it equals $2k - 1$ in a (p, k) ShuffleNet). Hence, ShuffleNet performs well when its diameter and consequently the number of nodes is small. ShuffleNet and de Bruijn graph are related in the sense that, if the second and subsequent columns in ShuffleNet are the same as the first, then one obtains a de Bruijn graph with degree p, diameter k, and p^k nodes [SiRa94].

An undesirable characteristic of the de Bruijn graph is that, even if the offered traffic to the network is fully symmetric, the link loadings can be unbalanced. This is due to the inherent asymmetry in the structure, e.g., in the (2,3) de Bruijn graph in Fig. 4.3, the self-loops on nodes "000" and "111" carry no traffic (and hence are wasted), and the link "1000" only carries traffic destined to node "000" while link "1001" carries all remaining traffic generated by or forwarded through node "100." Expressions for the average link loading and the maximum link loading for uniform offered traffic have been obtained in [SiRa94]. As a result of the link-load asymmetry, the maximum throughput supportable by a de Bruijn graph is lower than that supportable by an equivalent ShuffleNet structure with the same number of nodes and the same nodal degree.

A simplified delay analysis based on M/M/1 queueing models for links and independence assumptions indicates that, for uniform loading, the average packet delay in a de Bruijn graph can be slightly lower than that in an equivalent ShuffleNet for low to moderate loads [SiRa94].

A longest-path routing scheme by which one can achieve load-balanced routing in the de Bruijn graph and get throughputs higher than ShuffleNets has also been developed [SiRa94].

4.4.3 Torus (Manhattan Street Network)

An $N \times M$ Manhattan Street Network (MSN) is a regular mesh structure of degree 2 with its opposite sides connected to form a torus. Unidirectional communication links connect its nodes into N rows and M columns, with adjacent row links and column links alternating in direction. An example 4×6 MSN is shown in Fig. 4.4. The MSN structure was originally proposed as a metropolitan-area network in [Maxe85, Maxe87]. A locally implementable, adaptive deflection routing algorithm was proposed for the MSN in [Maxe85], and since then considerable additional research has been conducted on this structure. Another advantage of the MSN is that it is highly modular and easily growable. In addition, architectures for optical deflection switches applicable in MSNs have been investigated [ChFu91]. Extension of the torus' hop-distance characteristics to higher dimensions has been reported in [BaMS94]. For a comparison of MSN and ShuffleNet performance, please see [Mukh92b, Mukh97].

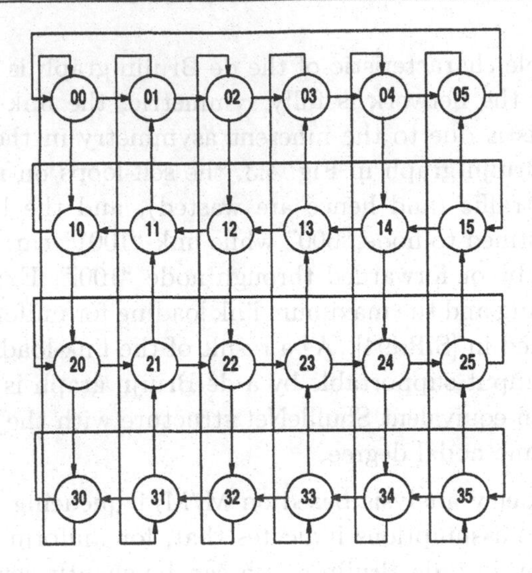

Figure 4.4 A 4×6 Manhattan Street Network (MSN) with unidirectional links.

4.4.4 Hypercube

The hypercube interconnection pattern has been actively investigated for multiprocessor architectures, and it has received attention as a virtual topology for multihop lightwave networks as well [Dowd91, Dowd92, LiGa92]. Below, we first discuss the binary hypercube, followed by the generalized hypercube.

Binary Hypercube

The simplest form of the hypercube interconnection pattern is the binary hypercube [LiGa92]. A p-dimensional binary hypercube has $N = 2^p$ nodes, each of which have p neighbors. A node requires p transmitters and p receivers, and it employs one transmitter-receiver pair to communicate directly and bidirectionally with each of its p neighbors. Any node i with an arbitrary binary address will have as its neighbors those nodes whose binary address differs from node i's address in exactly one bit position. An 8-node binary hypercube is shown in Fig. 4.5.

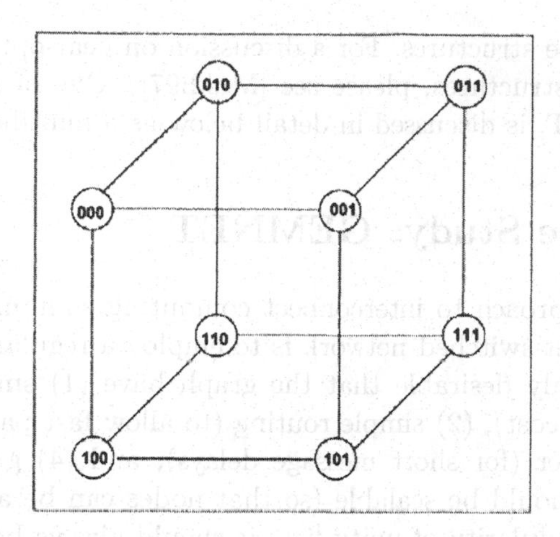

Figure 4.5 An eight-node binary hypercube.

The merits of this structure are its small diameter ($\log_2 N$) and short average hop distance [$(N \log_2 N)/(2(N-1))$]. Its disadvantage is that the nodal degree increases logarithmically with N.

Generalized Hypercube

The radix in the nodal address notation can be generalized to arbitrary integers, thereby resulting in the generalized hypercube structure. This structure can employ a mixed radix system to represent the node addresses. Let the number of nodes be given by $N = \prod_{i=1}^{p} n_i$, where the n_i are positive integers. A node's address P $(0 \leq P \leq N-1)$ is represented by the p-tuple $(m_p m_{p-1} \cdots m_1)$ where $0 \leq m_i \leq n_i - 1$. Thus, we have $P = \sum_{i=1}^{p} m_i w_i$ where $w_i = \prod_{j=1}^{i-1} n_j$.

The generalized hypercube shares similar merits and demerits as its binary version, except that it is more flexible in accommodating different numbers of nodes and their interconnection patterns.

4.4.5 Other Regular Multihop Topologies

Several other regular multihop topologies with nice properties have been examined in the literature, such as GEMNET [InBB95], Kautz graph [PaSe95] and CayleyNet [Tang94]. Please see the references for the descriptions and properties of these structures. For a discussion on near-optimal node placement in regular structures, please see [Mukh97]. One of these structures, namely GEMNET, is discussed in detail below as a multihop case study.

4.5 Multihop Case Study: GEMNET

An attractive approach to interconnect computing equipment (nodes) in a high-speed, packet-switched network is to employ a regular interconnection graph. It is highly desirable that the graph have (1) small nodal degree (for low network cost), (2) simple routing (to allow fast packet processing), (3) small diameter (for short message delays), and (4) growth capability, viz., the graph should be scalable (so that nodes can be added to it at all times) with a modularity of unity [i.e., it should always be possible to add one node to (or delete one node from) an existing (regular) graph while maintaining regularity]. We examine such a new network structure, the *GENeralized shuffle-exchange-based Multihop NETwork* architecture (called GEMNET) [InBB95].

GEMNET can serve as a logical (virtual), multihop topology for constructing the next generation of lightwave networks using WDM. GEMNET is a regular multihop network architecture, it is a generalization of shuffle exchange networks, it represents a family of network structures, and it includes the well-studied ShuffleNet [HlKa91] and de Bruijn graph [SiRa94] as members of its family. Figure 4.6(b) shows a multihop (logical) network (a 10-node GEMNET) embedded on the physical star topology network of Fig. 4.6(a). In general, by using wavelength-routing switches, one can construct wide-area, multihop optical networks as well (see Chapter 8).

In a (K, M, P) GEMNET, $K \times M$ nodes – each of degree P – are arranged in a cylinder of K columns and M nodes per column so that nodes in adjacent columns are arranged according to a generalization of the shuffle-exchange connectivity pattern using directed links [RaSi95a]. The generalization allows any number of nodes in a column as opposed to the constraint of P^K nodes in a column [HlKa91]. The logical topology in Fig. 4.6(b) is a (2,5,2) GEMNET.

In GEMNET, there is no restriction on the number of nodes as opposed to the cases in ShuffleNet and de Bruijn graph which can support only KP^K and P^D nodes, respectively, where $K, D = 1, 2, 3, \ldots$ and $P = 2, 3, 4, \ldots$. That is, GEMNET can represent arbitrary-sized networks in a regular graph; conversely, for any network size, at least two GEMNET configurations exist

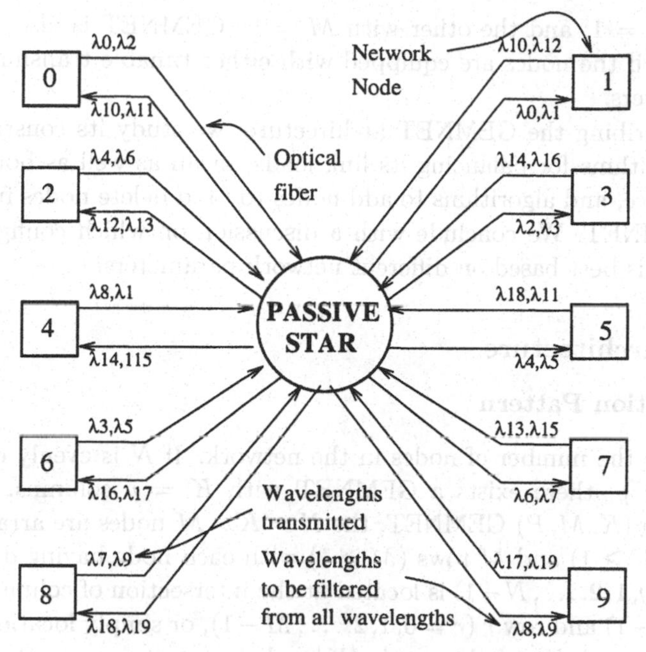

(a) Physical topology & transciever tuning pattern.

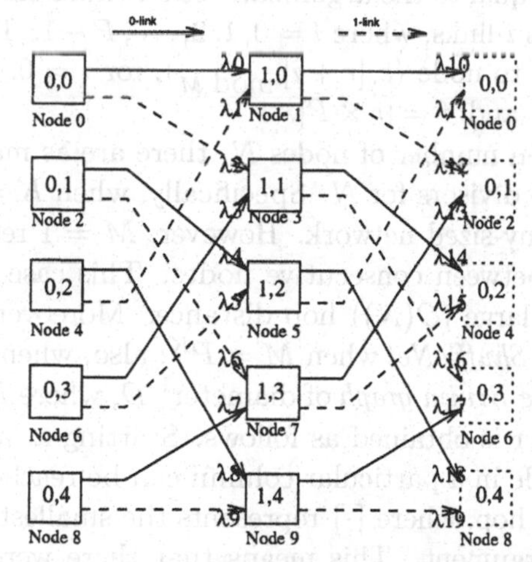

i-link from Node at (c,r) connects to Nodes at ((c+1) mod 2 , (2r+i) mod 5), i=0,1.

(b) Logical structure (virtual topology) corresponding to the transceiver assignments in (a).

Figure 4.6 A 10-node (2,5,2) GEMNET.

- one with $K = 1$, and the other with $M = 1$. GEMNET is also scalable in units of K if the nodes are equipped with either tunable transmitters or tunable receivers.

After describing the GEMNET architecture, we study its construction, routing, algorithms for balancing its link loads, mean as well as bounds on its hop distance, and algorithms to add nodes to (and delete nodes from) an existing GEMNET. We conclude with a discussion on which configuration of GEMNET is best based on different network parameters.

4.5.1 GEMNET Architecture

Interconnection Pattern

Let N denote the number of nodes in the network. If N is evenly divisible by an integer y, there exists a GEMNET with $K = y$ columns. In the corresponding (K, M, P) GEMNET, the $N = K \times M$ nodes are arranged in K columns ($K \geq 1$) and M rows ($M \geq 1$) with each node having degree P. Node a ($a = 0, 1, 2, \ldots, N-1$) is located at the intersection of column c ($c = 0, 1, 2, \ldots, K-1$) and row r ($r = 0, 1, 2, \ldots, M-1$), or simply location (c, r), where $c = (a \bmod K)$ and $r = \lfloor a/K \rfloor$, where $\lfloor \cdot \rfloor$ represents the largest integer smaller than or equal to the argument. The P links emanating out of a node are referred to as i-links, where $i = 0, 1, 2, \ldots, P-1$. The i-link from node (c, r) is connected to node $(\hat{c}, [\hat{r} + i]_{\text{mod } M})$, for $i = 0, 1, 2, \ldots, P-1$ where $\hat{c} = [c+1]_{\text{mod } K}$ and $\hat{r} = [r \times P]$.

Note that, for a given number of nodes N, there are as many (K, M, P) GEMNETs as there are divisors for N. Specifically, when $K = 1$ or $M = 1$, we can accommodate any-sized network. However, $M = 1$ results in a ring with P parallel paths between consecutive nodes. This case is not considered further due to its large ($O(N)$) hop distance. Moreover, a (K, M, P) GEMNET reduces to a *ShuffleNet* when $M = P^K$; also, when $M = P^D$ and $K = 1$, it reduces to a *de Bruijn graph* of diameter[1] D, where $D = 2, 3, 4, \ldots$.

GEMNET's diameter is obtained as follows. Starting at any node, note that each and every node in a particular column can be reached for the first time on the $\lceil \log_P M \rceil^{\text{th}}$ hop, where $\lceil \cdot \rceil$ represents the smallest integer larger than or equal to the argument. This means that there were one or more nodes not covered in the previously visited column. Due to the cylindrical

[1]The diameter of a network is defined as the longest distance it takes to get from an arbitrary node to another arbitrary node while taking the shortest possible path between them.

nature of GEMNET, the nodes in this column will be finally covered in an additional $K-1$ hops. Thus, $D = \lceil \log_P M \rceil + K - 1$.

Routing

Let (c_s,r_s) and (c_d,r_d) be the source node and the destination node, respectively. We define the "column distance" δ as the minimum hop distance in which the source node touches (covers) a node (not necessarily the destination) in the destination node's column. When $c_d \geq c_s$, we have $\delta = c_d - c_s$ because $(c_d - c_s)$ forward hops from any node in column c_s will cover a node in column c_d. When $c_d < c_s$, δ is given by $\delta = (c_d + K) - c_s$ because, after "sliding" c_d forward by K (i.e., $c_d + K$), due to wraparound, the situation becomes the same as when $c_d \geq c_s$. Thus, δ can be generalized as: $\delta = [(K + c_d) - c_s]_{\text{mod } K}$.

The hop distance from source node (c_s, r_s) to destination node (c_d, r_d) is given by the smallest integer h of the form $(\delta + jK)$, $j = 0, 1, 2, \ldots$, satisfying the following expression:

$$R = \left[M + r_d - \left(r_s \times P^h\right)_{\text{mod } M}\right]_{\text{mod } M} < P^h \qquad (4.2)$$

R, called the *route code*, specifies a shortest route from the source node to the destination node when it is expressed as a sequence of h base-P digits. For example, if $R = (11)_{\text{base } 10}$, $P = 3$, and $h = 4$, then R can be represented as $(0102)_{\text{base } 3}$. In general, if $R = [\alpha_1 \ \alpha_2 \ \cdots \ \alpha_h]_{\text{base P}}$, then the node about to send the packet on its jth hop will route the packet to its α_j^{th} outgoing link. The maximum number of iterations needed to solve for R is just $\lceil D/K \rceil$, where D is the diameter of the network. When $K = 1$, the number of iterations is maximum and the complexity in computing R is $O(\log_P N)$. Such a simplified routing scheme is one of the main advantages of a regular structure.

To explain Eqn. (4.2), we first note that δ is the minimum number of hops required to reach a node in column c_d from a node in column c_s. However, multiple passes around the GEMNET may be required to reach the destination node, i.e., h will be of the form $\delta + jK$, where $j = 0, 1, 2, \ldots$. Define an all-0-link path to be the path traced, from a particular source node, by taking the 0-link out of every intermediate node (including the source node) for an arbitrary number of hops. Now, note that $[r_s \times P^h]_{\text{mod } M}$ is the row index of the node in column c_d reachable from the source node in h hops, by following the all-0-link path. Then, $(h-1)$ 0-links followed by a 1-link leads to the node with row index $[r_s \times P^h + 1]_{\text{mod } M}$ in column c_d, and so on. However, on the hth hop, a maximum of P^h nodes can be *covered.* The node reached on the hth hop from the source node by following the all-$(P-1)$-link path (defined similar to the all-0-link path) will be $[r_s \times P^h + (P^h - 1)]_{\text{mod } M}$. Thus, if R is less than P^h [which means that the destination node falls somewhere between the all-0-link path and the all-$(P-1)$-link path], then the destination node is reachable in h hops and its route code is given by R. In Eqn. (4.2), the addition of M and the *mod* operations are required to accommodate the wraparound of row indices.

Often, the P^h nodes covered on the hth hop could be *greater* than the number of nodes in that column. This means that multiple shortest paths may exist to some nodes in that column. Having calculated R, if $(R+jM) < P^h$ for $j = 1, 2, 3, \ldots$, then $(R+jM)$ is also a routing code with path-length h for any j that satisfies this inequality. Thus, if the shortest path from node a to node b is h hops, the number of shortest paths is given by:

$$Y = \left\lceil \frac{P^h - R}{M} \right\rceil \tag{4.3}$$

Hence, for a given N, the number of alternate shortest paths increases as M decreases. The larger the number of shortest paths, the more opportunity there is to route a packet along a less-congested path and the greater is the network's ability to route a packet along a minimum-length path when a link or node failure occurs. The trade-off is that decreasing M will increase K, which, in general, will cause the average hop distance to increase.

4.5.2 GEMNET Properties

This section examines different shortest-path routing algorithms to balance GEMNET's link loads, analyze GEMNET's hop distance properties, and study how different GEMNET configurations perform relative to one another.

Routing Algorithms for Balancing Link Loading

In GEMNET, as in any other multihop network, an important goal is to balance the traffic on different links as much as possible. If a link's utilization is heavy, the corresponding link queue could become long and the delay for packets traversing that link could become significantly high. We study link loading properties of GEMNET by considering a uniform load which requires one unit of traffic to be moved between every source-destination pair.

So far, only shortest-length paths between source-destination pairs have been chosen for routing. However, if multiple shortest paths exist, then traffic can be routed over that shortest path which balances link flows as much as possible. The routing algorithm of Section 4.5.1 calculates R. If we choose to always route with the base-R route code of Eqn. (4.2) (henceforth referred to as the "unbalanced" routing scheme), the link loads tend to become unbalanced because, whenever there are multiple shortest paths, the "base-R" value is used.

Two alternative routing schemes, called "partially balanced" and "random," are found to perform better than the "unbalanced" scheme. Both of these schemes first calculate the base-R code as under the "unbalanced" scheme. Then, if only one shortest path to the destination exists, the base-R value is used to route the message. However, if multiple shortest paths exist, the "partially balanced" scheme will choose the route code R' such that, if $R + [(c_d \bmod P) \times M] < P^h$, then

$$R' = R + [(c_d \bmod P) \times M] \tag{4.4}$$

else set $R' = R$. When multiple paths exist, this approach spreads the traffic across different links, based on the destination node number. However, if the number of shortest paths exceeds P, this approach limits its selection to the first P such paths. The "random" scheme simply computes the number of alternate shortest paths from Eqn. (4.3) and assigns the route code:

$$R'' = R + (M \times Z) \tag{4.5}$$

where Z is a uniformly distributed random integer in the range $[0, Y-1]$ and Y is the number of shortest paths given by Eqn. (4.3). Some representative link loading statistics will be discussed later.

Bounds on the Average Hop Distance

A closed form solution for the average hop distance has not been found for all GEMNETs. However, based on the nature of the interconnection pattern, some fairly tight bounds can be placed on its average hop distance. In a GEMNET, beginning at a node, all of the nodes reachable on a certain hop count belong to a specific column, and are contiguous within that column (since row indices wraparound). This property can lead to some fairly tight bounds on the minimum and the maximum average hop distance. Below, these bounds and their maximum difference are analyzed. The interested reader is referred to [InBB95] for details of some of the derivations.

To determine the maximum average hop distance, note that if the set of nodes reached on hop i were to overlap all of the previously visited nodes, then the number of new nodes reached would be minimized. This pattern of visiting nodes would place an upper bound on the average hop distance for the network. This observation implies that, except for hops $D-K+1$ through D, the number of new nodes covered on hop i equals P^i when $i < K$, and $P^i - P^{(i-K)}$ when $i \geq K$. For hop $D-K+1$, we might cover fewer than P^i new nodes, and for hops $D-K+2$ through D, we are guaranteed to cover fewer than P^i new nodes. Then, C, the last column in which we are guaranteed to cover P^i new nodes on hop i (as outlined above), equals:

$$C = \begin{cases} D-K & \text{if } D-K \leq K-1 \\ K-1 & \text{otherwise} \end{cases} \tag{4.6}$$

Define the function $\delta(x, y) = 1$ if $x \leq y$, and $\delta(x, y) = 0$ otherwise. Then, it is a straightforward matter to verify that the maximum average hop distance bound for GEMNET equals:

$$\begin{aligned} \overline{h}_{\max} &= \frac{1}{M \times K - 1} \left\{ \sum_{i=0}^{C} iP^i + \delta(C+1, D-K) \sum_{i=C+1}^{D-K} i(P^i - P^{i-K}) \right. \\ &+ \left. \sum_{i=D-K+1}^{D} i[M - \delta(K, i)P^{i-K}] \right\} \end{aligned} \tag{4.7}$$

which, upon simplification, reduces to (see [InBB95] for details):

$$\overline{h}_{\max} = \frac{MK(D + \frac{1}{2} - \frac{1}{2}K) - K(\frac{P^{D-K+1}-1}{P-1})}{M \times K - 1} \tag{4.8}$$

When $C = D-K$, all nodes in the network are symmetric, so Eqn. (4.8) gives the *actual* average hop distance. Equation (4.8) also matches the average hop distance formula for ShuffleNet [HlKa91] and the upper bound for de Bruijn graph [SiRa91] when appropriate values for P, N, and K are used.

To determine the minimum average hop distance, we only want to visit new nodes on hop i (i.e., nodes which have not been visited before) until the contiguous property of the network forces some of the nodes reached to be previously visited nodes. This would place a lower bound on the average hop distance. So, P^i nodes, the maximum number of nodes one can possibly cover on hop i, will be covered until either hop $D-K$ or $D-K-1$. Let L be the last column in which P^i *new* nodes could be covered (as stated above). As described previously, the two possible values for L are $D-K$ and $D-K-1$. If the total number of nodes reached in a column on the $(D-K)^{\text{th}}$ hop, assuming P^i nodes covered on each hop is $\leq M$, then $L = D-K$, otherwise $L = D-K-1$. This is encompassed in the following definition of L:

$$L = \begin{cases} D-K & \text{if } P^{(D-K) \bmod K} \times \frac{(P^K)^{F+1}-1}{P^K-1} \leq M \\ D-K-1 & \text{otherwise} \end{cases} \tag{4.9}$$

where $F = \lfloor (D-K)/K \rfloor$. Now, the minimum average hop distance can be written as:

$$\overline{h}_{\min} = \sum_{i=0}^{L} iP^i + \sum_{i=L+1}^{L+K} i \left(M - P^{i \bmod K} \sum_{j=0}^{\lfloor \frac{i}{K} \rfloor - 1} P^{Kj} \right) \tag{4.10}$$

whose closed form is:

$$\begin{aligned}
\overline{h}_{\min} \;=\; & \frac{1}{N-1} \Bigg(P \left[\frac{(L+1)P^L(P-1) - (P^{L+1}-1)}{(P-1)^2} \right] \\
& + \frac{MK(2L+K+1)}{2} - \delta\left(0, \left\lfloor \frac{L+1}{K} \right\rfloor - 1\right) \frac{(P^K)^G - 1}{P^K - 1} \\
& \times \Bigg\{ (L+1)\frac{P^K-1}{P-1} + P^{Q+1} \frac{(K-Q)P^{K-Q-1}(P-1) - (P^{K-Q}-1)}{(P-1)^2} \\
& + \frac{P^Q \delta(K-Q, K-1)}{(P-1)^2} \left[(P-1)(K-(K-Q)P^{-Q}) - P(1-P^{-Q}) \right] \Bigg\} \\
& - P^{G\times K} \delta\left(\left\lfloor \frac{L+1}{K} \right\rfloor \times K + K, L+K \right) \Bigg[(G \times K + K) \frac{P^{L-G\times K+1}-1}{P-1} \\
& + P \frac{(L - G \times K + 1)P^{L-G\times K}(P-1) - (P^{L-G\times K+1}-1)}{(P-1)^2} \Bigg] \Bigg)
\end{aligned} \tag{4.11}$$

where $Q = (L+1) \bmod K$ and $G = \lfloor (L+1)/K \rfloor$. Equation (4.11) matches the formula for ShuffleNet [HlKa91] and the lower bound for de Bruijn graph [SiRa91] when appropriate values for P, N, and K are used [InBB95].

The largest potential difference between the maximum and the minimum average hop distance occurs when $P = 2$, $K = 1$, and $N(= M \times K)$ is large. When $P = 2$ and $K = 1$, the difference between the max and the min hop bounds equals:

$$\Delta = \frac{ND - 2^D - 1}{N-1} - \frac{[(L+1)P^{L+1} - P^{L+2} + 2] + N(L+1) - (2^{L+1}-1)(L+1)}{N-1} \tag{4.12}$$

where the first and the second terms correspond to the max and the min bounds, respectively. It can be verified that, for both cases of L, $\Delta < 1$.

Which Configuration of GEMNET Is the Best?

Multiple GEMNET configurations exist for a given number of nodes, so the question of which one has the lowest average hop distance or optimum link loading naturally arises.

Figure 4.7 demonstrates how the average hop distance in different GEMNET configurations compare with one another. This figure considers a 64-node GEMNET with nodal degree of 2. In general, the larger the number of columns (i.e., the "fatter" the GEMNET) is beyond an "optimal" configuration, the higher is the mean hop distance. In this example, the two-column GEMNET exhibits the lowest hop distance, and it is superior, in this sense, to the corresponding one-column GEMNET (a de Bruijn graph) because the latter has two nodes (top and bottom nodes) with links that transmit back to themselves (self-loops), and these two nodes have larger individual mean hop distances (at most 1 greater) than the other nodes.

In general, for a GEMNET with $P = 2$ and an even number of nodes, $K = 2$ gives the minimum average hop distance. For a GEMNET with $P = 2$, the minimum hop distance for a given N always has $K \leq 3$ columns. For $P = 2$ and an odd number of nodes, in general, the lowest average hop distance occurs for GEMNETs with $K = 1$. Also, for the GEMNETs with $P \geq 3$, in general, the best average hop distance is achieved by a GEMNET with $K = 1$. Most of these results can be seen in Table 4.1.

In terms of link loading, we observe from Table 4.1 that, in general, the "random" routing scheme performs better than the "unbalanced" and "partially balanced" schemes. Under the "unbalanced" scheme, the 0-links out of a node would be utilized more often when multiple paths exists. For the "partially balanced" scheme, the higher-numbered routes would not be chosen due to the mod-by-P operation (e.g., if there were four shortest paths and $P = 2$, only the two lowest-numbered routes would be considered). "Random" routing eliminates this bias toward the lowest-numbered routes; hence, it performs better. The "random" scheme also has the important advantage of being able to balance nonuniform traffic unlike fixed-path-routing schemes (like "unbalanced" and "partially balanced"). Henceforth, we limit our discussion to the "random" scheme only, but for additional results, please see Table 4.1.

Two opposing factors compete when we try to minimize the maximum

Table 4.1 Comparison of GEMNET (GN), ShuffleNet (SN) and de Bruijn (DB) graph. (Link loads are computed under the assumption of one unit of flow between every source-destination pair and different routing schemes. Also in this table, $\overline{h}$ is averaged over all the individual nodes' average hop distance ($\overline{h_i}$).)

P	N	K	Network GN	SN	DB	Avg. Hop dist. ($\overline{h}$)	D	Std. dev. of $\overline{h_i}$	Avg. link load ($\overline{l}$)	Fixed Routing Std. dev. of l_i	Fixed Routing $(l_i)_{max}$:$(l_i)_{min}$	Part. Bal. Rounting Std. dev. of l_i	Part. Bal. Rounting $(l_i)_{max}$:$(l_i)_{min}$	Random Routing Std. dev. of l_i	Random Routing $(l_i)_{max}$:$(l_i)_{min}$
2	8	1	X		X	2.107	3	0.211	7.375	3.08	11:7	3.08	11:7	3.08	11:7
		2	X	X		*2.0*	3	0	*7.0*	2.83	11:3	1.0	8:6	2.09	*10*:4
		4	X			2.286	4	0	8.0	6.67	19:1	4.12	15:5	2.18	12:4
2	24	1	X			3.406	5	0.257	39.167	17.23	67:15	17.23	67:15	17.23	67:15
		2	X			3.348	5	0.043	38.5	15.45	65:15	15.26	63:21	15.43	65:16
		3	X	X		*3.261*	5	0	*37.5*	16.13	67:14	4.50	44:31	6.67	*50*:26
		4	X			3.478	6	0	40.0	18.49	64:12	9.0	49:25	6.27	58:31
2	64	1	X		X	4.532	6	0.224	142.766	38.78	208:63	38.78	208:63	38.78	208:63
		2	X			*4.448*	6	0.090	*140.125*	40.53	227:31	34.92	191:39	34.28	189:44
		4	X	X		4.635	7	0	146.0	78.34	331:45	33.88	220:100	15.42	*188*:116
		8	X			5.714	10	0	180.0	236.16	1023:17	168.22	772:70	11.09	216:152
2	256	1	X		X	6.417	8	0.221	818.164	192.73	1151:255	192.73	1151:255	192.73	1151:255
		2	X			*6.336*	8	0.093	*807.844*	203.73	1259:127	177.44	1126:159	174.71	1111:183
		4	X			6.514	9	0.020	830.5	384.11	1755:245	201.94	1285:433	99.98	*980*:562
		8	X			7.561	12	0	964.0	1241.58	8535:129	845.93	5676:370	24.08	1028:888
2	1024	1	X		X	8.377	10	0.220	4284.929	907.59	6309:1000	907.59	6309:1000	907.59	6309:1000
		2	X			*8.297*	10	0.093	*4243.875*	961.10	6469:511	842.87	5797:639	835.53	5701:749
		4	X			8.474	11	0.022	4334.625	1790.35	10593:1000	948.28	6840:1000	525.89	*5065*:1000
		8	X			9.517	14	0	4868.0	5972.13	55953:769	4001.38	33755:1000	55.43	5101:1000
3	81	1	X		X	*3.386*	4	0.066	*90.296*	26.46	138:40	26.46	138:40	26.46	138:40
	82	2	X			3.523	5	0.039	95.130	28.59	153:32	17.29	128:45	14.08	127:49
	81	3	X	X		3.562	5	0	95.0	62.48	265:31	21.26	147:60	11.78	*121*:69
	81	9	X			5.625	10	0	150.0	372.79	2000:7	284.11	1552:13	11.39	180:125
4	256	1	X		X	*3.599*	4	0.031	*229.406*	58.02	313:85	58.02	313:85	58.02	*313*:85
		2	X			3.828	5	0.017	244.062	162.53	917:31	90.29	533:43	45.31	333:92
		4	X			4.188	6	0	267.0	362.81	3165:57	166.04	1185:103	13.29	*313*:218
		8	X			5.867	10	0	374.0	1344.05	14407:9	941.53	9882:17	18.33	432:314

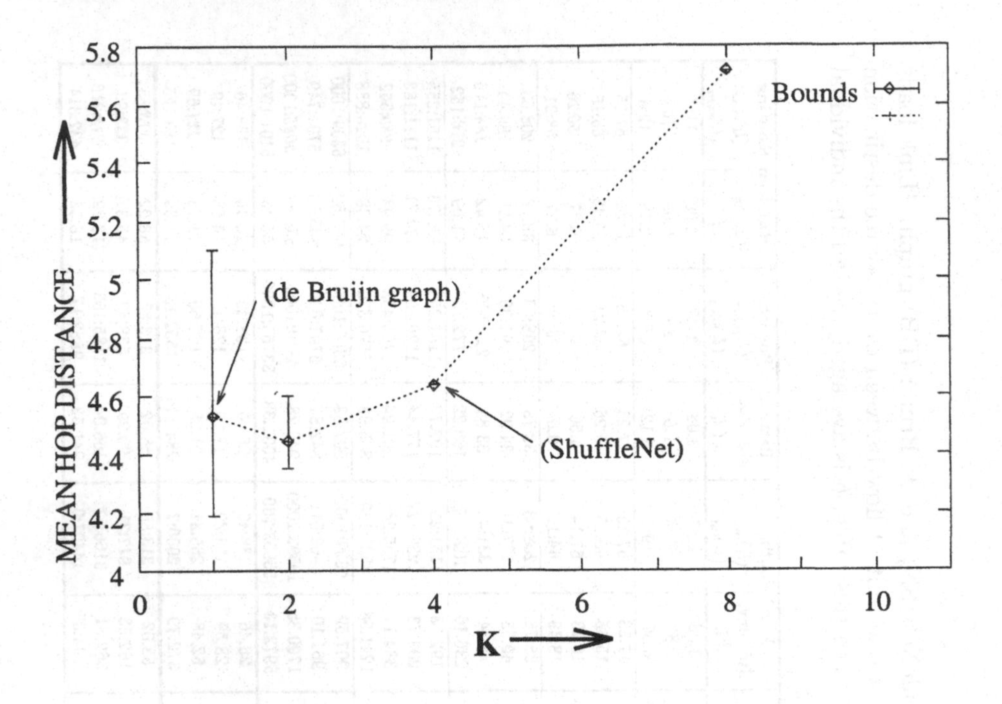

Figure 4.7 Bounds and average hop distance for a $P = 2$, 64-node GEMNET with different values of K.

link load. One is that, as a GEMNET is widened (by increasing K), a larger number of multiple shortest paths exist, allowing traffic to be better balanced, thereby decreasing the load on the most-congested link. On the other hand, as a GEMNET is widened, its average hop distance increases which will proportionally increase its average link load $\{= [(N-1) \times \overline{h}]/P\}$. For the "random" routing scheme, Table 4.1 reveals that an increase in the average link loading tends to increase the load on the most-congested link. For small N, the value of K that minimizes $\overline{h}$ also results in the minimum max-link load. [The optimum values of $\overline{h}$, $\overline{l_i}$ and $(l_i)_{\max}$ (for "random" routing), for a given (N, P) combination are italicized in Table 4.1.] However, as N gets larger, this trend does not hold. Due to various factors on which link loading depends, we can only state that, for large N, the value of K that minimizes the maximum link loading is generally slightly larger than the K which minimizes $\overline{h}$.

4.5.3 Scalability – Adapting the Size of a GEMNET

This section investigates how to scale the size of a GEMNET. Specifically, it examines different ways to grow a GEMNET. Approaches to decrease the size of a GEMNET are similar; see [InBB95] for more information. In addition, our approaches to scaling a GEMNET apply to its implementation based on a single passive star, because a topology reconfiguration under this implementation can be easily performed by retuning some of the transmitters and/or receivers at the various nodes. Reconfiguring a multistar or a switched (wavelength-routed) implementation of GEMNET would be more involved.

Since a one-column GEMNET can accommodate any number of nodes, there is always at least one interesting GEMNET (besides the one with P parallel rings corresponding to $M = 1$). Section 4.5.2 revealed that, for $P = 2$, the one-column and the two-column GEMNETs in general had the best average hop distance for an odd and even number of nodes, respectively. This property is very desirable for scalability since the one-column GEMNET can easily be grown by one node at a time (i.e., by adding an extra row), while the two-column GEMNET can be grown by two nodes at a time; also one node can be added to a two-column GEMNET with $2j$ nodes ($j = 1, 2, 3, \ldots$) by setting up the new structure as a one-column GEMNET with $2j + 1$ nodes. Since, in general, the best average hop distance for $P \geq 3$ is achieved by a one-column GEMNET, there is a dual benefit in that *the most scalable structure also possesses the best average hop distance.*

The easiest way to grow a GEMNET is to add one node at the bottom of each of its columns. Thus, a GEMNET can grow by K nodes at a time (i.e., with modularity K). The best location to add K nodes turns out to be the bottom row because the interconnection pattern is interrupted at the farthest point from the top-most row. Adding nodes closer to the top would have caused the interconnection pattern to be interrupted earlier. Thus, to add one row of nodes, each node must determine what are the row numbers of the nodes that it needs to connect to in the new *larger* network and then retune its transmitters or receivers accordingly. An example of growing a (1,6,2) GEMNET by one node to a (1,7,2) GEMNET in this fashion is shown in Fig. 4.8.

By examining the structure of a GEMNET, observe that, when adding a row of nodes, approximately the first M/P nodes in each column need not retune. Approximately the next M/P nodes perform one retuning, the next M/P nodes perform two retunings, and so on. To be exact, when adding K

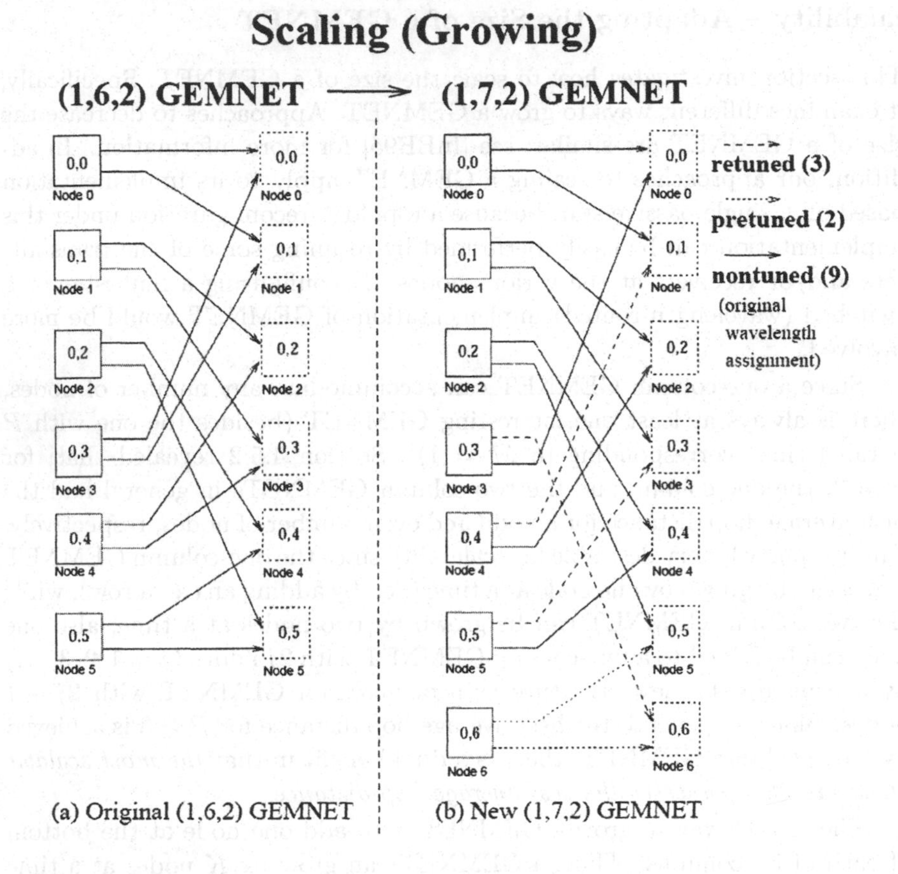

Figure 4.8 Growing a (1,6,2) GEMNET by one node.

nodes to an N-node GEMNET in the fashion described above, we obtain:

$$\text{number of retunings} = \sum_{i=2}^{P} \left\lceil \frac{N}{P} \right\rceil \times (i-1) \geq \frac{N \times (P-1)}{2} \tag{4.13}$$

with the equality holding when N is divisible by P. Since the total number of links (and hence transmitters/receivers) equals NP, for a $P = 2$ GEMNET, this means that approximately one-fourth of the total number of transmitters or receivers in the network need to be retuned, while for a $P = 3$ GEMNET, approximately one-third of the transmitters or receivers need to be retuned.

Adding more than one row to a GEMNET can be performed in stages by adding one row at a time until all rows are added. Alternatively, the network can be scaled from the original setup to the final setup in one massive retuning operation, by "optimally" "renumbering" the individual nodes in order to minimize the number of retunings. Deletion of a row of nodes from GEMNET can also be performed similarly. The interested reader is referred to [InBB95] for more information.

4.6 Shared-Channel Multihop Systems

Consider a local lightwave network based on a passive-star physical topology. The number of channels, w, in such a WDM network is limited by technology and is usually less than the number of nodes, N, in the network. This section provides a general method using *channel-sharing* to construct practical multihop networks under this limitation. *Channel-sharing* may be achieved through *time-division multiplexing (TDM)*. The method is demonstrated by application to GEMNET. Interesting characteristics of the channel-sharing approach are also highlighted.

4.6.1 Channel-Sharing in ShuffleNet

In the original ShuffleNet work in [Acam87], (p, k) ShuffleNets with $p = 2$ are considered, and the following channel-sharing strategy is used. Recall that there are k columns of nodes and p^k nodes per column in a (p, k) ShuffleNet. All nodes in the same row share their transmissions on p (=2) channels. That is, although node i $(0 \leq i \leq p^k - 1)$ in an arbitrary column transmits to nodes $j, j+1, \ldots, j+p-1$ in the next column [where $j = (i \bmod p^{k-1})p$] via p distinct channels, all k of the nth transmitters $(n = 0, 1, \ldots, p)$ from the ith nodes in all columns must share the same channel.

Due to channel-sharing, a packet can reach its destination in fewer hops, on the average, i.e., *channel-sharing reduces the hop distance.* For $p = 2$, an upper bound on the expected number of hops is obtained to be [Acam87]

$$\overline{h'} = \frac{1}{2^k}[2 + (k-1)2^k] \qquad (4.14)$$

The above result is not directly comparable to the mean hop distance for the dedicated-channel case in Eqn. (4.1) since Eqn. (4.14) includes self-hopping also (which was not included in Eqn. (4.1)).

Channel-sharing in generalized ShuffleNets is treated more rigorously in [HlKa91], but a different shared-channel allocation mechanism is employed. Specifically, it is required that, for $i = 0, 1, \ldots, p$, nodes i, $i+p^{k-1}$, $i+2p^{k-1}$, $\ldots$, and $i+(k-1)p^{k-1}$ in a column transmit on a shared channel which, in turn, is received by nodes $j, j+1, j+2, \ldots, j+p-1$ in the next column where $j = (i \bmod p^{k-1})p$. For other properties of shared-channel ShuffleNets, see [HlKa91].

4.6.2 Channel-Sharing in GEMNET

This section illustrates our sharing approach and its effect on the nodal degree in an example multihop network, viz., GEMNET.

Recall that, in a (K, M, P) GEMNET, N nodes – each of degree P – are arranged in a cylinder of K columns and M nodes per column so that nodes in adjacent columns are arranged according to a generalization of the shuffle-exchange connectivity pattern using directed links (see Fig. 4.6). Without channel-sharing, the number of channels required in a (K, M, P) GEMNET equals $N \times P = K \times M \times P$.

When each node can have only one fixed transmitter-receiver pair and $w < N$, *the degree of the network takes on a new meaning.* Channel-sharing through TDM implies a *higher logical nodal degree* (viz., ability to reach more nodes directly), albeit with a *lower capacity* on each of the *logical links.* The nodal degree is redefined as a logical quantity P, where $P = N/w$. For analytical convenience, N is assumed to be an integral multiple of w to ensure equal and fair sharing by all logical links. (Note that channel-sharing in this manner also leads to multicasting, for further information on which we refer the interested reader to [Mukh97, TrMu96, TrMu97].)

Thus, we define a (K, M, P)–SC_GEMNET (Shared-Channel GEMNET) as a GEMNET having K columns and M rows, where M is an integral multiple, n, of P (i.e., $M = nP$) and $K = N/M = N/(nP)$.

As examples, Fig. 4.9 shows several 12-node ($N = 12$) SC_GEMNETs with different values of w. Figure 4.9(a) shows the trivial 12-channel case. Since each node has only one fixed transmitter and one fixed receiver, the only SC_GEMNET that can be constructed is a ring (i.e., $K = 12, M = 1$). For the other trivial case, viz., the one-channel case, which is not shown in the figure, the resulting topology is a fully connected network, with all 12 nodes time-sharing the bandwidth available on the single channel. Parts (b) through (e) of Fig. 4.9 show the other four cases which allow for "perfect sharing," viz., $w = 6, 4, 3, 2$. In each of these figures, both a logical topology and the corresponding TDM frames on the various channels are shown. With reference to the portions of the figures representing the TDM frames, the numbers in the "slots" indicate the transmitting nodes and the numbers in parentheses indicate the nodes which have their receivers tuned to the channel which the frame represents. Note that, for each value of w, multiple SC_GEMNETs may exist. For example, in part (c) for the $w = 4$ case, it is also possible to construct a $K = 4, M = 2$ SC_GEMNET.

In deriving the average hop distance $\bar{h}$ of the (K, M, P)-GEMNET, its diameter $D = \lceil \log_p(M) \rceil + K - 1$ is used. Given D, an upper bound for the average hop distance $\bar{h}_{\max}$ was derived in Section 4.5 to be:

$$\bar{h}_{\max} = \frac{MK\left(D + \frac{1}{2} - \frac{K}{2}\right) - K\left(\frac{P^{D-K+1} - 1}{P-1}\right)}{M \times K - 1} \tag{4.15}$$

In determining a lower bound for the average hop distance, we need to define L to be the last column in which P^i *new* nodes could be covered starting from any arbitrary node. With this definition and also defining $Z = \lfloor (D - K)/K \rfloor$, we have $L = D - K$ if $P^{(D-K) \bmod K} \times [(P^K)^{Z+1} - 1]/[P^K - 1] \leq M$; otherwise, we have $L = D - K - 1$. The average value of the minimum hop distance can now be defined as follows:

$$\bar{h}_{min} = \sum_{i=0}^{L} iP^i + \sum_{i=L+1}^{L+K} i \left(M - P^{i \bmod K} \sum_{j=0}^{\lfloor \frac{i}{K} \rfloor - 1} P^{Kj} \right) \tag{4.16}$$

It has been shown in Section 4.5 that, in general, two-column GEMNETs (i.e., $K = 2$) have the best performance in terms of the average hop distance. Therefore, in constructing logical topologies for the multihop WDM network, it is, in general, best to consider $K = 2$ wherever possible and use the next higher value of K as dictated by the requirement $K = N/M = N/(nP) = w/n$, where n is an integer which equals the number of channels required to provide connectivity between two successive columns of the SC_GEMNET.

It is interesting to consider what happens when the available number of channels, w, is not an exact divisor of N. Such a system can be treated in two ways: (1) $w' < w$ channels may be used such that N/w' nodes share a channel with all nodes getting an equal share of the bandwidth, and $w - w'$ channels are wasted; (2) alternatively, all of the available channels may be used by employing different TDM frame lengths on different channels, i.e., with unequal sharing. For example, consider a system with $N = 100$. If

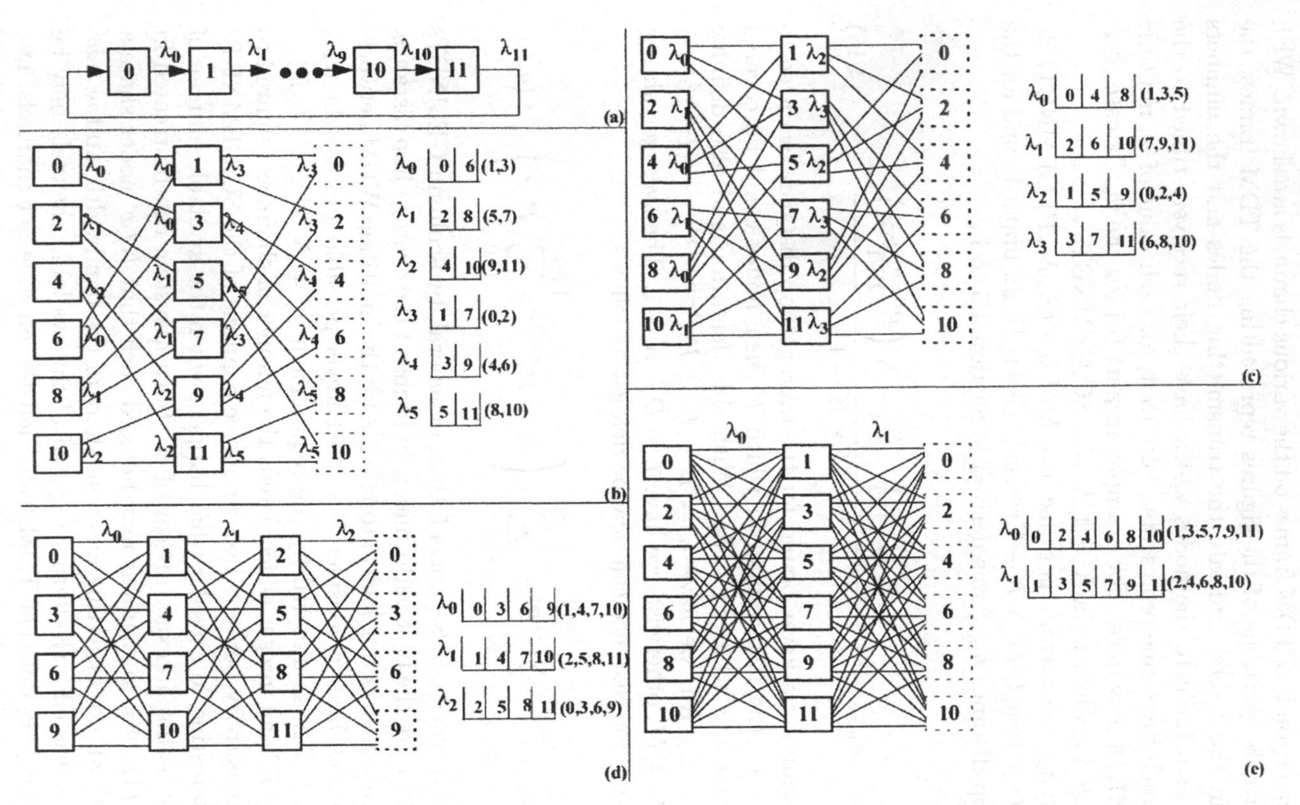

Figure 4.9 Twelve-node SC_GEMNETs, along with corresponding timing diagrams, for various values of w: (a) $w = 12$; (b) $w = 6$; (c) $w = 4$; (d) $w = 3$; and (e) $w = 2$. (Note: unless otherwise shown, all links are directed from left to right.)

$w = 50$, each channel is shared by two nodes, and the TDM frame length in each channel is two slots - one slot allocated to each of the two nodes that transmit on that channel. If $w = 51$, there are two possibilities. First, only $w' = 50$ channels may be used, and the fifty-first channel may be wasted resulting in a loss of efficiency of $1/51 \approx 2\%$. Alternatively, 98 of the nodes may share 49 channels - 2 nodes per channel, and the last two nodes may use the remaining two channels exclusively. The unfairness problem becomes more pronounced if $w = 99$. The first solution of using only 50 channels is quite wasteful in this case resulting in a loss of efficiency of approximately 50%. On the other hand, 98 nodes may be allowed exclusive use of one channel, and the remaining two nodes may share the remaining channel. For the remainder of this chapter, we will try to consider only parameters which lead to perfect sharing of channels. For cases where this is not possible, we will approximate the number of channels in the system to the closest integer value which leads to perfect sharing.

The SC_GEMNET lends itself to multicasting applications, because sharing at the receiving ends reduces the total number of transmissions required for multicasting (see [Mukh97, TrMu96, TrMu97] for details).

4.7 Summary

Figure 4.10 provides a summarized classification of multihop systems. Materials not discussed here can be found in [Mukh92b, Mukh97].

The modularity of regular structures, viz., how can one add nodes to or delete nodes from existing regular structures, is an important issue. One can refer to such structures as "injured" regular structures. How these multihop structures can reorganize their nodal connectivity pattern to restore optimality when the traffic pattern changes is another question which needs to answered.

An in-depth case study of a multihop network, GEMNET, was provided. We also presented a general method of constructing shared-channel, multihop WDM LANs. The method was illustrated using the GEMNET architecture as the target logical topology. Intuitively, it appears that channel-sharing may be particularly effective in the presence of multicast traffic.

Although this chapter's focus was on local lightwave networks employing the broadcast-and-select mechanism, work on optical WDM wide-area networks (WANs) has also been initiated based on wavelength-routing switches at intermediate nodes, so that an arbitrary multihop virtual topology may

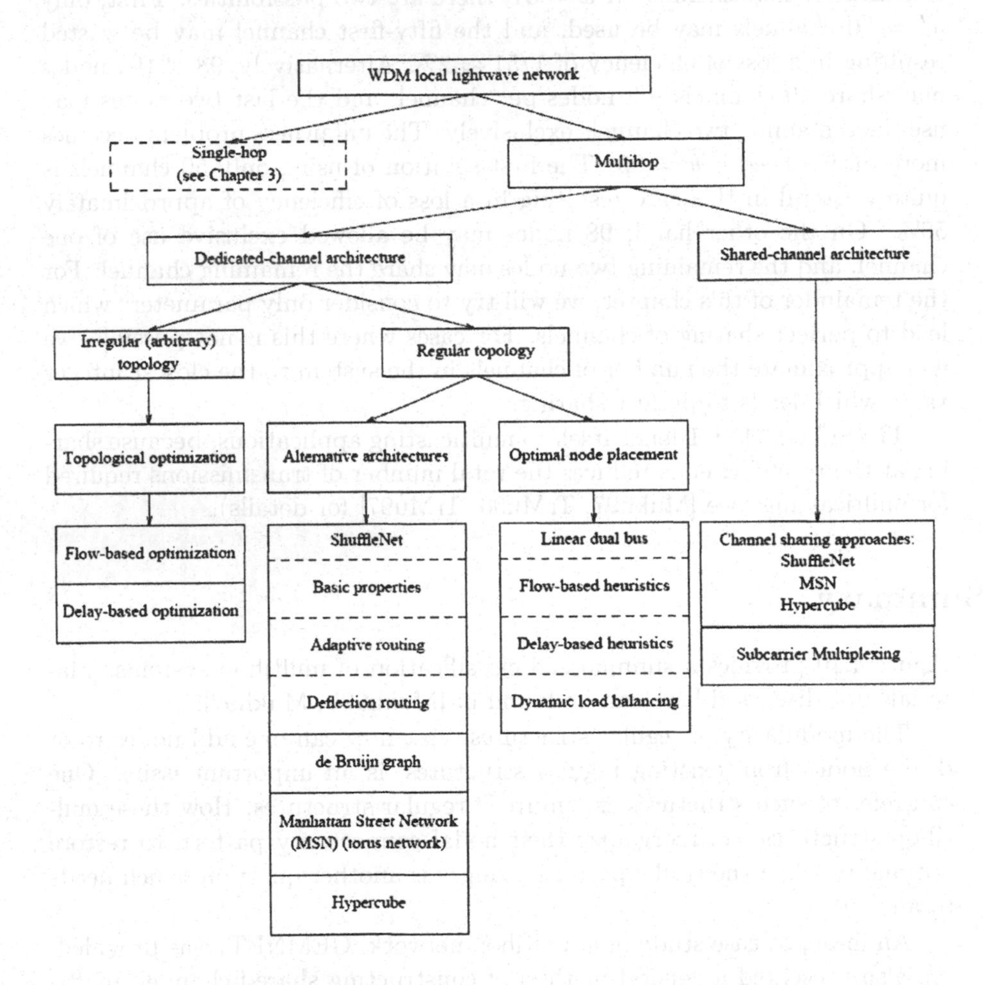

Figure 4.10 Classification of multihop network architectures [Mukh92b].

be embedded on an arbitrary physical fiber network (see Chapter 8). This category of networks does not need a centralized hub as local lightwave networks do. Hence, WDM WANs can employ spatial reuse of wavelengths in different parts of the network, thereby increasing the amount of concurrency in the network.

Exercises

4.1. What are some of the advantages and disadvantages of passive-star coupler multihop networks compared to single-hop networks?

4.2. Consider an eight-node (2,2) ShuffleNet. Calculate the average delay for a uniform traffic matrix (the arrival rate for each source-destination pair is λ packets per second). Assume that the service time for a packet at a node is exponentially distributed with mean $1/\mu$ seconds. Assume shortest-path routing.

4.3. Describe how a packet gets routed from node 0 to node 7 in a $(2, 2)$ ShuffleNet.

4.4. Show how a packet gets routed from node 0 to node 6 in a $(2, 3)$ de Bruijn graph. What is the maximum hop distance in this graph?

4.5. Draw a (3,2) de Bruijn graph. Compute the average hop distance.

4.6. Compare and contrast the following topologies and calculate the average hop distance for each:

(a) (2,2) ShuffleNet

(b) (2,3) de Bruijn graph

(c) 8-node binary hypercube

(d) 8-node 4×2 Manhattan Street Network

4.7. Find the average hop distance for a 4×4 Manhattan Street Network.

4.8. Derive the average hop distance and diameter for a p-dimensional binary hypercube.

4.9. Derive the average hop distance of ShuffleNet [Eqn. (4.1)].

4.10. Suppose we wish to build a multihop network in which each node has a nodal degree of two, and the diameter of the network is three. We consider only ShuffleNet and Manhattan Street Network. Give the possible logical topologies and compare (i) the maximum number of nodes that can be supported, and (ii) the average hop distance.

4.11. A de Buijn graph topology is chosen by a major internet network service provider. The network must contain at least one million nodes.

(a) If a binary ($\Delta = 2$) de Bruijn graph is used, what is the minimum number of bits required to represent the label of a node?

(b) How many self-loops does the graph contain?

(c) Find a bound for $\overline{h}$.

(d) A routing algorithm finds the next node to send a packet to so that this next node is on a shortest path from node a to b. The computer used to implement the algorithm can perform operations on machine words of size 32 bits in 1 nanosecond. Operations include bit shift, bitwise and, bitwise or, add, subtract, equality test, and inequality test. Outline, or code, a fast routing algorithm or program that runs on node a where $a = (a_1, a_2, \ldots, a_D)$, and outputs the bit which is appended to a, after shifting, i.e., output is a_{D+1}, where $(a_2, \ldots, a_D, a_{D+1})$ is the label of the next node in the shortest path toward $b = (b_1, b_2, \ldots, b_D)$. Attempt to make use of low-level (bit) parallelism in your code, possibly packing a and b as unsigned integers, into machine words, as much as possible. Explain how the bits are stored in the words, including the order.

(e) Assuming that branches also take 1 nanosecond, find a bound on the running time of your routine. Do not consider compiler optimization.

4.12. In the following topologies, the presence of an undirected link is understood to be bidirectional, i.e., it may be replaced by two links going in opposite directions. Compute the diameter and average hop distance for the following logical topologies (graphs). Also, compute the total number of wavelengths (channels) required to implement these logical multihop topologies with a passive-star coupler as the physical topology.

(a) a 10-node unidirectional ring

(b) a 10-node bidirectional ring

(c) a 10-node complete graph

(d) the (10-node) Peterson graph (see Fig. 4.11 below)

4.13. Consider an 8-node (2,3) de Bruijn graph. Determine the total number of paths from node 0 to node 5 that have at most 8 hops.

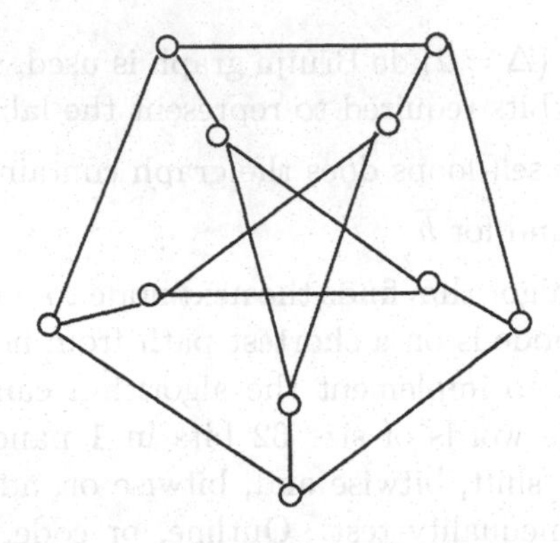

Figure 4.11 The (10-node) Peterson graph.

4.14. Find the transfer function between points A and B in the following graph. How many paths of distance 10 hops are there from A to B? How many paths of distance 15 hops are there from A to B?

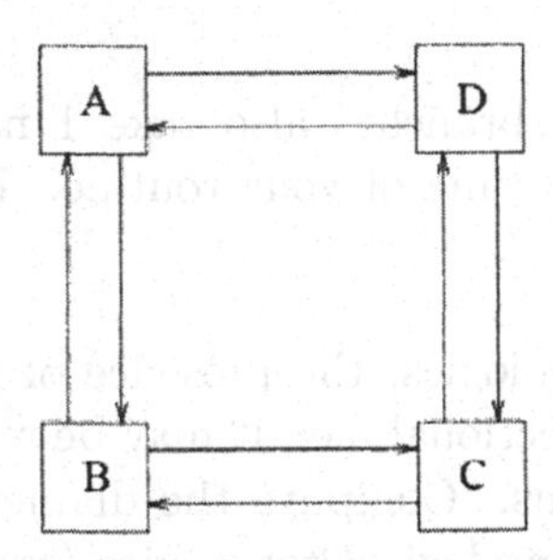

Figure 4.12 A bidirectional ring network.

4.15. Draw a two-dimensional radix-3 hypercube. Is the resulting graph isomorphic to a 3×3 torus? Will a two-dimensional radix-4 hypercube be isomorphic to a 4×4 torus? What is the number of edges in a d-dimensional radix-r hypercube?

4.16. Consider a multidimensional torus, with all links being bidirectional, and with all dimensions (M) being of the same size, i.e., an $N \times N \times \ldots \times N$ structure, of size N^M. Now consider the generalized hypercube with bidirectional links. Is it possible for a graph (logical topology) to be both? Does one class contain the other? In other words, can you find two graphs such that one is a generalized hypercube, but not a multidimensional torus, and the other is a torus, but not a generalized hypercube? Give a simple, one-sentence explanation of the difference the essential difference between these two structures.

4.17. Draw a generalized hypercube with 12 nodes. Explain how channel-sharing can be used in this hypercube network.

4.18. One of the disadvantages of a binary hypercube is that the degree of any node is unbounded (also true for generalized hypercubes). That is, for a hypercube of size $2^N = M$, the degree of a node is $N = \log_2 M$ for a binary hypercube. One solution to keep the nodal degree in check so that the structure can be scalable is to replace each node by a ring of N nodes. Each of the new nodes can then handle one of the connections required by the hypercube node. Such a graph is called a *cube-connected cycle.* It is also important to mention that the cycle must be ordered, i.e., hypercube nodes are connected when their labels differ by only one bit. The cycle must occur in bit order, so that connected cycles (hypercube nodes) are connected by corresponding cycle nodes. More precisely, if hypercube nodes a and b are connected, then their labels differ in one bit position, say the jth position. Thus, cycle node a_j and cycle node b_j must be connected. Also, for all cycles a, cycle nodes a_j and $a_{j+1 \bmod N}$ are connected.

When routing within the cube-connected cycle, a pair of nodes within a cycle must follow the cycle to form a shortest path. Outside of a cycle, routing can be accomplished without being much worse than the hypercube distance. Explain how. In other words, describe an efficient routing algorithm. Also, find a bound on the diameter of the graph.

Hint: Consider the routing algorithm of a hypercube.

4.19. Consider a six-dimensional binary hypercube.

(a) How would you represent the root of this hypercube?

(b) Consider node (010110_2). List the members of the 3-cube to which this node belongs.

(c) Consider node (101011_2). Which node is its "partner" in the 6-cube? 4-cube?

4.20. Given three nodes and traffic matrix A, use the Min-Max flow-based heuristic to find a near-optimal node placement on a linear dual bus.

$$A = \begin{bmatrix} 0 & 2 & 3 \\ 5 & 0 & 7 \\ 4 & 3 & 0 \end{bmatrix}$$

4.21. List the advantages of channel-sharing.

4.22. Discuss some of the salient features of GEMNET.

4.23. Show that a $(1, 2^N, 2)$ GEMNET is equivalent (having a one-to-one correspondence between all nodes and links) to a $(2, N)$ de Bruijn graph.

4.24. Show that a $(1, \Delta^N, \Delta)$ GEMNET is equivalent (having a one-to-one correspondence between all nodes and links) to a (Δ, N) de Bruijn graph.

4.25. Show that a (K, P^K, P) GEMNET is equivalent (having a one-to-one correspondence between all nodes and links) to a ShuffleNet.

4.26. Given nine nodes, find a GEMNET topology for which $K > 1$ and $M > 1$.

4.27. Under what conditions do we obtain the maximum and minimum hop distances in a GEMNET?

4.28. Explain Eqn. (4.2).

4.29. Find the average hop distance for a (2,5,2) GEMNET (see Fig. 4.6). Assume shortest path routing. Show how a packet is routed from node 0 to node 9.

4.30. Consider a (3,7,2) GEMNET. Find the route code R from source node (0,4) to destination node (2,1).

4.31. Find the number of shortest paths in a (3,27,3) GEMNET between nodes $s = (1, 25)$ and $d = (0, 10)$.

4.32. Consider a lightly but uniformly loaded (K, M, P) GEMNET. Suppose one link fails. How will hop distance be affected? If shortest path routing is being used, how many connections will have to be rerouted?

4.33. Compute the diameter and average hop distance for a 3×3 GEMNET. Compute the base route code from node (1,0) to node (1,2). Find all shortest paths from node (1,0) to node (1,2).

4.34. We wish to build a 10-node network with a GEMNET topology, trying to achieve the best performance. Come up with a GEMNET topology and explain your choice. Assume each node has a degree of two.

4.35. Draw all possible GEMNETs having six nodes and a nodal degree of two. Which of these will have the shortest average hop distance? The longest?

4.36. If we are trying to minimize the maximum link load in the network, which GEMNET design is better: one with a higher number of columns or one with a higher number of rows?

4.37. A (4,4,2) GEMNET is to be upgraded into a (4,5,2) GEMNET by adding a row of nodes. What is the number of retunings required?

4.38. Consider a network of six nodes. Construct a SC_GEMNET logical topology for each of the following cases:
(a) Number of channels = 6
(b) Number of channels = 3
(c) Number of channels = 2
Assume a $FT - FR$ system in each case.

4.39. Construct a (2,4,2) SC_GEMNET with $w = 3$, and show a possible TDM frame structure.

4.40. Find all possible channel-sharing configurations for a network with 10 nodes. Consider the situation in which w is an exact divisor of 10.

4.41. Consider a (3,4,2) SC_GEMNET with unicast traffic. Packets arrive at each node according to a Poisson process with rate $\lambda = 2 \cdot 10^5$ pkt/s. The distance between each node and the star coupler is 10 km. There are six wavelengths used in the system. Slot length is equal to 1 μs. Find the average packet delay. (Approximate the delay on each wavelength channel as a M/M/1 queueing system.)

4.42. Consider a 12-node SC_GEMNET with $w = 3$. The source node is node 0, and it has a multicast group consisting of nodes 2, 4, 6, and 10. Provide an efficient design of this SC_GEMNET. Justify your design. Compare your design with the nonshared case.

Chapter 5

Optical Access Networks

5.1 Introduction

This chapter discusses broadband access network architectures employing Passive Optical Networks (PONs). The potential of PONs to deliver high bandwidths to users in access networks and their advantages over current access technologies have been widely recognized. PONs have made strong progress in terms of standardization and deployment over the past few years. In this chapter, we first review the Ethernet PON (EPON), which has been standardized in the IEEE 802.3ah. We discuss the ATM-based PON (APON), and its enhanced versions, the Broadband PON (BPON) and the Generalized Framing Procedure (GFP) PON (GFP-PON). We then review the technologies available for introducing wavelength-division multiplexing (WDM) in PONs, and the progress of research in this area. Finally, we examine the issues related to deploying PONs in access networks.

The access network, also known as the "*first-mile*" network, connects the service provider central offices (COs) to businesses and residential subscribers. This network is also referred to in the literature as the "last-mile" network, the *subscriber access network*, or the *local loop*. Subscribers demand first-mile access solutions that have high bandwidth, offer media-rich Internet services, and are comparable in price with existing networks. Similarly, corporate users demand broadband infrastructure through which they can connect their local-area networks to the Internet backbone.

5.1.1 Challenges in Access Networks

Much of the emphasis over the past few years has been on developing high-capacity backbone networks. Backbone network operators currently provide high-capacity OC-192 (10 Gbps) links. However, current-generation access network technologies such as Digital Subscriber Loop (DSL) provide 1.5 Mbps of downstream bandwidth and 128 kbps of upstream bandwidth at best. The access network is, therefore, truly the bottleneck for providing broadband services such as video-on-demand, interactive games, video conference, etc., to end users.

The predominant broadband access solutions deployed today are the Digital Subscriber Line (DSL) and Community Antenna Television (CATV) (cable TV) based networks. However, both of these technologies have limitations because they are based on infrastructure that was originally built for carrying voice and analog TV signals, respectively; but their retrofitted versions to carry data are not optimal.

In addition, DSL has a limitation that the distance of any DSL subscriber to a central office must be less than 18000 feet because of signal distortions. Typically, DSL providers do not provide services to distances more than 12000 feet. Therefore, only an estimated 60% of the residential subscriber base in the United States can avail of DSL. Although variations of DSL such as very high bit-rate DSL (VDSL), which can support up to 50 Mbps of downstream bandwidth, are gradually emerging, these technologies have much more severe distance limitations. For example, the maximum distance which VDSL can be supported over is limited to 1500 feet.

Cable television (CATV) networks provide Internet services by dedicating some Radio Frequency (RF) channels in co-axial cable for data. Cable networks are mainly built for delivering broadcast services, so they don't fit well for distributing access bandwidth. At high load, the network's performance is usually frustrating to end users.

Faster access-network technologies are clearly desired for next-generation broadband applications. The next wave of access networks promises to bring fiber closer to the home. The FTTx model — Fiber-to-the-Home (FTTH), Fiber-to-the-Curb (FTTC), Fiber-to-the-Building (FTTB), etc. — offers the potential for unprecedented access bandwidth to end users. These technologies aim at providing fiber directly to the home, or very near the home, from where technologies such as VDSL can take over. FTTx solutions are mainly based on the Passive Optical Network (PON). The comparison between different access mechanisms is presented in Fig. 5.1.

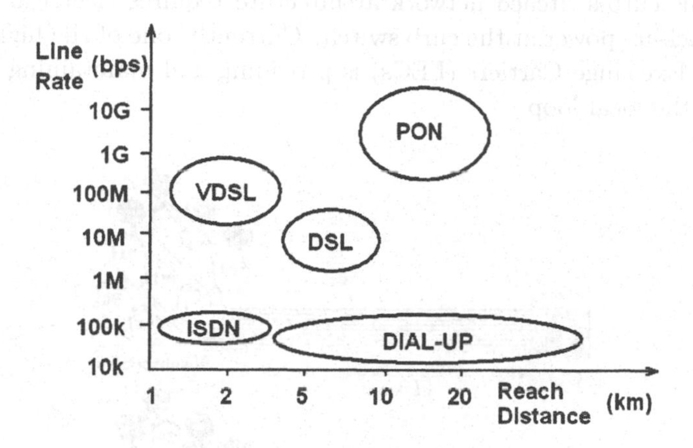

Figure 5.1 The comparison between different access mechanisms.

In this chapter, we shall review major developments in PON in recent years – EPON, APON, GFP-PON and WDM-PON. Finally, we review various issues related to deployment of PONs.

5.1.2 Next-Generation Access Networks

Optical fiber is capable of delivering bandwidth-intensive, integrated, voice, data and video services at distances of 20 kilometers or beyond in the subscriber access network. A logical way to deploy optical fiber in the local access network is to use a point-to-point (PtP) topology, with dedicated fiber runs from the CO to each end-user subscriber (Fig. 5.2(a)). While this is a simple architecture, in most cases it is cost prohibitive due to the fact that it requires significant outside plant fiber deployment as well as connector termination space in the Local Exchange (or Central Office (CO)). Considering N subscribers at an average distance L km from the central office, a PtP design requires 2N transceivers and N $\times$ L total fiber length (assuming that a single fiber is used for bi-directional transmission).

To reduce fiber deployment, it is possible to deploy a remote switch (concentrator) close to the neighborhood. This will reduce the fiber consumption to only L km (assuming negligible distance between the switch and customers), but will actually increase the number of transceivers to 2N+2, as there is one more link added to the network (Fig. 5.2(b)). In addition, this curb-switched network architecture requires electrical power as well as back-up power at the curb switch. Currently, one of the highest costs for Local Exchange Carriers (LECs) is providing and maintaining electrical power in the local loop.

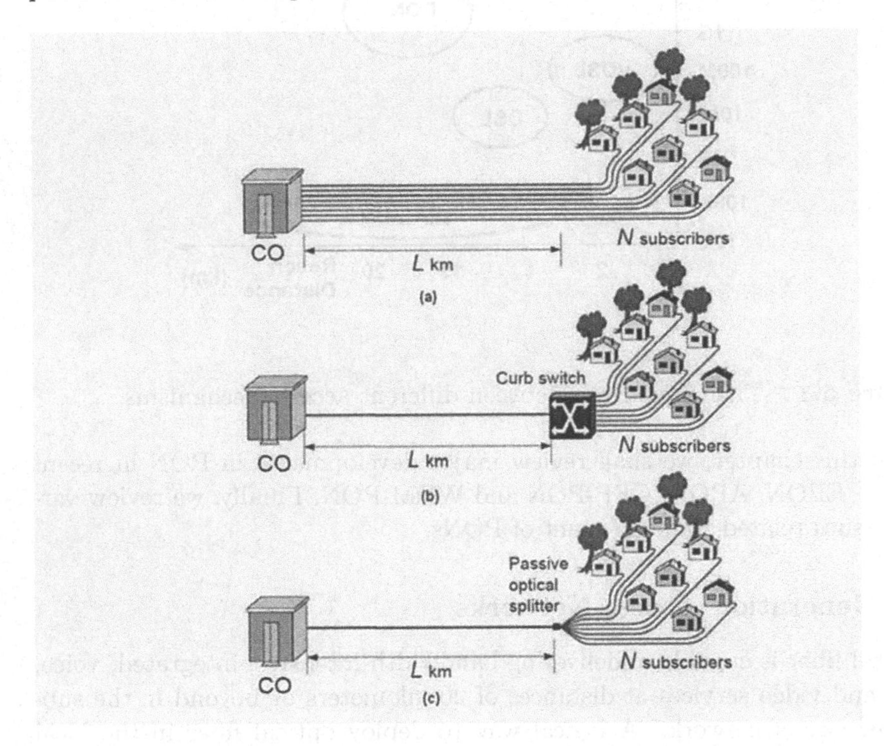

Figure 5.2 Fiber-to-the-home (FTTH) deployment scenarios.

Therefore, it is logical to replace the hardened (environmentally protected) active curb-side switch with an inexpensive passive optical splitter. Passive Optical Network (PON) is a technology viewed by many as an attractive solution to the first-mile problem [PeKe99, Lung99]; a PON minimizes the number of optical transceivers, CO terminations, and fiber deployment. A PON is a point-to-multipoint (PtMP) optical network with no active elements in the signal path from source to destination. The only interior elements used in a PON are passive optical components, such as optical fiber, splices, and splitters. An access network based on a single-fiber PON only require N + 1 transceivers and L km of fiber (Fig. 5.2(c)).

5.2 Overview of PON Technologies

5.2.1 Optical Splitters/Couplers

A PON employs a passive (not requiring any power) device to split an optical signal (power) from one fiber into several fibers and reciprocally, to combine optical signals from multiple fibers into one. This device is an optical coupler. In its simplest form, an optical coupler consists of two fibers fused together. Signal power received on any input port is split between both output ports. The splitting ratio of a splitter can be controlled by the length of the fused region and therefore is a constant parameter. $N \times N$ couplers are manufactured by staggering multiple 2×2 couplers (see Section 2.2.6) or by using planar waveguide technology.

Couplers are characterized by the following parameters:

1. *Splitting Loss*: Power level at the coupler's output vs. power level at its input, measured in dB. For an ideal 2×2 coupler, this value is 3 dB.

2. *Insertion Loss*: Power loss which results from imperfections of the coupler's manufacturing process. Typically, this value ranges from 0.1 dB to 1 dB.

3. *Directivity*: Some amount of input power leaked from one input port to another input port. Couplers are highly directional devices with the directivity parameter reaching 40–50 dB.

Often, couplers are manufactured to have only one input or one output. A coupler with only one input is referred to as a splitter. A coupler with only one output is called a combiner. Sometimes, 2×2 couplers are made highly asymmetric (with splitting ratios 5/95 or 10/90). This kind of coupler is used to branch off a small portion of the signal power, for example, for monitoring purposes. Such devices are called tap couplers.

5.2.2 PON Topologies

Logically, the first mile is a PtMP network, with a CO servicing multiple subscribers. There are several multipoint topologies suitable for the access network, including tree, tree-and-branch, ring, or bus (Fig. 5.3). Using 1:2 optical tap couplers and 1:N optical splitters, PONs can be flexibly deployed in any of these topologies. In addition, PONs can be deployed in redundant configurations such as double rings or double trees; or redundancy may be added to only a part of the PON, say the trunk of the tree (Fig. 5.3(d)) (see [EfIY01] for more redundant topologies).

All transmissions in a PON are performed between an Optical Line Terminal (OLT) and Optical Network Units (ONUs) (Fig. 5.3). The OLT resides in the CO and connects the optical access network to the metropolitan-area network (MAN) or wide-area network (WAN), which is also known as backbone or long-haul network. The ONU is located either at the end-user location (FTTH and FTTB), or at the curb, resulting in fiber-to-the-curb (FTTC) architecture.

The advantages of using PONs in access networks are numerous:

1. PONs allow for long reach between the CO and customer premises, operating at distances up to 20 km.

2. PONs minimize fiber deployment in both the CO and the local loop.

3. PONs provide higher bandwidth due to deeper fiber penetration, offering Gbps solutions.

4. Operating in the downstream as a broadcast network, PONs allow for video broadcasting either as IP video, or analog video.

5. PONs eliminate the necessity of installing active multiplexers at the splitting locations, thus relieving network operators from the task of maintaining active curbside units and providing power to them.

6. Being optically transparent end-to-end, PONs allow upgrades to higher bit rates or additional wavelengths.

5.2.3 Burst-Mode Transceivers

Due to unequal distances between the CO and the ONUs, optical signal attenuation in the PON may not be the same for each ONU. The power

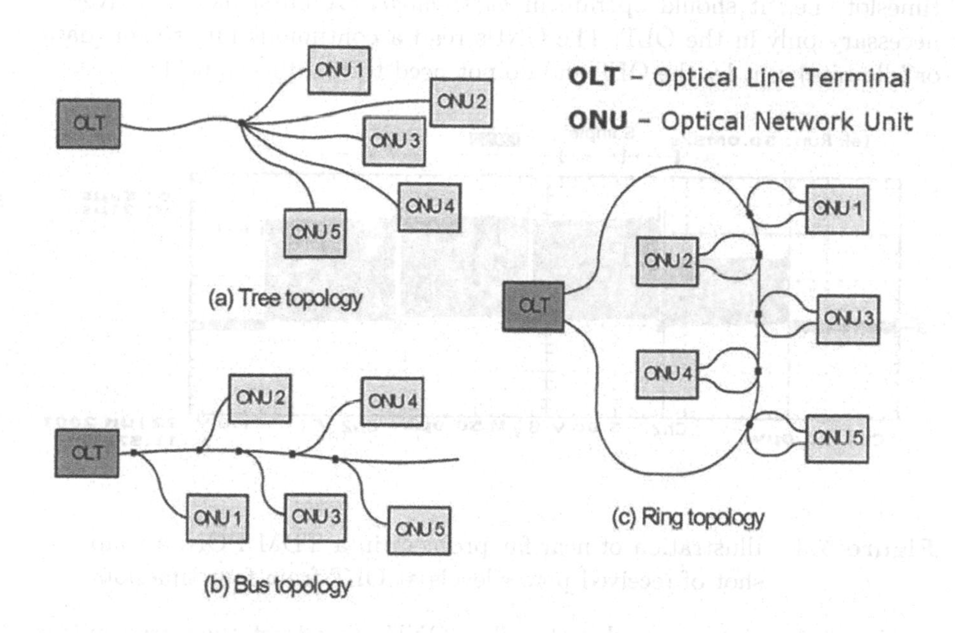

Figure 5.3 PON topologies.

level received at the OLT may be different for different ONUs (called the near-far problem). Figure 5.4 depicts power levels of four timeslots[1] received by the OLT from four different ONUs in a TDM-PON. As shown, one ONU's signal strength is lower at the OLT, most likely due to its longer distance. If the receiver at the OLT is adjusted to properly receive high-power signal from a close ONU, it may mistakenly read ones as zeros when receiving a weak signal from a distant ONU. In the opposite case, if the receiver is trained on a weak signal, it may read zeros as ones when receiving a strong signal.

To properly detect the incoming bit stream, the OLT receiver must be able to quickly adjust its zero-one threshold at the beginning of each received

[1] Timeslot refers to a period of time, which is recognized and uniquely defined, and during which certain data are transmitted by specific regulations in the network. Timeslot is not a periodic time allocation concept as in time-division multiplexing (TDM) here. Timeslot is a "timeslice" that is allocated to a node for transmission of its backlogged packets.

timeslot, i.e., it should operate in *burst mode*. A burst-mode receiver is necessary only in the OLT. The ONUs read a continuous bit stream (data or idle bits) sent by the OLT and do not need to re-adjust quickly.

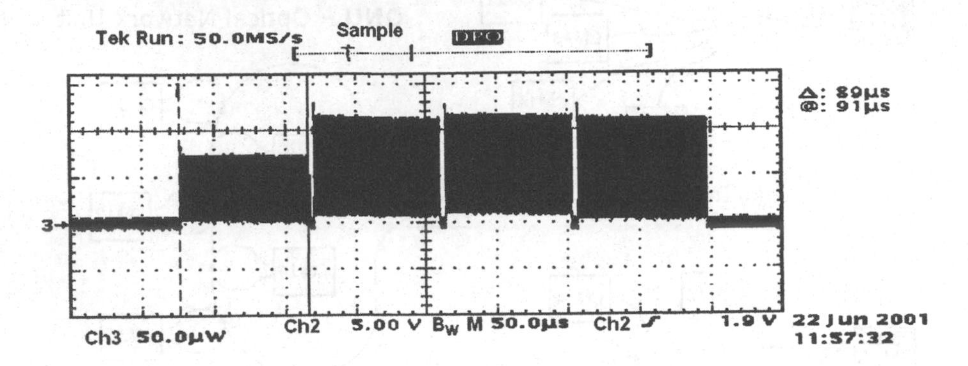

Figure 5.4 Illustration of near-far problem in a TDM PON: a snapshot of received power levels at OLT from four timeslots.

An alternative approach is to allow ONUs to adjust their transmitter powers such that power levels received by OLT from all ONUs become the same. This method is not particularly favored by transceiver designers as it makes the ONU hardware more complex, requires special signaling protocol for feedback from the OLT to each ONU, and most importantly, may degrade the performance of all ONUs to that of a most-distant unit.

Another issue is that it is not enough just to disallow ONUs to send any data. The problem is that, even in the absence of data, lasers generate spontaneous emission noise. Spontaneous emission noise from several ONUs located close to the OLT can easily obscure the signal from a distant ONU (called the capture effect). Thus, an ONU must shut down its laser between the timeslots. Because a laser cools down when it is turned off, and warms up when it is turned on, its emitted power may fluctuate at the beginning of a transmission. It is important that the laser be able to stabilize quickly after being turned on.

5.3 Ethernet PON (EPON) Access Network

Ethernet PON (EPON) carries data traffic encapsulated in Ethernet frames (defined in the IEEE 802.3 standard). It uses a standard 8b/10b line coding (in which 8 user bits are encoded as 10 line bits), and it operates at standard Ethernet data rates.

5.3.1 Ethernet Gaining in Prominence

The first-generation PON standardized by ITU–T G.983 employed Asynchronous Transfer Mode (ATM) as the medium-access control (MAC) protocol. When its standardization effort was started in 1995, the telecom community believed that ATM would be the prevalent technology in backbone networks. ATM had the advantages of streamlining voice and data services while providing operational and performance guarantees. However, since then, Ethernet has grown vastly popular. Ethernet line cards are cheap, and they are widely deployed in LANs today. Since access networks are focused towards end users and LANs, ATM has turned out to be not the best choice to connect to Ethernet-based LANs. High-speed Gigabit Ethernet deployment is widely accelerating and 10-Gigabit Ethernet products are becoming available. Ethernet is a much more efficient MAC protocol to use compared to ATM which imposes a considerable amount of overhead on variable-length Internet Protocol (IP) packets. Newly-adopted quality-of-service (QoS) techniques have made Ethernet networks capable of efficiently supporting voice, data, and video. These techniques include full-duplex transmission mode, prioritization (802.1p), and virtual LAN (VLAN) tagging (802.1Q). 802.1p is a specification which allows for prioritization of traffic into different priority classes at the MAC layer. 802.1Q defines an architecture for VLANs. Although 802.1Q doesn't directly define any QoS support, it defines a frame-format extension allowing Ethernet frames to carry priority information. EPONs, therefore, have much more promise in future access networks compared to ATM PONs (APONs). The following subsections describe the operation of the EPON, as stated in the IEEE 802.3ah [EPON02].

5.3.2 Principle of Operation

In the downstream direction (OLT to ONUs), Ethernet frames transmitted by the OLT pass through a 1:N passive splitter and reach each ONU. Typical values of N are between 8 and 32. Because EPON operation is broadcast in the downstream direction, it fits perfectly with the Ethernet philosophy. Packets are broadcast by the OLT and extracted by their destination ONU based on a Logical Link Identifier (LLID), which the ONU is assigned when it registers with the network. Figure 5.5 shows the downstream traffic in EPON.

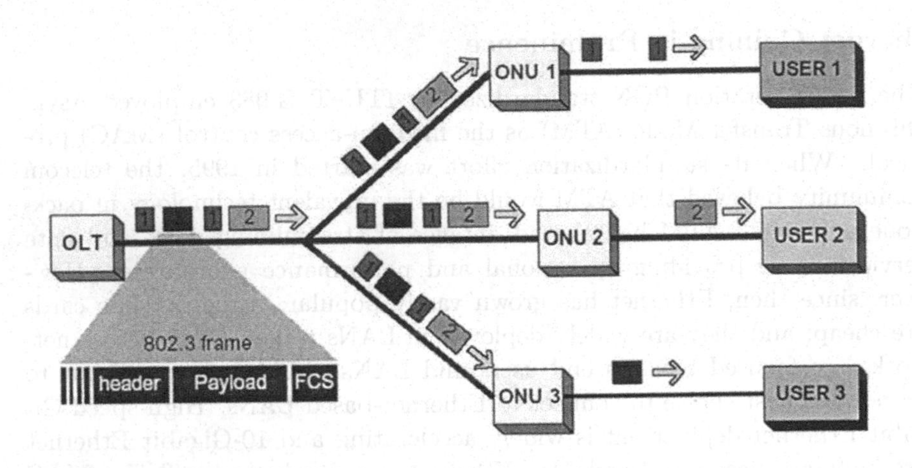

Figure 5.5 Downstream operation in EPON.

In the upstream direction, data frames from any ONU will only reach the OLT and will not reach any other ONU due to the directional properties of a passive optical combiner. Therefore, in the upstream direction, the behavior of EPON is similar to that of a point-to-point architecture. However, unlike in a true point-to-point network, in EPON, data frames from different ONUs transmitted simultaneously may collide. Thus, in the upstream direction, the ONUs need to employ some arbitration mechanism to avoid data collisions and fairly share the channel capacity. A contention-based media-access mechanism (similar to Carrier Sense Multiple Access/Collision Detection (CSMA/CD)) is difficult to implement because ONUs cannot detect a collision in the fiber from the combiner to the OLT due to directional properties of the combiner. An OLT could detect a collision and inform ONUs by sending a jam signal; however, the relatively large propagation delay in a PON (where the typical distance from the OLT to ONUs is 20 km) greatly reduces the efficiency of such a scheme. To introduce determinism in frame delivery in the upstream direction, different non-contention schemes have been proposed. Figure 5.6 illustrates the operation of upstream, time-shared, data flow in an EPON.

All ONUs are synchronized to a common time reference, and each ONU

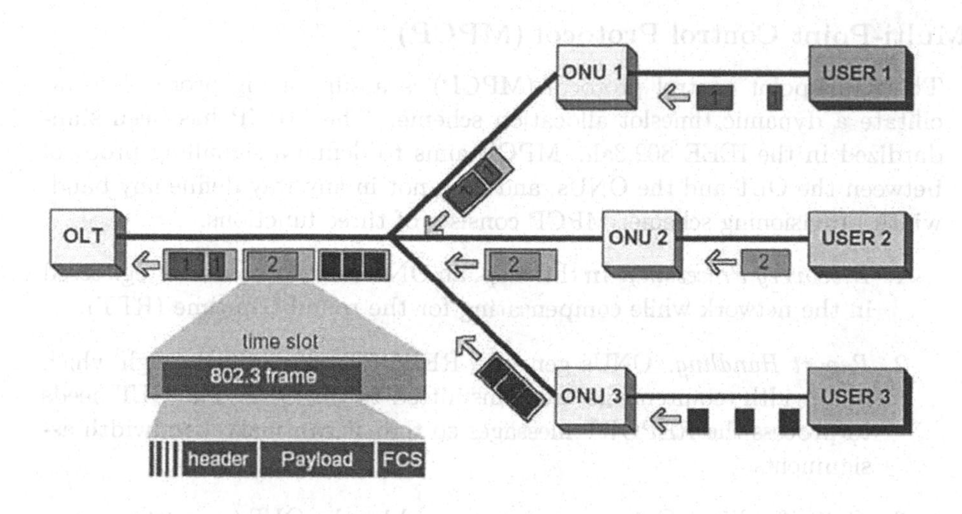

Figure 5.6 Upstream operation in EPON.

is allocated a timeslot in which to transmit. Each timeslot is capable of carrying several Ethernet frames. An ONU should buffer frames received from a subscriber until its timeslot arrives. When its timeslot arrives, the ONU would burst all stored frames at full channel speed. If there are no frames in the buffer to fill the entire timeslot, an idle pattern is transmitted.

Thus timeslot assignment is a very crucial step. The possible timeslot allocation schemes could range from a static allocation (fixed time-division multiple access (TDMA)) to a dynamically adapting scheme based on instantaneous queue size in every ONU (statistical multiplexing scheme). In the dynamic scheme, the OLT can play the role of collecting the queue sizes from the ONUs and then issuing timeslots. Although this approach leads to significant signalling overhead between the OLT and the ONUs, the centralized intelligence may lead to more efficient use of bandwidth. More advanced bandwidth-allocation schemes are also possible, including schemes utilizing notions of traffic priority, Quality-of-Service (QoS), Service-Level Agreements (SLAs), over-subscription ratios, etc. [KrMP02].

5.3.3 Multi-Point Control Protocol (MPCP)

The multi-point control protocol (MPCP) is a supporting protocol to facilitate a dynamic timeslot allocation scheme. The MPCP has been standardized in the IEEE 802.3ah. MPCP aims to define a signalling protocol between the OLT and the ONUs, and does not in any way define any bandwidth provisioning scheme. MPCP consists of three functions.

1. *Discovery Processing*: In this step, an ONU is discovered and registered in the network while compensating for the round-trip time (RTT).

2. *Report Handling*: ONUs generate REPORT messages through which bandwidth requirements are transmitted to the OLT. The OLT needs to process the REPORT messages so that it can make bandwidth assignments.

3. *Gate Handling*: Gate messages are used by the OLT to grant a timeslot at which the ONU can start transmitting data. Timeslots are computed at the OLT while making bandwidth allocation.

Discovery Processing

Discovery is the process in which newly-connected or offline ONUs register in the network. The steps are shown in Fig. 5.7.

1. OLT: The OLT periodically makes available a discovery time window during which the offline ONUs are given the opportunity to register themselves with the OLT. A DISCOVERY_GATE message is broadcast to all ONUs, this message includes the starting and the ending time of the discovery window.

2. ONU: Any offline ONU, which wishes to register, waits for a random amount of time within the discovery window, and then transmits a REGISTER_REQ message. The REGISTER_REQ message contains the ONU's MAC address. The random wait is required to reduce the probability of REGISTER_REQ messages transmitted by multiple ONUs from colliding.

3. OLT: The OLT, after receiving a valid REGISTER_REQ message, registers the ONU and allocates to it a Logical Link Identifier (LLID). The OLT now transmits a REGISTER message to the newly-discovered ONU which contains the ONU's LLID.

4. OLT: The OLT now transmits a standard GATE message, indicating a timeslot to transmit data.

5. ONU: Upon receiving the GATE message, the ONU responds with a REGISTER_ACK message in the assigned timeslot. Upon receipt of the REGISTER_ACK, the discovery process is complete and now normal operation may start.

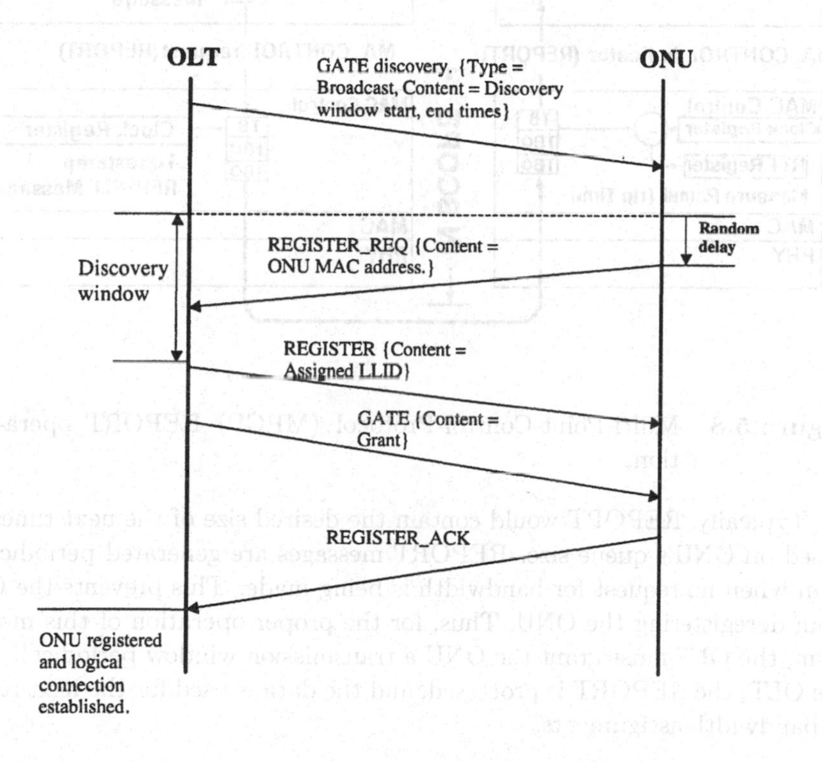

Figure 5.7 Discovery-phase message exchanges.

At each step, timeouts are maintained at the ONU and the OLT. If an expected message is not received before a timeout, then the OLT issues a DEREGISTER message, which makes the ONU register again.

REPORT Handling

REPORT messages are sent by ONUs in their assigned transmission windows along with data frames. A REPORT message is generated in the MAC control client layer and is time-stamped in the MAC controller, as shown in Fig. 5.8.

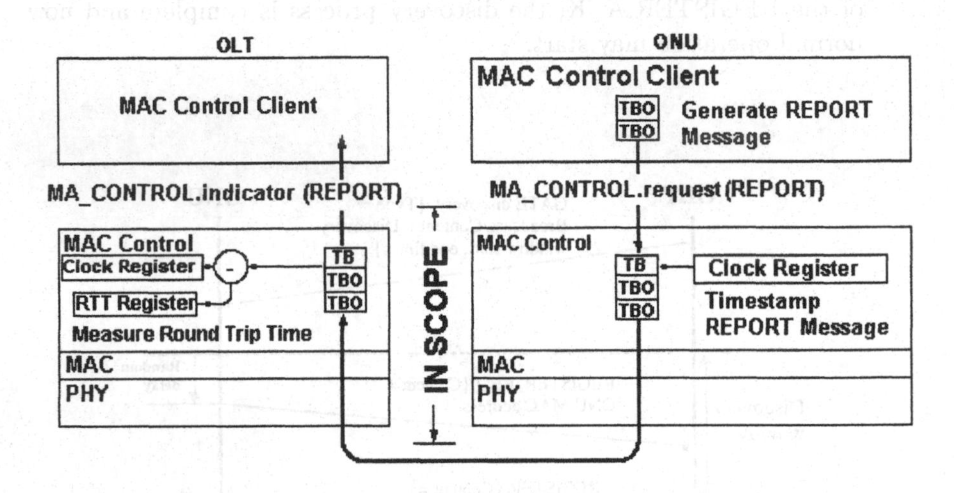

Figure 5.8 Multi-Point-Control-Protocol (MPCP) REPORT operation.

Typically, REPORT would contain the desired size of the next timeslot, based on ONU's queue size. REPORT messages are generated periodically, even when no request for bandwidth is being made. This prevents the OLT from deregistering the ONU. Thus, for the proper operation of this mechanism, the OLT must grant the ONU a transmission window periodically. At the OLT, the REPORT is processed, and the data is used for the next round of bandwidth assignments.

GATE Handling

The transmitting window of an ONU is indicated in the GATE message from the OLT. The *transmission start* and *transmission length* times are specified. Upon receiving a GATE message matching the ONU's LLID, the ONU will program its local registers with the *transmission start* and *transmission length* times, as shown in Fig. 5.9.

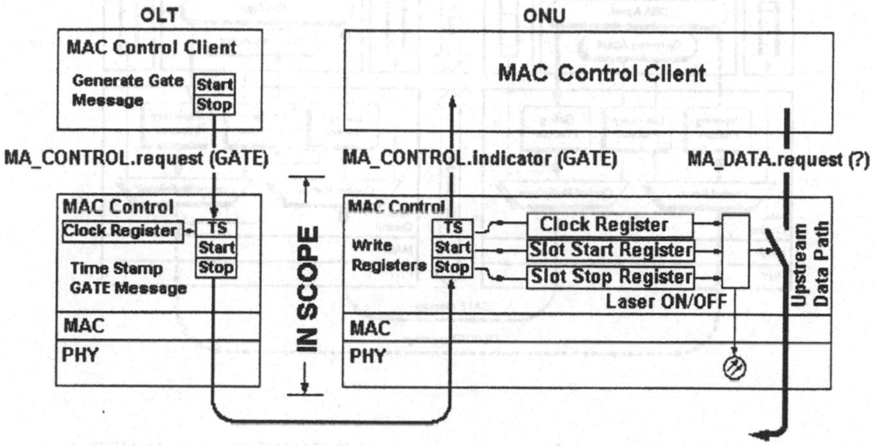

Figure 5.9 Multi-Point-Control-Protocol (MPCP) GATE operation.

The ONU will also verify that the time (according to the local clock at the ONU) when the GATE message arrived is close to the timestamp value contained within the message. If the difference in values exceeds some predefined threshold, the ONU will assume that it has lost its synchronization and will switch itself into offline mode. The ONU will then attempt to register again using the next discovery process.

Figure 5.10 shows the message exchange process of REPORT and GATE handling.

When the time at the local clock of the ONU reaches *transmission start*, the ONU starts transmitting data.

Clock Synchronization

The correct operation of MPCP depends on clock synchronization between the OLT and the ONU, which compensates for the RTT. RTT is expected to be different for each ONU as they may be located at different distances from the OLT. Clock synchronization to compensate for RTT is important because the OLT does not have to keep track of the different RTTs of different ONUs, when it issues timeslots in GATE messages.

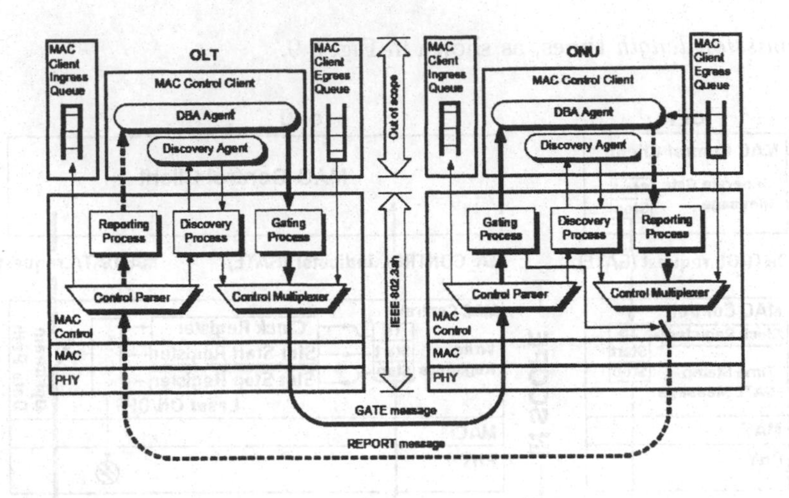

Figure 5.10 Message exchange process of REPORT and GATE handling.

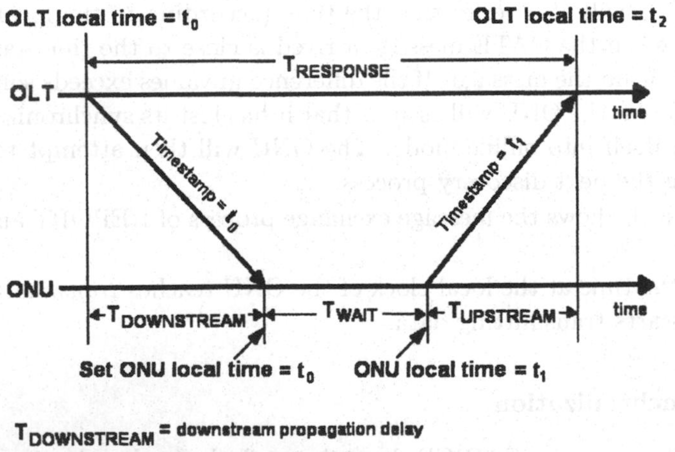

Figure 5.11 Calculation of round-trip time (RTT).

The above clock synchronization is achieved as follows. Whenever the ONU receives a MPCP message, it sets its local time from the time-stamp of that message. When the OLT receives a MPCP message, it calculates the RTT as the difference between its local time and time-stamp of the message (Fig. 5.11). Any significant change in RTT implies that the OLT and the ONU clocks are not in synchrony any more, and the OLT now issues a DEREGISTER message for that particular ONU. The ONU will then attempt to register in the network again through the discovery process.

5.3.4 Dynamic Bandwidth Allocation (DBA) Algorithms in EPON

EPON operation is broadcast in the downstream direction (from OLT to ONU): packets are broadcast by the OLT and extracted by their destination ONU based on the media-access control (MAC) address. In the upstream direction (from ONU to OLT), the ONUs should share the channel capacity and resources. Since the ONUs cannot communicate with one another, the OLT must assign timeslots in which the ONUs are allowed to transmit data. One method is to assign static timeslots for each ONU, which achieves synchronous Time-Division Multiple Access (TDMA). This results in a very cost-effective solution, since the OLT no longer has to poll the ONUs and schedule the timeslots, which avoids the need for REPORT messages in the MPCP protocol altogether. However, the TDMA approach has a major limitation: the lack of statistical multiplexing. The burstiness of network traffic results in a situation where some timeslots overflow even under very light load, resulting in packets being delayed for several timeslot periods, while a large number of slots remain underutilized

Hence, the need of Dynamic Bandwidth Allocation (DBA) algorithms, in which the OLT schedules the timeslots in which the ONUs may transmit, is to achieve statistical multiplexing. One of the first protocols proposed was the Interleaved Polling with Adapatative Cycle Time (IPACT, pronounced *eye-pact*). We describe IPACT based on MPCP below.

IPACT

The IPACT protocol [KrMP02] can be summarized as follows:

1. The OLT keeps track of the earliest scheduling time by a variable $T_{schedule}$. Thus, $T_{schedule}$ is changed after each allocated timeslot.

2. Whenever a REPORT message containing the requested timeslot from

the ONU arrives at the OLT, the DBA agent at the OLT is invoked to calculate the start time for the next transmission timeslot for that ONU. To maintain high utilization in the upstream channel, the DBA agent allocates the next timeslot immediately adjacent to the already allocated timeslot with only a guard time interval separation. The start time is therefore computed as:

$$T_{start} = T_{schedule} + T_{guard}$$

The DBA agent must also ensure that the ONU has enough time to process the GATE message before the granted timeslot is scheduled to begin. Let T_{local} denote the local time, and $T_{processing}$ denote the processing time. The start time T_{start} is therefore updated as:

$$if \quad T_{start} < T_{local} + T_{processing},$$

$$T_{start} = T_{local} + T_{processing}$$

3. *Maximum Scheduling Timeslot*: If the OLT authorizes each ONU to send its entire buffer contents in one transmission, ONUs with high data volume could monopolize the entire bandwidth, and subsequently the average delay in the network could become very large. To avoid this situation, the OLT must limit the maximum transmission size. We define this as a *limited-service scheme*, in which every ONU is allocated a timeslot to send as many bytes as it has requested, but no more than some upper limit which is defined as the *Maximum Scheduling Timeslot*. There could be various schemes for specifying the limit. It can be fixed, for example, based on a Service-Level Agreement (SLA) for each ONU, or dynamic based on network conditions. In the following section, we consider a few other service disciplines: *fixed*, *gated*, *constant credit*, *linear credit*, and *elastic*.

   Let the constant timeslot to transmit a REPORT message be T_{REPORT}, and *maxLength* denote the *Maximum Scheduling Timeslot*. Ignoring other overheads, such as laser on/off times and synchronization times, the length of the timeslot under the *limited-service scheme* is computed as:

$$length = REPORT.length + T_{REPORT}$$

$$if \quad length > maxLength,$$

$$length = maxLength$$

4. Once the above are computed, the corresponding GATE message is transmitted by the OLT. $T_{schedule}$ is now modified as:

$$T_{schedule} = T_{start} + length$$

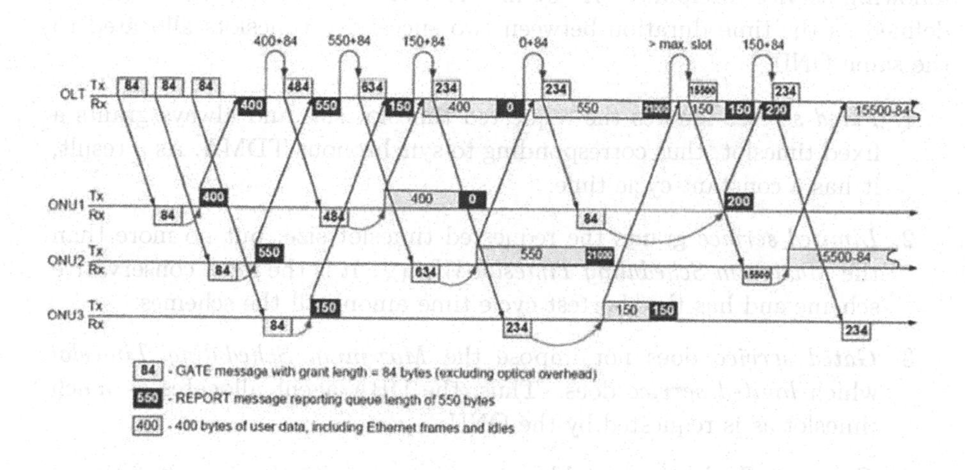

Figure 5.12 Timing diagram of the limited-service scheme in IPACT.

Figure 5.12 presents a timing diagram for the limited-service scheme in IPACT. For simplicity, only three ONUs are shown in the example. Upon completion of the discovery procedure in MPCP, the OLT issues individual GATE messages to all the ONUs. The ONUs send their REPORT messages in their respective timeslots. When the REPORT from an ONU arrives, the OLT schedules timeslots as described before. It may be observed in Fig. 5.12 that, when ONU2 requests a timeslot of 21,000 bytes, the *Maximum Scheduling Timeslot* of 15,500 bytes is granted. If the ONU emptied its buffer completely, it shall report 0 bytes to the OLT. Correspondingly, in the next cycle, the ONU is allocated only a small timeslot to send a REPORT message (which is 84 bytes in Fig. 5.12).

It may be observed that the upstream channel is almost 100% utilized (barring REPORT messages and guard times). Idle ONUs are allocated very small timeslots. Hence, if the system is under low load, then ONUs will be polled very frequently.

Service Disciplines

Now, we consider a few service disciplines for IPACT [KBSM04]. The size of the buffer (requested timeslot size) at the ONU is conveyed to the OLT using a REPORT message as defined in MPCP. The DBA agent at the OLT may decide the size of the next timeslot allocated to the ONU based on the following service disciplines. A key metric is the *cycle time*, which may be defined as the time duration between two successive timeslots allocated to the same ONU.

1. *Fixed service* ignores the requested timeslot size and always grants a fixed timeslot, thus corresponding to synchronous TDMA. As a result, it has a constant cycle time.

2. *Limited service* grants the requested timeslot size, but no more than the *Maximum Scheduling Timeslot* W_{MAX}. It is the most conservative scheme and has the shortest cycle time among all the schemes.

3. *Gated service* does not impose the *Maximum Scheduling Timeslot* which *limited service* does. Thus, the DBA agent allocates as much timeslot as is requested by the ONU.

4. *Constant-Credit service* adds a constant credit to the requested timeslot size. The idea behind adding the credit is the following: assume that x bytes arrived at an ONU between the time the ONU sent a REPORT and the time it received the corresponding GATE message with its timeslot assignment. If the granted window size equals the requested window + x (i.e., it has a credit of size x), these x bytes will not have to wait for the next GRANT to arrive; they will be transmitted with the current granted timeslot, and therefore the average packet delay is expected to be shorter.

5. *Linear-Credit service* uses a similar approach as the *Constant-Credit service* scheme. However, the size of the credit is proportional to the requested window. The reasoning here is the following: Network traffic possesses a certain degree of predictability [PaWi00]; specifically, if we observe a long burst of data, this burst is likely to continue for some time into the future. Correspondingly, the arrival of more data during the last cycle may signal that we are observing a burst of packets.

6. *Elastic service* is an attempt to get rid of a fixed *Maximum Scheduling Timeslot* W_{MAX} limit. The only limiting factor is the maximum

cycle time. The maximum window is granted in such a way that the accumulated size of last N Grants (including the one being granted) does not exceed $N \times W_{MAXbytes}$ (where N is the number of ONUs). Thus, if only one ONU has data to send, it may get a Grant of size up to $N \times W_{MAX}$.

Simulation Results for the Various Service Schemes

Consider a PON consisting of an OLT and N ONUs. Every ONU is assigned a downstream propagation delay (from the OLT to the ONU) and an upstream propagation delay (from the ONU to the OLT). To keep the model general, we assume independent upstream and downstream propagation delays and select them randomly (uniformly) over the interval [50 ms, 100 ms]. These values correspond to distances between the OLT and ONUs ranging from 10 to 20 km.

The transmission speeds of the PON and user access link may not necessarily be the same. In the model, we consider R_D Mbps to be the data rate of the access link from a user to an ONU, and R_U Mbps to be the rate of the upstream link from an ONU to the OLT. Note that, if $R_U \geq N \times R_D$, the bandwidth-utilization problem does not exist, since the system throughput is higher than the peak aggregated load from all ONUs. In this study, we consider a system with N = 16 and R_D and R_U being 100 Mbps and 1000 Mbps, respectively. Every ONU has a finite memory buffer of size Q (set to 10 Mbytes in the simulations). The simulation was performed using synthetic traffic traces that exhibit the properties of self-similarity and long-range dependence (LRD).

Figure 5.13 presents the mean packet delay for different service schemes as a function of an ONU's offered load. In this simulation, all ONUs have identical load. As can be seen, all service schemes except *fixed service* have almost coinciding plots. We will discuss *fixed service* results below. As for the rest of them, no other method gives a detectable improvement in packet delay. The explanation of this lies in the fact that all these methods are trying to send more data by way of increasing the granted timeslot. While this may clear the queue in fewer polling cycles, the polling cycle duration itself may increase. Overall, all these services have negligible effect on packet delay. The *fixed service* plot is interesting as an illustration of the traffic long-range dependence. Even at a very light load of 5 percent, the average packet delay is already very high (nearly 15 ms). This is because most packets arrive in very large packet trains. In fact, the packet trains

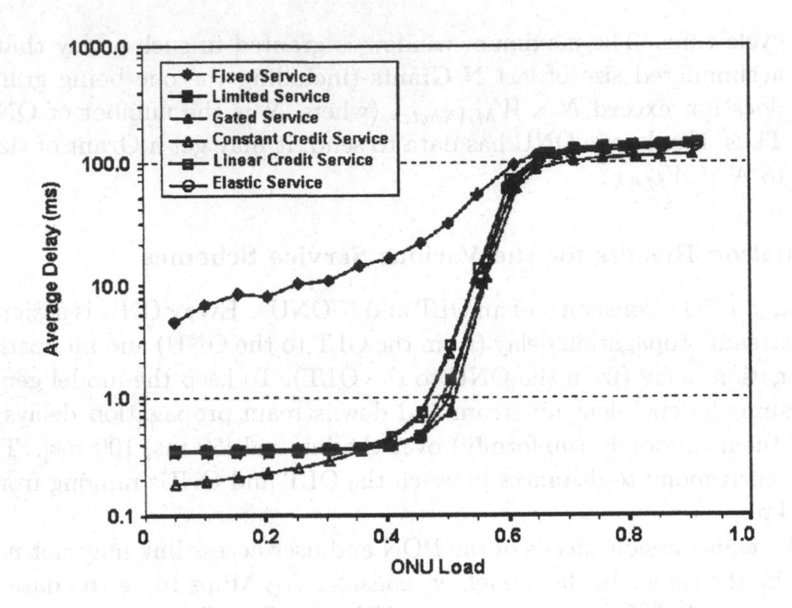

Figure 5.13 Average packet delay for different service schemes.

were so large that the 10-Mbyte buffers overflowed and about 0.14 percent of packets were dropped. Why do we observe this anomalous behavior only with fixed service? The reason is that all other services have much shorter cycle times; there is just not enough time in a cycle to receive more bytes than W_{MAX}; thus, the queue never builds up. In fixed service, on the other hand, the cycle is large (and fixed) from the very beginning, and several bursts that arrive close to each other can easily overflow the buffer.

Figure 5.14 shows the cycle time for the different service schemes. It may be observed that *fixed service* has a constant cycle time, while *gated service* has very high cycle time at a high load as the *Maximum Scheduling Timeslot* is not imposed by the gated service.

Derivatives of IPACT

Many other DBA algorithms have been proposed in the literature by researchers by enhancing the basic principles of IPACT. Derivatives of IPACT include supporting bandwidth guarantees for high priority traffic [MaZC03], deterministic effective bandwidth (DEB) admission control [Zhan03], and

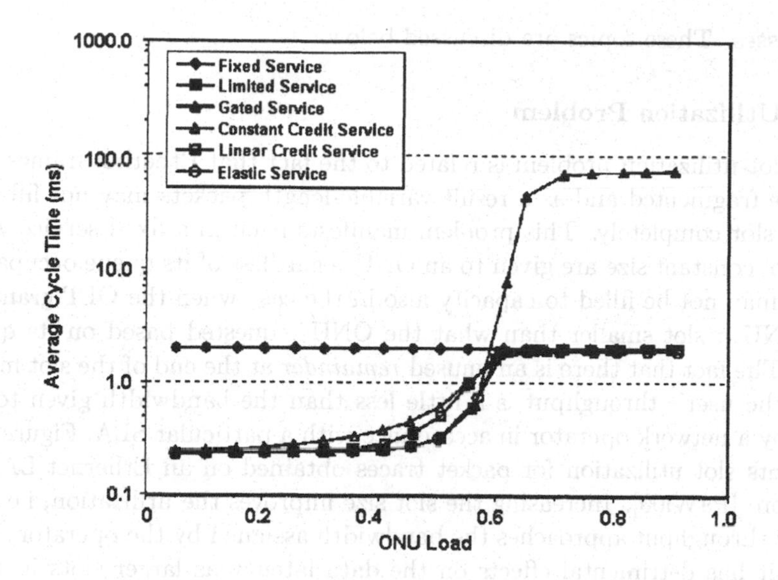

Figure 5.14 Average cycle times for various service schemes.

many others. A good survey of DBAs reported in the literature may be found in [McMR04].

5.3.5 Considerations for IP-based Services over EPON

The driving force behind extending Ethernet into the subscriber access area is Ethernet's efficiency for delivering IP packets. Data and telecom network convergence will lead to more and more telecommunication services migrating to variable-length-packet-based data networks. To ensure successful convergence, this migration should be accompanied by implementation of specific mechanisms traditionally available in telecom networks only.

Being designed with IP layer in mind, EPON is expected to seamlessly operate with IP-based traffic flows, similarly to any switched Ethernet network. One distinction with the typical switched architecture is that, in an EPON, the user's throughput is slotted (gated), i.e., packets cannot be transmitted by an ONU at any time. This feature results in two issues unique to EPONs: (a) potential slot under-utilization problem due to variable-length packets and (b) slot scheduling to support real-time and controlled-load traffic classes. These topics are discussed below.

Slot-Utilization Problem

The slot-utilization problem is related to the fact that Ethernet frames cannot be fragmented and as a result variable-length packets may not fill up a given slot completely. This problem manifests itself in a fixed service when slots of constant size are given to an ONU regardless of its queue occupancy. Slots may not be filled to capacity also in the case when the OLT grants to an ONU a slot smaller than what the ONU requested based on its queue size. The fact that there is an unused *remainder* at the end of the slot means that the user's throughput is a little less than the bandwidth given to the user by a network operator in accordance with a particular SLA. Figure 5.15 presents slot utilization for packet traces obtained on an Ethernet LAN in Bellcore [PaWi00]. Increasing the slot size improves the utilization, i.e., the user's throughput approaches the bandwidth assigned by the operator; however, it has detrimental effects on the data latency as larger slots increase the overall cycle time.

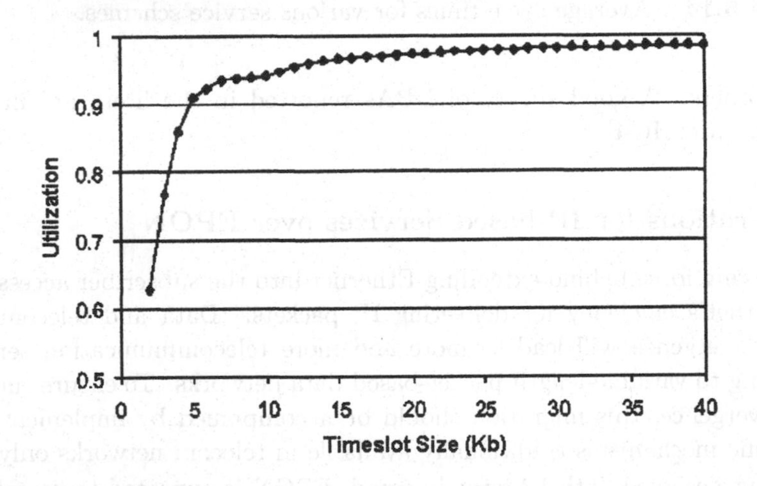

Figure 5.15 Slot utilization for various slot sizes.

Slot utilization can also be improved by employing smarter packet scheduling (e.g., the bin-packing problem). Rather than stopping the transmission when the head-of-the-queue frame exceeds the remainder of the slot, the algorithm may look ahead in the buffer and pick a smaller packet for immediate transmission (first-fit scheduling). However, as it turns out, first-fit scheduling is not such a good approach. To understand the problem, we need to look at the effects of packet reordering from the perspective of Transmission Control Protocol/Internet Protocol (TCP/IP) payloads carried by Ethernet frames. Even though TCP will restore the proper sequence of packets, an excessive reordering may have the following consequences:

1. According to the fast retransmission protocol, the TCP receiver will send an immediate acknowledgement (ACK) for any out-of-order packet, whereas for in-order packets, it may generate a cumulative acknowledgement (typically for every other packet) [Stev94]. This will lead to more unnecessary packets being placed in the network.

2. Second, and more importantly, packet reordering in ONUs may result in a situation where n later packets are being transmitted before an earlier packet. This would generate n ACKs (n-1 duplicate ACKs) for the earlier packet. If 'n' exceeds a predefined threshold, it will trigger packet retransmission and reduction of TCP's congestion window size (the cwnd parameter). Currently, the threshold value in most TCP/IP protocol stacks is set to 3 (see Fast Retransmission Protocol in [Stev94] or elsewhere).

Even if special care is taken at the ONU to limit out-of-order packets to only 1 or 2, the rest of the end-to-end path may contribute additional reordering. While true reordering typically generates less than 3 duplicate ACKs and is ignored by the TCP sender, together with reordering introduced by the ONU, the number of duplicate ACKs may exceed 3, thus forcing the sender to retransmit a packet. As a result, the overall throughput of the user's data may decrease.

So, what is the solution? It is reasonable to assume that the traffic entering the ONU is an aggregate of multiple flows. In the case of business users, it would be the aggregated flows from multiple workstations. In the case of a residential network, we may still expect multiple connections at the same time. This is because, as a converged access network, PON will carry not only data, but also voice-over-IP (VoIP) and video traffic. Also, home appliances are becoming network plug-and-play devices. The conclusion is that, if we have multiple connections, we can reorder packets that belong to different connections, and never reorder them if they belong to the same connection. Connections can be distinguished by examining the source/destination address pairs and source/destination port numbers. This will require an ONU to look up layer-3 and layer-4 information in the packets. Thus, the important tradeoff decision that EPON designers have to make is whether it makes sense to considerably increase the required processing power in an ONU to improve the bandwidth utilization.

Circuit Emulation (TDM over IP)

The migration of TDM circuit-switched networks to IP packet-switched networks is progressing at a rapid pace. But, even though the next-generation access network will be optimized for IP data traffic, legacy equipment (RF set-top boxes, analog TV sets, TDM private branch exchanges (PBXs), etc.) and legacy services (T1/E1, Integrated Services Digital Network (ISDN), Plain Old Telephone Service (POTS), etc.) will remain in use in the foreseeable future. Therefore, it is critical for next-generation access networks, such as Ethernet PONs, to be able to provide both IP-based services and jitter-sensitive and time-critical legacy services that have traditionally not been the focus of Ethernet.

The issue in implementing a circuit-over-packet emulation scheme is mostly related to clock distribution. In one scheme, users provide a clock to their respective ONUs, which in turn is delivered to the OLT. But, since the ONUs cannot transmit all the time, the clock information must be delivered in packets. The OLT will regenerate the clock using this information. It is somewhat trivial to impose a constraint that the OLT should be a clock master for all downstream ONU devices. In this scenario, an ONU will recover the clock from its receive channel, use it in its transmit channel, and distribute it to all legacy devices connected to it.

Real-Time Video and Voice Over IP

Performance of a packet-based network can be conveniently characterized by several parameters: bandwidth, packet delay (latency) and delay variation (jitter), and packet-loss ratio. Quality of Service (QoS) refers to a network's ability to provide bounds on some or all of these parameters. It is useful to further differentiate statistical QoS from guaranteed QoS. Statistical QoS refers to a case when parameters can exceed the specified bounds with some small probability. Correspondingly, guaranteed QoS refers to a network architecture where parameters are guaranteed to stay within the specified bounds for the entire duration of a connection. A network is required to provide QoS (i.e., bounds on performance parameters) to ensure proper operation of real-time services such as video-over-packets (digital video conference, VoD), voice-over-IP (VoIP), real-time transactions, etc. To be able to guarantee QoS for higher-layer services, QoS must be maintained in all traversed network segments, including the access network portion of the end-to-end path. Our current discussion only focuses on QoS in the EPON access network.

The original Ethernet standard based on the CSMA/CD MAC protocol was never concerned with QoS. All connections (traffic flows) were treated equally and were given best-effort service from the network. The first major step in allowing QoS in Ethernet was an introduction of the full-duplex mode. Full-duplex MAC can transmit data frames at any time; this eliminated non-deterministic delay in accessing the medium. In a full-duplex link (segment), once a packet is given to a transmitting MAC layer, its delay, jitter, and loss probability are known or predictable all the way to the receiving MAC layer. Delay and jitter may be affected by head-of-line blocking when the MAC port is busy transmitting the previous frame at the time when the next one becomes available. However, with a 1-Gbps channel, this delay variation becomes negligible since the maximum-sized Ethernet frame is transmitted in only about 12 ms. It is important to note that the full-duplex MAC does not make the Ethernet a QoS-capable network: switches located in junction points may still provide non-deterministic, best-effort services.

The next step in enabling QoS in Ethernet was the introduction of two new standards extensions — (a) 802.1p Supplement to MAC Bridges: Traffic Class Expediting and Dynamic Multicast Filtering (later merged with 802.1D), and (b) 802.1Q Virtual Bridged Local Area Networks. 802.1Q defines a frame-format extension allowing Ethernet frames to carry priority information. 802.1p specifies the default bridge (switch) behavior for different priority classes; specifically, it allows a queue in a bridge to be serviced only when all higher-priority queues are empty. The standard distinguishes the following traffic classes:

1. *Network Control* — characterized by a must-get-there requirement to maintain and support the network infrastructure.

2. *Voice* — characterized by less than 10-ms delay, and hence maximum jitter (one way transmission through the LAN infrastructure of a single campus).

3. *Video* — characterized by less than 100-ms delay.

4. *Controlled Load* — important business applications subject to some form of admission control, be that pre-planning of the network requirement at one extreme to bandwidth reservation per flow at the time the flow is started at the other.
5. *Excellent Effort or CEO's best effort* — the best-effort-type services that an information services organization would deliver to its most important customers.
6. *Best Effort* — LAN traffic as we know it today.
7. *Background* — bulk transfers and other activities that are permitted on the network but that should not impact the use of the network by other users and applications.

Both full-duplex and P802.1p/P802.1Q standards extensions are important but not sufficient QoS enablers. The remaining part is admission control. Without it, each priority class may intermittently degrade to best-effort performance. Here, EPON can provide a simple and robust method for performing admission control. Earlier, we mentioned that MPCP relies on GATE messages sent from the OLT to ONUs to allocate the transmission window. A very simple protocol modification may allow a single GATE message to grant multiple windows, one for each priority class. The REPORT message may also be extended to report queue states for each priority class. Alternatively, admission control can be left to higher-layer intelligence in ONUs. In this case, the higher layer will know when the next transmission window will arrive and how large it will be, and will schedule packets for transmission of different classes of service (CoS) accordingly.

Performance of CoS-Aware EPON

Now, we shall investigate how priority queuing may allow us to provide a delay bound for some services. Below, we describe a simulation setup in which data arriving from the user is classified in three priority classes (instead of seven) and directed to different queues in the ONU. The queues are then serviced in order of their priority; a lower-priority queue is serviced only when all higher-priority queues are empty. In this experiment, the tagged ONU has a constant load. We investigate the performance of each class as the ambient network load varies [KMDY02].

Best Effort (BE) class has the lowest priority. This priority level is used for non-real-time data transfers. There is no delivery or delay guarantees in this service. The BE queue in the ONU is served only if higher-priority queues are empty. Since all queues in the system share the same buffer, the packets arriving at higher-priority queues may displace the BE packets that are already in the BE queue. In the experiment, the tagged source has the BE traffic with an average load of 0.4 (40 Mbps).

Assured Forwarding (AF) class has higher priority than the BE class. The AF queue is served before the BE queue. In our experiment, the AF traffic consisted of a VBR stream with average bit rate of 16 Mbps. This corresponds to three simultaneous MPEG-2-coded video streams. Since the AF traffic is also highly bursty (LRD), it is possible that some packets in long bursts will be lost. This will happen if the entire buffer is occupied by AF or higher-priority packets.

Guaranteed Forwarding (GF) priority class was used to emulate a T1 line in the packet-based access network. The GF class has the highest priority and can displace the BE and AF data from their queues if there is not enough buffer space to store the GF packet. A new GF packet will be lost only if the entire buffer is occupied by GF packets. The GF queue is served before the AF and BE queues. The T1 data arriving from the user is packetized in the ONU by placing 24 bytes of data in a packet. Including Ethernet and User Datagram Protocol (UDP)/IP headers results in a 70-byte frame generated every 125 ms. Hence, the T1 data consumed bandwidth equal to 4.48 Mbps. Of course, we could put 48 bytes of T1 data in one packet and send one 94-byte packet every 250 ms. This would consume only 3.008 Mbps, but will increase the packetization delay.

Figures 5.16 and 5.17 show the average and the maximum packet delay for each type of traffic. The average load of the tagged ONU was set to 40 Mbps of BE data, 16 Mbps of AF data, and 4.48 Mbps of GF data, or to a total of approximately 60 Mbps. The horizontal axis shows the effective network load. These figures show how the traffic parameters depend on the overall network load.

We can observe that the BE traffic suffers the most when the ambient load increases. Its delay increases, and the simulation results showed that some packets were discarded when the network load exceeded 80%. The AF data also experienced an increased delay, but no packet losses were observed. The increased delay in the AF traffic can be attributed to long bursts of data. Clearly, applying some kind of traffic shaping/policing to limit the burst size at the source would improve the situation. The GF data experiences a very slight increase in both average and maximum delays. This is due to the fact

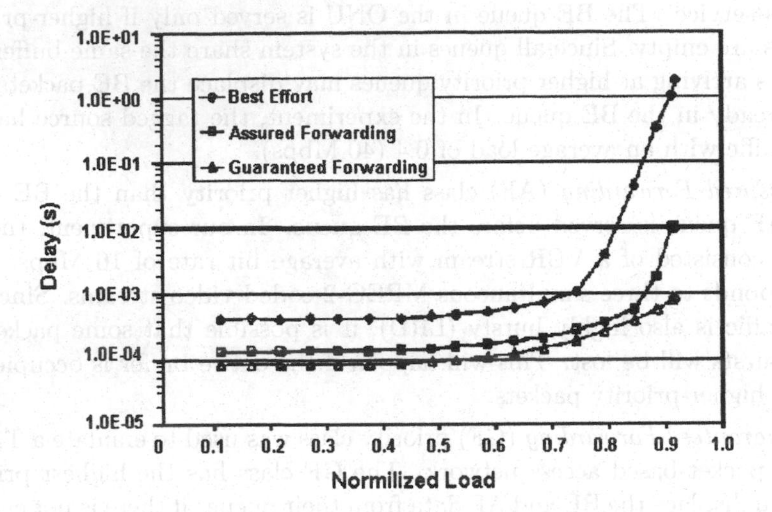

Figure 5.16 Average packet delay for various classes of traffic as a function of the effective network load.

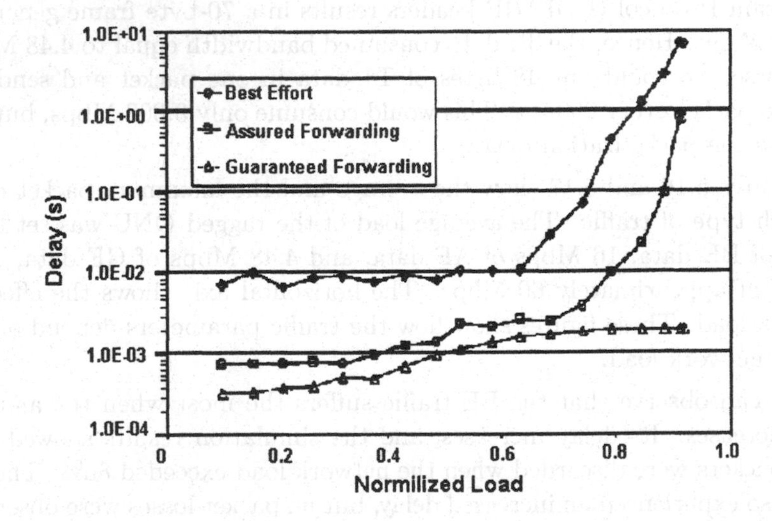

Figure 5.17 Maximum packet delay for various classes of traffic as a function of the effective network load.

that the packets were generated with a constant rate, i.e., no data bursts. The average delay in this case exactly followed the average cycle time, being one half of that. The maximum delay is equal to the maximum observed cycle time and for any effective network load is bounded by the maximum cycle time T_{MAX} (which was chosen to be 2 ms in the experiments).

Of course, to restore the proper T1 rate, a shaping buffer (queue with constant departure rate) should be employed at the receiving end (in the OLT). After receiving a packet (or a group of packets) from an ONU, the shaping buffer should have at least 2 ms worth of data, i.e., 384 bytes of T1 data. This is because the next packet from the same ONU may arrive after the maximum delay of 2 ms. When such a delayed packet arrives, it should still find the non-empty buffer. Let us say the minimum buffer occupancy is 24 bytes which is 125 ms of T1 transmission time (we call it a buffer underrun protection time). In this case, the overall latency experienced by T1 data will consist of:

1. 125 μs of packetization delay in ONU
2. Polling delay in ONU (up to T_{MAX})
3. Up to 100 μs of propagation delay (assuming maximum distance of 20 km), and
4. wait time in the shaping buffer.

Light-Load Penalty

One interesting observation is that combining default priority queuing with a simple polling mechanism in an EPON results in a very interesting phenomenon: As the load decreases from moderate (0.25) to very light (0.05), the average delay for the lowest priority class (P2) increases significantly. The average packet delay at a load of 0.05 (or 5 Mbit/s) is 17.8 ms, more than 1200% higher than the 1.4-ms delay at a load of 0.25 (25 Mbit/s). Similar behavior is observed for the maximum packet delay for P1 and P2 classes. We refer to this phenomenon as *light-load penalty*.

The *light-load penalty* may be explained as follows. At the end of every timeslot, an ONU generates a REPORT message containing the number of bytes that remain in the queue (residual queue occupancy). The residual queue occupancy is almost always less than the *Maximum Scheduling Timeslot* W_{MAX}, because the light-load penalty occurs only at light loads.

This means that whatever the slot size an ONU requested in the REPORT message, the OLT will grant the requested slot size through the next GATE message to that ONU. However, during the time lag between the ONU's sending a REPORT and the arrival of its assigned timeslot (i.e., between sending a REPORT and transmission of the reported data in the next allocated timeslot), more packets may arrive to the queue. Newly-arrived packets may have higher priority than some packets already stored in the queue, and they will be transmitted in the next transmission slot before the lower-priority packets. Since these new packets were not reported to the OLT before, the given slot cannot accommodate all the stored packets. This causes some lower-priority packets to be left in the queue. This situation may repeat over many cycles, causing some lower-priority packets to be delayed for multiple cycles. A lower-priority packet will finally be transmitted when more lower-priority packets (bytes) accumulate (and are reported to the OLT) behind a given packet than higher-priority packets cut in front of it. But, P0 traffic is periodic (CBR) and P2 traffic is bursty (i.e., a new P2 packet may not arrive for a long time), we observe in the experiments that, on average, at a load of 0.05, P2 packets are delayed by approximately 80 cycles. As the load increases, the queue behind a lower-priority packet grows faster and the light-load penalty decreases. At a load of 0.25, the average delay for P2 packets is only about 4.6 cycles. Since Ethernet packets cannot be fragmented (according to IEEE 802.3), packet preemption also results in an unused slot remainder (unless an added higher-priority packet is the same size as a preempted lower-priority packet, which is rare).

Several schemes have been suggested to avoid the light-load penalty in [KMDY02]. One way to eliminate the light-load penalty is to implement a two-stage queue in an ONU. In a two-stage system, stage I consists of multiple-priority queues and stage II consists of one First-Come-First-Serve (FCFS) queue. When a timeslot arrives, data packets from stage II are transmitted to the OLT, thereby vacating the queue; simultaneously, data packets from stage I are advanced into vacant spaces in the stage II queue. At the end of the current timeslot, the ONU reports to the OLT the occupancy of the stage II queue (to get a corresponding slot size in the next cycle). The total size of the stage-II buffer can be made exactly equal to W_{MAX} bytes, so that the ONU never requests a slot greater than W_{MAX} bytes. This configuration will ensure that the given slot is always 100% utilized, i.e., that the unused remainder is always zero because no packets are preempted. The biggest demerit of the two-stage scheme is that the average delay of all packets is increased because each packet must spend an average amount of half-cycle time in the stage I queues and one cycle time in the stage II queues.

Another interesting solution to the light-load penalty (without increasing the delay of the highest-priority class beyond one cycle time as in a two-stage scheme) is to predict the amount of high-priority packets that are expected to arrive at the ONU and to adjust the granted timeslot size accordingly. Of course, to predict the traffic with any reasonable accuracy, we need to have some knowledge about the traffic behavior. In the above example, we have this knowledge about the P0 traffic; namely, we know that this is a constant bit rate (CBR) flow with a given data rate. Therefore, when deciding on the size of the next timeslot for an ONU, the OLT can estimate the time of the next transmission and increase the timeslot size by the amount of CBR data it anticipates. We may call this scheme CBR credit, since the additional timeslot size increment (credit) is based on the known CBR arrival rate. Although this scheme performs relatively well, a major limitation is that external knowledge of the arrival process is necessary. Even though, for some time-critical applications, we may have such knowledge, it is by no means a universal case, especially for VBR traffic.

5.4 Other Types of PONs

Besides EPON, a few other alternative technologies have been also developed for PON, e.g., APON (or BPON), GFP-PON, and WDM-PON. We discuss them below.

5.4.1 APON/BPON

ATM PON (APON) is based on Asynchronous Transfer Mode (ATM) as the MAC layer protocol. The downstream frame, shown in Fig. 5.18(a), consists of 56 ATM cells (53 bytes each) for the basic rate of 155 Mbps, scaling upto 224 cells for 622 Mbps. There are two dedicated Physical Layer Operation, Administration, and Maintenance (PLOAM) cells, one at the beginning of the frame, and one in the middle. The remaining 54 cells are data ATM cells.

The upstream transmission (Fig. 5.18(b)) occurs in the form of bursts of ATM cells, with a 3-byte physical overhead appended to each 53-byte ATM cell to allow for burst-mode receivers. Burst-mode receivers are required at the OLT to synchronize to the different ONUs which may be located at different distances from the OLT; and, hence, the power received at the OLT from different ONUs may be different. The ATM cell may be either an ATM data cell, or a PLOAM cell.

In downstream direction, the PLOAM cells are used to carry grants from the OLT to the ONU. Each grant is a one-time permission for the ONU to transmit payload data in a ATM cell. 53 grants for the 53 upstream frame cells, are mapped into the PLOAM cells. During its operation, the OLT sends a continuous stream of grants to all the ONUs in the PON. Thus, the OLT can moderate the portion of upstream bandwidth assigned to each ONU. In upstream direction, the PLOAM cells are used by the ONUs to transmit their queue sizes to the OLT.

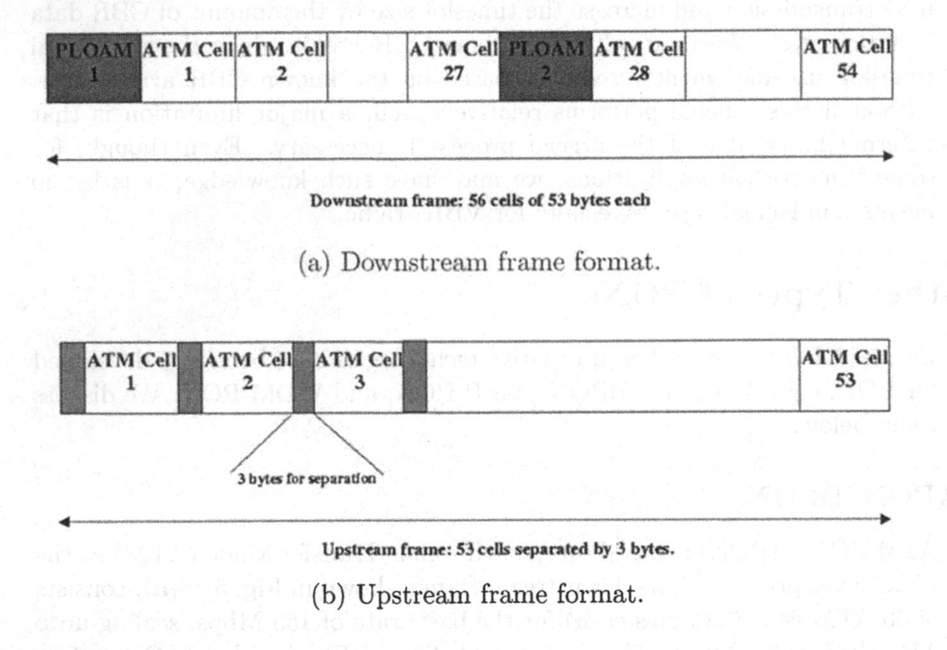

(a) Downstream frame format.

(b) Upstream frame format.

Figure 5.18 APON/BPON frame format.

Initial work on ATM PONs was launched in the mid 1990s by the Full Service Access Network (FSAN) [FSAN05] initiative which was started by service providers. Because the name APON led users to believe that only ATM-based services could be supported, the terminology was changed to Broadband PON (BPON). BPON has been standardized by the International Telecommunication Union (ITU) specification G.983.1. BPON provides overlay capabilities for services such as video and Ethernet traffic.

5.4.2 Generalized Framing Procedure (GFP) PON (GFP-PON)

The GFP-PON is being standardized by the ITU in specification G.984.x. It proposes bit rates of upto 2.5 Gbps. It also aims towards providing higher efficiency while carrying multiple services over the PON. It proposes a protocol using Generic Framing Procedure (GFP) [VaSZ02]. GFP provides a generic mechanism to adapt traffic from higher layers (Ethernet MAC/IP), over a transport layer such as Synchronous Optical Network/Synchronous Digital Hierarchy (SONET/SDH). Other functionalities such as dynamic bandwidth assignment, operation and maintenance, etc. are borrowed from APON.

As discussed earlier, both APON and GFP-PON have the disadvantage of a complex protocol and implementation, relative to EPON, because of which they have not gained much technical popularity amongst users and equipment vendors.

5.5 WDM-PON

This section reviews research contributions to WDM-PONs. We first emphasize the importance of WDM in PON. We then discuss various WDM-PON architectures which have been proposed in the literature.

5.5.1 Need for WDM in PONs

Although the PON is a significant step towards providing broadband access to the end user, it is not very scalable. Since the basic form of PON employs only a single optical channel, the available bandwidth is limited to the maximum bit rate of an optical transceiver, which, under current technologies is, 1 Gbps. The attenuation due to splitting limits the maximum number of ONUs to 64. This limits the network's scalability. The deployment cost of laying fiber in the access network being high, it is important to consider technologies which may help scale the PON capacity in future.

Many telecom operators are considering to deploy PONs using a FTTx model to support converged Internet Protocol (IP) video, voice, and data services — defined as "triple-play" — at a cheaper subscription cost than the cumulative of the above services deployed separately [BPCM05]. PONs are in the initial stages of deployment in many parts of the world. Although the

PON provides higher bandwidth than traditional copper-based access networks, there exists the need for further increasing the bandwidth of the PON by employing Wavelength-Division Multiplexing (WDM) so that multiple wavelengths may be supported in either or both upstream and downstream directions. Such a PON is known as a WDM-PON [BPCM05][2]. A WDM-PON is a point-to-point access network (as opposed to point-to-multipoint in PON), in which there exists a separate wavelength, between the OLT and each ONU.

Each wavelength is routed by a passive Arrayed Waveguide Grating (AWG). The AWG is discussed briefly in Section 5.5.2. In a WDM-PON, different ONUs can be supported at different bit rates, if necessary. Each ONU can operate at a rate upto the full bit rate of a wavelength channel; therefore, it does not have to share the available bandwidth with any other ONU in the network. Moreover, unlike the basic PON, the WDM-PON does not suffer power-splitting losses. Use of individual wavelengths for each ONU also facilitates privacy and reduces security concerns which the PON has. Finally, because of the periodic routing pattern of a AWG (described below), the WDM-PON is easily scalable. Keeping in view such advantages that WDM-PONs have, WDM has been recommended as an upgrade to the PON in the ITU-T G.983 [EfTY01].

5.5.2 Arrayed Waveguide Grating (AWG)

Wavelength routing in a WDM-PON may be implemented through an AWG. The AWG is a passive device with a fixed routing matrix, so it fits well with the PON philosophy. An AWG provides a fixed routing of an optical signal from a given input port to a given output port, based on the wavelength of the signal. Signals of different wavelengths coming into an input port will each be routed to a different output port. Similarly, different signals of the same wavelength may be directed to different input ports, and they shall be routed to different output ports. One of the main advantages of the AWG is its periodic routing behavior, shown in Fig. 5.19.

Consider a broad-spectrum optical source entering the input port x. For the optical signals entering port x and routed to a given output port y,

[2] This view of WDM-PON where each ONU has its own unique dedicated wavelength is what exists in the literature today, and our review will be conducted accordingly. However, this is not a view we share for the future mode of operation of WDM-PONs, in which we expect that multiple ONUs may share a wavelength or an ONU may be allocated multiple wavelengths because different ONUs may source and/or sink different amounts of traffic.

the AWG routes wavelengths which are separated by a fixed wavelength interval called the free spectral range (FSR). Therefore, considering a base wavelength λ_0, the output wavelengths at port y are λ_0, λ_{0+FSR}, λ_{0+2FSR}, and so on. For output port $y + 1$, the wavelengths routed from port x are shifted by a wavelength interval $\Delta\lambda$ compared to y. Thus, the output wavelengths at port $y+1$ are $\lambda_{0+\Delta\lambda}$, $\lambda_{0+\Delta\lambda+FSR}$, $\lambda_{0+\Delta\lambda+2FSR}$, and so on. This periodic routing property of the AWG helps immensely in scaling the network as described in Section 5.5.4.

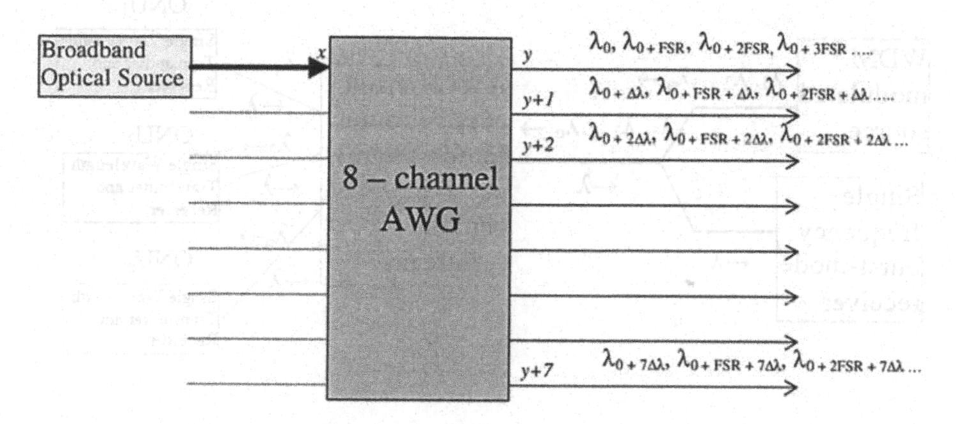

Figure 5.19 The periodic routing behavior of an AWG.

5.5.3 WDM-PON Architectures

All WDM-PON architecture proposed in the literature so far typically employ a separate wavelength channel for each ONU in the downstream direction (from OLT to ONU). However, the various architectures proposed in the literature differ in the amount of resources used in the upstream direction (from ONU to OLT). Upstream communication differs from downstream communication due to two main reasons. ONU equipment (transmitters) must be inexpensive if ONUs are to deployed in a large scale. It is preferable to not have wavelength-specific equipment at the ONU, because it is difficult to manage and maintain different kinds of inventory.

One of the earliest WDM-PON architecture proposals employed WDM in the 1550-nm band in downstream and a single upstream wavelength in the 1300-nm band shared through time-division multiple access (TDMA)

[LiSp89]. The upstream and downstream transmissions can be supported on a single fiber through coarse WDM (CWDM). This architecture has been referred to as Composite PON (CPON) in the literature [Zirn98]. A single-wavelength, burst-mode receiver is used at the OLT to receive the upstream signal. A burst-mode receiver is required to synchronize to the clock signals of different transmitting ONUs, which may be at different distances from the OLT. Figure 5.20 shows the layout of a CPON.

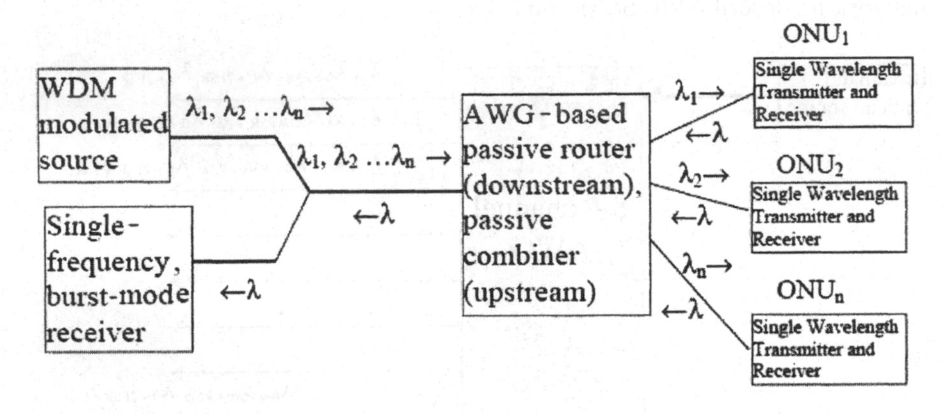

Figure 5.20 Composite PON (CPON) [Zirn98].

A limitation of the CPON architecture is that a single-frequency laser, such as a DFB laser, at the ONU may be economically prohibitive. Moreover, it may be difficult to control the wavelength changes that may arise due to temperature fluctuations at the remote (ONU) end.

The LARNET (Local Access Remote Network) architecture [ZJSD95] attempts to work around the above limitations by using a broad-spectrum source at the ONU such as an inexpensive edge-emitting LED, whose spectrum is sliced by the AWG-based router in the upstream direction. As illustrated in Section 5.5.2, when a broad-spectrum source is directed into one input port of the AWG, the various constituent wavelengths are directed to different output ports. The edge-emitting LED emits a broad spectrum of wavelengths centered around a single wavelength, as compared to DFB lasers which emit only one wavelength of light. The OLT employs a broadband burst-mode receiver (unlike a single-wavelength burst-mode receiver employed in CPON), which can receive any spectral component of the LED. TDMA is used to share the upstream channel. Figure 5.21 shows the architecture of LARNET.

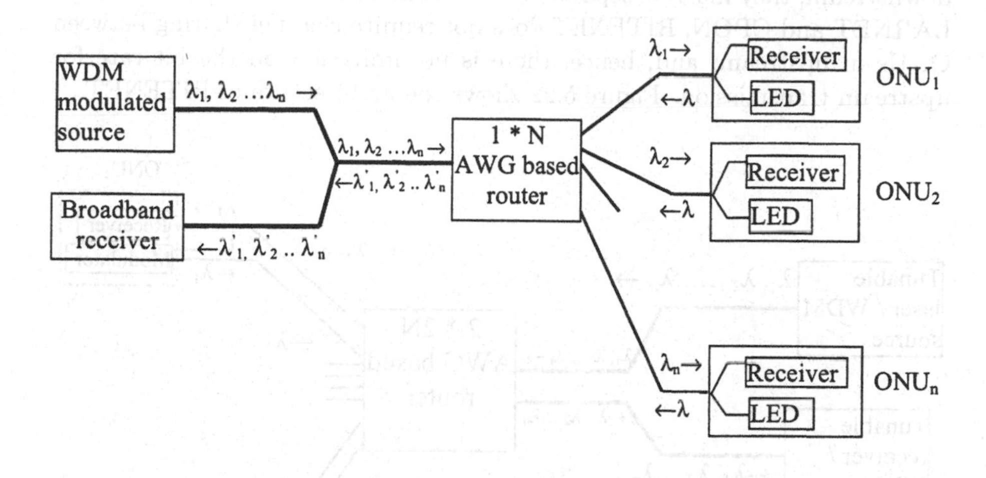

Figure 5.21 LARNET architecture [ZJSD95].

LARNET is attractive because the technology for edge-emitting LEDs is quite mature, and such devices have been commercially available for some time. Edge-emitting LEDs are much cheaper compared to DFB lasers, so they help in the economics of the ONU equipment. The limitation is that spectrally slicing a broad-spectrum source by a AWG leads to a very high power loss. Therefore, the distance from the OLT to the ONU is considerably reduced in LARNET.

Recently, variations to the LARNET architecture have been suggested in which the upstream signal from an ONU can be looped back downstream to all other ONUs from the AWG, through appropriate wiring at the AWG [Desa01] based on the periodic routing property of the AWG. The AWG is typically deployed very close to the ONU. Because the propagation delay from the ONU to the AWG is very small, a CSMA/CD MAC protocol such as Ethernet can now be used for contention resolution of upstream traffic.

The RITENET (Remote Integrated Terminal Network) architecture [FIMD94] aims at avoiding the transmitter at the ONU by modulating the downstream signal from the OLT and sending it back. The signal from the OLT is shared for downstream and upstream through time-sharing. A frame is split into two parts, one used for downstream transmission, and the other for upstream transmission. A 2 × N AWG-based router is used to route the wavelengths. Since the same optical channel is used for both upstream and downstream, they must be separated on two different fibers. However, unlike LARNET and CPON, RITENET does not require channel sharing between ONUs in upstream; and, hence, there is no limitation on the bit rate for upstream transmission. Figure 5.22 shows the architecture of RITENET.

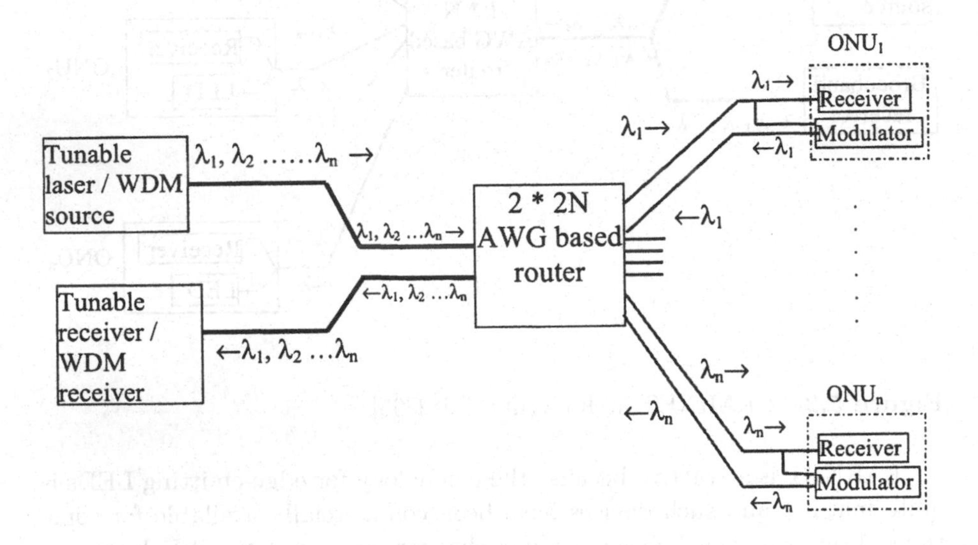

Figure 5.22 RITENET architecture [FIMD94].

While RITENET helps in reducing the end-terminal cost at the ONU, the distance from the OLT to the ONU is now much less as the signal at the OLT now has to travel double the distance. Also, since the signal is now shared between the two ends, the bit rate of the PON must be doubled. Moreover the number of fibers employed is also doubled, which doubles the cost of deployment and maintenance. A WDM receiver has to be deployed at the OLT (unlike LARNET or CPON), which increases the cost of OLT terminal equipment.

All of the above architectures employ a single multi-wavelength laser source at the OLT. Commercial products, which produce a multi-wavelength optical spectrum composed of many stable individual optical frequencies that can be locked into the standard ITU grid, are now available. The multi-wavelength source can then be modulated with independent modulators as shown in Fig. 5.23. A multi-wavelength laser source also implies greater wavelength stability in the network compared to using numerous DFB lasers in an array, because the single laser source can be very easily and efficiently controlled for temperature variations.

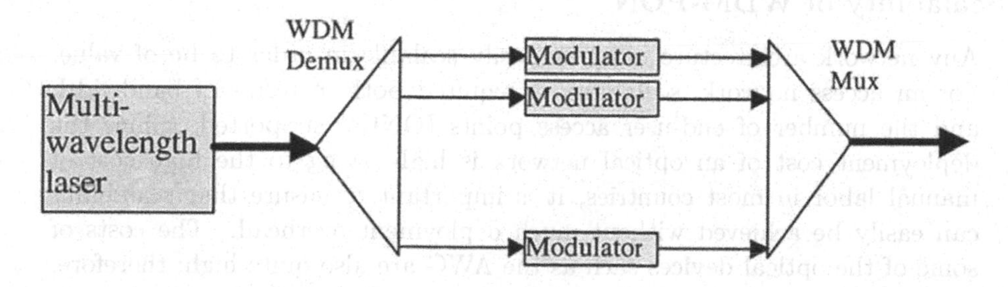

Figure 5.23 Modulating a multi-wavelength laser source.

AWG-based routers are the building blocks of many WDM-PON architectures. Integrated-optics technology has matured over the years and the number of channels supported has also scaled very well. 40-channel AWGs are available commercially today. As more of these devices are deployed, their cost is expected to reduce. Since the AWG needs outdoor deployment in an access network, thermal stability is a very important issue. Outdoor temperatures may vary between -40^oC to 85^oC. Temperature variations may cause the passbands at which the AWG operates to drift. Various measures have been suggested in the literature to improve the thermal performance of the AWG. Some proposed solutions such as keeping the AWG in a constant-temperature environment are not suited for passive networks, as the router will then have to be supplied with power. Other solutions are based on drifting the input and output wavelengths as temperature changes the refractive index, to accommodate for the passband drifts of the AWG.

Various field trials based on the above architectures have been reported in the recent literature. An experiment on an optical access network providing Gigabit Ethernet access to over 100 users was demonstrated in [KaTA03]. A variation of the RITENET architecture is used, a difference being that, instead of using one wavelength per ONU, two are used, one for upstream, the other for downstream. This eliminates the need for time-sharing. Thus, 256 wavelengths are used to support 128 users. An Optical Carrier Supply Module (OCSM) is used to generate 256 wavelengths with 25-GHz spacing.

A testbed using a variation of the LARNET architecture has been described in [OHTP02]. A variation of the CPON architecture, using optical amplifiers, called the SuperPON architecture, has been described in [VMVO00].

5.5.4 Scalability of WDM-PON

Any network architecture must be easily scalable in order to be of value. For an access network, scalability is required both in terms of bandwidth and the number of end-user access points (ONUs) supported. Since the deployment cost of an optical network is high owing to the high cost of manual labor in most countries, it is important to ensure that scalability can easily be achieved without much deployment overhead. The costs of some of the optical devices such as the AWG are also quite high; therefore, it must be ensured that such devices can be reused and they do not have to be replaced when scalability is desired. Similarly, since a large number of ONUs are deployed and such devices are typically located in end-user homes, buildings, or communities, it is desired that replacements, if needed due to scalability, should be minimum. Since all end users may not wish to upgrade to higher bandwidths at the same time, it should be ensured that legacy users can be supported while scaling the network.

The combination of all the above makes scalability in WDM-PON architectures a challenging issue. A novel solution in [MMPE00] proposes exploiting the periodic routing pattern of the AWG by deploying AWGs in series. Figure 5.24 shows how additional AWGs may be deployed to scale from a 8-wavelength, 8-ONU WDM-PON architecture to a 32-wavelength 32-ONU WDM-PON architecture. The subscript of the wavelength denotes the wavelength number while the superscript denotes the source (e.g., λ_1^2 denotes wavelength 1 from laser 2). This idea has several merits. The legacy ONUs - ONU_1 to ONU_8 - remain unaffected and continue to use wavelengths λ_1^1 to λ_4^1 and λ_1^2 to λ_4^2. The legacy 2 × 8 channel AWG is maintained and 8 new 1 × 4 channel AWGs are used to scale the network. This architecture also has the benefit of wavelength re-use. For example, in Fig. 5.24, laser 1 and laser 2 share the same wavelength domain λ_1 to λ_{16}, but cater to different ONUs.

5.6 Deployment of WDM-PONs

Various models have been proposed for the deployment of fiber in the access network. While Fiber-To-The-Home (FTTH) is the ultimate objective, Fiber-To-The-Curb (FTTC), Fiber-To-The-Building (FTTB), Fiber-

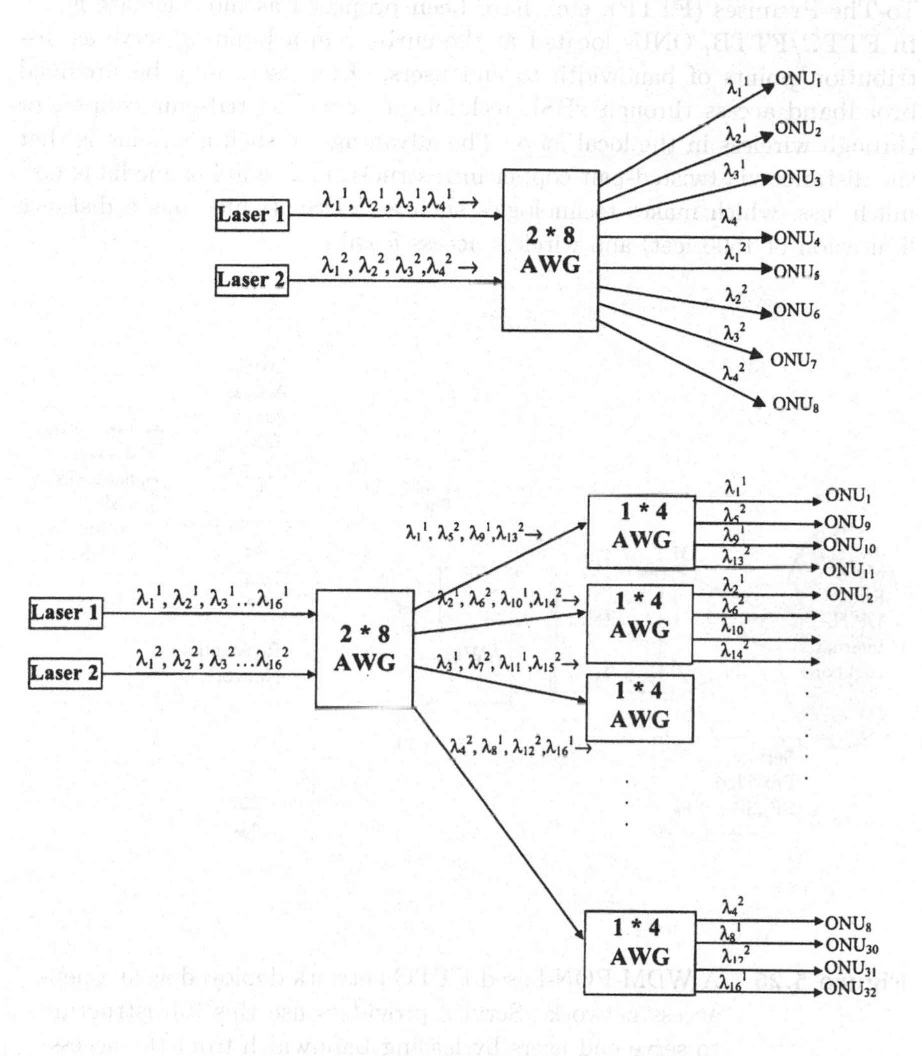

Figure 5.24 Scalability of a WDM-PON architecture [MMPE00]. A 8-ONU WDM-PON architecture is scaled to 32 ONUs. The subscript of the wavelength denotes the wavelength number while the superscript denotes the source of the wavelength.

To-The-Premises (FTTP), etc., have been proposed as intermediate goals. In FTTC/FTTB, ONUs located at the curb or in a building serve as distribution points of bandwidth to end users. End users may be provided broadband access through xDSL technologies over twisted-pair copper, or through wireless in the local loop. The advantage of such a scheme is that the distance via twisted-pair copper infrastructure, or wireless media is now much less, which makes technologies such as VDSL (which has a distance limitation of 1500 feet) and wireless access feasible.

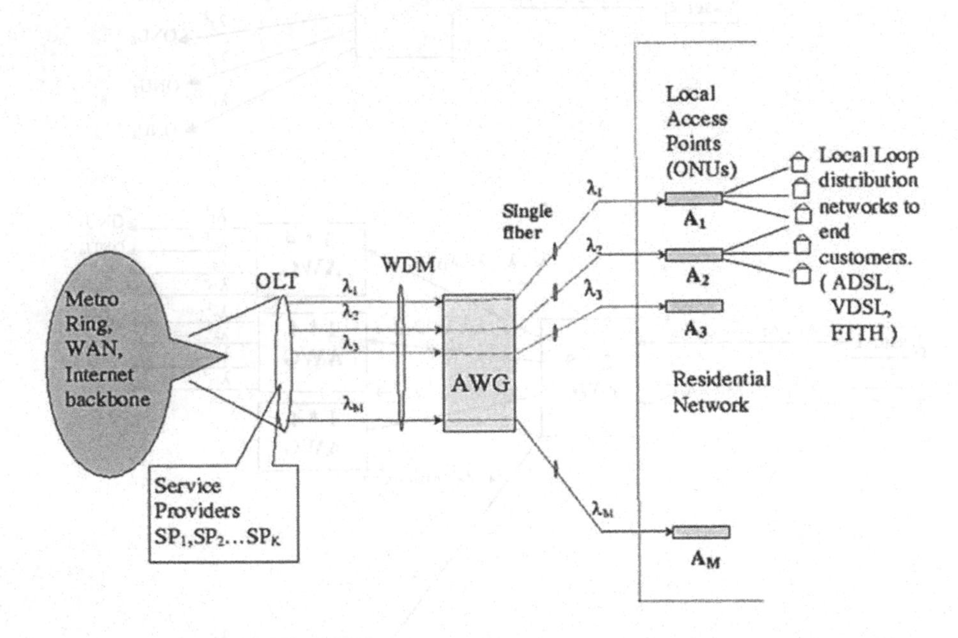

Figure 5.25 A WDM-PON-based FTTC network deployed as an open-access network. Service providers use this infrastructure to serve end users by leasing bandwidth from the access-network operator (ANO).

Thus, a FTTC network would act as a single broadband access infrastructure through which different service providers can provide many application services to the end users. The OLT may be connected to the broader Internet via a metro ring network, a wide-area LAN, or a long-haul optical network. The ONUs are defined as Local Access Points (LAPs) which act as distribution centers for bandwidth to end users (see Fig. 5.25). The access network needs to be shared because it is not feasible for every service provider to deploy its own access network because of deployment and operational costs and right-of-way issues. We call such an access network an *open access network*. This access network could be deployed and maintained by an *access-network operator (ANO)*. We describe the concept of *open access* in the next section.

5.6.1 Open Access

In the context of a broadband access network, the term *open access* implies the ability of multiple service providers (SPs) to share the deployed access-network infrastructure to make services available to the end users. Multiple services may thereby be delivered over a shared access channel. A PON deploying such a model of *open access* is shown in Fig. 5.26.

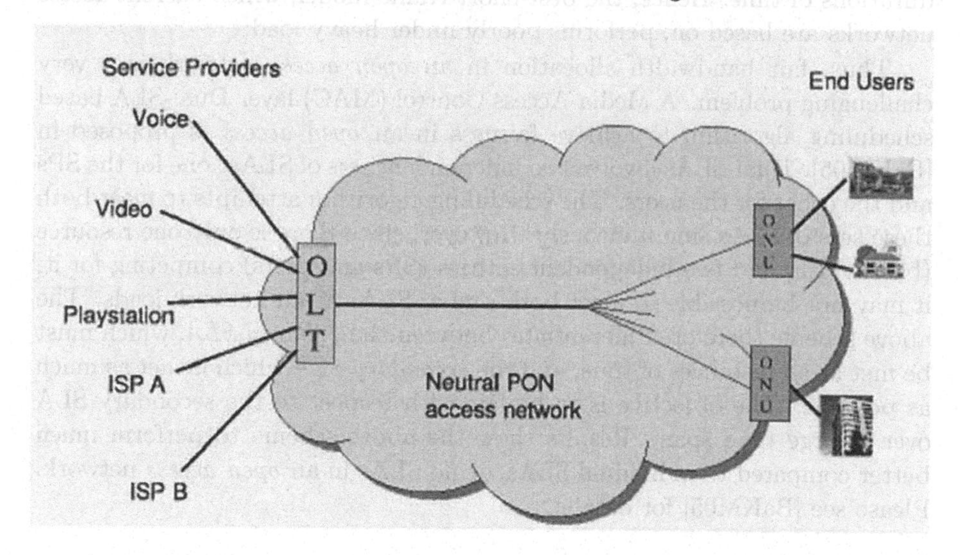

Figure 5.26 A PON deploying *open access.*

A predominant characteristic in an access network is that a single channel (e.g., a 1- Gbps access channel in EPON) is shared across multiple users. An important difference between an access network and a LAN is that an access network serves independent, non-cooperating, and bandwidth-competing users. Moreover, the propagation delay between the OLT and

ONU in EPON can be significantly larger than in a LAN. Hence, it is important to ensure some minimal degree of performance for each user, so that a bandwidth-intensive user does not adversely affect the performance of other users in the network. Similarly, if two competing SPs are providing the same category of services, e.g., IP video, then it is required that the bandwidth access to them be non-discriminatory. Most services are available on-demand; and, hence, users may avail of different services from different SPs at different intervals of time. Developing a set of fairness requirements for such a scenario, and meeting them, is an important problem. Such fairness requirements may be in terms of network throughput, delay, jitter, and other network parameters.

Moreover, traffic in an access network may be described as the multiplexing of several independent self-similar streams, which lead to very bursty traffic. The network may thus be subjected to very heavy load for brief durations of time. Hence, the best-effort traffic model, which current access networks are based on, performs poorly under heavy load.

Thus, fair bandwidth allocation in an *open access* network is a very challenging problem. A Media-Access Control (MAC) layer Dual-SLA based scheduling algorithm to achieve fairness in an *open access* is proposed in [BaKM05]. Dual-SLAs involve two independent sets of SLAs: one for the SPs and the other for the users. The scheduling algorithm attempts to meet both these sets of SLAs simultaneously. However, since there is only one resource (bandwidth) and two independent entities (SPs and users) competing for it, it may not be possible to meet both sets of SLAs at all network loads. The above scheme therefore differentiates between the *primary SLA*, which must be met at all instances of time, and the *secondary SLA* which is met as much as possible. The objective is to be fair with respect to the secondary SLA over a large time span. Results show the above scheme to perform much better compared to individual SLAs, or no SLAs in an *open access* network. Please see [BaKM05] for details.

5.7 Summary

In this chapter, we first reviewed the Ethernet PON (EPON) which is standardized in the IEEE 802.3ah. We reviewed the multi-point control protocol (MPCP) standardized for EPON. We then discussed dynamic bandwidth allocation (DBA) algorithms for EPON, and the IPACT protocol in particular. Several commercial initiatives on EPON are undergoing and products are expected as soon as standardization efforts are complete.

Then, we studied the ATM-PON (APON) (also called Broadband PON (BPON)) and the GFP-PON. We discussed the WDM-PON, which is gaining attention as an attractive solution for solving the broadband access needs in the future. WDM-PONs are still in the process of test trials by several organizations and research communities. These two technologies (EPON and WDM-PON) will clearly go a long way towards meeting future end-user broadband requirements. Finally, we discussed various aspects related to the scalability and deployment of PONs and the concept of *open access* in broadband access networks which is gaining a lot of ground.

Exercises

5.1. Let us consider the costs of deploying a point-to-point network, a curb-switched network, and a passive optical network (PON) shown in Fig. 5.2. Ten homes in a residential network wish to be connected to a central office (CO) via fiber, so as to receive a shared bandwidth connection of 1 Gbps.

Assume the following estimated costs (numbers are for example only):

(a) Cost of fiber installation: $200/meter

(b) Cost of curb switch: $15000 + $500 × number of ports

(c) Cost of passive optical splitter: $750 + $100 × number of ports

(d) Cost of a transceiver (either at CO/OLT or at ONU): $750

(e) Distance of homes from CO: 5 km

(f) Possible location of curb switch/splitter: 50 meters away from each home

Compare the costs of network deployments using each of the three approaches. Please try to justify the importance of a PON.

5.2. Assume that the PON solution was chosen after analyzing the data in Problem 5.1. Several years after the deployment, the homes wish to upgrade their connection from a 1-Gbps shared line to a 10-Gbps shared line. Let us consider the following possibilities for upgrading the connection:

(a) *Rate upgrade*: Upgrade all the transceivers from 1 Gbps to 10 Gbps.

(b) *Spatial upgrade*: Deploy nine additional fibers, so that each ONU can now have one fiber dedicated to it, thus giving it a bandwidth of 1 Gbps, and making the cumulative bandwidth 10 Gbps.

(c) *WDM upgrade*: Use WDM to support 10 channels of 1 Gbps each on the same fiber. A passive AWG routes each channel to a different ONU.

In addition to the costs in Problem 5.1, assume the following additional cost estimates:

(a) Cost of AWG router: $4000 + $300 × number of ports

(b) Cost of a 10-Gbps transceiver: $2000

(c) Cost of multiplexing 10 channels at OLT for WDM: $1000

Compare the costs for each of the three types of upgrades. Analyze the advantages and disadvantages of each approach.

5.3. Consider the IPACT protocol with the *limited service scheme*, and a maximum scheduling timeslot W_{MAX} of 2500 bytes. Draw a timing diagram for a 4-ONU EPON system, similar to the timing diagram is in Fig. 5.12.

Assume the following sequence of requests as shown in Table 5.1. Assume RTTs are constant but different for different ONUs and assume a 5-μs guard timeslot between data transmissions of different ONUs. The polling order starts with ONU 1.

Table 5.1 Sequence of requests from ONUs

ONU No.	Request 1	Request 2	Request 3	RTT
1	1800	1000	3000	100
2	1500	1200	4500	200
3	3000	3000	1000	250
4	2700	2500	2500	400

5.4. Assume that, in an EPON, the downstream delay is equal to the upstream delay. When a REPORT message arrives from ONU_i, the OLT's local time is 10. The timestamp on the REPORT message is 2. What is the current local time at ONU_i? Assume negligible processing time at the OLT.

5.5. Consider IPACT with *limited service scheme*. Consider a maximum scheduling timeslot W_{MAX} of 5000 bytes. The number of ONUs in the EPON is 16. Assume a guard time interval of 1 μs. Compute:

(a) the maximum cycle time,

(b) average throughput for each ONU, when all ONUs are active,

(c) maximum throughput when only one ONU is active, while all the other ONUs are idle,

(d) average throughput for active ONUs, when 8 ONUs are active, and the remaining 8 are idle.

5.6. Once a window is granted to an ONU, there are two approaches for transmitting packets in this window. The first approach is to transmit packets in order. This means that, if the first packet in the queue is smaller than the remaining timeslot, this packet will not be transmitted, which is also known as *head-of-line blocking*. Another approach is to use techniques such as bin-packing algorithms to ensure that as many packets may be transmitted in the allocated timeslot as possible. Contrast and compare these two approaches. Write an algorithmic description for the second approach.

5.7. In a PON, the downstream traffic is point-to-multipoint, whereas the upstream traffic is point to-point. Discuss the implications of such an architecture from a network security point of view.

5.8. Draw a transition state diagram for discovery handling, GATE message handling, and REPORT message handling, as defined in the MPCP protocol for an ONU.

5.9. Consider the PON network architecture. We would like to modify the PON architecture to build a (1 + 1) protected PON, so that if any component, namely, the OLT or ONU transceivers, fiber, or splitter fails, the PON network is not disrupted at all. Draw a suitable architecture for such a (1 + 1) protected PON.

5.10. Consider a PON which has 32 ONUs. Let the transmission power of the OLT transceiver be 0.01 mW, and the receiver be able to detect signals of at least 0.0001 mW for a reasonable bit-error rate. Assume that the splitter is 10% lossy. The optical fiber has an attenuation of 0.2 dB/km. What is the maximum possible distance of an ONU from the OLT. What is the RTT corresponding to this distance.

5.11. Consider the *near-far problem* in an EPON with four different ONUs. The distances of different ONUs from the OLT are: 1 km for ONU_1, 5 km for ONU_2, 10 km for ONU_3, and 20 km for ONU_4.

Assume that each ONU's transmission power is 0.01 mW. Also assume that the splitter is lossless, and the attenuation in the fiber is 0.2 dB/km.

Let the adjustment time for the OLT's transceiver to determine the 0-1 threshold for a signal to be linearly dependent on the difference between the power of the current signal and the previous signal, and let it be defined by the following equation:

$$t_{0-1} = 500 \times [P_{new}[\mu W] - P_{old}[\mu W]] \quad ns$$

What is the minimum value of the guard time interval, in order to allow for the adjustment time of the OLT's transceiver?

5.12. What is the *light-load penalty*. Discuss the possible solutions for avoiding the light-load penalty.

5.13. Identify whether the following statements are *true* or *false*. Justify your answer.

(a) The Ethernet PON is based on the CSMA/CD protocol.

(b) Fixed timeslot assignment (synchronous TDMA) is an efficient scheme in EPON.

(c) Dynamic slot assignment in EPON is based on a distributed (decentralized) approach.

(d) The GFP-PON uses ATM cells for the MAC frames.

Chapter 6

Optical Metro Networks

6.1 Introduction

Telecommunications networks are normally segmented in a three-tier hierarchy: access, metropolitan, and long-haul (and further delineations are also possible). Long-haul/backbone networks span inter-regional/global distances (1000 km or more) and are optimized for transmission and related costs. At the other end of the hierarchy are access networks, providing connectivity to a plethora of customers within close proximity. Straddled in the middle are metropolitan (metro) networks, averaging regions between 10–100 km and interconnecting access and long-haul networks.

Metro networks today are based on synchronous digital hierarchy (SDH)/synchronous optical network (SONET) ring architectures[1]. Namely, smaller tributary rings, e.g., OC-3/STM-1 (155 Mbps) or OC-12/STM-4 (622 Mbps), aggregate traffic onto larger core inter-office (IOF) rings that interconnect central office (CO) locations at higher bit rates, e.g., OC-48/STM-16 (2.5 Gbps)[2]. Overall, SONET/SDH has been very successful in delivering the

[1]The work in [GhPC02] serves as an excellent motivation for optical metro networks. Our introduction summarizes these motivations. We refer the interested reader to [GhPC02] for more details.

[2]Synchronous Transport Signal (STS) is the electrical-level transmission frame struc-

early wave of end-user connectivity, namely voice.

Internet data traffic growth has significantly altered the networking domains bordering the metro. Long-haul networks felt the Internet crunch first and have undergone large-scale expansions using optical wavelength division multiplexing (WDM) technology. WDM yields the best backbone cost/capacity tradeoff, and many backbone networks now have terabit capacities. Meanwhile, access networks have also seen their share of progress. Residential cable and digital subscriber line (DSL) modems have increased user access rates from kbps to Mbps, and other advanced technologies (such as PONs, see Chapter 5) promise to further this trend. Also, large corporate customers are now deploying advanced switching/routing gear, capable of direct line-rate inputs to the metro core, e.g., concatenated OC-48c/192c, and 10 Gbps Ethernet interfaces. Collectively, these increased access rates are blurring traditional access boundaries and beginning to stifle legacy "voice-centric" metro architectures. Many SONET/SDH rings are experiencing capacity exhaust at even OC-48/192 rates. Furthermore, as market expansion and deregulation take form, increased competition is forcing operators to support a highly-dynamic and an increasingly-diverse mixture of client protocols, both legacy and asynchronous data. Examples include IP (Internet Protocol), ATM (Asynchronous Transfer Mode), SONET/SDH, Ethernet (10/100 Mbps, 1.0/10 Gbps), multiplexed TDM voice (DS-n), and other more specialized data protocols such as ESCON (Enterprise System Connectivity), FICON (Fiber Connectivity Channel), Fiber Channel, cable video, etc. (see Fig. 6.1).

Clearly, new metro solutions are required to offer superior price/ performance alternatives for legacy SONET/SDH expansion, and from the above, a host of necessary features can be derived. For example, new platforms must offer high bandwidth scalability and carry multiple protocols over a common infrastructure to reduce costs. Rapid, intelligent service provisioning and survivability are also crucial, i.e., advanced service-level agreements (SLA). Moreover, given current infrastructure investments, new schemes must provide backward compatibility, i.e., legacy support, to enable more cost-effective, staged migrations. WDM technology meets many of these requirements, and is being increasingly deployed in the metro domain. WDM component technologies (such as amplifiers, switches, filters,

---

ture, while Optical Carrier (OC) is the optical level. STS-N frame is usually carried by an OC-N transmission link. Synchronous Transport Module (STM) is the counterpart of STS in SDH terminology.

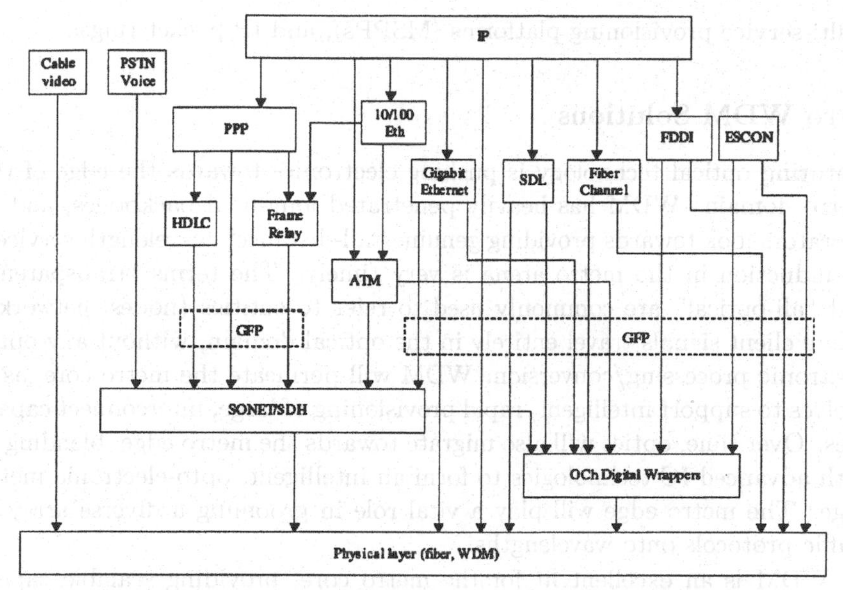

Figure 6.1 Sample protocol mappings for metro edge [GhPC02].

lasers, fibers) are enabling network-level wavelength routing and protection over fiber topologies, such as rings or meshes. These capabilities, coupled with intelligent control architectures, will allow operators to provision large amounts of capacity with an advanced degree of service definition/differentiation.

Even though WDM technology offers many benefits, its deployment within metro arena is quite complex [GhPC02]. WDM is best suited for larger metro core networks, where scalable large granularity "lambda" tributary provisioning is required, in Gbps range. At the metro edge, where operators must interface with increased protocol heterogeneities, namely, interfaces and bit rates, the need to cost-effectively handle finer "sub-wavelength" capacity increments is paramount. This contrasts sharply with long-haul networks where input signals comprise a few protocols (SONET/SDH or digital wrappers) and interface bit rates (2.5, 10, and possibly 40 Gbps). As a result, the metro edge requires integrated, intelligent opto-electronic solutions to perform multi-protocol aggregation/grooming on to larger WDM tributaries, with a particular focus on data protocol efficiency. Various "data-centric" solutions are emerging, driven by advances in high-density electronic integrated circuit (IC) technology, e.g., "next-generation" SONET/SDH, multi-service provisioning platforms (MSPPs), and IP packet rings.

6.1.1 Metro WDM Solutions

Maturing optical technology is pushing electronics towards the edge of the metro domain. WDM has heavily penetrated long-haul backbones, and as operators look towards providing genuine "all-distance" wavelength services, its induction in the metro arena is very timely. The terms "transparent" and "all-optical" are commonly used to refer to entities (nodes, networks) where client signals travel entirely in the optical domain, without any opto-electronic processing/conversion. WDM will permeate the metro core, as it evolves to support intelligent, rapid provisioning of large, interconnect capacities. Over time, optics will also migrate towards the metro edge, blending in with advanced IC technologies to form an intelligent, opto-electronic metro edge. The metro edge will play a vital role in grooming a diverse array of traffic protocols onto wavelengths.

WDM is an excellent fit for the metro core, providing scalable capacity (Tbps/fiber), reducing signal regeneration needs, and supporting format transparency.

Optical WDM Ring Networks

WDM rings have evolved from SONET/SDH concepts, essentially replacing timeslots with wavelengths and performing "optically equivalent" channel operations, i.e., add-drop, pass-through, protection, etc. However, unlike SONET/SDH, these rings offer excellent bandwidth scalability, data transparency, and multiple data rates. In particular, optical bypass eliminates the need for detailed electronic knowledge of, or access to, client signals and yields significant cost savings over traditional ADM/DXC nodes[3]. WDM ring architectures have been proposed, varying from simple static setups (i.e., fixed nodes) to advanced sharing schemes (i.e., dynamic nodes).

The basic building block of an optical WDM ring is the optical ADM (OADM) node. Fixed OADMs, which can operate on static or factory-tuned wavelengths, are usually used in static rings for static traffic. With increasing customer dynamics, static rings will become very limiting, requiring complex manual wavelength planning and yielding reduced wavelength

[3]ADM = add-drop multiplexer; DXC = digital crossconnect, which is essentially a TDM switch.

efficiencies. Here, reconfigurability is a key issue (in addition to scalability), and dynamic OADM rings are very amenable solutions, where reconfigurable OADMs (ROADMs), which is also known as dynamic wavelength ADMs (DWADMs), are the building blocks. We will explore such WDM ring network architecture in Sections 6.5 and 6.6.

6.1.2 Metro-Edge Solutions

The metro edge represents a merging between the core inter-office and the client access spaces. It is becoming increasingly evident that SONET/SDH is not the best unifying layer. Conversely, since WDM is bandwidth-inefficient for sub-gigabit line-rates, advanced electronic multiplexing technologies are needed to aggregate diverse end-user protocols onto large-granularity optical (WDM) tributaries. Dense IC technologies are finding particular favor here, helping collapse legacy multiplexing hierarchies and further blurring traditional access boundaries. Many new metro-edge solutions have been proposed, including next-generation SONET/multi-service provisioning and IP routing/packet rings.

Next-Generation SONET/Multi-Service Provisioning Paradigms

Even though legacy TDM technology has many shortcomings, it will continue to play a significant role in the convergence of data and optical networks at the metro edge. Demand for short-haul SONET/SDH gear is still high and may continue to grow for the next several years. A large part of this market comprises of larger OC-48/STM-16 and OC-192/STM-64 systems, although smaller OC-3/STM-1 and OC-12/STM-4 systems will also see increased deployments (see [GhPC02]). Moreover, many existing routers/switches have SONET/SDH interfaces, and recent efforts to define a broader generic framing protocol (GFP) (for mapping "non-standard" data protocols) may further propagate the ubiquity of such framing. In light of the above, many proposals have sought to "enhance" SONET/SDH paradigms to better suit data traffic needs. Overall, all these solutions share two main features, namely, efficient data-tributary mappings and integrated higher-layer protocol functionalities. Concurrently, these solutions also leverage ubiquitous SONET/SDH performance monitoring, protection switching, and network-management capabilities.

SONET/SDH mapping of smaller packet interfaces (10, 100 Mbps Ethernet) is usually done in "coarse" STM-1 increments and the resultant bit rate incongruencies usually yield large amounts of stranded bandwidth, e.g., 10 Mbps Ethernet allocated a full STS-1 yields 80% unused capacity. Bursty data profiles can further exacerbate bandwidth inefficiencies. By combining finer tributaries, e.g., by using virtual concatenation (VCAT), native packet interface rates can be matched much more closely, e.g., 1-Gbps Ethernet via seven STS-3c's. Multiple "matched" tributaries can then be more efficiently packed into existing standardized tributaries, and this will help collapse multiplexing (equipment) hierarchies. This topic will be discussed more in Chapter 14.

Packet-Based Solutions

Recently, the concept of "packet rings" has been proposed, aiming to combine the salient features of "TDM" with the advantages of packet switching (statistical multiplexing and finer QoS), namely, resilient packet ring (RPR, IEEE 802.17). Specifically, a new Ethernet-layer media-access control (MAC) protocol is defined to statistically multiplex multiple IP packets onto Ethernet frames. The MAC protocol itself is "media-independent", and will be capable of running over various underlying network infrastructures, including SONET/SDH, WDM, or dark fiber.

RPR designs can provide separate "layer-two" packet bypass capabilities at coarser granularities, and this will relieve packet loads at the IP (layer-three) routing level and improve QoS provisioning. Additionally, packet rings can also provide a rapid "layer-two" protection signaling protocol, designed to match the 50-ms protection-switching time of SONET/SDH.

The current RPR framework focuses on two (i.e., dual) counter-propagating "rings" which can both carry working traffic, i.e., no reserved protection bandwidth for bandwidth efficiency. All control messages are carried "in-stream", making the control strictly in-band. Additionally, (layer-two) destination stripping is performed for unicast flows unlike earlier source-stripping FDDI rings, thereby permitting spatial re-use of bandwidth (note that multicast and broadcast still require source stripping, however). Collectively, the above features significantly improve ring capacity utilization/throughput. Currently, a spatial re-use protocol (SRP) framework has been tabled for standardization and is commercially available (amongst others), aiming to provide all RPR features, e.g., protection switching, topology discovery, bandwidth fairness, etc.

The rest of the chapter is organized as follows. Section 6.2 presents a research overview of traffic grooming in SONET/WDM ring networks. Section 6.3 investigates the traffic-grooming problem in SONET ring networks. Interconnected ring networks are explored in Section 6.4. Section 6.5 studies packet communications in WDM ring networks by using ROADMs, i.e., tunable (or reconfigurable) wavelength add-drop multiplexers. Section 6.6 investigates online connection provisioning using ROADMs. Section 6.7 summerizes this chapter.

6.2 Overview of Traffic Grooming in SONET Ring Networks

6.2.1 Node Architecture

SONET/SDH ring, hereafter referred to as SONET ring only, is the most widely used optical network infrastructure today. In a SONET ring network, WDM is mainly used as a point-to-point transmission technology. Each wavelength in such a SONET/WDM network is operated at OC-N line rate, e.g., $N = 192$. The SONET system's hierarchical TDM schemes allow a high-speed OC-N channel to carry multiple OC-M channels (where M is smaller than or equal to N). The ratio of N and the smallest value of M carried by the network is called "*grooming ratio*". Electronic add-drop multiplexers (ADMs) are used to add/drop traffic at intermediate nodes to/from the high-speed channels.

In a traditional SONET ring network, one ADM is needed for each wavelength at every node to perform traffic add/drop on that particular wavelength. With the progress of WDM, over a hundred wavelengths can now be supported simultaneously by a single fiber. It is, therefore, too costly to put the same amount of ADMs (each of which has a significant cost) at every network node since a lot of traffic is only bypassing an intermediate node. With the emerging optical components such as optical add-drop multiplexers (OADM) (also referred to as wavelength add-drop multiplexers (WADM)), it is possible for a node to bypass most of the wavelength channels optically and only drop the wavelengths carrying the traffic destined to the node. Figure 6.2 shows the architecture of a typical node in a SONET/WDM ring network. For some wavelengths (λ_1 in this example), since there is no need to add or drop any of its timeslots, they can be optically bypassed at the node. For other wavelengths (λ_2 and λ_3) where at least one timeslot needs to be added or dropped, an electronic ADM is used.

Compared with the wavelength channel resource, ADMs form the domi-

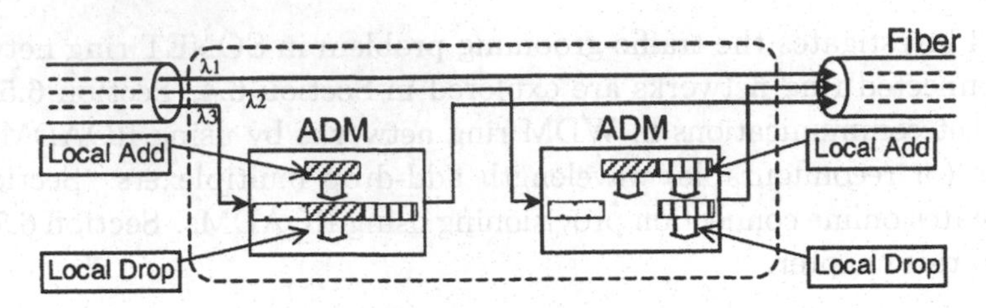

Figure 6.2 Architecture of a WDM ring network node.

nant cost in a SONET/WDM ring network. Hence, carefully arranging these optical bypasses can reduce a large amount of the network cost. It is clear that using OADMs can decrease the number of SONET ADMs needed in the network and eventually bring down the network cost. Then, the problems are, for a given low-speed set of traffic demands, which low-speed demands should be groomed together, which wavelengths should be used to carry the traffic, which wavelengths should be dropped at a local node, and how many ADMs are needed at a particular node?

6.2.2 Single-Hop Grooming in SONET/WDM Ring

As shown in Fig. 6.2, (1) ADMs do not have the timeslot interchange function, and (2) wavelength conversion is not possible without additional equipment, so there are timeslot-continuity and wavelength-continuity constraints at nodes where only ADMs are used. We refer to these rings that are built only with ADMs as single-hop rings since all the connections are "direct" connections. Low-speed OC-M connections are groomed on to OC-N wavelength channels. Based on this network model, for a given traffic matrix, satisfying all the traffic demands as well as minimizing the total number of ADMs is a network design/optimization problem and has been studied extensively in the literature.

Figure 6.3 shows a five-node network with uniform traffic requests. In this example, we assume a bi-directional ring with grooming ratio 2 (which may not be practical in a real SONET ring but it is used here for illustration purposes). The total number of bi-directional requests is 10 and each request is for 1 unit of sub-channel capacity. Figure 6.3(a) illustrates all the 10 requests. Figures 6.3(b) and 6.3(c) illustrate two ways of organizing the connections on two wavelengths (one and half are actually in use). In Fig. 6.3(b), there is only bypassing traffic at node 2 of the second ring (dashed line) so nine ADMs are needed to carry all the requests. The num-

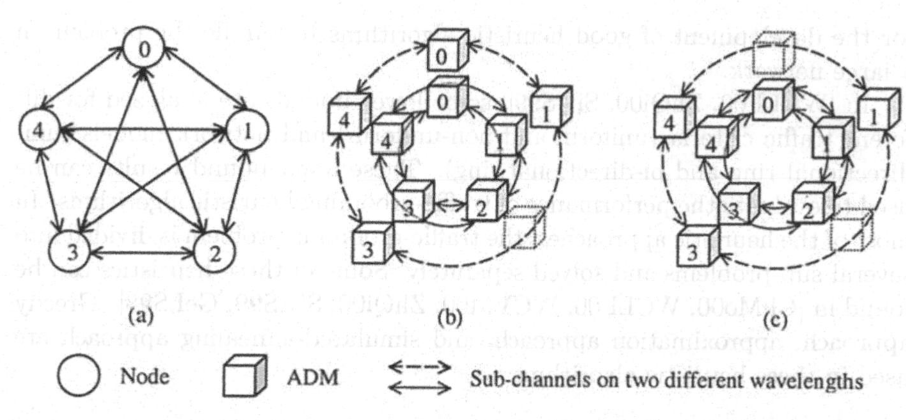

Figure 6.3 An example of traffic grooming on a 2-wavelength network showing that different strategies can lead to different network cost. Total number of traffic connections is 10 and each request has capacity of one unit. (a) Ten connection requests. (b) One way of putting the 10 requests on two wavelengths. Nine ADMs are needed. (c) Another way of putting the 10 requests on two wavelengths. Eight ADMs are needed.

ber of ADMs can be further reduced by simply reconfiguring the rings as shown in Fig. 6.3(c), where the positions of the connections between nodes 1 and 4 on the second ring ($1 \leftrightarrow 4$) and part of the first ring ($4 \leftrightarrow 0$, $0 \leftrightarrow 1$) are swapped.

It has been proven in [ChMo00, WCLF00] that the general traffic-grooming problem is *NP*-Complete. The authors in [WCVM00] formulate the optimization problem as an integer linear program (ILP). When the network size is small, some commercial software can be used to solve the ILP equations to obtain an optimal solution. The formulation in [WCVM00] can be applied to both uniform and non-uniform traffic demands, as well as to unidirectional and bi-directional SONET/WDM ring networks. The limitation of the ILP approach is that the numbers of variables and equations increase explosively as the size of network increases. The computation complexity makes it hard to be useful on networks with practical size. By relaxing some of the constraints in the ILP formulation, it may be possible to get some results, which are close to the optimal solution for reasonable-size networks. The results from the ILP may give some insights and intuition for the development of good heuristic algorithms to handle the problem in a large network.

In [WCLF00, ZhQi00, SiGS99], some lower bounds are analyzed for different traffic criteria (uniform and non-uniform) and network models (unidirectional ring and bi-directional ring). These lower-bound results can be used to evaluate the performance of traffic-grooming heuristic algorithms. In most of the heuristic approaches, the traffic-grooming problem is divided into several sub-problems and solved separately. Some of these heuristics can be found in [ChMo00, WCLF00, WCVM00, ZhQi00, SiGS99, GeLS98]. Greedy approach, approximation approach, and simulated-annealing approach are used in these heuristic algorithms.

6.2.3 Multi-Hop Grooming in SONET/WDM Ring

In single-hop (a single-lightpath hop) grooming, traffic cannot be switched between different wavelengths.

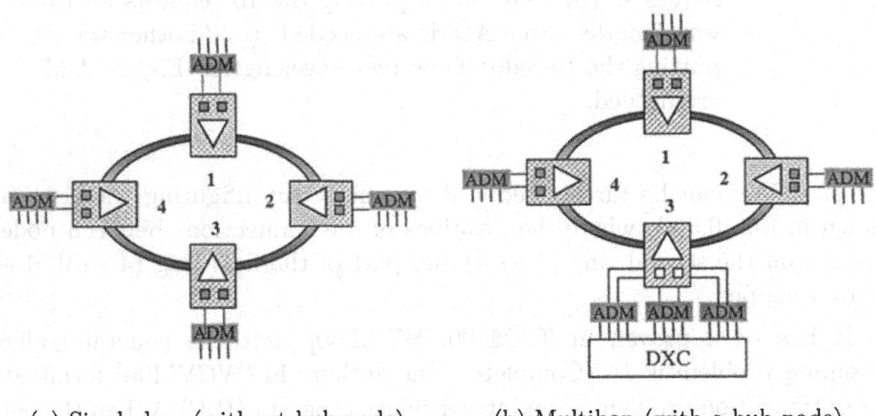

(a) Single-hop (without hub node) (b) Multihop (with a hub node)

Figure 6.4 SONET/WDM ring with/without a hub node.

Figure 6.4(a) shows this kind of a network configuration. Another network architecture has been proposed in [SiGS99, GeRS00], in which there are some nodes equipped with Digital Crossconnects (DXCs). In Fig. 6.4(b), node 3 has a DXC installed. This kind of node is called a hub node. Traffic from one wavelength/timeslot can be switched to any other wavelength/timeslot at the hub node. Because the traffic needs to be converted from optical to electronic at the hub node when wavelength/timeslot exchange occurs, this grooming approach is called multihop (multi-lightpath hops) grooming. Depending on the implementation, there can be a single hub node or multiple hub nodes in the network. A special case is that every node is a hub node, i.e., there is a DXC at every node. This kind of network is called point-to-point WDM ring network (PPWDM ring) [GeRS00].

The work in [GeRS00] provides some excellent theoretical analysis on comparing network cost of PPWDM ring, a SONET/WDM ring without hub node, a SONET/WDM ring with one or multiple hub nodes, etc. The authors of [WCVM00] have compared the single-hop grooming with multihop grooming (with one hub node) network performance using simulation. The results indicate that, when the grooming ratio is large, the multihop approach tends to use fewer ADMs, but when the grooming ratio is small, the single-hop approach tends to use fewer ADMs, and in general, the multihop approach uses more wavelengths than the single-hop approach.

Table 6.1 gives a brief summary of previous work reported in this area. The various terms used in the table are described below:

- Static or dynamic traffic: The traffic pattern will not or will change with time, respectively.
- Uniform (non-uniform) traffic: Traffic demands between any node pair are the same (not the same).
- Single-hop ring: No virtual connection on the ring will be terminated electronically at any intermediate node.
- Multihop ring: Some or all of the virtual connections on the ring may be regenerated electronically at some intermediate nodes.
- Unidirectional (bidirectional) ring: All the traffic on the ring can go along one (both) direction.
- Circle: The virtual ring that is established on one timeslot of a wavelength. It has the capacity of one unit.
- PPWDM: Point-to-point (opaque) WDM ring, in which signals get crossconnected and regenerated at every node.
- Distance-dependent traffic (used in [SiGS98]): The nodes farthest apart exchange one unit of traffic, and the inter-node traffic demand increases by one unit as the inter-node distance decreases by one link.

Table 6.1 Comparison of previous work on traffic grooming on SONET/WDM ring networks.

	Traffic assumptions	Ring architecture	Main result
Gerstel et al. [GeLS98] [GeRS00]	• Static, uniform • Non-statistic, dynamic, and fixed wavelength	• PPWDM (multihop) • Fully-optical ring (single-hop) • Single-hub (multihop) • Double-hub (multihop) • Hierarchical ring (multihop) • Incremental ring (multihop)	• First paper that tries to minimize transceiver cost. • Study of dynamic traffic and fixed lightpath. • Different architectures are compared.
Chiu & Modiano [ChMo00]	• Egress traffic • Static, uniform	• Unidirectional ring with egress node (single-hop) • Bidirectional ring (single-hop)	• Proof of NP-completeness. • Optimal solution for uniform traffic on egress ring. • Lower bound on uniform all-to-all traffic.
Simmons et al. [SiGS98] [SiGS99]	• Static, uniform • Static, distance-dependent traffic	• Bidirectional ring with odd number of nodes (single-hop)	• How to group timeslots. • Maximal savings for some special cases. • Super node model for distance-dependent traffic.
Zhang & Qiao [ZhQi00]	• Static, uniform traffic • Static, non-uniform traffic	• Unidirectional and bidirectional ring (single-hop)	• Greedy heuristic for grooming arbitrary traffic. • Heuristic for circle construction for non-uniform traffic.
Wang et al. [WCVM00]	• Static, uniform traffic • Static, non-uniform traffic	• Unidirectional and bidirectional ring (single-hop) • Single hub (multihop)	• Formal mathematical problem specification (ILP). • Simulated-annealing-based heuristic for traffic grooming • Greedy heuristic for single-hop and multihop grooming.

- Egress node: A special node, like the central office of an access network, on the ring where all the traffic terminates at or originates from.

6.2.4 Dynamic Grooming in SONET/WDM Ring

Instead of using a single static traffic matrix to characterize the traffic requirement, it is also possible to describe it by a set of traffic matrices. The traffic pattern may change within this matrix set over a period of time, say throughout a day or a month. The network needs to be reconfigured when the traffic pattern transits from one matrix to another matrix in the matrix set. The network design problem for supporting any traffic matrix in the matrix set (in a non-blocking manner) as well as minimizing the overall cost is known as a dynamic-grooming problem in a SONET/WDM ring [BeMo00].

The dynamic-grooming problem proposed in [BeMo00] is like a network design problem with reconfiguration consideration. The authors of [BeMo00] have formulated the general dynamic-grooming problem in a SONET/WDM ring as a bipartite graph-matching problem and provided several methods to reduce the number of ADMs. A particular traffic matrix set is then considered and the lower bound on the number of ADMs is derived. They also provide the necessary and sufficient conditions so that a network can support such a traffic pattern. This kind of traffic matrix set is called a t-allowable traffic pattern. For a given traffic matrix, if each node can source at most t duplex circuits, we call this traffic matrix a t-allowable traffic matrix. The traffic matrix set, which only consists of t-allowable traffic matrices, is called a t-allowable matrix set or a t-allowable traffic pattern. We use an example from [MoLi01] to illustrate dynamic traffic grooming for a t-allowable traffic pattern in a SONET/WDM ring.

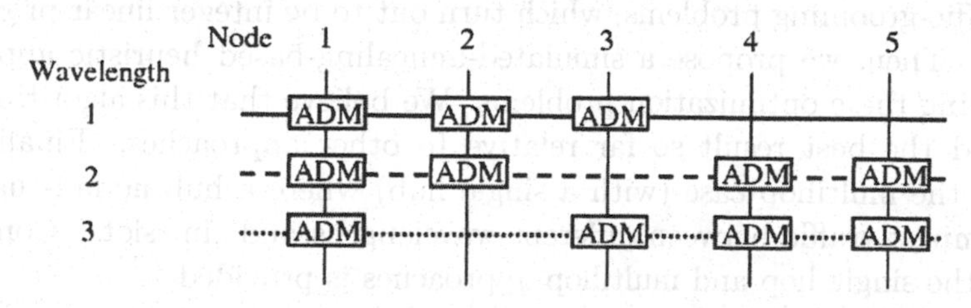

Figure 6.5 Network design for 2-allowable traffic.

Figure 6.5 shows a 5-node SONET/WDM ring network. Three wavelengths are supported in the network. Assume that each wavelength can support 2 low-speed circuits. The network configuration in Fig. 6.5 is a

2-allowable configuration, i.e., it can support any 2-allowable traffic matrix (set). For instance, consider a traffic matrix with request streams 1–2, 1–3, 2–3, 2–4, 3–4, 4–5, and 4–5. The traffic matrix can be supported by assigning 1–3, 2–3 on wavelength 1, assigning 1–2, 2–4, 4–5, 4–5 on wavelength 2, and assigning 3–4 on wavelength 3. Note that, for a particular traffic matrix, there may be some redundant ADMs in the configuration. However, the configuration is able to support other potential t-allowable traffic matrices. Designing such configurations to support any t-allowable traffic matrix while minimizing the network cost is a very interesting problem with some practical utility. The authors in [BeMo00] provide an excellent analysis on t-allowable traffic pattern. The study of dynamic-traffic grooming in a SONET/WDM ring with other generic traffic pattern can potentially be challenging research.

6.2.5 Grooming in Interconnected SONET/WDM Rings

Most traffic-grooming studies in SONET/WDM ring networks have assumed a single-ring network topology. The authors of [WaMu02] have extended the problem to an interconnected-ring topology. Today's backbone networks are mainly constructed as a network of interconnected rings. Extending the traffic-grooming study from a single-ring topology to the interconnected-ring topology will be very useful for a network operator to design their network and to engineer the network traffic. We will explore this topic in Section 6.4.

6.3 Traffic Grooming in WDM Ring Networks

In this section, we first provide formal mathematical definitions for the various traffic-grooming problems, which turn out to be integer linear programs (ILPs). Then, we propose a simulated-annealing-based heuristic approach for solving these optimization problems. We believe that this algorithm has achieved the best result so far relative to other approaches. Finally, we discuss the multihop case (with a single hub) where a hub node is used to crossconnect traffic between different wavelengths and timeslots. Comparison of the single-hop and multihop approaches is provided.

6.3.1 Problem Definition

Here are the common notations that are used throughout this study.

N: Number of nodes in the network.

W: Number of wavelengths in the network (each wavelength can transmit several circles in time-division fashion).

C: Grooming ratio, which is the number of circles a wavelength can carry.

$T(t_{ij})$: Non-uniform traffic matrix, in which t_{ij} represents the traffic from node i to j. The traffic matrix is given, and these traffic elements need to be carried by the ring at minimum cost.

${}^{d}V_{ij}^{cw}$: Virtual connection from node i to node j on circle c, wavelength w. d represents the direction of a connection and it can be either clockwise or counter-clockwise.

O_i, I_i: In the multihop case, O_i represents the virtual connection that starts from node i and terminates at the hub node. Similarly, I_i represents the virtual connection that starts from the hub node and terminates at node i.

ADM_i^w: Number of ADMs at node i on wavelength w.

e: A link on the physical ring.

Formulation 6.1: Single-Hop Connections

The traffic-grooming problem on a single-hop bi-directional ring can be mathematically formulated as shown in Formulation 6.1.

In Formulation 6.1, ${}^{d}V_{ij}^{cw}$ represents whether there is a virtual connection (one unit of capacity) from node i to node j along direction d (which can be clockwise or counter clockwise) on circle c and wavelength w. The t_{ij} represents the total amount of traffic from node i to node j. ADM_i^w represents whether there is an ADM on wavelength w at node i. Both ${}^{d}V_{ij}^{cw}$ and ADM_i^w can be either 0 or 1.

The traffic-load constraint simply states that the number of links from node i to node j on all circles is equal to the traffic specified in the traffic matrix. In the channel-capacity constraint, ${}^{d}e^{cw}$ represents a d-directional link on wavelength w and sub-channel c. If the virtual connection ${}^{d}V_{ij}^{cw}$ uses it, we say that ${}^{d}e^{cw} \in {}^{d}V_{ij}^{cw}$. The channel-capacity constraint requires that a circle carry only one connection on any given link. The last two constraints specify that the number of connections that start and terminate at a node of a ring is bounded by the capacity of the electronic ADM at that node.

- Objective function:

$$Minimize \quad \sum_{i}\sum_{w} ADM_i^w$$

- Subject to:

$$\sum_{w}\sum_{c}\sum_{d} {}^dV_{ij}^{cw} = t_{ij} \qquad \forall i,j \qquad \text{(Traffic-load constraint)}$$

$$\sum_{{}^de^{cw} \in {}^dV_{ij}^{cw}} {}^dV_{ij}^{cw} \leq 1 \qquad \forall d,e,c,w \quad \text{(Channel-capacity constraints)}$$

$$\sum_{c}\sum_{j} {}^dV_{ij}^{cw} \leq C \cdot ADM_i^w \qquad \forall d,i,w \quad \text{(Transmitter constraint)}$$

$$\sum_{c}\sum_{i} {}^dV_{ij}^{cw} \leq C \cdot ADM_j^w \qquad \forall d,j,w \quad \text{(Receiver constraint)}$$

- Bounds:

$${}^dV_{ij}^{cw} \text{ and } ADM_i^w \text{ are both binary numbers.}$$

Formulation 6.1: Mathematical problem formulation for traffic grooming in a single-hop network.

If there is an ADM, at most C connections can start and terminate there; otherwise, no add/drop can occur.

The unidirectional-ring case can be viewed as a special case of the above formulation, where d can only be either clockwise or counter-clockwise. Sometimes, shortest-path routing is required in the bidirectional ring. This requirement can also be accommodated in the above formulation by specifying that ${}^dV_{ij}^{cw}$ will be zero if the distance from node i to j in the specified direction exceeds $\lceil N/2 \rceil$.

Formulation 6.2: Multihop Method

A multihop ring uses a DXC to do sub-channel consolidation or segregation at a hub node. In this study, we simplify the cost calculation by using the following equivalent architecture instead. The hub node has as many ADMs as there are wavelengths. All connections that are passing through it can be terminated and switched to any wavelength and timeslot. A connection can go through the hub node at most once. The formulation for the unidirectional, single-hub ring case is shown in Formulation 6.2. Bi-directional formulation for the single-hub-based multihop network is also available. We omit it here since it is a straightforward extension of Formulation 6.2.

- Objective function:

$$Minimize \sum_i \sum_w ADM_i^W$$

- Subject to:

$$\sum_w \sum_c \Big(\sum_{j(j>i)} V_{ij}^{cw} + O_i^{cw} \Big) = \sum_j t_{ij} \quad \forall i \qquad \text{(Traffic-load constraint)}$$

$$\sum_w \sum_c \Big(\sum_{i(j>i)} V_{ij}^{cw} + I_j^{cw} \Big) = \sum_i t_{ij} \quad \forall j \qquad \text{(Traffic-load constraint)}$$

$$\sum_{e^{cw} \in V_{ij}^{cw}} V_{ij}^{cw} + \sum_{i<e^{cw}} O_i^{cw} + \sum_{j>e^{cw}} I_j^{cw} \leq 1 \qquad \forall e, c, w$$

(Channel-capacity constraints)

$$\sum_c \sum_j V_{ij}^{cw} + \sum_c O_i^{cw} \leq C \cdot ADM_i^w \quad \forall i, w \ \text{(Transmitter constraint)}$$

$$\sum_c \sum_i V_{ij}^{cw} + \sum_c I_i^{cw} \leq C \cdot ADM_j^w \quad \forall j, w \ \text{(Receiver constraint)}$$

- Bounds:

$$V_{ij}^{cw},\ O_i^{cw},\ I_i^{cw} \text{ and } ADM_i^w \text{ are binary numbers.}$$

Formulation 6.2: Mathematical problem formulation for traffic grooming in a single-hub multihop network.

In Formulation 6.2, O_i^{cw} represents the virtual connection from node i to the hub node on circle c, wavelength w. Similarly, the notation I_i^{cw}

represents the connection that starts from the hub node and terminates at node i. The condition $(j > i)$ means the virtual connection that starts from node i and ends at node j without going through the hub node. The condition $i < e^{cw}$ means the following: if s is the start node of link e^{cw} and t is the end node, then i is at upstream of t. Similarly, $j > e^{cw}$ means that j is at downstream of s. The traffic-load constraint needs to be broken into two parts for the multihop case. The first part in the multihop formulation specifies the following fact: any virtual connection that starts from node i will either terminate before it reaches the hub node or will terminate at the hub node. Similarly, the second traffic-load constraint states that any virtual connection that terminates at node i is either coming from the hub node or from some node downstream of the hub node. The explanations of the other constraints are the same as in the single-hop case.

6.3.2 Solving the ILPs Directly

We attempted to solve the ILPs directly under the current restriction that a SONET ring has a maximal size of 16 nodes. Uniform traffic on single-hop, unidirectional ring was initially chosen to understand how fast the ILPs can be solved although the formulation is capable of solving non-uniform traffic cases (examples of which will be given in Section 6.3.4). A commercially-available ILP solver (CPLEX) was used. Table 6.2 shows the computation time and the optimal solution. For small networks, i.e., 6 nodes or less, the ILP solver can find the optimal solution in a reasonably short time of a few seconds to a few hours. When the network size grows beyond 6 nodes, the solver was found to take more than 6 hours to discover the optimal solution for some cases (shown as question marks in Table 6.2). The results shown in Table 6.2 were obtained from a HP-Visualize B132l machine running UNIX operating system. For the cases N=4, 5 and C=12 (shaded cells), no grooming is needed since one wavelength can carry all the traffic. When the network size is larger than 7 or 8 nodes, we need to turn to heuristics. Table 6.3 provides some additional information. When we give the ILP solver a time limit of half-hour for each problem, it usually fails to find even one feasible solution when the network size is larger than 8 nodes. Notice that, when the grooming ratio is large enough so that one wavelength is enough to carry all the traffic, traffic grooming does not have practical meaning any more since an ADM is needed at each node. Also, in Table 6.3, note that the simulated-annealing algorithm reaches better results than the greedy algorithm, even reaching the lower bound sometimes.

Question marks represent the cases where the ILP solver could not find a feasible solution within a half-hour time limit. Shaded data means that all traffic can be carried on one wavelength, so no "traffic grooming" is necessary. Lightly-shaded data indicates the only situation where Greedy was found to outperform simulated annealing in our experiments.

Table 6.2 Computation time (in seconds) and optimal solution for the single-hop case.

	C=3		C=4		C=12	
	$T(s)$	N_{ADM}	$T(s)$	N_{ADM}	$T(s)$	N_{ADM}
N=4	1	7	2	7	0	4
N=5	471	12	45	10	0	5
N=6	?	?	9783	15	191	9
N=7	?	?	?	?	549	12

6.3.3 Heuristics

The general traffic-grooming problem is NP-complete [ChMo00]. Solving the ILP directly is not practical even for moderate-size networks because of the long solution time. Although a simple greedy heuristic has been provided in [ZhQi00], we propose alternative approaches to discover improved results. Specifically, we propose a simulated-annealing algorithm for single-hop connections and a simple heuristic for multihop connections.

Simulated Annealing for Single-Hop Connections

To reduce its computational complexity, the traffic-grooming problem is usually divided into two components (see [SiGS98, SiGS99, ZhQi00]). In the first step, the traffic demands are assigned to "circles". In the second step, a traffic-grooming algorithm is employed to reorganize the circles or connections on wavelengths. We adopted the same strategy and our first-step heuristic is built upon the wavelength-assignment algorithm proposed in [ZhQi00] for reducing the wavelength consumption. An important practical concern is whether the heuristic is a good foundation for the second-step heuristic. It has been shown in [GeLS98] that the number of wavelengths and the number of ADMs cannot always be minimized simultaneously. Both the

Table 6.3 Results from different approaches to solve the traffic-grooming problem in an unidirectional ring with uniform traffic and single-hop connections. LB: lower bound; SA: simulated annealing [WCVM00]; GD: greedy algorithm [ZhQi00].

		N=4	N=5	N=6	N=7	N=8	N=9	N=10	N=11	N=12	N=13	N=14	N=15	N=16
C=3	LB	6	11	15	21	29	36	45	56	66	78	92	105	120
	ILP	7	12	18	28	46	?	?	?	?	?	?	?	?
	SA	7	12	17	21	31	36	48	57	69	78	95	105	124
	GD	7	12	17	26	35	44	56	67	81	98	113	131	152
C=4	LB	6	10	15	21	28	36	45	55	66	78	91	105	120
	ILP	7	10	15	27	31	?	?	?	?	?	?	?	?
	SA	7	10	15	21	28	36	45	55	66	78	91	105	120
	GD	7	11	17	23	30	37	48	60	72	84	99	112	130
C=12	LB	4	5	9	11	15	18	23	28	33	39	46	53	60
	ILP	4	5	9	13	18	?	?	?	?	?	?	?	?
	SA	4	5	9	12	16	18	24	30	36	39	49	57	64
	GD	4	5	10	13	18	19	27	33	41	49	56	69	73
C=16	LB	4	5	6	10	12	16	18	23	28	33	37	42	48
	ILP	4	5	6	11	14	18	?	?	?	?	?	?	?
	SA	4	5	6	11	14	18	20	26	33	37	42	46	57
	GD	4	5	6	13	15	20	21	30	35	42	48	58	65
C=48	LB	4	5	6	7	8	9	10	15	17	19	21	26	29
	ILP	4	5	6	7	8	9	10	?	?	?	?	?	?
	SA	4	5	6	7	8	9	10	16	19	22	24	31	37
	GD	4	5	6	7	8	9	10	19	23	24	25	31	34
C=64	LB	4	5	6	7	8	9	10	11	15	18	20	22	24
	ILP	4	5	6	7	8	9	10	11	?	?	?	?	?
	SA	4	5	6	7	8	9	10	11	15	19	22	25	28
	GD	4	5	6	7	8	9	10	11	15	25	26	27	28

ILPs (discussed in Section 6.3.1) and heuristic (to be discussed in this section) have the potential to find the true optimal (number of ADMs) when there are more than enough wavelengths. Furthermore, the emphasis of this study is on comparing different second-step heuristics. Our second-step heuristic randomly chooses some virtual connections in the network and changes their position, by using the simulated-annealing technique to help accelerate the process of branch-and-bound to find a good solution. The implementation of simulated annealing follows the Monte Carlo method [Laar88] referred to as the *Metropolis algorithm*. The algorithm is depicted in Algorithm 6.1.

Algorithm 6.1 Simulated-annealing for single-hop traffic grooming

repeat {iterate around all states}
  repeat {accept $ANN\_CONST \times C$ times}
    $dcost \Leftarrow perturb()$ {randomly pick two circles, swap part or all of them}
    if $\Delta cost < 0$ or ($\Delta cost > 0$ and $exp(-\Delta cost/control) > rand[0,1)$) **then**
      $accept\_change()$ {accepted the change}
      $chain \Leftarrow chain + 1$
    else
      $reject\_change()$
    end if
  until $chain < ANN\_CONST \times C$
  $control \Leftarrow control \times DEC\_CONST$
until $control > END$

In Algorithm 6.1, "perturb" means to randomly pick two circles on different wavelengths, swapping part of them or the whole circles (illustrated in the comment of the pseudo-code). The chance for doing partial swapping is selected to be very small. Whenever the criterion for doing partial swapping is satisfied, we go ahead to check if there are segments in the two circles that can be swapped (if swapped, they do not break any existing connections). If the result is no, we simply swap the whole circles. The consequence of the perturbation may or may not be helpful to bring down the ADM usage. If the perturbation helps, we will accept the perturbation; otherwise, we will calculate $exp(-\Delta cost/control)$ and compare it with a random number. The perturbation will still be accepted if the random number is smaller. After repeating the above process for a certain number of times, we consider that the system has reached the equilibrium state, and then we go on to decrease the control variable (the temperature). The process will be terminated when the control variable satisfies some predefined criterion.

In the above implementation, as the computation progresses and the control variable goes down, the chance of having a "good perturbation" and the chance of accepting a "bad perturbation" will decrease, so the time spent in lower temperature is much longer than in higher temperature. The constants ANN_CONST and DEC_CONST are critical for the algorithm's performance. ANN_CONST decides how long we have to wait before we consider that the system has reached its equilibrium. DEC_CONST decides how fast we decrease the temperature. After experimenting with these parameter values, we finally adopted ANN_CONST to be between 4 and 20 depending on the size of the search space, and DEC_CONST = 0.95 for best results for our numerical examples. The start temperature is found to be important for some cases. We will not recommend very high starting temperature. The circle-assignment heuristic in the first step serves as a pre-grooming heuristic, which puts similar connections as close to one another as possible; thus, a very high starting temperature can counter-balance the effort of the first step and prolong the convergence process. Some special techniques are also needed when implementing the perturb() function, details of which are skipped here. These techniques are mainly used to increase the convergence speed of the algorithm.

Although the simulated-annealing-based heuristic adopted the two-step strategy, it is intrinsically different from the greedy heuristic [ZhQi00] in that it always has the potential to find the true optimal. For the greedy heuristic, the second step can only regroup the circles but not change the existing circles. This is one of the major reasons for getting a sub-optimal solution in [ZhQi00]. The simulated-annealing-based, second-step heuristic provides a chance to change the circles so that it can jump out of "traps."

Heuristic for the Multihop Method

A greedy heuristic is proposed for the multihop method [ChWM01]. It puts all the traffic on circles sequentially, and then it applies Algorithm 6.2.

In Algorithm 6.2, the wavelength-combining function checks two wavelengths link by link. If the total load on any link will not exceed the wavelength capacity, then combine them. The segment-swapping function finds the underutilized links in different wavelengths and combines them into one wavelength through segment swapping.

Algorithm 6.2 Traffic-grooming algorithm for multihop method

while *the number of ADMs and the number of wavelengths can be reduced* **do**
  Establish connection on the shortest path
  Wavelength_Combining()
  Segment_Swapping()
end while

6.3.4 Illustrative Numerical Results and Comparisons

Single-Hop Approach

Uniform Traffic. The uniform-traffic case has been well studied before in the literature so there exist good references to evaluate our algorithms. Table 6.3 shows the required number of ADMs from different algorithms for the single-hop case (Formulation 6.1) for an unidirectional ring. For each value of the grooming ratio (C), the first row shows the lower bound calculated by the algorithm proposed in [ZhQi00]. The second row shows the result from solving our ILP formulation, given half an hour for each value (a question mark means that the ILP solver failed to obtain a feasible solution within half an hour). The third row shows the result from our simulated-annealing heuristic, and the fourth row shows the result from the greedy heuristic in [ZhQi00]. The simulated-annealing algorithm was run for 30 trials and the best result was chosen.

We noticed that, most of the time, the simulated-annealing approach achieves better results than the greedy heuristic. Sometimes, it even reaches the lower bound. Even for the one case where greedy showed better result ($N = 16$, $C = 48$), we believe that the simulated-annealing algorithm might have done a better job if it was given more time. In this study, all algorithms started from the same network configurations (step-1 heuristic results), but the simulated-annealing algorithm can always be started from the greedy result if we choose to do so and this will guarantee equal or better results. To calculate all the simulated-annealing results in Table 6.3, the computational time for all 30 trials was a little less than 100 minutes on a 200-MHz Pentium machine running Windows NT. In [ChMo00, SiGS98], the authors provided a more elaborate algorithm for the special case of uniform traffic on bidirectional rings. Although claimed to be optimal, it does need one more ADM than our simulated-annealing-based heuristic for the case $N = 9$ and $C = 4$. The reason for this difference is that the algorithm in [ChMo00, SiGS98] will always try to fill partially-occupied wavelengths first before it examines a new wavelength. For this special case ($N = 9$, $C = 4$), 3 wavelengths are needed to carry all the traffic and the optimal happens to occur when two wavelengths are filled with only 3 circles and one with 4 circles, which is out of the search space for the algorithm in [ChMo00, SiGS98]. For our algorithm, this is a natural result.

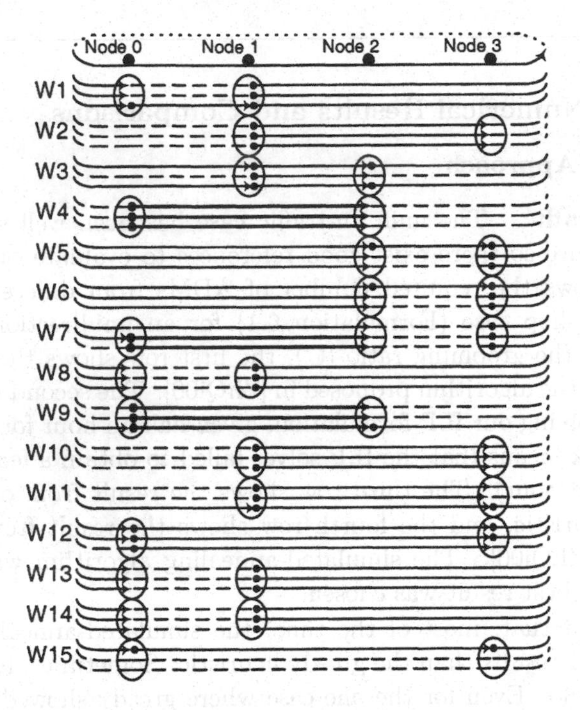

Figure 6.6 Non-uniform traffic example. Dots and arrows represent the start and the end of a connection. Circles represent ADMs.

Non-Uniform Traffic. For handling non-uniform traffic, we first show an example, which is small enough to be solved by an ILP solver in a reasonable amount of time. We picked a 4-node unidirectional-ring network with a random traffic matrix: {{0, 1, 8, 4}, {12, 0, 3, 9}, {1, 2, 0, 2}, {4, 1, 7, 0}}. In this traffic matrix, if nodes are numbered 0 through 3, then the traffic demand from node 1 to node 3 is 9 units. The grooming ratio C

(or wavelength capacity) of this network is chosen to be 3. We run all the three programs, i.e., ILP, simulated annealing, and greedy. After a 6-hour time limit, the ILP does not reach its final conclusion and its best result was: 15 wavelengths and 31 ADMs; simulated annealing gives the same result in about 2 seconds; while greedy gives 33 ADMs in negligible time. Our experiment indicates that the ILP usually cannot handle networks larger than 6 nodes when given non-uniform traffic. Results from simulated annealing are shown in Fig. 6.6. In this figure, each wavelength is drawn as three lines (recall that $C = 3$) representing the three circles. The four nodes on the ring are denoted as nodes 0, 1, 2, 3. As an example, the 8 units of traffic from node 0 to node 2 are carried by wavelength 4, wavelength 9, and part of wavelength 7 (timeslots 2 and 3) (shown in bold in Fig. 6.6).

Notice that the traffic matrix is highly asymmetric which is very practical in today's data networks and is very different from most previous work on traffic grooming. In the above example, suppose the wavelength capacity is OC-3, then traffic demand from node 0 to node 2 equals OC-8, which can be two OC-3 and two OC-1 connections (as was shown in Fig. 6.6). One demand may have to be transmitted on different wavelengths if needed, e.g., the OC-8 traffic from a single user may be transmitted over three OC-3 wavelengths.

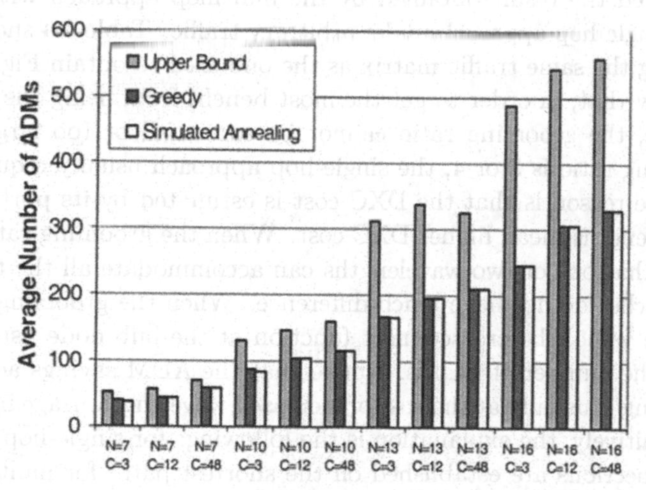

Figure 6.7 Statistical result comparison for non-uniform traffic on a bidirectional ring.

Figure 6.7 compares statistical results for the bidirectional rings. The results are calculated in the following way: for each ring scenario (described by its size and wavelength capacity), 30 random traffic matrices were generated. The traffic demand between any node pair is uniformly distributed between 0 and twice the wavelength capacity (t_{ij} = rand(0, $2 \times C$)). Three values are calculated for each traffic matrix, namely, the upper bound (no grooming), result from the greedy approach, and result from the simulated-annealing approach. Then, the average number of ADMs is shown in the figure for each approach.

The results show that both traffic-grooming algorithms achieve very good savings. Although still visible in some cases, the benefit of the simulated-annealing-based approach is not that significant for this example. The reason is that the traffic load given to the networks is heavy and unregulated. Unregulated traffic causes "bubbles" (unusable segments) in single-hop networks. The two reasons combined lead to the number of circles to tens and even hundreds, which in turn increases the search space dramatically. Even after increasing the computation time, the simulated-annealing result does not improve significantly because the search space is very large.

Multihop approach

We compared the result obtained by the multihop approach with the one from the single-hop approach under arbitrary traffic. Table 6.4 shows an example using the same traffic matrix as the one used to obtain Fig. 6.6. Our results show that, in order to get the most benefit from using the switching (hub) node, the grooming ratio cannot be too small or too large. When the grooming ratio is 3 or 4, the single-hop approach usually requires fewer ADMs. The reason is that the DXC cost is estimated by its port count, so more wavelengths mean higher DXC cost. When the grooming ratio is large enough so that one or two wavelengths can accommodate all the traffic, the two approaches do not have much difference. When the grooming ratio has a moderate value, the crossconnect function at the hub node usually helps to reduce the number of ADMs. Notice that the ADM savings achieved by multihopping is usually at the cost of increased wavelength usage in our algorithm. Intuitively, the explanation is the following: for single-hop networks, all the connections are established on the shortest path; for multihop case, some connections may travel more than one hop and traverse the ring more than once in order to take advantage of the switching function of the hub node. Although switching may help to fill the "bubbles" around the switching node, with just one switching node, it is not easy to counter-balance the waste caused by the "longer" connections. Currently, our multihop heuristic is only designed for minimizing the number of ADMs. An open problem is to incorporate minimizing the number of wavelengths as well.

Table 6.4 Comparison of single-hop and multihop approaches on non-uniform traffic ($N = 4$).

	Single-hop		Multihop	
	W	ADM	W	ADM
C=3	15	31	19	38
C=12	4	14	5	11
C=48	1	4	2	5

6.4 Interconnected WDM Ring Networks

In order to provide large geographical coverage, multiple SONET rings can be interconnected together. The optical crossconnect (OXC) technology provides a convenient way for these interconnections. Two problems, viz. (1) how to interconnect WDM ring networks with OXCs and (2) how to groom traffic in interconnected rings will be addressed in this section. We present four different strategies for interconnecting WDM ring networks with OXCs. Then, we provide mathematical formulations of the traffic-grooming problems for each of the four architectures. Next, we discuss improvements to an existing heuristic algorithm [ZhQi00, WCVM00] in detail to solve the traffic-grooming problems. Finally, numerical results from the traffic-grooming problems are presented for the four different architectures.

6.4.1 Interconnected Rings

Node Architecture

In an interconnected SONET/WDM ring, there are non-intersection nodes and intersection nodes. A non-intersection node has two interfaces to connect it with its two neighbors and a local interface bank for adding or dropping traffic. Figure 6.8 shows the architecture of a non-intersection node in a unidirectional ring. A bi-directional ring comprises of two unidirectional rings and this requires extra hardware. Most of our study here will assume bi-directional rings and bi-directional SONET ADMs (SADMs) (back-to-back double SADMs for both directions assembled as one unit).

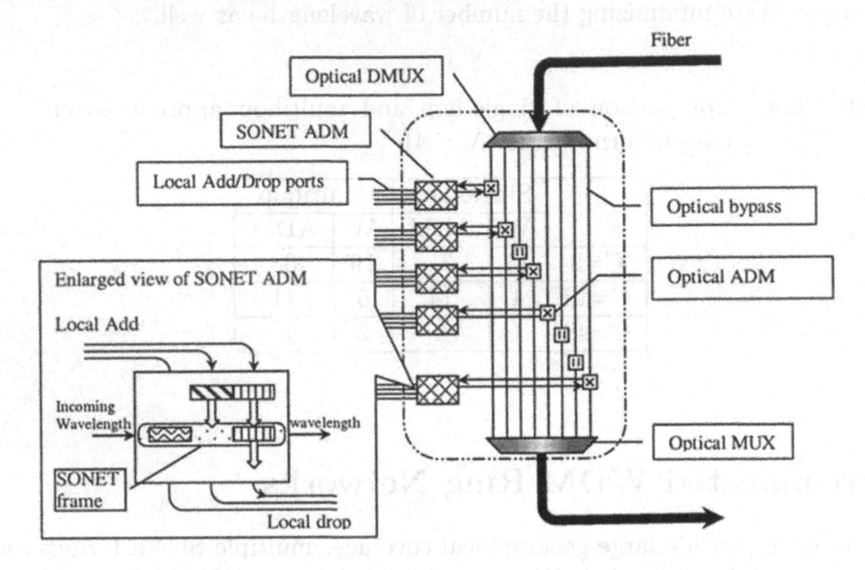

Figure 6.8 Simplified node architecture of a SONET/WDM ring network.

The architecture of the intersection node varies depending on the availability of hardware. Current deployment is based on optical add/drop multiplexer (OADM) and SADMs. A digital crossconnect (DXC) or a manual patch panel may be used to connect low-speed streams between two rings. Figure 6.9(a) illustrates the way this mechanism works. Traffic, either going to local ports or to another ring, will be dropped by OADMs and SADMs, which are then relayed by DXC to their desired destination.

Newly-developed hardware technologies allow OXCs to be built so that they can switch an entire wavelength (not sub-channels as in the case of DXC). There are transparent and opaque technologies to build these OXCs. Transparent here refers to all-optical switching. It is still unclear when all-optical wavelength converters will be available, so, like most other studies, we assume absence of wavelength-conversion capability in transparent crossconnects. By opaque technology, we refer to electronic switching in a crossconnect. The signal has to be converted from optical to electronic domain before it is switched and converted back to optical. Opaque technology implies full wavelength conversion, so it potentially has advantage in deliver-

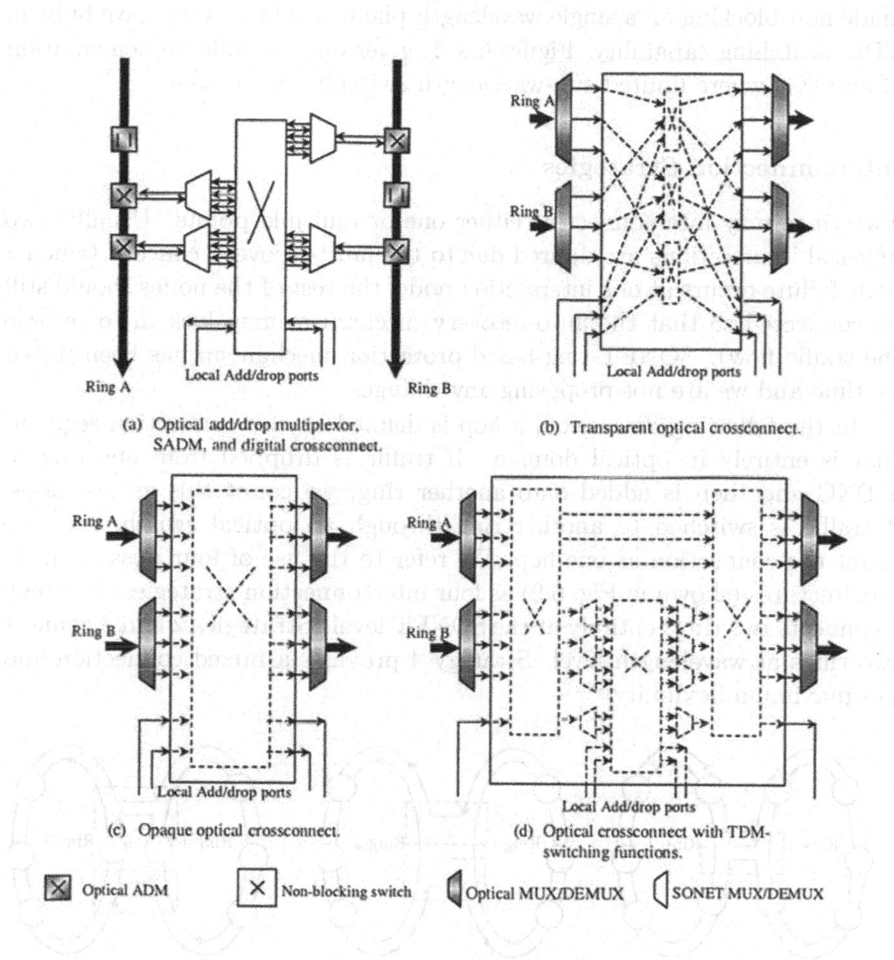

Figure 6.9 Simplified architectures of different optical crossconnects (OXCs).

ing higher network utilization when compared with a wavelength-continuous switch. Figures 6.9(b) and 6.9(c) compare these two technologies. Conceptually, a transparent OXC may not be fully non-blocking although it can be made non-blocking on a single wavelength plane. An OXC may have built-in TDM-switching capability. Figure 6.9(d) gives one possible implementation of an OXC where limited sub-wavelength switching is possible.

Interconnection Strategies

Two rings may interconnect at either one or multiple points. Usually, two physical intersections are desired due to the fault-recovery concern (when a node failure occurs at one intersection node, the rest of the nodes should still be connected so that the auto-recovery mechanism may kick in to resume the traffic flow). SONET-ring-based protection mechanism has been tested by time and we are not proposing any changes.

In the following discussion, a hop is defined as any connection segment that is entirely in optical domain. If traffic is dropped from one ring to a DXC and then is added onto another ring, we count this as two hops; if traffic is switched to another ring through an optical switch, then we count the connection as one hop. We refer to the use of four possible node architectures (shown in Fig. 6.9) as four interconnection strategies. Strategy 1 connects two rings entirely at the SONET level. Strategies 2 and 3 connect two rings at wavelength level. Strategy 4 provides a mixed connection and has maximum flexibility.

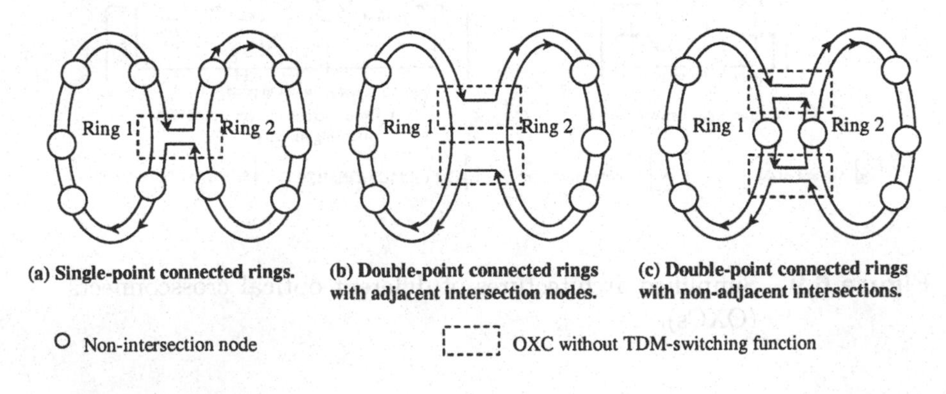

Figure 6.10 Three forms of interconnections that can be formed by using strategies 2 and 3.

In strict sense, strategies 2 and 3 do not "crossconnect" two rings together if they are used alone. The two strategies can only form one (or multiple) logical ring(s) that covers all the non-intersection nodes. Figure 6.10 shows the three cases for the single-point (one intersection node between the two rings) and double-point (two intersection nodes) connected rings. Because inter-ring traffic cannot be truly bridged to the local rings, logical rings need to have TDM add/drop ports. This may imply a scalability problem since no more than 16 nodes can be on a single SONET ring. Strategies 2 and 3 have to be used with other ones for interconnecting multiple rings.

6.4.2 Traffic Grooming in Interconnected Rings

Problem Specifications for Interconnected Double Rings

In Section 6.3, we formulated the single-ring traffic-grooming problem mathematically as an integer linear program (ILP), and an improved heuristic algorithm for general traffic pattern was proposed. Results from the research efforts in single-ring networks serve as good references and provide a strong base for the study of traffic grooming in interconnected rings. The mathematical problem specification we are going to provide below has been extended from the work done for single-ring networks.

We assume arbitrary (static) traffic pattern for our discussion in this section. The traffic pattern is represented by a $N \times N$ traffic matrix, where N is the total number of nodes in the rings. The traffic matrix is denoted by T and each of its element t_{ij} is the total amount of traffic demand from node i to node j measured in units of sub-channel capacity. All our formulations are designed for bi-directional rings (unidirectional formulations are straightforward and are not shown here to conserve space). The formulation also assumes that all wavelengths have same capacity, i.e., they can accommodate equal number of sub-channels, which is assumed to be C and is called the grooming ratio. For simplicity, we only explain the formulation for single-point-connected double rings.

In our formulations, we assume that the minimum number of wavelengths that are needed to achieve maximum SADM savings (while carrying all the given traffic) is known. Actually, finding this minimum number is almost as challenging as the traffic-grooming problem itself. Fortunately, there exists a heuristics that comes in handy [ZhQi00]. In our study, we used the result obtained from this heuristic.

Let us start with the problem formulation for the crossconnect architecture shown in Figure 6.9(a), namely, strategy 1. In this case, we use a DXC to interconnect two rings, and all the inter-ring traffic is added/dropped at the intersection node regardless of how they are switched. The optimization problem on one ring is independent of the one on the other ring. To each ring, all traffic to or from the other ring is the traffic to or from the intersection node. The unified traffic matrix can easily be broken into two parts, each part belonging to a ring. We use the notation of $T' = \{t'_{ij}\}$ to represent decomposed traffic matrix on each ring. The mathematical formulation for the traffic-grooming problem on each ring is the same as Formulation 6.1, except that t_{ij} is replaced with t'_{ij}.

In the modified Formulation 6.1, ${}^{d}V_{ij}^{cw}$ represents whether there is a virtual connection (one unit of capacity) from node i to node j along direction d (which can be clockwise or counter clockwise) on circle c and wavelength w. The t'_{ij} represents the total amount of traffic from node i to node j on a ring. Notice that t'_{ij} is equal to t_{ij} when neither i nor j is the intersection node. When one of the two is the intersection node, t'_{ij} will include both the intra-ring and inter-ring traffic. The intra-ring traffic is t_{ij}, and the inter-ring traffic part will be the summation of all traffic from (to) nodes on another ring to (from) the non-intersection node. ADM_i^w represents whether there is an SADM on wavelength w at node i. Both ${}^{d}V_{ij}^{cw}$ and ADM_i^w can be either 0 or 1.

The traffic-load constraint states that the number of virtual links from node i to node j is equal to the traffic specified in the traffic matrix T'. For each TDM sub-channel of a wavelength on a certain link, the number of connections that can use it is no more than 1. In the channel-capacity constraint, ${}^{d}e^{cw}$ represents a d-direction (d can be either 0 or 1) link on wavelength w and sub-channel c. If the virtual connection ${}^{d}V_{ij}^{cw}$ uses it, we say that ${}^{d}e^{cw} \in {}^{d}V_{ij}^{cw}$. The transmitter and receiver constraints specify that the number of virtual connections that start or terminate at any node is bounded by the electronic multiplexing (demultiplexing) capacity, which is equal to the number of transmitters or receivers. Sometimes, shortest-path routing is required in a bi-directional ring. This requirement can also be accommodated in the formulation by specifying ${}^{d}V_{ij}^{cw}$ to be zero if the distance from node i to j in a specified direction exceeds half the ring size. It is worth mentioning that some redundant constraints were added to the basic formulation shown in Formulation 6.1 in our implementation to reduce the search space. Just to name one, we specified that the capacity occupied by all connections from node i to node j on all the intermediate links along a path remains the same.

When compared with strategy 1, strategy 2 has one more logical ring, which covers all the non-common nodes. How to break the traffic matrix into the three rings is not that obvious and inappropriate decomposition may lead to a sub-optimal solution. One way to take this into consideration is the following. The new logical ring will take the responsibility of carrying all the inter-ring traffic, as well as part of the local traffic if this can help in bringing down the total network cost. A local ring takes care of all or part of the local traffic. We use the notation $\overline{T} = \{\overline{t}_{ij}\}$ to represent traffic on the logical ring, and $T' = \{t'_{ij}\}$ for the traffic on the local rings. Substituting $\overline{T}$ and the two T''s into constraints in the formulation provided before, we get three sets of constraints. By adding the constraints shown in Formulation 6.3, the new constraints for strategy 2 are complete. The new objective function is to minimize the summation of all SADMs on all the three rings.

$$\begin{aligned} \overline{t}_{ik} &= t_{ik} && \textit{(i, k belong to different rings)} \\ t'_{ij} + \overline{t}_{ij} &= t_{ij} && \textit{(i, j belong to same ring)} \end{aligned}$$

Formulation 6.3: Additional traffic-load constraints for transparent-OXC-connected double rings (strategy 2).

$$\begin{aligned} \overline{t}_{ik} &\leq t_{ik} && \textit{(i, k belong to different rings)} \\ t'_{ij} + \overline{t}_{ij} &= t_{ij} + \sum_k (t_{ik} - \overline{t}_{ik}) && \textit{(i, j belong to same ring, i is the intersection node)} \\ t'_{ij} + \overline{t}_{ij} &= t_{ij} + \sum_k (t_{kj} - \overline{t}_{kj}) && \textit{(i, j belong to same ring, j is the intersection node)} \\ t'_{ij} + \overline{t}_{ij} &= t_{ij} && \textit{(i, j belong to same ring and none of them is intersection node)} \end{aligned}$$

Formulation 6.4: Strategy 4's additional traffic-load constraints for OXC with TDM-switching capability.

For the crossconnect architecture shown in Fig. 6.9(c), which is referred to as strategy 3, wavelength conversion is available for inter-ring traffic at the intersection node. One way to formulate the problem is to enumerate all possible wavelength-conversion patterns and to apply the same formulation of strategy 2 for each pattern. A slight modification to replace the notion of ${}^{d}V_{ij}^{cw}$ with ${}^{d}V_{ij}^{cw \to w'}$ is needed. A program needs to be written to figure out which ${}^{d}V_{ij}^{cw \to w'}$ exists (does not matter whether it will turn out to be 0 or 1 eventually) when vector (i, j) goes across the intersection node for a given wavelength-conversion pattern. For example, if wavelength conversion from w_1 to w_2 is available at node s for connections along direction d, then ${}^{d}V_{ij}^{cw_1 \to w_2}$ exists. Wavelength conversion may lead to long connections that go around the ring more than once. To ensure that the channel-capacity constraint holds, we have to enforce that no connection longer than a full circle is allowed.

Strategy 4 differs from strategy 1 in that part of the inter-ring traffic can be switched without electronic multiplexing and demultiplexing. When compared with strategies 2 and 3, not all the inter-ring traffic has to be carried by a logical ring in strategy 4. These differences make the traffic decomposition a little more complex. To incorporate all traffic-decomposition possibilities, the additional traffic-load constraint for strategy 2 is amended in the way shown in Formulation 6.4. In the formulation, $\overline{T} = \{\overline{t}_{ij}\}$ represents the traffic on the wavelengths that are optically switched across the border of the two rings, and $T' = \{t'_{ij}\}$ represents the traffic carried by the wavelengths that traverse through only one ring.

All formulations can be modified for two-point-connected double rings. Figure 6.11 shows an illustrative example problem solved by using the above formulations. The grooming ratio is assumed to be 1. Each ring has three nodes. Uniform (one unit of request per node pair), full-mesh traffic demand is carried by the network.

Figures 6.11(a)–(d) give an optimal solution obtained from strategies 1 through 4. Figure 6.11(a) shows the most expensive solution under this particular traffic demand and network configuration. Fourteen SADMs are used. In Fig. 6.11(b), we observe that, in order to achieve maximum savings, two directions of some bi-directional connections are routed along different paths (e.g., connection between node 0 and node 2). This result also shows that a SADM is used at some nodes even when it is used in only one direction (e.g., wavelength 2 of node 2). In Fig. 6.11(c), wavelength conversion is used. This strategy achieves the minimum of 10 SADMs in this case. In Fig. 6.11(d),

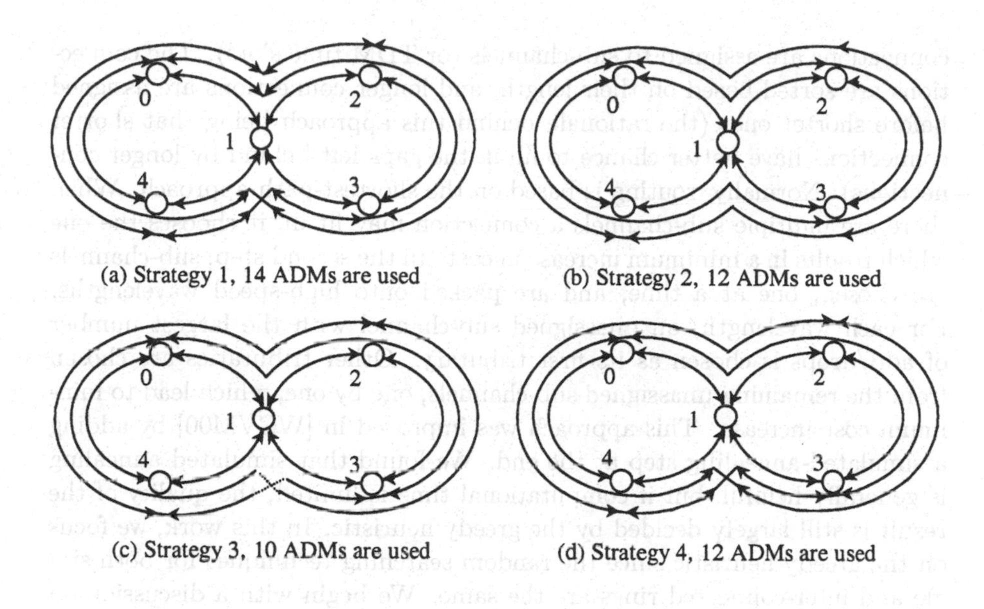

Figure 6.11 Illustrative results of traffic grooming.

where we assume wavelength-continuous OXC with TDM-switching capability, 12 SADMs are used. Figure 6.11(d) differs from Fig.6.11(c) in that wavelength conversion is done by the combination of drop and add.

Heuristic Design

Mathematical problem specifications are useful in providing insights into the nature of these problems. These formulations can be translated into computer programs, which perform a detailed search to find an optimal solution. It has been proven that the traffic-grooming problem in a single-ring network is NP-complete [ChMo00]. The problem of grooming traffic in a multi-ring scenario is also NP-complete. Our experience has shown that even an industrial-strength MILP solver, "CPLEX", may be incapable of solving the problem specified by Formulation 6.1 on a single 7-node bi-directional network with uniform traffic within a reasonable period of time. Hence, while constructing large interconnected rings efficiently, we look for heuristics.

A greedy heuristic developed in [ZhQi00] has been found to achieve fairly good results. This greedy approach consists of two steps. In the first step, all connections are assigned to sub-channels (or TDM time slots). The connections are sorted based on their length, and longer connections are assigned before shorter ones (the rationale behind this approach being that shorter connections have better chance to fill in the gaps left behind by longer connections). Normally, routing is based on the shortest-path approach. When there are multiple sub-channels a connection may fit in, it chooses the one which results in a minimum increase in cost. In the second step, sub-channels are chosen, one at a time, and are packed onto high-speed wavelengths. For each wavelength, an unassigned sub-channel with the largest number of add/drops is chosen as its first tributary. Other tributaries are chosen from the remaining unassigned sub-channels, one by one, which lead to minimum cost increase. This approach was improved in [WCVM00] by adding a simulated-annealing step at the end. We found that simulated annealing is generally helpful, but if computational time is limited, the quality of the result is still largely decided by the greedy heuristic. In this work, we focus on the greedy heuristic since the random searching techniques for both single and inter-connected rings are the same. We begin with a discussion on different traffic-pattern characteristics in interconnected rings.

Characteristics of Inter-Ring Traffic in Different Interconnection Strategies

Consider interconnecting two rings, with N nodes on one and M nodes on the other, carrying traffic uniformly distributed in the network. In this network, it can be shown, statistically, that the internal traffic on each ring is proportional to N^2 and M^2, respectively. The inter-ring traffic is roughly proportional to $2 \times N \times M$. If N and M are comparable, the inter-ring traffic is about twice the intra-ring traffic on each ring.

When strategy 1 (with DXC at intersection node) is used, each inter-ring connection is broken down into two parts and is mixed with intra-ring traffic on the local rings. The above-mentioned greedy heuristic needs to be re-optimized because all the inter-ring traffic (most of the time the dominant component of all traffic on a ring) has to go through the intersection node. This kind of traffic is usually referred to as the egress/ingress traffic. If the egress/ingress traffic alone is considered, then each sub-channel can carry at most two bi-directional connections since the intersection node has only two interfaces. Shortest-path routing, which can help save network resources for a ring with normal traffic pattern, does not have the same advantage here, i.e., in case of only egress/ingress traffic. Instead, since no connection will take a route longer than half the ring, most SADMs in the network cannot be exploited to their maximum capacity. An algorithm for computing the optimal solution on a unidirectional egress ring has been proposed in [ChMo00]. We propose and discuss a similar approach for bi-directional rings below.

As discussed earlier, there is no traffic exchanged between the logical ring and local rings in strategies 2 and 3. This necessitates decomposition of the unified traffic representation, so that all inter-ring traffic and part of the intra-ring traffic can be carried by the logical ring. All the inter-ring traffic (usually the dominant component on the logical ring) has to go through the intersection node(s) and this creates a capacity bottleneck at physical intersection node(s). Global shortest-path routing is likely to result in most nodes using one side of their interface (closer to the intersection node) more often than their other side. As a result, most SADMs will only be used for half of their capacity and the other half is left idle. A solution to this problem will be addressed later.

Strategy 4 uses an OXC with TDM-switching function to interconnect the two rings, a design which provides maximum flexibility in traffic grooming. If none of the TDM-switching ports is used, then this strategy is identical to strategies 2 or 3, depending on whether or not wavelength conversion is allowed. If all the traffic is sent to TDM switches, then it is identical to strategy 1. The key question is which portion of inter-ring traffic (using traffic-matrix decomposition in grooming algorithm) to switch at TDM level to minimize the total number of SADMs used.

In a real network, several rings are inter-connected and minimizing SADMs to carry all the traffic in such a network is important. If DXC-based switching is used at all intersection nodes, a long connection may have to travel multiple hops (rings) to reach its destination. This not only increases the multiplexing and de-multiplexing cost at each intermediate intersection node, but also increases the switching complexity. An alternative approach is to deploy OXCs at intersection nodes to build one or multiple large rings (we refer them as *hyper-rings*), which can cover all local rings. Hyper-rings serve as "highway" to carry inter-ring traffic. Because they crossconnect with each local ring (at TDM level) at a limited number of nodes, a hyper-ring can connect several local rings. This approach can be repeatedly used to build a hierarchical network to increase geographical coverage. If a network is carefully designed, traffic from one ring to its peer ring will go through as few as 3 hops. In the first hop, traffic moves from its source to an intersection node where it is digitally crossconnected to the hyper-ring. In the second hop, traffic traverses along the hyper ring. In the last hop, traffic moves to the local ring on which the destination node resides. In this study, we focus only on the simplest two-layered architecture with one large hyper ring.

For multiple rings connected by the layered strategies, the traffic-grooming heuristic for each local ring remains the same as the one developed for strategy 1. The traffic on a hyper ring has the same characteristics as that of a regular ring. This is easy to see when every local ring has only one intersection with the hyper ring. If there are two intersections, we can treat the two nodes together as a super node. So, no new heuristics are needed for the hyper ring.

Improved Traffic-Grooming Algorithms

In order to groom the egress/ingress traffic that appear on the rings, interconnected with strategy 1, we form full circles using sub-channels and thus save on the number of wavelengths. Without losing generality, we use a single-point-connected network as an example to explain our heuristic. Since there is only one intersection node in each ring, a sub-channel can carry at most two bi-directional traffic demands. If there is a pair of traffic from one node to the intersection node (or vice versa), they use a common sub-channel in opposite directions. We assign connections for each node sequentially, a procedure which is described in detail below.

We route the egress/ingress traffic for each node one by one. For the first node, all outgoing traffic from it is routed alternately in two different directions, and then all the incoming traffic is routed alternately from two different directions. Because there may be a mismatch in the number of outgoing and incoming connections, there can be some free spots left in the sub-channels which are used by the first node. For the second node, we start with some new sub-channels to form full circles, and come back to check if there are free spots to accommodate some newly-generated unbalanced (more outgoing or incoming) traffic. The procedure for node 2 is repeated for all other nodes. After we finish the assignment of all the traffic into sub-channels, we use the same greedy heuristic (proposed in [ZhQi00]) to pack them onto as few wavelengths as possible.

When strategy 1 is used to interconnect two rings, a local ring carries not only the egress/ingress type of traffic, but also its local traffic. After we have used the above heuristic to groom the inter-ring traffic, there are some free spots available in the used sub-channels, which can be used to carry local traffic. When we groom the local traffic, a concern is whether or not we should mix it with the groomed inter-ring traffic. We compared the following two approaches: The first approach separates the two types of traffic (inter-ring and intra-ring), and the local traffic starts from an unused sub-channel; while the second approach inserts as much local traffic as possible onto the existing egress/ingress sub-channels before grooming the remaining traffic. Our simulation results show that the first approach is usually economical than the second. The reason is that the uniform distribution of traffic within the local ring is distorted (mostly the connections that are far away from the intersection nodes have a good chance to be carried on by the logical ring) and the heuristic (proposed in [ZhQi00]) is not optimized for this unbalanced traffic pattern.

The logical ring formed when strategies 2 and 3 are used has two points that are potential capacity bottlenecks. In case of double rings connected at a single point, the two bottlenecks are physically located in the same OXC. In case of two-point interconnected double rings, the two bottleneck nodes are at the two intersections. These bottlenecks are formed because all the inter-ring traffic has to be crossconnected at one intersection node, which restricts a sub-channel to carry at most two (bi-directional) inter-ring connections. The heuristic we propose for strategy 2 is similar to that in strategy 1, in which it tries to form a full circle on each sub-channel. When we assign connections to sub-channels, we first pick bi-directional connections and route each of the two directions along different half circles of a unidirectional sub-channel. Then, we use the greedy heuristic (from [ZhQi00]) for these circles and unidirectional connections. We should be careful while carrying local traffic onto the logical ring (even if there is space) for reasons similar to those described for strategy 1.

The example in Fig. 6.11(c) shows that wavelength conversion may save SADMs in interconnected ring networks. For strategy 3, we treat the hyper-ring to be wavelength-continuous (with no wavelength-conversion capability at intersection node) and apply our above traffic-grooming algorithm designed for strategy 2. Then, we check if there are any unused wavelength segments, adjacent to the common interconnection node, so that they can be connected by utilizing the wavelength-conversion capability of OXCs. If we discover such wavelength segments, we search for the connections that fit in these spaces and test if, by moving any of these connections, we can reduce the total number of SADMs. In our study, since each wavelength carries several tens of sub-channels, there is a slim chance of having a wavelength segment free of traffic and adjacent to the intersection node. Experiments (using our heuristics designed for strategy 2) confirm that we seldom exploit the wavelength-conversion capability of OXCs at intersection nodes. So, we do not differentiate between the heuristics for strategies 2 and 3 while conducting experiments to obtain the numerical results presented in the next section.

We assume that a wavelength-continuous OXC with TDM-switching capability is used for strategy 4, and it attempts to shift different portions of the inter-ring traffic back and forth between the logical ring and local rings to fully exploit the switching capability. The results from these experiments are shown in the following section.

6.4.3 Numerical Results and Analysis

Interconnected Double Rings

To show the results from our simulation experiments, we use the following example. Two bi-directional rings are interconnected at two points. The first ring (local ring 1) consists of 7 nodes and the second ring (local ring 2) has 8 nodes. The nodes are numbered as shown in Fig. 6.12. They are interconnected at two consecutive nodes (0 and 6). We run the experiment under three grooming ratio (C) assumptions of 4, 16, and 64. Grooming ratio here refers to the number sub-channels each wavelength can carry. The traffic pattern is represented by a random traffic matrix. Each element in the traffic matrix is a uniformly-distributed random integer between 0 and 15. The traffic between nodes 0 and 6 is equally distributed between the two local rings. For strategies 1 and 4, inter-ring traffic (or part of it) is switched at one of the intersection nodes which is decided by the shortest-path routing algorithm. If there are two equal-cost paths, traffic is bifurcated. For strategies 2 and 3, the logical ring follows the node-number sequence.

Figures 6.13(a)–(c) show the average number of SADMs used by the four interconnection strategies and Fig. 6.13(d)–(f) show the average number of wavelengths used. The results are obtained from the average of 100 trials. In each trial, a different random traffic matrix is used. For all SADM-usage figures, the y-axis represents the total number of SADMs used. For all the wavelength-usage figures, the y-axis represents the number of wavelengths times the number of links. The number of wavelengths used on the logical ring is large because the ring is large. All the results here are obtained by

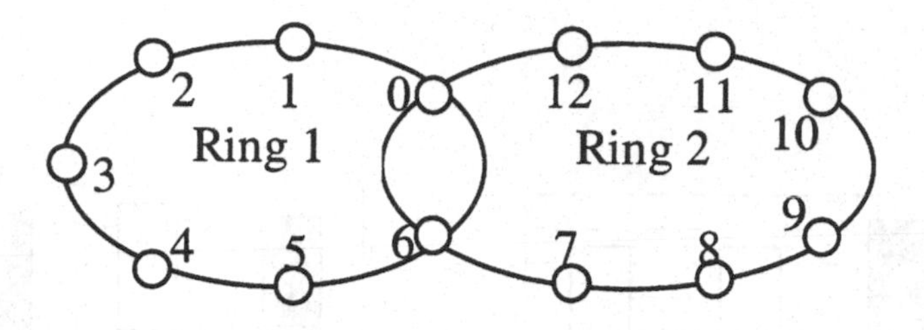

Figure 6.12 Interconnected ring used for experiments.

using our heuristic described earlier.

We compared how the relative performance of each interconnection strategy changes under different grooming-ratio parameters. The results match with those we obtained in our single-ring traffic-grooming research. When grooming ratio is low (say $C = 4$ in Fig. 6.13(a)), the optical-switched strategies (2 and 3) yield cheaper solutions. In this low grooming-ratio case, the TDM-level switching at the intersection node(s) plays an important role in bringing down the number of SADMs at the non-intersection nodes. However, this benefit is not enough to compensate for the extra cost that has to be spent at the switching node. With increase of the grooming ratio (C=16 and C=64), the advantage of using TDM-level switching can be gradually seen (in Figs. 6.13(b) and 6.13(c)). Whatever the grooming ratio is, the OXC-connected network always uses fewer wavelengths (note that we are not trying to minimize wavelength usage).

For C=4, strategies 2 and 3 show about 11% savings on SADM cost and about 9.5% savings on wavelength cost when compared with strategy 1. For C=16, strategy 1 shows about 4% savings on SADM cost with the addition of 10% more wavelengths when compared with strategies 2 and 3. For C=64, the differences in SADM usage become smaller, with about 1% savings achieved by strategy 1 when compared with strategies 2 and 3. The difference between the wavelength cost is still about 10%. The reason for not seeing even bigger savings on SADM cost when compared with the C=16 case is the following. The grooming ratio is so large here that we only need a few wavelengths to carry all the traffic. For strategy 1, an average of 3.86 and 4.39 wavelengths were used for each local ring. For strategies 2 and 3, the average of 2.41, 1.42, and 1.95 wavelengths are needed for the logical ring and the two local rings. As a result, every ring tends to have SADMs everywhere. For strategy 1, an average of 5.02 (out of 7 possibilities) and 6.02

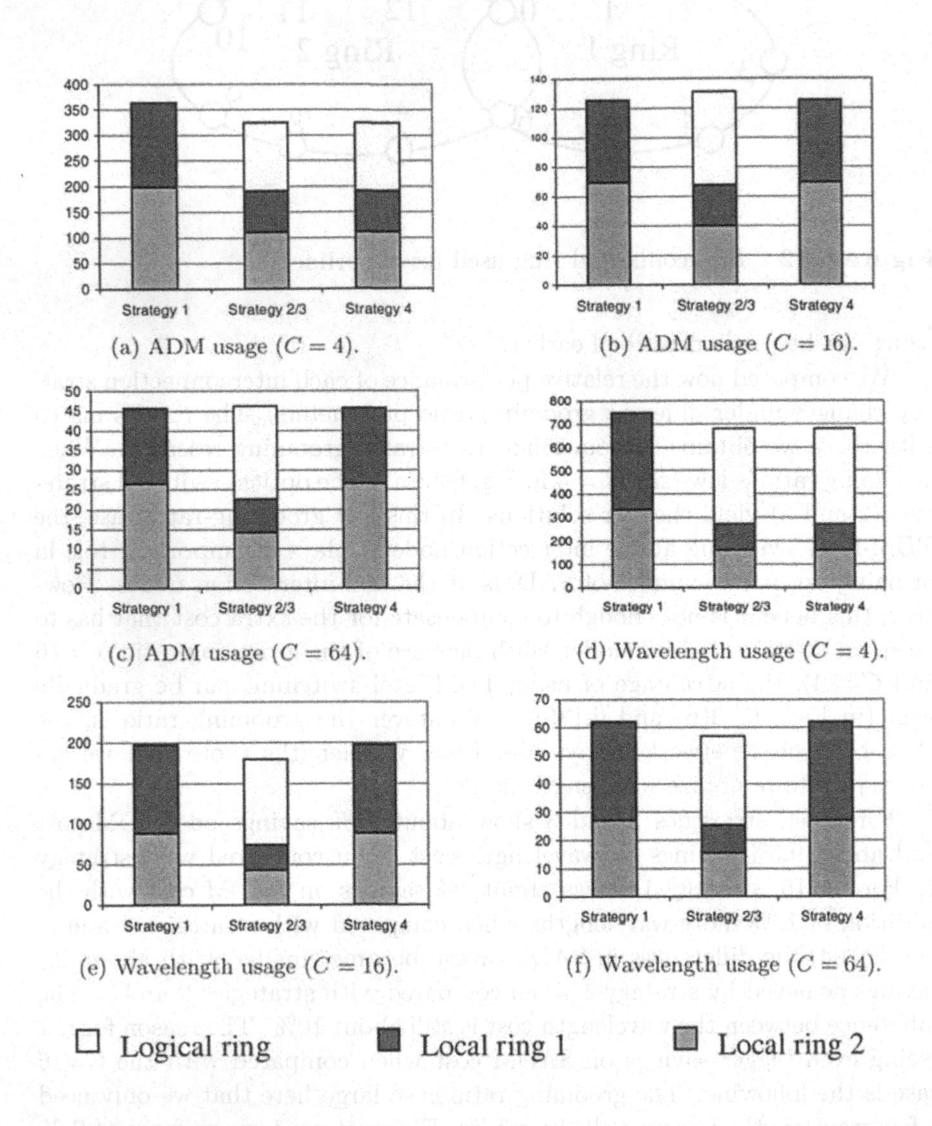

Figure 6.13 ADM and wavelength usage comparison.

(out of 8) SADMs are used for each ring. For strategies 2 and 3, an average of 5.89 and 7.31 SADMs are used for each local ring. The only exception is the logical ring where 9.83 (out of 13) SADMs are used, which is close to the lower grooming-ratio cases. For networks with very large grooming ratio, different traffic-grooming strategies do not have considerable difference in terms of savings on the number of SADMs.

It is interesting to observe that strategy 4 can always achieve the best SADM savings by adopting a winning strategy. We compared several ways to decompose the inter-ring traffic for strategy 4 and tried to find the best operating point. For example, we place only very regulated traffic on the logical ring but all remaining traffic on the local rings. We were able to achieve very low cost on the logical ring but this increased the local ring cost too much so that the overall network cost increased. To conclude, if the grooming ratio is small or very large, switch all the inter-ring traffic optically. For networks with medium grooming ratio, consider using TDM-level switching. Thus, in order to design an OXC that has TDM-switching functionality and is intended only for interconnecting rings, the grooming ratio should be between 8 and 64.

Interconnected Multi-Ring

We now show the traffic-grooming results for another network topology with 15 nodes interconnected by four rings. The physical topology of the network is shown in Fig. 6.14. Such a network is a typical one deployed by a telecom network operator. Thirty randomly-generated traffic matrices are used to obtain the average SADM usage. The hyper-ring architecture is used. The hyper-ring (chosen arbitrarily for illustration) consists of the following eight nodes: 11, 9, 7, 6, 15, 13, 13, and 1. Notice that the nodes 7 and 6 only have connection with ring 2, while node 13 is connected to both rings 3 and 4. Figure 6.15 shows the average number of SADMs used on each ring.

In the above experiment, traffic demand between any node pair is assumed to be a uniformly-distributed random number between 0 and 15. The grooming ratio is chosen to be 4. The intersection nodes between the hyper ring and a local ring are randomly chosen. Because the traffic on the hyper ring is usually much heavier than any local ring, we need to reduce this part of the network cost as much as possible. In the experiment, we simply try to divide the traffic to and from the hyper ring evenly between the two intersections on each local ring. Results show that the traffic on the hyper ring is more regulated, so the traffic-grooming algorithm gives good

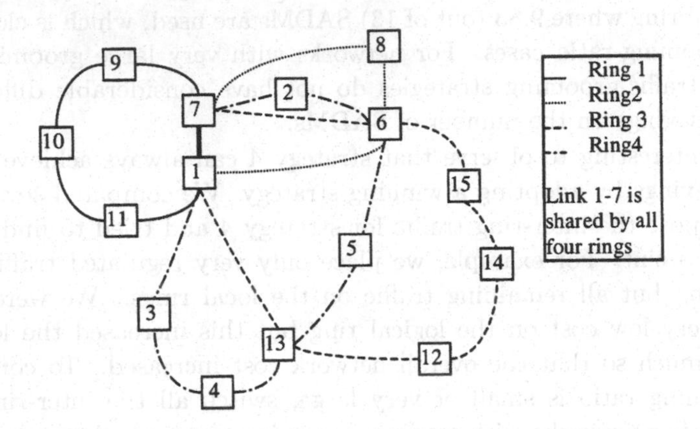

Figure 6.14 An interconnected multi-ring network topology.

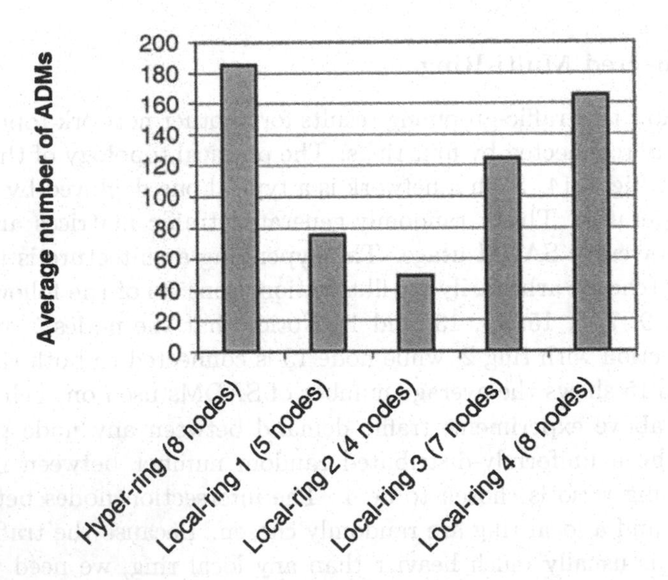

Figure 6.15 Average ADM usage for the multi-ring network.

results. When compared with local ring 4, the hyper ring used a little more SADMs but carried about 2 times more traffic.

6.5 Packet Communication using Tunable Wavelength Add-Drop Multiplexers

6.5.1 Introduction

Using several SONET Add-Drop Multiplexers (ADMs) which equal the number of wavelengths, Wavelength Add-Drop Multiplexers (WADMs) are most applicable in fiber-based high-speed ring networks. A WADM (also referred to as an OADM) consists of a demux, followed by a set of 2 × 2 switches — one switch per wavelength — followed by a mux. The WADM can be essentially "inserted" on a physical fiber link. If all of the 2 × 2 switches are in the "bar" state, then all of the wavelengths flow through the WADM "undisturbed." However, if one of the 2 × 2 switches is configured into the "cross" state via electronic control, then the signal on the corresponding wavelength is "dropped" locally, and a new data stream can be "added" on to the same wavelength at this WADM location.

In [MBLM96], novel proposals were illustrated for medium-access control (MAC) protocols in all-optical packet networks based on WDM multichannel ring topologies where nodes are equipped with one fixed-wavelength receiver and one wavelength-tunable transmitter. Three access protocols based on local status information are described. Global fairness control algorithms derived from those adopted in the Metaring high-speed metropolitan-area network [CiOf93] were also proposed. Access delays and throughputs were taken as performance indices for a simulation-based comparison of the proposed protocols, in the case of a 16-node multiring with either balanced or unbalanced traffic. Simulation results showed that the throughput limitations and the fairness problems inherent in the network topology could be overcome with relatively simple protocols.

In this section, we consider multi-wavelength ring networks where each node is equipped with one dynamic WADM (DWADM) and one SONET ADM. Rings allow for slot synchronization, even at high data rates; hence, they offer an efficient use of the available optical bandwidth for packet communications. The ADM is used for cost-effective traffic grooming in WDM ring networks. Thus, if a DWADM, an ADM, and the advantage of WDM can be used together, the designed ring network should show good performance. Since all source nodes that need to transmit to a destination must share the channel, MAC protocols are needed for packet communication to arbitrate access to the shared channels [ZhQi00].

DWADM

A DWADM (also referred to as a ROADM) can add and drop signals on a selected wavelength to and from a main fiber bus like a tunable transmitter and a tunable receiver. It can operate over a range of speeds, with high selectivity and low loss. Figure 6.16 shows the characteristics of DWADM and its difficulty in developing a protocol by comparing with the architecture proposed in [MBLM96].

In a general protocol design of WDM networks, if each node has one or more tunable (or fixed) transmitter and receiver, it can independently receive packets on some wavelengths while transmitting some other packets on some other wavelength. However, the case of a DWADM is quite different since the input and output wavelength must be the same, i.e., if the wavelength to receive at some node s is λ_i, the wavelength to transmit must be the same wavelength λ_i; and, furthermore, if the node has a packet to send to another node d, then node d should also use the wavelength λ_i to receive and to send a packet. Thus, as one can observe in Fig. 6.16(b), a virtual loop is created, and such loops can change dynamically over time, based on new packet arrivals at the various nodes. Let us consider the case in Fig. 6.16(d). This is the best method to increase the channel utilization and prohibit the channel collision. Since the ring network can have up to 16 nodes, we can consider a maximum of 8 wavelengths for transmission. Naturally, by this limitation in wavelength usage, the performance in the ring network equipped with DWADMs is restricted than that in a general WDM ring network. However, the cost and simplicity of an ADM can be a significant advantage to build such a ring network.

6.5.2 Protocol

System Architecture

We introduce the following assumptions.

- There are N (≤ 16) stations.
- There are $W + 1$ wavelengths of which W wavelengths ($\lambda_1 \sim \lambda_W$) are for data channels and one wavelength (λ_0) is for a control channel.

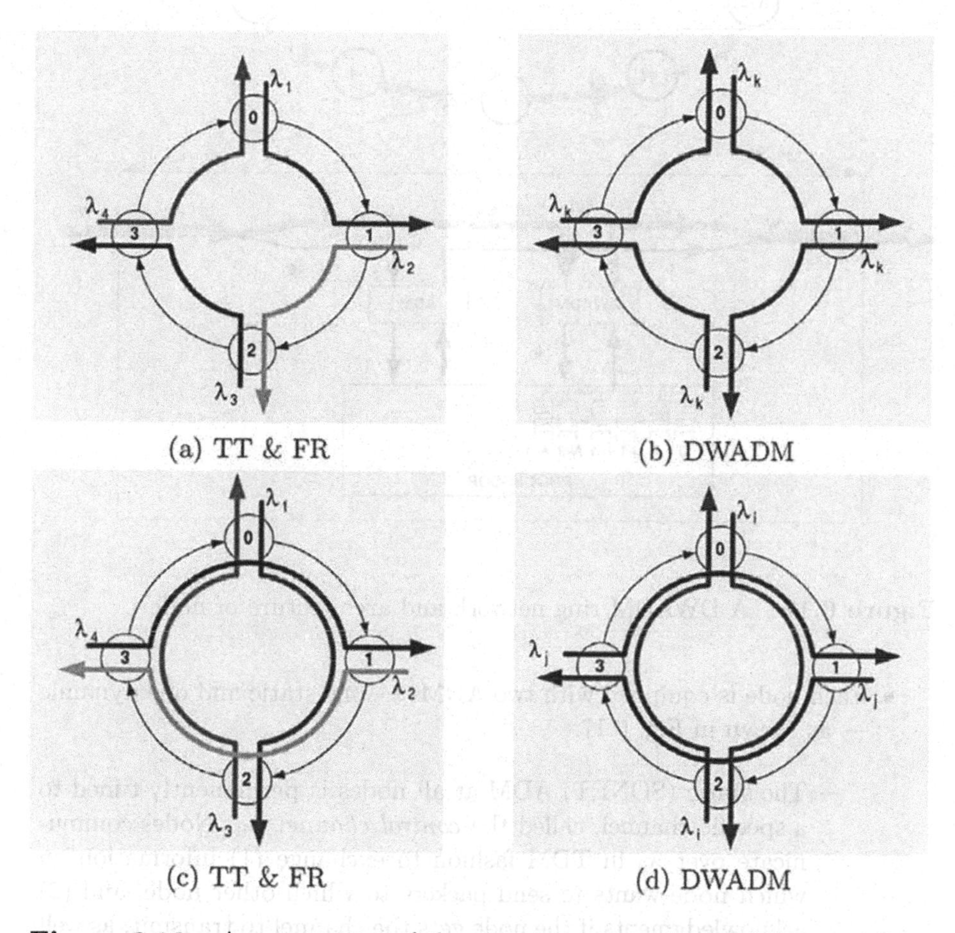

(a) TT & FR (b) DWADM

(c) TT & FR (d) DWADM

Figure 6.16 A comparison of the ring network with tunable transmitters and a fixed receiver at each node and the ring network with a DWADM and a SONET ADM at each node.

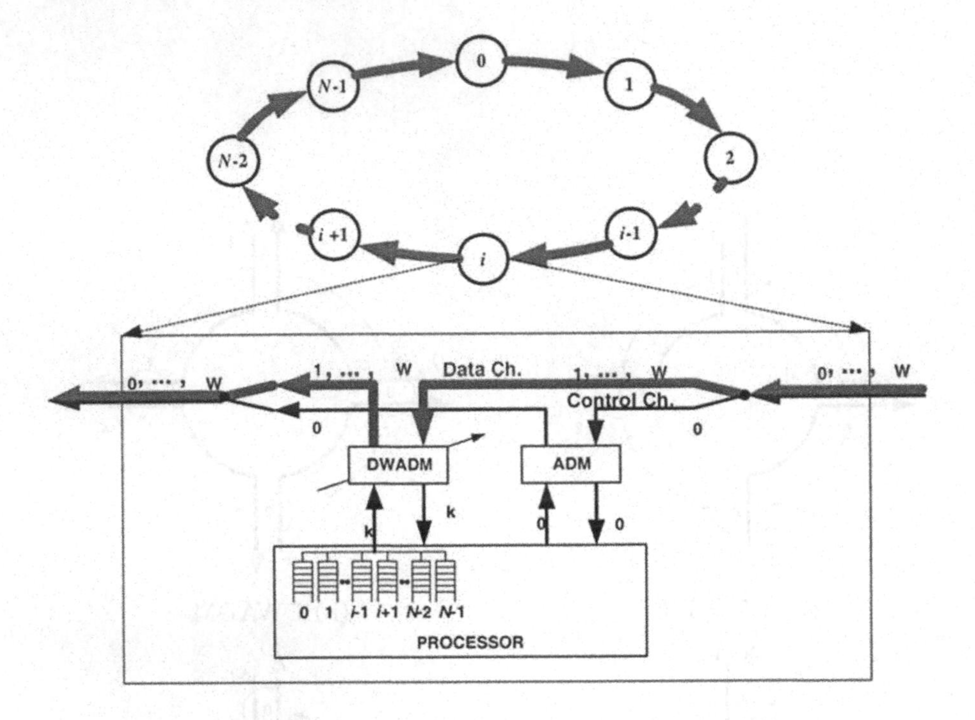

Figure 6.17 A DWADM ring network and architecture of node i.

- Each node is equipped with two ADMs — one static and one dynamic — as shown in Fig. 6.17.
  - The static (SONET) ADM at all nodes is permanently tuned to a specific channel, called the *control channel*, λ_0. Nodes communicate over λ_0 in TDM fashion to exchange (1) information on which node wants to send packets to which other node, and (2) acknowledgments if the node gets the channel to transmit, as well as the wavelength channel number.
  - The DWADM at each node is used for actual data transfer over the *data channels*, $\lambda_1 \sim \lambda_W$. Note that the information transfer now takes just a single hop over a data channel.

- A *slot* is defined as the time duration to transmit a packet from node i to the next node $i+1$. Each slot has the same time duration.

- A *cycle* is defined as a sequence of N consecutive slots, as shown in Fig. 6.18. The duration of a cycle must be equal to or longer than the propagation delay around the ring.

- All packets are of the same fixed length, and the slot size is such that one packet exactly fits into one slot.

- The ring network is globally synchronized.

- Nodes must keep one separate logical queue for every destination. Thus, $N-1$ packet queues are necessary at each node. Every queue handles packets in a first-in-first-out (FIFO) order. The queues are scanned by a pre-determined method to select a packet to transmit in a cyclic fashion.

- A packet that cannot be transmitted has no priority in the next cycle (the notion of a cycle will become clear later) and is queued in the destination queue, with newly-arrived packets, for later selection by a packet-selection protocol.

The challenge is to design an efficient protocol, which is also easy to implement, viz. how to perform a data-channel selection after a control-packet transmission?

Notations

We employ the following notations.

- D_i: Destination node of a packet to send from node i.

- C_i: Channel number for a packet to transmit from node i.

- $P_{i,j}$: A packet with destination node j generated at source node i.

- p_{ij}: Probability that a packet originating at node i is directed to node j.

- $i \oplus j = (i + j) \bmod N$

- $i \ominus j = (i - j) \bmod N$

Traffic Type

We assume that the traffic at node i is generated by an exponential distribution with mean α_i. But we should consider how to choose the destination node because the traffic characteristic is different depending on how the network is used. If the network acts as a peer-to-peer one, the destination node will be chosen with the same probability, but if the network is a client-server one, the server node will be chosen with a larger probability by client nodes. For analysis, we consider the following cases.

1. Random selection of a destination node with the same probability: All source nodes generate the same amount of traffic and all $N-1$ destinations are selected with equal probability. We call this *Balanced traffic.*

2. The pre-determined server node chooses the client destination node in $N-1$ client nodes with the same probability. The client node's destination can be either only the server node or the server node with a higher probability and the other nodes with a lower probability. We call this *Unbalanced traffic.*

6.5.3 Algorithm for Creating Virtual Paths and Assigning Wavelengths

At any given time, the channels to add a packet and to drop a packet at a node must be the same wavelength in a DWADM, like in an ADM. This may be both an advantage and a disadvantage for designing a protocol for a DWADM ring network.

Let us elaborate using the following example. If node i was scheduled to send a packet to node j via wavelength λ_k in cycle t; and if node j, the destination node of node i's packet transmission, has a packet to send to node m, how and when can node j send node m its packet to achieve the best performance?

Two cases can be considered. First, node j sends node m its packet via wavelength λ_k in cycle t in which node i also sends node j a packet. Node j should be easily able to send or receive a packet via the *same wavelength* λ_k simultaneously, like an ADM. This method yields good channel utilization, and is also easy to implement. In the second case, node j has to send node m its packet via some other wavelength in the next cycle $t+1$. Since we use a DWADM, after node i sends a packet to node j via wavelength λ_k in cycle t, node j has to send the packet to node m via wavelength λ_k or some other wavelength in cycle $t+1$. Naturally, this method wastes the channel and may lead to a larger packet delay. Finally, if node j receives a packet from node i and node j also send a packet to node i or some node l which, in turn, has a packet to send to node i, these packet-transmission paths can be combined into a circle, and we can get good performance from the DWADM ring network.

Therefore, given a traffic matrix, we create virtual circles (or paths) as much as possible by connecting as many paths as possible between source nodes and destination nodes, and then assigning them wavelengths to get good performance such as low packet delay and high channel utilization.

For sending a control packet, gathering information about traffic (packet backlog) between nodes, executing the algorithm to create virtual circle paths and assign wavelengths, and sending acknowledgments with the results, a specific node should be determined first. We call this specific node *a server node.* But, if we assume that the server node is fixed, e.g., at node 0, then all packets from node 0 will always have higher priority on sending, and this will create a fairness problem. Thus, we need to introduce another special node, *a priority node*, to solve this fairness problem. This priority node is considered first when the virtual circle path is computed by the server node. Although all of the operations are started and done by the server node, the starting node for the computation to create virtual circle paths is the priority node. Several priority-node-selection protocols are shown below.

Priority-Node-Selection Protocols

- Fixed-priority at node i (FiPi): The priority is always given to node i. If node i is node 0, this is the same when we do not consider the priority node. Since the computation to create virtual circle paths is started at node i, node i has *the best access chance* and node $i-1$ has *the worst access chance.* If node i has a packet to node $i-1$ and node $i-2$ also has a packet to the same node $i-1$, since node i has the best access chance, node $i-2$ cannot send a packet to $i-1$ in the same cycle. This protocol can be used to decrease the access delay at the special node like a server node on a client-server network.

- Random priority (RanP): The priority node is determined randomly among the N nodes.

- Round-robin priority (RRP): To solve the fairness problem that occurs when node i always has priority, priority passes from a node to the next node continuously along a ring path, even if the node has no packet to send. Since each node has even priority now, each node sends packets with equal opportunity, resulting in a good solution to the fairness problem in a DWADM ring network.

- Special-round-robin priority (SRRP): This is the same as round-robin priority except that the priority node is set only when it has a packet to send. If a node has no packet to send, priority passes to the next node which has a packet to send.

Packet-Selection Protocols

Since each node has $N-1$ destination queues, we have to design a protocol to select one destination queue from which to transmit a packet. There are seven choices.

- FIFO (FIFOS): The packets are chosen according to the order in which they were placed in the queues. This will decrease the access delay. In a strict sense, this method does not need $N-1$ queue at each node.

- Random (RanS): The packet to be transmitted is chosen at random among the non-empty queues. This protocol is easy to implement, but each node has a fairness problem when selecting destination node.

- Round-robin (RRS): The packet to be transmitted is chosen in a round-robin manner. This is a good protocol for solving the fairness problem in destination node queues.

- Counterclockwise-round-robin (CRRS): Let us say that the destination queues are scanned in a clockwise direction by the other protocols. This protocol scans the destination queues in counter-clockwise direction.

- Longest-waiting-queue-first (LWQFS): The packet to be transmitted is chosen from the longest queue. We can achieve a shorter packet delay when the system is assumed to have infinite size queues. However, we cannot guarantee fairness.

- Nearest-node-queue-first (NNQFS): The packet to be transmitted is chosen from the destination queue of the nearest node. This protocol was considered to get a better throughput by transmitting as many packet as possible by avoiding collisions. However, the access delays at the farthest nodes will be increased significantly.

- Farthest-node-queue-first (FNQFS): The packet to be transmitted is chosen from the destination queue of the farthest node. The virtual circle has smaller number of paths between the nodes because this protocol makes a long virtual path.

Thus, the number of protocols considered to create virtual paths and assign wavelengths is equal to the number of the priority-node-selection protocols times the number of the packet-selection protocols (please see Algorithm 6.3). We will show just some protocols which can explain the merits of these ideas.

Control Packet Format and Packet-Transmission Procedure

The control packet format and the packet-transmission procedure are shown in Fig. 6.18. The control packet has N fields, one for each node, and each field is subdivided into three mini-fields which are *source node address*, *destination node address*, and *channel number*. The number of slots in one cycle is equal to the number of nodes. Three cycles are needed to transmit a packet: *control-packet transmission cycle*, *acknowledgment reception cycle*, and *data-packet transmission cycle*.

Let us assume that the server node is node 0 and the priority node is node i. The control packet at node 0 includes the predetermined destination node of node 0 by the packet-selection protocol and is sent to the next node 1. Each source node address mini-field is filled at the server node when the control packet leaves the server node. Node 1 receives it and writes the destination node information in the destination node address mini-field, if node 1 has a packet to send. Then, the control packet propagates to the next node, and this procedure continues until node $N-1$ has had an opportunity to write its control information. Finally, server node 0 will get the control packet with information from all the nodes about their respective destination nodes.

After receiving the control packet, by using the algorithm, server node 0 creates virtual circle paths and assigns wavelengths to each of them. Now, the priority node should have the best packet-transmission chance relative to the other nodes.

After the computation is done, if a wavelength is assigned successfully to a node to transmit a packet, the results, i.e., the assigned wavelengths,

Algorithm 6.3 Creating virtual paths and assigning wavelengths

// **Priority-node-selection protocol**
1. $s = k$; or // Fixed-priority at node k.
2. $s = i$; or // Round-robin priority.
3. if (there is a packet to transmit at node i), $s = i$;
// Special round-robin priority.
{
// **Packet-selection protocol from the destination queues**
Select the packet P_{s,D_s}
a. first entered; or // FIFO
b. randomly; or // Random
c. in round-robin order; or // Round-robin
d. in counterclockwise round-robin; or // Counterclockwise round-robin
e. at the longest-waiting queue; or // Longest-queue-first
f. at the nearest-destination-node queue; or // Nearest-node-queue first
g. at the farthest-destination-node queue; // Farthest-node-queue first
while ((if $P_{D_s,D_{D_s}}$ exists) &&
(the path from D_s to D_{D_s} was not already assigned a wavelength)) {
Create a path from s to D_{D_s} by adding the path from D_s to D_{D_s};
$D_s = D_{D_s}$;}
Find an available channel and assign the path a channel;
if (there is no available channel to assign), Collision;
if (D_{D_s} was already assigned a wavelength), Collision;
$i = i \oplus 1$;
if (every selected packets are considered), end;
}

are included in the channel number mini-field in the control packet and are sent again to each node via the ACK reception cycle (Fig. 6.18). If a node fails to be assigned a wavelength, its corresponding field is left empty. The appropriate data packets will be sent in the next data-packet transmission cycle via the assigned wavelengths by each node simultaneously.

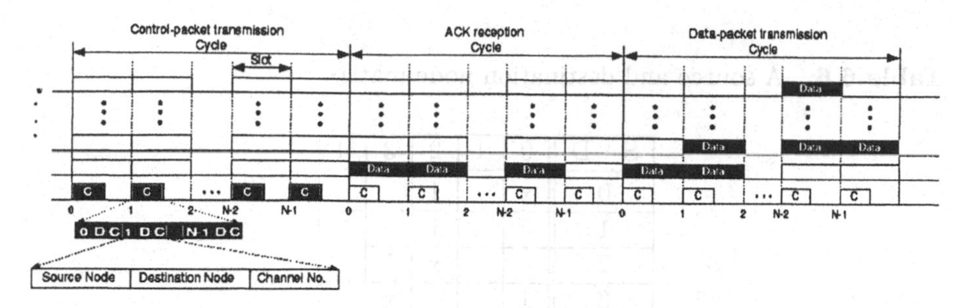

Figure 6.18 Data and control channel structure, and control packet format.

A Simple Example

Let us consider a simple example to understand these protocols. In this example, let us assume that $N = 5$, and node 0 is the server node and the priority node. By the packet-selection protocol, node 0 chose a packet to send node 2, node 1 chose node 2, node 2 chose node 3, node 3 chose node 0, and node 4 chose node 1. The control packet is shown in Table 6.5 when it arrives at node 0, the server node, after traveling around the ring.

Table 6.5 Control packet with destination node information of each node.

0	2		1	2		2	3		3	0		4	1	

And then, the server node, node 0, computes a source and destination node matrix like the one shown in Table 6.6.

Since we assumed the server node is node 0, the task of creating virtual circle paths and assigning wavelengths is done at node 0. As one can see from Table 6.6, a collision occurs between node 0 and node 1 because their packets have the same destination. Since node 0, the priority node, has priority, node 1 cannot have a chance to transmit a packet in this cycle. After the virtual path from node 0 to node 2 is created, since the destination of node 2 is node 3, this path can be added to the virtual circle path previously created from node 0 to node 2. So, the resulting virtual path is from node 0 to node 3. Fortunately, since node 3 wants to send a packet to node 0, one long virtual

Table 6.6 A source and destination node matrix.

S \ D	0	1	2	3	4
0	x		1		
1		x	1		
2			x	1	
3	1			x	
4		1			x

circle path from node 0 to node 2, and to node 3, and to node 0 is created, and it is assigned as wavelength λ_1. Since node 4 has a packet to node 1, and node 1 cannot send or receive any packet in this cycle, node 4 can send a packet to node 1 using some other wavelength such as λ_2 if this system has more than two wavelengths. Thus, only node 1 cannot access the network. If node 1 has a priority in the next cycle or there are no packets to send from other nodes, node 1 will be able to send the packet in the next cycle. This configuration can be drawn as shown in Figs. 6.19(a) and (b). In addition to this result, if we change the priority node to node 1, the result will be also change as shown in Figs. 6.19(c) and (d). After assigning the channels, node 0 will send the control packet with the computed channel number as an acknowledgment shown in Table 6.7. As shown in Fig. 6.19, the results can be changed significantly by the priority-node selection.

Table 6.7 Control packet with assigned channel number of a simple example.

0	2	1	1	2	x	2	3	1	3	0	1	4	1	2

6.5.4 Simulation and Illustrative Results

This subsection presents numerical results (via simulation) for a 16-node DWADM ring network where each node has 15 destination queues to store packets awaiting transmission. The packet delay is the delay encountered

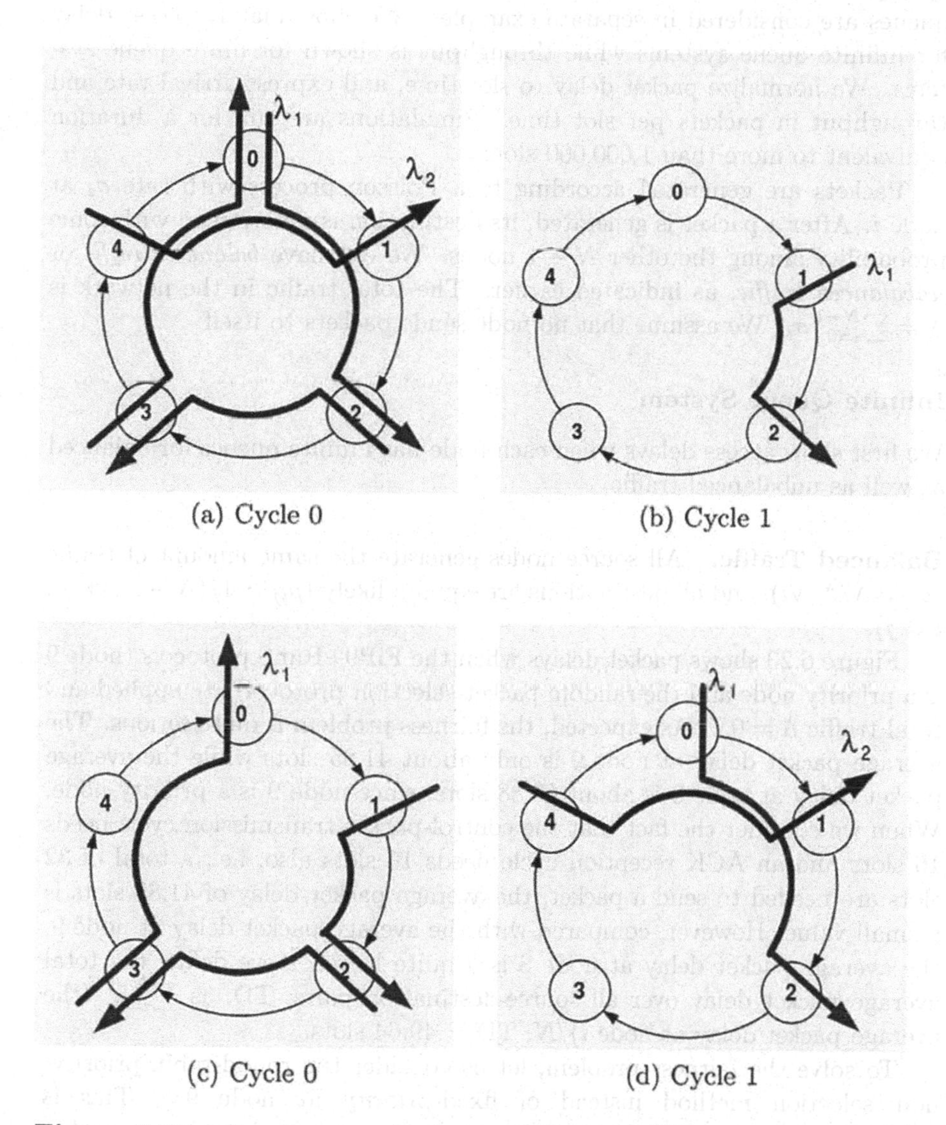

Figure 6.19 Results of a simple example with different priority-node selection.

by a packet prior to accessing the network. Both infinite queues and finite queues are considered in separate examples. We show mainly access delay for infinite queue systems while throughput is shown for finite queue systems. We normalize packet delay to slot time, and express arrival rate and throughput in packets per slot time. Simulations are run for a duration equivalent to more than 1,000,000 slots.

Packets are generated according to a Poisson process with rate σ_i at node i. After a packet is generated, its destination is determined with some probability among the other $N-1$ nodes. We can have *balanced traffic* or *unbalanced traffic*, as indicated earlier. The total traffic in the network is $\Lambda = \sum_{i=0}^{N-1} \sigma_i$. We assume that no node sends packets to itself.

Infinite Queue System

We first show access delays when each node has infinite queues for balanced as well as unbalanced traffic.

Balanced Traffic. All source nodes generate the same amount of traffic ($\sigma_i = \Lambda/N, \forall i$), and all destinations are equally likely ($p_{ij} = 1/(N-1), \forall i,j : i \neq j$).

Figure 6.20 shows packet delays when the FiP9+RanS protocols (node 9 is a priority node and the random packet-selection protocol) are applied and total traffic $\Lambda = 0.1$. As expected, the fairness problem is quite serious. The average packet delay at node 9 is only about 41.85 slots while the average packet delay at node 8 is about 60.88 slots, since node 9 is a priority node. When we consider the fact that the control-packet transmission cycle needs 16 slots and an ACK reception cycle needs 16 slots also, i.e., a total of 32 slots are needed to send a packet, the average packet delay of 41.85 slots is a small value. However, compared with the average packet delay at node 9, the average packet delay at node 8 is a quite large. If we define the total average packet delay over all source-destination pairs, TD, as $\sum_{i=0}^{N-1}$ (the average packet delay at node i)$/N$, TD = 49.64 slots.

To solve the fairness problem, let us consider the round-robin priority-node-selection method instead of fixed-priority at node 9. This is RRP+RanS protocols. Figure 6.21 shows that, now, the average packet delays for nearer nodes from node i such as $i+1$ are smaller than those farther downstream from node i such as $i-1$ (assuming cyclic symmetry). This is because there are more collisions in access from node i to node $i-1$ than from node i to node $i+1$. Thus, because there are the best-access-

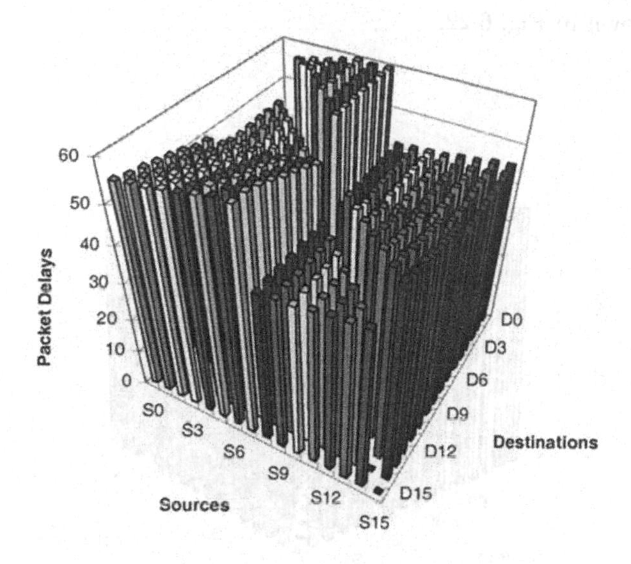

Figure 6.20 Packet delays for FiP9+RanS protocols with balanced traffic, $\Lambda = 0.1$. (S0 - S15 are source nodes and D0 - D15 are destination nodes.)

chance destination node and the worst-access-chance destination node at each source node, a slight fairness problem on destination nodes still exists. However, when we compute the average packet delay at each source node, we find that there is no fairness problem between the source nodes, as expected, because of cyclic symmetry, as shown in Fig. 6.22. When $\Lambda = 0.1$, TD = 49.1 slots (=3.07 cycles), which is smaller than that for the previous FiP9+RanS case (49.64 slots).

Like the round-robin priority-node-selection protocol, if we select a packet from destination queues sequentially (i.e., round-robin packet selection), i.e., RRP+RRS protocols, we get results similar to those in Fig. 6.21 but with smaller values of packet delay. TD is only 46.42 slots which turns out to be the lowest average packet delay for the three cases using the RanS packet-selection protocol. The fairness problem is also less significant now, as shown in Fig. 6.22.

Using the RanP+RRS protocols, the packet delays are very similar to those in Fig. 6.21; however, now, we have the best result, with TD = 46.19

slots, as shown in Fig. 6.22.

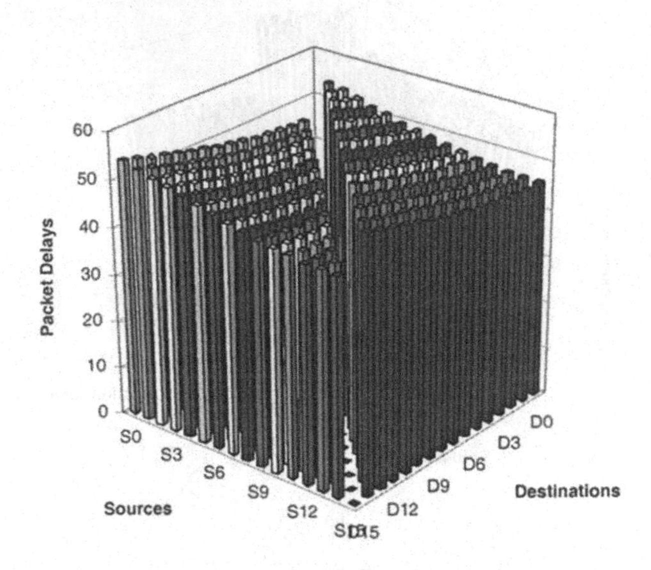

Figure 6.21 Packet delays for RRP+RanS protocols with balanced traffic, $\Lambda = 0.1$.

When we increase the total balanced traffic Λ to 1, we get the result of RanP+RRS protocols shown in Fig. 6.23(a). Compared with Fig. 6.21, Fig. 6.23(a) shows a significant delay difference between the packet delay on the destination queue for node $i-1$ and the packet delay on the destination queue for node $i+1$ at node i, although the fairness problem on each source node does not show up much in Fig. 6.23(b). Therefore, by increasing the total balanced traffic, the packet delays also increase significantly, while the fairness issues remain the same as before.

Unbalanced Traffic. We use the following model for unbalanced traffic. One server node (say node 0) is present in the network, and the remaining $N-1$ client nodes direct half of their traffic to the server node, while the other half of their traffic is uniformly distributed among the other client nodes. In this study, we assume that all nodes generate the same amount of traffic ($\sigma_i = \Lambda/N, \forall i$), but destinations are distributed differently ($p_{0j} = 1/(N-1), \forall j \neq 0$, $p_{ij} = [1/2(N-2)]$, $\forall i$, $j \neq 0$, $i \neq j$, and $p_{i0} = 1/2$,

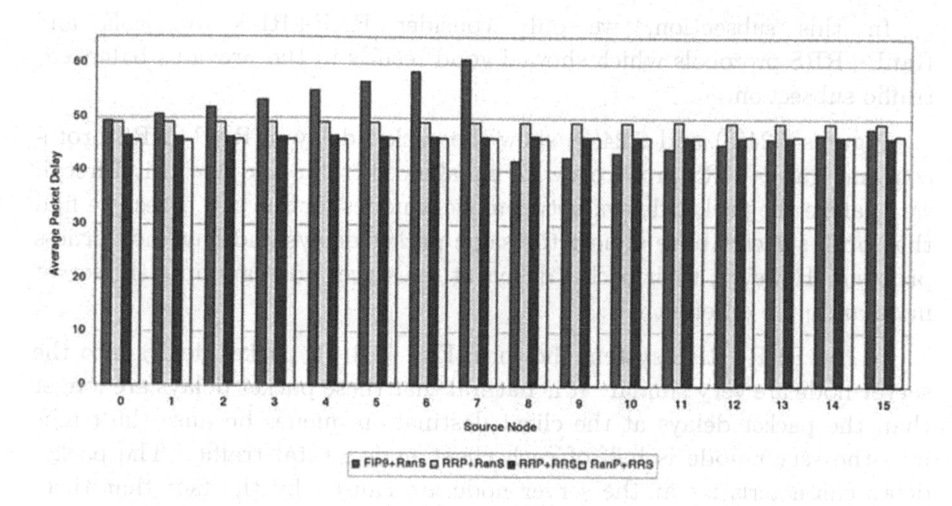

Figure 6.22 Comparison of average packet delays of protocols with balanced traffic, $\Lambda = 0.1$.

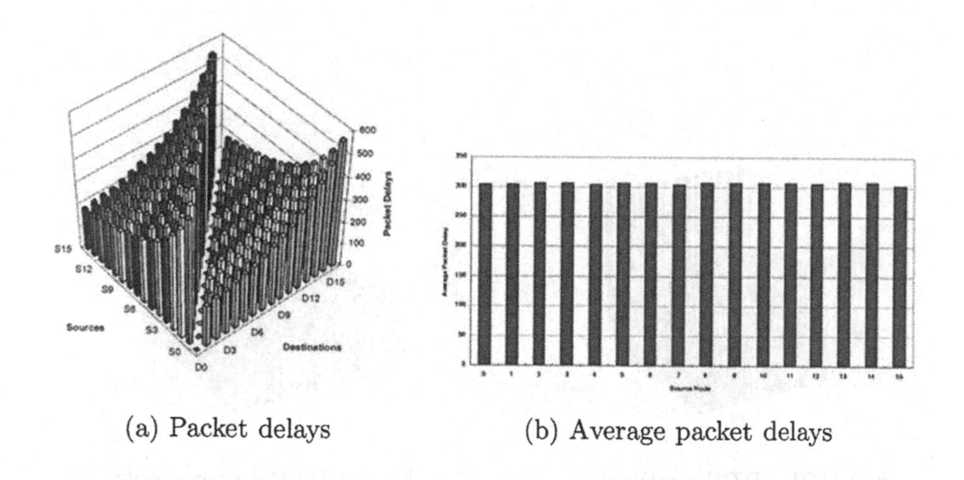

(a) Packet delays (b) Average packet delays

Figure 6.23 Delays for RanP+RRS protocols with balanced traffic, $\Lambda = 1$.

$\forall i \neq 0$).

In this subsection, we only consider RRP+RRS protocols and RanP+RRS protocols which showed good results in the previous balanced-traffic subsection.

Figures 6.24(a) and 6.24(b) show the packet delay of RRP+RRS protocols and RanP+RRS protocols with unbalanced traffic and $\Lambda = 0.1$. First, if we analyze the packet delays between client nodes for the two cases, we find that both systems have almost the same packet delays and have no fairness problem. However, their packet delays at the server node and into the server node are quite different.

In the RRP+RRS system shown in Fig. 6.24(a), packet delays into the server node are very similar. It is natural that these packet delays are longer than the packet delays at the client destination queues because the traffic into the server node is half of each client node's total traffic. The packet delay characteristics at the server node are caused by the fact that there are the best-access-chance destinations and the worst-access-chance destinations. Moreover, the traffic at the server node is very high, thereby resulting in the large difference between the delay for the best-access-chance destination and that for the worst-access-chance destination. At the server node, node 0, the best chance node is node 1 and the worst chance node is node 15.

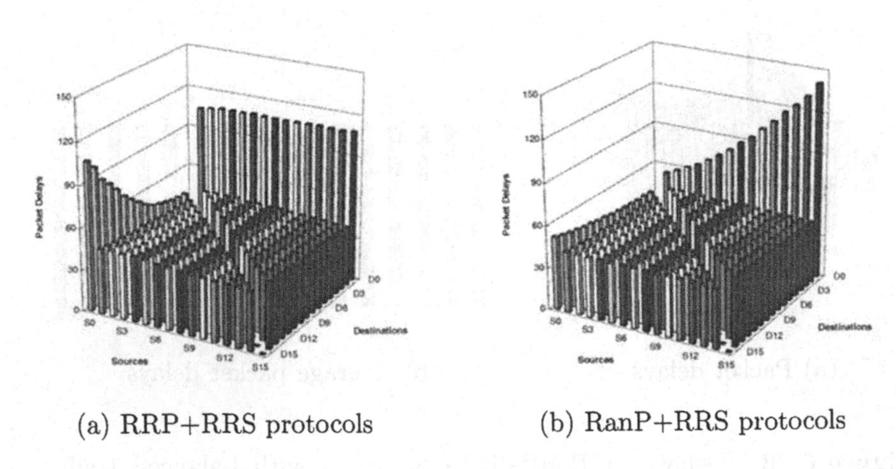

(a) RRP+RRS protocols (b) RanP+RRS protocols

Figure 6.24 Packet delays with unbalanced traffic, $\Lambda = 0.1$.

Finite Queue System

Let us now show the results when each destination queue is finite. We show throughputs now. We assume that each node can transmit at most N packets in each cycle. For this assumption, each destination queue size is greater than N. We simulate systems with a destination queue size of $2N$ system and a destination queue size of $3N$ under balanced and unbalanced traffic, respectively.

Balanced Traffic. Let us define total throughput (TT) as the sum of the throughput between each node-pair. The simulation results, TTs, of both balanced traffic and unbalanced traffic are shown in Table 6.8. We vary $\Lambda = 1$, 2, and 4, when the queue size is 32 and 48. Just like the result for the previous infinite queue system, the RanP protocol shows good TT for balanced traffic, while the FiP9 protocol has the worst TT result. The RanS protocol has better TT than RRS. So, below, we only consider the RanP+RanS protocols with finite buffer size.

Figure 6.25 shows throughputs of RanP+RanS protocols with buffer size 32 and balanced traffic when $\Lambda = 1$ and 4. While Fig. 6.25(a) (at low load) shows good fairness between nodes, by increasing the load, the worst-access-chance nodes, $i-1$, $i-2$, $\cdots$, at node i have worse throughput than the best-access-chance nodes, $i+1$, $i+2$, $\cdots$, at node i as Fig. 6.25(b) shows. Thus, increasing load should make this throughput difference larger between the best-access-chance node and the worst-access-chance node.

Unbalanced Traffic. Let us consider protocols of unbalanced traffic with finite size buffer. As both Table 6.8 and the previous infinite-buffer subsection show, the RRP protocol has better performance than the RanP protocol and the FiPi protocol. Although the RRP protocol and the RanP protocol show similar TT, these two protocols have distinct characteristics as shown in Fig. 6.24. We consider only these two protocols in this subsection.

Figure 6.26 shows throughputs of RanP+RanS protocols with buffer size 32 and unbalanced traffic when $\Lambda = 1$ and 4. There is not much fairness problem in both cases except that nodes near the server node have a better throughput at the destination queue from a client node to the server node. This means that the priority-node-selection protocol is not affected greatly at the destination queue from a client node to the server node as in Fig. 6.24(b) about the RanP+RRS protocols. By increasing Λ to 4, fairness is not a problem, as shown in Fig. 6.26(b). However, throughputs from the server

Table 6.8 Comparison of throughputs of protocols with balanced and unbalanced traffic.

Λ		1	2	4	1	2	4
Queue size		32	32	32	48	48	48
Balanced	FiP9+RanS	0.9949	1.8825	3.1849	0.9984	1.9248	3.2692
	FiP9+RRS	0.9972	1.8663	2.6869	0.9987	1.9121	2.8628
	RanP+RanS	1.0000	1.9943	3.7794	1.0000	1.9992	3.8854
	RanP+RRS	0.9990	1.9932	3.3799	0.9990	1.9980	3.4760
	RRP+RanS	0.9991	1.9943	3.7807	0.9991	1.9980	3.8847
	RRP+RRS	0.9999	1.9932	3.2688	0.9999	1.9989	3.3474
Unbalanced	FiP9+RanS	0.7133	1.3349	2.3804	0.7165	1.3458	2.4329
	FiP9+RRS	0.7638	1.4815	2.3428	0.7697	1.4984	2.4074
	RanP+RanS	0.9109	1.5062	2.5232	0.9393	1.5201	2.5470
	RanP+RRS	0.8782	1.3918	2.3568	0.9054	1.3959	2.3719
	RRP+RanS	0.9923	1.7507	2.7578	0.9982	1.7666	2.7809
	RRP+RRS	0.9727	1.5929	2.5172	0.9881	1.6051	2.5307

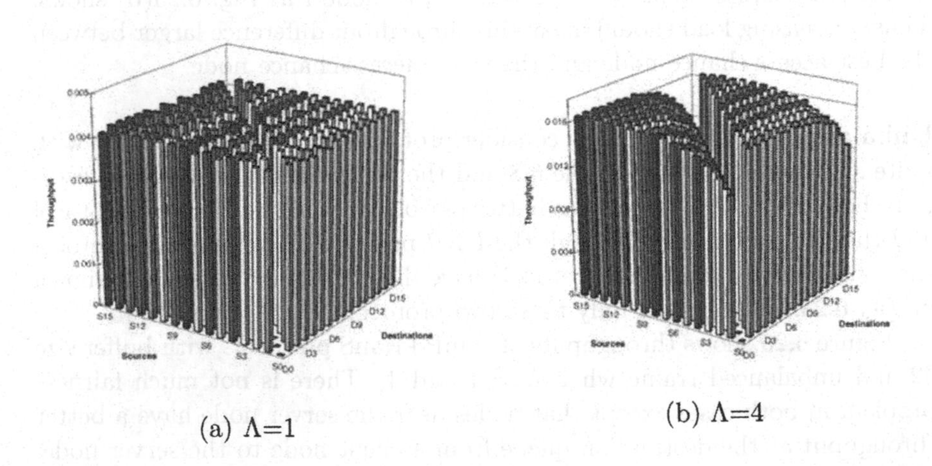

(a) Λ=1 (b) Λ=4

Figure 6.25 Throughputs of RanP+RanS protocols with buffer size 32 and balanced traffic.

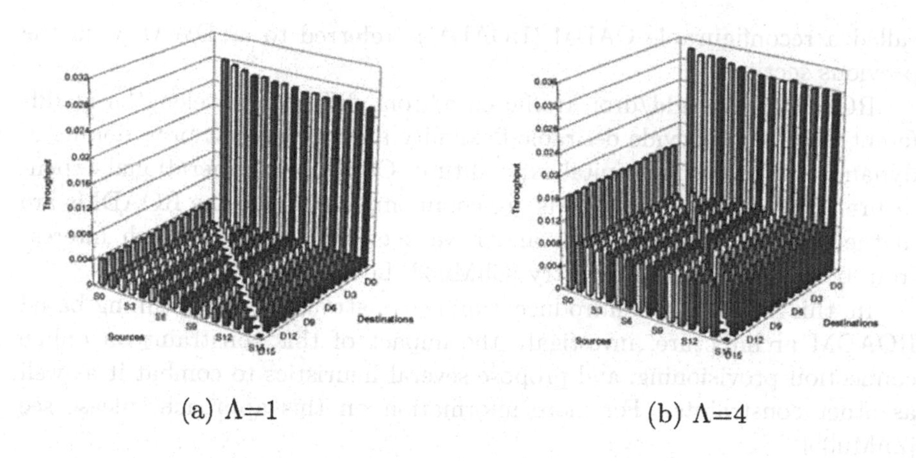

(a) $\Lambda=1$ (b) $\Lambda=4$

Figure 6.26 Throughputs of RanP+RanS protocols with buffer size 32 and unbalanced traffic.

node to client nodes are increased about four times while throughputs from client nodes to the server node are increased a little.

6.6 Online Connection Provisioning using Reconfigurable OADMs (ROADMs)

6.6.1 Introduction

A simple and important building block for a WDM network is the Optical Add-Drop Multiplexer (OADM). Given that most networks deployed by telecom service providers were implemented as (single-wavelength) fiber-optic SONET rings, it is relatively straightforward to upgrade them to operate over multiple wavelengths by employing OADMs and appropriate terminal equipment at the network nodes, i.e., the same fiber cable plant can be reused. In such a W-wavelength ring network (where W can be as high as 32 or 64 today), an OADM "drops" information from and "adds" information to the ring on a given wavelength on which it is set to operate. OADMs can significantly reduce the network cost by allowing traffic to bypass intermediate nodes without expensive O-E-O conversion [GiSp99]. An OADM can be made to operate permanently (statically or factory tuned) on a particular wavelength, in which case it is called a fixed OADM (FOADM); or it can be dynamically tuned using some control mechanism and then it is called a reconfigurable OADM (ROADM) (referred to as DWADM in the previous section).

ROADMs can add/drop traffic onto/from different wavelengths at different time, and provide desirable flexibility and enable fast provisioning of dynamic traffic, saving capital expenditure (CapEx) and operational expenditure (OpEx). While FOADMs are commonplace today, the ROADMs are in the early technology stages and have been attracting research interest from both academe and industry [ChMu03, LeSi03].

In this section, we introduce tuning constraint for the tuning-based ROADM architecture, investigate the impact of this constraint on online connection provisioning, and propose several heuristics to combat it as well as other constraints. For more information on this approach, please see [ZhMu05].

6.6.2 Tuning Constraint

ROADMs can have several different architectures. One architecture employs switches to achieve add/drop functionality, and we call it switching-based architecture (shown in Fig. 6.27). Another ROADM architecture uses tunable devices, e.g., fiber Bragg grating (FBG) or thin film filter (TFF), to selectively add/drop a designated wavelength, and we call it tuning-based architecture (shown in Fig. 6.28). The tuning-based architecture is more cost-effective than the switching-based one, and hence we focus on the tuning-based architecture.

In the tuning-based architecture in Fig. 6.28, a ROADM adds/drops different wavelengths by tuning the tuning head to the corresponding wavelengths. If a ROADM is not adding/dropping, the tuning head can stay or park on any free wavelength, or between any two adjacent wavelengths.

To set up a connection, ROADMs at the source and destination nodes must be tuned to the same free wavelength. As the tuning process is virtually continuous, the tuning head will pass through all the wavelengths between the current position and chosen wavelength. However, if the tuning process passes through some *working wavelengths*, it will cause a service interruption on the traffic carried by those wavelengths, which is unacceptable and should be avoided.

Consider the case where the current positions of the tuning heads of ROADM 1 and ROADM 2 are on wavelengths λ_1 and λ_4, respectively, as shown in Fig. 6.29, and a connection needs to be set up between these two ROADMs. Suppose there is traffic on wavelength λ_3 bypassing these two

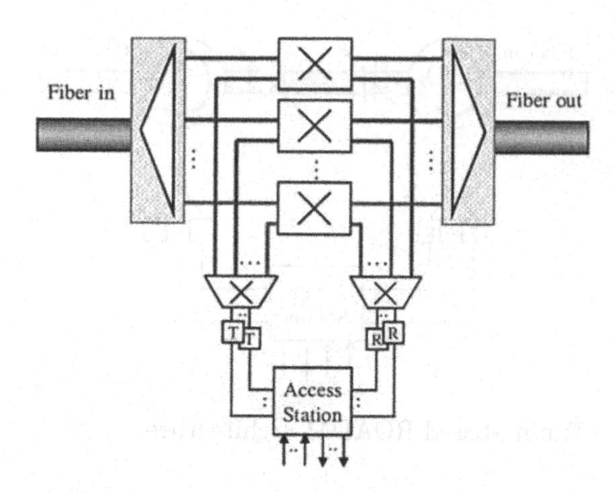

Figure 6.27 Switching-based ROADM architecture.

ROADMs. Since the tuning heads of the ROADMs are on the different sides of wavelength λ_3, the two ROADMs cannot tune to the same wavelength without crossing wavelength λ_3 and the connection has to be blocked. This imposes another constraint on setting up a connection, and we call it the *tuning constraint.* Since the tuning head of a ROADM is not allowed to cross working wavelengths, it can only tune within a certain *tuning range.* Note that the tuning range varies dynamically as new connections are set up and existing connections are terminated.

6.6.3 Problem Statement

A network node can have multiple ROADMs in order to add/drop multiple wavelengths simultaneously. Two or more ROADMs at a node may be connected serially, so adding/dropping at one ROADM may also affect the tuning range of the other ROADMs at the same node. When setting up a connection, we need to determine which ROADMs should be used at the source and destination nodes, as well as the wavelength and direction, if the ring is bidirectional. The chosen wavelength must be in the tuning range of both the ROADMs. This is called *direction, wavelength and ROADM assignment* (DWRA) subproblem.

Moreover, since the connection may affect the tuning range of the bypassed ROADMs, it should be decided to which positions the tuning heads of those ROADMs should be tuned before the connection is established. This

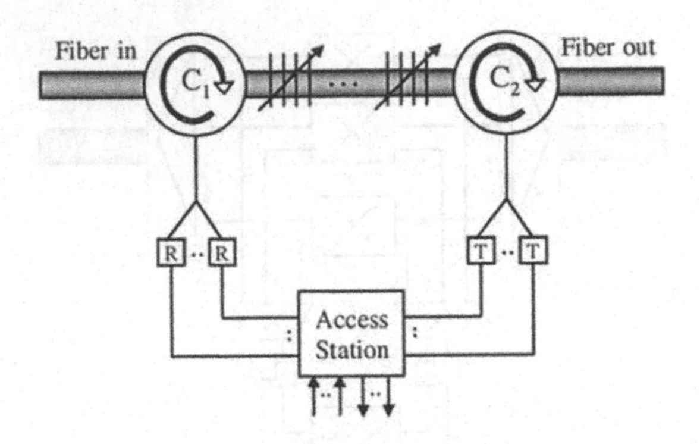

Figure 6.28 Tuning-based ROADM architecture.

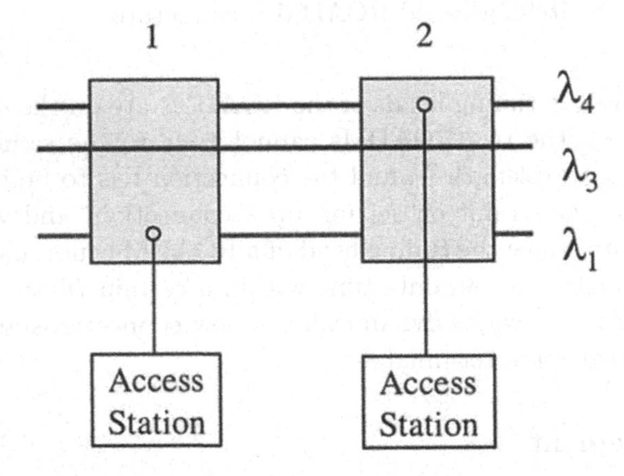

Figure 6.29 An illustration of tuning constraint.

decision will determine the tuning ranges of the ROADMs, and hence has significant impact on the establishment of future connections. This is called the *tuning-head positioning* (TP) subproblem.

Thus, the online provisioning problem can be formulated as follows.

- Given: a WDM ring network using tuning-based node architecture with tuning constraint, and dynamic traffic, which arrives one by one.
- Determine: direction, wavelength and ROADM assignment (DWRA) for the connection, and tuning-head positioning (TP) for the affected

ROADMs.

- Objective: minimize the overall connection blocking probability.

6.6.4 Heuristics

The tuning constraint is a very important factor when setting up connections. A connection cannot be set up if the ROADMs at the source and destination nodes cannot tune to the same wavelength, no matter how many ROADMs are free at both end nodes and how many wavelengths are available along the path.

When solving the DWRA problem, several factors need to be considered:

- *Hop distance (bidirectional rings).* If the hop distance is shorter, fewer wavelength-links are used. So, it is more likely that fewer ROADMs will be affected by the connection.

- *Utilization of ROADM.* A ROADM has two ports: add-port and drop-port. A ROADM can use one port without using the other port (since connections are assumed to be unidirectional). However, using only one port does not fully utilize the capability of the ROADM. In addition, it limits the flexibility of the ROADM in the sense that the ROADM cannot tune to other wavelengths. If the connection can be set up by using the unused port of a working ROADM, i.e., two connections share one ROADM, then a ROADM can be saved.

- *Number of affected ROADMs.* Different solutions for the DWRA subproblem may affect different number of ROADMs along the path. The connection will partition the tuning range of the affected ROADMs, causing the tuning range to shrink, so the flexibility of the ROADMs will be reduced.

All of these goals may not be achievable at the same time. If we prioritize these goals, we can get various solutions. Below, we propose four heuristics for the DWRA subproblem.

- *Least-Hop-First-Fit (LHFF)* chooses the solution that has fewest hop count and first-fit wavelength assignment for tie-breaking.

- *Least-Affected-ROADMs-First-Fit (LAFF)* chooses the solution that affects the least number of ROADMs, and first-fit wavelength assignment for tie-breaking.

- *Least-Affected-ROADMs-Maximum-Sharing (LAMS)* chooses the solution that affects the least number of ROADMs. If there are multiple solutions, the one that can share more ROADMs with existing connections will be chosen.
- *Least-Hop-Least-Affected-ROADMs (LHLA)* chooses the solution that has the fewest hop count, and if there is a tie, the one that affects the least number of ROADMs will be chosen.

We study three heuristics for tuning-head positioning (TP) subproblem.

- *Tune-to-Large (TL).* When the tuning range is divided into two parts, this heuristic always positions the tuning head in the part whose range is larger.
- *Proportional-Tuning (PT).* PT uses a probabilistic method to determine the tuning. The probability that the tuning head will fall in a part is proportional to the size of the tuning range of that part.
- *Max-Cover-Tuning (MCT).* In the above two heuristics, the tuning head of a ROADM is positioned independently. MCT tries to position the tuning heads of the ROADMs at the same node in a more coordinated way. It positions the tuning heads such that the number of wavelengths that can be tuned to by at least one ROADM at the node after setting up the connection is maximum. If there are multiple solutions, it will choose the one with the maximum number of wavelengths that can be tuned to by at least two ROADMs at the node. It repeats this procedure for tie-breaking until it finds the solution.

6.6.5 Illustrative Numerical Examples

In our simulation experiments, the network is a bidirectional WDM ring with 8 nodes. Each node has 4 ROADMs for each direction, and each unidirectional link has 8 wavelengths. Traffic (consisting of unidirectional connection requests) is uniformly distributed among all the node pairs, and each connection requires the entire bandwidth of a wavelength. The traffic arrival is a Poisson process and the connection-holding time is exponentially distributed (whose average value is normalized to unity in our studies reported here). We simulated 100,000 connection requests to obtain the network performance using different heuristics.

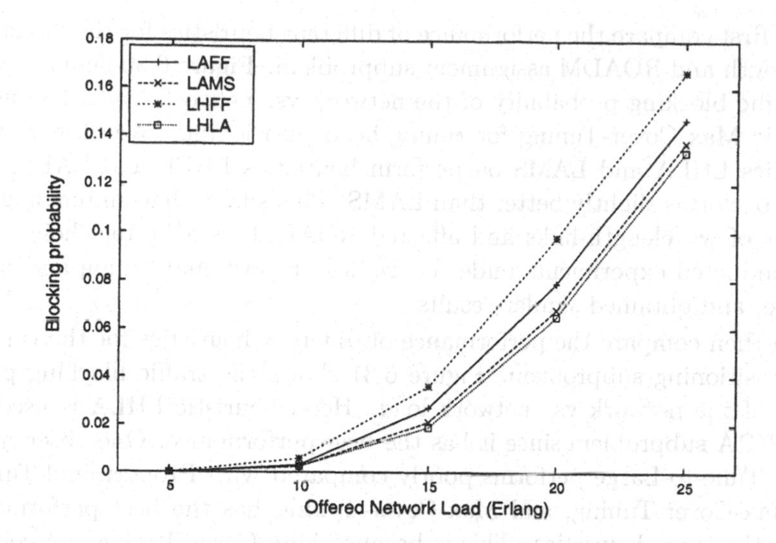

Figure 6.30 Comparison of heuristics for DWRA subproblem with MCT tuning.

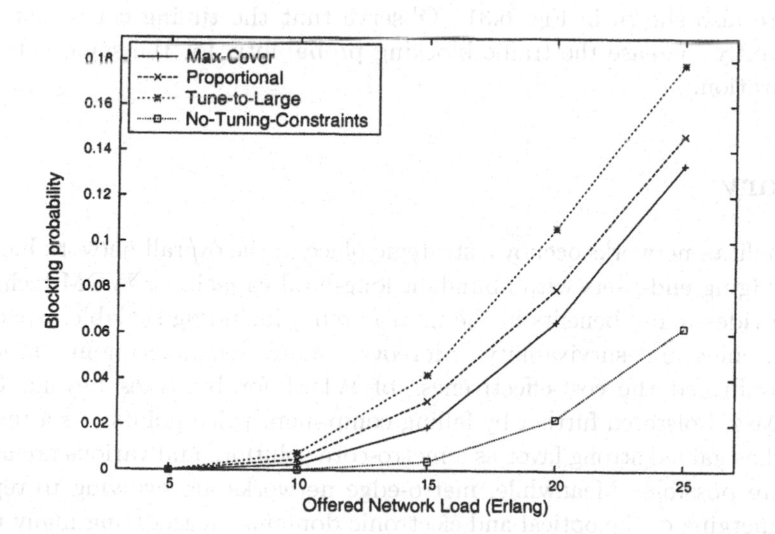

Figure 6.31 Comparison of heuristics for TP subproblem with LHLA for DWRA subproblem.

We first compare the performance of different heuristics for the direction, wavelength and ROADM assignment subproblem. Figure 6.30 demonstrates the traffic blocking probability of the network vs. network load when using heuristic Max-Cover-Tuning for tuning-head positioning. We observe that heuristics LHLA and LAMS outperform heuristics LHFF and LAFF, and LHLA performs slightly better than LAMS. This shows that minimizing the number of wavelength-links and affected ROADMs is a better choice. We also conducted experiments under heuristics proportional-tuning and tune-to-large, and obtained similar results.

We then compare the performance of different heuristics for the tuning-head positioning subproblem. Figure 6.31 shows the traffic blocking probability of the network vs. network load. Here, heuristic LHLA is used for the DWRA subproblem since it has the best performance. One observation is that Tune-to-Large performs poorly compared with Proportional-Tuning and Max-Cover-Tuning, and Max-Cover-Tuning has the best performance among the three heuristics. This is because Max-Cover-Tuning can coordinate the tuning heads of the ROADMs at a node to achieve larger tuning range for the node. To evaluate how the tuning constraint will impact the blocking probability, we conducted experiments on the WDM ring network without tuning constraint using first-fit wavelength assignment, and the results are also shown in Fig. 6.31. Observe that the tuning constraint will significantly increase the traffic blocking probability, for the same network configuration.

6.7 Summary

Metropolitan networks occupy a strategic place in the overall network hierarchy, bridging end-users with abundant long-haul capacities. WDM technology provides many benefits in the metro arena, including scalable capacity, transparency, and survivability. Moreover, many techno-economic studies have confirmed the cost-effectiveness of WDM for bit rates beyond OC-12/STM-4, bolstered further by falling component price points. As a result, WDM has gained strong favor as a metro-core solution, and various architectures are possible. Meanwhile, metro-edge networks are evolving to represent a merging of the optical and electronic domains, aggregating many user protocols onto large metro-core wavelength tributaries. Many metro-edge solutions have been proposed, ranging from next-generation SONET/SDH and multi-service provisioning platforms, to "IP-based" packet rings. The choice of edge solution will clearly depend on an individual operator's needs, but over time, those incorporating WDM technology will be most beneficial and hence will likely gain prominence.

In Section 6.2, we gave an overview of research on traffic grooming in SONET/WDM ring networks, including node architecture, single-hop grooming, multihop grooming, dynamic grooming, and traffic grooming in interconnected rings.

In Section 6.3, we first provided formal mathematical specifications of the traffic-grooming problem in several ring networks, i.e., single-hop and multihop (with a single hub) cases of unidirectional and bidirectional rings. Then, we proposed a simulated-annealing-based traffic-grooming algorithm for the single-hop case and a greedy heuristic for the multihop case. The simulated-annealing-based heuristic overcomes the sub-optimal problem caused by the two-step strategy in previous greedy heuristic. It was also shown that, for non-uniform traffic, the greedy approach usually is good enough when compared with the simulated-annealing algorithm. The multihop approach could achieve more ADM savings when the grooming ratio is neither too small nor too large, but it usually results in more wavelength usage due to the prolonged connection length.

In Section 6.4, we compared four interconnection strategies ranging from using traditional digital crossconnects to various optical crossconnects. Both mathematical formulations and heuristic approaches were presented for solving the traffic-grooming problem in interconnected rings. For this particular problem, ILP formulations are useful in understanding the nature of the problem, but not efficient for solving practical-size problems. The heuristics discussed in this study improved the existing ones by adding full-circle-forming strategies which improve the performance of the interconnected-ring environment. Our results show that TDM-switching capability will be useful in reducing the total network cost if the grooming ratio is neither too small nor too large. The pure optically-crossconnected strategies provide better SADM cost savings when the grooming ratio is not too large; plus, they always save more wavelengths and switch port count. The OXCs with TDM-switching function can always take advantage of either side. A true mixed-switching strategy is difficult to design, so a more practical solution is to either use pure optical switching or TDM-level switching depending on which one can yield more benefit.

In Section 6.5, a new ring architecture and protocols were introduced that use a SONET ADM and a DWADM at each node which have the advantage of WDM and an ADM. The DWADM is an emerging device which could turn out to be very inexpensive, so we explored its usage in multi-wavelength ring networks for dynamic packet transport. Packet delays and throughputs of proposed protocols were computed by simulation when traffic is balanced and unbalanced and when the queue size is infinite and finite. Although the DWADM has the limitation that a node must transmit and receive on the same wavelength at the same time, we find that, if both the priority-node-selection protocol and the packet-selection protocol are chosen well, then shorter packet delays can be obtained, and the fairness problem can also be solved.

ROADMs (another name for DWADMs) provide a new dimension of flexibility to the network, enable online connection provisioning, and reduce the cost of the network. When a tuning-based ROADM tunes from one wavelength to another wavelength, it is not allowed to cross a working wavelength because it will interfere with the traffic carried by the wavelength. The tuning constraint of tuning-based ROADMs plays an important role in the online connection provisioning environment. In Section 6.6, we investigated the online provisioning problem and divided it into two subproblems: direction, wavelength, and ROADM assignment (DWRA) subproblem and tuning-head positioning (TP) subproblem. Several heuristics were proposed for each subproblem, and their performances were compared via simulation.

Exercises

6.1. For interconnected double rings, what switching strategy should we adopt for small, medium, and large traffic-grooming ratio? Explain why.

6.2. Consider an optical metro ring networks consisting of five nodes, where node 0 is a hub node. OC-48 rings can be supported, which can groom four OC-12 connections. Traffic originates only from node 0 and is transmitted to the other nodes based on the following 1×5 traffic vector:

$$\begin{pmatrix} 0 & 2 & 3 & 3 & 4 \end{pmatrix}$$

The entry (i, j) in the vector represents the number of OC-12 connections of traffic demand from node i to node j. There are three wavelengths available. Design the rings so that the minimum number of ADMs is employed in the network.

6.3. For the network topology solution obtained in Problem 6.2, draw the node architecture of node 3. Indicate which switches are in the cross state and which are in the bar state.

6.4. Consider the network in Fig. 6.32. The network is an unidirectional ring with two wavelengths (λ_1 and λ_2) and two timeslots on each wavelength channel (C_1 and C_2). Assume that the traffic matrix $T = [t_{ij}]$ is 3×3.
(a) Write the objective function for node 1.
(b) Write the traffic-load constraints for the connection between node 1 and node 2 (assume $t_{12} = 3$).
(c) Write the channel-capacity constraints for the physical link $L_{12}^{C_1\lambda_1}$
(d) Write the transmitter and receiver constraints for node 1. What is the grooming ratio?

6.5. The bidirectional ring network in Fig. 6.33 has uniform traffic request, with grooming ratio 3. Each request is one unit of sub-channel capacity. Find a feasible solution to put all the connection requests on two wavelengths such that the minimum number of ADMs are used. Verify the minimum ADMs used with the tabular result given by the ILP solver using the simulated-annealing-based traffic-grooming algorithm.

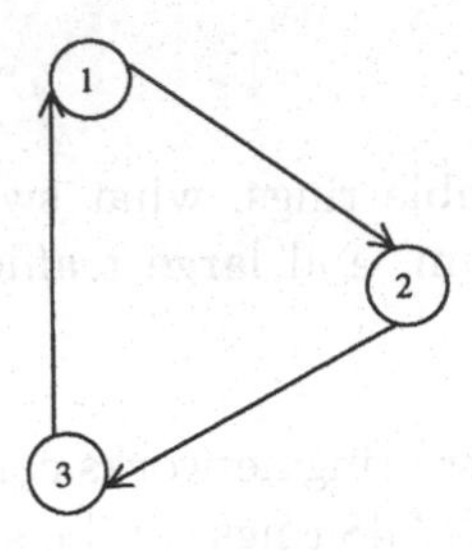

Figure 6.32 3-node unidirectional network.

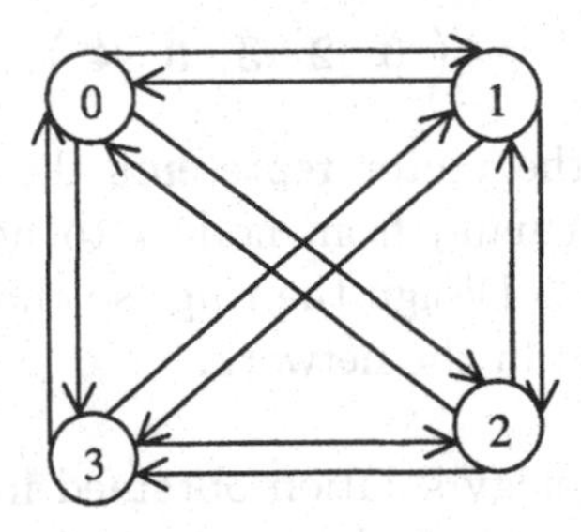

Figure 6.33 4-node bidirectional network.

6.6. Consider the bidirectional single-point-connected double-ring network shown in Fig. 6.34. Given the following traffic matrix

$$A = \begin{pmatrix} 0 & 2 & 4 & 6 & 8 & 10 & 12 \\ 1 & 0 & 4 & 3 & 2 & 1 & 7 \\ 3 & 8 & 0 & 5 & 7 & 9 & 3 \\ 5 & 7 & 11 & 0 & 1 & 4 & 6 \\ 7 & 6 & 3 & 8 & 0 & 12 & 2 \\ 9 & 5 & 8 & 2 & 13 & 0 & 9 \\ 11 & 4 & 14 & 7 & 6 & 5 & 0 \end{pmatrix}$$

Show:
(a) the decomposed traffic matrix on each ring for strategy 1 (DXC), and
(b) one reasonable decomposition of the traffic matrix on each ring for strategy 2 (OXC).

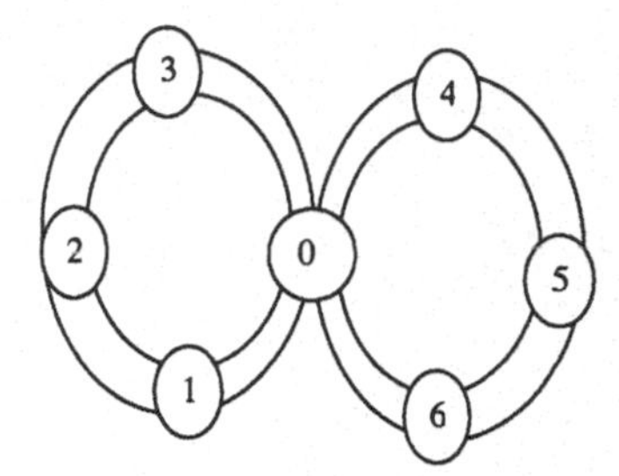

Figure 6.34 Single-point-connected double-ring network.

6.7. Given the following traffic matrix for an unidirectional ring:

$$A = \begin{pmatrix} 0 & 1 & 3 & 0 \\ 2 & 0 & 0 & 0 \\ 1 & 1 & 0 & 1 \\ 1 & 0 & 1 & 0 \end{pmatrix}$$

Assume $c = 1$ and transmission in the clockwise direction only. Give a virtual connection assignment. Calculate the number of ADMs and the number of wavelengths required.

6.8. Describe the relation between grooming ratio and the required ADMs of strategy 1 (DXC) compared to the strategies 2 and 3 (OXC) under uniform traffic. Explain why this is so.

6.9. For the mathematical problem formulation for traffic grooming in a single-hub multihop networks, extend it to the bidirectional formulation for uniform traffic, and analyze its complexity in terms of big-O notation.

6.10. Same question as Problem 6.9, for the mathematical problem formulation for traffic grooming in a single-hop unidirectional ring, extend it to the bidirectional formulation for uniform traffic, and analyze its complexity in terms of big-O notation.

Part III

Wavelength-Routed (Wide-Area) Optical Networks

Chapter 7

Routing and Wavelength Assignment

7.1 Introduction

This chapter (as well as subsequent chapters) will consider long-haul optical networks in which nodes employ wavelength-routing switches (or optical cross-connects (OXCs)), which enable the establishment of wavelength-division multiplexed (WDM) channels, called lightpaths, between node pairs. It develops a practical approach to solve the routing and wavelength assignment (RWA) of lightpaths in such networks. A large RWA problem is partitioned into several smaller subproblems, each of which may be solved independently and efficiently using well-known approximation techniques. A multicommodity flow formulation, combined with randomized rounding, is employed to calculate the routes for lightpaths. Wavelength assignments for lightpaths are performed based on graph-coloring techniques. Representative numerical examples indicate the accuracy of the algorithms.

Wavelength-routed optical networks are being deployed mainly as backbone networks for large regions, e.g., for nationwide or global coverage. Access stations (or nodes) – to whom the architecture and operation of the backbone will be transparent – will attach to the network through a wavelength-sensitive switching/routing node, as shown, for example, in Fig. 7.1. An access station in this context need not necessarily be a terminal equipment, but the aggregate activity from a collection of terminals – including those that may possibly be feeding in from other regional and/or local subnetworks – so that the aggregate activity on any of its transmitters is close to the peak electronic transmission rate on a wavelength channel.

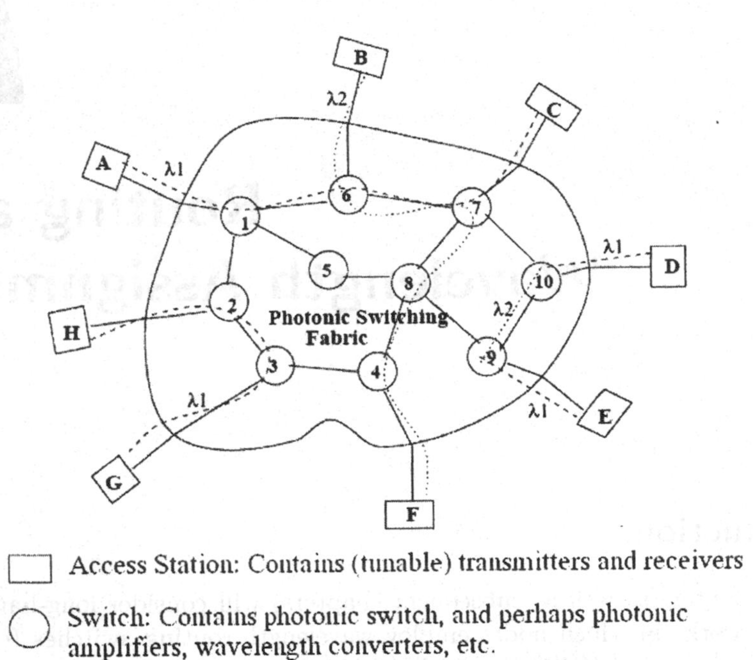

Figure 7.1 Lightpath routing in an all-optical network.

In fact, multiple edge devices, e.g., IP routers, ATM switches, etc., may attach themselves to the same switching/routing node, and the collection of these edge devices is referred to as an access station in our present context. Also, frequently, for modeling purposes, a switching/routing node and its associated access station is considered to be an integrated unit, and referred to as a "node".

Access stations communicate with one another via *optical (WDM) channels*, called *lightpaths* [ChGK92]. The terms *lightpath* and *connection* are used interchangeably, since, in order to establish a "connection" between a source-destination (sd) pair, we need to set up a "lightpath" between them.

A lightpath may span multiple fiber links, e.g., to provide a "circuit-switched" interconnection between two nodes which may have a heavy traffic flow between them and which may be located "far" from each other in the physical fiber network topology. Each intermediate node in the lightpath essentially provides a circuit-switched optical bypass facility to support the lightpath.

In an N-node network, if each node is equipped with $N-1$ transceivers [transmitters (lasers) and receivers (filters)] and if there are enough wavelengths on all fiber links, then every node pair could be connected by an optical lightpath, and there is no networking problem to solve. However, it should be noted that the network size (N) should be scalable, transceivers are expensive so that each node may be equipped with only a few of them, and technological constraints dictate that the number of WDM channels that can be supported in a fiber be limited to W (whose typical value is 64 to 160 today, but is expected to improve with time and technological breakthroughs). Thus, only a limited number of lightpaths may be set up on the network.

The complete set of lightpaths will be considered to form a *virtual topology* in Chapters 8 and 9, and packet traffic will be routed over it. In the present chapter, the offered traffic is itself "circuit-oriented," i.e., the offered traffic consists of a set of connections such that each connection requires the full bandwidth of a lightpath in order for it to be routed between its corresponding source-destination pair. Thus, this problem corresponds to that of Business Model #2 (see Section 1.3) where the network operator owns the layer-one infrastructure, and sells its services to its customers (ISPs) (Business Model #3) who generate circuit requests as the offered traffic for Business Model #2.

Once a set of lightpaths has been chosen or determined, we need to route each such lightpath in the network and assign a wavelength to it. This is referred to as the *routing and wavelength assignment (RWA)* problem.

Formally, the RWA problem can be stated as follows. Given a set of lightpaths that need to be established on the network, and given a constraint on the number of wavelengths, determine the routes over which these lightpaths should be set up and also determine the wavelengths which should be assigned to these lightpaths so that the maximum number of lightpaths may be established. While shortest-path routes may often be preferable, note that this choice may have to be sometimes sacrificed, in order to allow more lightpaths to be set up. Thus, RWA algorithms generally allow several alternate routes for each lightpath that needs to be established. Lightpaths that cannot be set up due to constraints on routes and wavelengths are said to be blocked, so the corresponding network optimization problem is to minimize this blocking probability.

In this regard, note that, normally, a lightpath operates on the same wavelength across all fiber links that it traverses, in which case the lightpath is said to satisfy the *wavelength-continuity constraint.* Thus, two lightpaths that share a common fiber link should not be assigned the same wavelength. However, if a switching/routing node is also equipped with a wavelength converter facility, then the *wavelength-continuity constraints* disappear, and a lightpath may switch between different wavelengths on its route from its origin to its termination.

For such a network, given a set of connection demands to be established, an integer linear programming (ILP) formulation of the RWA problem can be developed as a multicommodity flow problem [RaSi95a]. Then, one can derive any generic RWA algorithm's performance bounds – an upper bound on the carried traffic (number of lightpaths established), or, equivalently, a lower bound on the lightpath blocking probability. The approach can be extended to accommodate wavelength conversion as well. The wavelength-assignment problem is equivalent to the graph-coloring problem [ChGK93] and is hence NP-complete. A number of heuristics exist on how to obtain good solutions to the RWA problem [ZaJM00a].

In this chapter, we employ well-known approximation algorithms to solve the RWA problem. Our approach consists of practical algorithms which enable us to solve the RWA problem for large network sizes. Our final objective is to minimize the number of wavelengths that will be needed to carry a certain number of connections in the network, given a certain physical topology.

The RWA problem is decomposed into four different subproblems, and each subproblem is solved independently with the results of one stage fed in as the input to the next stage. We first formulate a linear program (LP) relaxation (using the idea of multicommodity flow in a network), which is based on the physical topology as well as the set of connections to be routed, and use a general-purpose LP solver to derive solutions to this problem. Since the general form of the LP can easily overwhelm the capabilities of today's state-of-the-art computing facilities, even for moderate sized networks (of, say, 10 nodes with 4 connections per node), we develop simple specialized techniques to drastically prune the size of the LP (in terms of both the number of variables and the number of equations it needs to handle) so that solutions to large instances of the problem (e.g., a few hundred nodes) can be obtained. Specifically, this pruning approach is based on tracking only a limited number of alternate breadth-first paths between source-destination pairs to reduce the size of the LP formulation. We then use a technique called *randomized rounding* (explained in Section 7.2.3) to convert the fractional flows provided by the LP solution to integer flows through the physical fiber links. Once the flows through the fiber links have been determined, *sequential coloring algorithms* are used to assign wavelengths to the lightpaths by taking into account the wavelength-continuity constraints. This method of subdividing the overall problem into smaller subproblems, each of which may be solved efficiently, allows practical solutions to the RWA problem for networks with a large number of nodes.

The solution to our LP formulation gives a lower bound on the number of wavelengths we would need in the network to route the given set of connections.[1] How close the final result is to this value determines the efficacy of the algorithms proposed. The proposed solutions turn out to be very close to the lower bound for large networks.

This chapter initially examines the problem for a known set of connections that need to be routed – referred to as the static lightpath establishment (SLE) problem [ChGK93]. Then, simple heuristics are employed to extend our approach and obtain good results for the case of dynamic lightpath establishment (DLE).

The remainder of the chapter is organized as follows. Section 7.2 formulates the RWA problem and describes our tools and techniques to solve this problem for large networks. Section 7.3 describes the results for an experimental network with 100 nodes; the results are given for both the SLE and the DLE problems. Because the RWA problem is a complex one, the Routing Subproblem and the Wavelength-Assignment Subproblem are treated separately in Sections 7.4 and 7.5. Illustrative examples and comparisons relevant to these subproblems are discussed in Sections 7.6 and 7.7. Section 7.8 concludes this chapter.

[1] Because of our pruning the search space, there might be a small discrepancy between the absolute lower bound and our result; however, we can make the difference arbitrarily small by increasing the number of alternate paths.

7.2 Problem Formulation of RWA

Significant work has been reported on combinatorial formulations, in terms of using mixed-integer linear programs, for solving the RWA problem [RaSi95a, ZaJM00a]. However, these formulations, when applied to solving large problems, become computationally expensive in spite of using sophisticated techniques such as branch-and-bound methods. The algorithms in the next few sections elaborate on how it is possible to use approximation techniques to arrive at optimal results for the RWA problem in large networks.

7.2.1 Solution Approach

Our objective is to minimize the number of wavelengths needed to set up a certain set of lightpaths for a given physical topology. Our approach employs approximation results based on a combinatorial formulation to arrive at optimally close results to the lower bound for the number of wavelengths required in the network. The RWA problem, without the wavelength-continuity constraint, can be formulated as a straightforward multicommodity flow problem with integer flows in each link. This corresponds to an *integer linear program* (ILP) with the objective function being to minimize the flow in each link, which, in turn, corresponds to minimizing the number of lightpaths passing through a particular link.

Let λ_{sd} denote the traffic (in terms of a lightpath) from any source s to any destination d. We consider at most one lightpath from any source to any destination; hence $\lambda_{sd} = 1$ if there is a lightpath from s to d; otherwise $\lambda_{sd} = 0$. (However, this condition can easily be generalized.) Let F_{ij}^{sd} denote the traffic (in terms of number of lightpaths) that is flowing from source s to destination d on link ij. The linear programming formulation is written as follows:

$$\textbf{Minimize}: \quad F_{max} \tag{7.1}$$

such that:

$$F_{max} \geq \sum_{s,d} F_{ij}^{sd} \;\; \forall \;\; ij \tag{7.2}$$

$$\sum_i F_{ij}^{sd} - \sum_k F_{jk}^{sd} = \begin{cases} \lambda_{sd} & \text{if } s = j \\ -\lambda_{sd} & \text{if } d = j \\ 0 & \text{otherwise} \end{cases} \tag{7.3}$$

$$\lambda_{sd} = 0, 1 \tag{7.4}$$

$$F_{ij}^{sd} = 0, 1 \tag{7.5}$$

This problem is NP-complete [EvIS76] but it can be approximated successfully using *randomized rounding*, which is outlined below in Section 7.2.3.

7.2.2 Problem-Size Reduction

If we consider the general multicommodity formulation, the number of equations and the number of variables in the formulation grow rapidly with the size of the network. For example, let us assume that there are 10 nodes, 30 links (ij pairs), and an average of 4 connections per node, i.e., 40 connections (sd pairs) need to be set up on the network.

In the simplest and most general formulation, the number of λ_{sd} variables is $10 \times 9 = 90$, since there are 90 sd pairs. The number of F_{ij}^{sd} variables will be 90 sd pairs $\times$ 30 ij pairs = 2,700. The number of equations will be 3,721.[2] Thus, even for a small problem, we observe that the number of variables and the number of equations are very large, and these numbers grow proportionally with the square of the number of nodes.

A smarter solution can be obtained by only considering the λ_{sd} variables that are 1, thus reducing the number of λ_{sd} variables from 100 to 40. This can eliminate all of Eqns. (7.4). Also, this approach will reduce the number of F_{ij}^{sd} variables to be $40 \times 30 = 1{,}200$. This approach is more specific to the particular instance of lightpaths that need to be set up, since it has to take into account the lightpaths that need to be established.

Further reduction of the number of variables can be achieved if we assume that a particular lightpath will not pass through all of the ij links. If we can determine the links which have a good probability of being in the path through which a lightpath may pass, we can only consider those links as the F_{ij}^{sd} variables for that particular sd pair. Thus, if on an average, a lightpath sd passes through seven links, there will be approximately $40 \times 7 = 280$ F_{ij}^{sd} variables. We employ an *extended* breadth-first search (BFS) to obtain

[2] There are one instance of Eqn. (7.1), 30 instances of Eqns. (7.2), 900 instances of Eqns. (7.3), 90 instances of Eqns. (7.4), and 2,700 instances of Eqns. (7.5).

a set of alternate, shortest paths between a given source-destination pair. The links constituting these alternate paths are then used as part of the LP formulation. Also, as part of the randomized rounding algorithm (to be discussed in Section 7.2.3 below), we can relax the integrality constraints, hence getting rid of all of Eqns. (7.5). Using this approach, we will be left with a total of 351 equations. This will comprise of Eqn. (7.1), 30 instances of Eqns. (7.2), and 320 instances of Eqns. (7.3). Since there are on an average of seven links considered per connection, we need to enumerate Eqns. (7.3) for eight nodes (six intermediate nodes and the two end nodes) per connection, for each of 40 connections.

Hence, using knowledge which is specific to a particular set of lightpaths, we can drastically reduce the size of the LP problem formulation and make it tractable for large networks.

7.2.3 Randomized Rounding

Randomized rounding has been studied extensively in [RaTh87]. The relation of an ILP to its fractional relaxation has been the subject of considerable interest. Such efforts fall into two categories: (1) showing existence results for feasible solutions to an ILP in terms of the solution to its fractional relaxation and (2) using the information derived from the solution of the relaxed problem in order to construct a provably-good solution to the original ILP.

The *randomized rounding* technique is applicable to a class of 0-1 ILPs and yields results in both of the categories listed above. The technique is probabilistic, i.e., with high probability, the algorithm will provide an integer solution in which the objective function takes on a value close to the optimum of the rational relaxation. This is a sufficient condition to show the near-optimality of our 0-1 solution since the optimal value of the objective function in the relaxed version is better than the optimal value of the objective function in the original 0-1 integer program. This technique has been effectively used in multicommodity flow problems.

In a general, undirected, multicommodity flow problem, we are given an undirected graph $G(V, E)$. Let there be k commodities that need to be routed. In an instance of the problem, various vertices are the sites of *sources* and *sinks* for a particular commodity. One unit of flow is to be conveyed from each source s_t to each destination d_t through the edges in E. Each edge $e \in E$ has a capacity $c(e)$ which is an upper limit on the total amount of flow in E. The flow of each commodity in each edge must be 0 or 1. The objective function is to minimize the common capacity in each link, while still realizing unit flows for all commodities. The general integral problem is known to be NP-complete [EvIS76], although the nonintegral version can be solved using linear programming methods [Karp72] in polynomial time.

In our formulation of the problem, each commodity corresponds to a lightpath from a source node to a destination node, the capacity is the number of wavelengths supported in each fiber, and the objective function is to minimize the number of wavelengths needed to accommodate all of the requests, so as to maximize the spare capacity in the network.

The algorithm consists of the following three major phases:

1. solving a nonintegral multicommodity flow problem,
2. path stripping, and
3. randomized path selection.

1. *Nonintegral Multicommodity Flow.* We relax the requirement of the 0-1 flows to allow fractional flows in the interval [0,1]. The relaxed capacity-minimization problem can be solved by a suitable linear programming method. If the flow for each commodity i on edge $e \epsilon E$ is denoted by $f_i(e)$, a capacity constraint of the form:

$$\sum_{i=1}^{k} f_i(e) \leq C$$

is then satisfied for each edge in the network, where C is the optimal solution to the nonintegral, edge-capacity optimization problem. The value of C is also a lower bound on the best possible integral solution.

2. *Path Stripping.* The main idea in this phase is to convert the edge flows for each commodity i into a set τ_i of possible paths which could be used to realize the flow of that commodity. Initially, τ_i is empty. For each commodity i, the following steps need to be performed:

a. Discover a loop-free, depth-first, directed path $e_1, e_2, \ldots, e_p$ from the source to the destination.

b. Let $f_m = \min f_i(e_j)$, where $1 \leq j \leq p$. For $1 \leq j \leq p$, replace $f_i(e_j)$ by $f_i(e_j) - f_m$. Add the path $e_1, e_2, \ldots, e_p$ to τ_i along with its weight f_m.

c. Remove any edges with zero flow from the set of edges that carry any flow for commodity i. If there is nonzero flow leaving s_i, repeat Step b. Otherwise, continue for next commodity i.

Upon termination, the sum of the weights of all the paths in τ_i is 1. Path stripping gives us a set of paths τ_i that may carry the flow for commodity i in the optimal case.

3. *Randomization.* For each i, cast a $\mid \tau_i \mid$ die with face probabilities equal to the weights of the paths in τ_i. Assign to commodity i the path whose face comes up.

It has been shown in [RaTh87] that, provided $C \geq 2 \ln \mid E \mid$, the integer capacity of the solution produced by the above procedure does not exceed

$$C + \sqrt{3C \ln \frac{\mid E \mid}{\epsilon}}$$

where $0 \leq \epsilon \leq 1$ with probability at least $1 - \epsilon$.

The formulation of the problem allows the F_{ij}^{sd} variables to take on fractional values. These values are then used to find the fractional flow through each of a set of alternate paths. A coin-tossing experiment is used to select the path over which to route the lightpath λ_{sd} based on the probability of the individual paths. This technique can be used to solve large problems for which solving the original ILP would have been computationally very expensive.

7.2.4 Graph Coloring

Once a path has been chosen for each connection, the number of lightpaths going through any physical fiber link defines the *congestion* on that particular link. Now, we need to assign wavelengths to each lightpath such that any two lightpaths that pass through the same physical link are assigned different wavelengths.

If the intermediate switches do not have the capability to perform wavelength conversion, a lightpath is constrained to operate on the same wavelength throughout its path. This *wavelength-continuity constraint* may reduce the effective utilization of the wavelengths in the network, because a lightpath that needs to be set up may not find a free wavelength *of the same color* in *all* of the physical fiber links it passes through, even though these link may have free wavelengths. This problem may be alleviated by the use of wavelength converters in a switching node. (Recall that a wavelength converter may optically switch a signal arriving on a wavelength on an input fiber to a different free wavelength on an output fiber corresponding to the next physical link.)

Assigning wavelength colors to different lightpaths, so as to minimize the number of wavelengths (colors) used under the wavelength-continuity constraint, reduces to the graph coloring problem, as stated below.

1. Construct a graph $G(V, E)$, so that each lightpath in the system is represented by a node in graph G. There is an undirected edge between two nodes in graph G if the corresponding lightpaths pass through a common physical fiber link (see Fig. 7.2).

2. Color the nodes of the graph G such that no two adjacent nodes have the same color. Figure 7.2 shows an example use of an auxiliary graph, where three wavelengths are needed to color the eight lightpaths.

This problem has been shown to be NP-complete, and the minimum number of colors needed to color a graph G (called the chromatic number $\chi(G)$ of the graph) is difficult to determine. However, there are efficient *sequential graph coloring* algorithms which are optimal in the number of colors used.

In a *sequential graph coloring* approach, vertices are sequentially added to the portion of the graph already colored, and new colorings are determined to include each newly adjoined vertex. At each step, the total number of colors necessary is kept at a minimum. It is easy to observe that some particular sequential vertex coloring will yield a $\chi(G)$ coloring. To see this, let A_i be the vertices colored i by a $\chi(G)$ coloring of G. Then, for any ordering of the vertices $V(G)$, which has all members of A_i before any member of A_j for $1 \leq i \leq j \leq \chi(G)$, the corresponding sequential coloring will be a $\chi(G)$ coloring.

It is easy to note that, if $\Delta(G)$ denotes the maximum degree in a graph, then $\chi(G) \leq \Delta(G)+1$. However, intuitively, if a graph has only a few nodes of very large degree, then coloring these nodes early will avoid the need for using a very large set of colors. This gives rise to the following theorem:

Theorem: Let G be a graph with $V(G) = v_1, v_2, \ldots, v_n$ where $\deg(v_i) \geq \deg(v_{i+1})$ for $i = 1, 2, \ldots, n-1$. Then, $\chi(G) \leq \max_{1 \leq i \leq n} \min \{i, 1 + \deg(v_i)\}$. Determination of a sequential coloring procedure corresponding to

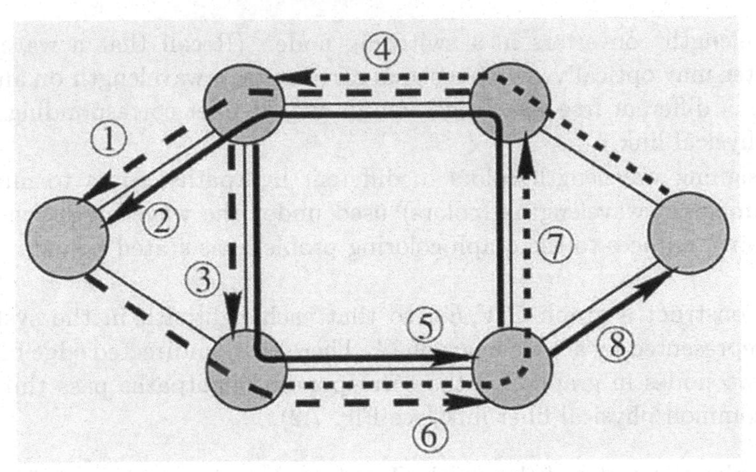

(a) A network with eight routed lightpaths.

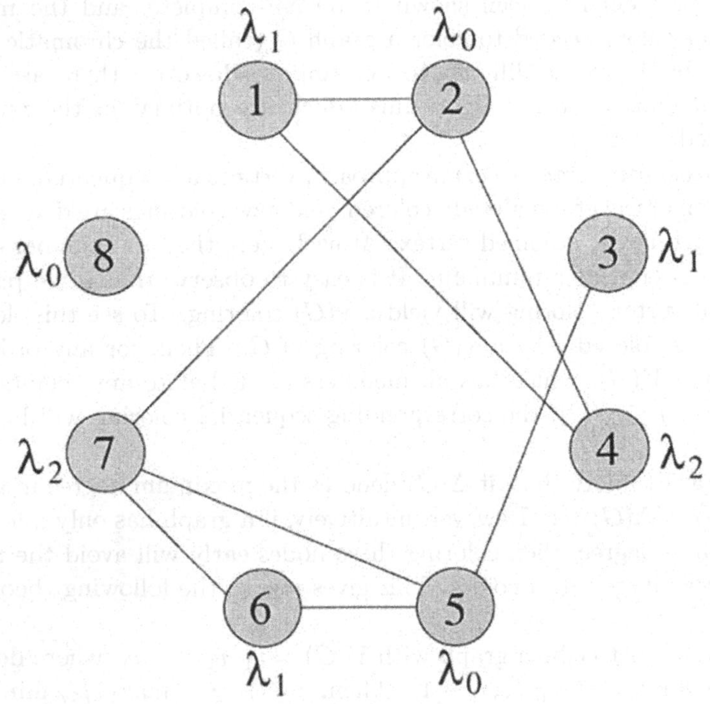

(b) The auxiliary graph, $G(V, E)$, for the lightpaths in the network.

Figure 7.2 A network and a corresponding auxiliary graph.

such an ordering will be termed the *largest-first* algorithm. The proof is straightforward and can be found in [MaMI72].

A closer inspection of the sequential coloring procedure shows that, for a given ordering $v_1, v_2, \ldots, v_n$ of the vertices of a graph G, the corresponding sequential coloring algorithm could never require more than k colors where

$$k = \max_{1 \le i \le n} \{1 + \deg_{<v_1, v_2, \ldots, v_n>}(v_i)\}$$

and $\deg_{<v_1, v_2, \ldots, v_n>}(v_i)$ refers to the degree of node v_i in the vertex-induced subgraph denoted by $< v_1, v_2, \ldots, v_n >$. The determination of a vertex ordering that minimizes k was derived in [Matu72] and can be found in the following procedure:

1. For $n = \mid V(G) \mid$, let v_n be chosen to have minimum degree in G.
2. For $i = n-1, n-2, \ldots, 2, 1$, let v_i be chosen to have minimum degree in:
$< V(G) - v_n, v_{n-1}, \ldots, v_{i+1} >$.

For any vertex ordering $v_1, v_2, \ldots, v_n$ determined in this manner, we must have:

$$\deg_{<v_1, v_2, \ldots, v_i>}(v_i) = \min_{1 \le j \le i} \deg_{<v_1, v_2, \ldots, v_i>}(v_j)$$

for $1 \le i \le n$, so that such an ordering will be termed a *smallest-last* (SL) vertex ordering. The fact that any smallest-last vertex ordering minimizes k over the $n!$ possible orderings is shown in [Matu72].

We employ the *smallest-last* coloring algorithm to color the lightpaths. There may exist better algorithms for graph coloring; however, we chose this algorithm for the sake of simplicity and because our contribution is a methodology on how we can solve large RWA problems efficiently rather than to propose the best possible method to solve each subproblem.

7.3 Illustrative Examples from ILP

For illustration purposes, consider a randomly-generated physical topology consisting of 100 nodes, with each node having a physical nodal degree uniformly distributed between two and five. All links are unidirectional, and it turns out that there are 357 directed links in our simulation of this network.

The model of offered traffic is the following. A set of lightpaths needs to be established between randomly-chosen source-destination (*sd*) pairs, so that an *sd* pair can have zero or one lightpath, and all *sd* pairs are treated equally. Associated with each (source) node, we identify a quantity d' which is the average number of lightpath connections the node will source. Thus, in an N-node network, the probability that a node will have a lightpath with each of the remaining $(N-1)$ nodes equals $d'/(N-1)$. Note that d' can also be considered to be the average "logical degree" (hereafter, referred to simply as "degree") of a node.

We assume – as do most RWA algorithms – that there are enough transceivers at the access nodes to accommodate all of the lightpath requests that need to be established. That is, no lightpath request will be blocked due to lack of transceivers at the access nodes.

7.3.1 Static Lightpath Establishment (SLE)

First, let us consider the static lightpath establishment (SLE) problem. Given the physical topology and the set of connections to be routed, generate the LP to obtain a lower bound on the number of wavelengths required for the RWA problem. To reduce the size of the LP formulation, we consider a set of K alternate, shortest paths between a given *sd* pair. Only the links which constitute these alternate paths are used as the F_{ij}^{sd} variables. An extended breadth-first search (BFS) algorithm is used to derive the set of K alternate shortest paths between a *sd* pair.

The LP formulation is solved by a LP solver called *lpsolve* [Berk94], and the resultant output (containing the flow values for the F_{ij}^{sd} variables) is used as input for the randomized rounding algorithm. The value of the objective function denotes the lower bound on congestion that can be achieved by any RWA algorithm. Using the values of the individual variables F_{ij}^{sd}, we use the path-stripping technique and the randomization technique outlined in Section 7.2.3 to assign physical routes for the different lightpaths. Once this procedure is completed, we obtain the congestion on the different links in the network.

Once the lightpaths have been assigned physical routes, we need to assign a wavelength to each individual lightpath. This is performed by generating the conflict graph for a lightpath; each lightpath corresponds to a node in a conflict graph G, and lightpaths that pass through a common physical link are adjacent nodes in the graph G. Coloring the nodes of graph G, such that adjacent nodes get different colors, corresponds to assigning wavelengths correctly to the lightpaths. We use the smallest-last vertex coloring as discussed earlier in Section 7.2.4.

A set of numerical results for the 100-node random network is shown in Table 7.1. The LP experiments were run on an unloaded DEC-Alpha workstation and the time taken for each experiment provides relative guidance as to the time complexity of this approach.

In Table 7.1, note that, when the number of alternate paths is one, the lower bound exactly matches the congestion. This is expected because this case corresponds to shortest-path routing. For larger values of K, observe that the number of wavelengths needed to color all of the lightpaths is a little higher but very close to the maximum congestion in the network. The maximum network congestion gives the number of wavelengths we would need to have in the network, if the intermediate switching nodes were equipped with wavelength converters, so that there was no wavelength-continuity constraint for a lightpath.

The time taken to solve the LP increases rapidly as the number of connections increase (corresponding to larger problem formulations). The table entries which are *nil* correspond to the case when the LP solver failed to give a solution due to lack of memory.

Note that, as the number of connections in the network increases (as in the case where there are 20 connections per node and $K = 2$ alternate paths for each connection), the difference between the LP lower bound and the maximum network congestion continues to stay small; however, the difference between the network congestion and the number of wavelengths needed to color the lightpaths has increased. This indicates that we might have to employ better coloring algorithms which use backtracking techniques to achieve better performance. Figure 7.3 shows the effect of the nodal degree d' (which is proportional to the number of connections in the network) on the wavelength congestion, if we consider $K = 2$ alternate paths. The correspondence is linear in this case.

As expected, there is a sharp decrease in the number of wavelengths needed when we change from pure shortest-path routing to considering $K = 2$ alternate paths. This is because we transit from a situation in which there can be no congestion balancing when $K = 1$, to a situation where we can balance the flow of lightpaths by using alternate paths ($K > 1$). As we increase from two alternate paths to three or more alternate paths, we would still expect some improvement, although the rate of improvement reduces. The LP lower bound is a monotonically decreasing function as

Table 7.1 Sample numerical results for static lightpath establishment (SLE) on a 100-node random network.

Avg. Deg. d'	*Alt. Paths K*	*Connections*	*LP vars.*	*LP eqns.*	*LP time (sec)*	*LP Lower Bound*	*Congestion per link*	*Wavelengths needed*
1	1	100	319	776	0.5	4.00	4	4
2	1	205	684	1246	1.4	8.00	8	8
3	1	300	1024	1681	2.7	10.00	10	10
4	1	400	1362	2119	4.4	11.00	11	11
10	1	984	3334	4675	22.0	22.00	22	22
20	1	1958	6692	9007	90.6	38.00	38	41
1	2	100	647	1004	2.7	3.00	3	4
2	2	205	1353	1710	28.0	4.33	5	6
3	2	300	2001	2358	100.7	6.00	7	7
4	2	400	2678	3035	271.9	7.00	8	8
10	2	984	6618	6975	2585.1	12.00	14	17
20	2	1958	13219	13576	9113.7	22.00	24	29
1	3	100	962	1219	18.8	2.50	3	4
2	3	205	2003	2155	215.2	3.75	5	6
3	3	300	2931	2988	545.2	5.00	7	8
4	3	400	3923	3880	1205.6	6.50	8	10
10	3	984	9655	9028	-	-	-	-
20	3	1958	19270	17669	-	-	-	-
1	4	100	1257	1414	52.4	2.50	4	4
2	4	205	2607	2555	420.0	3.67	5	6
3	4	300	3811	3569	1225.2	4.67	7	8
4	4	400	5102	4661	2253.7	5.50	8	9
10	4	984	12521	10915	-	-	-	-

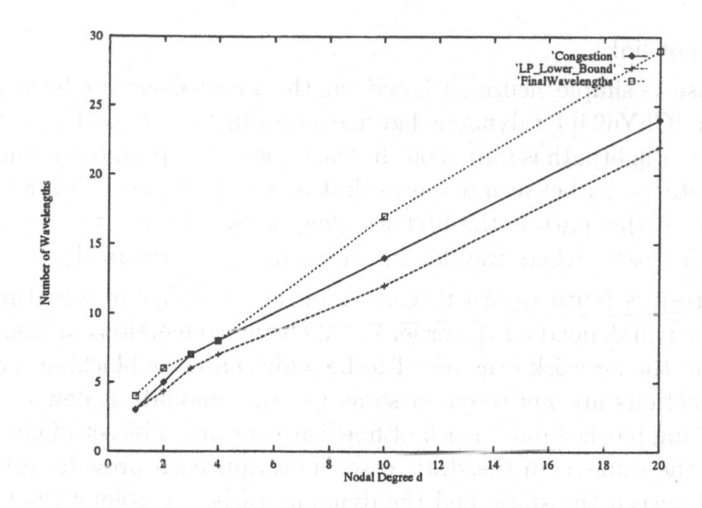

Figure 7.3 Effect of nodal degree d' (for $K = 2$ alternate paths) on wavelength routing.

the number of paths is increased. However, when K is large, with a lot of choices, randomized rounding may choose a nonoptimal path with nonzero probability. This may explain why the congestion and maximum wavelengths needed when we use four alternate paths is slightly higher than the case when we use two or three alternate paths. In large networks with higher number of connections, we would expect such occurrences to be rare due to averaging effects. Because of the large time complexity required for solving larger LP problems, and because of the minimal improvement to be gained from having a large number of alternate paths, the recommended choice is two or three alternate paths.

7.3.2 Dynamic Lightpath Establishment (DLE)

The previous subsection discussed the static problem in which lightpath requests are known in advance. The LP formulation and subsequent algorithms enabled us to determine how many wavelengths would be needed to accommodate all of the lightpath requests. However, if the lightpath requests are dynamic, we cannot make use of the globally optimal algorithms and we would need to have a dynamic algorithm which adaptively routes the incoming traffic connections over different paths based on congestion on the different links.

We use a simple heuristic based on the Least-Congested-Path (LCP) algorithm [ChYu94] for dynamic lightpath establishment (DLE). In the LCP algorithm, a lightpath is routed on the least-congested path from among a set of alternate paths between a source-destination (*sd*) pair. The wavelength allocated on this path is the first wavelength that is free among all of the links in this path, where wavelengths are numbered arbitrarily.

The results (with regard to congestion and number of wavelengths) in this study will depend on the order in which the connections arrive. For the DLE case, the network is assumed to be wide-sense nonblocking, i.e., existing connections are *not* rerouted so as to accommodate a new connection that is being blocked due to lack of free wavelengths. The set of connections are kept the same as in the static case. This approach provides a fair comparison between the static and the dynamic cases. A point to note is that the congestion results using LCP routing are very close to the optimal results obtained in the static optimization case, after running the randomized rounding algorithm. This is understandable since the LCP algorithm tries to adaptively minimize the congestion as each connection arrives.

It is expected that, as the congestion increases in the network, the number of wavelengths needed in the dynamic case will be more than the wavelengths needed in the static case. In the static case, the coloring algorithm assigned wavelengths to the lightpaths in a specific order so as to minimize the number of wavelengths needed. This method of assigning a wavelength to a lightpath is not possible in the dynamic case where lightpaths arrive in a random order. Thus, under the wavelength-continuity constraint, optimal allocation of wavelengths may not be possible in the dynamic case.

To alleviate the above problem, there are known techniques by which it may be possible to *reassign* existing connections to unused wavelengths in order to optimally allocate resources for new connections. This problem has been studied in the graph-theoretical case in [MaMI72]. The algorithm uses a fundamental operation called *color interchange* which corresponds to two lightpaths interchanging the wavelengths they use so as to create spare capacity in the network. This approach corresponds to global synchronization across four nodes (the endpoints of these two lightpaths) as well as the intermediate switches, and may be difficult to achieve in wide-area networks.

Illustrative results for the dynamic case are given in Table 7.2. The set of lightpaths is the same as those in the static case; however, the order in which the lightpaths arrive is changed randomly. Each experiment is analyzed over

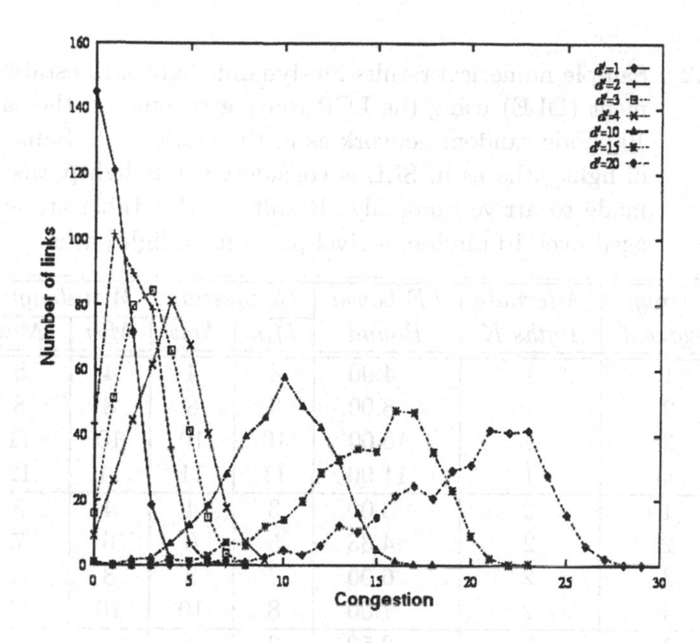

Figure 7.4 Effect of number of connections on link congestion.

10 random arrival patterns for the lightpaths.

Observe in Table 7.2 that the results in the DLE problem are very close to the lower bound for number of wavelengths achievable, for the given set of connections. Another important fact is that the variance in the results is very low (over the 10 sample runs that were conducted for each set of parameters). This leads us to believe that the LCP routing algorithm is quite a suitable choice for dynamic routing of lightpaths, because of its simplicity and adaptive properties.

Figure 7.4 shows the distribution of the congestion on the network's links, viz., the number of links that have a particular congestion value, with increase in the nodal degree d'. As expected, the congestion in the links increases with increasing nodal degree, i.e., the distribution shifts to the right with increasing number of connections.

Table 7.2 Sample numerical results for dynamic lightpath establishment (DLE) using the LCP routing scheme on the same 100-node random network as in the static case. Same set of lightpaths as in SLE is considered, but lightpaths are made to arrive randomly. Results in this table are averaged over 10 random arrival patterns of lightpaths.

Average Degree d'	*Alternate Paths K*	*LP Lower Bound*	*Congestion*		*Wavelengths*	
			Min	*Max*	*Min*	*Max*
1	1	4.00	4	4	4	5
2	1	8.00	8	8	8	8
3	1	10.00	10	10	10	11
4	1	11.00	11	11	11	12
1	2	3.00	3	4	4	5
2	2	4.33	5	6	6	7
3	2	6.00	7	8	8	10
4	2	7.00	8	10	10	11
1	3	2.50	3	4	4	4
2	3	3.75	5	6	6	7
3	3	5.00	6	8	8	9
4	3	6.50	8	9	10	11
1	4	2.50	3	4	4	4
2	4	3.67	5	6	5	7
3	4	4.67	6	7	8	9
4	4	5.50	7	8	10	11

7.4 Routing Subproblem

Although combined routing and wavelength assignment is a hard problem (see Section 7.2), it can be simplified by decoupling the problem into two separate subproblems: the routing subproblem and the wavelength-assignment subproblem. In this section, we focus on various approaches to routing connection requests by following the ILP for SLE and DLE in Section 7.2. The wavelength-assignment subproblem will be treated separately in Section 7.5.

7.4.1 Fixed Routing

The most straightforward approach to routing a connection is to always choose the same fixed route for a given source-destination pair. One example of such an approach is fixed shortest-path routing. The shortest-path route for each source-destination pair is calculated off-line using standard shortest-path algorithms, such as Dijkstra's algorithm or the Bellman-Ford algorithm, and any connection between the specified pair of nodes is established using the pre-determined route.

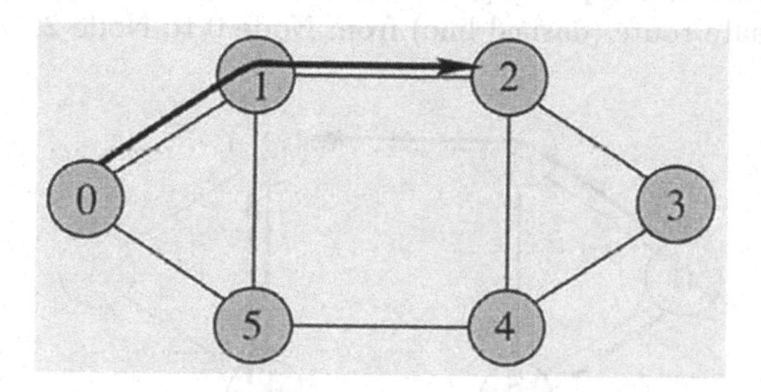

Figure 7.5 Fixed shortest-path route from Node 0 to Node 2.

In Fig. 7.5, the fixed shortest-path route from Node 0 to Node 2 is illustrated. This approach to routing connections is very simple; however, the disadvantage of such an approach is that, if resources (wavelengths) along the path are tied up, it can potentially lead to high blocking probabilities in the dynamic case, or may result in a large number of wavelengths being used in the static case. Also, fixed routing may be unable to handle fault situations in which one or more links in the network fail. To handle link faults, the routing scheme must either consider alternate paths to the destination, or must be able to find the route dynamically. Note that, in Fig. 7.5, a connection request from Node 0 to Node 2 will be blocked if a common wavelength is not available on both links in the fixed route, or if either of the links in the fixed route is cut.

7.4.2 Fixed-Alternate Routing

An approach to routing that considers multiple routes is fixed-alternate routing. In fixed-alternate routing, each node in the network is required to maintain a routing table that contains an ordered list of a number of fixed routes to each destination node. For example, these routes may include the shortest-path route, the second-shortest-path route, the third-shortest-path route, etc. A primary route between a source node s and a destination node d is defined as the first route in the list of routes to node d in the routing table at node s. An alternate route between s and d is any route that does not share any links (is link-disjoint) with the first route in the routing table at s. The term "alternate routes" is also employed to describe all routes (including the primary route) from a source node to a destination node. Figure 7.6 illustrates a primary route (solid line) from Node 0 to Node 2, and an alternate route (dashed line) from Node 0 to Node 2.

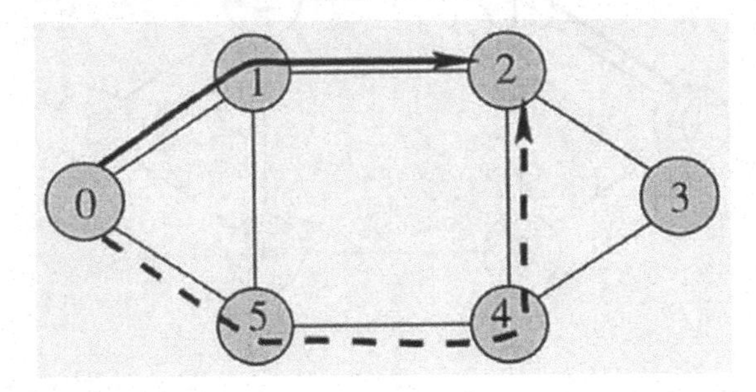

Figure 7.6 Primary (solid) and alternate (dashed) routes from Node 0 to Node 2.

When a connection request arrives, the source node attempts to establish the connection on each of the routes from the routing table in sequence, until a route with a valid wavelength assignment is found. If no available route is found from the list of alternate routes, then the connection request is blocked and lost. In most cases, the routing tables at each node are ordered by the number of fiber link segments (hops) to the destination. Therefore, the shortest path to the destination is the first route in the routing table. When there are ties in the distance between different routes, one route may be selected at random. Fixed-alternate routing provides simplicity of control for setting up and tearing down lightpaths, and it may also be used to provide some degree of fault tolerance upon link failures (as discussed Section 7.4.4). Another advantage of fixed-alternate routing is that it can significantly reduce the connection blocking probability compared to fixed routing. It has also been shown that, for certain networks, having as few as two alternate routes provides significantly lower blocking probabilities than having full wavelength conversion at each node with fixed routing [Rama98, RaMu02].

7.4.3 Adaptive Routing

In adaptive routing, the route from a source node to a destination node is chosen dynamically, depending on the network state. The network state is determined by the set of all connections that are currently in progress. One form of adaptive routing is adaptive shortest-cost-path routing, which is well-suited for use in wavelength-converted networks. Under this approach, each unused link in the network has a cost of 1 unit, each used link in the network has a cost of ∞, and each wavelength-converter link has a cost of c units. If wavelength conversion is not available, then $c = \infty$. When a connection arrives, the shortest-cost path between the source node and the destination node is determined. If there are multiple paths with the same distance, one of them is chosen randomly. By choosing the wavelength-conversion cost c appropriately, we can ensure that wavelength-converted routes are chosen only when wavelength-continuous paths are not available. In shortest- cost adaptive routing, a connection is blocked only when there is no route (either wavelength-continuous or wavelength- converted) from the source node to the destination node in the network. Adaptive routing requires extensive support from the control and management protocols to continuously update the routing tables at the nodes. An advantage of adaptive routing is that it results in lower connection blocking than fixed and fixed-alternate routing. For the network in Fig. 7.7, if the links (1,2) and (4,2) in the network are busy, then the adaptive-routing algorithm can still establish a connection between Nodes 0 and 2, while both the fixed-routing protocol and the fixed-alternate routing protocols with fixed and alternate paths as shown in Fig. 7.6 would block the connection.

Another form of adaptive routing is least-congested-path (LCP) routing [ChYu94]. Similar to alternate routing, for each source-destination pair, a sequence of routes is pre-selected. Upon the arrival of a connection request, the least-congested path among the pre-determined routes is chosen. The congestion on a link is measured by the number of wavelengths available on the link. Links that have fewer available wavelengths are considered to be more congested. The congestion on a path is indicated by the congestion on the most congested link in the path. If there is a tie, then shortest-path routing may be used to break the tie. An alternate implementation is to always give priority to shortest paths, and to use LCP only for breaking

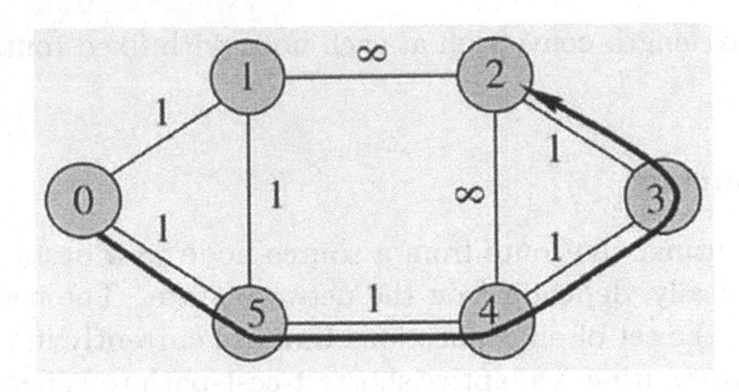

Figure 7.7 Adaptive route from Node 0 to Node 2.

ties. Both combinations are examined through simulation in [LiSo99], and it has been shown that using shortest-path routing first and LCP second works better than using LCP alone.

A disadvantage of LCP is its computational complexity. In choosing the least-congested path, all links on all candidate paths have to be examined. A variant of LCP is proposed in [LiSo99] which only examines the first k links on each path (referred to as the source's neighborhood information), where k is a parameter to the algorithm. It has been shown that, when $k = 2$, this algorithm can achieve similar performance to fixed-alternate routing. It is also shown in [LiSo99] that LCP performs much better than fixed-alternate routing.

7.4.4 Fault-Tolerant Routing

When setting up connections in a wavelength-routed optical WDM network, it is often desirable to provide some degree of protection against link and node failures in the network by reserving some amount of spare capacity [Rama98, RaMu99a]. A common approach to protection is to set up two link-disjoint lightpaths (the routes for the lightpaths do not share any common links) for every connection request. One lightpath, called the primary lightpath, is used for transmitting data, while the other lightpath is reserved as a backup in the event that a link in the primary lightpath fails. This approach can be used to protect against any single-link failures in the network (a situation in which any one physical fiber link in the network fails). To further protect against node failures, the primary and alternate paths may also be node-disjoint.

Fixed-alternate routing provides a straightforward approach to handling protection. By choosing the alternate paths such that their routes are link-disjoint from the primary route, we can protect the connection from any single-link failures by allocating one of the alternate paths as a backup path.

In adaptive routing, a protection scheme may be implemented in which the backup path is set up immediately after the primary path has been established. The same routing protocol may be used to determine the backup path, with the exception that a link cost is set to ∞ if that link is being used by the primary path on any wavelength. The resulting route will then be link-disjoint from the primary path. An alternative is to implement restoration, in which the backup path is determined dynamically after the failure has occurred. Restoration will only be successful if sufficient resources are available in the network. Note also that, when a fault occurs, dynamic discovery and establishment of a backup path under the restoration approach might take significantly longer than switching over to the pre-established backup path using the protection approach.

The static formulation in Section 7.2 may also be extended to provide for fault protection in the network. The modified formulation would include additional constraint equations requiring that two lightpaths be set up for each connection (one primary lightpath and one backup lightpath), and that the routes for these two lightpaths do not share any links. For further information regarding protection and restoration, the reader is referred to Chapter 11.

7.5 Wavelength-Assignment Subproblem (Heuristics)

By following the studies of the wavelength-assignment problem in Sections 7.2 and 7.3, we turn to discuss several wavelength-assignment heuristics for the dynamic wavelength-assignment problem. These heuristics may also be applied to the static wavelength-assignment problem by ordering the lightpaths and then sequentially assigning wavelengths to the ordered lightpaths.

For the case in which lightpaths arrive one at a time (either incremental or dynamic traffic), heuristic methods must be used to assign wavelengths to lightpaths. For the dynamic problem, instead of attempting to minimize the number of wavelengths as in the static case, we assume that the number of wavelengths is fixed (this is the practical situation), and we attempt to minimize connection blocking.

The following heuristics have been proposed in the literature: (1) Random, (2) First-Fit, (3) Least-Used/SPREAD, (4) Most-Used/PACK, (5) Min-Product, (6) Least Loaded, (7) MAX-SUM, (8) Relative Capacity Loss, (9) Distributed Relative Capacity Loss, (10) Wavelength Reservation, and (11) Protecting Threshold. These heuristics can all be implemented as on-line algorithms and can be combined with different routing schemes. The first nine schemes attempt to reduce the overall blocking probability for new connections, while the last two approaches aim to reduce the blocking probability for connections that traverse more than one link. In our discussions, we use the following notation and definitions:

- L: Number of links.
- M_l: Number of fibers on link l.
- M: Number of fibers per link if all links contain the same number of fibers.
- W: Number of wavelengths per fiber.
- $\pi(p)$: Set of links comprising path p.
- S_p: Set of available wavelengths along the selected paths p.
- D: L-by-W matrix, where D_{lj} indicates the number of assigned fibers on link l and wavelength j. Note that the value of D_{lj} varies between 0 and M_l.
- Load: For dynamic traffic, the holding time is exponentially distributed with a normalized mean of one unit, and connection arrivals are Poisson; thus, load is expressed in units of Erlangs.

We describe the wavelength-assignment heuristics below.

1. **Random Wavelength Assignment (R).** This scheme first searches the space of wavelengths to determine the set of all wavelengths that are available on the required route. Among the available wavelengths, one is chosen randomly (usually with uniform probability).
2. **First-Fit (FF).** In this scheme, all wavelengths are numbered. When searching for available wavelengths, a lower-numbered wavelength is considered before a higher-numbered wavelength. The first available

wavelength is then selected. This scheme requires no global information. Compared to Random wavelength assignment, the computation cost of this scheme is lower because there is no need to search the entire wavelength space for each route. The idea behind this scheme is to pack all of the in-use wavelengths toward the lower end of the wavelength space so that continuous longer paths toward the higher end of the wavelength space will have a higher probability of being available. This scheme performs well in terms of blocking probability and fairness, and is preferred in practice because of its small computational overhead and low complexity. Similar to Random, FF does not introduce any communication overhead because no global knowledge is required.

3. **Least-Used (LU)/SPREAD.** LU selects the wavelength that is the least used in the network, thereby attempting to balance the load among all the wavelengths. This scheme ends up breaking the long wavelength paths quickly; hence, only connection requests that traverse a small number of links will be serviced in the network. The performance of LU is worse than Random, while also introducing additional communication overhead (e.g., global information is required to compute the least-used wavelength). The scheme also requires additional storage and computation cost; thus, LU is not preferred in practice.

4. **Most-Used (MU)/PACK.** MU is the opposite of LU in that it attempts to select the most-used wavelength in the network. It outperforms LU significantly [SuBa97]. The communication overhead, storage, and computation cost are all similar to those in LU. MU also slightly outperforms FF, doing a better job of packing connections into fewer wavelengths and conserving the spare capacity of less-used wavelengths.

5. **Min-Product (MP).** MP is used in multi-fiber networks [JeAy96]. In a single-fiber network, MP becomes FF. The goal of MP is to pack wavelengths into fibers, thereby minimizing the number of fibers in the network. MP first computes:

$$\prod_{l \in \pi(p)} D_{lj} \tag{7.6}$$

for each wavelength j, i.e., $1 \leq j \leq W$. If we let X denote the set of wavelengths j that minimizes the above value, then MP chooses the lowest-numbered wavelength in X. As shown in [SuBa97], MP does not perform as well as the multi-fiber version of FF in which the fibers, as well as the wavelengths, are ordered. MP also introduces additional computation costs.

6. **Least-Loaded (LL).** The LL heuristic, like MP, is also designed for multi-fiber networks [KaAy98]. This heuristic selects the wavelength that has the largest residual capacity on the most-loaded link along route p. When used in single-fiber networks, the residual capacity is either 1 or 0; thus, the heuristic chooses the lowest-indexed wavelength with residual capacity 1. Thus, it reduces to FF in single-fiber networks. LL selects the minimum indexed wavelength j in S_p that achieves:

$$\max_{j \in S_p} \min_{l \in \pi(p)} (M_l - D_{lj}) \tag{7.7}$$

   In [KaAy98], it is shown that LL outperforms MU and FF in terms of blocking probability in a multi-fiber network.

7. **MAX-SUM ($M\sum$).** $M\sum$ [BaSu97, SuBa97] was proposed for multi-fiber networks but it can also be applied to the single-fiber case. $M\sum$ considers all possible paths (lightpaths with their preselected routes) in the network and attempts to maximize the remaining path capacities after lightpath establishment. It assumes that the traffic matrix (set of possible connection requests) is known in advance, and that the route for each connection is pre-selected. These requirements can be achieved since the traffic matrix is assumed to be stable for a period of time, and routes can then be computed for each potential path on the fly.

   To describe the heuristic, we introduce the following notation. Let Ψ be a network state that specifies the existing lightpaths (routes and wavelength assignments) in the network. In $M\sum$, the link capacity on link l and wavelength j in state Ψ, $r(\Psi, l, j)$, is defined as the number of fibers on which wavelength j is unused on link l, i.e.,

$$r(\Psi, l, j) = M_l - D(\Psi)l_j, \tag{7.8}$$

where $D(\Psi)$ is the D matrix in state Ψ.

The path capacity $r(\Psi, p, j)$ on wavelength j is the number of fibers on which wavelength j is available on the most-congested link along the path p, i.e.,

$$r(\Psi, p, j) = \min_{l \in \pi(p)} r(\Psi, l, j). \tag{7.9}$$

The path capacity of path p in state Ψ is the sum of path capacities on all wavelengths, i.e.,

$$R(\Psi, p) = \sum_{j=1}^{\max} \min_{l \in \pi(p)} c(\Psi, l, j). \tag{7.10}$$

Let

- $\Omega(\Psi, p)$ be the set of all possible wavelengths that are available for the lightpath that is routed on path p, and
- $\Psi'(j)$ be the next state of the network if wavelength j is assigned to the connection.

$M\sum$ chooses the wavelength j that maximizes the quantity

$$\sum_{p \in P} R(\Psi'(j), p), \tag{7.11}$$

where P is the set of all potential paths for the connection request in the current state. Once the lightpath for the connection has been established, the network state is updated and the next connection request may be processed.

8. **Relative Capacity Loss (RCL).** RCL was proposed in [ZhQi98] and is based on $M\sum$. $M\sum$ can also be viewed as an approach that chooses the wavelength j that minimizes the capacity loss on all lightpaths, which is:

$$\sum_{p \in P} (R(\Psi'(j)) - (R(\Psi'(j)), p)), \tag{7.12}$$

where Ψ is the network state before the lightpath is set up. Since only the capacity on wavelength j will change after the lightpath is set up on wavelength j, $M\sum$ chooses wavelength j to minimize the total capacity loss on this wavelength, i.e.,

$$\sum_{p \in P} (r(\Psi'(j)) - (r(\Psi'(j)), p)). \tag{7.13}$$

On the other hand, RCL chooses wavelength j to minimize the relative capacity loss, which can be computed as:

$$\sum_{p \in P} (R(\Psi'(j)) - (r(\Psi'(j)), p))/r(\Psi, p, j). \tag{7.14}$$

RCL is based on the observation that minimizing total capacity loss sometimes does not lead to the best choice of wavelength. When choosing wavelength i would block one lightpath p_1, while choosing wavelength j would decrease the capacity of lightpaths p_2 and p_3, but not block them, then wavelength j should be chosen over wavelength i, even though the total capacity loss for wavelength j is greater than the total capacity loss for wavelength i. Thus, RCL calculates the Relative Capacity Loss for each path on each available wavelength and then chooses the wavelength that minimizes the sum of the relative capacity loss on all the paths.

Both $M\sum$ and RCL can be used for non-uniform traffic by taking a weighted sum over the capacity losses. RCL has been shown to perform better than $M\sum$ in most cases.

9. **Distributed Relative Capacity Loss (DRCL).**

   There are additional costs in implementing algorithms LU, MU, MP, LL, $M\sum$, and RCL that involve global knowledge of the network state in a distributed-controlled network. Information on the network state must be exchanged frequently to ensure accurate calculations, similar to what must be done in implementing the link-state routing protocol. $M\sum$ and RCL perform well but are difficult and expensive to implement in a distributed environment. Furthermore, $M\sum$ and RCL both require fixed routing, which makes it difficult to improve network performance. In order to implement an effective wavelength-selection policy in a distributed adaptive routing environment, two problems have to be solved:

   - how is information of network state exchanged? and

- how can we reduce the amount of calculation upon receiving a connection request?

To speed up the wavelength-assignment procedure, each node in the network stores information on the capacity loss on each wavelength so that only a table lookup and a small amount of calculation are required upon the arrival of a connection request. To maintain a valid table, the related values should be updated as soon as the network state has changed. To simplify the computation, we propose an algorithm called Distributed Relative Capacity Loss (DRCL). The routing is implemented using the Bellman-Ford algorithm [Acev92]. In Bellman-Ford, each node exchanges routing tables with its neighboring nodes and updates its own routing table accordingly. We introduce an RCL table at each node and allow the nodes to exchange their RCL tables as well. The RCL tables are updated in a similar manner as the routing table. Each entry in the RCL table is a triple of (wavelength w, destination d, $rcl(w,d)$). When a connection request arrives and more than one wavelength is available on the selected path, computation is carried out among these wavelengths. Similar to the manner in which $M\sum$ and RCL consider a set of potential paths for future connections, DRCL considers all of the paths from the source node of the arriving connection request to every other node in the network, excluding the destination node of the arriving connection request. DRCL then chooses the wavelength that minimizes the sum of $rcl(w,d)$ over all possible destinations d.

The $rcl(w,d)$ at node s is calculated as follows:

- If there is no path from node s to node d on wavelength w, then $rcl(w,d) = 0$; otherwise,
- If there is a direct link from node s to node d, and the path from s to d on wavelength w is routed through this link (note that it is possible for a direct link to exist between two nodes, but for the path to be routed around this link), then $rcl(w,d) = 1/k$, where k is the number of available wavelengths on this link through which s can reach d; otherwise,
- If the path from node s to node d on wavelength w starts with node n (n is s's next node for destination d on wavelength w), and there are k wavelengths available on link $s \rightarrow n$ through which s

can reach d, then $rcl(w, d)$ at node s is set to $(1/k, rcl(w, d)$ at node n).

Thus far, the wavelength-assignment schemes that we have described attempt to minimize the blocking probability. However, considering that longer lightpaths have a higher probability of getting blocked than shorter paths, some schemes attempt to protect longer paths. These schemes are wavelength reservation (Rsv) and protecting threshold (Thr) [BiKe95]. Rsv and Thr differ from other wavelength-assignment schemes in two ways: First, they do not specify which wavelength to choose, but instead specify whether or not the connection request can be assigned a wavelength under the current wavelength-usage conditions. Hence, they can not work alone and must be combined with other wavelength-assignment schemes. Second, other schemes aim at minimizing the overall blocking probability for all connection requests, while the Rsv and Thr schemes attempt to protect only the connections that traverse multiple fiber links (multihop connections). Therefore, when these two schemes are used, the overall blocking probability performance in the network may be higher, but a greater degree of fairness can be achieved, in that connections that traverse multiple fiber links will not have significantly higher blocking probabilities than connections that traverse only a single fiber link.

10. **Wavelength Reservation (Rsv).** In Rsv, a given wavelength on a specified link is reserved for a traffic stream, usually a multihop stream. For example, in Fig. 7.5, wavelength λ_1 on link $(1, 2)$ may be reserved only for connections from Node 0 to Node 3; thus, a connection request from Node 1 to Node 2 cannot be set up on λ_1 link (1,2), even if the wavelength is idle. This scheme reduces the blocking for multihop traffic, while increasing the blocking for connections that traverse only one fiber link (single-hop traffic) [BiKe95].

11. **Protecting Threshold (Thr).** In Thr, a single-hop connection is assigned a wavelength only if the number of idle wavelengths on the link is at or above a given threshold [BiKe95].

The above wavelength-assignment schemes can work online since they make use of the current network state information. It is straightforward to show that they can also work off-line for static network traffic by handling the static set of lightpaths sequentially. An additional issue when applying these heuristics to the static problem is how to order the lightpaths when assigning wavelengths. Approaches similar to those in Section 7.2 may be applied.

7.6 Illustrative Examples from RWA Subproblem

We use an example to illustrate how the above wavelength-assignment schemes work in a single-fiber network. This example was borrowed from and was initially used to illustrate the $M\sum$, RCL, and DRCL heuristics.

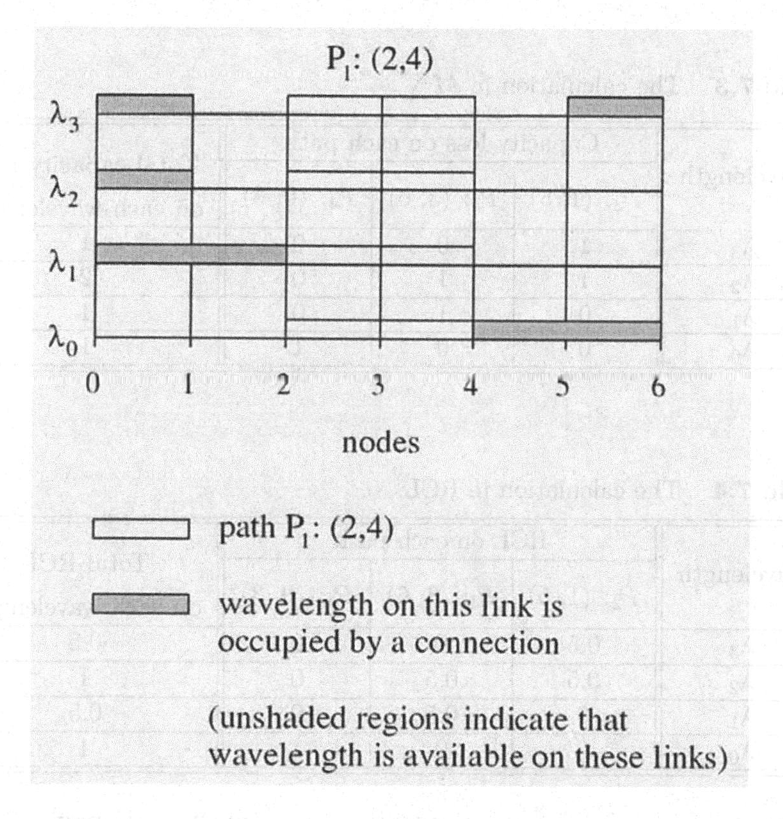

Figure 7.8 Wavelength-usage pattern for a network segment consisting of six fiber links in tandem.

Consider a six-link segment of a single-fiber network spanning a tandem sequence of seven nodes (numbered 0 through 6) with a current wavelength-usage pattern as shown in Fig. 7.8. If we want to set up a lightpath P_1: (2, 4), we observe that four wavelengths (λ_0 through λ_3) are available.

In the Random scheme, any of the four wavelengths can be chosen with equal probability.

If First-Fit is used, λ_0 will be assigned. λ_0 will also be assigned in Min-Product and Least-Loaded, since they reduce to First-Fit in single-fiber networks. Since λ_0, λ_1, and λ_3 are each used on two out of the six links in the network and λ_2 is used only on one link, Least-Used will choose λ_2 and Most-Used will choose either λ_0, λ_1, or λ_3 with equal probability.

Table 7.3 The calculation in $M\sum$.

wavelength	Capacity loss on each path			Total capacity loss on each wavelength
	P_2: (1, 5)	P_3: (3, 6)	P_4: (0, 3)	
λ_3	1	0	0	1
λ_2	1	1	0	2
λ_1	0	1	0	1
λ_0	0	0	1	1

Table 7.4 The calculation in RCL.

wavelength	RCL on each path			Total RCL on each wavelength
	P_2: (1, 5)	P_3: (3, 6)	P_4: (0, 3)	
λ_3	0.5	0	0	0.5
λ_2	0.5	0.5	0	1
λ_1	0	0.5	0	0.5
λ_0	0	0	1	1

When performing the calculations using $M\sum$ and RCL, a pre-determined traffic matrix that specifies a set of connections and their paths must be assumed. We consider the case in which there are only three other potential paths that share common links with P_1, and that no other paths need to be considered. These paths are P_2: (1, 5), P_3: (3, 6), and P_4: (0, 3).

The total capacity loss (for $M\sum$) and total relative capacity loss (for RCL) are calculated in Tables 7.3 and 7.4, respectively. For $M\sum$, we observe that setting up lightpath P_1 on wavelength λ_0 will block path P_4 on λ_0, setting up P_1 on λ_1 will block P_3, setting up P_1 on λ_2 will block both P_2 and P_3, and setting up P_1 on λ_3 will block P_2. Choosing λ_2 will result in the highest total capacity loss, or the highest amount of blocking for possible future calls; thus, any of λ_0, λ_1, and λ_3, which have equal total capacity loss, may be chosen by $M\sum$. However, note that, by choosing λ_0, path P_4 will be blocked on all wavelengths, whereas if we choose λ_1 or λ_3, each of P_2, P_3, and P_4 would still have at least one wavelength on which they would not be blocked. RCL attempts to improve on $M\sum$ by also taking into consideration the number of available alternate wavelengths for each potential future connection. We observe that path P_2 may choose either of two wavelengths, λ_2 or λ_3; thus, if P_1 is established on either of these wavelengths, then the relative capacity loss for P_2 is 1/2. Similarly, P_3 has two wavelengths on which a connection can be established; therefore, its relative capacity loss on these wavelengths is also 1/2. However, a connection on P_4 can only be established on wavelength λ_0; thus, its relative capacity loss is 1 for wavelength λ_0. Summing the relative capacity loss for each wavelength over all paths yields the total relative capacity loss on a given wavelength. Choosing the wavelength with the smallest total relative capacity loss results in either λ_1 or λ_3 being chosen, but not λ_0 or λ_2.

Table 7.5 The calculation in DRCL.

wavelength	RCL from source to each destination					Total RCL on each wavelength
	(2, 0)	(2, 1)	(2, 3)	(2, 5)	(2, 6)	
λ_3	0	1/3	1/4	1/3	0	11/12
λ_2	0	1/3	1/4	1/3	1/2	17/12
λ_1	0	0	1/4	1/3	1/2	13/12
λ_0	1	1/3	1/4	0	0	19/12

Table 7.5 shows the computation carried out by DRCL. If we are attempting to set up a connection on path (2,4), we must then calculate the RCL for each of the paths (2, 0), (2, 1), (2, 3), (2, 5), and (2, 6) on each wavelength. The path (2, 0) can only be established on one possible wavelength, λ_0; thus, its RCL on wavelength λ_0 is 1, and its RCL on the other

wavelengths is 0. Path (2, 1) can be established on one of three wavelengths, yielding RCL values of 1/3 for each of these wavelengths. Path (2, 3) can be established on any of the four wavelengths, leading to RCL values of 1/4 for all wavelengths. Path (2, 5) can be established on three wavelengths, giving RCL values of 1/3, and path (2, 6) can be established on two wavelengths, yielding RCL values of 1/2. Note that these RCL values can be calculated at Node 2 only using the RCL tables from the adjacent nodes 1 and 3. Also, the calculations for RCL can be done prior to the arrival of a connection request, reducing the time required to select a wavelength and set up the lightpath. When a connection request arrives, DRCL simply has to sum the RCL values for each wavelength over all destinations excluding the destination of the connection request itself. The wavelength that yields the lowest total relative capacity loss is then selected. In the above example, wavelength λ_3 is chosen (see Table 7.5).

Since Wavelength Reservation and Protecting Threshold must work together with other protecting schemes, their operation is not discussed here.

7.7 Simulation Results and Comparison from RWA

7.7.1 Comparison among Wavelength-Assignment Heuristics

We compare the first nine wavelength-assignment heuristics via simulation. Fixed routing is used in all simulations as required by $M\sum$ and RCL. Simulations are carried out on the network shown in Fig. 7.9, and each link in the network contains M fibers, and each fiber supports W wavelengths. Results are shown in Figs. 7.10 through 7.13 for different values of M and W.

A distributed link-state control protocol [ZaMu01] is used in the simulations, and the results depend on the propagation delays in the network. This approach differs significantly from that in [SuBa97], in which centralized control is used and no propagation delays are assumed. In the link-state protocol, each node has full information regarding the network state. When a lightpath is set up, the appropriate information is broadcast to all nodes, which then update their state information. Since it takes a certain amount of time for the information to reach all of the nodes, some nodes may make routing and wavelength-assignment decisions based on outdated information if connection requests are arriving at a high rate. These decisions based on outdated information can lead to higher blocking probabilities; thus, heuristics that rely on more state information may potentially have higher blocking

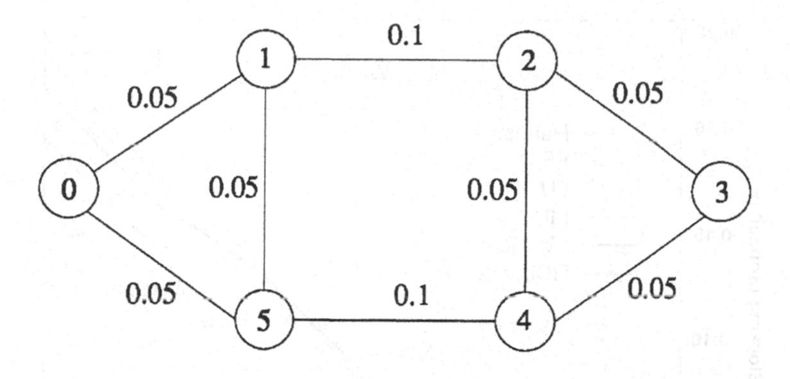

Figure 7.9 Simulation network.

probabilities if the propagation delay is high and connection-arrival rates are high.

In the single-fiber case (Fig. 7.10), MU is found to achieve the best performance under low load while $M\sum$ and RCL work well when the load is high (≥ 50 Erlangs), with the other approaches not that far behind. When the number of fibers per link is two ($M = 2$), MU, MP, and RCL perform well under low load, while LL and $M\sum$ offer better performance under a higher load (Fig. 7.11). When the number of fibers per link is four ($M = 4$), LL appears to give the best performance (Fig. 7.12). In each case, we observe that the difference among the various heuristics is not too significant.

DRCL works well with adaptive routing for distributed-controlled networks because it distributes the computation load among all the network nodes, and each node can thus utilize information from other nodes. DRCL's performance is compared to those of RCL (with fixed routing) and FF (with fixed and adaptive routing) in Fig. 7.13. Note that RCL cannot be implemented with adaptive routing. We observe that DRCL slightly outperforms FF (with adaptive routing) in the reasonable region (the region in which the network performs well in terms of blocking probability, which is 45-65 Erlangs in this network), and they both perform better than RCL and FF with fixed routing. Overall, it appears that the routing scheme has much more of an impact on the performance of the system than the wavelength-assignment scheme. This "routing is more significant" conclusion is consistent with the findings in previous studies [RaMu02]. Therefore, it is important to first decide on a good routing mechanism, and then to choose a wavelength-

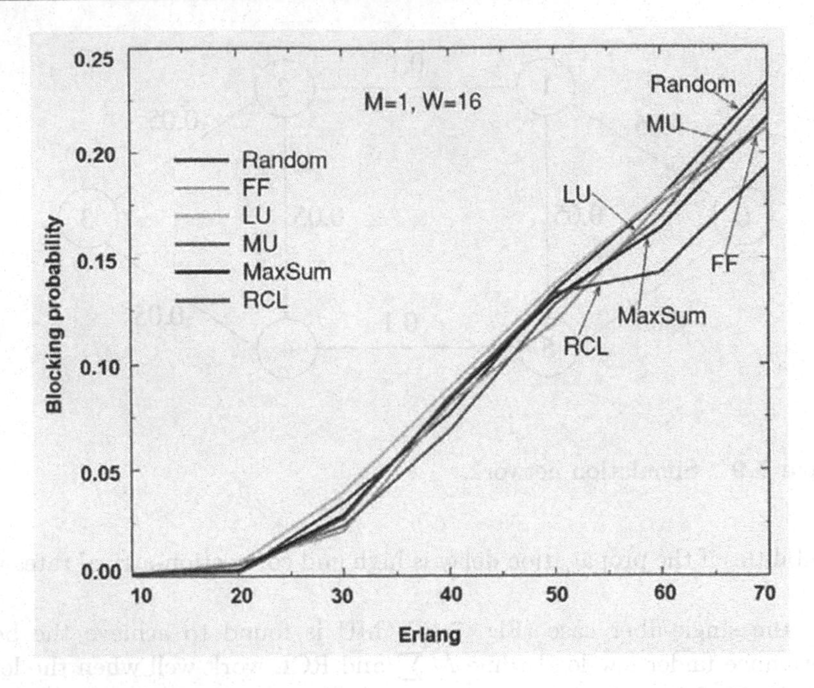

Figure 7.10 Comparison of Random, FF, LU, MU, Max-Sum, and RCL for single-fiber network with 16 wavelengths.

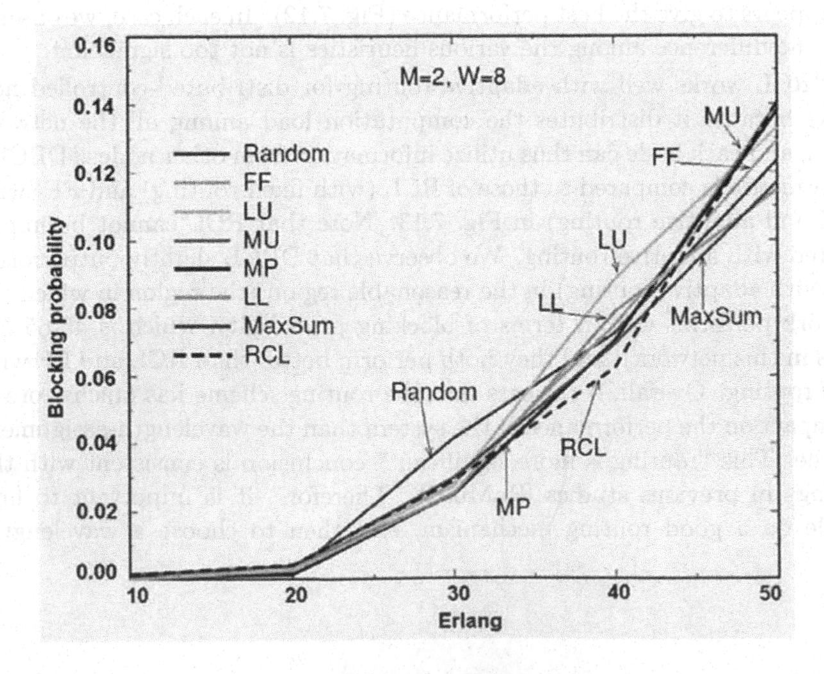

Figure 7.11 Comparison of Random, FF, LU, MU, MP, LL, Max-Sum, and RCL for two-fiber network with 8 wavelengths.

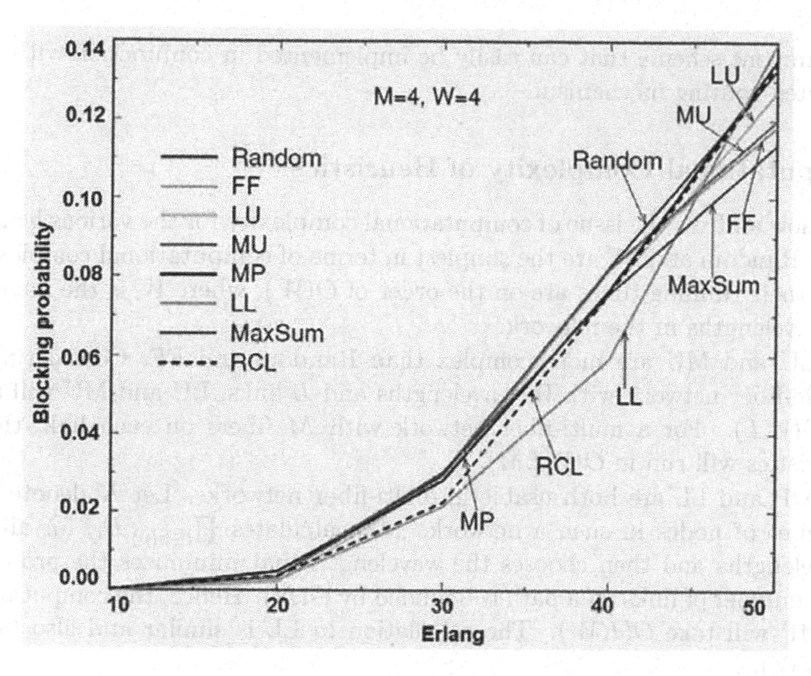

Figure 7.12 Comparison of Random, FF, LU, MU, MP, LL, Max-Sum, and RCL for four-fiber network with 4 wavelengths.

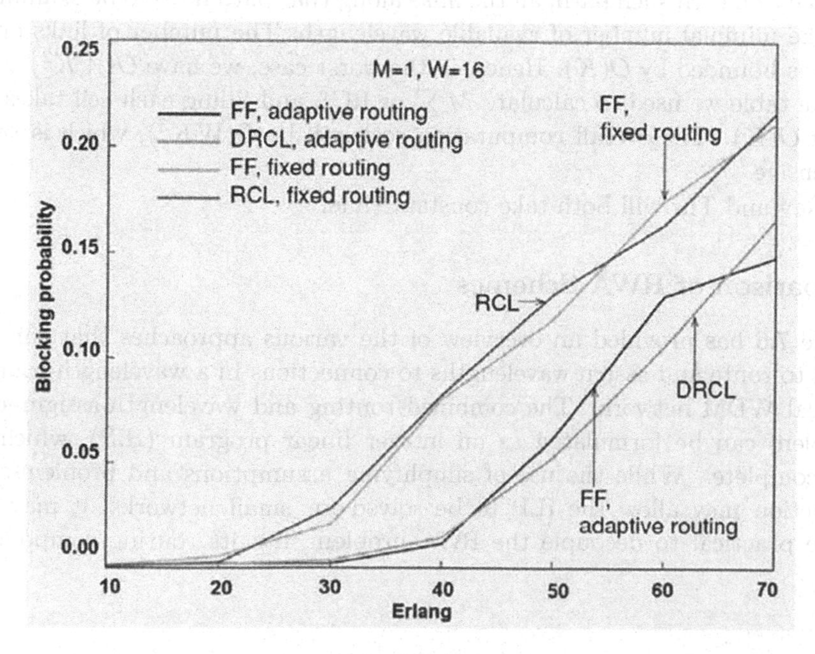

Figure 7.13 Comparison of DRCL, FF (with or not adaptive routing) and RCL for single-fiber network with 16 wavelengths.

assignment scheme that can easily be implemented in conjunction with the selected routing mechanism.

7.7.2 Computational Complexity of Heuristics

We now address the issue of computational complexity for the various heuristics. Random and FF are the simplest in terms of computational complexity and their running times are on the order of $O(W)$, where W is the number of wavelengths in the network.

LU and MU are more complex than Random and FF. Considering a single-fiber network with W wavelengths and L links, LU and MU will run in $O(WL)$. For a multi-fiber network with M fibers on each link, these heuristics will run in $O(WLM)$.

MP and LL are both usable in multi-fiber networks. Let K denote the number of nodes in such a network. MP calculates $\prod_{l\in\pi(p)} D_{lj}$ for all W wavelengths and then chooses the wavelength that minimizes the product. The number of links on a path is bounded by $O(N)$. Hence, the computation in MP will take $O(KW)$. The calculation in LL is similar and also takes $O(KW)$.

$M\sum$ and RCL are relatively expensive. If we consider the number of paths in a single-fiber network that share common links with a given path, the worst-case running time is on the order of $O(K^2)$. To calculate the capacity on each such path, all the links along that path have to be examined for the minimal number of available wavelengths.The number of links on a path is bounded by $O(K)$. Hence, in the worst case, we have $O(WK^2)$ cells in the table we used to calculate $M\sum$ or RCL and filling each cell takes at most $O(K)$. The overall computation cost will be $O(WK^3)$, which is very expensive.

Rev and Thr will both take constant time.

7.7.3 Comparison of RWA Schemes

Table 7.6 has provided an overview of the various approaches that can be used to route and assign wavelengths to connections in a wavelength-routed optical WDM network. The combined routing and wavelength-assignment problem can be formulated as an integer linear program (ILP), which is NP-complete. While the use of simplifying assumptions and problem-size reduction may allow the ILP to be solved for small networks, it may be more practical to decouple the RWA problem into its routing component

Table 7.6 Summary of RWA schemes.

Problems		Approaches	On/Off line	Comments	References
Static RWA		ILP formulation	off line	NP-complete	[RaSi95a, Mukh97]
Routing		ILP formulation	off line	NP-complete	[RaSi95a, Mukh97]
		fixed routing alternate routing	on/off line	in order increasing performance & complexity	[ChYu94, Rama98] [LiSo99, RaMu02]
		adaptive routing	on line		
WA	WA (connections and routes are known)	graph coloring	off line	NP-complete	[ChGK92, Mukh97]
	WA+ fixed routing	LU (SPREAD) Random MP (multi-fiber) FF MU (PACK) LL (multi-fiber) $M\sum$ RCL	on/off line performance	heuristics approximately in order of increasing	[ChGK89, JeAy96, BaSu97] [SuBa97, KaAy98, ZhQi98]
	WA+ adaptive routing	DRCL	on line	combined with adaptive routing	[ZaJM00a]
	Other WA algorithms	Rsv Thr	on line	combined with other WA algorithms	[BiKe95]

and its wavelength-assignment component for larger networks. The static routing problem, in which the set of connections which need to be routed is known in advance, may also be formulated as an ILP which is NP-complete. A more traditional approach to routing is to use shortest-path algorithms; however, relying on a single fixed shortest path may lead to high blocking probabilities. It has been shown that techniques such as fixed-alternate routing and adaptive routing provide significant benefits over fixed-shortest-path routing, and often, these routing approaches even provide improved performance over wavelength conversion [RaMu02]. In networks that require protection, having either fixed-alternate or adaptive routing is a requirement [Rama98, RaMu99a].

Table 7.6 also gives an overview of some of the proposed wavelength-assignment algorithms. Graph-coloring heuristics can be applied to the static case in which all connections and their routes are known in advance. For dynamic traffic in which connection requests arrive one at a time, there are a number of possible heuristics. It is found that more complex heuristics, such as Max-Sum and RCL, provide lower blocking probability than simpler heuristics; however, the difference in performance among the various heuristics is not very large (as shown in Section 7.7.1). Also, these two heuristics rely on fixed routing and cannot be directly applied within an adaptive-routing environment. Distributed Relative Capacity Loss (DRCL), which is based on RCL, is well suited for a distributed-controlled network in which adaptive routing is utilized. For more detailed information, we refer the interested reader to [ZaJM00a].

7.8 Summary

This chapter first studied large instances of the routing and wavelength assignment (RWA) problem as applied to a wavelength-routed optical network. The objective was to minimize the number of wavelengths needed, given a certain set of lightpath requests that were to be satisfied on a given physical topology. The problem was partitioned into two parts: (1) routing of a lightpath over the physical fiber links, and (2) assigning wavelengths to each lightpath so that no two lightpaths passing through the same physical fiber link are assigned the same wavelength. Algorithms were proposed which can solve the RWA problem for large networks. The problem was studied for the static case (where all lightpath requests are available in advance) as well as the dynamic case (where lightpath requests arrive and need to be established one by one). A linear programming (LP) formulation was used along with good approximation techniques to solve the static lightpath establishment (SLE) problem. Our algorithms provided results which were very close to the lower bound for the number of wavelengths that will be needed to establish a given set of lightpaths.

Some heuristics were further studied, and used for the dynamic case; the results obtained were also close to the aforementioned lower bound. These algorithms can be used for the design and analysis of large wavelength-routed optical networks. Specifically, it is shown that DRCL performs as well as the other wavelength-assignment heuristics in an adaptive-routing environment.

Since routing decisions play a significant role in determining the blocking performance of a network, it is critical to choose a wavelength-assignment scheme, not based solely on its blocking performance relative to other wavelength-assignment schemes, but also based on its compatibility with the chosen routing protocol.

Exercises

7.1. In the linear programming formulation in Section 7.2.1, we minimize $F_{\max}$. What is $F_{\max}$? Why should we minimize $F_{\max}$?

7.2. Let G be a graph with $V(G) = v_1, v_2, \ldots, v_n$, where $\deg(v_i) \geq \deg(v_{i+1})$ for $i = 1, 2, \ldots, n-1$. Prove that $\chi(G) \leq \max_{1 \leq i \leq n} \min(i, 1 + \deg(v_i))$.

7.3. Given a graph $G = (V, E)$. Define

$$k = \max_{1 \leq i \leq n} 1 + \deg_{<v_1, v_2, \ldots, v_n>}(v_i) \tag{7.15}$$

where $v_1, v_2, \ldots, v_n \in V$ and $< v_1, v_2, \ldots, v_n >$ is a vertex ordering. Define *smallest-last (SL)* vertex ordering $< v_1, v_2, \ldots, v_n >$ to be the vertex ordering such that $\deg(v_{i+1}) \leq \deg v_i$, for $1 \leq i \leq n$. Show that, over all the $n!$ possible vertex orderings, the SL vertex ordering minimizes the value of k.

7.4. We know that the static lightpath establishment (SLE) problem is NP-complete. Show a simple transformation that transforms a SLE problem into a graph coloring problem.

7.5. Consider the network shown in Fig. 7.1. Let the connection requests be as follows:
B-H, A-E, B-D, D-F, B-F, C-E, C-H, A-G, A-C
Set up lightpaths to satisfy the above connection requests using at most three wavelengths per link. Assume no wavelength conversion.

7.6. Consider the network in Fig. 7.1, and the following lightpaths:

(a) C-7-8-9-E

(b) A-1-5-8-9-E

(c) H-2-1-5-8-7-C

(d) B-6-7-8-9-E

(e) A-1-6-7-10-D

(f) G-3-2-1-6-B

(g) H-2-3-4-F

Color the lightpaths using the minimum number of wavelengths.

7.7. Consider the NSFNET physical topology shown in Fig. 9.1. Remove the nodes WA, CO, NE, and GA. Suppose the connection requests are as shown in Fig. 7.14. What is the minimum number of wavelengths needed to satisfy all the connection requests?

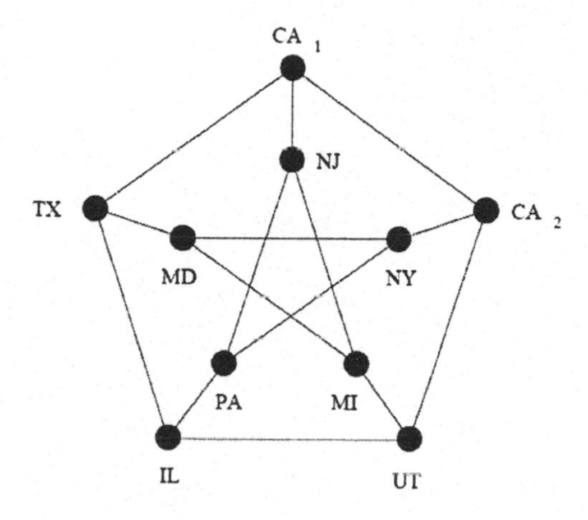

Figure 7.14 Connection requests.

7.8. Compare the characteristics of various routing schemes.

7.9. Compare the characteristics of various wavelength-assignment schemes.

7.10. Consider the addition of a ninth unidirectional lightpath in Fig. 7.2 between the left-most node (as source) and the right-most node (as destination). Let the route of this lightpath follows the lower three links. Draw the corresponding auxiliary graph, and determine the number of wavelengths needed to solve the problem.

7.11. Repeat the above problem where the ninth lightpath is routed over the top three links.

7.12. Choose a different route for the ninth lightpath from the ones chosen in the previous two problem. Repeat the analysis.

Chapter 8

Elements of Virtual-Topology Design

8.1 Introduction

This chapter explores design principles for next-generation optical wide-area networks, employing wavelength-division multiplexing (WDM) and targeted to nationwide and global coverage. This optical network exploits wavelength multiplexers and optical switches (crossconnects) in routing nodes, so that an arbitrary virtual topology may be embedded on a given physical fiber network. The virtual topology, which is operated as a packet-switched network and which consists of a set of "lightpaths," is set up to exploit the relative strengths of both optics and electronics – viz., packets of information are carried by the virtual topology "as far as possible" in the optical domain using *optical circuit switching*, but packet forwarding from lightpath to lightpath is performed via *electronic packet switching*, whenever required[1].

Each lightpath in the virtual topology is set up using the routing and wavelength assignment (RWA) techniques (discussed in Chapter 7). The

[1]The problem treated in this chapter corresponds to that of Business Model #1, where the business owns the entire network infrastructure from the ground up and offers packet-based services to its customers (see Section 1.3)

collection of lightpaths, i.e., the virtual topology, is also referred to as a "*Lamda Grid*", or just *Grid*, for short.

In this chapter, we do not consider the number of available wavelengths to be a constraint. This constraint as well as additional relaxations are studied in a companion chapter (Chapter 9) so that the entire virtual-topology design problem can be *linearized*, and it can, hence, be *solved optimally*. Also, in this chapter, we ignore the routing of lightpaths and wavelength assignment for these lightpaths, a topic which was examined in Chapter 7.

In a wavelength-routed optical network, optical circuit switching at a node is achieved by using an optical crossconnect (OXC), which is capable of switching a lightpath from an input fiber to an output fiber[2]. If there is no wavelength conversion in the OXC (as assumed in this chapter), the wavelength of the lightpath stays the same in the output fiber as it was in the input fiber.

This network architecture is a combination of the well-known "single-hop" and "multihop" approaches, and it attempts to exploit the characteristics of both. A "lightpath" in this architecture provides "single-hop" communication between any two nodes, which could be far apart in the physical topology. However, by employing a limited number of wavelengths, it may not be possible to set up "lightpaths" between all user pairs; as a result, "multihopping" between "lightpaths" may be necessary. In addition, when the prevailing traffic pattern changes, a different set of "lightpaths" forming a different "multihop" virtual topology may be more desirable. A networking challenge is to perform the necessary reconfiguration with minimal disruption to the network's operations [LaHA94, RoAm95]. In this architecture, using wavelength multiplexers provides the advantage of much higher aggregate system capacity due to spatial reuse of wavelengths and supports a large number of users, given a limited number of wavelengths. Specifically, we investigate the overall design, analysis, and optimization of a nationwide WDM network consistent with device capabilities, e.g., aimed at upgrading the NSFNET.

We formulate the virtual-topology design problem as an optimization problem with one of two possible objective functions: (1) for a given traffic matrix (intensities of packet flows between various pairs of nodes), mini-

[2]If the OXC is implemented using an all-optical technology, then there is no electronic processing at the intermediate node. However, an OXC can be implemented using an opaque optical-electronic-optical (OEO) technology as well, in which case the electronic processing also is a circuit-switched operation. An OEO-based OXC, however, generally has a built-in property that is logically equivalent to that of wavelength conversion.

mize the network-wide average packet delay (corresponding to a solution for *present traffic demands*), or (2) maximize the scale factor by which the traffic matrix can be scaled up (to provide the maximum capacity upgrade for *future traffic demands*). Since simpler versions of this problem have been shown to be NP-hard, we shall also explore heuristic approaches. Specifically, we shall employ an iterative approach which combines "simulated annealing" (to search for a good virtual topology) and "flow deviation" (to optimally route the traffic – and possibly bifurcate its components – on the virtual topology). We illustrate our approaches by employing experimental traffic statistics collected from NSFNET.

Section 8.2 explains the system architecture including motivation, general problem statement, and an illustrative example. The problem is formulated as a combinatorial optimization in Section 8.3. Since the problem is NP-hard, heuristic solutions are developed in Section 8.4. Results of applying experimental data (collected from NSFNET) to our algorithms are discussed in Section 8.5.

8.2 System Architecture

8.2.1 Motivation

Consider the NSFNET T1 backbone in Fig. 8.1[3].

Information is transferred over this backbone as packets (of possibly variable sizes) at a rate of 1.544 Mbps per link. The backbone consists of over a dozen nodes, each containing one or more computers operating as electronic packet switches. These switches are connected with one another via optical fibers to form an irregular mesh structure[4]. Store-and-forward packet switching is performed at the network nodes. That is, a packet traveling from a source node to a destination node may have to pass through zero or more intermediate nodes, and at each such intermediate node, the packet has

[3]The NSFNET has been upgraded from time to time by adding new nodes and/or links, and the backbone we show in Fig. 8.1 existed in the early 1990s. It was replaced with a faster T3 backbone in the 1992-93 time frame, with faster backbones subsequently, and eventually passed on to commercial entities for operation. However, we proceed with the T1 backbone since we have obtained a significant amount of traffic statistics from this configuration, and we will employ these statistics to illustrate our proposed "optical" solutions to the design of a wide-area WDM network.

[4]A structure or graph employed to interconnect a number of nodes, possibly computers or processing elements, is known as a mesh, and it is referred to as irregular if it does not possess any well-defined connectivity pattern.

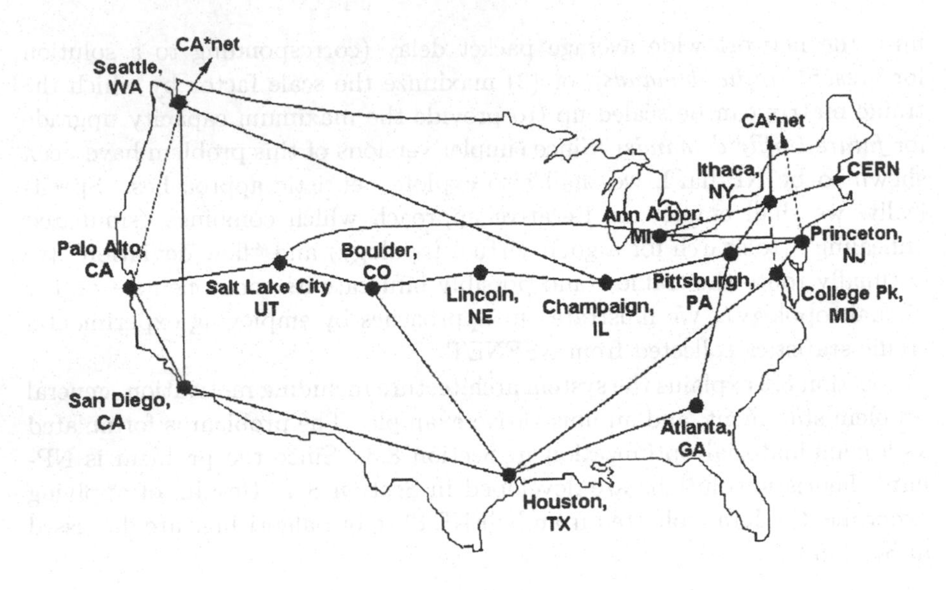

Figure 8.1 NSFNET T1 backbone.

to be completely received (stored in memory), its header has to be processed by the intermediate node to determine on which of the node's outgoing links this packet must be forwarded, and the packet may have to wait at this node longer if that corresponding outgoing link is busy due to the transmission of other packets. Although fiber is employed to connect the nodes, the fiber's tremendous transmission bandwidth is not exploited since data transmission on each fiber link is performed only at T1 (1.544 Mbps) rate, and on a single wavelength.

While WDM networks are being investigated, advances in packet switching such as Asynchronous Transfer Mode (ATM), Gigabit Ethernet (GigE), Multi-Protocol Label Switching (MPLS), etc., for high-speed networking solutions have also been making rapid progress. Thus, we expect that nationwide and global backbone networks such as NSFNET should also support these solutions. Although many networking issues will evolve with time, we expect that the underlying physical network will continue to be a fiber plant on which our WDM solutions will still be applicable.

Accordingly, our examination of optical networks is fueled by the following criteria and observations: (1) Any future technology must be incrementally deployable. That is, instead of scrapping an existing and operational wide-area fiber-based network such as NSFNET, we examine how such networks can be upgraded to support (and exploit the capabilities of) WDM. In this regard, we show later how one might replace (or upgrade) the existing packet switches in NSFNET with wavelength-routing switches (WRS), or alternately, how one might embellish an electronic switch with an optical component to transform it to a WRS. (2) We must also explore how the WDM solution can be used to upgrade an existing ATM solution. (3) In addition, we must explore how the WDM solution can support a variety of electronic interfaces or services, based on WDM's data and protocol transparency property.[5] However, the last point is beyond the scope of this chapter. In the following subsections, we provide a general problem statement, followed by an illustrative example using the NSFNET T1 topology.

8.2.2 General Problem Statement

The problem of embedding a desired virtual topology on a given physical topology (fiber network) is formally stated below. We are given the following inputs to the problem:

1. A physical topology $G_p = (V, E_p)$ consisting of a weighted undirected graph, where V is the set of network nodes, and E_p is the set of links connecting the nodes, where the subscript p is used to denote physical network links, as opposed to v for virtual links (or lightpaths) in a virtual topology. Undirected means that each link in the physical topology is bidirectional. Nodes correspond to network nodes (packet switches) and links correspond to the fibers between nodes; since links are undirected, each link may consist of two fibers or two channels multiplexed (using any suitable mechanism) on the same fiber. Links are assigned weights, which may correspond to physical distances between nodes. A network node i is assumed to be equipped with a $D_p(i) \times D_p(i)$ wavelength-routing switch (WRS), where $D_p(i)$, called the physical degree of node i, equals the number of physical fiber links

[5] Since multiple WDM channels on the same fiber can be operated independently, they can carry data at different rates and formats (including some analog and some digital channels, if desired); also, the protocols controlling the data transfers over different channels can be different, so that one can establish independent subnetworks operating on different sets of WDM channels over the same fiber plant. This is referred to as WDM's *transparency* property.

emanating out of (as well as terminating at) node i.[6]

2. Number of wavelength channels carried by each fiber $= M$.
3. An $N \times N$ traffic matrix, where N is the number of network nodes, and the (i,j)-th element is the average rate of packet traffic flow from node i to node j. Note that the traffic flows may be asymmetric, i.e., flow from node i to node j may be different from the flow from node j to node i.
4. The number of wavelength-tunable lasers (transmitters) and wavelength-tunable filters (receivers) at each node.

Our goal is to determine the following:

1. A virtual topology $G_v = (V, E_v)$ as another graph where the out-degree of a node is the number of transmitters at that node and the indegree of a node is the number of receivers at that node. The nodes of the virtual topology correspond to the nodes in the physical topology[7]. Each link between a pair of nodes in the virtual topology corresponds to a "lightpath" directly between the corresponding nodes in the physical topology. (Noting that each such link of the virtual topology may be routed over one of several possible paths on the physical topology, an important design issue is "optimal routing" of *all lightpaths* so that the constraint on having a limited number of wavelengths per fiber is satisfied. This RWA problem was treated in Chapter 7.)
2. A wavelength assignment for lightpaths, such that if two lightpaths share a common physical link, they must necessarily employ different wavelengths.
3. The sizes and configurations of the WRSs at the intermediate nodes. Once the virtual topology is determined and the wavelength assignments have been performed, the switch sizes and configurations follow directly.

[6] Note that $D_p(i)$ includes the fiber(s) corresponding to local connections, i.e., for attaching an electronic router to the WRS (shown later in this chapter). There are wavelength-related and cost-related issues which affect the decision on the number of fibers chosen to connect the local node to the local switch, and these issues are discussed in Section 8.2.3.

[7] In a more general version of the problem, the set of nodes in the virtual topology could be a subset of the nodes in the physical topology.

Communication between any two nodes now takes place by following a path (a sequence of lightpaths) from the source node to the destination node on the virtual topology. Each intermediate node in the path must perform (1) an opto-electronic conversion, (2) electronic routing (or packet switching in the electronic domain), and (3) electrooptic forwarding onto the next lightpath.

8.2.3 An Illustrative Example

We employ an illustrative example to demonstrate how WDM can be used to upgrade an existing fiber-based network. Using a slightly modified version of the NSFNET in Fig. 8.1 as an example (see Fig. 8.2), we demonstrate how a hypercube (Fig. 8.3) can be embedded as a virtual topology over this physical topology (to be discussed shortly). We also assume an undirected virtual topology comprising bidirectional lightpaths in this example. In general, the virtual topology may be a directed graph, as our formulation in Section 8.3 will assume.

The NSFNET backbone in Fig. 8.1 has 14 nodes, with some links connecting these nodes to one another and some links connecting to the "outside world." For illustration purposes, we supplement this physical topology by adding two fictitious nodes, AB and XY, to capture the effect of NSFNET's connections to Canada's communication network, CA*net. Node XY is connected to Ithaca (NY) and Princeton (NJ) nodes of NSFNET, while node AB is connected to the Seattle (WA) and Salt Lake City (UT) nodes, where we have employed the last link as a fictitious one to render the physical topology richer and fault tolerant. (Note that, to provide fault tolerance, each node in the NSFNET is connected to at least two other nodes, and we did not want to violate this policy for node AB.) Thus, we get the modified physical topology, hereafter referred to as the physical topology, of Fig. 8.2.

For each node, the nodal switching architecture consists of an optical component and an electronic component. The optical component is a wavelength-routing switch (WRS), which can switch some lightpaths, and which can locally terminate some other lightpaths by directing them to node's electronic component. The electronic component is an electronic packet router (which may be an IP router), which serves as a store-and-forward electronic overlay on top of the optical virtual topology. Figure 8.4 provides a schematic diagram of the architecture of the Utah node (UT) in our illustrative example.

The design of a WRS can take several forms. An attractive choice is

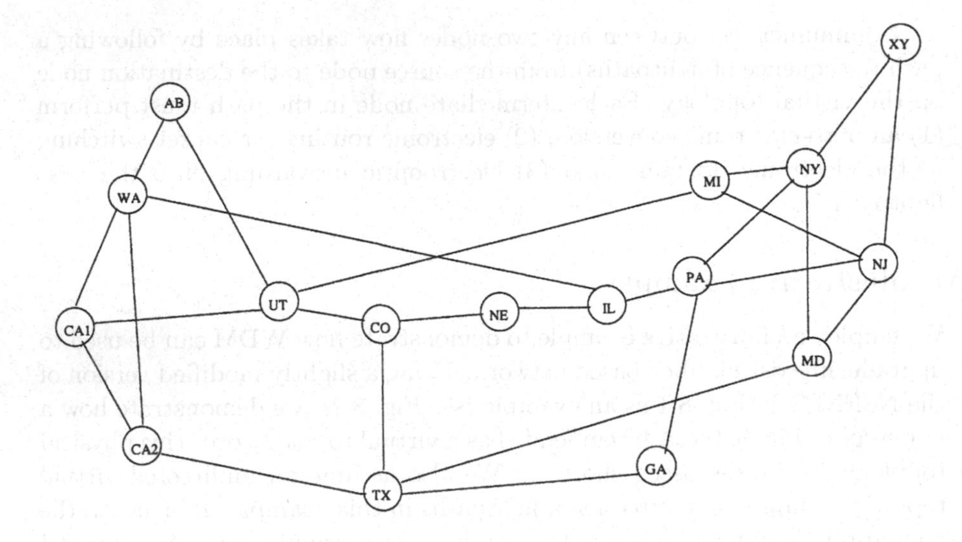

Figure 8.2 Modified physical topology.

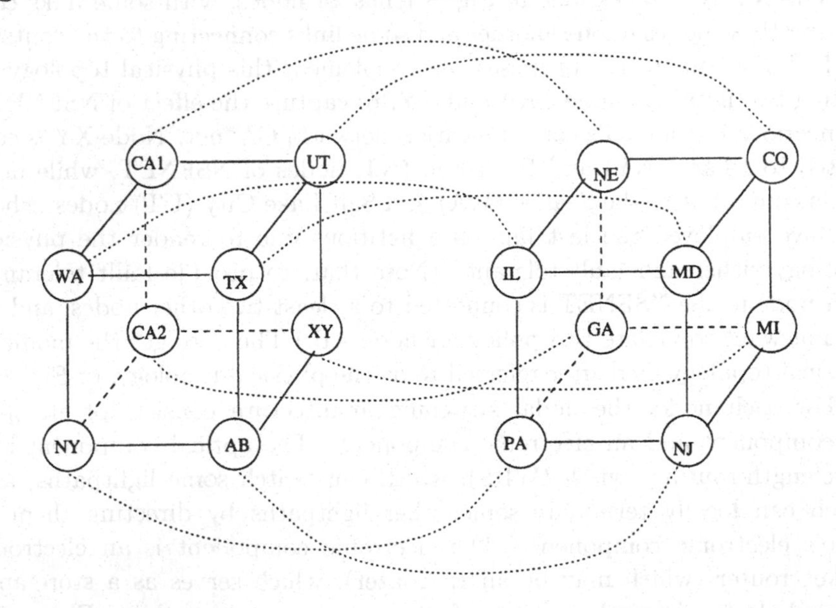

Figure 8.3 A 16-node hypercube virtual topology embedded on the NSFNET physical topology.

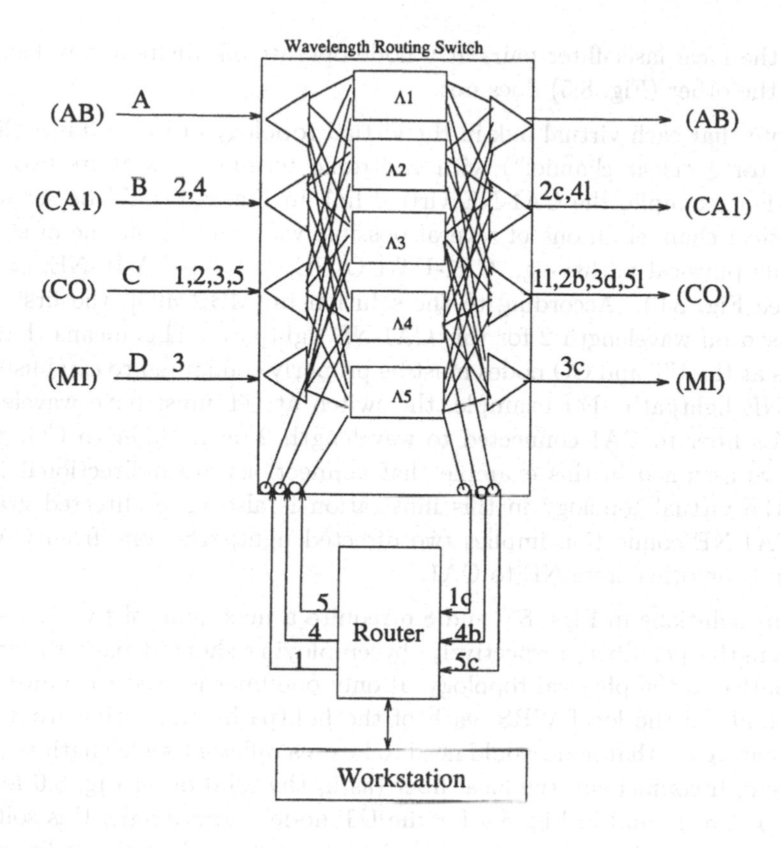

Figure 8.4 Details of the Utah (UT) node.

to employ an array of optical space-division switches, one per wavelength, between the demux and mux stages (see Fig. 8.4). These switches can be reconfigured under electronic control, e.g., to adapt the network's virtual topology on demand. One approach would be to have the local lasers/filters normally operated on fixed wavelengths, but a facility to tune them to different wavelengths must be provided.

The virtual topology chosen for our illustration is a 16-node hypercube (e.g., Fig. 8.3), although algorithms for arbitrary virtual-topology embedding will be studied later in this chapter. Two embeddings which result from the optimization criteria for hypercube embedding studied in [MBRM96] are shown in Figs. 8.5 and 8.6. One of these embeddings (Fig. 8.6) assumes that all of the local laser-filter pairs at a node operate on different wavelengths, while the other (Fig. 8.5) does not.

Note that each virtual link in the virtual topology of Fig. 8.3 is a "lightpath" (or a "clear channel") with electronic terminations at its two ends only. For example, the CA1-NE virtual link in Fig. 8.6 could be set up as an optical channel on one of several possible wavelengths on one of several possible physical paths, e.g., CA1-UT-CO-NE, or CA1-WA-IL-NE, or others (see Fig. 8.6). According to the solution in [MBRM96], the first path is chosen on wavelength 2 for this CA1-NE lightpath. This means that the WRSs at the UT and CO nodes must be properly configured to establish this CA1-NE lightpath. For example, the switch at UT must have wavelength 2 on its fiber to CA1 connected to wavelength 2 on its fiber to CO. Since we have assumed in this example that connections are bidirectional (note that the virtual topology in this illustration is also an undirected graph), the CA1-NE connection implies two directed lightpaths, one from CA1 to NE and the other from NE to CA1.

The solutions in Figs. 8.5 and 8.6 require a maximum of five and seven wavelengths per fiber, respectively, by employing shortest-path routing of lightpaths on the physical topology. If only one fiber is used to connect the local node to the local WRS, each of the lightpaths emanating from (and terminating at) that node would need to be on a different wavelength to avoid wavelength conflicts on the local fiber (as in the solution in Fig. 8.6 for the entire network, and in Fig. 8.4 for the UT node); accordingly, this solution needs more wavelengths to embed a virtual topology. If multiple fibers are used to connect the local node to the local WRS, multiple lightpaths may emerge from a node on the same wavelength, and hence fewer wavelengths would be needed, e.g., see the corresponding solution in Fig. 8.5 for the same hypercube virtual topology. However, in both solutions, not all wavelengths on all fibers may be utilized, e.g., in Fig. 8.6, only wavelengths 2 and 4 are used on the CA1-UT fiber, while wavelengths 1, 2, 3, and 5 are used on the UT-CO fiber.

For the solution in Fig. 8.6, the details of one of the nodes, viz., the one at UT, are shown in Fig. 8.4. Note that this switch has to support four incoming fibers plus four outgoing fibers, one each to nodes AB, CA1, CO, and MI, as dictated by the physical topology. In general, each switch also interfaces with four lasers (inputs) and four filters (outputs), with each laser-filter pair dedicated to accommodate each of the four virtual links which a node has to support on the virtual topology. The labels "$1l$ $2b$ $3d$ $5l$" on the

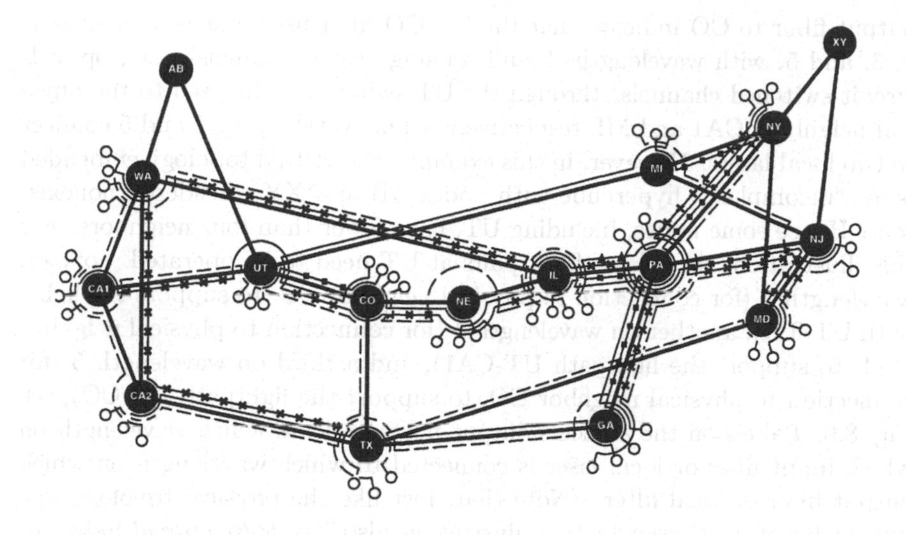

Figure 8.5 The physical topology with embedded wavelengths corresponding to an optimal solution (more than one transceiver at any node can tune to the same wavelength).

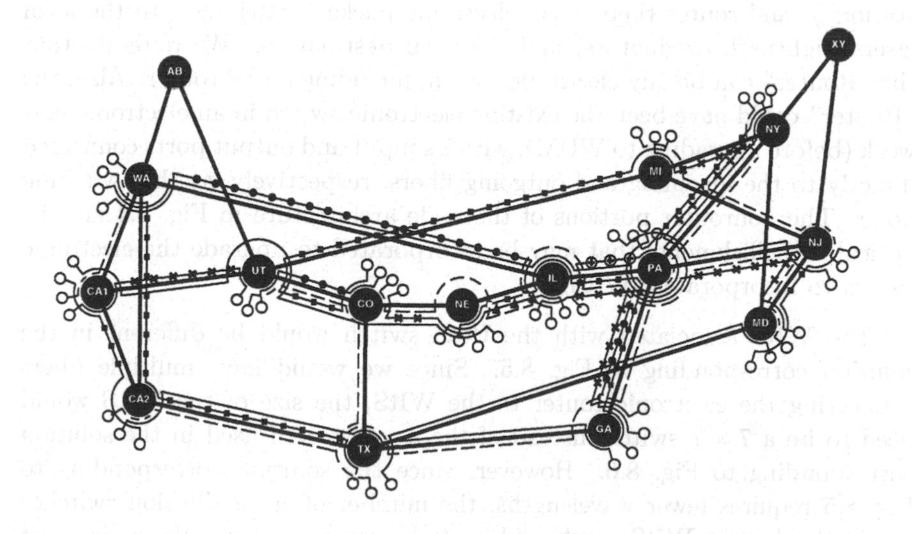

Figure 8.6 The physical topology with embedded wavelengths corresponding to an optimal solution (all transceiver pairs at any node must be tuned to different wavelengths).

output fiber to CO indicate that the UT-CO fiber uses four wavelengths 1, 2, 3, and 5, with wavelengths 2 and 3 being "clear channels" (i.e., optical-circuit-switched channels) through the UT switch and directed to the physical neighbors CA1 and MI, respectively, while wavelengths 1 and 5 connect to two local lasers. However, in this example, the virtual topology embedded is an "incomplete" hypercube with nodes AB and XY considered nonexistent. Hence some nodes, including UT, have fewer than four neighbors. For this illustration, three laser-filter pairs at UT need to be operated – one on wavelength 1 (for connection to physical neighbor CO, to support the lightpath UT-TX); another on wavelength 4 (for connection to physical neighbor CA1, to support the lightpath UT-CA1); and a third on wavelength 5 (for connection to physical neighbor CO, to support the lightpath UT-CO); see Fig. 8.6. Labels on the switch's output ports indicate which wavelength on which input fiber or local laser is connected to which wavelength on which output fiber or local filter. (Note that, just like the physical topology, the virtual topology chosen in this illustration also has *bidirectional* links, although this is not a requirement, i.e., a virtual topology with unidirectional links may also be embedded.)

Figure 8.4 shows a few more interesting issues of this architecture. The box labeled "Router" is an electronic switch which takes information from terminated lightpaths (1*c* 4*b* 5*c*) as well as a local source (labeled "Workstation"), and routes them – via electronic packet switching – to the local lasers (lightpath originators) and the local destination. We reiterate that the "Router" can be any electronic switch, including an IP router. Also, the "Router" could have been the existing electronic switch in an electronic network (before upgrading to WDM), with its input and output ports connected directly to the incoming and outgoing fibers, respectively, at the switching node. The nonrouter portions of the node architecture in Fig. 8.4 are the optical embellishments that may be incorporated to upgrade the electronic switch to incorporate a WRS.

The WRS associated with the Utah switch would be different in the solution corresponding to Fig. 8.5. Since we would have multiple fibers connecting the electronic router to the WRS, the size of the WRS would need to be a 7×7 switch instead of the 4×4 switch used in the solution corresponding to Fig. 8.6. However, since the solution corresponding to Fig. 8.5 requires fewer wavelengths, the number of space-division switches inside the largest WRS would reduce from seven to five. There are cost implications associated with the WRS design and these costs may influence the solution that is adopted.

8.3 Formulation of the Optimization Problem

We formulate the problem as an optimization problem, using principles from multicommodity flow for physical routing of lightpaths and traffic flow on the virtual topology, and using the following notation:

1. s and d used as subscript or superscript denote *source* and *destination* of a packet, respectively,
2. i and j denote *originating* and *terminating nodes, respectively, in a lightpath*, and
3. m and n denote *endpoints of a physical link that might occur in a lightpath.*

8.3.1 Formulation Equations

- **Given:**
  - Number of nodes in the network = N.
  - Maximum number of wavelengths per fiber = M (a system-wide parameter).
  - Physical topology P_{mn}, where $P_{mn} = P_{nm} = 1$ if and only if there exists a direct physical fiber link between nodes m and n, where $m, n = 1, 2, 3, \ldots, N$; $P_{mn} = P_{nm} = 0$ otherwise (i.e., fiber links are assumed to be bidirectional). (The problem can be generalized to accommodate multi-fiber networks, where $P_{mn} > 1$ also.)
  - Distance matrix, viz., fiber distance d_{mn} from node m to node n. For simplicity in expressing packet delays, d_{mn} is expressed as a propagation delay (in time units). Note that $d_{mn} = d_{nm}$ since fiber links are bidirectional, and $d_{mn} = 0$ if $P_{mn} = 0$.
  - Number of transmitters at node $i = T_i$ ($T_i \geq 1$).
    Number of receivers at node $i = R_i$ ($R_i \geq 1$).
  - Traffic matrix λ_{sd} which denotes the average rate of traffic flow from node s to node d, with $\lambda_{ss} = 0$ for $s, d = 1, 2, \ldots, N$. [Additional assumptions are that packet interarrival durations at node s and packet lengths are exponentially distributed, so standard

M/M/1 queueing results can be applied to each network link (or "hop") by employing the independence assumption on interarrivals and packet lengths due to traffic multiplexing at intermediate hops. Also, by knowing the mean packet length (in bits per packet), the λ_{sd} can be expressed in units of packets per second.]

– Capacity of each channel = C (normally expressed in bits per second, but converted to units of packets per second by knowing the mean packet length).

- **Variables:**
  - Virtual topology: The variable $V_{ij} = 1$ if there exists a lightpath from node i to node j in the virtual topology; $V_{ij} = 0$ otherwise. Note that this formulation is general since lightpaths are not necessarily assumed to be bidirectional, i.e., $V_{ij} = 1 \nRightarrow V_{ji} = 1$. (Further generalization of the problem can be performed by allowing multiple lightpaths between node pairs, i.e., $V_{ij} > 1$.)
  - Traffic routing: The variable λ_{ij}^{sd} denotes the traffic flowing from node s to node d and employing V_{ij} as an intermediate virtual link. Note that traffic from node s to node d may be "bifurcated" with different components (or fractions) taking different sets of lightpaths.
  - Physical-topology route: The variable $p_{mn}^{ij} = 1$ if the fiber link P_{mn} is present in the lightpath for virtual link V_{ij}; $p_{mn}^{ij} = 0$ otherwise.
  - Wavelength color: The variable $c_k^{ij} = 1$ if a lightpath from originating node i to terminating node j is assigned the color k, where $k = 1, 2, \ldots, M$; $c_k^{ij} = 0$ otherwise.

- **Constraints:**
  - On virtual-topology connection matrix V_{ij}:

$$\sum_j V_{ij} \leq T_i \quad \forall i \tag{8.1}$$

$$\sum_i V_{ij} \leq R_j \quad \forall j \tag{8.2}$$

The equalities in Eqns. (8.1) and (8.2) hold if all transmitters at node i and all receivers at node j are in use.

– On physical route variables p_{mn}^{ij}:

$$p_{mn}^{ij} \leq P_{mn} \tag{8.3}$$

$$p_{mn}^{ij} \leq V_{ij} \tag{8.4}$$

$$\sum_m p_{mk}^{ij} = \sum_n p_{kn}^{ij} \quad if\ k \neq i, j \tag{8.5}$$

$$\sum_n p_{in}^{ij} = V_{ij} \tag{8.6}$$

$$\sum_m p_{mj}^{ij} = V_{ij} \tag{8.7}$$

– On virtual-topology traffic variables λ_{ij}^{sd}:

$$\lambda_{ij}^{sd} \geq 0 \tag{8.8}$$

$$\sum_j \lambda_{sj}^{sd} = \lambda_{sd} \tag{8.9}$$

$$\sum_i \lambda_{id}^{sd} = \lambda_{sd} \tag{8.10}$$

$$\sum_i \lambda_{ik}^{sd} = \sum_j \lambda_{kj}^{sd} \quad if \quad k \neq s, d \tag{8.11}$$

$$\sum_{s,d} \lambda_{ij}^{sd} \leq V_{ij} * C \tag{8.12}$$

– On coloring of lightpaths c_k^{ij}:

$$\sum_k c_k^{ij} = V_{ij} \tag{8.13}$$

$$\sum_{ij} p_{mn}^{ij} . c_k^{ij} \leq 1 \quad \forall m, n, k \tag{8.14}$$

- **Objective: Optimality Criterion**

(1) Delay minimization (**minimizing**):

$$\sum_{ij} \left[\sum_{sd} \lambda_{ij}^{sd} \left(\sum_{mn} p_{mn}^{ij} d_{mn} + \frac{1}{C - \sum_{sd} \lambda_{ij}^{sd}} \right) \right] \tag{8.15}$$

(2) Maximizing offered load (equivalent to minimizing maximum flow in a link):

$$\min \left[\max \left(\sum_{sd} \lambda_{ij}^{sd} \right) \right] \equiv \max \frac{C}{\left[\max \left(\sum_{sd} \lambda_{ij}^{sd} \right) \right]} \quad \forall i, j \quad (8.16)$$

8.3.2 Explanation of Equations

The above equations are based on principles of conservation of flows and resources (transceivers, wavelengths, etc.) as well as on conflict-free routing, e.g., two lightpaths that share a fiber should not be assigned the same wavelength. Equations (8.1) and (8.2) ensure that the number of lightpaths emanating out of and terminating at a node are at most equal to that node's out-degree and in-degree, respectively. Equations (8.3) and (8.4) constrain the problem so that p_{mn}^{ij} can exist only if there is a physical fiber and a corresponding lightpath present. Equations (8.5) through (8.7) are the multicommodity equations that account for the routing of a lightpath from its origin to its termination. Equations (8.8) through (8.12) are responsible for the routing of packet traffic on the virtual topology (assuming no packet dropping in the router), and they take into account the fact that the combined traffic flowing through a channel cannot exceed the channel capacity. Equation (8.13) requires that a lightpath be of one color only. Equation (8.14) ensures that the colors used in different lightpaths are mutually exclusive over a physical link.

Equations (8.15) and (8.16) represent two possible objective functions. In Eqn. (8.15), in the innermost brackets, the first component corresponds to the propagation delays on the links mn which form the lightpath ij, while the second component corresponds to delay due to queueing and packet transmission on lightpath ij (using a M/M/1 queueing model for each lightpath). If we assume shortest-path routing of the lightpaths over the physical topology, then the p_{mn}^{ij} values become deterministic. If, in addition, we neglect queueing delays, the optimization problem in Eqn. (8.15) reduces to minimizing $\sum_{sd} \sum_{ij} \sum_{mn} \lambda_{ij}^{sd} p_{mn}^{ij} d_{mn}$ which leads to a mixed-integer linear program (MILP) in which the V_{ij} and the c_k^{ij} variables need to have integer solutions, while the λ_{ij}^{sd} variables do not.

The objective function in Eqn. (8.16) is also nonlinear and it minimizes the maximum amount of traffic that flows through any lightpath. This corresponds to obtaining a virtual topology which can maximize the offered load to the network if the traffic matrix is allowed to be scaled up. We choose this optimization for our algorithms in Section 8.4 because our purpose is to demonstrate how to upgrade the capacity of existing fiber-based networks by employing WDM.

8.4 Algorithms

8.4.1 Subproblems

The optimization problem in Section 8.3 is NP-hard, since several subproblems of this problem are NP-hard. The problem of optimal virtual-topology design can be partitioned into the following four subproblems, which are not necessarily independent:

1. determine a good virtual topology, viz., which nodal transmitter should be directly connected to which nodal receiver,
2. route the lightpaths over the physical topology,
3. assign wavelengths optimally to the various lightpaths (this problem has been shown to be NP-hard in [ChGK93]), and
4. route packet traffic on the virtual topology (as in any packet-switched network).

Subproblem 1 addresses how to properly utilize the limited number of available transmitters and receivers. Subproblems 2 and 3 deal with proper usage of the limited number of available wavelengths, and was addressed as a RWA problem in Chapter 7. Subproblem 4 minimizes the effect of store-and-forward (queueing plus transmission) delays at intermediate electronic hops. The remainder of this chapter will address Subproblems 1 and 4.

Previous Work

The problem of designing optimal virtual topologies has been studied before [BaFG90, LaAc91]. Our formulation is more general in the sense that we accommodate many of the physical connectivity constraints which were not considered earlier. In general, the optimal virtual-topology problem has been conjectured to be NP-hard, which means that the problem cannot be solved optimally for large problem sizes, unless one resorts to some form of exhaustive search. One instance of this problem has been formulated as a

MILP which gets difficult to solve with increasing problem size [LaAc91]. Accordingly, heuristic approaches have been employed to solve these problems [BaFG90, LaAc91].

Related work on these problems can be found in [ChGK93, MBRM96, RaSi95b, ZhAc94, DuRo00, RaRa00, GeMu03]. Embedding of a packet-switched virtual topology on a physical fiber plant in a switched network was first introduced in [ChGK93], and this network architecture was referred to as a *lightnet.* Some algorithms to embed a hypercube virtual topology were provided in [ChGK93, MBRM96]. The work in [RaSi95b] proposes a virtual-topology design for packet-switched networks. The average hop distance is minimized, which automatically increases the network traffic supported. The work in [RaSi95b] uses the physical topology as a subset of the virtual topology, employing algorithms for maximizing the throughput, subject to bounded delay characteristics.

The work in [GeMu03] proposes an adaptation mechanism to follow the changes in traffic without *a priori* knowledge of the future traffic pattern by utilizing the measured Internet backbone traffic characteristics. The work differs from most previous studies on this subject which redesign the virtual topology according to an expected (or known) traffic pattern, and then modify the connectivity to reach the target topology. The key idea of the approach is to adapt the underlying optical connectivity by measuring the actual traffic load on lightpaths continuously (periodically based on a measurement period) and reacting promptly to the load imbalances caused by fluctuations on the traffic, by either adding or deleting one or more lightpath at a time. When a load imbalance is encountered, it is corrected either by tearing down a lightpath that is lightly loaded or by setting up a new lightpath when congestion occurs. Specifically, this method adapts very well to the changes in the offered traffic.

Our Solution Approach

To obtain a thorough understanding of the problem, we concentrate on Subproblems 1 and 4 above, i.e., *for the purposes of this chapter, we do not consider the number of available wavelengths to be a constraint.* In the expanded problem, both the number of wavelengths and their exact assignments are critical, and they are accommodated in Chapter 9. Specifically, we employ an iterative approach consisting of "simulated annealing" to search for a good virtual topology (Subproblem 1), in conjunction with the "flow-deviation" algorithm for optimal (possibly "bifurcated") routing of packet traffic on the virtual topology (Subproblem 4). Also, although the virtual topology can be an undirected graph, we consider lightpaths to be bidirectional in our solution here since most (Internet) network protocols rely on bidirectional paths and links. In addition, we consider Optimization Criterion (2) of Eqn. (8.16) (maximizing offered load) for our illustrative solution below, mainly because we are interested in upgrading an existing fiber-based network to a WDM solution.

We start with a random configuration (virtual topology) and try to find a good virtual topology through simulated annealing by using node-exchange (similar to branch-exchange [LaAc91]) techniques. Then, we *scale up* the traffic matrix to ascertain the *maximum throughput* that can be accommodated by the virtual topology, using flow deviation for packet routing over the virtual topology. For a given traffic matrix, the flow-deviation algorithm minimizes the network-wide packet delay by properly distributing the flows on the virtual links (to reduce the effect of large queueing delays).

We have used measured data over the NSFNET backbone as our sample traffic matrix. We scale up each entry in the traffic matrix by a constant scaleup factor and verify if the offered load from the scaled-up traffic matrix can be accommodated by the virtual topology. Our goal is to design the virtual topology that can accommodate the maximum traffic scaleup. This provides an estimate of the maximum throughput we can expect from the current fiber network if it were to support WDM, and if future traffic characteristics were to model present-day traffic characteristics except for the traffic intensities to grow by a constant scale factor. While it is difficult to predict future traffic characteristics, we believe that our approach provides a reasonable framework for analysis and design.

8.4.2 Simulated Annealing

Simulated annealing (along with genetic algorithms) has been found to provide good solutions for complex optimization problems [AaKo89]. In the simulated annealing process, the algorithm starts with an initial random configuration for the virtual topology. Node-exchange operations are used to arrive at neighboring configurations. In a node-exchange operation, adjacent nodes in the virtual topology are examined for swapping, e.g., if node i is connected to nodes j, a, and b, while node j is connected to nodes p, q, and i in the virtual topology, after the node-exchange operation between nodes i and j, node i will be connected to nodes p, q, and j, while node j will be connected to nodes a, b, and i. Neighboring configurations which give better results (lower average packet delay) than the current solution are accepted automatically. Solutions which are worse than the current one are accepted with a certain probability which is determined by a system control parameter. The probability with which these failed configurations are chosen, however, decreases as the algorithm progresses in time so as to simulate the "cooling" process associated with annealing. The probability of acceptance is based on a negative exponential factor and is inversely proportional to the difference between the current solution and the best solution obtained so far.

The initial stages of the annealing process examine random configurations in the search space so as to obtain different initial starting configurations without getting stuck at a local minimum as in a greedy approach. However, as time progresses, the probability of accepting bad solutions goes down, and the algorithm settles down into a minimum after several iterations. The state become "frozen" when there is no improvement in the objective function of the solution even after a large number of iterations. For further information on simulated annealing, see [AaKo89].

8.4.3 Flow-Deviation Algorithm

By properly adjusting link flows, the flow-deviation algorithm [FrGK73] provides an optimal algorithm for minimizing the network-wide average packet delay. However, traffic from a given source to a destination may be bifurcated, i.e., different fractions of it may be routed along different paths in order to minimize the packet delay. If the flows are not balanced, then excessively loading of a particular channel may lead to large delays on that channel and thus have a negative influence on the network-wide average packet delay. The algorithm is based on the notion of *shortest-path flows* which first calculates the linear rate of increase in the delay with an infinitesimal increase in the flow on any particular channel. These "lengths" or "cost rates" are used to pose a shortest-path flow problem (which can be solved using one of several well-known algorithms such as Dijkstra's algorithm, Bellman-Ford algorithm, etc.) and the resulting paths represent the "cheapest" paths on which some of the flow may be deviated. An iterative algorithm determines *how much* of the original flow needs to be deviated. The algorithm continues until a certain performance tolerance level is reached.

8.5 Experimental Results

The simulated-annealing algorithm as well as the flow-deviation algorithm were both used to derive results for the virtual-topology design problem, viz., to study Subproblems 1 and 4 outlined in Section 8.4.1. The traffic matrix employed for this mapping is an actual measurement of the traffic on the NSFNET backbone for a 15-minute period (11:45 pm to midnight on January 12, 1992, EST).

The raw traffic matrix showing traffic flow in bytes per 15-minute intervals between network nodes is shown in Table 8.1.[8] Nodal distances used are the actual geographical distances and they are not shown here to conserve space. Initially, it was assumed that each node could set up at most four bidirectional lightpath channels, but later more experiments were conducted to study the effect of having higher nodal degree. The number of wavelengths per fiber was assumed to be large enough so that all possible virtual topologies could be embedded[9].

Some of our results are tabulated in Table 8.2. For each experiment, the maximum scaleup achieved is tabulated along with the corresponding individual delay components, the maximum and minimum link loading as well as the average hop distance[10]. Since the aggregate capacity for the carried traffic is fixed by the number of links in the network, reducing the average hop distance can lead to higher values of load that the network can carry. The queueing delay was calculated using a standard M/M/1 queueing system with a mean packet length calculated from the measured traffic

[8]NSFNET backbone data was collected by the **nnstat** program and made available to us by the National Science Foundation (NSF) through its Merit partnership. The traffic matrix shown in Table 8.1 is not exactly the same as the one used in our previous work [MBRM96]. Discrepancies can be attributed to different ways of filtering the raw data. The matrix we are currently using is more accurate. Also, the traffic matrix in Table 8.1 shows that several nodes have "self-traffic." This is due to the fact that nodes in the NSFNET are actually gateways connecting to regional networks, so some of the intra-regional traffic at each node also showed up in our measurements.

[9]If we limit the number of wavelengths supported on each fiber, it might not be possible to set up all possible lightpaths in a virtual topology. There are established bounds on the minimum number of wavelengths required to set up arbitrary virtual topologies for a given number of nodes [Pank92].

[10]The average hop distance for packets is an important figure of merit for the topology chosen. It not only has a direct bearing on the queueing delays that a packet suffers, but more importantly, it determines the maximum offered load for the network. The product of the average hop distance and the offered load in the network equals the aggregate traffic load on all the links.

Table 8.1 Traffic matrix (in bytes per 15-minute interval).

	Traffic Matrix (multiply by 10 to get bytes per 15-minute interval)													
	WA	*CA1*	*CA2*	*UT*	*CO*	*TX*	*NE*	*IL*	*PA*	*GA*	*MI*	*NY*	*NJ*	*MD*
WA	53065	268225	117095	27185	196575	8795	53825	249035	34220	18515	311785	96725	44195	191410
CA1	719135	39095	610070	301290	586365	261795	398780	1549725	114465	214105	799325	1031395	552420	775865
CA2	109160	475695	360	466130	85075	363705	86575	856715	100325	46200	516350	62095	139155	215780
UT	70165	62050	136445	0	19090	6090	7015	28815	20030	32600	131120	121575	23925	69745
CO	1227715	1599940	190245	34365	3595	40365	107790	622270	240235	179245	721075	1185620	131820	217565
TX	18415	165355	34265	55215	34030	0	26145	26850	8780	38720	60580	48245	15425	69555
NE	370065	620060	1023115	44755	220365	79000	0	1141795	198280	219565	1540270	933255	236695	1638750
IL	149525	2345475	2103480	85225	282150	26675	970820	3185	439515	330060	900570	711585	202005	889010
PA	849300	199420	373500	60070	249915	68140	250690	610200	0	396225	1106855	1476115	456745	631390
GA	18630	419260	102640	37395	223405	94815	49890	570845	68455	1400	363215	261715	126950	143675
MI	111680	376125	582980	50665	94450	129940	187900	378915	204770	251215	454995	596745	322770	371920
NY	312275	1318380	198710	146225	429985	71520	173240	573215	396000	294260	2116365	742465	279960	659740
NJ	393745	553420	186005	75395	84165	8515	44905	243960	1176820	356930	691765	792130	70690	521995
MD	819050	2270095	542870	229610	892785	318230	327035	918520	306075	16635	1296980	1375980	627470	1216290

(133.54 bytes per packet) and a link speed of 45 Mbps. *We assume infinite buffers at all nodes.* The "cooling" parameter for the simulated annealing is updated after every 100 acceptances using a geometric parameter of value 0.9. A state is considered "frozen" when there is no improvement over 100 consecutive trials.

Table 8.2 Summary of experimental results.

Parameter	*Physical Topology (No WDM)*	*Multiple Pt-to-Pt Links (No WRS)*	*Arbitrary Virtual Topology (Full WDM)*
Maximum scale up	49	57	106
Avg. Pkt. Delay	11.739 ms	12.5542 ms	17.8146 ms
Avg. Prop. Delay	10.8897 ms	10.9188 ms	14.4961 ms
Queueing Delay	0.8499 ms	1.6353 ms	3.31849 ms
Avg. Hop Distance	2.12186	2.2498	1.71713
Max. Link Loading	98%	99%	99%
Min. Link Loading	32%	23%	71%

8.5.1 Physical Topology as Virtual Topology (No WDM)

Our goal was to obtain a fair estimate of what optical hardware can provide in terms of extra capabilities. In this experiment, we start off with just the existing hardware (as in any electronic, packet-switched network), comprising fiber and point-to-point connections using a single bidirectional lightpath channel per fiber link, i.e., *no WDM is used.* Using flow deviation, the maximum scaleup that could be achieved was found to be 49 (only integer values of the scaleup were considered). The link with the maximum traffic (WA-IL) was loaded at 98%, while the link with the minimum traffic (NY-MD) was at 32%. (These numbers are truncated to show their integer part only.) These values serve as a basis for comparison as to what can be gained in terms of throughput by adding extra WDM optical hardware, viz., tunable transceivers and wavelength routing switches at nodes.

8.5.2 Multiple Point-to-Point Links (No WRS)

The goal of the next experiment was to determine how much throughput we could obtain from the network without adding any optical switching capability at a node, but by adding extra transceivers (up to four) at each node, i.e., *WDM is used on some links, but no WRS capability is employed at any node.* The initial network had 21 bidirectional links in the physical topology (see Fig. 8.2). Using extra transceivers at the nodes, we set up extra links on the paths NE-CO, NE-IL, WA-CA2, CA1-UT, MI-NJ, and NY-MD. These lightpaths are chosen manually. Different combinations were considered and the choice of channels which provided the maximum scaleup was chosen. Given 14 nodes, each with a nodal degree of four, we should have been able to have 28 channels. However, the GA node is connected only to TX and PA, both of which are physically connected to four nodes already; hence, they do not have any free transceivers to create an extra channel to GA. Thus, we could add only six new channels. In this case, the maximum scaleup was found to be 57. We found that the two NY-MD channels had a minimum load of only 23%, while the UT-MI channel had a maximum load of 99%.

8.5.3 Arbitrary Virtual Topology (Full WDM)

In the final experiment, we assumed *full WDM with all nodes equipped with WRSs*, i.e., it may now be possible to set up lightpaths between any two nodes. We did not constrain the number of wavelengths supported in each fiber, so that all graphs of degree four were candidates for possible virtual topologies; also, all lightpaths were assumed to be routed over the shortest path on the physical topology. Starting off with a random initial topology, we used simulated annealing to get the best virtual topology. The experiment ran on an unloaded Sparc 10 for approximately three days. The best virtual topology, which is shown in Table 8.3, provided a maximum scaleup of 106. Clearly, the increased scaleup demonstrates the benefits of the WDM-based virtual-topology approach. Now, the minimum loading was on link UT-TX at 71%, while all the other links were above 98% loading.

8.5.4 Comparisons

The various delay characteristics [viz., overall average packet delay (FD), average propagation delay encountered by each packet (PD), average queueing delay experienced by each packet (QD), and the mean hop distance (HD)], as functions of the scaleup (throughput) for the above three experiments are shown in Figs. 8.7 through 8.9. The scaleup provides an estimate of the throughput in the network. We note from these figures that the propagation

Table 8.3 Virtual topology for nodal degree $P = 4$ and best scaleup (106).

Source	*Neighbors*
WA	CA1, CA2, MI, UT
CA1	WA, CO, IL, TX
CA2	WA, PA, NE, GA
UT	WA, TX, IL, MD
CO	CA1, MD, NE, GA
TX	CA1, UT, GA, NJ
NE	CA2, CO, IL, MI
IL	CA1, UT, NE, PA
PA	CA2, IL, NY, NJ
GA	CA2, CO, TX, NY
MI	WA, NE, NY, NJ
NY	PA, GA, MI, MD
NJ	TX, PA, MI, MD
MD	CO, NY, NJ, UT

delay is the dominant component of the packet delay. Also, at light loads, the average propagation delay faced by packets in NSFNET is a little over 9 ms (for the given traffic matrix), and this serves as a lower bound on the average packet delay. As a basis for comparison, the coast-to-coast, one-way propagation time in the U.S. is nearly 23 ms. Thus, on an average, each packet travels about 40% of the coast-to-coast distance.

Note that the average queueing delay increases slightly with increasing traffic until the scaleup nearly reaches its maximum value, after which there is a sharp increase in queueing delay. The propagation delay also increases slightly with increasing scaleup as more traffic is deviated away from shortest path routes by the flow-deviation algorithm. One interesting feature is that the average hop distance decreases as the traffic load is increased; this is again because the flow-deviation algorithm will deviate traffic onto longer links, which might increase the propagation delay encountered by a packet (compared to shortest-path routing), but helps to decrease the average hop distance. Note that our target network in Fig. 8.1 has only 14 nodes and is not very dense; it has a hop distance of a little over 2.0 under the no-

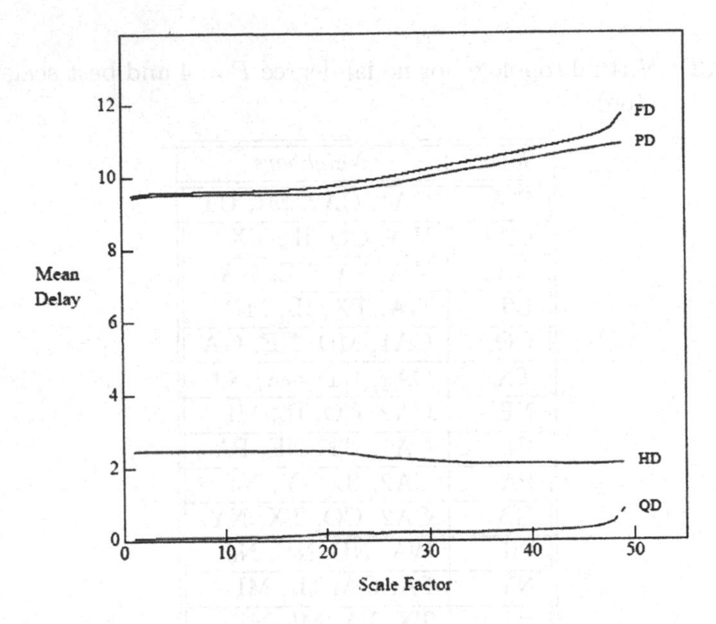

Figure 8.7 Delay vs. throughput (scaleup) characteristics with no WDM, i.e., physical topology as virtual topology.

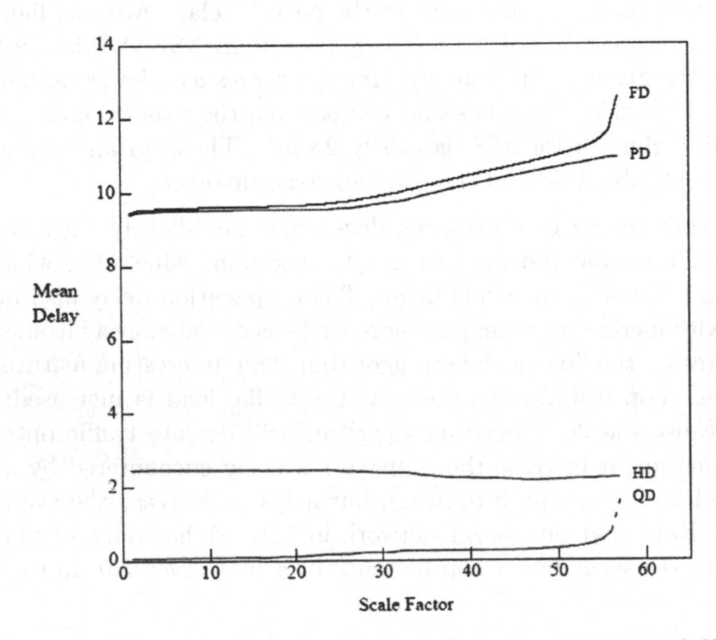

Figure 8.8 Delay vs. throughput (scaleup) characteristics with WDM used on some links, but no WRSs, i.e., multiple point-to-point links are allowed on the physical topology.

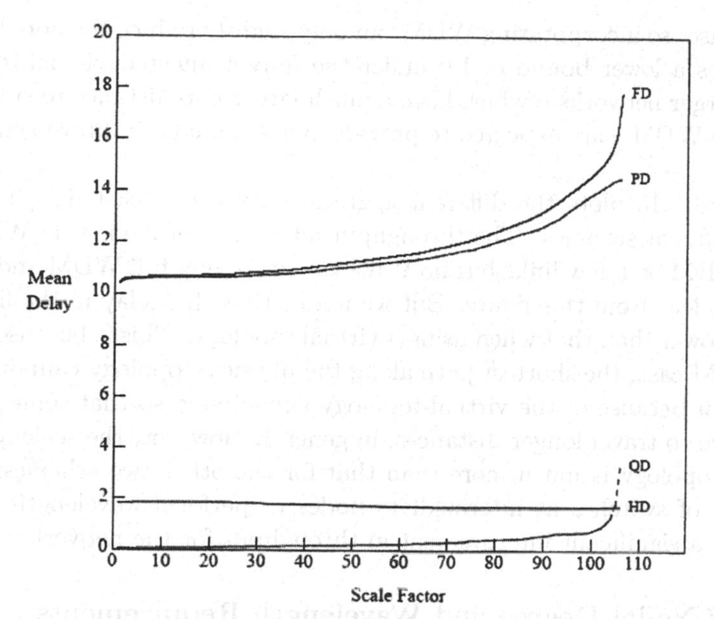

Figure 8.9 Delay vs. throughput (scaleup) characteristics with full WDM on some links and a WRS at each node, i.e., arbitrary virtual topologies are allowed.

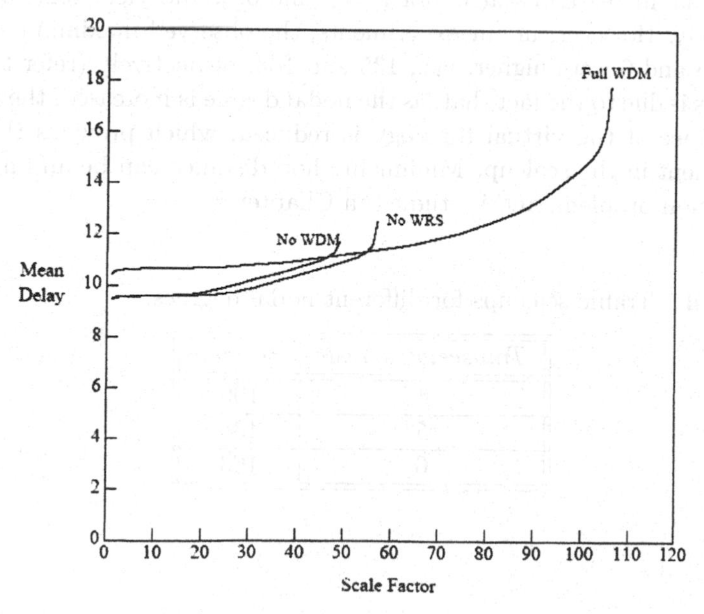

Figure 8.10 Delay vs. throughput (scaleup) characteristics for different virtual topologies.

WDM case, so incorporating WDM can only slightly reduce the hop distance which has a lower bound of 1.0 under the fully-connected virtual-topology case. Larger networks – which have a much larger hop distance to start with under no WDM – are expected to provide more-dramatic improvements with WDM.

Figure 8.10 plots the different aggregate average packet delays for the three different schemes. The throughput advantage of having no WDM vs. using WDM on a few links but no WRS vs. employing full WDM and WRSs is again clear from this figure. But we notice that the delay in the first two cases is lower than that when using a virtual topology. This is because, in the full WDM case, the shortest path along the physical topology cannot always be chosen because of the virtual-topology embedding, so that some packets may have to travel longer distances, in general. However, the scaleup in the virtual topology is much more than that for the other two schemes; so the addition of switches at intermediate nodes to perform wavelength routing provides a significant improvement in throughput for the network.

8.5.5 Effect of Nodal Degree and Wavelength Requirements

So far, the nodal degree (P) was four. Now, we consider full WDM (with a WRS at each node), and increase the nodal degree to five and six. We find that the maximum scaleup increased nearly proportionally with increasing nodal degree. Actually, with the scaleup of 106 for $P = 4$ as a baseline, proportional increase in scaleup for $P = 5$ and 6, would yield 132.5 and 159, respectively. However, in our experiments, the observed maximum scaleups for $P = 5$ and 6 were higher, viz., 135 and 163, respectively (refer to Table 8.4). This is due to the fact that, as the nodal degree is increased, the average hop distance of the virtual topology is reduced, which provides the extra improvement in the scaleup. Minimizing hop distance can be an important optimization problem, and is studied in Chapter 9.

Table 8.4 Traffic scaleups for different nodal degrees.

Transceivers/node	*Scaleup*
4	106
5	135
6	163

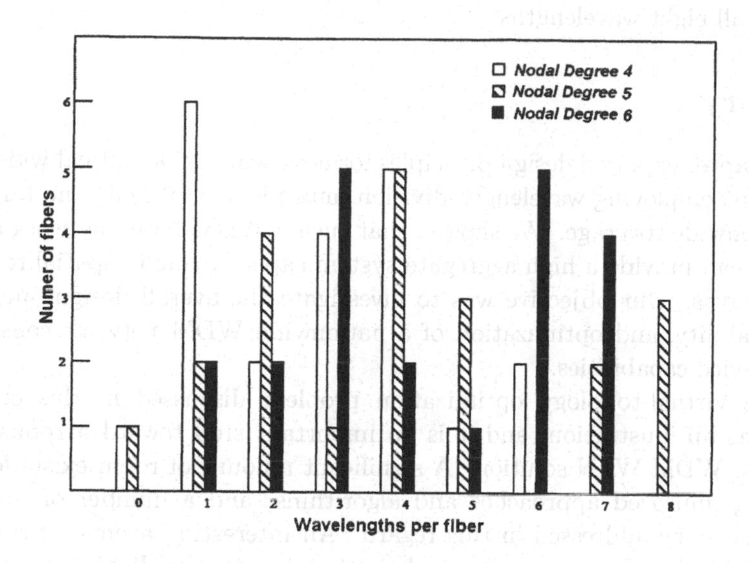

Figure 8.11 Distributions of the number of wavelengths used in each of the 21 fiber links of the NSFNET for the virtual topology approach with nodal degree $P = 4$, 5, and 6.

Although no constraints on wavelengths per fiber were imposed in this study, we also examined the wavelength requirements to set up a virtual topology using shortest-path routing of lightpaths on the physical topology. Assuming no limit on the supply of wavelengths, but with the wavelength constraints as outlined before (Section 8.3), the maximum number of wavelengths required for embedding the best virtual topology (which provided the maximum scaleup) with degree $P = 4$, 5, and 6 were found to be 6, 8, and 8 wavelengths, respectively. The corresponding distributions of the number of wavelengths used in each of the 21 fiber links of the NSFNET (see Fig. 8.1) are plotted in Fig. 8.11. We find that, with increasing nodal degree, i.e., with an increasing number of lightpaths to be supported, the average number of wavelengths a fiber needs to support increases. However, due to the combination of reasons such as desired virtual topology, shortest-path routing of lightpaths, and wavelength constraints, it may so happen that there is no link on the physical topology that employs all of the required wavelengths. This happened for our $P = 6$ experiment, i.e., although eight wavelengths were required to embed the virtual topology, no physical link carried all eight wavelengths.

8.6 Summary

This chapter explored design principles for next-generation optical wide-area networks, employing wavelength-division multiplexing (WDM) and targeted to nationwide coverage. We showed that such a WDM-based network architecture can provide a high aggregate system capacity due to spatial reuse of wavelengths. Our objective was to investigate the overall design, analysis, upgradability, and optimization of a nationwide WDM network consistent with device capabilities.

The virtual-topology optimization problem discussed in this chapter serves as an illustration, and it is an important step toward a robust and versatile WDM WAN solution. A significant amount of room exists for developing improved approaches and algorithms, and a number of research issues must be addressed in this regard. An interesting avenue of research is to study how routing and wavelength assignment of lightpaths can be combined with the choice of virtual topology and its corresponding packet routing in order to arrive at an optimum solution. Dynamic establishment and reconfiguration of lightpaths is an important issue which needs to be thoroughly studied. Some of these topics are studied in Chapters 9 and 7.

Exercises

8.1. What are the advantages of embedding a virtual topology on a physical topology?

8.2. What is the wavelength-continuity constraint? Which of the constraint equations relate to the wavelength-continuity constraint?

8.3. Consider the NSFNET physical topology shown in Fig. 8.1. Remove nodes WA, CO, NE, and GA.

(a) Draw the new physical topology.

(b) Set up lightpaths on the new topology so that the resulting virtual topology is the Petersen graph. In your virtual topology, what is the maximum number of wavelengths used on any link in the network?

(c) Show the details of the UT (Utah) switch (similar to the one in Fig. 8.4.

(d) Assume a uniform traffic matrix, i.e., equal amount traffic between any two nodes. Assume that packets are routed via the shortest path. Calculate the average packet hop distance when packets are routed over the physical topology and when packets are routed over the virtual topology.

8.4. Give a set of lightpaths which embeds a 4×4 Manhattan Street Network with bidirectional links onto the NSFNET physical topology shown in Fig. 8.2. How many wavelengths are required for this embedding?

8.5. For the physical NSFNET topology shown in Fig. 8.2, find a logical ring embedding which uses only one wavelength.

8.6. The embedding shown in Fig. 8.6 assumes that all of the local laser-filter pairs at a node operate on different wavelengths. What are the advantages of this design? What are the disadvantages?

8.7. In Fig. 8.10, at low loads, the average delay in a network employing full WDM is more than the average delay in a network with no WDM. Why?

Chapter

9

Advanced Topics in Virtual-Topology Optimization

9.1 Introduction

For the same *wavelength-routed network* setting as in Chapter 8, this chapter presents the design of a lightpath-based optical network as a *linear* optimization problem, and uses the problem formulation to derive an *exact* optimal network design. The formulation presented here is a modified version of the problem formulation given in Chapter 8, which – though complete – contained nonlinear equations, and was difficult to solve exactly. We simplify the objective function to minimize the average packet hop distance[1] (which is inversely proportional to the network throughput under balanced network flows and which is a linear objective function). By making a shift in the objective function to minimal hop distance and by relaxing the wavelength-continuity constraints (i.e., assuming wavelength converters at all nodes), we demonstrate that the entire optical network design problem can be *linearized* and hence solved *optimally*.

If the channel (lightpath) capacity is C, the number of lightpaths is L,

[1] The average packet hop distance is defined as the number of lightpaths that a packet has to traverse on average, and is a function of the virtual topology.

and the average packet hop distance is H, then the network throughput is bounded by:

$$T \leq \frac{CL}{H} \tag{9.1}$$

Therefore, minimizing H and maximizing the network throughput are equivalent in an asymptotic sense when the equality can be satisfied.[2] A linear program (LP) formulation is developed to minimize H, the average packet hop distance, for a virtual-topology-based, wavelength-routed network. The LP provides a complete specification to the virtual-topology design, routing of the constituent lightpaths, and intensity of packet flows though the lightpaths.

The LP formulation can be used to design a balanced network, such that the utilization of both transceivers and wavelengths is maximized, thus reducing cost of the network equipment. We analyze the trade-offs in budgeting of resources (transceivers and switch sizes) in the optical network, and demonstrate how an improperly designed network may have low utilization of one of these resources.

Section 9.2 presents the mathematical problem formulation as an LP. This section also discusses some of our underlying assumptions that can drastically reduce the size of the problem, and consequently minimize the running time of the solution. Section 9.3 presents two simple heuristics with fast running times whose performance compares favorably with the performance of the optimized solution (obtained from the LP solution). Heuristics become important when the size of the problem becomes larger than what an LP solver can handle, or when the optimization needs to be achieved in real time; the proposed heuristics will be demonstrated to perform well with respect to the optimal bound (viz., LP output). Section 9.4 examines trade-offs that affect the quality of the solution, and discusses resource-budgeting strategies for transmission/reception as well as switching equipment. This approach allows the design of a balanced network, which can provide "optimal" network throughput along with high utilization of transceivers and wavelengths. Section 9.5 proposes an algorithm which can be used for virtual-topology reconfiguration. The proposed algorithm computes a new virtual topology from an existing virtual topology, such that the new virtual topology is optimal with respect to the changing traffic patterns;

[2] In a non-WDM network, the number of channels is given by M, the number of fiber links in the network. Therefore, the non-WDM network throughput is given by $T_p \leq CM/H_p$, where H_p denotes the average packet hop distance in the non-WDM network.

among all such optimal virtual topologies, the algorithm selects the topology which is "closest" in structure to the current virtual topology. Section 9.6 provides numerical simulation results obtained from solving the above formulation for the NSFNET topology (see Fig. 9.1). From our LP output, two sample embeddings of a virtual topology over the NSFNET for two and five wavelengths are shown in Fig. 9.2; they indicate that, even though full wavelength-conversion capabilities are assumed at all nodes, in reality, only a few wavelength converters are needed in the network anyway. In Section 9.7, we use the LP formulation to provide a reconfiguration methodology in order to adapt the virtual topology to dynamic traffic conditions.

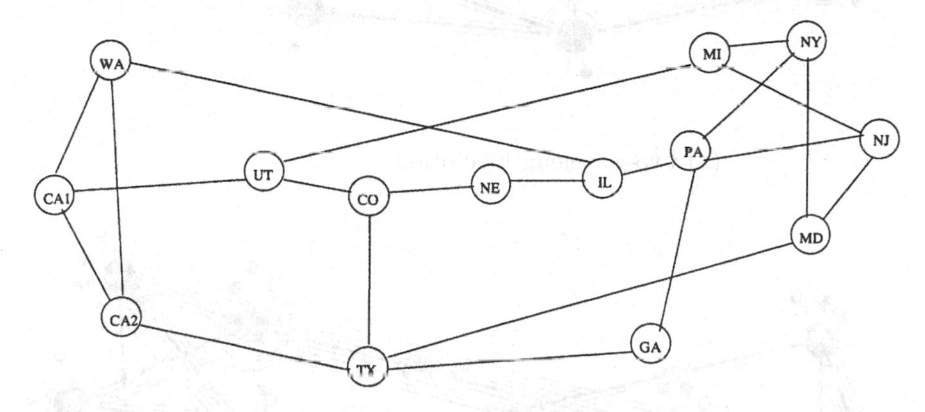

Figure 9.1 NSFNET T1 backbone.

For clarity of exposition, this chapter deals in detail with LP approaches to virtual-topology design and adaptation.

9.2 Problem Specification of Linear Program (LP)

Much of the notation and the constraints used in this section are borrowed from Chapter 8. They are repeated here for completeness of the problem specification. New material, relative to that in Chapter 8, include a new *linear* objective function (Eqn. (9.2)), a constraint to bound the lightpath length (Eqn. (9.17)), a constraint to bound the maximum loading per channel (Eqn. (9.26)), a constraint to incorporate the physical topology as part of the virtual topology (Eqn. (9.16)), all of the simplifying assumptions in Section 9.2.2 (to make the problem linear and tractable), and several other

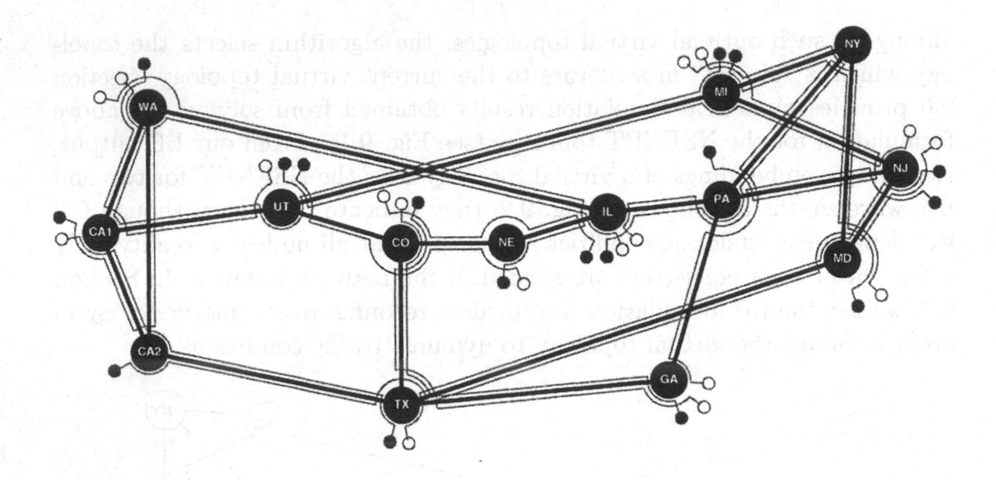

(a) two-wavelength solution

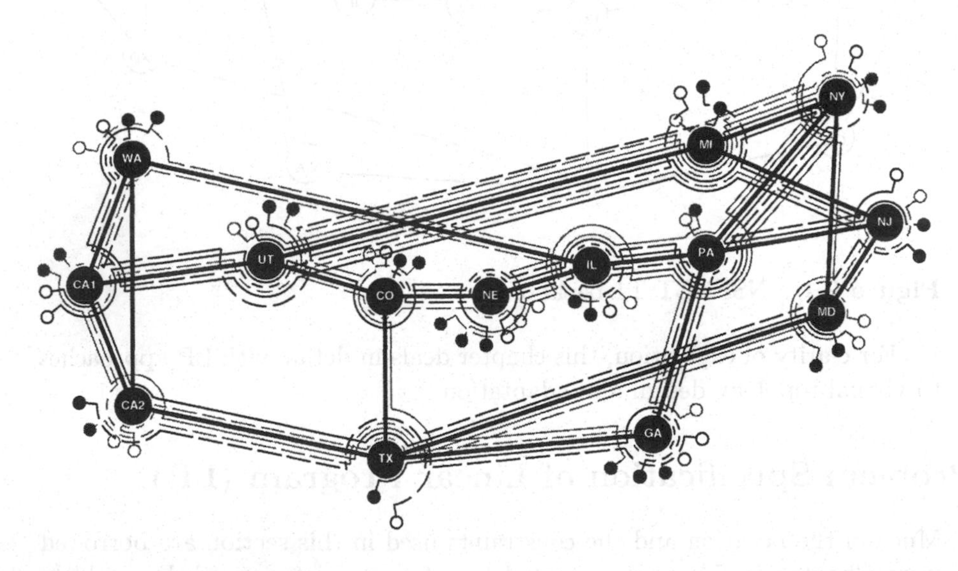

(b) five-wavelength solution

Figure 9.2 Optimal solution for a two-wavelength and a five-wavelength network. Each physical link consists of two unidirectional fibers carrying transmissions in opposite directions (hence, each wavelength may appear twice on any link in the diagrams; their signal propagation directions are opposite to each other in such cases). Wavelength 0 is used to embed the physical topology over the virtual topology, so the Wavelength-0 lightpaths are not shown explicitly in these diagrams to preserve clarity. Note: ∘ = transmitter: • = receiver.

generalizations (multiple fibers between nodes, multiple lightpaths between node pairs, etc.).

9.2.1 Linear Formulation

We formulate the problem as an optimization problem, using principles from multicommodity flow for physical routing of lightpaths and traffic flow on the virtual topology, and using the following notation:

- s and d used as subscript or superscript denote *source* and *destination* of a packet, respectively,
- i and j denote *originating* and *terminating* nodes, respectively, in a lightpath, and
- m and n denote *endpoints of a physical link* that might occur in a lightpath.

Given:

- Number of nodes in the network = N.
- Maximum number of wavelengths per fiber = W (a system-wide parameter).
- Physical topology P_{mn} denotes the number of fibers interconnecting node m and node n. $P_{mn} = 0$ for nodes which are not physically adjacent to each other. $P_{mn} = P_{nm}$ indicates that there are an equal number of fibers joining two nodes in different directions. Note that there may be more than one fiber link connecting adjacent nodes in the network. $\sum_{m,n} P_{mn} = M$ denotes the total number of fiber links in the network.
- Fiber-length matrix, viz., fiber distance d_{mn} from node m to node n. For simplicity in expressing packet delays, d_{mn} is expressed as a propagation delay (in time units). Note that $d_{mn} = d_{nm}$, and $d_{mn} = \infty$ if $P_{mn} = 0$.
- Shortest-path delay matrix D where D_{sd} denotes the delay (sum of propagation delays only) over the shortest path between nodes s and d.

- Lightpath length bound $\alpha, 1 \leq \alpha < \infty$, bounds the delay over a lightpath (and hence the length of the lightpath) between two nodes i and j, with respect to the shortest-path delay D_{ij} between them, i.e., the maximum permissible propagation delay over the lightpath between the two nodes i and j is αD_{ij}.

- Number of transmitters at node $i = T_i$ ($T_i \geq 1$). Number of receivers at node $i = R_i$ ($R_i \geq 1$). In general, we would assume that $T_i = R_i, \forall i$, although this is not a strict requirement.

- Traffic matrix Λ_{sd} which denotes the average rate of traffic flow (in packets per second) from node s to node d, with $\Lambda_{ss} = 0$ for s, $d = 1, 2, \ldots, N$.

- Capacity of each channel = C (normally expressed in bits per second, but converted to units of packets per second by knowing the mean packet length).

- Maximum loading per channel $= \beta, 0 < \beta < 1$. β restricts the queueing delay on a lightpath from getting unbounded by avoiding excessive link congestion. *We do not incorporate queueing delays explicitly in the problem formulation, under the assumption that they are negligible for suitably chosen values of* β. Also, previous results in Chapter 8 indicate that queueing delays are negligibly small compared to propagation delays for a large network setting as in Fig. 9.1, except under *extremely heavy loading.*

Variables:

- Virtual topology: The variable V_{ij} denotes the number of lightpaths from node i to node j in the virtual topology. Note that the current formulation is general, since lightpaths are not necessarily assumed to be bidirectional, i.e., $V_{ij} = 0$ does not imply that $V_{ji} = 0$. As an example, Fig. 9.2 contains a lightpath from CA1 to TX, but not in the reverse direction. Therefore, $V_{CA1,TX} = 1$. Moreover, there may be multiple lightpaths between the same source-destination pair, i.e., $V_{ij} > 1$, for the case when traffic between nodes i and j is greater than a single lightpath's capacity (C).

- Traffic routing: The variable λ_{ij}^{sd} denotes the traffic flowing from node s to node d and employing V_{ij} as an intermediate virtual link. Note that traffic from node s to node d may be "bifurcated," with different components flowing through different sets of lightpaths.

- Physical-topology route: The variable p_{mn}^{ij} denotes the number of lightpaths between nodes i and j being routed though fiber link mn. For example, since the lightpath from CA1 to TX passes through CA2, the variables $p_{CA1,CA2}^{CA1,TX} = 1$ and $p_{CA2,TX}^{CA1,TX} = 1$.

Objective: Optimality Criterion

$$\textbf{Minimize}: \quad \frac{1}{\sum_{s,d} \Lambda_{sd}} \sum_{i,j} \sum_{s,d} \lambda_{ij}^{sd} \tag{9.2}$$

The objective function minimizes the average packet hop distance in the network. This is a linear objective function because $\sum_{i,j} \sum_{s,d} \lambda_{ij}^{sd}$ is a linear sum of variables, while $\sum_{s,d} \Lambda_{sd}$ is a constant for a given traffic matrix. The two objective functions used in Chapter 8 were (1) minimization of the average packet delay over the network, and (2) minimization of the maximum traffic flow in a lightpath; both were nonlinear.

Constraints:

- On virtual-topology connection matrix V_{ij}:

$$\sum_j V_{ij} \leq T_i \quad \forall i \tag{9.3}$$

$$\sum_i V_{ij} \leq R_j \quad \forall j \tag{9.4}$$

$$V_{ij} \quad \text{int} \tag{9.5}$$

The above equations ensure that the number of lightpaths emerging from a node is constrained by the number of transmitters at that node, while the number of lightpaths terminating at a node are constrained by the number of receivers at that node. V_{ij} variables can only hold integer values. If V_{ij} has a value greater than 1, it means that there is more than one lightpath between the particular source-destination pair. These lightpaths may follow the same route or different routes through the network.

- On physical route variables p_{mn}^{ij}:

$$\sum_m p_{mk}^{ij} = \sum_n p_{kn}^{ij} \quad if\ k \neq i, j \tag{9.6}$$

$$\sum_n p_{in}^{ij} = V_{ij} \tag{9.7}$$

$$\sum_m p_{mj}^{ij} = V_{ij} \tag{9.8}$$

$$\sum_{i,j} p_{mn}^{ij} \leq WP_{mn} \tag{9.9}$$

$$p_{mn}^{ij} \quad \text{int} \tag{9.10}$$

Equations (9.6) through (9.8) are multicommodity-flow-based equations governing the routing of lightpaths from source to destination. Equations (9.9) ensure that the number of lightpaths flowing through a fiber link does not exceed W. Note, however, that the equations do not ensure the wavelength-continuity constraint (under which the lightpath is assigned the same wavelength on all the fiber-links through which it passes). In the absence of wavelength-continuity constraints in the equations, the solution obtained from our current formulation may require the network to be equipped with wavelength converters.

- On virtual-topology traffic variables λ_{ij}^{sd}:

$$\sum_j \lambda_{sj}^{sd} = \Lambda_{sd} \tag{9.11}$$

$$\sum_i \lambda_{id}^{sd} = \Lambda_{sd} \tag{9.12}$$

$$\sum_i \lambda_{ik}^{sd} = \sum_j \lambda_{kj}^{sd} \quad if \quad k \neq s, d \tag{9.13}$$

$$\lambda_{ij}^{sd} \leq \Lambda_{sd} V_{ij} \tag{9.14}$$

$$\sum_{s,d} \lambda_{ij}^{sd} \leq \beta V_{ij} C \tag{9.15}$$

Equations (9.11) through (9.13) are multicommodity-flow equations governing the flow of traffic through the virtual topology. Note that the routing of traffic from a given source to a given destination may be "bifurcated." Equations (9.14) ensure that traffic can only flow through an existing lightpath, while Eqns. (9.15) specify the capacity constraint in the formulation.

- Optional constraints:

1. Physical topology as a *subset* of the virtual topology:

$$P_{mn} = 1 \Rightarrow V_{mn} = 1, p_{mn}^{mn} = 1 \tag{9.16}$$

2. Bound lightpath length:

$$\sum_{m,n} p_{mn}^{ij} d_{mn} \leq \alpha D_{ij}, \; for\ K\ alternate\ paths \tag{9.17}$$

Equations (9.16) and (9.17) are optional, and may be incorporated to ensure bounded packet delays in the network. These equations reduce the solution space of the problem, and could theoretically affect the optimality of the solution. However, in general, their effect on the solution quality is found to be minimal for most networks of interest.

Equations (9.16) embed the physical topology as a subset of the virtual topology, i.e., every link in the physical topology is also a lightpath in the virtual topology, in addition to which there are lightpaths which span multiple fiber links, e.g., Fig. 9.2(a) demonstrates a virtual-topology embedding for a two-wavelength solution. This approach for choosing lightpaths can satisfy packets with the tightest delay constraints [RaSi96]. The lightpaths corresponding to the physical topology may also be used to route network-control messages efficiently, and this approach can simplify network management. For these equations to be valid, $T_n \geq \delta_n^o$ and $R_n \geq \delta_n^i$, where δ_n^o and δ_n^i denote the physical number of fibers emerging from and terminating at node n, respectively.

Equations (9.17) may be used to restrict the enumerated p_{mn}^{ij} variables to be only among those present in K alternate shortest paths from i to j, where $K \geq 1$. These equations prevent long and convoluted lightpaths, i.e., lightpaths with an unnecessarily long route instead of a much shorter route, from occurring. The value of K may be selected by the network designer. We choose $K = 2$ in our experiments reported in Section 9.6, and found this choice to work well (relative to higher values of K).

9.2.2 Simplifying Assumptions

This section outlines some simplifying assumptions to make the problem more tractable.

- *Wavelength-continuity constraints for a lightpath are intentionally ignored* in the current formulation, which only ensures that the total number of lightpaths routed through a fiber is less than or equal to W. Adding wavelength-continuity constraints to the above set of equations significantly increases the complexity of the problem[3], e.g., if the variable $c_k^{ij} = 1$ signifies that a lightpath from node i to node j is assigned the wavelength k (where $k = 1, 2, \ldots, W$), the relevant equations are as follows:

$$\sum_k c_k^{ij} = V_{ij} \tag{9.18}$$

$$\sum_{ij} p_{mn}^{ij} \cdot c_k^{ij} \leq 1 \quad \forall m, n, k \tag{9.19}$$

- Equations (9.19) are nonlinear because they involve the product of two variables. Therefore, we intentionally ignore the wavelength assignment of lightpaths in the current problem formulation[4], assuming that the wavelength-assignment problem will be solved separately, based on the lightpath routes obtained through the current formulation, or that wavelength converters are available at the routing nodes.

- *Queueing delays are also intentionally ignored*, partly to simplify (linearize) the objective function, and also because it has been observed that the propagation delay dominates the overall network delay in nationwide optical networks, as in Fig. 9.1. The exact optimization function for delay minimization is as follows (see Chapter 8):

$$\textbf{Minimize}: \quad \sum_{ij} \left[\sum_{sd} \lambda_{ij}^{sd} \left(\sum_{mn} p_{mn}^{ij} d_{mn} + \frac{1}{C - \sum_{sd} \lambda_{ij}^{sd}} \right) \right] \tag{9.20}$$

[3] The wavelength-assignment problem has been demonstrated to be NP-complete [ChGK93].

[4] The efficacy of using wavelength converters in the routing nodes is currently an active area of research. It has been shown that, in many situations, networks with sparse wavelength conversion have performance nearly equivalent to that of networks with full wavelength conversion (see Chapter 10).

This is a nonlinear equation because it involves the product of two variables, λ_{ij}^{sd} and p_{mn}^{ij}, and also because the term $1/(C - \sum_{sd} \lambda_{ij}^{sd})$ is nonlinear in $\{\lambda_{ij}^{sd}\}$.

- *The number of variables and equations in the formulation are reduced.* The number of variables and equations in the original problem formulation grows as $O(N^4)$, and can very easily overwhelm today's state-of-the-art computing facilities. To make the problem more tractable, we reduce the number of constraints by pruning the search space. Pruning is based on tracking a limited number of alternate shortest paths, denoted by K, between source-destination pairs, such that the selected routes are within a constant factor ($\alpha \geq 1$) of the shortest-path distance between the given source-destination pair. We assume that traffic flow will only use the lightpaths which interconnect nodes present in these alternate paths, i.e., all values of λ_{ij}^{sd} are not enumerated. Likewise, lightpaths may only be routed through one of a few permissible routes, i.e., all possible values of p_{mn}^{ij} are not enumerated. Since these assumptions are incorporated during the generation of the problem formulation, it helps reduce the total number of equations and variables. The amount of pruning (hence, the value of K) required is a function of the size of the problem that can be solved in "reasonable time" by the chosen LP solver. In our experiments, we used the *lpSolve* package [Berk94], running on an unloaded DEC Alpha, but were still restricted to using two alternate shortest paths for the network in Fig. 9.1 in order to get reasonable running times.

To understand the pruning process, let us consider the NSFNET topology in Fig. 9.1. If we consider two alternate shortest paths between any source-destination pair, then the two alternate paths from node (CA) to node (IL) may be CA1-UT-CO-NE-IL and CA1-WA-IL. Then, the enumerated variables for lightpath routing are as follows:

$$p_{CA1,UT}^{CA1,IL}, p_{UT,CO}^{CA1,IL}, p_{CO,NE}^{CA1,IL}, p_{NE,IL}^{CA1,IL}, p_{CA1,WA}^{CA,IL}, p_{WA,NE}^{CA1,IL}$$

Likewise, the enumerated variables for packet routing are as follows:

$$\lambda_{CA1,UT}^{CA1,IL}, \lambda_{CA1,CO}^{CA1,IL}, \lambda_{CA1,NE}^{CA1,IL}, \lambda_{CA1,IL}^{CA1,IL}, \lambda_{UT,CO}^{CA1,IL}, \lambda_{UT,NE}^{CA1,IL}, \lambda_{UT,IL}^{CA1,IL},$$

$$\lambda_{CO,NE}^{CA1,IL}, \lambda_{CO,IL}^{CA1,IL}, \lambda_{NE,IL}^{CA1,IL}, \lambda_{CA1,WA}^{CA1,IL}, \lambda_{WA,IL}^{CA1,IL}.$$

- The current formulation allows bifurcated routing of packet traffic. To specify nonbifurcated routing of traffic, we can use new variables γ_{ij}^{sd} which are only allowed to take binary values, and the equations are suitably modified. Under nonbifurcated routing, Eqns. (9.11) through (9.15) become:

$$\gamma_{ij}^{sd} \in 0,1 \tag{9.21}$$

$$\sum_j \gamma_{sj}^{sd} = 1 \tag{9.22}$$

$$\sum_i \gamma_{id}^{sd} = 1 \tag{9.23}$$

$$\sum_i \gamma_{ik}^{sd} = \sum_j \gamma_{kj}^{sd} \quad if \quad k \neq s,d \tag{9.24}$$

$$\gamma_{ij}^{sd} \leq V_{ij} \tag{9.25}$$

$$\sum_{s,d} \gamma_{ij}^{sd} \Lambda_{sd} \leq \beta V_{ij} C \tag{9.26}$$

The objective function becomes:

$$\textbf{Minimize}: \quad \frac{1}{\sum_{s,d} \Lambda_{sd}} \left(\sum_{i,j} \sum_{s,d} \gamma_{ij}^{sd} \Lambda_{sd} \right) \tag{9.27}$$

We used bifurcated routing in our experiments, since it was found that nonbifurcated routing of packet traffic significantly increased the running time of the optimization solution. The increase in running time is primarily due to the computation of the product terms in Eqns. (9.26) and in the modified objective function. Note, however, that these equations are also strictly linear.

9.3 Heuristic Approaches

This section presents two heuristic approaches that allow us to solve large problem instances of the virtual-topology design problem, in order to minimize the average packet hop distance. Heuristics become important when the problem formulation becomes large due to an increase in the physical size of the network, and becomes difficult to solve by traditional LP methods due to computational constraints. Results from these heuristics compare favorably with the optimal result obtained by solving the exact problem formulation, for small to medium-sized networks that can be solved exactly by the LP method.

To ensure that the results from the heuristics are comparable with those from the optimization formulation, we do not impose wavelength-continuity constraints on the lightpath routing, although the heuristics can easily accommodate this feature without any sacrifice in their running time.

(1) **Maximizing Single-Hop Traffic.** This simple heuristic attempts to establish lightpaths between source-destination pairs with the highest Λ_{sd} values, subject to constraints on the number of transceivers at the two end nodes, and the availability of a wavelength in some path connecting the two end nodes. Pseudo-code for this heuristic follows.

```
procedure MaxSingleHop(void)
  while (not done)
    Find Λs'd' = Max (Λsd)
      if ((free transmitter available at s') AND
        (free receiver available at d') AND
        (free wavelength available in any alternate path from s' to d'))
        begin
            Establish lightpath between s' and d'
            Λs'd' = Λs'd' − C
        end
      endif
  endwhile
```

(2) **Maximizing Multihop Traffic.** In a packet-switched network, the traffic carried by a link may include forwarded traffic as well as traffic originating from that node. Intuitively, it seems that any lightpath-establishment heuristic which accounts for the forwarded traffic that the lightpath will carry should provide better performance than a heuristic which only tries to maximize the single-hop traffic. This intuition led to the derivation of the current heuristic. The heuristic starts with the physical topology as the initial virtual topology, and attempts to add more lightpaths one by one. The performance of this heuristic is found to be slightly better than that of the previous heuristic (see Section 9.6).

Let H_{sd} denote the number of electronic hops needed to send a packet from source s to destination d. The heuristic attempts to establish lightpaths in decreasing order of $\Lambda_{sd}(H_{sd} - 1)$, subject to constraints on the number of transceivers at the two end nodes, and the availability of a wavelength in some path connecting the two end nodes.[5] After each lightpath is established, H_{sd} values are recalculated, as traffic flows might have changed due to the new lightpath, in order to minimize the average packet hop distance. This algorithm allows only a single lightpath to be established between any source-destination pair (but this can be generalized). Pseudo-code for this heuristic is provided below.

procedure MaxMultiHop(void)
Initial Virtual Topology = Physical Topology
while (not done)
Compute $H_{sd}\ \forall s, d$
Find $\Lambda_{s'd'}(H_{s'd'} - 1) = \text{Max}\ (\Lambda_{sd}(H_{sd} - 1))$
if ((free transmitter available at s') AND
(free receiver available at d') AND
(free wavelength available in any alternate path from s' to d'))
begin
Establish lightpath between s' and d'
end
endif
endwhile

9.4 Network Design: Resource Budgeting and Cost Model

This section discusses some of the network-design principles that can be derived from the LP formulation. We present a cost model for the network design, in terms of the costs for the transmission as well as the switching equipment, which may be used in conjunction with the optimization formulation, to derive a minimum-cost solution.

[5] This approach is similar to the longest-lightpath-first strategy proposed in [ChGK93]; however, in that case, the lightpaths to be established were known a priori. In the current problem specification, we need to select the lightpaths in addition to routing them.

9.4.1 Resource Budgeting

It is intuitive that, in a network with a very large number of transceivers per node, but with very few wavelengths per fiber and few fibers between node pairs, a large number of transceivers may be unused because some lightpaths may not be establishable due to wavelength constraints. Similarly, a network with few transceivers but a large number of available wavelengths may have a large number of wavelengths unutilized because the network is transceiver-constrained.

This mismatch in transceiver vs. wavelength utilization has a direct impact on the cost of the network. The number of wavelengths supported in the network determines the cost of the switching equipment.[6] Likewise, the number of transceivers per node determines the cost of the terminating equipment. We would like to balance these network resources, in order to maximize the utilizations of both the transceivers and the wavelengths in the network. Resource budgeting becomes an important issue when we attempt to optimize network design with constrained total cost.

A very simple analysis leads to some insights into the resource-budgeting problem. Given a physical topology, and a routing algorithm for lightpaths, we can asymptotically determine the average length of a lightpath (in terms of the number of fiber links traversed by a lightpath, averaged over all source-destination pairs in the network); let the average length of a lightpath be denoted by H_P. If there are M fiber links in the network, each supporting W wavelengths, then the maximum number of lightpaths that can be supported is MW/H_P, assuming uniform utilization of wavelengths on all fiber links.[7] Therefore, the number of transceivers per node should be approximately:

$$T_i = R_i \approx \frac{MW}{NH_P} \tag{9.28}$$

in order to get a balanced network. Our optimization-based and heuristic-based network simulations on the NSFNET reinforce this conjecture (see Section 9.6).

[6] A WRS (see Fig. 9.3) with δ input ports and δ output ports (including the ports used to connect to the local node used for electronic termination of lightpaths), supporting W wavelengths, requires W $\delta \times \delta$ wavelength-insensitive optical switches. Increasing the number of fibers interconnecting two nodes increases δ, and increasing W increases the number of crosspoint switching elements required; thus, in both cases, the cost of the switching equipment would increase.

[7] Note, however, that in many cases, the routing on the network is dependent on the wavelength congestion in the network, and any static routing policy may not yield very accurate results.

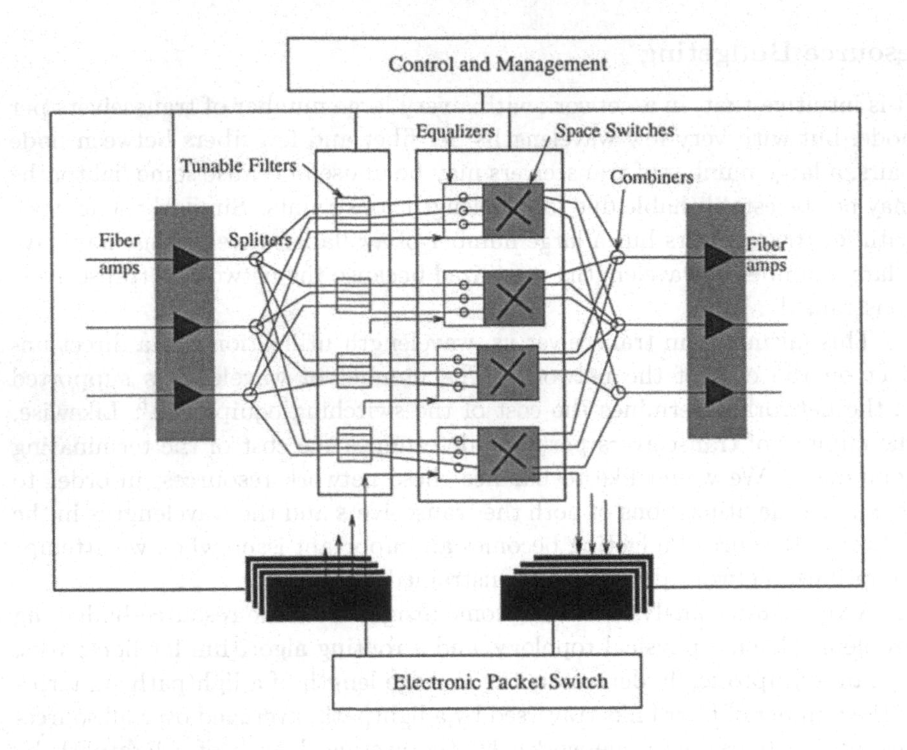

Figure 9.3 Transport node in the RACE WDM optical network architecture.

9.4.2 Network-Cost Model

Resource budgeting in the network has a direct impact on the cost of setting up the network. In this chapter, we only consider equipment costs[8]. Our model of the WRS is based on the prototype used in the RACE project (see Fig. 9.3) [Hill93]. This cost model is valid for $W \geq 2$ wavelengths. Also, this model assumes that OXCs and transmission equipment at a node are integrated together to form the corresponding WRS.

- Let C_t be the cost of a transceiver. Then, the aggregate network-wide equipment cost for transceivers is $C_t(\sum_i T_i + \sum_i R_i)$.
- Let C_m be the cost of a multiplexer or a demultiplexer. Then, the

[8]A lot of fiber has already been buried in the ground, and hence, for our purposes here, the cost of laying fiber is assumed to be zero.

aggregate cost of multiplexers/demultiplexers in the network is given by $C_m(2M + \sum_{i=1}^{N} \lceil T_i/W \rceil + \sum_{j=1}^{N} \lceil R_j/W \rceil)$, where $(\sum_{i=1}^{N} \lceil T_i/W \rceil + \sum_{j=1}^{N} \lceil R_j/W \rceil)$ denotes the cost of providing (de)multiplexers for the local access ports needed to launch or terminate lightpaths.

- Let C_x be the cost of a 2×2 optical crosspoint switching element. Then, the cost of a switch with δ input and δ output ports built from 2×2 optical switching elements arranged in a multistage interconnection network (MIN) is $C(\delta) = C_x \delta \log \delta / 2$. There is a MIN switch per wavelength in a WRS; hence, the cost for node m with degree $\delta_m = \sum_{n=1}^{N} P_{mn} + \lceil T_m/W \rceil$ is $WC(\delta_m)$. The total network cost is therefore $W \sum_{m=1}^{N} C(\delta_m)$.

- The cost of a wavelength converter may also be incorporated in the current model; we, however, choose to ignore the converter cost in this chapter, due to lack of a clear consensus regarding the exact architecture (and, hence, cost) of optical switches based on wavelength converters. Also, very few wavelength converters are needed anyway [SuAS96], and as our illustrative example in Section 9.6 will show.

To summarize, the total cost of a wavelength-routed optical network, excluding fiber-layout cost, may be expressed as:

$$\mathbf{C} = C_t \left(\sum_i T_i + \sum_i R_i \right) + C_m \left(2M + \sum_{i=1}^{N} \left\lceil \frac{T_i}{W} \right\rceil + \sum_{j=1}^{N} \left\lceil \frac{R_j}{W} \right\rceil \right) + WC_x \left(\sum_{m=1}^{N} \delta_m \log \delta_m \right) /2 \quad (9.29)$$

for $W \geq 2$.

Using the above equation, and assuming that $C_t = \$5,000, C_m = \$100$, and $C_x = \$1,000$, Table 9.1 demonstrates the equipment cost of building a wavelength-routed optical network embedded on the NSFNET of Fig. 9.1.[9]

[9]We remark that the costs shown in Table 9.1 are meant to provide only a comparative understanding of how the system cost scales with the number of transceivers and wavelengths, based on rough estimates of market values for C_t, C_m, and C_x. The actual cost of upgrading the network in Fig. 9.1 to accommodate WDM will probably be orders of magnitude higher than those shown in Table 9.1, when costs for the electronics, other supporting optics (such as preamplifiers and signal equalizers), installation, and maintenance are also taken into account.

The network cost does not increase monotonically with the number of wavelengths because of the $\lceil T/W \rceil$ and the $\lceil R/W \rceil$ components in the above equation. Also, since we use 2×2 crosspoint switching elements to build the MIN, it is not possible to build a MIN switch of odd degree, i.e., δ_m *odd*. Therefore, a node with $\delta_m = 5$ would require a three-stage MIN with three crosspoint elements per column. The costs given in Table 9.1 are independent of the utilizations of wavelengths and transceivers in the network, and in several improperly-designed cases, a significant amount of the resources may be underutilized. Thus, it is easy to observe from Table 9.1 that, given a cost constraint, it is possible to iteratively try different combinations of wavelengths and transceivers, in order to optimize the network performance for a given network cost.

Table 9.1 Cost of upgrading the NSFNET using WDM.

Network Equipment Cost (in $1000)							
	Wavelengths						
Transceivers	2	3	4	5	6	7	8
4	806.0	922.0	835.2	901.2	967.2	1033.2	1099.2
5	980.8	1062.0	1178.0	1041.2	1107.2	1173.2	1239.2
6	1120.8	1202.0	1318.0	1434.0	1247.2	1313.2	1379.2
7	1323.6	1392.8	1458.0	1574.0	1690.0	1453.2	1519.2
8	1463.6	1532.8	1598.0	1714.0	1830.0	1946.0	1659.2
9	1650.4	1672.8	1804.8	1854.0	1970.0	2086.0	2202.0
10	1790.4	1905.6	1944.8	1994.0	2110.0	2226.0	2342.0
11	2093.2	2045.6	2084.8	2216.8	2250.0	2366.0	2482.0
12	2233.2	2185.6	2224.8	2356.8	2390.0	2506.0	2622.0
13	2424.0	2394.4	2487.6	2496.8	2628.8	2646.0	2762.0
14	2564.0	2534.4	2627.6	2636.8	2768.8	2786.0	2902.0
15	2786.8	2674.4	2767.6	2776.8	2908.8	3040.8	3042.0
16	2926.8	3057.2	2907.6	3069.6	3048.8	3180.8	3182.0

9.5 Virtual-Topology Reconfiguration

A major advantage of an optical network is that it may be able to reconfigure its virtual topology to adapt to changing traffic patterns. Some reconfiguration studies on optical networks have been reported [BiGu92, LaHA94, RoAm95]; however, these studies assumed that the new virtual topology *was known a priori*, and were concerned with the cost and sequence of branch-exchange operations to transform from the original virtual topology to the new virtual topology. We propose a methodology to *obtain the new virtual topology*, based on optimizing a given objective function, as well as minimizing the changes required to obtain the new virtual topology from the current virtual topology [BaMu00].

The LP formulation in Section 9.2 can help us derive new virtual topologies from existing virtual topologies. In the ideal situation, given a small change in the traffic matrix, we would prefer for the new virtual topology to be largely similar to the previous virtual topology, in terms of the constituent lightpaths and the routes for these lightpaths, i.e., we would prefer to minimize the changes in the number of WRS configurations needed to adapt from the existing virtual topology to the new virtual topology. More formally, it would be preferable if a large number of the V_{ij} and the P_{mn}^{ij} variables retain the same values in the two solutions, without compromising the quality of the solution (in terms of minimizing the average packet hop distance).

Let us consider the snapshot of two traffic matrices, Λ_{sd}^1 and Λ_{sd}^2, taken at two not-too-distant time instants. We assume that there is a certain amount of correlation between these two traffic matrices. Given a certain traffic matrix, there may be many different virtual topologies, each of which has the same optimal value with regard to the objective function, i.e., Eqn. (9.2). Usually, an LP solver will terminate after it has found the first such optimal solution. Our reconfiguration algorithm finds the virtual topology corresponding to Λ_{sd}^2 which matches "closest" with the virtual topology corresponding to Λ_{sd}^1 (based on our above definition of "closeness") [BaMu00].

9.5.1 Reconfiguration Algorithm

We perform the following sequence of actions:

- Generate linear formulations $\mathcal{F}(1)$ and $\mathcal{F}(2)$, corresponding to traffic matrices Λ_{sd}^1 and Λ_{sd}^2, respectively, based on the formulation in Section 9.2.

- Derive solutions $\mathcal{S}(1)$ and $\mathcal{S}(2)$, corresponding to $\mathcal{F}(1)$ and $\mathcal{F}(2)$, respectively. Denote the variables' values in $\mathcal{S}(1)$ as $V_{ij}(1), P_{mn}^{ij}(1), \lambda_{ij}^{sd}(1)$, and those in $\mathcal{S}(2)$ as $V_{ij}(2), P_{mn}^{ij}(2), \lambda_{ij}^{sd}(2)$, respectively. Let the values of the objective function for $\mathcal{S}(1)$ and $\mathcal{S}(2)$ be OPT_1 and OPT_2, respectively.

- Modify $\mathcal{F}(2)$ to $\mathcal{F}'(2)$ by adding the new constraint:

$$\frac{1}{\sum_{s,d} \Lambda_{sd}} \sum_{i,j} \sum_{s,d} \lambda_{ij}^{sd} = OPT_2 \tag{9.30}$$

  This ensures that all the virtual topologies generated by $\mathcal{F}'(2)$ would be optimal with regard to the objective function.

- The new objective function for $\mathcal{F}'(2)$ is:

$$\textbf{Minimize}: \quad \sum_{ij} \sum_{mn} \mid p_{mn}^{ij}(2) - p_{mn}^{ij}(1) \mid \tag{9.31}$$

  We could also have used the following objective function:

$$\textbf{Minimize}: \quad \sum_{ij} \mid V_{ij}(2) - V_{ij}(1) \mid \tag{9.32}$$

  Note that the mod operation, $\mid x \mid$, is a nonlinear function. If we assume that p_{mn}^{ij} and V_{ij} can only take on binary values, then Eqns. (9.31) and (9.32) become linear, i.e., if $V_{ij}(1) = 1$, then $\mid V_{ij}(2) - V_{ij}(1) \mid \equiv (1 - V_{ij}(2))$, else if $V_{ij}(1) = 0$ then $\mid V_{ij}(2) - V_{ij}(1) \mid \equiv V_{ij}(2)$. Hence, $\mathcal{F}'(2)$ may be solved directly using an LP solver.

  Also, note that $\mid p_{mn}^{ij}(2) - p_{mn}^{ij}(1) \mid$ also implies that $\mid V_{ij}(2) - V_{ij}(1) \mid$. Therefore, Eqn. (9.31) is a stronger condition than Eqn. (9.32). Hence, we chose Eqn. (9.31) for our simulation studies on reconfiguration at the end of Section 9.6.

9.6 Illustrative Examples

This section presents numerical examples of the virtual-topology network-design problem, using the NSFNET illustrative backbone (Fig. 9.1) as our physical topology. The NSFNET consists of 14 nodes connected in a mesh network. Each of its links are bidirectional, i.e., for each link, there is a pair of unidirectional fibers which carry transmissions in opposite directions and which join physically adjacent nodes, i.e., $P_{mn} = P_{nm} = 1$. Each node consists of a WRS along with multiple transceivers for optical origination and termination of lightpaths. The number of transmitters is assumed to be equal to the number of receivers, and is the same for all nodes.

The traffic matrix is randomly generated, such that a certain fraction F of the traffic is uniformly distributed over the range $[0, C/a]$ and the remaining traffic is uniformly distributed over $[0, C\Upsilon/a]$, where C is the lightpath channel capacity, a is an arbitrary integer which may be 1 or greater, and Υ denotes the average ratio of traffic intensities between node-pairs with high traffic values and node-pairs with low traffic values. This model allows us to generate traffic patterns with varying characteristics.

Figures 9.4, 9.5, and 9.6 plot system characteristics averaged over 25 different virtual topologies, each corresponding to an independent traffic matrix, obtained with the parameters $C = 1250, a = 20, \Upsilon = 10, F = 0.7, \beta = 0.8, K = 2$, and $\alpha = 2$. T_i and R_i were assumed to be equal for all nodes, and were allowed values between 4 and 8. W was allowed to take values between 1 (no WDM) and 7.

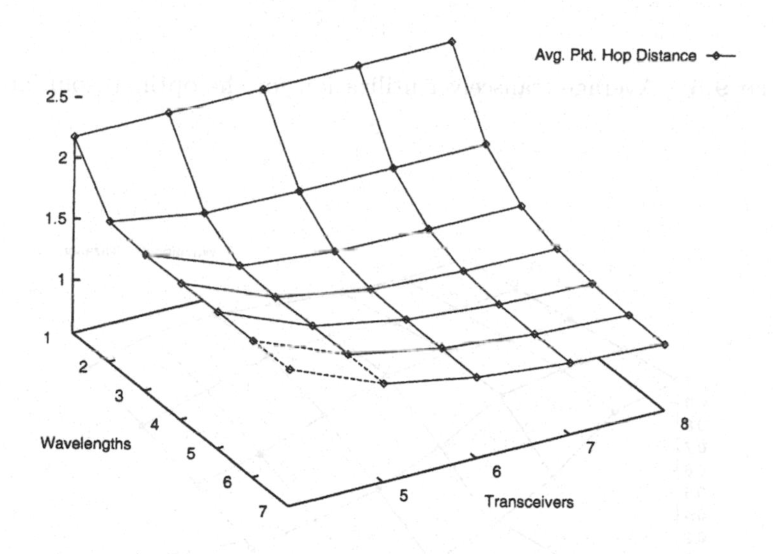

Figure 9.4 Average packet hop distance for the optimal solution.

Figure 9.4 plots the average packet hop distance for optimal virtual topologies, given a different number of transceivers per node, and a different number of supported wavelengths in the system. The average hop distance in the network is a function of the number of lightpaths set up in the network, which directly depends on the number of transceivers and wavelengths supported. The case corresponding to one wavelength in the

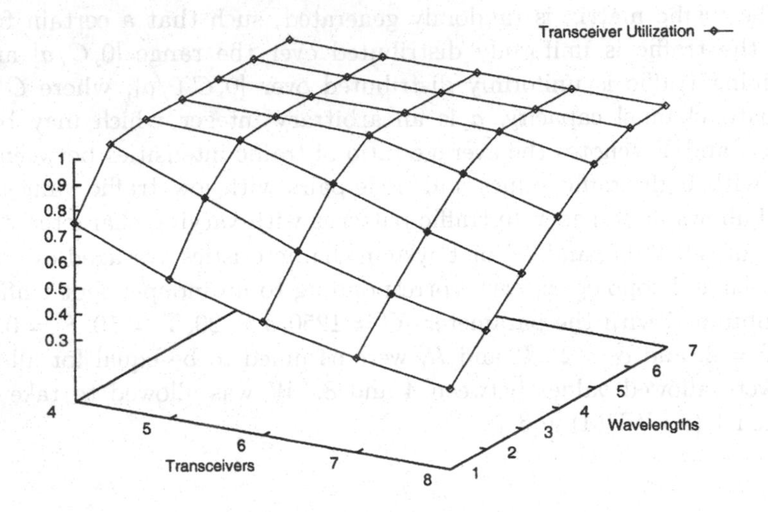

Figure 9.5 Average transceiver utilization for the optimal solution.

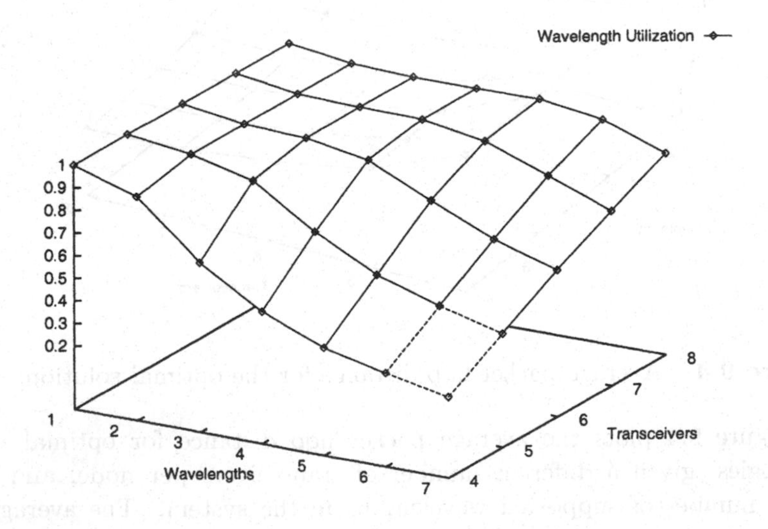

Figure 9.6 Average wavelength utilization for the optimal solution.

system corresponds to a point-to-point network (no WDM). As expected, the average hop distance decreases with a balanced increase in the number of transceivers and wavelengths in the network. Increasing transceivers without adding extra wavelengths marginally improves the quality of the solution. For more than six transceivers, and more than four wavelengths, the performance improvement is marginal for the network in Fig. 9.1. Note that, ideally, we would like the average hop distance to be as close as possible to its lower bound of unity.

Figure 9.5 plots the transceiver utilization for different values of the number of wavelengths in the system, and number of transceivers at a node. Figure 9.6 plots the wavelength utilization for the same set of experiments. As one would expect, Fig. 9.5 quantitatively demonstrates that the transceiver utilization decreases as the number of wavelengths is reduced and/or the number of transceivers is increased. Similarly, Fig. 9.6 demonstrates that the wavelength utilization decreases when the number of wavelengths is increased and/or the number of transceivers is reduced. These results confirm our original hypothesis that it is necessary to obtain the correct balance between transceivers and wavelengths in the system in order to properly utilize both of these expensive resources. Given a cost constraint, resource-budgeting trade-off becomes an important issue, in order not to underutilize transceivers and wavelengths in the system. Again, note that, ideally, both transceiver utilization and wavelength utilization should be as close as possible to their upper bounds of unity.

Based on the results in Figs. 9.4, 9.5, and 9.6, for the example considered here, a good operating point for the chosen system parameters appears to be *four wavelengths and six transceivers* since one would like to keep the system cost down as well as achieve low average hop distance and high utilizations (all close to unity!) of transceivers and wavelengths.

Figure 9.2 demonstrates virtual-topology solutions for the NSFNET, given a specific traffic matrix, and for $T_i = R_i = 5$, and for $W = 2$ and 5. Since there are 14 nodes in the network, and five transceivers per node, a maximum of $14 \times 5 = 70$ lightpaths may be established in the network. In the two-wavelength solution, only 59 lightpaths could be established, out of which 42 lightpaths constituted the physical topology embedding over the virtual topology. In the five-wavelength solution, all of the 70 lightpaths could be established, so that the transceiver utilization is 100% as opposed to a transceiver utilization of less than 85% for the two-wavelength case. It is also evident that the wavelength utilization in the two-wavelength case is much higher than that in the five-wavelength case (96% vs. 57%). Therefore, increasing the number of wavelengths in the system might increase the transceiver utilization, but tends to decrease the wavelength utilization in the system.

The wavelength assignment of lightpaths in the two solutions was done arbitrarily, and was not based on any optimal algorithm. In the current wavelength assignment, *only two wavelength converters are needed at the NY node*, in order to establish all the lighpaths. This vindicates our original assumption that sparse wavelength conversion may be sufficient to ensure good virtual topologies. Both of these solutions also assume that there may be multiple lightpaths on the same wavelength, emerging from a particular node. These solutions may require larger switch sizes (specifically, there may need to be multiple fibers from the local node to the local WRS) (see Chapter 8), and can increase the cost of the WRS beyond that presented in the cost model in Section 9.4.2.

Table 9.2 tabulates the average hop distance for the two heuristic approaches as compared to the optimal solution obtained in Fig. 9.4. The same sample of 25 traffic matrices were used to evaluate the performance of the heuristics. Figure 9.7 also plots these performance results for a four-wavelength system. As expected, the average hop distance decreases with an increase in the number of transceivers in the system, with the heuristics performing a little poorly relative to the LP's optimal solution which can be treated as a lower bound. Also, the heuristic which maximizes the multihop traffic is found to perform slightly better than the heuristic which maximizes single-hop traffic for smaller number of transceivers.

We demonstrate the reconfiguration capabilities of the problem formulation in Section 9.5.1. We generate two sequences of 25 traffic matrices each, with the same set of statistical parameters as used before. In the first sequence, exactly 20% of the entries in successive traffic matrices in the sequence are forced to differ. In the second sequence, 80% of the entries differ. The traffic sequence is created by generating an initial traffic matrix, and then swapping a fraction (either 20% or 80%, depending on the chosen sequence) of nondiagonal entries in the traffic matrix. The algorithm in Section 9.5.1 was applied to this traffic sequence, in order to generate virtual topologies in a network with eight transceivers per node, and eight wavelengths per fiber. Figure 9.8 plots the fraction of lightpath additions and deletions as observed over the sequence of 25 traffic matrices, for 20% and 80% changes in the traffic matrix. The fraction of common lightpaths be-

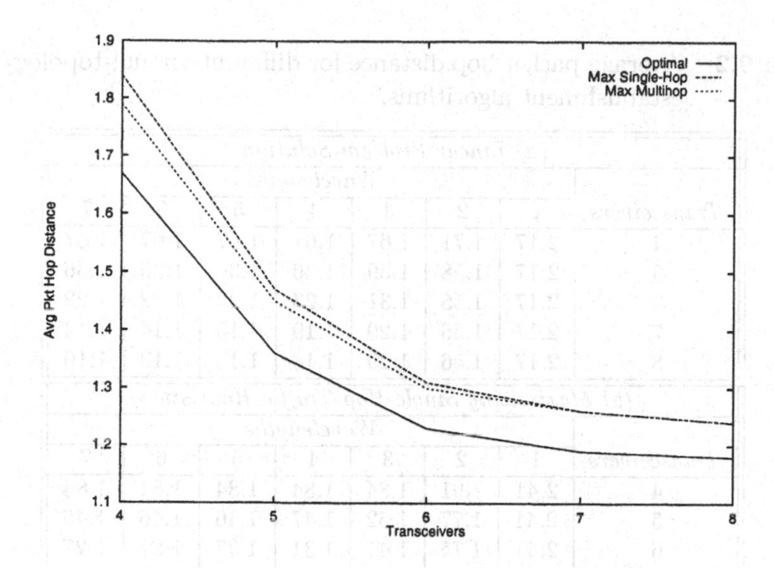

Figure 9.7 Comparison of heuristic algorithms for a four-wavelength network.

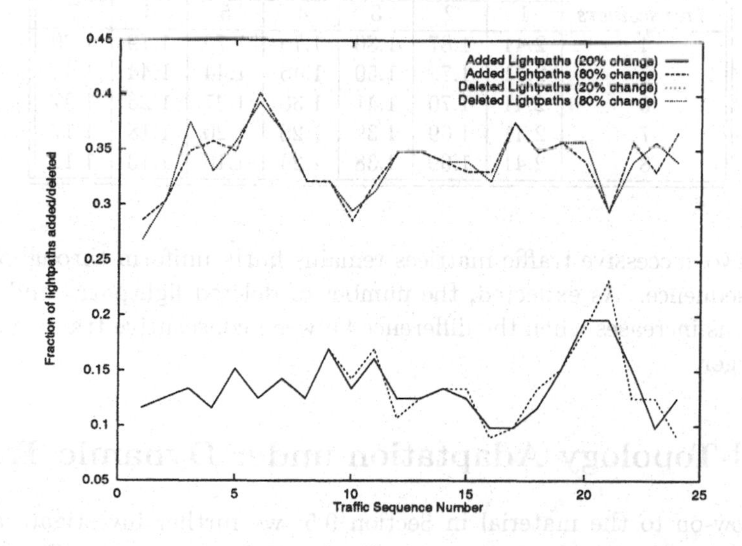

Figure 9.8 Reconfiguration statistics.

Table 9.2 Average packet hop distance for different virtual-topology establishment algorithms.

(a) Linear Problem Solution							
	Wavelengths						
Transceivers	1	2	3	4	5	6	7
4	2.17	1.71	1.67	1.67	1.67	1.67	1.67
5	2.17	1.58	1.39	1.36	1.36	1.36	1.36
6	2.17	1.56	1.31	1.23	1.22	1.22	1.22
7	2.17	1.56	1.29	1.19	1.15	1.14	1.14
8	2.17	1.56	1.29	1.18	1.13	1.10	1.10
(b) Maximizing Single-Hop Traffic Heuristic							
	Wavelengths						
Transceivers	1	2	3	4	5	6	7
4	2.41	1.91	1.84	1.84	1.84	1.84	1.84
5	2.41	1.77	1.52	1.47	1.46	1.46	1.46
6	2.41	1.75	1.41	1.31	1.37	1.27	1.27
7	2.41	1.74	1.39	1.26	1.20	1.20	1.17
8	2.41	1.74	1.39	1.24	1.17	1.14	1.12
(c) Maximizing Multihop Traffic Heuristic							
	Wavelengths						
Transceivers	1	2	3	4	5	6	7
4	2.41	1.87	1.80	1.79	1.79	1.79	1.79
5	2.41	1.72	1.50	1.45	1.44	1.44	1.44
6	2.41	1.70	1.41	1.30	1.27	1.26	1.27
7	2.41	1.69	1.39	1.26	1.20	1.18	1.17
8	2.41	1.69	1.38	1.24	1.16	1.13	1.12

tween two successive traffic matrices remains fairly uniform throughout the entire sequence. As expected, the number of deleted lightpaths and added lightpaths increases when the difference between consecutive traffic matrices gets larger.

9.7 Virtual-Topology Adaptation under Dynamic Traffic

As follow-on to the material in Section 9.5, we further investigate an advanced topic in virtual-topology reconfiguration in this section. By utilizing the measured Internet backbone traffic characteristics, we propose an adaptation mechanism to follow the changes in traffic without *a priori* knowledge of the future traffic pattern. Our work differs from most previous studies on this subject which redesign the virtual topology according to an expected (or known) traffic pattern, and then modify the connectivity to reach the target topology. The key idea of our approach is to adapt the underlying optical connectivity by measuring the actual traffic load on lightpaths continuously (periodically based on a measurement period) and reacting promptly to the load imbalances caused by fluctuations on the traffic, by either adding or deleting one or more lightpath at a time.

9.7.1 Problem Definition

A virtual topology is defined to be the set of all such lightpaths in a network. Such a virtual topology can be employed by an Internet service provider (ISP) or a large institutional user of bandwidth (hereafter referred to as an ISP) to connect its end equipment (e.g., IP routers) by leasing bandwidth (wavelength channels) from the network operator who owns the fiber plant and OXCs. In fact, multiple virtual topologies, possibly belonging to different ISPs, may coexist on the same fiber plant. The ISP in this context represents Business Model #3, while the network operator who owns the layer-one fiber infrastructure represents Business Model #2 (see Section 1.3).

In real networks, however, the traffic rates between node pairs fluctuate distinguishably over time [GeMu03, ANTS05, Caid05], which is an important obstacle in virtual-topology design. (An exemplary traffic measurement can be seen in Fig. 9.9 [ANTS05]. In this example, the measurements for both directions of a link on the Abilene network are displayed as two profiles over a 33-hour period, starting at 9:00 a.m. on one day and ending a little after 6:00 p.m. the next day). A virtual topology which is optimized for a specific traffic demand may not be able to respond with equal efficiency to a different traffic demand. Thus, in such a situation, a reconfiguration of the virtual topology may be needed to match it with the changing traffic.

In this section, we need to investigate the problem of on-line reconfiguration of the virtual topology in WDM mesh networks when the traffic load changes dynamically over time. Reconfiguration of optical networks has been studied, both for broadcast optical networks [LaHA94, BaRo01], and for wavelength-routed networks [Bala96a, BaMu00, SMGM01]. In these studies, the problem is generally treated as a two-phase operation where the first phase is virtual-topology design for the new traffic conditions and the second phase is the transition operation from the old virtual topology to the newly designed one. A major difficulty with this view is that

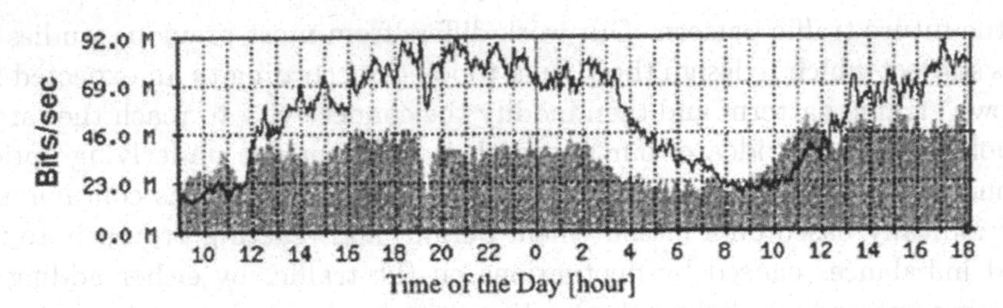

Figure 9.9 Traffic measurements on a link in the Abilene network during a 33-h period from 9:00 a.m. on Day 1 to 6:00 p.m. on Day 2. The two profiles correspond to the two directions of traffic on a link.

the future traffic demand is assumed to be known. With this information in hand, the design of a new virtual topology is practicable, and in many studies considerable effort was spent for the design of the next virtual topology so as to reduce the topology changes during the transition phase [BaMu00, DuRo00, SMGM01, KrSi01]. In practice, the assumption on the future traffic may decrease the value of the newly designed virtual topology if the traffic changes are predicted inaccurately. What we need is a reconfiguration method that can update the virtual topology without relying on traffic forecasts.

Another problem originates from the transition process. During the second phase, lightpaths involved in transition cannot be used by the ongoing traffic. Most earlier studies developed techniques to minimize the disruption to the ongoing traffic [LaHA94, RoAm95, NaMo00]. These studies dealt with this problem either by performing the reconfiguration on all network elements concurrently [RoAm95], or by applying step-by-step changes until the new virtual topology is settled upon [LaHA94, NaMo00]. In both cases, the traffic disruption because of the transition phase cannot be eliminated completely.

In another earlier study, hitless reconfiguration was defined as a reconfiguration process without the loss of any data [Bala96a]. According to the methodology proposed in [Bala96a], the transition between topologies was achieved by first establishing all new links without removing any link. The links of the old topology were removed only when the traffic was rerouted through the links of the new topology.

In this study, our problem definition is different. As traffic fluctuates over time, it will be monitored systematically, and the virtual topology will be changed accordingly, but we do not make any other assumptions on future traffic pattern. The reconfiguration process is seen as a continuous measurement plus adaptation system where small adjustments are made, instead of waiting for a noticeable drop in system efficiency and changing the entire topology.

Basically, in this adaptation method, a new lightpath is added when a congestion is encountered, and a lightpath is deleted if it is being underutilized. Thus, an ISP can optimize the operational cost of its virtual topology by leasing only the appropriate amount of lightpaths as it considers to be really necessary. In previous studies on virtual-topology design, the methods established as many lightpaths as possible and reconfiguration did not change the number of lightpaths [LaHA94, NaMo00, GeMu03]. On the contrary, our method keeps only the necessary number of lightpaths that can grow during peak traffic hours and fall when the overall network traffic decreases. This type of operation is more cost effective for ISPs. We introduce two system parameters to detect the link-usage (in) efficiencies: high watermark W_H and low watermark W_L. At the end of an observation period (typically hundreds of seconds), if the load of one or more lightpaths is higher than W_H, a new lightpath is established to decrease that load. When the load of a link drops below W_L, that link will be torn down if alternate paths exist for diverting the traffic using that lightpath. Please see [GeMu03] for details.

Network Model

We consider a network of nodes connected by bidirectional optical links forming an arbitrary physical topology. Each optical link supports W wavelengths, and any node i is assumed to have T_i transmitters and R_i receivers. We assume that each node is equipped with an OXC with full wavelength-conversion capability, so that a lightpath can be established between any node pair if the resources (an optical transmitter at source, an optical receiver at destination, and at least a wavelength on each fiber link) are available along the path. Mechanisms to accommodate no wavelength conversion and different numbers of wavelengths on different links are straightforward.We consider unidirectional lightpaths, since the traffic between two nodes is not necessarily symmetric (as can be seen in Fig. 9.9).

Each OXC is connected to an edge device, e.g., an IP router, which can be a source or a destination of a traffic flow and which can provide routing for multihop traffic passing by that node. We assume that each router is capable of processing all packet traffic flowing through it and of observing the amount of traffic on its outgoing lightpaths. In this chapter, for ease of explanation, we consider a centralized approach to the virtual-topology reconfiguration problem. A central manager will collect the virtual-link usage information from routers at the end of every observation period. Specifically, the link-usage information needed to make a reconfiguration decision consists of which links are overloaded, which links are underloaded, and what are the end-to-end packet-traffic intensities flowing through the overloaded links. The decision for a topology change will then be made by the central manager, and a signaling mechanism will be started if a lightpath addition or deletion is required as a result of the decision algorithm (in this chapter, for simplicity, we ignore the details of the signaling protocol). An implicit assumption here is that the observation period is much longer (typically hundreds of seconds or longer) than the time it takes for control signals to propagate from various nodes to the central manager. We expect that it is possible to design a decentralized protocol to do this job as well, but this is outside the scope of our present discussion.

In the optical layer, we use shortest-path routing for routing lightpaths on the physical topology and the first-fit scheme for wavelength assignment [ZaJM00a, GeMu03]. For packet routing, we consider a shortest-path (minimum-hop) routing scheme, since it provides better usage of network links and is frequently used by existing routing protocols.

Problem Statement

- **Given:**
  - A network graph $P(V, E_P)$ where V is the set of nodes and E_P is the set of links connecting the nodes. Graph nodes correspond to network nodes with OXCs and links correspond to the fibers between nodes.
  - Number of wavelength channels carried by each fiber.
  - Number of transmitters and receivers at each node.
  - Current virtual topology $V(V, E_V)$ as another graph where the nodes correspond to the nodes in the physical topology. Each link in E_V corresponds to a direct lightpath between the nodes.
  - Current traffic load carried by each lightpath.
- **Determine:**

  - Whether the current virtual topology of the network is efficient for the current traffic.
  - Whether a change in the virtual topology should be made.
  - If a change is necessary, which lightpaths should be added and/or deleted.

- **One can identify the following important steps in solving such a problem:**
  - Traffic should be monitored continuously to provide adequate information to the reconfiguration system.
  - A decision mechanism is needed to trigger a virtual-topology change if the current topology is not convenient.
  - Finally, the exact modification to the topology should be determined.

9.7.2 Local Optimization for Virtual-Topology Adaptation

This problem can be formally stated as a mixed-integer linear program (MILP). This formulation is defined for one step of the adaptation, so it defines a local optimization. Observations on several backbone networks show that the amount of traffic between nodes changes in a smooth and continuous manner [ANTS05, Caid05] (see Fig. 9.9). Long-term variations have time-of-the-day characteristics where traffic intensities change in terms of hours. Therefore, the adaptation mechanism would be beneficial for backbone networks of ISPs since the traffic characteristics in these environments favor slow adaptation.

A. Formulation of Adaptation as a MILP

We assume full wavelength-conversion capability at each node in this formulation. We use the following notations:

- s and d denote *source* and *destination* of a traffic flow when used as a superscript or subscript.
- i and j denote originating and terminating nodes of a lightpath, respectively.
- m and n denote the end points of a physical link.

- W_H and W_L denote the high and low watermarks which are used to detect the link-usage efficiencies in a network.

At any step of the adaptation, one of these three decisions can be made: addition of a lightpath, deletion of a lightpath, or no change to the virtual topology. The choice of the decision is related to the highest and the lowest lightpath loads and the watermark values. Selecting the proper action and the best lightpath which keeps the maximum link load as low as possible is a local optimization problem ("local" with respect to time). This problem turns out to be a MILP. We give the MILP formulation for one step of the adaptation method.

- **Given:**
  - Number of nodes in the network is equal to N.
  - Physical topology of the network $P = P_{mn}$, where P_{mn} indicates the number of fibers between nodes m and n, and $P_{mn} = P_{nm}$ for $m = 1, 2, 3, ..., N$ and $n = 1, 2, 3, ..., N$.
  - Current traffic matrix $\Lambda = \Lambda_{sd}$ denotes the average traffic rate (in bits/s) measured during the last observation period between every node pair, with $\Lambda_{ss} = 0$ for $s = 1, 2, 3, ..., N$.
  - Current virtual topology $V = V_{ij,q}$, where $V_{ij,q}$ is a binary value denoting the qth lightpath between nodes i and j, and $V_{ii,q} = 0$. $V_{ij,0} = 0$ if there is no lightpath from node i to node j. $V_{ij,0} = V_{ij,1} = ...V_{ij,k-1} = 1$ and $V_{ij,k} = 0$ if there are k lightpaths from node i to node j. Since lightpaths are not necessarily assumed to be bidirectional, $V_{ij,q} = 0 \not\Rightarrow V_{ji,q} = 0$.
  - Number of wavelengths on each fiber is equal to W.
  - Capacity of each wavelength channel is equal to C bps.
  - Number of transmitters and receivers at node i: T_i and R_i respectively.
  - High watermark value is equal to W_H, where $W_H \in (0, 1)$, e.g., $W_H = 0.8$ implies that a lightpath is considered to be overloaded when its load exceeds $0.8 * C$.
  - Low watermark value is equal to W_L, where $W_L \in (0, 1)$.
  - Highest and lowest lightpath loads measured during the observation period: L^P_{Max} bps and L^P_{Min} bps, respectively.

- **Variables:**
  - Physical routing binary variable $p_{mn}^{ij,q} = 1$ if the qth lightpath from node i to node j is routed through the physical link (m, n).
  - New virtual topology: $V' = V'_{ij,q}$, where $V'_{ij,q}$ is defined similar to $V_{ij,q}$. Note that V' is at most one lightpath different from V.
  - Traffic routing: The binary variable $\Upsilon_{ij,q}^{sd}$ is 1 when the traffic flowing from node s to node d traverses lightpath $V'_{ij,q}$, and 0 otherwise. $\Upsilon_{ii,q}^{sd} = 0$ by definition. The traffic from s to d is not bifurcated, i.e., all traffic between s and d will flow through the same path.
  - Load of maximally-loaded lightpath in the network: L_{Max}.

- **Objective:**
  - Minimize L_{Max}.

The objective function minimizes the load of the maximally-loaded lightpath in the network. This objective allows us to balance the network load in the new virtual topology, by addition or deletion of the best possible lightpath.

- **Constraints:**
  - On physical topology:

$$\forall i, j, q, \quad \sum_n p_{in}^{ij,q} = V'_{ij,q} \qquad (9.33)$$

$$\forall i, j, q, \quad \sum_n p_{nj}^{ij,q} = V'_{ij,q} \qquad (9.34)$$

$$\forall i, j, q, l, \quad \sum_n p_{nl}^{ij,q} - \sum_n p_{ln}^{ij,q} = 0, \; i \neq l \text{ and } j \neq l \qquad (9.35)$$

$$\forall m, n, \quad \sum_i \sum_j \sum_q p_{mn}^{ij,q} \leq W * P_{mn} \qquad (9.36)$$

$$\forall m, n, i, j, q, \quad p_{mn}^{ij,q} \leq V'_{ij,q} \qquad (9.37)$$

Equation (9.33) ensures that only one outgoing physical link of the source node will be assigned to a lightpath. Equation (9.34) ensures that only one incoming physical link at the destination node will be assigned to a lightpath. Equation (9.35) guarantees that the number of incoming and outgoing links reserved for a lightpath at any intermediate node will be equal. The total number of wavelengths used between two nodes is limited to (*the number of fiber links*)$*W$ by Eqn. (9.36). Note that we assume wavelength conversion capability on network nodes, and we use the wavelength channels on different fibers as nondistinguishable entities. To capture wavelength continuity in a nonwavelength-conversion network, some additional constraints will be needed. Equation (9.37) states that a physical link is assigned only if the lightpath exists.

– On virtual-topology connections:

$$\sum_i \sum_j \sum_q V'_{ij,q} = \sum_i \sum_j \sum_q V_{ij,q} + k_H - (1 - k_H) * k_L \quad (9.38)$$

where $k_H = \lceil \frac{L^P_{Max}}{C} - W_H \rceil$ and $k_L = \lceil W_L - \frac{L^P_{Min}}{C} \rceil$.

$$\forall i, j, q, \quad [1 + 2 * (k_H - 1) * k_L] * (V'_{ij,q} - V_{ij,q}) \geq 0 \quad (9.39)$$

Note that the values of k_H and k_L are binary, and they are calculated by using the maximum and the minimum lightpath loads measured in the last observation period, watermark values, and channel capacity. Therefore, these values are constant for the MILP. Note that k_H is unity when one or more lightpaths are experiencing heavy load. This will ensure that a new lightpath will be added to the virtual topology. k_L is unity when one or more lightpaths has a load below the low watermark. In this case, if $k_H = 0$ (i.e., none of the lightpaths in the virtual topology is heavily loaded), a lightpath will be deleted. Thus, we are giving a higher priority to a lightpath addition than to a lightpath deletion to better accommodate the traffic. Equation (9.38) specifies the total number of lightpaths in the new virtual topology, i.e., it decides whether a change should be made, and if the answer is affirmative, whether a lightpath should be added or deleted. Equation (9.39) guarantees that the new virtual topology consists of the same set of lightpaths of the old virtual topology except that one lightpath is added or deleted.

– On virtual-topology traffic variables:

$$\forall s,d,l,q, \sum_i \Upsilon_{il,q}^{sd} - \Upsilon_{li,q}^{sd} = \begin{cases} 1, & l = d \\ 0, & l \neq s \text{ and } l \neq d \\ -1, & l = s \end{cases} \quad (9.40)$$

$$\forall s,d,i,j,q, \quad \Upsilon_{ij,q}^{sd} \leq V'_{ij,q} \quad (9.41)$$

$$L_{Max} \leq C * W_H \quad (9.42)$$

$$\forall i,j,q, \quad \sum_s \sum_d \Lambda_{sd} * \Upsilon_{ij,q}^{sd} \leq L_{Max} \quad (9.43)$$

Equation (9.40) is a multicommodity-flow equation controlling the routing of packet traffic on virtual links. Equation (9.41) ensures that traffic can only flow through an existing lightpath, and Eqn. (9.42) specifies the capacity constraint for any lightpath. Equation (9.43) constrains the load on any lightpath to be lower than or equal to the maximum load L_{Max}.

– On transceivers:

$$\forall i, \quad \sum_j \sum_q V'_{ij,q} \leq T_i \quad (9.44)$$

$$\forall j, \quad \sum_i \sum_q V'_{ij,q} \leq R_j \quad (9.45)$$

Equations (9.44) and (9.45) limit the total number of lightpaths originating from and terminated at a node to the total number of transmitters and receivers at that node.

The above formulation gives the best selection for a virtual-topology adjustment of one lightpath. By solving the MILP, let us compare this adaptation scheme with the earlier reconfiguration method proposed in Section 9.5.

B. Adaptation With Minimal Lightpath Change

Now, we focus on minimizing the number of lightpaths (i.e., resource usage cost) and the number of changes made for reconfiguring the topology (i.e., operation cost). We show that the new adaptation approach is cost effective in both sense by comparing our results with the earlier reconfiguration study (Section 9.5). The comparison is based on solving MILP formulations of both methods by using the standard solver CPLEX [CPLE05].

The reconfiguration method in Section 9.5 (which we call full reconfiguration) can be summarized as follows:

Start with initial virtual topology.

Every $\triangle$ seconds do:

- Find the optimal virtual topology for the new traffic pattern.
- Find the virtual topology such that: it requires minimum number of changes from the previous topology.

We solve this MILP by substituting the objective function with:

$$\text{Minimize } \sum_i \sum_j \sum_q V_{ij,q}$$

to minimize the total number of lightpaths in the network. We also limit the maximum load L_{Max} to W_H. When the optimum virtual topology is obtained from this formulation, this solution gives the minimum number of lightpaths that any virtual topology must contain to carry the given traffic demand. Let this number be η. This MILP can then be modified as follows:

(1) The objective function guarantees that the new virtual topology will be as close as possible to the previous one.

(2) A new constraint is added to the formulation to guarantee that the new virtual topology will have η lightpaths.

This second step selects the closest virtual topology among the feasible topologies having exactly η lightpaths and able to carry the given traffic demand. We use two metrics to compare our adaptation method and the full reconfiguration method: the total number of lightpaths in the network and the number of lightpath additions and deletions.

Since finding the optimal topology using full reconfiguration is computationally tractable only for small networks, our comparison is based on a six-node topology (see Fig. 9.10). (For details of the traffic model, and larger

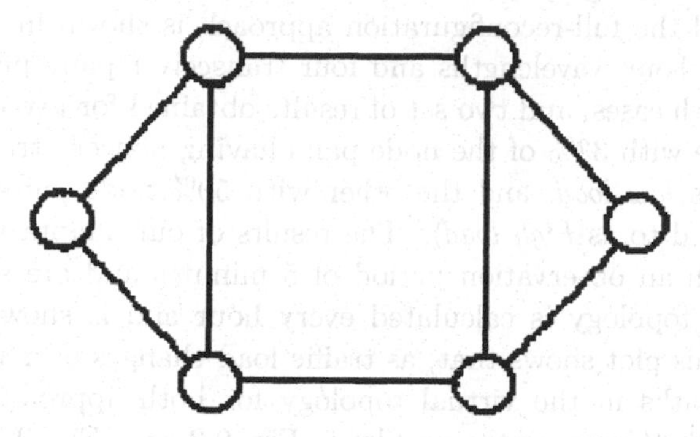

Figure 9.10 Six-node network used in experiments.

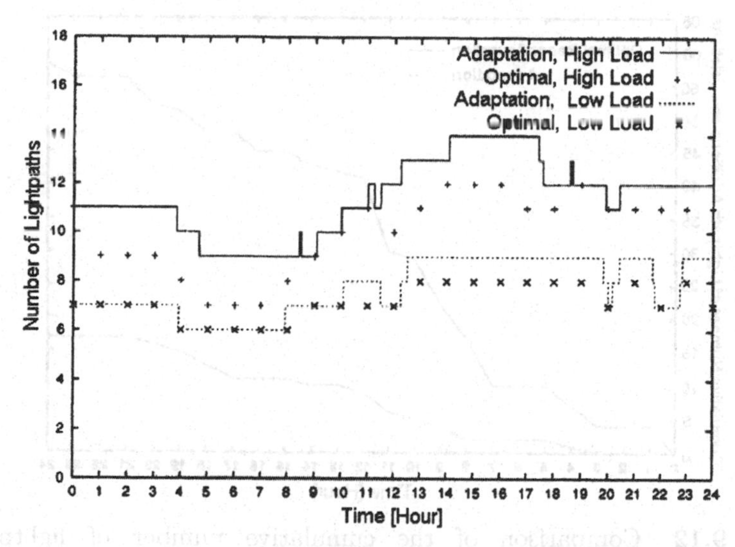

Figure 9.11 Number of lightpaths in virtual topologies obtained by solving our proposed adaptation MILP and the optimal reconfiguration MILP.

network examples employing these approaches, see [GeMu03].) Comparison of the number of lightpaths as a function of time used by our proposed approach and the full-reconfiguration approach is shown in Fig. 9.11 for a 24-hour run. Four wavelengths and four transceiver pairs per node are assumed for both cases, and two set of results obtained for two different traffic matrices (one with 33% of the node pairs having nonzero traffic intensities, referred to as *low load*, and the other with 50% node pairs with nonzero traffic, referred to as *high load*). The results of our adaptation scheme are obtained with an observation period of 5 minutes and are shown as lines. The optimal topology is calculated every hour and is shown as points in Fig. 9.11. This plot shows that, as traffic load changes over time, the number of lightpaths in the virtual topology for both approaches changes in unison as well. (Compare the profiles in Fig. 9.9 with Fig. 9.11, noting that the measurements in Fig. 9.9 starts at 9:00 a.m.). However, as also expected, the adaptation method establishes a few more lightpaths than those in the optimal topology, and the difference in the number of lightpaths increases slightly with the traffic load.

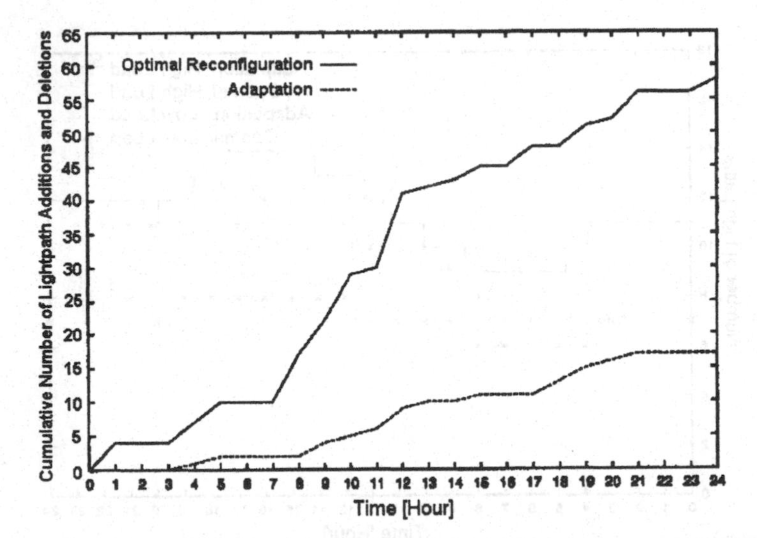

Figure 9.12 Comparison of the cumulative number of lightpath changes (sum of lightpath additions and deletions) between our adaptation scheme and optimal reconfiguration.

Fig. 9.12 plots the cumulative number of lightpath additions and deletions for both methods during the same 24-hour period (results from only the high-traffic-load experiment are shown in this figure). We observe that the adaptation scheme can reconfigure the topology using far fewer lightpath changes compared with the optimal reconfiguration; thus, the adaptation scheme is a very efficient one.

We remark that the complexity of the MILP limits its use for large networks. In following section, we design an efficient heuristic method to adapt the virtual topology in response to traffic changes in large networks.

9.7.3 Heuristic Adaptation Algorithm for Larger Networks

Since the MILP does not scale, an efficient heuristic adaptation algorithm is needed for solving larger instances of this problem. For such a heuristic, we refer the interested reader to [GeMu03].

We show some results from [GeMu03] by applying its heuristic to a larger network with 19 nodes and 31 bidirectional links, where each link has 16 wavelengths and each node has 8 transmitters and 8 receivers.

The operation of the system is demonstrated by measuring the maximal and minimal lightpath loads in the network at the end of every observation period. High and low watermarks will define a *balance region* in which the lightpath loads are allowed to fluctuate as long as they do not exceed the high watermark or drop below the low watermark. Every time a link load goes out of the balance region, the system will try to adapt to the new traffic conditions.

An example operation of the system for a three-day period with $W_H = 70$ and $W_L = 10$ is shown in Fig. 9.13. Note that, in this section we represent the high and low watermark values as $0 \leq W_H, W_L \leq 100$, normalized to 100, so that W_H and W_L can be interpreted as percentage loads. Figure 9.13(a) plots the maximal and minimal loads in the network observed at the end of each observation period (300 s for this experiment). Figure 9.13(b) shows the times of topology adjustments, where a positive impulse indicates a lightpath addition and a negative impulse indicates a lightpath deletion. The same average traffic rate functions were repeated for every day, but slight differences can be seen for the same time of different days related to randomly generated traffic and to the dynamic nature of the algorithm. For more illustrative numerical examples, we refer the reader to [GeMu03].

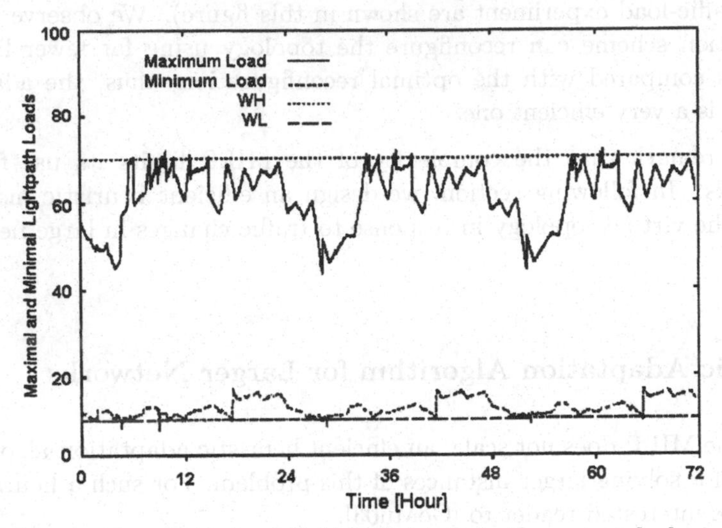

(a) Maximal and minimal lightpath loads in the network during a three-day run ($W_H = 70, W_L = 10$).

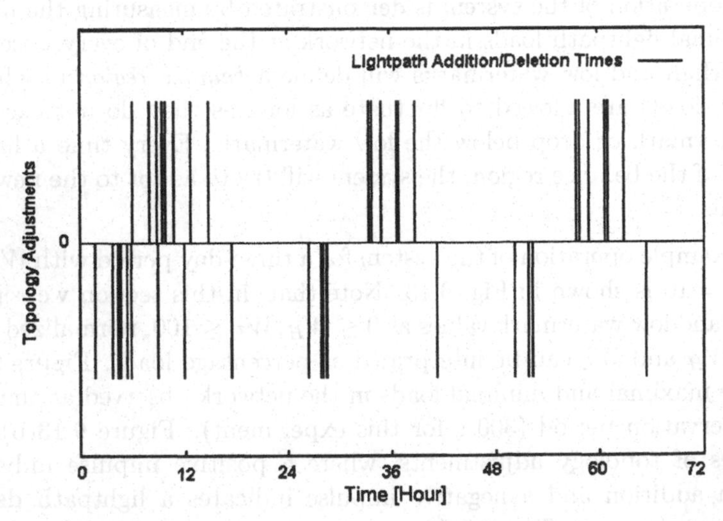

(b) Impulse plot indicating times of lightpath addition or deletion.

Figure 9.13 Example operation of the system for a three-day period.

9.8 Summary

This chapter presented a linear program (LP) formulation to derive an exact, minimal-hop-distance solution to the virtual-topology design problem in a wavelength-routed optical network, in the absence of wavelength-continuity constraints. The problem formulation is general, and can be used to derive a complete virtual-topology solution, including choice of the constituent lightpaths, routes for these lightpaths, and intensity of packet flows through these lightpaths. We showed that adding wavelength-continuity constraints and queueing delays makes the problem nonlinear. We used simplifying assumptions to make the problem tractable. We also proposed two simple heuristics and demonstrated that these heuristics perform well with respect to the optimal solution.

We studied resource-budgeting trade-offs in the allocation of transceivers per node, and wavelengths per fiber. A simple analysis in Section 9.4.1 [Eqn. (9.28)] provided an approximate bound regarding the number of transceivers that can be supported in a network with W wavelengths. We demonstrated how we can equip the network with an optimal balance of transceivers and wavelengths, in order to derive minimal-hop-distance solutions, along with high utilization of both transceivers and wavelengths.

We proposed an exact reconfiguration procedure which, for a changed traffic matrix, searches through all possible *optimal* virtual topologies, in order to obtain a solution which shares the maximum number of lightpaths with the previous virtual topology. The solution to the reconfiguration algorithm generates a virtual topology which minimizes the amount of reconfiguration that needs to be performed, in order to adapt the virtual topology to the new traffic matrix.

Thus, based on the new view of the virtual-topology reconfiguration problem, an adaptation scheme for WDM mesh networks under dynamic traffic is further studied in Section 9.7. We defined the problem as tracking the long-term traffic fluctuations by adapting the topology in a measurement-adaptation cycle with the following constraints: no assumption should be made on future traffic rates and the ongoing traffic should not be interrupted by a transition phase. A MILP formulation is presented for the selection of the lightpath to be added or deleted to minimize the maximum link load in the network for the adaptation algorithm which allows one lightpath change at a time. By solving this MILP, it is shown that the performance of the adaptation scheme is comparable to the optimal reconfiguration in Section 9.5 in terms of number of lightpaths and much better in terms of cumulative number of changes. The development of a distributed version of the adaptation algorithm is an open problem for further study.

Exercises

9.1. In this chapter, average packet hop distance is used as the objective function. Why?

9.2. Heuristics `MaxSingleHop` and `MaxMultiHop` assign lightpaths based on traffic flows. `MaxSingleHop` adjusts the traffic matrix after assigning a lightpath, while `MaxMultiHop` does not. Why?

9.3. For this exercise, consider the physical topology shown in Fig. 9.14 and the traffic matrix shown below. Assume two channels per link, where each channel has a capacity of five units. Use the `MaxSingleHop` heuristic to set up lightpaths. Assume that each node has sufficient number of transceivers. The traffic matrix is:

$$T = \begin{bmatrix} 0 & 4 & 5 & 0 & 0 \\ 0 & 0 & 0 & 0 & 0 \\ 1 & 0 & 0 & 2 & 3 \\ 0 & 0 & 0 & 0 & 0 \\ 0 & 0 & 0 & 0 & 0 \end{bmatrix}$$

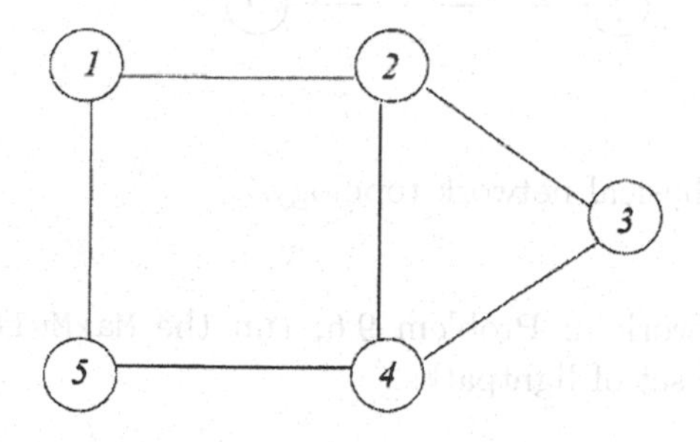

Figure 9.14 Physical network topology.

9.4. Draw the virtual topology of the network shown in Problem 9.3. Calculate V_{ij}, λ_{ij}^{sd}, and p_{mn}^{ij} for this network.

9.5. Derive Eqn. (9.1). Show that the inequality holds for the network shown in Problem 9.3. When will the equality hold?

9.6. Consider the network shown Fig. 9.15 with two transmitters and two receivers per node, two wavelengths, capacity of each wavelength equal to 10 units, and the following traffic matrix:

$$\Lambda = \begin{bmatrix} 0 & 2 & 2 & 1 & 4 & 3 \\ 1 & 0 & 4 & 4 & 3 & 2 \\ 3 & 2 & 0 & 6 & 2 & 1 \\ 1 & 1 & 2 & 0 & 1 & 7 \\ 3 & 5 & 2 & 1 & 0 & 1 \\ 2 & 4 & 5 & 3 & 2 & 0 \end{bmatrix}$$

Determine a set of lightpaths using the `MaxSingleHop` heuristic.

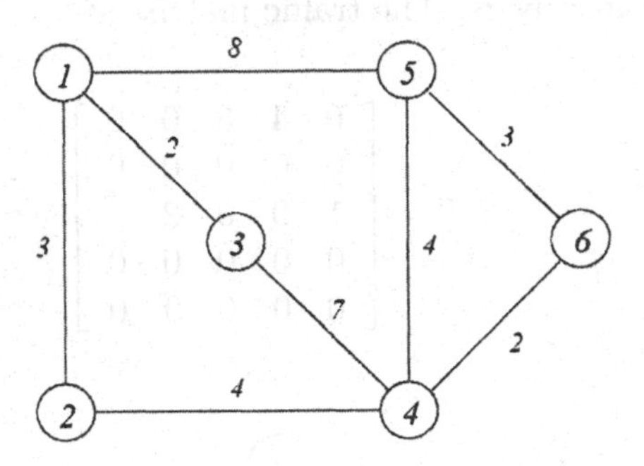

Figure 9.15 Physical network topology.

9.7. For the network in Problem 9.6, run the `MaxMultiHop` heuristic to determine a set of lightpaths.

9.8. Assume that the network equipment cost budget is $1,000,000. Using Table 9.1, find the network configurations (i.e., number of transceivers and wavelengths) that can be supported. Which network configuration maximizes the total network throughput?

9.9. Assume the Petersen graph as the physical topology. Assume that the network can support 10 wavelengths. Now assume that we embed a complete graph on 10 nodes as the virtual topology. (Use shortest-path routing to embed the virtual topology). Using the cost model described in Section 9.4, compute the cost to build the network. Which type of component has the highest networkwide cost?

Chapter 10

Wavelength Conversion

10.1 Introduction

Wavelength conversion can be used in WDM networks to improve efficiency. The enabling technologies for building wavelength converters were reviewed in Section 2.7. After briefly re-reviewing what wavelength conversion (WC) is, its objectives, and the corresponding switch designs, this chapter will examine the network performance issues when wavelength conversion is incorporated. Various analytical models that can be employed to assess the performance benefits in a wavelength-convertible network will reviewed.

Consider the network in Fig. 10.1. It shows a wavelength-routed network containing two *WDM* nodes with OXCs (S1 and S2) and five *access stations* (A through E). Figure 10.1 shows that more than one "access station" may be attached to an OXC. Thus, this model can perform a clear demarkation between the "optical cloud" and the electronic portion of the network, as can be seen in Fig. 10.1.

In Fig. 10.1, three lightpaths have been set up (C to A on wavelength λ_1, C to B on λ_2, and D to E on λ_1). To establish a lightpath, we may prefer that the *same* wavelength be allocated on all of the links in the path. Recall that this requirement is known as the *wavelength-continuity constraint.* This constraint distinguishes the wavelength-routed network from a "circuit-

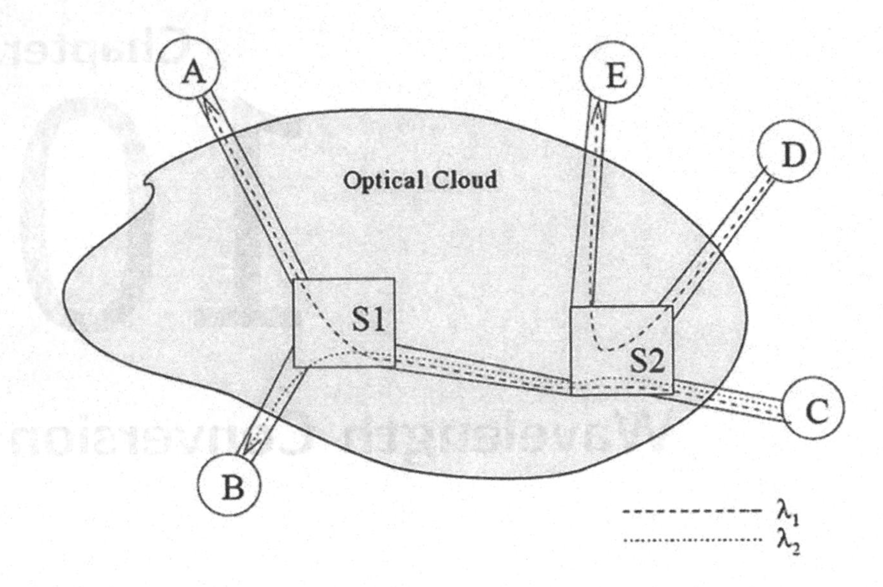

Figure 10.1 An optical wavelength-routed network.

switched" network which has no such constraints since the latter blocks a connection only when there is no capacity along any of the links in the path assigned to the connection.

Now, consider the example in Fig. 10.2(a) with $W = 2$ wavelengths per link. Two lightpaths have been established in the network: (1) between node 1 and node 2 on wavelength λ_1 and (2) between node 2 and node 3 on wavelength λ_2. Now, suppose a new lightpath between node 1 and node 3 needs to be set up. Establishing such a lightpath is impossible even though there is a free wavelength on each of the links along the path from node 1 to node 3. This is because the available wavelengths on the two links are *different.* Thus, a wavelength-routed network *with wavelength-continuity constraint* may suffer from higher blocking as compared to a circuit-switched network.

It is easy to eliminate the wavelength-continuity constraint, if we can *convert* the data arriving on one wavelength along a fiber link into another wavelength at an intermediate node and forward it along the next fiber link. Such a technique is feasible and is referred to as *wavelength conversion* (WC). In Fig. 10.2(b), a wavelength converter at node 2 is employed to convert data

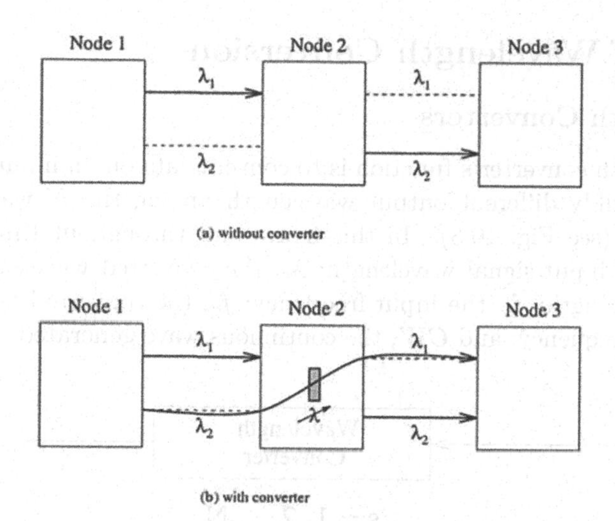

Figure 10.2 Wavelength-continuity constraint in a wavelength-routed network.

from wavelength λ_2 to λ_1. The new lightpath between node 1 and node 3 can now be established by using wavelength λ_2 on the link from node 1 to node 2 and then by using wavelength λ_1 to reach node 3 from node 2. Notice that a single lightpath in such a *wavelength-convertible* network can use a different wavelength along each of the fiber links in its path. Thus, wavelength conversion may improve the efficiency in the network by resolving the wavelength conflicts of the lightpaths. A wavelength continuous light-path has also been referred to as a wavelength path in the literature [MPBC04].

This chapter will examine the effects of wavelength converters in wavelength-routed networks. A note on terminology: wavelength converters have been referred to in the literature as wavelength shifters, wavelength translators, wavelength changers, and even frequency converters. Throughout this book, we refer to these devices as wavelength converters.

Section 10.2 reviews the basic principles of wavelength conversion and the corresponding switch designs. Section 10.3 highlights the network design, control, and management issues for effectively using wavelength conversion. The approaches taken to tackle some of these issues are highlighted and new problems in this area are introduced. Section 10.4 discuss various approaches that can be used to quantify the benefits of wavelength conversion.

10.2 Basics of Wavelength Conversion

10.2.1 Wavelength Converters

A wavelength converter's function is to convert data on an input wavelength onto a possibly different output wavelength among the N wavelengths in the system (see Fig. 10.3). In this figure and throughout this section, λ_s denotes the input signal wavelength; λ_c, the converted wavelength; λ_p, the pump wavelength; f_s, the input frequency; f_c, the converted frequency; f_p, the pump frequency; and CW, the continuous wave generated as the signal.

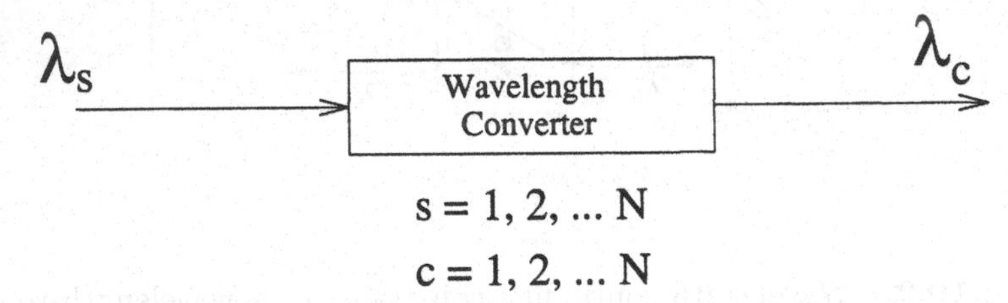

Figure 10.3 Functionality of a wavelength converter.

An ideal wavelength converter should possess the following characteristics [DMJD96]:

- transparency to bit rates and signal formats,
- fast setup time of output wavelength,
- conversion to both shorter and longer wavelengths,
- moderate input power levels,
- possibility for same input and output wavelengths (i.e., no conversion),
- insensitivity to input signal polarization,
- low-chirp output signal with high extinction ratio[1] and large signal-to-noise ratio, and
- simple implementation.

[1]The *extinction ratio* is defined as the ratio of the optical power transmitted for a bit "0" to the power transmitted for a bit "1."

A classification of wavelength conversion schemes was provided in Section 2.7.

10.2.2 Switches

As wavelength converters become readily available, a vital question comes to mind: where do we place them in the network? An obvious location is in the switches (i.e., crossconnects) in the network. A possible architecture of such a wavelength-convertible switching node is the dedicated wavelength convertible switch (see Fig. 10.4, from [LeLi93]). In this architecture, each wavelength along each output link in a switch has a *dedicated* wavelength converter, i.e., an $M \times M$ switch in an N-wavelength system requires MN converters. The incoming optical signal from a link at the switch is first wavelength demultiplexed into separate wavelengths. Each wavelength is switched to the desired output port by the nonblocking optical switch. The output signal may have its wavelength changed by its wavelength converter. Finally, various wavelengths combine to form an aggregate signal coupled to the outbound link.

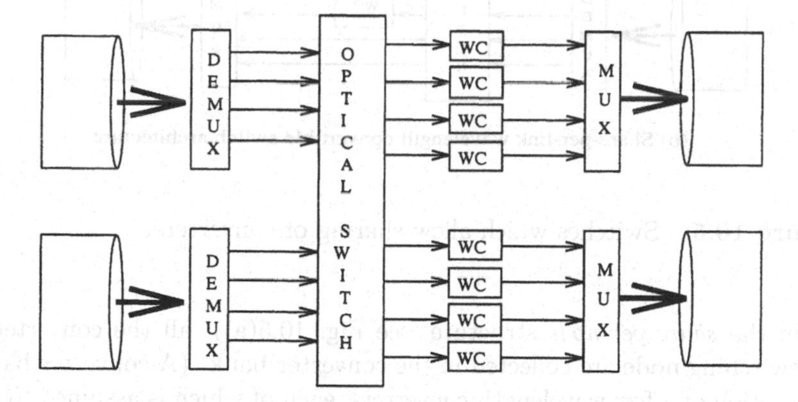

Figure 10.4 A switch with dedicated converters at each output port for each wavelength.

However, the dedicated wavelength-convertible switch is not very cost efficient since all the wavelength converters may not be required all the time. An effective method to cut costs is to share the converters. Two architectures have been proposed for switches sharing converters [LeLi93].

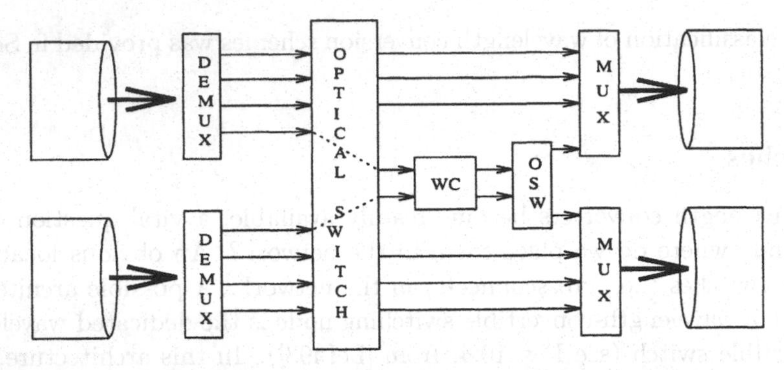

(a) Share-per-node wavelength-convertible switch architecture

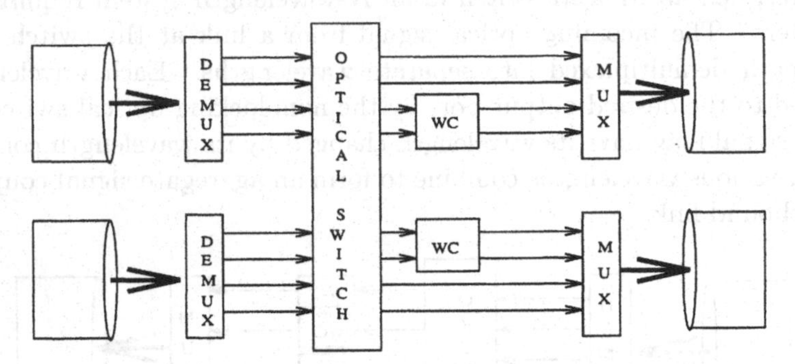

(b) Share-per-link wavelength-convertible switch architecture

Figure 10.5 Switches which allow sharing of converters.

In the *share-per-node* structure (see Fig. 10.5(a)), all the converters at the switching node are collected in the converter bank. (A converter bank is a collection of a few wavelength converters, each of which is assumed to have identical characteristics and can convert any input wavelength to any output wavelength.) This bank can be accessed by any of the incoming lightpaths by appropriately configuring the larger optical switch in Fig. 10.5(a). In this architecture, only the wavelengths which require conversion are directed to the converter bank. The converted wavelengths are then switched to the appropriate outbound link by the second optical switch.

In the *share-per-link* structure (see Fig. 10.5(b)), each outgoing link is provided with a dedicated converter bank which can be accessed only by those lightpaths traveling on that particular outbound link. The optical switch can be configured appropriately to direct wavelengths toward a particular link, either with conversion or without conversion.

When opto-electronic wavelength conversion is used, the functionality of the wavelength converter can be performed at the access stations instead of at the switches. The *share-with-local* switch architecture in the simplified network access station architecture fall under this category (see Chapter 2).

10.3 Network Design, Control, and Management Issues

10.3.1 Network Design

Network designs must evolve to incorporate wavelength conversion effectively. Network designers must choose not only among the various conversion techniques described in Section 2.7, but also among the several switch architectures described in Section 10.2.2. An important challenge in the design is to overcome the limitations in using wavelength conversion technology. These limitations fall into the following three categories:

1. *Limited availability of wavelength converters at the nodes.* As long as wavelength converters remain expensive, it may not be economically viable to equip all the nodes in a WDM network with them. Some effects of sparse conversion (i.e., having only a few converting switches in the network) have been examined [SuAS96]. An interesting question is *where* (optimally?) to place these few converters in the network.

2. *Sharing of converters.* Even among the switches capable of wavelength conversion, it may not be cost effective to equip all the output ports of a switch with this capability. Designs of switch architectures have been proposed (see Section 10.2.2) which allow sharing of converters among the various signals at a switch. It has been shown in [LeLi93] that the performance of such a network saturates when the number of converters at a switch increases beyond a certain threshold. An interesting problem is to quantify the dependence of this threshold on the routing algorithm used and the blocking probability desired.

3. *Limited-range wavelength conversion.* Four-wave-mixing-based all-optical wavelength converters provide only a limited-range conversion capability. If the range is limited to k, then an input wavelength λ_i can only be converted to wavelengths $\lambda_{\max(i-k,1)}$ through $\lambda_{\min(i+k,W)}$,

where W is the number of wavelengths in the system (indexed 1 through W). Analysis shows that networks employing such devices, however, compare favorably with those utilizing converters with full-range capability, under certain conditions [YLES96].

Other wavelength converter techniques too have some limitations. As seen in Section 2.7, the wavelength converter using SOAs in XGM mode suffers greater degradation when the input signal is up-converted to a signal of equal or longer wavelength than when it is down-converted to a shorter wavelength. Moreover, since the signal quality worsens after multiple such conversions, the effect of a cascade of these converters can be detrimental. The implications of such a device on the design of the network need to studied further.

Apart from efficient wavelength-convertible switch architectures and their optimal placement, several other design techniques offer promise. Networks equipped with multiple fibers on each link have been considered for potential gains [JeAy96] in wavelength-convertible networks and suggested as a possible alternative to conversion. This work will be reviewed in greater detail in Section 10.4. Another important problem is the design of a fault-tolerant wavelength-convertible network. Such a network could reserve capacity on the links to handle disruptions due to link failure caused by a cut in the fiber. Quantitative comparisons need to be developed for the suitability of a wavelength-convertible network in such scenarios.

10.3.2 Network Control

Control algorithms are required in a network to manage its resources effectively. An important task of the control mechanism is to provide routes to the lightpath requests while maximizing a desired system parameter, e.g., throughput. Such routing schemes can be classified into *static* and *dynamic* categories, depending on whether the lightpath requests are known a priori or not. These two categories are described below.

1. *Dynamic Routing.* In a wavelength-routed optical network, lightpath requests between source-destination pairs arrive at the source node at random, and each lightpath has a random holding time after which it is torn down. These lightpaths need to be set up dynamically between source-destination pairs by determining a route through the network connecting the source to the destination and assigning a free wavelength along this path. Two lightpaths which have at least a link in

common cannot use the same wavelength. Moreover, the same wavelength has to be assigned to a path on all of its links. This is the wavelength-continuity constraint described in Section 10.1. This routing and wavelength assignment (RWA) problem was studied in Chapter 7.

However, if all switches in the network have full wavelength conversion (see Fig. 10.4), the network becomes equivalent to a circuit-switched telephone network [RaSi95b]. Routing algorithms have been proposed for use in wavelength-convertible networks. In [LeLi93], the routing algorithm approximates the cost function of routing as the sum of individual costs due to using channels and wavelength converters. For this purpose, an auxiliary graph is created [BaSB91] and the shortest-path algorithm is applied on the graph to determine the route. In [ChFZ96], an algorithm with provably optimal running time has been provided for such a technique. Algorithms have also been studied which use a *fixed path* or *deterministic* routing [RaSi95b]. In such a scheme, there is a fixed path between every source-destination pair in the network. The work in [RaMu02] demonstrates that most of the benefits of wavelength conversion can also be achieved by alternate routing algorithms. Several RWA heuristics have been designed based on which wavelength to assign to a lightpath along the fixed path [BaSB91, MoAz96, MoAz98] and which, if any, lightpaths to block selectively.

2. *Static Routing.* In contrast to the dynamic routing problem described above, the static RWA problem assumes that all the lightpaths that are to be set up in the network are known initially. The objective is to maximize the total throughput in the network, i.e., the total number of lightpaths which can be established simultaneously in the network. An upper bound on the carried traffic per available wavelength has been obtained (for a network with and without wavelength conversion) by relaxing the corresponding integer linear program (ILP) [RaSi95b]. Several heuristic-based approaches have been proposed for solving the static RWA problem in a network without wavelength conversion [ChBa96]. Again, efficient algorithms which incorporate the limitations in Section 10.3.1 for a wavelength-convertible network are still unavailable.

10.3.3 Network Management

Issues arise in network management regarding the use of wavelength conversion to promote interoperability across subnetworks managed by independent operators. Wavelength conversion supports the distribution of network control and management functionalities into smaller subnetworks by allowing flexible wavelength assignments within each subnetwork. As shown in Fig. 10.6, network operators 1, 2, and 3 manage their own subnetworks and may use wavelength conversion for communication *across* subnetworks.

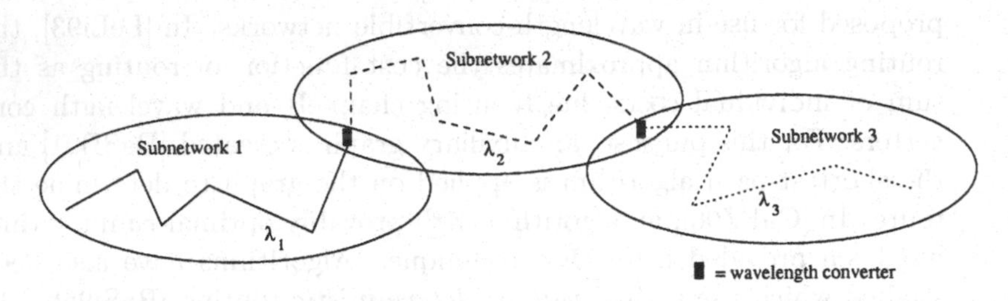

Figure 10.6 Wavelength conversion for distributed network management.

A related network interconnection problem has also been examined in [SoAz97]. This work shows that wavelength conversion for communication across subnetworks offers very little benefit when the subnets are broadcast stars.

10.4 Benefit Analysis

As mentioned in Section 10.1, wavelength conversion eliminates the wavelength-continuity constraint in wavelength-routed WDM networks. In fact, a wavelength-routed network with full conversion at all nodes behaves essentially like a circuit-switched telephone network. The wavelength assignment algorithm in such a network becomes trivial because all the wavelengths can be treated equivalently, and wavelengths used on successive links along a path can be independent of one another. In addition to reducing blocking, converters can improve fairness by allowing many long distance calls which would have been blocked otherwise due to the wavelength-continuity constraint.

Efforts have been made to quantify the benefits of wavelength conversion. Such attempts involve either probabilistic approaches or they employ deterministic algorithms on specific network topologies. These studies have shown that the benefits of wavelength conversion are greater in a mesh network than in a ring or a fully connected network [KoAc96]. In the following subsections, we first present a probabilistic approach to quantify the benefits (e.g., decrease in connection blocking probability) of using wavelength conversion, followed by a brief review of several of benefits-analysis studies.

10.4.1 A Probabilistic Approach to Wavelength-Conversion Benefits Analysis

This development is due to the work in [BaHu96], and is based on standard serial-independent-link assumptions, i.e., a connection (lightpath) request sees a network in which a wavelength's usage on a fiber link is statistically independent of other fiber links and other wavelengths. However, this model generally tends to overestimate the blocking probability because it ignores the correlation of usage of wavelength in successive links, especially for a multi-link lightpath.

Let there be W wavelengths per fiber link, and let ρ be the probability that a wavelength is used on any fiber link. (Since ρW is the expected number of busy wavelengths on any fiber link, ρ is also the "fiber utilization" of any fiber.) We will consider an H-link path for a connection from node A to node B that needs to be set up.

First, let us consider a network *with* wavelength converters. The probability P_b' that the connection request from A to B will be blocked equals the probability that, along this H-link path, *there exists a fiber link with all of its W wavelengths in use*, so that:

$$P_b' = 1 - \left(1 - \rho^W\right)^H \tag{10.1}$$

Defining q to be the achievable utilization for a given blocking probability in a wavelength-convertible network, we have:

$$q = \left[1 - (1 - P_b')^{1/H}\right]^{1/W} \approx \left(\frac{P_b'}{H}\right)^{1/W} \tag{10.2}$$

where the approximation holds for small values of P_b'/H, i.e., when the correlation of successive link utilizations are small.

Next, let us consider a network *without* wavelength converters. The probability P_b that the connection request from A to B will be blocked equals the probability that, along this H-link path, *each wavelength is used on at least one of the H links,* so that:

$$P_b = \left[1 - (1-\rho)^H\right]^W \tag{10.3}$$

Defining p to be the achievable utilization for a given blocking probability in a network without wavelength conversion, we have:

$$p = 1 - \left(1 - P_b^{1/W}\right)^{1/H} \approx -\frac{1}{H}\ln\left(1 - P_b^{1/W}\right) \tag{10.4}$$

where the approximation holds for large values of H, and for $P_b^{1/W}$ not too close to unity. Observe that the achievable utilization is inversely proportional to the "length of the lightpath connection" (H), as expected.

Define $G = q/p$ to be a measure of the benefit of wavelength conversion, which is the increase in (fiber or wavelength) utilization for the same blocking probability. From Eqns. (10.2) and (10.4), after setting $P_b = P_b'$, we get:

$$G \approx H^{1-(1/W)} \frac{P_b^{1/W}}{-\ln\left(1 - P_b^{1/W}\right)} \tag{10.5}$$

where the approximation holds for small P_b, large H, and moderate W so that $P_b^{1/W}$ is not too close to unity.

Observe that, if $H = 1$ or $W = 1$, then $G = 1$, i.e., there is no difference between networks with and without wavelength converters in these cases.

It is also reported in [BaHu96] that the gain increases as the blocking probability decreases, but this effect is small for small values of P_b. Also, as W increases, G also increases until it peaks around $W \approx 10$ (for $q \approx 0.5$), and the maximum gain is close to $H/2$. After peaking, G decreases, but very slowly. Generally, it is found that, for a moderate to a large number of wavelengths, the benefits of wavelength conversion increases with the "length" of the connection, and decreases (slightly) with an increase in the number of wavelengths.

While this was a simple analysis, more detailed and rigorous treatment of the benefits of wavelength conversion can be found in the references cited in the following subsection.

10.4.2 A Review of Benefit-Analysis Studies

1. *Bounds on RWA algorithms with and without wavelength converters* [RaSi95b]. Upper bounds on the carried traffic (i.e.,, lower bounds on the blocking probability) in a wavelength-routed WDM network are derived in [RaSi95b]. The generalized RWA problem (for both the static and the dynamic cases) are formulated as integer linear programs (ILP) with the objective of maximizing the number of lightpaths that are successfully routed. The formulation is similar to a multicommodity flow problem with integer flows through the links. The upper bound is obtained by relaxing the integrality constraints in the formulation. A similar bound was also obtained for networks with full wavelength conversion at all nodes. The bound is shown to be achievable asymptotically by a fixed RWA algorithm using a large number of wavelengths. A heuristic shortest-path RWA algorithm for dynamic routing is provided which employs a set of shortest paths and assigns the first available free wavelength to the lightpath requests. The wavelength reuse factor – which is defined as the maximum offered traffic per wavelength for which the blocking probability can be made arbitrarily small by using sufficiently large number of wavelengths – is found to increase by using wavelength converters in large networks.

2. *Probabilistic model <u>with</u> independent link-load assumption* [KoAc96]. An approximate analytical model is developed for a (deterministic) fixed-path wavelength-routed network with an arbitrary topology, both with and without wavelength conversion. This model is then used along with simulations to study the performance of three example networks: the nonblocking centralized switch, the two-dimensional torus network, and the ring network. The traffic loads on the different links in the network are assumed to be independent. The wavelength occupancy probabilities on the links are also assumed to be independent. A wavelength-assignment strategy is employed in which a lightpath is assigned a wavelength at random from among the available wavelengths in the path. The blocking probability of the lightpaths is used to study the performance of the network. The benefits of wavelength conversion are found to be modest in networks such as the nonblocking centralized switch and the ring; however, wavelength conversion is found to significantly improve performance of large two-dimensional-torus networks. The analytical model employed in this study cannot

be applied to a ring network because the very high load correlation along the links of a path in a ring network invalidates the independent link-load assumption.

3. *Probabilistic model without independent link-load assumption* [BaHu96]. A model which is more analytically tractable than the ones in [Birm96, KoAc96] is provided in this study; however, it uses more simplistic traffic assumptions. The link loads are not assumed to be independent; however, the assumption is retained that a wavelength is used on successive links independent of other wavelengths. The concept of *interference length* (L), i.e., the expected number of links shared by two sessions which share at least one link, is introduced. Analytical expressions for the link utilization and the blocking probability are obtained by considering an average path which spans H (*average hop distance*) links in networks with and without wavelength conversion. The gain (G) due to wavelength conversion is defined as the ratio of the link utilization with wavelength conversion to that without wavelength conversion for the same blocking probability. The gain is found to be directly proportional to the effective path length (H/L). A larger switch size (Δ) tends to increase the blocking probability in networks without wavelength conversion. The model used in [BaHu96] is applicable to ring networks unlike the work in [KoAc96], and correctly predicts the low gain in utilizing wavelength conversion in ring networks.

4. *Probabilistic model for a class of networks* [Birm96]. The work in [Birm96] provides an approximate method for calculating the blocking probability in a wavelength-routed network. The model considers Poisson input traffic and uses a Markov chain model with state-dependent arrival rates. Two different routing schemes are considered: *fixed routing*, where the path from a source to a destination is unique and is known beforehand; and *least-loaded routing* (LLR), an alternate-path scheme where the route from source to destination is taken along the path which has the largest number of idle wavelengths. Analysis and simulations are carried out using fixed routing for networks of arbitrary topology with paths of length at most three hops, and using LLR for fully-connected networks with paths of one or two hops. The blocking probability is found to be larger without wavelength conversion. However, this method is computationally intensive and is tractable only for

networks with a few nodes.

5. *Multifiber networks* [JeAy96]. The benefits of wavelength conversion in a network with *multiple* fiber links are studied in [JeAy96], by extending the analysis presented in [BaHu96] to multifiber networks. Multifiber links are found to reduce the gain obtained due to wavelength conversion, and the number of fibers is found to be more important than the number of wavelengths for a network. A heuristic is also provided to solve the capacity assignment problem in a wavelength-routed network without wavelength conversion where the multiplicity of the fibers is sought to be minimized. Multiple lightpaths between a source-destination pair is allowed in the network, with each lightpath utilizing a separate wavelength channel. It is concluded that a mesh network enjoys a higher utilization gain with wavelength conversion for the same traffic demand than a ring or a fully-connected network.

6. *Sparse wavelength conversion* [SuAS96, InMu99]. The works in [SuAS96, InMu99] quantifies the effects of sparse wavelength conversion (see Section 10.3.1), where only a few of the nodes in the network are capable of full wavelength conversion and the remaining nodes do not support any conversion at all. (But it must be noted that the performance model in [SuAS96] applies equally well to systems with full conversion and systems without conversion.) The analytical model in [SuAS96] improves on the model proposed in [KoAc96] by relaxing the wavelength-independence assumption while retaining the link-load independence assumption, thus incorporating, to a certain extent, the correlation between the wavelengths used in successive links of a multilink path. This work as well as that in [InMu99] shows that, in most cases, only a small fraction of the nodes has to be equipped with wavelength conversion capability for good performance. Also, in general, converters are more effective when the number of wavelengths is substantial and when the load in the network is low. This is especially true in networks with a high degree of connectivity, like the mesh-torus, but with fairly large average hop distances. It is also concluded that, in a wide-area network (WAN) with an irregular topology and a large number of links, the connectivity and the number of wavelengths are much more important than the availability of wavelength conversion.

7. *Limited-range wavelength conversion* [YLES96, InMu99].

The effects of limited-range wavelength conversion (see Section 10.3.1) on the performance gains achievable in a network are considered in [YLES96, InMu99]. The model used in these work captures the functionality of certain all-optical wavelength converters (e.g., those based on four-wave mixing) whose conversion efficiency drops with increasing range. The analytical model employs both link-load independence and wavelength-independence assumptions. The routing algorithm used in [YLES96] is a fixed one. The wavelength-assignment algorithm attempts to minimize the number of converters used and breaks ties by choosing the input wavelength with the lowest index at the converters. Simulations are conducted on a unidirectional ring and a mesh-torus network. The results obtained indicate that significant improvement in the blocking performance of the network is obtained when limited-range wavelength converters with as little as one-quarter of the full range are used. Moreover, converters with just half of the full-conversion range deliver almost all of the performance improvement offered by an ideal full-range converter. For additional results on sparse and limited-range conversion, please see [InMu99], a subset of which is discussed in the following section.

10.5 Benefits of Sparse Conversion

10.5.1 Goals

The goals of this study are summarized below. In most cases, it may be undesirable (e.g., due to budget constraints) to deploy "full" wavelength-conversion capabilities at all nodes, so some degree of sparse wavelength conversion may be desirable.

- Given that "full" wavelength conversion will be installed at a few nodes, what are the "best" nodes at which wavelength converters are placed?
- Different wavelength-converting-switch designs can utilize fewer wavelength converters effectively, so which switch design should be implemented in the network?
- In order to avoid under- or over-utilizing the wavelength converters at each node, how many wavelength converters should be placed at each node?

- Make a general determination as to whether wavelength converters offer significant benefits to optical networks, i.e., does the reduction in blocking probabilities justify the increased costs due to deploying wavelength converters?

- Analyze how different traffic loads affect the need and/or desirability for wavelength converters.

While this section provides a summary of our sparse-conversion findings, much more detailed results on analytical models as well as detailed analysis and simulation examples can be found in [InMu99, Ines97], including results on a "interconnected rings" network topology that is very typical of the telecom network environment.

Three Degrees of Sparseness

- *Sparse nodal conversion:* This design attempts to reduce costs by only giving a few nodes in the network "full" conversion capabilities.

- *Sparse switch-output conversion:* This design attempts to reduce costs by limiting the number of wavelength converters at each node.

- *Sparse- (or limited-) range conversion:* In this environment, costs (both monetary and power costs) are reduced by limiting the conversion capabilities (or range, distance) of the wavelength converters. For some wavelength converter design technologies, e.g., wave-mixing-based approaches, it is more efficient, in terms of power maintenance, to convert to wavelengths that are closer (in terms of nm) than to those that are farther away [YLES96].

10.5.2 Simulator

We have developed a network simulator to study the benefits of wavelength conversion (including sparse wavelength conversion) in a wavelength-routed optical network. The simulation also assumes that there are enough transceivers at each node such that they never become a limitation.

The simulator is designed to be flexible enough to test all possible aspects of wavelength converters, such as traffic model, (sparse) switch design, arbitrary network topology with arbitrary set of nodes with conversion capabilities, and arbitrary routing and wavelength assignment (RWA) algorithms [InMu99, Ines97].

Our present study will assume Poisson arrivals, exponential holding times, and uniform (symmetric, balanced) traffic; $W = 5$ wavelengths per fiber link; and fixed-path routing (shortest path with respect to hops) for each connection, with one chosen randomly when multiple shortest paths exist.

10.5.3 Single Optical Rings

A study on unidirectional ring networks with "dynamic" traffic was conducted to assess the benefits wavelength converters. A varying number of nodes were given conversion capabilities and these nodes were spread out across the ring as evenly as possible. Each node that was given any converters was given full wavelength-conversion capabilities (i.e., any wavelength entering a node can exit on any free wavelength on any output fiber). However, even with this "full-conversion" capability, our results tend to support earlier predictions based on "static" analysis [RaSa97] that wavelength converters have limited usefulness in a single optical ring (see Fig. 10.7).

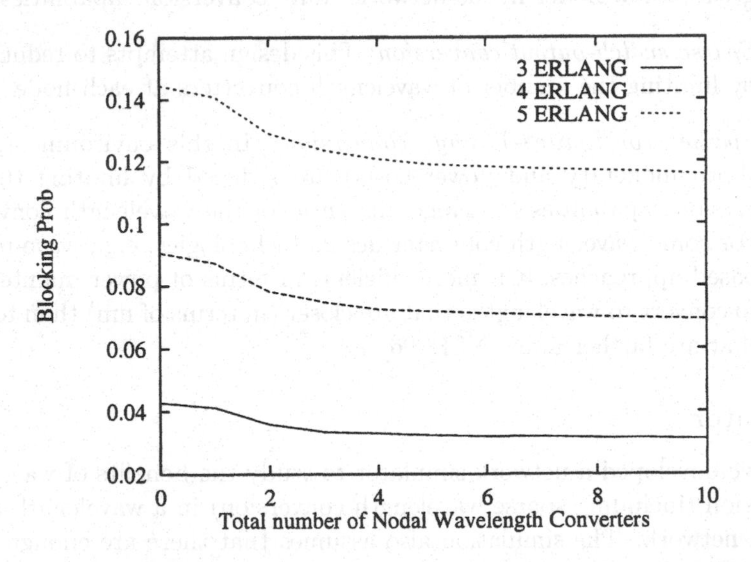

Figure 10.7 Blocking probabilities for different loads in a 10-node optical ring with sparse nodal conversion.

When all connections to be established are known in advance, the work in [RaSa97] showed that *only one node* in the ring having "full" conversion capabilities was enough to satisfy all possible sets of requests that could have been established if there was full conversion capabilities at all nodes. Our dynamic results (*where connections come and go*) in Fig. 10.7 show that one converter is not enough; however, for this 10-node case, more than two or three nodes having full conversion seems to yield only marginal benefits.

10.5.4 NSFNET

Our next network, the backbone of the NSFNET (see Fig. 10.8), allows wavelength converters to perform better than in a ring because it allows more "mixing" of traffic. (Additional results for a network of interconnected rings may be found in [InMu99, Ines97].

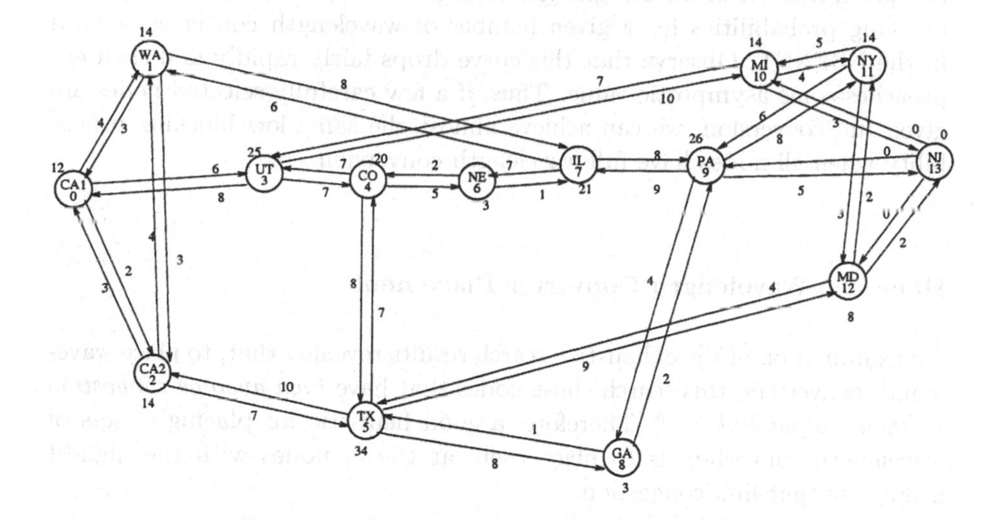

Figure 10.8 NSFNET with the number of convertible routes shown. A number on a link indicates how many source-destination paths passed through the previous node and possibly could have been converted. A number next to a node indicates how many source-destination paths pass through the node and can possibly be wavelength converted.

Sparse Nodal Conversion: Which Nodes Need Full Conversion?

Let us determine how much benefit can be achieved by giving a certain number of nodes in the NSFNET full conversion capabilities. We do this by determining the blocking probabilities, for certain traffic parameters, when one node is allowed full conversion, when two nodes are allowed full conversion, all the way up to allowing all nodes full conversion. The wavelength-assignment strategy used is the First-Fit (FF) algorithm, which orders the wavelengths in some manner, and whenever there is a choice of wavelengths, it chooses the first available wavelength according to its ordering of the wavelengths. This wavelength-assignment strategy has been shown to be fairly efficient (relative to a "Best-Fit (BF) Strategy" [InMu99, MoAz96]) at utilizing the wavelengths and is fairly simple to implement.

Figure 10.9 shows the results of exhaustively searching all combinations of a given number of wavelength-converter placements to determine the best blocking probabilities for a given number of wavelength converters placed in the NSFNET. Observe that this curve drops fairly rapidly and then approaches some asymptotic value. Thus, if a few carefully-selected nodes are given full conversion, we can achieve almost the same low blocking probability when all nodes have full wavelength conversion.

Heuristic Wavelength Converter Placement

An examination of the exhaustive search results revealed that, to place wavelength converters, this search chose nodes that have *high average congestion of their output links.*[2] [3] Therefore, a good heuristic for placing C sets of wavelength converters is to place them at the C nodes with the highest average output link congestion.

Figure 10.9 also shows that the heuristic placement compares extremely favorably with the optimal placement in the NSFNET.

[2]Two other, but less dominant, factors that seemed to influence converter placements were: the distributions of the congestion on the output links of a node, and the average length of paths that pass through a node.

[3]These statements are made with respect to traffic that does not start at that node. Any traffic that is sourced or sinked by a node will never need to be converted by that node.

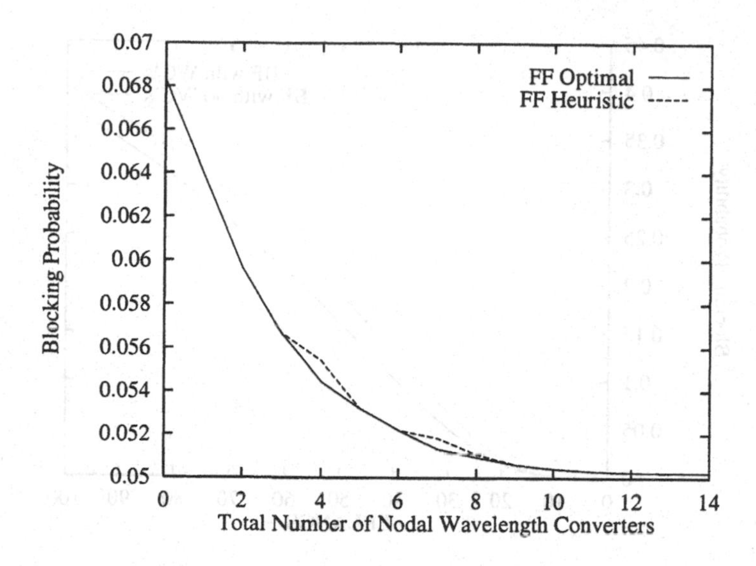

Figure 10.9 Blocking probabilities in the NSFNET for optimal and heuristic placement of wavelength converters (30 ERLANG load).

Traffic-Load Influences on the Benefit of Wavelength Converters

At light load, there is not much need for wavelength conversion since the few connections can find a route to their respective destinations. So, it was a common belief that the benefit of wavelength converters increases with increasing traffic. However, Figs. 10.10 and 10.11 demonstrate that this belief is only partially true.

Figure 10.10 reveals that the difference between the blocking probabilities with and without conversion becomes fairly constant after the traffic load reaches a certain value. This is because, at heavy load, the network is "capacity-limited," i.e., using wavelength converters can only squeeze through a limited number of additional connections beyond what no conversion allows. This also means that, as a percentage of total blocking, the benefit is decreasing. Thus, as can be seen from Fig. 10.11, there is an operating load at which the percentage gain from using wavelength converters is maximized.

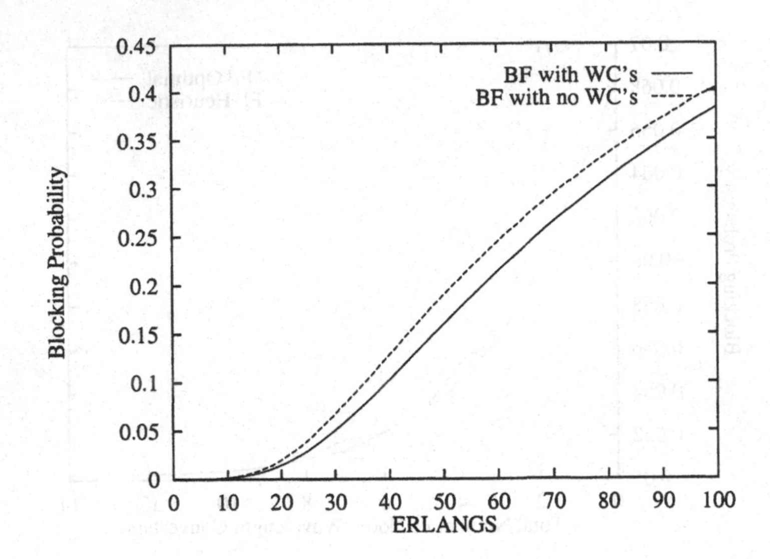

Figure 10.10 Comparison of blocking probabilities in the NSFNET when using full conversion and no conversion in the network with the Best-Fit algorithm.

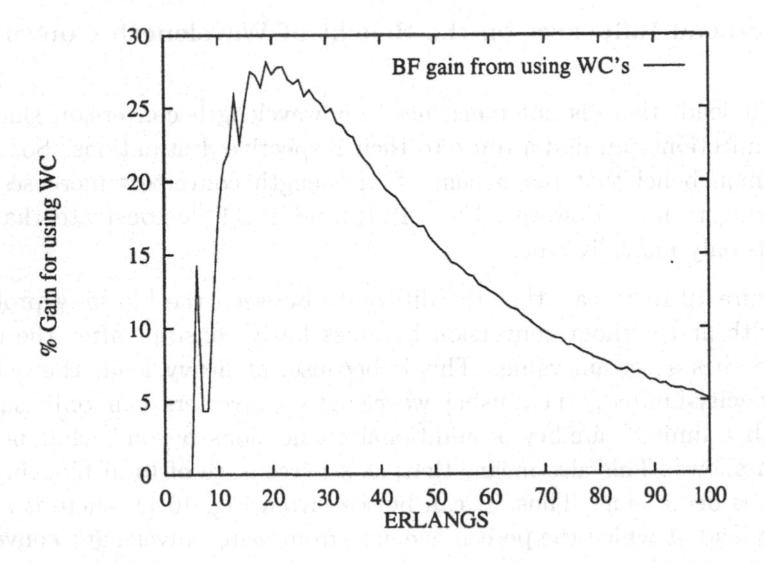

Figure 10.11 Percent gain in the NSFNET from using full-conversion at every node as opposed to no conversion in the network.

Limited Wavelength Converters at each Node

It may be possible to operate a switch with only a few wavelength converters (as in Fig. 10.5) and still achieve most of the benefits of a switch that has full conversion (see Fig. 10.4). Let us now examine how many wavelength converters are actually needed/utilized at a switch.

Let us focus on one of the nodes in NSFNET (node 2), and study how its wavelength converters were utilized (for the same offered traffic as in Fig. 10.9, with all nodes having full conversion). The distribution of node 2's converter utilization is shown in Fig. 10.12. Observe that node 2 spends

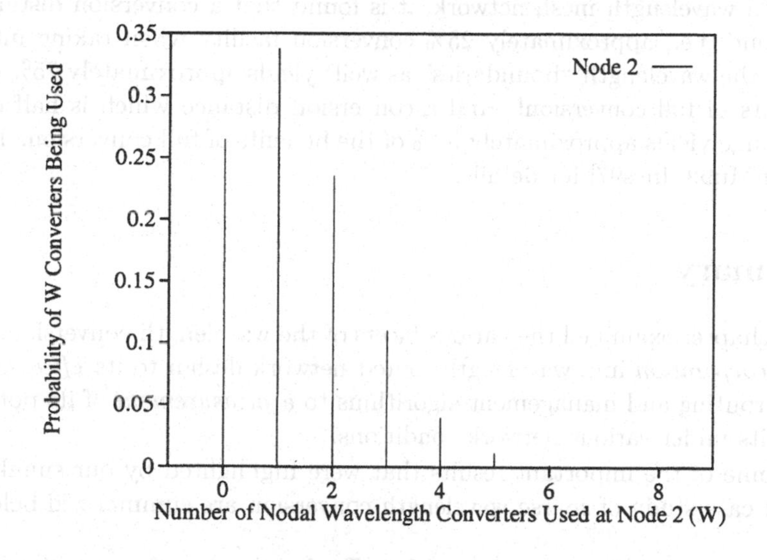

Figure 10.12 Distribution of the number of wavelength converters utilized at node 2 in the NSFNET (30 ERLANG load).

almost 95% of the time utilizing three or fewer wavelength converters. Since there are three output fibers and five wavelengths in the network, if node 2 had "full" conversion, it would have required 15 wavelength converters. Implementing a three-converter version of the switch in Fig. 10.5 (with three output fibers) would be very reasonable at node 2. Similarly, other nodes also utilize very few converters [InMu99, Ines97]. For additional related results, please see [InMu99, Ines97].

Sparse- (or Limited-) Range Wavelength Conversion

This section briefly deals with issues regarding how far (in terms of distance between wavelengths) a wavelength converter needs to convert in order to allow the network to operate efficiently. Previous theoretical studies [YLES96] indicate that most of the reduction in blocking probabilities, from using full-conversion-capable converters, can be achieved by using only limited-conversion-capable converters.

Detailed analytical and simulation models in [InMu99, Ines97], as applied to mesh networks such as the NSFNET (Fig. 10.8) and a network of interconnected rings, corroborate the above observations. Specifically, for a seven-wavelength mesh network, it is found that a conversion distance of only one (i.e., approximately 25% conversion facility when taking into account the wavelength "boundaries" as well) yields approximately 75% of the benefits of full conversion! And a conversion distance which is half of the full range yields approximately 90% of the benefits of full conversion. Please see [InMu99, Ines97] for details.

10.6 Summary

This chapter examined the various facets of the wavelength conversion: from its *incorporation* in a wavelength-routed network design to its *effect* on efficient routing and management algorithms to a *measurement* of its potential benefits under various network conditions.

Some of the important results that were highlighted by our simulation-based case study of sparse wavelength conversion are summarized below:

- A network needs a mixing of traffic for wavelength converters to be beneficial (i.e., single rings benefit little from wavelength converters, while graphs with higher connectivity benefit more).

- A network with sparse wavelength conversion, whether it is sparse nodal conversion or sparse output conversion, can achieve almost the same benefit as a network that has "full" conversion capabilities.

- Simple heuristics can be employed to efficiently place wavelength conversion capabilities.

- Traffic load can influence the benefit of wavelength conversion.

- A shared-output-wavelength-conversion switch appears to be the best switch based on its reduced cost (due to only having sparse-wavelength-conversion) and reasonable flexibility.

Exercises

10.1. What are the different methods in which we can increase the capacity of a WDM optical network? Also, mention the changes in the network that will be required to implement the method.

10.2. Suppose we have an optical network N_1 with one fiber between adjacent nodes in the physical topology and *four* wavelengths per fiber. Network N_1 does not allow wavelength conversion. Now consider another network N_2 with *four* fibers between adjacent nodes in the physical topology and one wavelength per fiber. Let N_3 be a network similar to N_1, but with full wavelength conversion. Assume that connection requests are set up dynamically. Let p_1, p_2, and p_3 be the average blocking probabilities of networks N_1, N_2, and N_3, respectively. What can we say about how p_1, p_2, and p_3 compare with one another?

10.3. For a network with *two* wavelengths, show an example of dynamic connection setup requests which can be satisfied with wavelength conversion and cannot be satisfied otherwise.

10.4. Given an optical network with the facility of recoloring existing lightpaths, show that such a network may block a connection request which could have been satisfied if wavelength conversion was allowed.

10.5. Explain why employing multiple fibers between nodes is better (i.e., results in lower blocking probabilities) than increasing the number of wavelengths?

10.6. Let N_1 and N_2 be two networks with the same physical topology and number of wavelengths per fiber. Wavelength conversion is not allowed in network N_1 while it is allowed in network N_2. Assume that we use the *least-congested path* routing scheme to satisfy dynamic connection requests. Let the blocking probabilities for a sequence of connection requests S be p_1 and p_2, for the networks N_1 and N_2, respectively. Is $p_1 > p_2$ for all S? If yes, prove it. If not, show an example of a network topology and sequence of connections S, such that $p_2 > p_1$.

10.7. Figure 10.11 shows a plot of the percent gain from using full conversion vs. the network load. Explain the local maximum in the plot.

10.8. Consider a node with two input fibers and two output fibers. There are four wavelengths that can be used in the system. For each of the following sets of connections, determine which node architecture – share-per-node, share-per-link, and dedicated converters – can support the connections.

(a) λ_1 from input 1 to output 1
λ_2 from input 1 to output 1
λ_3 from input 1 to output 2
λ_1 from input 2 to output 1
λ_2 from input 2 to output 1
λ_3 from input 2 to output 2
λ_4 from input 2 to output 2

(b) λ_1 from input 1 to output 2
λ_2 from input 1 to output 2
λ_3 from input 1 to output 2
λ_1 from input 2 to output 1
λ_2 from input 2 to output 2
λ_3 from input 2 to output 1
λ_4 from input 2 to output 1

10.9. Consider a path consisting of six links. Each link supports up to four wavelengths, and average link utilization is 0.5. Calculate the blocking probability with and without wavelength conversion. What is the gain for a blocking probability of 0.8?

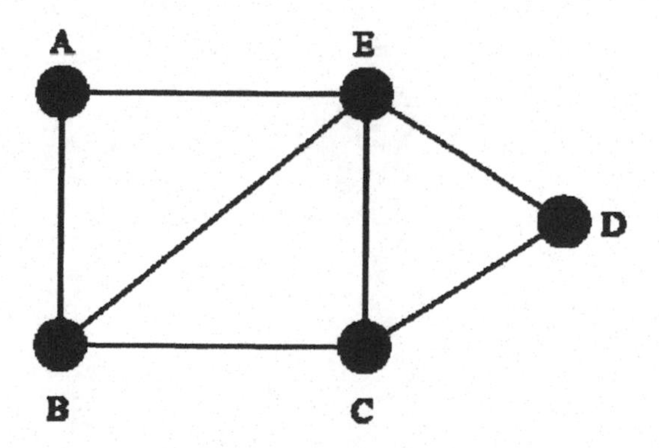

Figure 10.13 Network with uniform loading.

10.10. Given a path consisting of five links, suppose full wavelength conversion is allowed between the second and third links, but not at any other location along the path. There are five wavelengths in the system, and the average link utilization is 0.6. What is the blocking probability?

10.11. Calculate the pass-through traffic for each node in the network shown in Fig. 10.13). Assume uniform loading.

Chapter 11

Survivable WDM Networks

11.1 Introduction

In this chapter, we consider equipment failures which may occur in a network and disrupt traffic. Figure 11.1 shows different types of failures that may occur in an optical WDM network. A "duct" is a bidirectional physical pipe between two nodes. In practice, fibers are put into cables, which are buried into ducts under the ground (or hung on poles in the air). A fiber cut usually occurs due to a duct cut during construction or destructive natural events, such as earthquakes, etc. All the lightpaths that traverse the failed fiber will be disrupted so a fiber cut can lead to tremendous traffic loss. For example, if a fiber supports 160 wavelength channels and each wavelength operates at 10 Gbps (OC-192), a fiber cut can lead to 1.6 Tbps data loss. In addition, noting that fiber is laid in bundles (cables), with each cable carrying as many as 864 fiber strands (or higher), and each duct carrying many bundles (perhaps 10 or higher), a duct cut can lead to huge data (and revenue) loss.

A central office (CO) can also fail where OXCs are located, usually because of catastrophic events such as fire or flooding. This is referred to as node failure. Node failures are rare but the disruption will be very significant if it occurs. Besides node and link failures, another type of failure

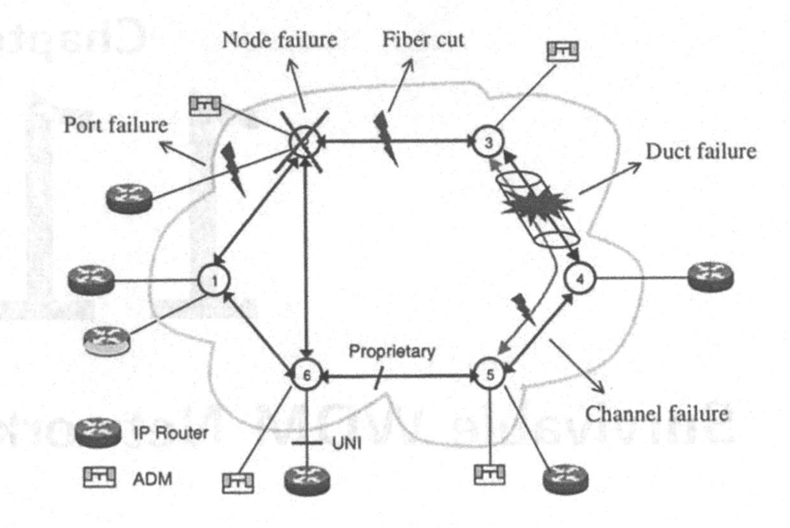

Figure 11.1 Different types of failures in an optical WDM network.

called channel failure is also possible in optical WDM networks. A channel failure is usually caused by the failure of transmitting and/or receiving equipment operating on that channel. Table 11.1 shows some typical data on network component (transmitter, receiver, fiber link (cable), etc.) failure rates and failure-repair times, according to Bellcore (Telcordia) [ToNe94]. In Table 11.1, FIT (failure-in-time) denotes the average number of failures in 10^9 hours, Tx denotes optical transmitter systems, Rx denotes optical receiver systems, and $MTTR$ means mean time to repair.

Table 11.1 Failure rates and repair times (Bellcore Statistics).

Metric	Bellcore Statistics
Equipment MTTR	2 hrs
Cable-Cut MTTR	12 hrs
Cable-Cut Rate	1-5 cuts/yr/1000 km
Tx failure rate	10867 FIT
Rx failure rate	4311 FIT

With the high frequency that a fiber cut may occur and the tremendous traffic loss a failure may cause, network survivability becomes a critical concern in network design and its real-time operation. We need to design effective methods to recover from failures of network links and nodes. An individual channel failure can be handled locally by quickly switching to another idle local channel, or it can be handled as a link failure when no idle channel is available. In addition, the software designed for control and management in today's intelligent network is becoming more and more complex. Software bugs may lead to unstable network states, but this problem is hard to protect against in the network. We rely on proper software design and testing to solve this problem. Besides the failures that may occur in the infrastructure of an optical network, other failures can also occur in other network layers, e.g., the port failure at an IP router, as shown in Fig. 11.1. In general, these failures will be handled by the fault-management schemes in the network layers they belong to. For instance, link failures in an IP network will be handled by IP rerouting.

Most of the research work on survivability in WDM networks focus on the recovery from a single link or node failure, where one failure is repaired before another failure is assumed to occur in the network, since they are the predominant form of failures in optical networks[1]. This is known as the assumption of single failure scenario. Nevertheless, multiple, near-simultaneous failures are also possible in a realistic network (see Fig. 11.2), and appropriate recovery methods can be designed [ZhMu04].

Shared Risk Groups (SRGs) express the risk relationship that associates all the optical channels with a single failure. An SRG may consist of all the optical channels in a single fiber, all of the optical channels through all the fibers wrapped in the same cable, or all of the optical channels traversing the same conduit. Since a fiber may run through several conduits, an optical channel may belong to several SRGs. The provisioning algorithms must exploit SRG maps to discover SRG-diverse routes so that, after any conduit

[1]Single-fiber failures are the dominant failures in communication networks, and they occur mainly due to construction equipment ploughing through buried cable. Node failures are relatively rare because the switch fabric and the switch-control unit in a carrier-class node are typically dedicated (1+1) – i.e., master/slave – protected. A node's port cards, however, are generally not 1+1 protected since they take up the bulk of the space (perhaps over 80%) and cost of a node (switch); also a port-card failure can be handled as link failure(s). Thus, nodes are more robust than links. But node failures can be handled by calculating node-disjoint routes. Node failures are important to protect against in scenarios where an entire node (or a collection of nodes in a network) may be taken down, possibly by a natural disaster or a malicious attacker.

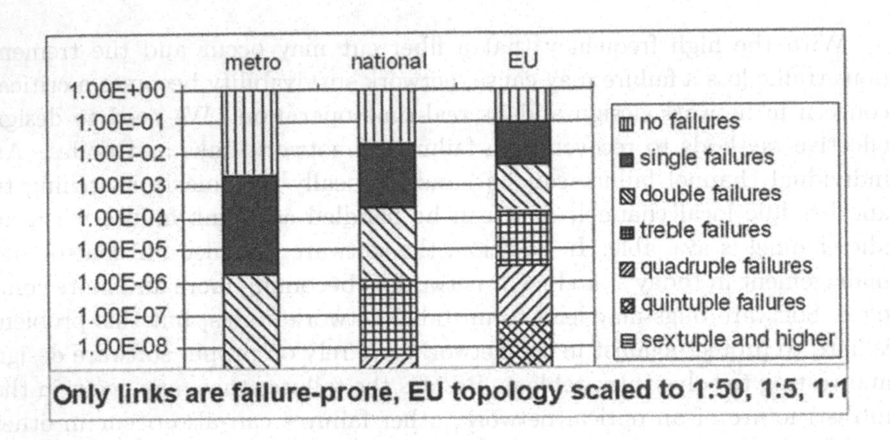

Figure 11.2 Multiple failures: significance (based on [ToNe94]).

is cut, there is always at least one viable route remaining. This constraint is known as the SRG constraint. The SRG concept can be generalized to include a group of nodes and links that are in close proximity so that a large-scale disaster covering a wide geographical region may disrupt all members of the SRG simultaneously. Since link failure is the dominant failure scenario, shared-risk link group (SRLG) is a commonly-used form of SRG.

Survivability can be provided in many layers in the network, e.g., ATM, IP, SONET/SDH, etc. The fault-management schemes in each layer have their own functionalities and characteristics. For instance, in an IP network, the Internet can be viewed as a collection of autonomous systems (ASs), each of which can range in size from a small corporate network to a large backbone network. An AS consists of a set of routers that belong to the same administrative domain. Routers within an AS exchange routing information by employing an interior gateway protocol (IGP). (An example of an IGP employed by routers in an AS is the open shortest path first (OSPF) protocol.) By using an IGP, an AS can combat a link failure, i.e., when a link fails along a primary path between two nodes/routers in the AS, the IGP can dynamically find an alternate path between the two nodes. The routing table at each router within the domain is updated in a distributed manner. In practice, it can take seconds after the failure is detected before the routing tables at all the routers converge and have consistent routing information. During this process, packets may be routed incorrectly since the current versions of the routing tables at the routers may be inconsistent and incorrect. Another problem for fault management in IP network is the long failure-detection time. Usually, adjacent routers exchange "hello" packets periodically. A router declares a link to have failed and initiates rerouting if it misses a certain number of "hello" packets. In general, this recovery procedure may take upto tens of seconds.

In an optical network, whenever there is a loss of signal on an optical link or the bit-error rate (BER) on a link exceeds a threshold, line terminals can detect the failures in milliseconds. And the optical layer can handle some faults more efficiently than the upper (client) layers. For instance, a fiber cut results in the loss of all the traffic streams carried by the fiber. Without optical-layer protection, each traffic stream will be restored independently by the client layers. Thus, the network-management system may be flooded with a large number of messages (failure notification, traffic rerouting, etc.) for this single failure. However, fewer entities need to be rerouted if the optical layer can quickly restore the traffic.

11.2 Terminology

Next, we introduce some terminology used in this chapter.

- *Restoration time (RT)* is measured from the instant a connection goes "down" due to a failure to the instant when the traffic is restored either through a predesigned protection scheme or via a dynamically discovered route. It is the exact disruption-holding time and should be minimized as much as possible.

- *Restoration success rate (RSR)* denotes the ratio between the number of successfully restored traffic streams and the number of disrupted traffic streams after a network failure occurs.

- *Service restorability* is usually a network-wide parameter representing the capability that a network can survive a specific failure scenario.

- *Availability* is defined as the asymptotic probability that a system (a connection in the case of this chapter) will be found in the operating state at a random time in the future. Availability of a system can be computed statistically, based on the failure frequency and failure repair rate of the underlying network components that the system is using, reflecting the percentage of time the system (or, a connection) is "alive" or "up" during its entire service period.

- *Reliability* is the probability that a system will operate without any disruption for a predefined period of time. Service reliability can be represented by the number of "hits" or disruptions in a unit of time. Availability and reliability are different measures of service quality. For example, consider a service (or, a connection) which is disrupted once during the period starting from time T_1 to time T_2. If the disruption holds for (i) 50 ms or (ii) 5 seconds, the availability of the service (or, the connection) will differ in two orders of magnitude for the two cases while the service (or, the connection) reliability is the same (i.e., one failure during period $T_2 - T_1$) for both cases.

11.3 Fault Management in SONET/SDH

SONET and SDH networks can provide high availability to end-to-end connections through the use of extensive protection techniques. In protection schemes, two paths are provided to a connection. The path that is used to carry traffic under normal operation is called the *working path* (or *primary path*) while the second path is called *protection path* (or *backup path*) and it carries traffic in case of failures. If the connection is unidirectional and traffic is transmitted simultaneously on the working and the protection paths, the receiver at the end of the paths simply selects the signal with better quality. However, if the connection is bidirectional, sometimes traffic in both direction needs to be switched from the working path to the protection path even if the failure is only on the working path of one direction. This requires a signaling protocol, called an *automatic protection-switching* (APS) protocol. For example, as shown in Fig. 11.3, a failure affects the working path from A to B without affecting the working path from B to A. The transmitter in A is not aware that there has been a failure. A simple APS protocol works as follows: after the receiver in B detects the failure on the working path, node B turns off its transmitter on the working path and then switches over the traffic to the protection path. Then, the receiver in A detects the signal loss, and switches the traffic to the protection path as well.

If traffic is transmitted in both directions over a single fiber, a fiber cut can be detected by both ends. An APS protocol is not needed in this case but it will still be needed to handle unidirectional equipment failures. APS

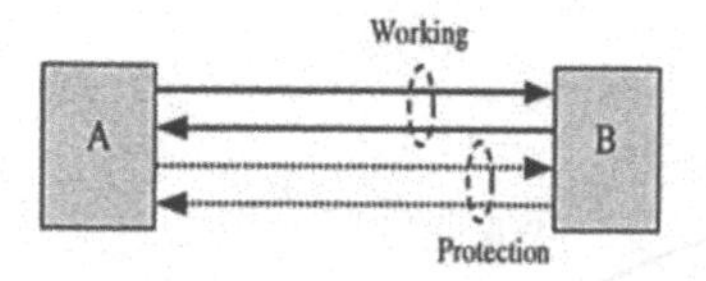

(a) A bidirectional connection.

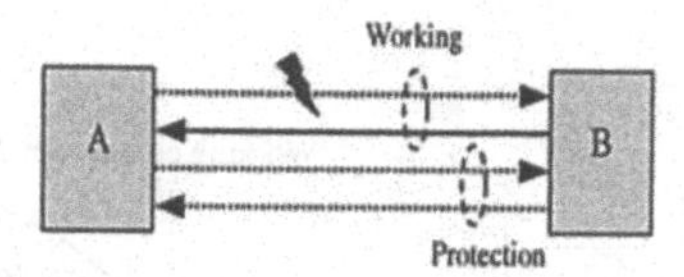

(b) Failure occurs on the working path from A to B.

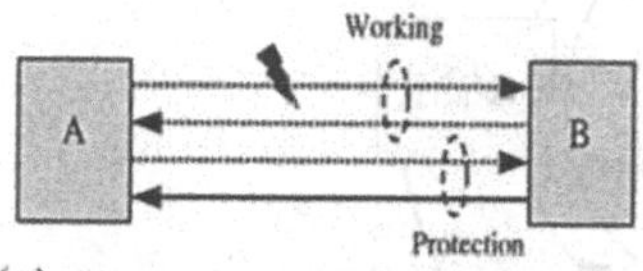

(c) B detects failure and switches traffic.

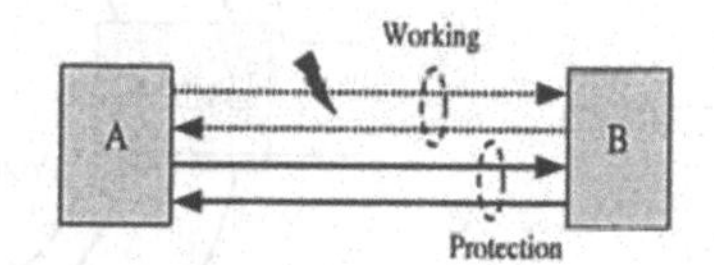

(d) A switches traffic to protection path.

Figure 11.3 An example of automatic protection-switching (APS) protocol.

protocols used in optical networks are far more complex since they need to handle many possible scenarios.

Much of the carrier infrastructure today uses SONET/SDH rings, where multiple nodes are interconnected with a single physical ring. A ring is a 2-connected topology, and provides two disjoint paths between any pair of nodes that do not have any nodes or links in common except the source and destination nodes. These rings are called *self-healing rings* (SHRs) because they can automatically detect failures and quickly reroute traffic away from failed links or nodes by using the other half of the ring.

Three ring architectures have been widely deployed: two-fiber unidirectional path-switched ring (UPSR), two-fiber bidirectional line-switched rings (BLSR/2), and four-fiber bidirectional link-switched rings (BLSR/4). Next, we introduce the architectures and protection schemes in UPSR, BLSR/2, and BLSR/4.

11.3.1 Unidirectional Path-Switched Ring (UPSR)

In a UPSR, the adjacent nodes are connected by two fibers. For example, the fiber in the clockwise direction may be used as the working fiber and the other in the counter-clockwise direction is used as the protection fiber. For each SONET connection, traffic is sent both on the working fiber and on

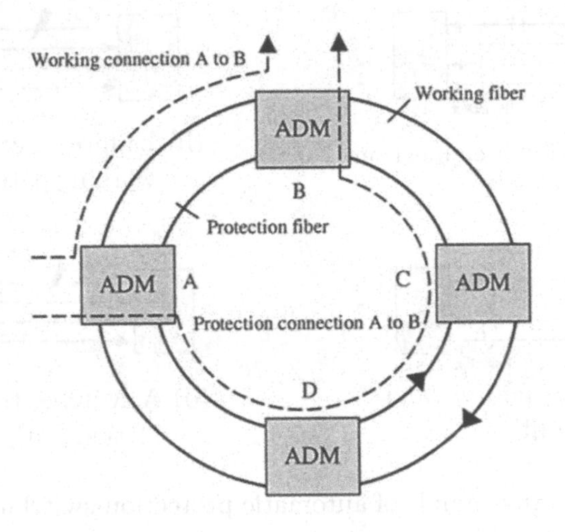

Figure 11.4 A unidirectional path-switched ring (UPSR).

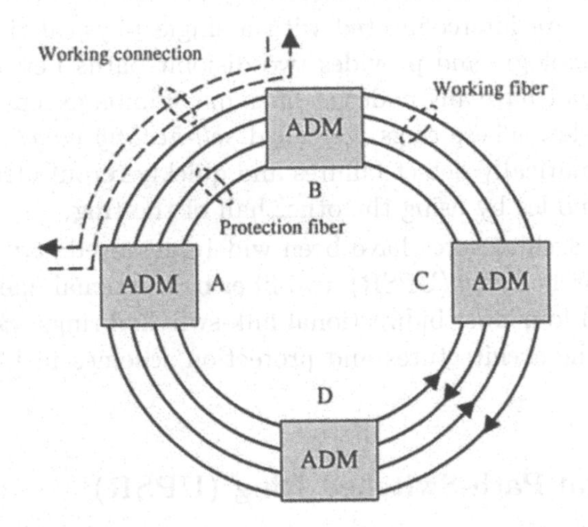

Figure 11.5 A four-fiber bidirectional line-switched ring (BLSR/4).

the protection fiber. The destination monitors both the working and protection fibers and selects the better signal between the two. Suppose that the destination receives traffic from the working fiber under normal operation. If there is a link or node failure on the working path, the destination will switch over to the protection fiber and continue to receive the data. This scheme is referred to as 1+1 (or *dedicated*) *path protection* scheme since it is operating at the end-to-end path layer, and protection resource is dedicated to one working connection at one time. Figure 11.4 shows a UPSR with four nodes.

The protection scheme in a UPSR is easy to implement, requires no communications between the nodes, and provides fast failure recovery. However, this architecture is not capacity efficient since half of the capacity is devoted to protection purposes. For example, if the connection is bidirectional and requires the bandwidth of one wavelength, one wavelength channel on each fiber in the ring will be dedicated to the connection request, either for working or for protection purpose. Thus, there is no sharing of the protection bandwidth between connections. This is the main drawback with the UPSR.

UPSR is a popular topology in low-speed local exchange and access networks due to its simplicity and, thus, low cost. Typical ring speeds today are OC-3 and OC-12. There is no special limit on the number of nodes in a UPSR. But the clockwise (working) and counter-clockwise (protection) paths taken by a signal will have different delays associated with them. The delay difference between the two paths is determined by the ring length and will affect the restoration time in the event of a failure. Thus, in practice, the ring length will be limited by the requirement of the restoration time. The SONET/SDH standards dictate that, in SONET/SDH rings, services must be restored within 60 ms after a failure.

11.3.2 Bidirectional Line-Switched Ring (BLSR)

Figure 11.5 shows a four-fiber BLSR. Two adjacent nodes in the ring are connected by four fibers, two for working traffic and two for protection traffic. Usually, the traffic between two nodes will be routed via the shortest path. For example, the traffic for node A to node B is routed clockwise and the traffic from B to A is routed counter-clockwise. When a working fiber fails, the traffic is routed onto the protection fiber between the two nodes on the same link. This scheme is referred to as *span switching.* If the working fibers and protection fibers between two nodes fail simultaneously, which is the common case since the working fibers and the protection fibers between two

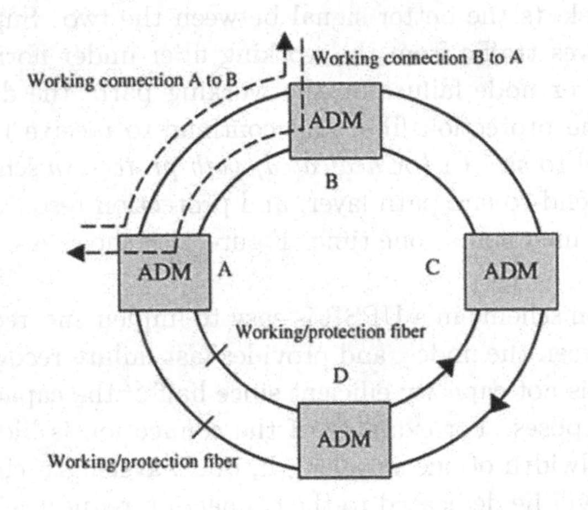

Figure 11.6 A two-fiber bidirectional line-switched ring (BLSR/2).

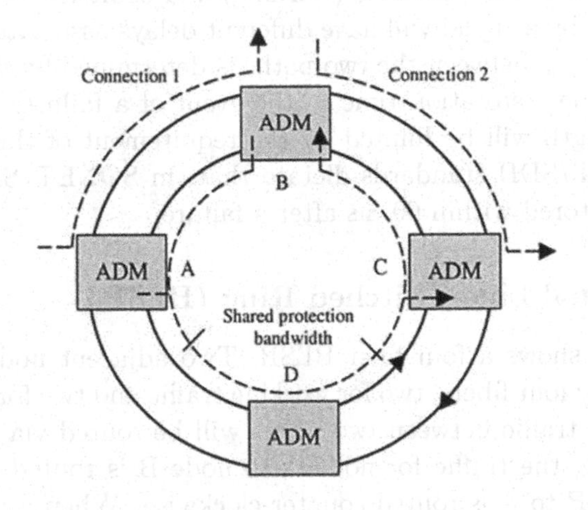

Figure 11.7 Protection bandwidth spatial reuse in a BLSR/2.

nodes are usually routed jointly (because they are part of the same bundle), or if there is a node failure in the network, the traffic between the two nodes is routed around the ring on the protection fiber. This scheme is referred to as *ring switching.*

Figure 11.6 shows a two-fiber BLSR. Two adjacent nodes in the ring are connected by two fibers, one in each direction. In each fiber, half the capacity is reserved for protection purpose. In this case, span switching is not possible but ring switching works in the same way as in a BLSR/4. In the event of a link failure, the traffic on the failed link is rerouted along the other part of the ring using the protection capacity. In a BLSR, the protection capacity is not utilized until the failure occurs. This scheme is called 1:1 protection.

A BLSR allows protection bandwidth to be shared between spatially separated connections. Figure 11.7 shows such a case, where protection capacity from node A to node D (and from node D to node C) on the fiber in the counter-clockwise direction is shared by connections 1 and 2. Furthermore, the protection bandwidth can be used to carry low-priority traffic during normal operation. This traffic is preempted when the bandwidth is needed for protection.

A BLSR can support upto 16 nodes due to the fact that a 4-bit addressing field is used for the node identifier. The ring length is limited by the requirement on the restoration time. When longer rings are needed, e.g., for undersea applications, the 60-ms restoration time needs to be relaxed. BLSRs are widely deployed in long-haul and interoffice networks. Most metro carriers have deployed BLSR/2s, while many long-haul carriers have deployed BLSR/4s. Today, these rings operate at OC-12, OC-48, and OC-192 rates.

BLSRs are more efficient than UPSRs in protecting distributed traffic pattern since the protection capacity in the ring is shared among all the connections. However, BLSRs are more complex than UPSRs in both implement and operation. BLSRs require extensive signaling between the nodes for coordinating multiple protection mechanisms, managing the low-priority traffic, etc. For more details on SONET/SDH rings and their protection characteristics, we refer the reader to [Gora02].

11.4 Fault Management in WDM Mesh Networks

As networks migrate from interconnected rings to meshes because of the poor scalability of interconnected rings and the excessive resource redundancy used in ring-based fault-management schemes, designing and operating a survivable WDM mesh network have been receiving increasing attention [ElHS00, RaSM03, MoMS01, RaBS01, GeRa00c, GeRa00b, LWKD02, ZhSu00, ZaMu01]. Our objective here is to present a broad overview of the fault-management issues involved in designing an optical mesh network employing optical crossconnects (OXCs) and in their real-time network operation, including dynamic connection provisioning. We introduce the basic mechanisms in fault-management schemes and the concept of ring cover and protection cycles. Then, we discuss various challenges in fault-management schemes, e.g., primary (working) and backup (protection) route computation, maximizing sharability for the shared-protection schemes, different considerations in dynamic restoration, etc., and we discuss the appropriate techniques to solve them (also see Chapter 16).

11.4.1 Basic Concepts

There are two types of fault-recovery mechanisms: *protection* and *restoration.* If backup resources (routes and wavelengths) are precomputed and reserved in advance, we call it a *protection* scheme. Otherwise, when a failure occurs, if another route and a free wavelength have to be discovered dynamically for each interrupted connection, we call it a *restoration* scheme.

Generally, dynamic restoration schemes are more efficient in utilizing network capacity because they do not allocate spare capacity in advance, and they provide resilience against different kinds of failures (including multiple failures). But protection schemes have faster recovery time, and they can guarantee recovery from disrupted services they are designed to protect against (a guarantee which restoration schemes cannot provide).

Protection schemes can be classified as ring protection and mesh protection. Ring-protection schemes include Automatic Protection Switching (APS) and Self-Healing Rings (SHR), as discussed in Section 11.3. Both ring protection and mesh protection can be further divided into two groups: *path protection* and *link protection.*

In *path protection*, the traffic is rerouted through a backup route (*backup path* or protection path) once a link failure occurs on its working path (*primary path*). The primary and backup paths for a connection must be link-disjoint so that no single link failure can affect both of these paths. In *link protection*, the traffic is rerouted only around the failed link. While path protection leads to efficient utilization of backup resources and lower end-to-end propagation delay for the recovered route, link protection provides faster protection-switching time. Figure 11.8 shows an example of path protection vs. link protection for the same connection from node 1 to node 4. Essentially, UPSR is path protection implemented in a ring, and BLSR is link protection implemented in a ring.

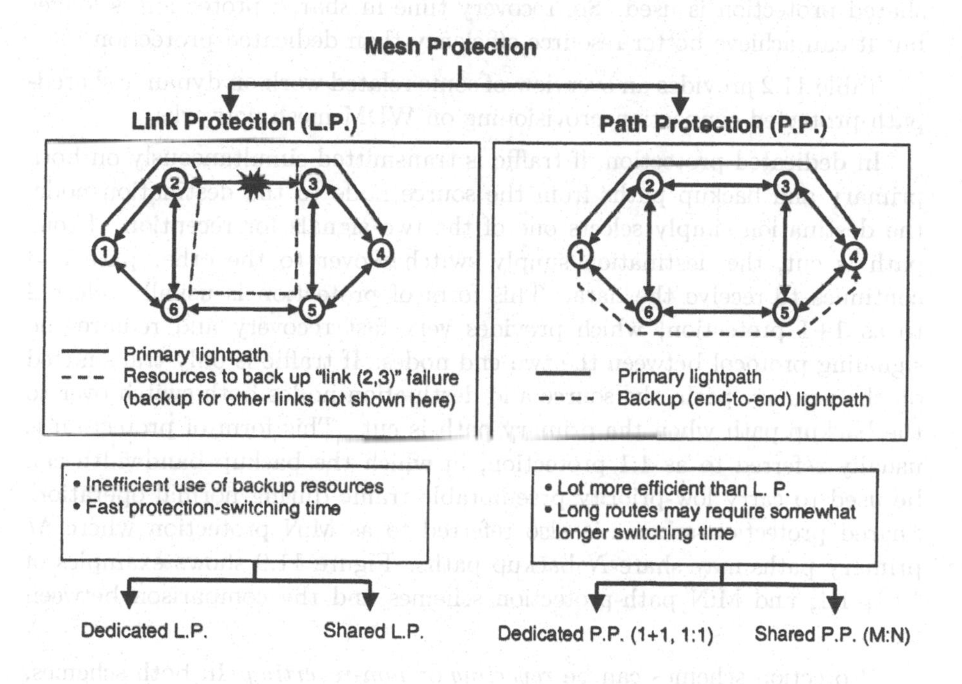

Figure 11.8 Link protection vs. path protection.

Recently, researchers have proposed the idea of *sub-path protection* in a mesh network by dividing a primary path into a sequence of segments and protecting each segment separately (or by dividing the whole network into different domains, and a lightpath segment in one domain must be protected by the resources in the same domain) [OZSM04, Mouf02, AnCQ01]. Compared with path protection, sub-path protection can achieve high scalability and fast recovery time but sacrifices some resource efficiency.

Protection schemes are usually designed to protect against a single link failure since it is the dominant failure scenario in current optical networks. But node failures can also be considered by calculating node-disjoint routes.

Protection schemes can be dedicated or shared. In *dedicated protection*, sharing is not allowed between backup bandwidth, while in *shared protection*, backup bandwidth can be shared on some links as long as their protected segments (links, sub-paths, paths) are mutually diverse or not in the same SRGs. OXCs on backup paths are not configured until the failure occurs if shared protection is used. So, recovery time in shared protection is longer but it can achieve better resource efficiency than dedicated protection.

Table 11.2 provides an overview of some related work on dynamic shared-path-protected connection provisioning on WDM mesh networks.

In dedicated protection, if traffic is transmitted simultaneously on both primary and backup paths from the source node to the destination node, the destination simply selects one of the two signals for reception. If one path is cut, the destination simply switches over to the other path and continues to receive the data. This form of protection is usually referred to as 1+1 protection, which provides very fast recovery and requires no signaling protocol between the two end nodes. If traffic is only transmitted on the primary path, the source and destination nodes both switch over to the backup path when the primary path is cut. This form of protection is usually referred to as 1:1 protection, in which the backup bandwidth can be used to carry low-priority preemptable traffic during normal operation. Shared protection scheme is also referred to as M:N protection where M primary paths may share N backup paths. Figure 11.9 shows examples of 1+1, 1:1, and M:N path-protection schemes and the comparison between them.

Protection schemes can be *reverting* or *non-reverting.* In both schemes, if a failure occurs, traffic is switched from the primary path to the backup path. In reverting, the traffic is switched back to its primary path after the failure on the primary path is repaired. In non-reverting, the traffic stays on the backup path for the remaining service time. Reverting allows the network to return to its original state once the failure is restored. Dedicated protection schemes can be either reverting or non-reverting but reverting may be applied for a shared protection scheme. Since multiple connections are sharing the common backup bandwidth, the backup bandwidth must be freed up as soon as possible after the original failure has been repaired, so that it can be used to protect other connections when another failure occurs.

Table 11.2 Comparison of related work on dynamic shared-path-protected lightpath/connection provisioning on WDM/MPLS mesh networks.

Research Work	Objective	P/L	C/D	Info.	WC	D/P	Contributions (in brief)
Bouillet et al. [BLER02, BLRC02]	Minimize the total cost of working and backup paths for each lightpath	P	C	A	Y	P & D	Stochastic approaches; cost model; K-shortest path routing.
Elie-Dit-Cosaque et al. [ElAT02]		P	D	F, P	Y	D	Protection-sharing table.
Mohan et al. [MoMS01]		P	C	A	N	D	Primary-backup sharing; Cost model for route computation.
Ramamurthy et al. [RaBS01]		P	D	F	Y	D	Performance comparison of different schemes.
Su et al. [SuSu01a, SuSu01b]		L	D	F	Y	D	Bucket-based link metric; ILP & two-step heuristic.
Xin et al. [XYDQ01]		P	C	F	Y	D	K-shortest path routing.
Xiong et al. [XiXQ03]		P	C & D	F & P	Y	D	ILP formulations.
Our work		P	C	A	Y	D	NP-complete proof; heuristic for optimization; heuristic for finding a feasible solution.
Kodialam et al. [KoLa00, KoLa01]	Minimize the total cost of working and backup paths for each connection	P & L	C & D	F, N, P	Y	D	ILPs for different scenarios & a heuristic based on primal-dual and LP-relaxation.
Li et al. [LWKD02]		P	C	F	Y	D	Two-step heuristic using a bucket-like link metric; distributed signaling.
Liu et al. [LiTS01]		P	D	A	Y	D	Aggregating per-flow information with a matrix; successively updating existing backups.
Qiao et al. [QiXu02]		P	D	P	Y	D	ILP & two-step heuristic; distributed signaling.

P/L = Path/Link: Path protection or link protection; C/D = Centralized/Distributed: Whether the algorithm is centralized or distributed;
Info.: The amount of information needed; A: aggregated lightpath/connection information; P: partial information about existing lightpaths/connections;
F: full per-lightpath/connection information; N: no information about existing lightpaths/connections; D/P = Deterministic/Probabilistic;
WC = Wavelength Conversion: Whether the work applies to wavelength-continuous network or wavelength-convertible network.
Please note that the basic ideas of the work in [KoLa00, KoLa01, LiTS01, LWKD02, QiXu02] (which is devoted to MPLS networks) are applicable to WDM networks.

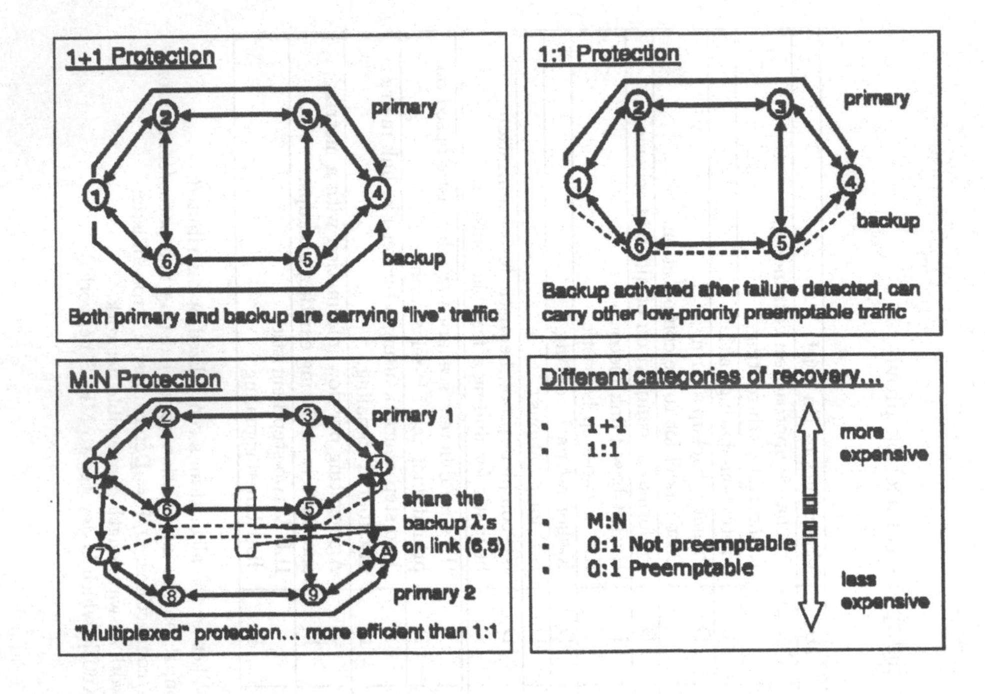

Figure 11.9 Examples of 1+1, 1:1, and M:N path-protection schemes and the comparison between them.

Reverting, however, will cause an additional "hit" on the data flow, which may not be appreciated by network operators for some specific applications.

Dynamic restoration can also be classified as link, sub-path, or path based, depending on the type of rerouting. In *link restoration*, the end nodes of the failed link dynamically discover a route around the link, for each connection (or "live" wavelength) that traverses the link. In *path restoration*, when a link fails, the source and the destination node of each connection that traverses the failed link are informed about the failure (possibly via messages from the nodes adjacent to the failed link). The source and destination nodes of each connection independently discover a backup route on an end-to-end basis. In *sub-path restoration*, when a link fails, the upstream node of the failed link detects the failure and discovers a backup route from itself to the corresponding destination node for each disrupted connection. Link restoration is fastest and path restoration is slowest among the above three schemes. Sub-path restoration time lies in between. Figure 11.10 summarizes the classification of protection and restoration schemes.

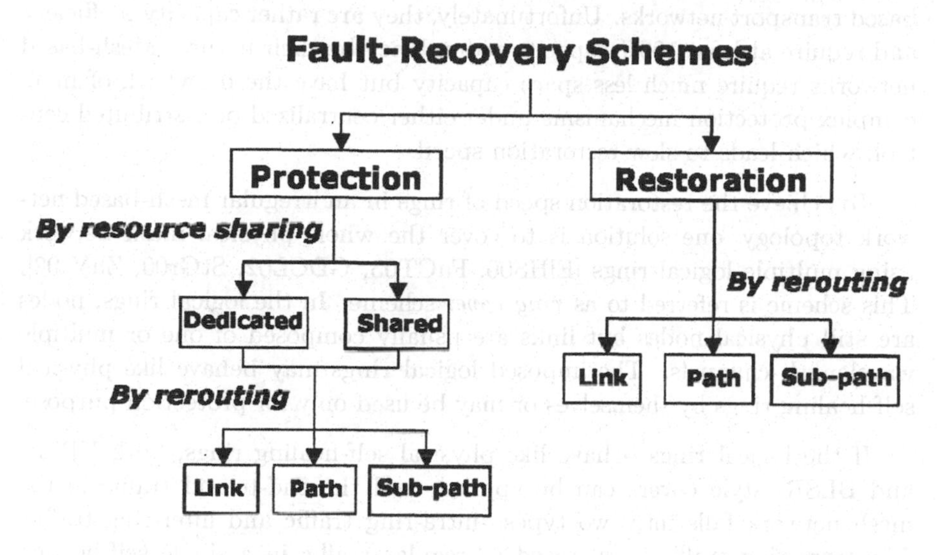

Figure 11.10 Different protection and restoration schemes in WDM mesh networks.

As we have discussed in Section 11.3, the ring network topology provides fast protection schemes, and it is easy to construct and operate. However, today's backbone networks are more densely connected than a ring, and traffic is becoming more and more distributed. For such networks, mesh protection schemes offer more bandwidth-efficient solution than rings. It has been reported that mesh protection schemes can achieve 20% - 60% efficiency improvements relative to ring protection schemes [OZSM04].

Next, we discuss various issues emerging from designing a survivable WDM mesh network and appropriate techniques to solve them. Network traffic can be static, dynamic, or incremental; and these techniques can be applied to different provisioning scenarios according to different network characteristics.

11.4.2 Ring Cover, Stacked Rings, and Protection Cycles

Optical networks based on a single physical ring are favored for their fast restoration since rings use a simple switching mechanism which permits restoration in about 50-60 ms. For example, self-healing ring (SHR) techniques have already been successfully applied to digital cross connect (DXC) based transport networks. Unfortunately, they are rather capacity-inefficient and require at least 100% capacity redundancy by their nature. Mesh-based networks require much less spare capacity but have the drawback of more complex protection mechanisms under either centralized or distributed control, which leads to slow restoration speed.

To achieve the restoration speed of rings in an irregular mesh-based network topology, one solution is to cover the whole physical mesh network using multiple logical rings [ElHS00, FuCT03, GDCL02, StGr00, ZhYa02]. This scheme is referred to as *ring cover* scheme. In the logical rings, nodes are still physical nodes but links are usually composed of one or multiple wavelength channels. The imposed logical rings may behave like physical self-healing rings by themselves or may be used only for protection purpose.

If the logical rings behave like physical self-healing rings, both UPSR- and BLSR- style covers can be applied, and the end-to-end traffic in the mesh network falls into two types: intra-ring traffic and inter-ring traffic. The intra-ring traffic is operated as regular traffic in a single self-healing ring but the inter-ring traffic is more expensive since extra switching and add/drop operations are needed at the node crossed over by two or more rings. The inter-ring traffic is segmented by the rings it traverses, and each segment is protected by the bandwidth in the same ring. Figure 11.11 shows a 10-node network covered by five logical BLSRs. Node A and node B are not covered by the same ring so the traffic from node A to node B needs to traverse multiple rings. In this example, traffic from node A to node B can be routed via rings 5 and 2. It can also take other routes through other rings, e.g., using ring 5 and 4. This type of interconnected logical rings is usually called *stacked rings.* In the horizontal dimension, adjacent rings provide extended geographic coverage compared to a single physical ring. In the vertical dimension, the rings are stacked on top of each other on some links, thereby providing increased network capacity for the area covered by the ring stack. This vertical stacking is often achieved through the use of WDM, wherein one fiber can carry multiple, stacked rings by using different wavelengths to transmit the data. Note, for example, that by using WDM, four wavelength on a fiber may be used to simulate the effect of a link in a BLSR/4 network. It is easy to see that 100% capacity redundancy is required for protection purpose in stacked rings since each ring is operated as a self-healing ring by itself.

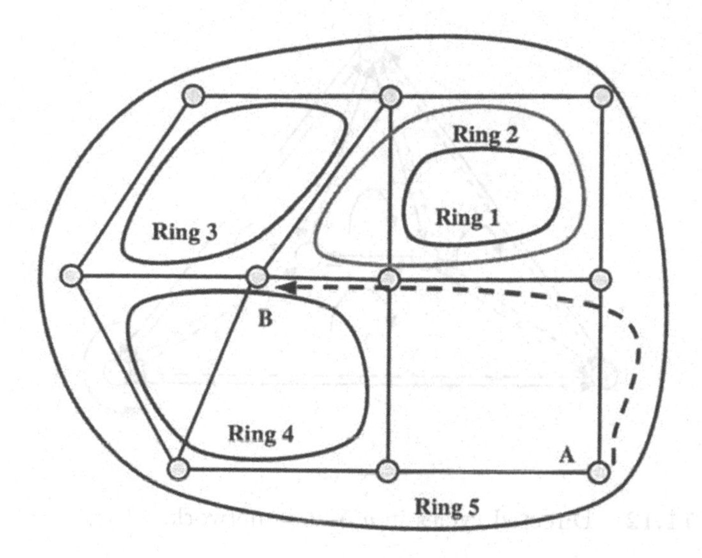

Figure 11.11 A 10-node network covered by five logical BLSRs.

If the logical rings are only used for protection purpose, we need to design a set of directed cycles which cover all the links in the network. Therefore, if any link goes "down", we can have a cycle to recover the traffic on the failed link, which can achieve automatic protection switching (APS) in a mesh network. Figure 11.12 shows such a case. The networks considered here have a pair of unidirectional working fibers (constituting a bidirectional working link) and a pair of unidirectional protection fibers (constituting a bidirectional protection ink) in each fiber link. So, half of the capacity is devoted for protection purpose.

We can also design a set of cycles which cover all the links that need to be protected. This usually applies in a static network design problem where a set of traffic demands are given a priori. The objective here is to minimize the total (working and protection) bandwidth required in a given network topology to support a given set of traffic demands. This problem can be decomposed into three subproblems:

- for every traffic demand, route the working lightpath in the mesh topology;

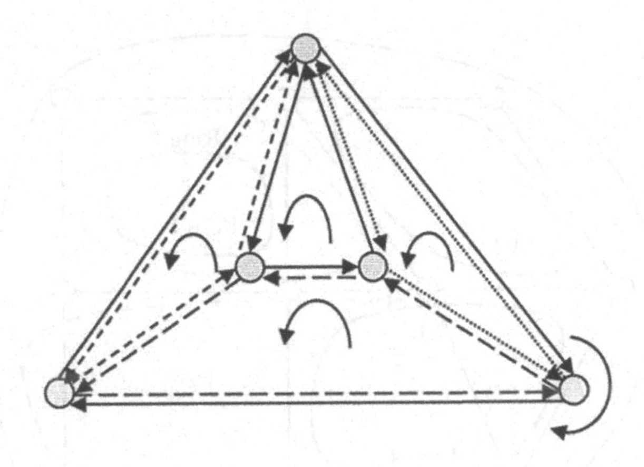

Figure 11.12 Directed cycles in a 5-node network.

- for every link carrying at least one working lightpath, identify the ring(s) covering the link and protecting the traffic; and

- for every ring in the cover, provision the spare bandwidths that are necessary to protect the working lightpath.

There are also some design constraints, e.g., the size of each ring (number of nodes) is bounded in order to achieve fast restoration [ZaOM03]. ILP approaches can be applied to solve this problem.

Protection cycle (*p*-cycle) has been investigated in recent years as a variation of ring-cover techniques [Grov03]. In a *p*-cycle, not only these on-cycle links could be protected by each other in a self-healing-ring manner, but the chordal (or straddling links) could also be protected. A straddling link is an off-cycle link having *p*-cycle nodes as end-points. In the case of a straddling-link failure, two node-disjoint paths around the *p*-cycle running between the two end nodes can be used to protect the failed link, as shown in Fig. 11.13. A *p*-cycle can be more capacity-efficient than a BLSR or UPSR since the on-cycle capacity is used to protect both on-cycle link failures and the failures of straddling links.

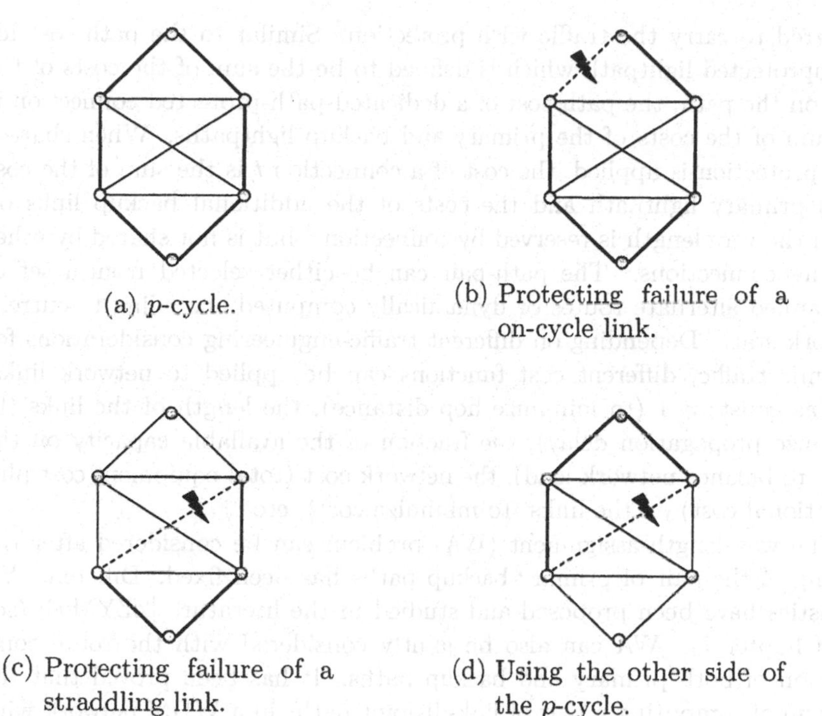

(a) p-cycle.

(b) Protecting failure of a on-cycle link.

(c) Protecting failure of a straddling link.

(d) Using the other side of the p-cycle.

Figure 11.13 A network with one p-cycle which can protect on-cycle links and straddling links.

11.4.3 Survivable Routing and Wavelength Assignment (S-RWA)

In a wavelength-continuous[2] WDM mesh network employing end-to-end path protection, the problem of finding a link-disjoint primary-backup path pair and assigning a proper wavelength channel to each path is known as the survivable routing and wavelength assignment (S-RWA) problem and has been extensively studied.

Usually, a path-pair with least cost from a source to a destination is

[2]If wavelength converters are equipped in OXCs, a lightpath can be assigned to different wavelengths on the links it traverses. Such a network is known as a *wavelength-convertible* network. If wavelength converters are not equipped in OXCs, we require that, when establishing a lightpath, the same wavelength be allocated on all the links in the path. This requirement is known as the *wavelength-continuity constraint*, and such a network is known as a *wavelength-continuous* network. This topic was discussed in Chapter 10.

preferred to carry the traffic with protection. Similar to the path cost for an unprotected lightpath which is defined to be the sum of the costs of the links on the path, the path cost of a dedicated-path-protected connection is the sum of the costs of the primary and backup lightpaths. When shared-path protection is applied, the cost of a connection t is the sum of the cost of t's primary lightpath and the costs of the additional backup links on which the wavelength is reserved by connection t but is not shared by other existing connections. The path-pair can be either selected from a set of preplanned alternate routes or dynamically computed according to current network state. Depending on different traffic-engineering considerations for dynamic traffic, different cost functions can be applied to network links, such as constant 1 (to minimize hop distance), the length of the links (to minimize propagation delay), the fraction of the available capacity on the links (to balance network load), the network cost (total equipment cost plus operational cost) on the links (to minimize cost), etc.

The wavelength-assignment (WA) problem can be considered after the routing of the pair of primary-backup paths has been fixed. Different WA heuristics have been proposed and studied in the literature [ZZYM03] (see also Chapter 7). WA can also be jointly considered with the route computation of both primary and backup paths. It has been proved that the problem of computing a pair of link-disjoint paths in a WDM network with wavelength-continuity constraint is NP-complete [ZZYM03]. When a network has full wavelength-conversion capability, such a problem is reduced to an optimal routing problem for a link-disjoint path pair, which can be solved using existing algorithms. Next, we describe two algorithms to compute a link-disjoint path pair.

Algorithms For Computing Link-Disjoint Paths

One algorithm to compute link-disjoint routes, which is referred to as the *two-step algorithm*, works as follows. In the first step, the primary path is computed using a shortest-path algorithm. In the second step, we compute the backup path by first removing the edges along the primary path, and then finding the shortest path in the reduced graph. The two-step algorithm has some potential weaknesses in that its computation is sequential and it does not allow backtracking to re-compute the primary path if necessary. As a result, it may not always find a pair of link-disjoint paths between the source and the destination nodes even if such paths exist. Also, the paths found by the two-step algorithm may not be optimal, i.e., the total cost of the two paths may not be minimal.

Figure 11.14 shows an example where the two-step algorithm fails to compute a pair of link-disjoint paths from node 0 to node 5, while such a pair of paths actually exists. (Numbers next to links in Fig. 11.14 represent link weights.) In the first step, route (0, 1, 4, 5) is computed to be the (least-cost) primary path (cost = 3), shown by dashed lines in Fig. 11.14(a). After removing the links along the primary path, node 0 and node 5 become disconnected, as shown in Fig. 11.14(b). Hence, the two-step algorithm fails. However, a pair of link-disjoint paths does exist, namely route (0, 1, 2, 5) and route (0, 3, 4, 5), as shown in Fig. 11.14(c).

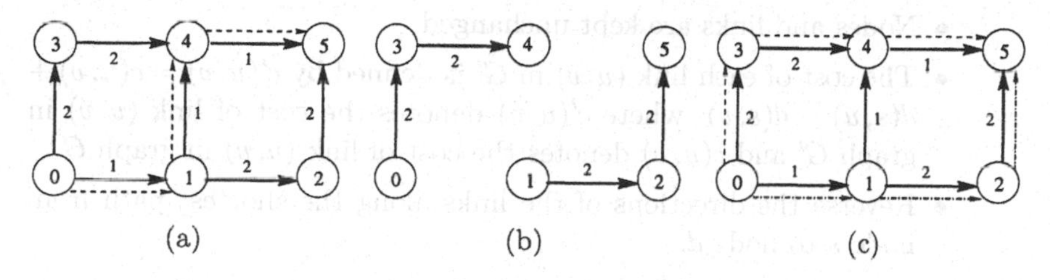

Figure 11.14 An example where the two-step algorithm fails.

A polynomial time ($O(N^2)$, where N = number of nodes) algorithm for simultaneously computing two link-disjoint paths between a node pair has been proposed [SuTa84], and it is referred to as the *one-step algorithm.* (This algorithm is referred to as Suurballe's algorithm in the literature, and a slight variation which achieves the same result is referred to as the Bhandari algorithm in [Bhan99].) The total cost of the two paths computed by this algorithm has been proved to be minimum, and we refer to this path pair as the Optimal Path Pair (OPP). The one-step algorithm is briefly described in Algorithm 11.1.

The authors in [SuTa84] have proved that the link-disjoint path pair found by this algorithm has the minimum cost.

Figure 11.15 gives an example to describe how the one-step algorithm works, in which a pair of link-disjoint paths from node 0 to node 5 needs to be computed. The example is the same as in Fig. 11.14 which showed how the two-step algorithm failed. Numbers next to links in Fig. 11.15(a) represent link costs and the number next to a node u represents the shortest-path distance between the nodes 0 and u (i.e., $d(0, u)$). The dashed line gives the shortest-path tree rooted at node 0, from which we can determine

Algorithm 11.1 One-Step Algorithm

We are given graph G, source node s, and destination node d. The OPP between nodes s and d is calculated as follows.

1. Compute the shortest-path tree rooted at node s using Dijkstra's shortest-path algorithm. Let $d(s,u)$ denote the shortest-path distance from node s to node u.
2. Transform the original graph G to an auxiliary graph G' as follows.
   - Nodes and links are kept unchanged.
   - The cost of each link (u,v) in G' is defined by $c'(u,v) = c(u,v) + d(s,u) - d(s,v)$, where $c'(u,v)$ denotes the cost of link (u,v) in graph G' and $c(u,v)$ denotes the cost of link (u,v) in graph G.
   - Reverse the directions of the links along the shortest path from node s to node d.
3. Compute the shortest path from node s to node d in graph G'.
4. The shortest path between nodes s and d in G (G') is denoted as T (T'). After removing the links appearing in both T and T' (in opposite direction), all the other links in T and T' form a cycle when ignoring their directions. Two link-disjoint paths between nodes s and d can be found from the cycle.

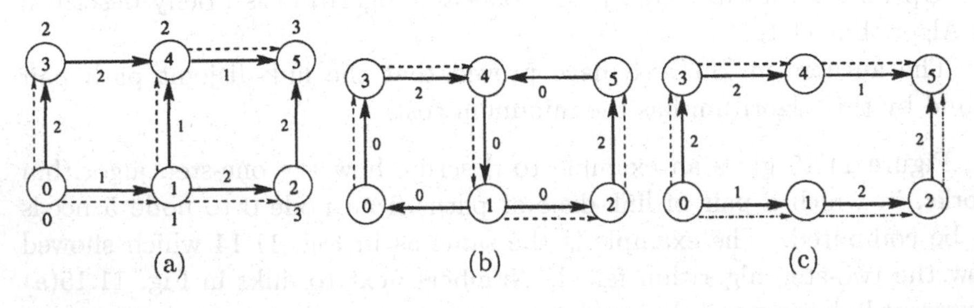

Figure 11.15 An example describing how the one-step algorithm works.

that the shortest path from node 0 to node 5 is $T = (0,1,4,5)$. After graph transformation, the shortest path between nodes 0 and 5 is $T' = (0,3,4,1,2,5)$, as shown in Fig. 11.15(b). After removing the interlacing link $(1,4)$, Fig. 11.15(c) shows the two link-disjoint paths $(0,3,4,5)$ and $(0,1,2,5)$.

Bhandari in [Bhan99] outlines a similar algorithm except that, in the graph transformation, he simply reverses the directions of the links along the shortest path from the source to the destination node in graph G, and negates the cost of those links. Then, he develops a modified Dijkstra's algorithm, by which the shortest-path problem can be solved even when a graph has some links with negative cost. The presence of negative links, however, slows down the shortest-path algorithm.

The one-step algorithm is composed of several runs of a standard shortest-path algorithm and a graph transformation procedure. The computational complexities of a standard shortest-path algorithm and the graph transformation procedure are $O(N^2)$, where N = number of nodes, and $O(E)$, where E = number of links, respectively. Hence, the overall computational complexity for the one-step algorithm is still $O(N^2)$, which is the same as that for the two-step algorithm.

ILP of S-RWA For Static Traffic Demands

Beside different S-RWA heuristics, linear program (LP) based approaches are also used to attack the problem. The LP approach can be used to precompute a set of candidate routes, to compute a pair of primary-backup paths according to current network state in an on-demand manner, or to optimize the total amount of bandwidth used on all the links for a given set of static traffic demands in an off-line manner. Although an LP-based scheme is not very scalable because it is computation intensive, such an approach can provide valuable insights for designing efficient heuristic algorithms. Different approximation schemes have been proposed for LP-based approaches which make them suitable for use in a practical network with a reasonable volume of traffic demands.

In this chapter, we provide ILP formulations of path-protection schemes to protect against single-link failures. We assume that the network topology and a demand matrix (consisting of the number of connections to be established between each node-pair) are given. We assume that the set of alternate routes[3] (that are used to satisfy any demand) between each node-pair can

[3]In alternate routing, each network node has a routing table which contains a list of a

be pre-computed or is given. Our objective is to minimize the total number of wavelengths used on all the links in the network (for both the primary paths and backup paths). The ILP solution also determines the routing and wavelength assignment (RWA) of the primary and backup paths. Generally, capacity efficiency can be measured in two ways (which serve as "dual" problems to each other): (a) given a certain capacity, maximize the protected carried demand, or (b) given a certain demand set, and given a 100% restoration requirement, minimize the total capacity used. In our formulations, we require that all demands should be protected, and we minimize the total capacity used. ILPs 1 and 2 minimize the capacity utilizations for dedicated-path protection and shared-path protection, respectively.

Notation

We define the notation employed to formulate the ILPs. We are given the following: (a) the network topology, represented as a directed graph G, (b) a demand matrix, i.e., the number of lightpath requests between node-pairs, and (c) alternate routing tables at each node. Also given are the following.

- N: Nodes in the network (numbered 1 through N). (Node-pairs are numbered 1 through $N \times (N-1)$.)
- E: Links in the network (numbered 1 through E).
- W: Number of wavelengths on a link.
- R^i: Set of alternate routes for node-pair i.
- $M^i = |R^i|$: Number of alternate routes between node-pair i. Let M be the maximum number of alternate routes between any node-pair, i.e., $M = \max_i M^i$.
- R^i_j: Set of eligible alternate routes between node-pair i after link j fails.
- *end-nodes(j)*: Set of alternate routes between the node-pair adjacent to link j.
- d_i: Demand for node-pair i, in terms of number of connection requests. (Each connection requires the bandwidth of a full wavelength channel.)

limited number of fixed routes to each destination node. The list of routes can be based on one or more constraints, such as shortest path, shared-risk link groups (SRLGs), etc. A connection request arriving at a node utilizes one of the routes to the destination node from the list of available routes.

We require the ILPs to solve for the following variables:

- w_j: Number of wavelengths used by primary lightpaths on link j.
- s_j: Number of *spare* wavelengths used on link j.
- $\gamma_w^{i,r}$ takes on the value of 1 if the rth route between node-pair i utilizes wavelength w before any link failure; 0 otherwise. These variables are employed in both ILPs.
- $\alpha_{w,p}^{i,b}$ takes on the value of 1 if the dedicated backup route b on wavelength w is employed for protecting a primary route p between node-pair i; 0 otherwise. These variables are employed only in ILP1.
- $\delta_{w,p}^{i,b}$ takes on the value of 1 if the shared backup route b on wavelength w is employed for protecting a primary route p between node-pair i; 0 otherwise. These variables are employed only in ILP2.
- m_w^j takes on the value of 1 if wavelength w is utilized by some back-up route r that traverses link j; 0 otherwise. These variables are employed only in ILP2.

ILP Formulations

1. ILP1: Dedicated-Path Protection

   Minimize the total capacity used:

$$Minimize \sum_{j=1}^{E}(w_j + s_j) \tag{11.1}$$

   Number of lightpaths on each link is bounded:

$$(w_j + s_j) \leq W \qquad 1 \leq j \leq E \tag{11.2}$$

   Demand between each node-pair i is satisfied:

$$d^i = \sum_{r=1}^{M_i} \sum_{w=1}^{W} \gamma_w^{i,r} \qquad 1 \leq i \leq N(N-1) \tag{11.3}$$

Number of primary lightpaths traversing link j:

$$w_j = \sum_{i=1}^{N(N-1)} \sum_{r \in R^i, j \in r} \sum_{w=1}^{W} \gamma_w^{i,r} \qquad 1 \le j \le E \tag{11.4}$$

Number of spare channels utilized for link j:

$$s_j = \sum_{i=1}^{N(N-1)} \sum_{b \in R^i, j \in b} \sum_{p \in R^i, p \ne b} \sum_{w=1}^{W} \alpha_{w,p}^{i,b} \qquad 1 \le j \le E \tag{11.5}$$

Wavelength-continuity constraint, i.e., only one primary or backup lightpath can use wavelength w on link j:

$$\sum_{i=1}^{N(N-1)} \sum_{r \in R^i : j \in r} \gamma_w^{i,r} + \sum_{i=1}^{N(N-1)} \sum_{b \in R^i : j \in b} \sum_{p \in R^i, p \ne b} \alpha_{w,b}^{i,p} \le 1 \tag{11.6}$$

for $1 \le w \le W, 1 \le j \le E$.

Due to a link failure, if route p fails between node-pair i, then the demand between node-pair i should still be satisfied:

$$\sum_{w=1}^{W} \gamma_w^{i,p} = \sum_{b \in R^i, b \ne p} \sum_{w=1}^{W} \alpha_{w,p}^{i,b} \qquad p \in R^i, 1 \le i \le N(N-1) \tag{11.7}$$

2. ILP2: Shared-Path Protection

Minimize the total capacity used:

$$Minimize \sum_{j=1}^{E} (w_j + s_j) \tag{11.8}$$

Number of channels on each link is bounded:

$$w_j + s_j \le W \qquad 1 \le j \le E \tag{11.9}$$

Demand between each node-pair is satisfied:

$$\sum_{r=1}^{M_i} \sum_{w=1}^{W} \gamma_w^{i,r} = d^i \qquad 1 \le i \le N(N-1) \tag{11.10}$$

Definition of the number of primary lightpaths traversing a link:

$$w_j = \sum_{i=1}^{N(N-1)} \sum_{r \in R^i, j \in r} \sum_{w=1}^{W} \gamma_w^{i,r} \qquad 1 \le j \le E \tag{11.11}$$

Definition of the spare capacity required on link k:

$$s_k = \sum_{w=1}^{W} m_w^k \qquad 1 \le k \le E \tag{11.12}$$

Constraints to indicate whether wavelength w is reserved for some restoration path on link k:

$$m_k^w \le \sum_{i=1}^{N(N-1)} \sum_{p,b \in R^i, k \in b} \delta_{w,p}^{i,b} \qquad 1 \le k \le E, 1 \le w \le W$$

$$N(N-1) \times E \times M \times m_k^w \ge \sum_{i=1}^{N(N-1)} \sum_{p,b \in R^i, k \in b} \delta_{w,p}^{i,b}$$

for $1 \le k \le E, 1 \le w \le W$.

Wavelength-continuity constraint, i.e., only one primary or backup lightpath can use wavelength w on link j:

$$\left(\sum_{i=1}^{N(N-1)} \sum_{r \in R^i : j \in r} \gamma_w^{i,r} \right) + m_w^j \le 1 \qquad 1 \le j \le E, 1 \le w \le W \tag{11.13}$$

Constraints to ensure that two backup lightpaths can share wavelength w on link k only if the corresponding primary paths are fiber disjoint:

$$\sum_{i=1}^{N(N-1)} \sum_{p \in R^i : f \in p} \sum_{b \in R^i : k \in b} \delta_{w,p}^{i,b} \le 1$$

for $1 \le f \le E, 1 \le k \le E, 1 \le w \le W$.

Constraints to ensure that every primary lightpath is protected by a back-up lightpath:

$$\sum_{w=1}^{W} \gamma_w^{i,p} = \sum_{b \in R^i, b \ne p} \sum_{w=1}^{W} \delta_{w,p}^{i,b}$$

for $1 \le i \le N(N-1), \forall p \in R^i, 1 \le w \le W$.

11.4.4 Maximizing Sharability For Shared-Protection Schemes

One of the key advantages of WDM mesh networks vs. SONET-based interconnected-rings networks is that WDM mesh networks are capable to support differentiated protection schemes and can be more efficient than those in SONET ring networks. Particularly, through path-based shared protection, optical WDM mesh networks may only require 40%-60% extra capacity to protect against any single failure in the network, compared with 100% spare-capacity requirement in SONET-ring-based protection schemes. In a shared-protection scheme, network resources along the backup path can be shared between different connections, as long as only one connection will switch its traffic from the primary path to the backup path when a network failure occurs. There are several investigations on how to maximize resource sharability for the shared-protection scheme in WDM mesh networks in order to optimize network resource efficiency [MoMS01, LWKD02]. It is generally assumed that: (a) link failure is the dominant network failure scenario, and (b) there is at most a single link failure at any time since the multiple-failure scenario is a relatively rare event in the network. Recently, approaches that can handle multiple link failures are beginning to appear in the literature (see [ZaOM03, SoZM05, ZhZM05]). Following approaches and considerations have been investigated for maximizing resource sharability with or without the wavelength-continuity constraint.

Backup Route Optimization

One way to achieve high resource sharability is to spread the primary path of different connections as much as possible, and simultaneously plan their backup paths such that they will share the same resources extensively. This joint optimization is a very hard problem, and we lack effective approaches. An alternative approach is to fix the primary path according to current network state (e.g., minimal-cost route) while optimizing the backup route for a connection request. This can be realized by adjusting the link costs based on current resource usage information of network links. For instance, let $l_j(p, b_1, b_2, ..., b_i, ..., b_N, b)$ denote the resource usage information of link j (i.e., link vector) in a fully wavelength-convertible network, where p denotes the number of wavelength channels allocated for the primary paths (connections) on link j; b_i denotes the number of wavelength channels allocated on link j to protect failure of link i ($1 \leq i \leq N$, and N is the number of links in the network), i.e., when link i fails, there are b_i connections which are originally supported by link i and will be reverted to link j; and b denotes the total number of allocated spare wavelength channels for protection purposes. Under the assumptions that there is a single link failure at any time and shared-path protection is employed in the network, b is equal to the maximal value of b_i $(1 \leq i \leq N)$. When the primary path of a connection traverses links $l_{m1}, l_{m2}, ..., l_{mn}$, the link cost of l_j can be adjust to:

$$Cost(l_j) = \begin{cases} \infty & \text{if } l_j \text{ is on the primary path,} \\ 0 & \text{if } b_{m1} < b, \text{ ..., and } b_{mn} < b, \\ 1 & \text{otherwise.} \end{cases} \quad (11.14)$$

Note that, using this link-cost adjustment function, the link cost is set to 0 if no new wavelength channel needs to be allocated; otherwise, link cost is set as 1.[4] After each network link has been assigned a proper link cost, the backup path can be computed using any shortest-path algorithm such as Dijkstra's algorithm.

Figure 11.16 illustrates an example on how to compute a backup path which can optimize resource sharability when the primary path is known. Figure 11.16(a) shows the state of a part of a network, where only a few network nodes and links are shown. There is a connection request between node pair (s, d), whose primary path traverses links l_1, l_2, and l_3 as the solid lines shown in Fig. 11.16(a). It is straightforward to see that the candidate backup path for the connection request either traverse links l_4 and l_5, or link l_6. (Note that each backup link in Fig. 11.16(a) can be a collection of links. To simplify the example, we just use one or two links on each candidate backup path.) Assuming that there are enough wavelength channels on each link, Fig. 11.16(b) shows the link vector for each network link. Based on these resource-usage information and the cost function shown in Eqn. (11.14), Fig. 11.16(c) shows the network state with adjusted link cost from which a resource-sharability-optimized backup route can be calculated for the given primary path using a standard shortest-path algorithm. Finally,

[4]We assume that minimizing the number of wavelength links used in the network is the optimization objective. An additional refinement to the link cost can be used, e.g., $\epsilon = 10^{-5}$ for each connection sharing a link for shared-path protection [ZhZM05, OZZM04], where ϵ is a small positive number and it is used to avoid loops and unnecessarily long paths when a shortest path is computed using the cost function. This refinement can enable us to control the number of backup paths sharing a link under shared-path protection.

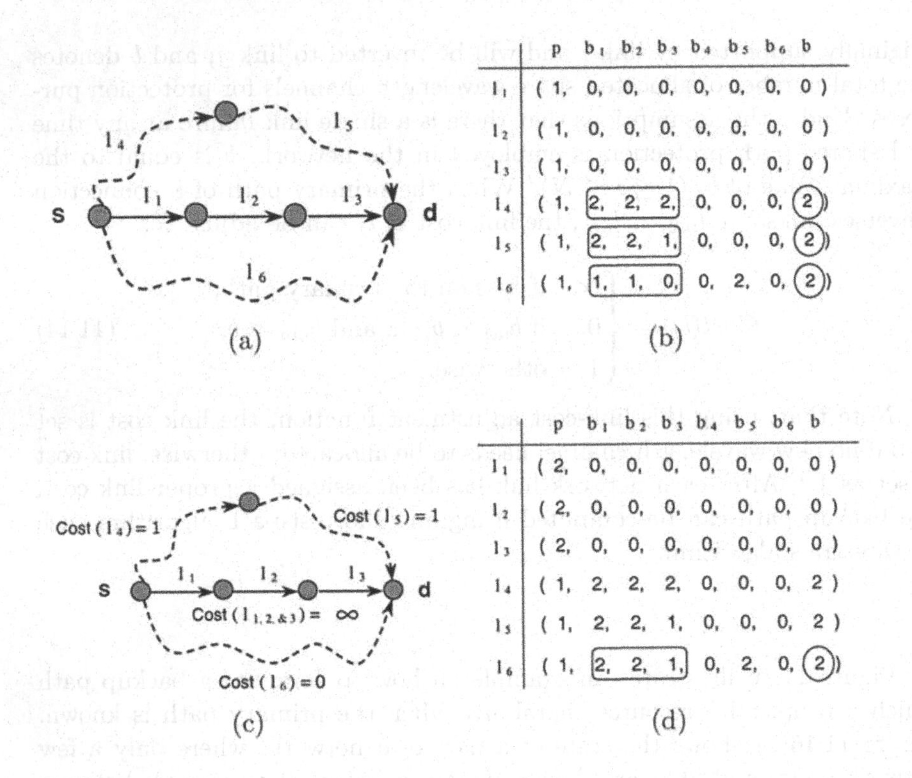

s
d
l_1
l_2
l_3
l_4
l_5
l_6
(a)
p b1 b2 b3 b4 b5 b6 b
l_1 (1, 0, 0, 0, 0, 0, 0, 0)
l_2 (1, 0, 0, 0, 0, 0, 0, 0)
l_3 (1, 0, 0, 0, 0, 0, 0, 0)
l_4 (1, 2, 2, 2, 0, 0, 0, 2)
l_5 (1, 2, 2, 1, 0, 0, 0, 2)
l_6 (1, 1, 1, 0, 0, 2, 0, 2)
(b)
Cost (l_4) = 1
Cost (l_5) = 1
Cost ($l_{1,2,\&3}$) = ∞
Cost (l_6) = 0
(c)
p b1 b2 b3 b4 b5 b6 b
l_1 (2, 0, 0, 0, 0, 0, 0, 0)
l_2 (2, 0, 0, 0, 0, 0, 0, 0)
l_3 (2, 0, 0, 0, 0, 0, 0, 0)
l_4 (1, 2, 2, 2, 0, 0, 0, 2)
l_5 (1, 2, 2, 1, 0, 0, 0, 2)
l_6 (1, 2, 2, 1, 0, 2, 0, 2)
(d)

Figure 11.16 An example of backup-sharing optimization.

Fig. 11.16(d) illustrates the updated link vectors after both primary and backup paths have been fixed for the connection request.

Such a backup-resource-optimization technique can be applied to different provisioning scenarios based on different network characteristics. Network traffic can be static, dynamic, or incremental. It may also be combined with different S-RWA schemes.

Physical Constraint on Backup-Route Optimization

Although the backup-route-optimization technique can greatly improve resource efficiency, one problem may arise. When this approach is extensively used, some connection may have a backup path traversing long (hop) distances even though the primary path is short. Long backup path may lead to a signal-quality degradation problem, especially in an all-optical WDM network. In such a network, transmission and switching impairments can accumulate along the lightpath, and they may affect the signal quality at the destination node. As a result, after an optical signal travels a long distance, the bit-error rate (BER) at the destination node may not be tolerable for services at upper network layers. Therefore, when a network failure occurs, even through a preplanned backup path can be used to restore the affected connections, the unexpected long backup paths can potentially degrade signal quality or even fail to restore the connections.

Recently, different research groups have started to investigate such a problem. The authors in [QiXX02] proposed an ILP-based model to jointly compute the shared-protected primary-backup path pair for dynamic traffic. The model takes both the network-resource usage and the backup-path distance into consideration. The idea of such a model is to incorporate one additional cost component μ to the link cost, such that the link cost reflects both the extra resources the backup path may use and the distance it may traverse.

Failure-Independent Backup Routing (FIBR) vs. Failure-Dependent Backup Routing (FDBR)

Another possible approach to improve backup-resource sharability is failure-dependent backup routing (FDBR). In such a scheme, one backup route can be computed according to a certain network failure on the primary path. That is, if the primary path traverses m links, there may exist m backup paths, one for each failure of the m links. In a failure-independent backup routing (FIBR) scheme, one single backup path will be used independent of the failed link. It is easy to see that FIBR is a special case for FDBR in the sense that the m backup paths are the same. The m backup paths in FDBR may share resources with other backup paths or even between themselves. The resources along the primary path may also be reused by the m backup paths. In this way, FDBR may further improve resource sharability among the backup paths and eventually increase overall network resource efficiency.

11.4.5 Dynamic Restoration

Beside protection schemes, traffic restoration schemes have also been an important area of research interest. The following performance metrics have been used to evaluate a restoration scheme:

- *Restoration success rate (RSR)* denotes the ratio between the number of successfully restored connections and the number of affected connections after a network failure occurs.
- *Restoration time (RT)* denotes the average time needed to successfully restore an disrupted connection request.

There are different considerations which are under study for restoration schemes in optical WDM mesh networks.

1. *Distributed Control vs. Centralized Control*: In a distributed control system, the source node of each interrupted connection can restore the service following either a precomputed route or a dynamically-computed route. Since the connections are restored in a distributed manner, it is possible that resource contention may occur at some network link. Although such contentions can be resolved through restoration retries, they may affect the RSR and RT performance. In a centralized control system, connections will be restored one by one so resource contention is avoided but this scheme may affect the RT performance of some connections. Compared with distributed control, a centralized-controlled restoration scheme may achieve better RSR since it can perform global optimization of network resource usage.
2. *Preplanned Restoration Routes vs. On-line Dynamically-Computed Restoration Routes*: In a distributed control system, the restoration routes can be preplanned or dynamically computed. In a preplanned scheme, a candidate restoration route set can be precomputed for each connection. When a connection fails, one route can be selected as the restoration path without on-line computation. This scheme may improve the RT performance. The route set may be periodically updated according to different network states in order to improve the probability of successful restoration.
3. *Path vs. Sub-path vs. Link Restoration*: As we mentioned before, restoration schemes can be classified into path-based, sub-path-based, and link-based schemes, according to which a node initializes the alternate path and how the new path is routed to bypass the failed link. The work in [WaSM02] has compared the performance tradeoff of these different restoration mechanisms under a distributed control and signaling system.

4. *IP Restoration vs. WDM Protection*: It has been well accepted that IP traffic is the dominant traffic in today's Internet. The IP-over-WDM network architecture has gained significant attention and has been widely studied. In such a network architecture, different network layers may employ different fault-management schemes. For example, it may not be cost-effective to employ all fault-management schemes at every network layer. The authors in [SaRM02, FuVa00] have investigated the tradeoffs of different fault-management schemes at different layers. It is reported that a network may have better performance if restoration schemes are employed at the IP layer and protection schemes are used at the optical WDM layer. There is a growing interest on this research topic, and more in-depth studies are needed to design and develop such IP-over-WDM networks.

11.5 Advanced Topics in Network Survivability

As our knowledge on resource management in survivable network design and its real-time operation keeps maturing, more and more researchers are shifting their attention to a service perspective [FPRT02, FTUF02, Hac94, ASLC02, WeMY02, Grov99]. Naturally, how to provide a certain quality of service (QoS) per a customer's requirement and how to guarantee the service quality become critical concerns. A WDM mesh network may provide services for IP network backbones, ATM network backbones, leased lines, virtual private networks (VPNs), etc. The QoS requirements for these services can be very different because of their diverse characteristics, e.g., on-line trading, military applications, and banking services will require stringent reliability while IP best-effort packet-delivery service may be satisfied without a special constraint on reliability. Service quality can be measured in many different ways such as signal quality, service availability, service reliability, restoration time, service restorability, etc. Signal quality is mainly represented by the optical signal-to-noise ratio (OSNR), bit-error rate (BER), etc., and it is affected by the transmission equipment characteristics. This is a problem in all-optical networks, and it will be addressed in Chapter 15.

11.5.1 Service Availability

It is clear that a protection scheme will help to improve a connection's availability since traffic on the failed primary segment (link/path/sub-path) will be quickly switched to the backup segment. For example, a path-protected connection will have 100% availability in the presence of any single failure if the contribution of the reconfiguration time from primary path to backup path towards unavailability is disregarded (since it is relatively small (usually on the order of milliseconds) with respect to the failure repair time (usually on the order of hours) and the connection's holding time (usually on the order of weeks or months)).

Nevertheless, when the more realistic scenario of multiple, near-simultaneous failures is considered, a path-protected connection may become unavailable in some failure scenarios, e.g., when two concurrent failures occur, one on the backup path and the other on the primary path. Therefore, when considering multiple failures, connection availability depends intimately on the precise details of the failures (locations, repair times, etc.), how much backup resources are reserved (i.e., single backup route or multiple backup routes), and how the backup resources are allocated (i.e., dedicated or shared). Intuitively, the more backup resources (paths) there are, the higher is the connection availability, while more backup sharing leads to lower connection availability. Therefore, instead of simply stating that a connection has been protected, we need to quantitatively evaluate how well the connection is protected, i.e., we need to have a relatively accurate estimation of the connection availability.

As discussed above, a customer of an optical network operator may lease some bandwidth (STS-1, STS-3, etc.) with certain service-quality requirements. Availability is one of them, which is usually defined in a Service-Level Agreement (SLA). The SLA is a contract between the network operator and a customer. (Normally, the customer pays for the services provided by the network operator.) A SLA violation may cause a certain amount of penalty to be paid by the network operator, according to the contract, e.g., providing free services for one additional month. Although over-provisioning may help a network operator to avoid such a penalty, extra resource (or cost) consumption will be introduced, which may not be necessary if the connection is provisioned properly. Thus, a cost-effective, availability-aware, connection-provisioning scheme is very desirable such that, for each customer's service request (static or dynamic), a proper protection scheme (dedicated, shared, or unprotected) is designed and the degree of sharing is consciously controlled (in shared-protection case) so that the SLA-defined availability requirement can be guaranteed and high overall resource efficiency can be achieved. Through such a scheme, differentiated protection services are also inherently provided in optical WDM mesh networks.

A network component's availability is a relatively static value since it is based on the component's failure rate and average time to fix a failure. One may notice that a connection's availability is also a static value as long as the routes of the connection's primary and backup paths have been fixed and there is no resource correlation between two connections (which means no resource sharing). Hence, some candidate routes can be predesigned, and the availability of each of them can be calculated. Then, in on-line provisioning, one of the routes will be picked to carry a new connection as long as its availability is larger than that required by the customer. This strategy can also be applied in off-line network design with a given set of traffic demands which need to be set up concurrently. We can formulate the problem into an ILP and solve it for moderate problem sizes.

However, connection availability becomes a dynamic value when the connection (t) is sharing some wavelengths on its backup path with others. Let S_t contain all the connections that share some backup wavelength on some link with t. We denote S_t as the sharing group of t. Each time a new connection joins S_t, the availabilities of all the connections in the group will be affected. Meanwhile, there are various backup-sharing-related operational decisions which also affect connection availability. For example, if there are multiple failures in the network which, unfortunately, affect more than one connection in S_t, some questions will arise such as which connection will be chosen to restore and how to deal with other failed connections. Usually, connection t's traffic can be switched back to its primary path after the failure on the primary path is repaired, which is called *reverting*; or the traffic can stay on the backup path for the remaining service time, which is called *non-reverting*. A network operator may choose their desired policies based on the operational cost and the service characteristics. Each of these policies will have an effect on the availability analysis.

Availability Analysis in WDM Mesh Networks

We present an availability analysis for a connection with different protection schemes (could be unprotected, dedicated-path protected, or shared-path protected) in a WDM mesh network. The availability of a system (which could be a component, path, connection, etc.) in a mesh network is analyzed with the following typical assumptions:

1. a system is either available (functional) or unavailable (experiencing failure);

2. different network components fail independently; and
3. for any component, the "up" time (or Time To Failure) and the "down" time (or Time To Repair) are independent exponentially distributed stochastic variables with known mean values. Thus, we assume that Mean Time To Failure (MTTF) and Mean Time To Repair (MTTR) are given.

The availability of a system is the fraction of time the system is "up" during the entire service time. If a connection t is carried by a single path, its availability (denoted by A_t) is equal to the path availability; if t is dedicated or shared protected, A_t will be determined by both the primary and the backup paths. Here, the contribution of the reconfiguration time for switching traffic from the primary path to the backup path (including signal propagation delay of control signals, processing time of control messages, and switching time at each node) towards unavailability is disregarded since it is relatively small, usually on the order of milliseconds, compared to the failure-repair time (usually on the order of hours) and the connection's holding time (usually on the order of weeks or months).

- **Assessing Network-Component Availability**

  A network component's availability can be estimated based on its failure characteristics. Upon the failure of a component, it is repaired and restored to be "as good as new". This procedure is known as an alternating renewal process. Consequently, the availability of a network component j (denoted as a_j) can be calculated as follows [Triv82]:

$$a_j = \frac{MTTF}{MTTF + MTTR} \tag{11.15}$$

  In particular, the MTTF of a fiber link is distance related and can be derived according to measured fiber-cut statistics. Table 11.1 shows some typical data on failure rates and failure-repair times of network components (transmitters, receivers, fiber links, etc.) according to Bellcore (now Telcordia).

- **End-to-End Path Availability**

  Given the route of a path i, the availability of i (denoted as A_i) can be calculated based on the known availabilities of the network components along the route. Path i is available only when all the network components along its route are available. Let a_j denote the availability of network component j. Let G_i denote the set of network components used by path i. Then, A_i can be computed as follows:

$$A_i = \prod_{j \in G_i} a_j \tag{11.16}$$

- **Availability of a Dedicated-Path-Protected Connection**
  In path protection, connection t is carried by one primary path p and protected by one backup path b that is link disjoint[5] with p. If b's backup wavelengths are dedicated to connection t, then, when primary path p fails, traffic will be switched to backup path b (or b may already be carrying the same traffic, and the receiver will switch its reception from path p to path b) as long as b is available; otherwise, the connection becomes unavailable until the failed component is replaced or restored. t is "down" only when both paths p and b are unavailable, so A_t can be computed straightforwardly as follows:

$$A_t = 1 - (1 - A_p) \times (1 - A_b) = A_p + (1 - A_p) \times A_b \tag{11.17}$$

  where A_p and A_b denote the availabilities of paths p and b, respectively. Note that a connection may employ multiple backup paths to increase its availability. Assuming that all backup paths are disjoint, the availability of a connection with multiple backup paths can be derived following the similar principle in Eqn. (11.17).
  In sub-path protection [OZSM04, AnCQ01, GuPM03], the primary path is usually divided into non-overlapping segments, and each segment is protected by a link-disjoint backup route. Then, the availability of each dedicated-protected segment can be computed according to Eqn. (11.17). The overall connection can be treated as a serial system composed of protected segments. Hence, the connection availability will be equal to the product of the availability of each protected segment. Link protection is a special case of sub-path protection where each segment is a link.

- **Availability of a Shared-Path-Protected Connection**

[5] By link disjoint, we mean that the backup path for a connection has no links in common with the primary path for that connection. Node failures can also be accommodated by making the primary and the backup paths node disjoint as well.

Issues Affecting Availability in Backup Sharing

In this section, we describe various issues or policies in backup sharing that will affect the availability of a shared-path-protected connection.

- **Reverting vs. Non-reverting**

  In shared-path protection, connection t is carried by primary path p, protected by a link-disjoint backup path b, and the reserved wavelength on each link of b can be shared by other connections as long as SRLG constraint can be satisfied. Let S_t contain all the connections that share some backup wavelength on some link with t. We denote S_t as the sharing group of t. Connection t's traffic will be switched to b when a failure occurs on p. After the failure is repaired, connection t's traffic can be switched back to p, an approach which is called *reverting*; or it can stay on b for the remaining service time (or till b fails), an approach which is called *non-reverting*. Both the reverting and non-reverting strategies have their pros and cons. For example, traffic may be disturbed twice in the reverting strategy, which may be undesirable for some services such as on-line trading. In the non-reverting strategy, the backup paths for the connections in S_t may need to be rearranged since some of the shared backup wavelengths on parts of their backup paths have been taken by t when t is switched to its backup path. These connections become vulnerable during their backup-recomputation and backup-resource-reservation processes; and, furthermore, their successful backup rearrangement is not guaranteed; so, non-reverting may result in unpreferred service degradation. A network operator may choose the desired policies based on the operational cost and the service characteristics. The reverting model is preferable when the operator wants simplicity in network control and management. We assume a reverting model in our analysis.

- **Active Recovery vs. Lazy Recovery**

  In the reverting model, after traffic is reverted back to p, the released backup resources on b can be actively used to recover the other connections in S_t that are experiencing failure and waiting for their backup resources to be released. We call this mechanism *active recovery*. On the contrary, if the backup resources wait to be activated when the next failure arrives, these currently-failed

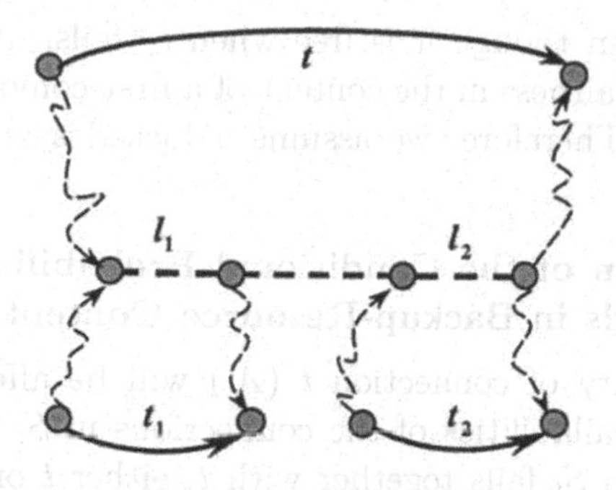

Figure 11.17 A general backup sharing example.

connections cannot be recovered even though their backup is free. This mechanism is called *lazy recovery.*

In active recovery, the released backup resources on b may be able to recover more than one connection. Figure 11.17 shows such an example, where t, t_1, and t_2 are three connections; t and t_1 share the same backup wavelength on link l_1; and t and t_2 share the same backup wavelength on link l_2. When t fails, both backup wavelengths on links l_1 and l_2 will be taken by t. Let t_1 and t_2 fail sequentially after t, and they are still in failed state when t is fixed. Then, both t_1 and t_2 can be recovered since they don't have backup resource contention. Note that neither of them (t_1 and t_2) could be recovered if the lazy recovery scheme is used. Obviously, backup resources are utilized more intelligently in the active-recovery model so we assume an active-recovery system in our study.

If active recovery is employed, another problem will arise, i.e., when a connection frees the backup resources and if there are multiple failed connections waiting for the backup resources, which connection should be chosen to recover next? Connections can be recovered in the exact order of their failure sequence, i.e., earliest failure recovered first. We call this a resource-locked system in the sense that a failed connection will "lock" all the free backup wavelengths it needs and wait for others to be released. For example, if the failure sequence in Fig. 11.17 is t_1, t, and t_2, then t will lock the backup wavelength on link l_2 so it cannot be used by t_2 even though it is free when t_2 fails. A locked system can provide fairness in the context of a first-come-first-served (FCFS) policy. Therefore, we assume a locked system in the following analysis.

- **Computation of the Conditional Probability that a Connection Succeeds in Backup-Resource Contention**

  The availability of connection t (A_t) will be affected by the size of S_t and the availabilities of the connections in S_t. When one or more connections in S_t fails together with t, either t or some of the failing connections in S_t can acquire the backup wavelengths. Hence, we need to compute the conditional probability (denoted as $\delta_t^{(k)}$) that t will successfully acquire the backup wavelengths when k connections in S_t fail concurrently with t.

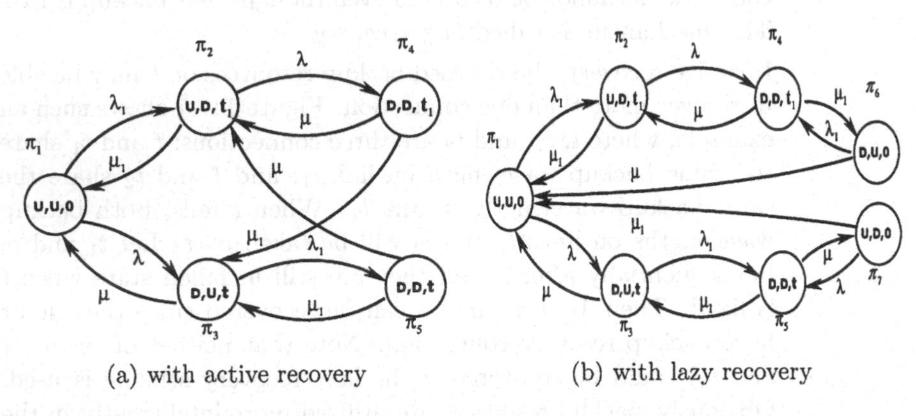

Figure 11.18 State-transition diagram for computing $\delta_t^{(1)}$ while assuming backup always available.

  We employ a continuous-time Markov chain to derive $\delta_t^{(k)}$. Figure 11.18(a) shows the corresponding state-transition diagram (when $k = 1$) for the Markov chain when an active-recovery, resource-locked system is applied. Let t_1 denote the other connection which shares backup resources with t. The label for each state in Fig. 11.18 is a 3-tuple (x, y, z), where x and y represent the state of the primary paths of connections t and t_1, respectively, and z represents which connection the backup path is recovering. Tuples x and y could be "Up" (U) or "Down" (D); and z could be "None" (0), "t", or "t_1". Let $MTTF = \frac{1}{\lambda}$

and $MTTR = \frac{1}{\mu}$ ($MTTF_1 = \frac{1}{\lambda_1}$ and $MTTR_1 = \frac{1}{\mu_1}$) to be the mean failure parameters for connection t (t_1). The state-transition probabilities can thus be represented by these parameters. Let π_i denote the long-run proportion of time the system is in state i. After solving for π_i, $\delta_t^{(1)}$ can be computed as follows for Fig. 11.18(a) (the details of these derivations are straightforward and not included here):

$$\delta_t^{(1)} = \frac{\pi_5}{\pi_4 + \pi_5} = \frac{\mu_1}{\mu + \mu_1} \tag{11.18}$$

One may notice that the value of $\delta_t^{(1)}$ is only determined by the repair rates of the concurrently failed connections (and not their failure rates)! If we assume that all the connections have the same repair rate[6], which is referred to as Assumption I in what follows, we have $\delta_t^{(1)} = \frac{1}{2}$ from Eqn. (11.18). Similarly, we can compute the conditional probability for one connection to acquire the backup wavelengths when k connections in S_t are experiencing failures concurrently, where $k \geq 2$. We find that, under Assumption I, $\delta_t^{(2)} = \frac{1}{3}$. (Figure 11.19 shows the state-transition diagram for $k = 2$, where the label for each state is similar to Fig. 11.18 and not all the transitions are included for sake of clarity, but it is straightforward for one to derive the full diagram from Fig. 11.19.) We expect that $\delta_t^{(k)} = \frac{1}{k+1}$ for any $k \geq 3$. Intuitively, each one of the $k+1$ failed connections (including t) should have equal chance to get the backup wavelengths since the conditional probability is only affected by the repair rate and under the assumption that all of them have the same repair rate. We can follow the same approach to derive $\delta_t^{(k)}$ for other recovery policies, e.g., lazy recovery, even though the Markov chain may be different. As an example, Fig. 11.18(b) shows the corresponding Markov chain to compute $\delta_t^{(1)}$ with the lazy-recovery policy.

With the value of $\delta_t^{(k)}$, we can compute the availability of a shared-path-protected connection t now. Connection t will be available if:

1. path p is available; or

[6] We can use the network-wide smallest repair rate among all components as the common repair rate for each connection. It is also expected that statistics to repair a failed link will be independent of the link, i.e., it will be the same network-wide.

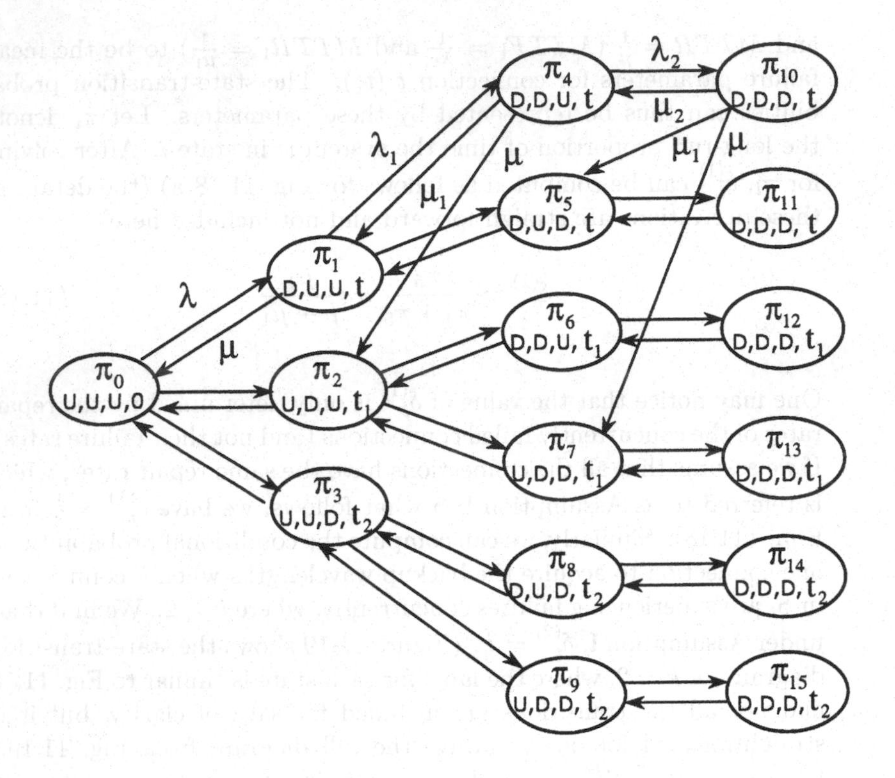

Figure 11.19 State-transition diagram (incomplete) for computing $\delta_t^{(2)}$ while assuming backup always available.

2. p is unavailable, b is available, and p can get the backup wavelengths when other connections in the sharing group S_t have also failed.

Therefore, A_t can be computed as follows:

$$A_t = A_p + (1 - A_p) \times A_b \times \sum_{k=0}^{N} \delta_t^{(k)} \times p_k \tag{11.19}$$

where A_p and A_b denote the availability of p and b, respectively; N is the size of S_t; $\delta_t^{(k)}$ is the probability that t can get the backup resources when both p and the other k primary paths in S_t fail; and p_k is the probability that exactly k primary paths in S_t are unavailable.

We can enumerate all the possible k connection failures to compute p_k. Note that it may not be necessary for us to enumerate all the possible multiple failure cases (up to N) since the probability of k simultaneous failures decreases drastically as k increases. Hence, such failure scenarios will have little effect on the connection availability. In a practical network, instead of enumerating all possible failure scenarios, we may only consider up to B simultaneous connection failures, where B is known as the *approximation bound.* By properly choosing the value of B, we can get a very tight lower bound on the connection availability. The value of B depends on the network failure characteristics, i.e., the more fragile the network is, the larger the value of B should be, and vice verse. The computational complexity of Eqn. (11.19) depends on:

1. the size of the sharing group (N); and
2. the approximation bound (B).

We find that it will only take several seconds to compute Eqn. (11.19) when several tens of connections are in the sharing group and B is around 10 using a computer with a 1.4-GHz Pentium processor and 512 Mbytes RAM; thus, the computation is feasible in a practical network.

- **Provisioning Strategies Based on the Analytical Model**

  We propose some availability-aware connection-provisioning approaches in which an appropriate level of protection is provided to each connection according to its predefined availability requirement. Both formal optimization techniques and heuristic strategies are studied for a given set of traffic demands. Our goal is to determine the route for each connection request and protect them (if necessary) while satisfying their availability requirements and minimizing the total network cost (wavelength links, particularly).

  To optimize network-resource usage, we first classify the connection requests into two categories: T_1 (containing *one-path-satisfiable* connections whose availability requirements can be satisfied without using any backup paths) and T_2 (containing *protection-sensitive* connections for which backup paths are necessary); and then, we provide different treatments to different connection sets, as follows.

  1. For a connection in T_1, one path is needed to carry each of them. We use an ILP to find the routes which can satisfy the connec-

tions' availability requirements while minimizing the consumed resources (wavelength links).

2. Protection schemes are necessary for connections in T_2. The problem of providing dedicated-path protection while satisfying the connections' availability requirements is mathematically formulated. Due to the nonlinearity of the formulations, we propose two schemes to approximately solve them. Several heuristic algorithms are studied since mathematical formulations are not scalable when the network size and the number of connection requests increase. To further improve resource efficiency, we incorporate a failure-independent shared-path protection scheme into the heuristics.

Our numerical results demonstrate that availability-aware provisioning strategies are promising; hence, further research is very encouraging to undertake.[7]

11.5.2 Service Reliability and Restorability

Service Reliability — Disruption Rate

Service disruption rate is not only affected by the failure rate but also by the operation policies. For example, traffic will be disturbed twice in the reverting strategy, which may be undesirable for some services such as online trading. However, if non-reverting strategy is employed, the backup paths for the connections which are sharing backup resources with connection t may need to be rearranged since some resources on parts of their backup paths may have been taken by t after t is switched to its backup path. These connections become vulnerable during their backup recomputation and resource reservation; and, furthermore, their successful backup rearrangement is not guaranteed, especially when the network load is high, so non-reverting may result in unpreferred service degradation. Hence, the operation policies need to be carefully selected according to the customers' requirements and network resource utilization.

[7] Please see [ZZZM03] for the connection-classification technique, detailed mathematical formulations, heuristic algorithms, and results from ILP and heuristics.

Restoration Time

Service restoration time varies according to different fault-management schemes. It is usually called protection-switching time when traffic is restored through predesigned protection resources. In dedicated (link, path, or sub-path) protection, OXCs on the backup paths can be preconfigured when the connection is set up. Then, no OXC configuration is necessary when the failure occurs. This type of recovery can be very fast. If traffic is not transmitted on both paths, the destination node needs to wait until the source node is notified and it switches traffic to the backup path. So, the protection-switching time will include the time for failure detection, failure notification, and the propagation delay. In shared-path protection, the OXCs on the backup paths cannot be configured until the failure occurs. The protection-switching time is longer in this scheme. For dynamic restoration, the service restoration time includes the time for route computation and resource discovery besides failure detection, notification, OXC reconfiguration, and propagation delay.

Network partitioning has been proposed to achieve high network scalability and fast fault-restoration time. The ideas are (1) to partition a large network into several smaller domains, and then (2) to protect each connection such that an intra-domain lightpath does not use resources of other domains and the primary and backup paths of an inter-domain lightpath exit a domain (and enter another domain) through a common egress (or ingress) domain-border node [OZSM04]. When a failure occurs, only the affected domain will activate its protection *sub-path* so that the restoration time is reduced due to the reduced path length.

Service Restorability

Service restorability is usually a network-wide parameter representing the capability that a network can survive a specific failure scenario. The restorability $R_f(i)$ of a network for a specific f-order ($f \geq 1$) failure scenario (i) is defined as the fraction of failed working capacity that can be restored by a specified mechanism within the spare capacity that is provided in a network [ClGr02]. The restorability R_f of a network as a whole is the average value of $R_f(i)$ over the set of f-order failure scenarios. For example, the network-wide ratio of restorable capacity to failed capacity over all single-failure (dual-failure) scenarios is called the single-failure (dual-failure) network restorability, R_1 (R_2).

Network restorability is an important criteria in network design which can be used to evaluate the quality of a specific mechanism, e.g., a protection scheme. Suppose we have two protection schemes P_1 and P_2, both of which consume the same amount of spare capacities to provide 100% R_1 to the network. We can essentially compare R_2 of the network under P_1 and P_2 to distinguish between their qualities for the 2-failure scenario. Obviously, the one with higher R_2 is preferred since it can provide higher network restorability when dual failures occur. The work in [ClGr02] showed that R_1-designed mesh-restorable networks inherently have high levels of dual-failure restorability (R_2) using an adaptive restoration process. However, how to efficiently design a network with $R_1 = 100\%$ and how to restore connections for dual failures play important roles in R_2 evaluation, which needs further study.

11.6 Summary

Fault management in WDM networks is an important and exciting research area. This chapter was devoted to studying the fault-management mechanisms involved in deploying a survivable optical network in both SONET/SDH rings and mesh-based architectures using optical crossconnects (OXCs). Specifically, we discussed various protection and restoration schemes, the concepts of stacked rings and protection cycles, routing algorithms for computing a pair of link-disjoint paths, ILP formulations of path-protection schemes to optimize the total number of bandwidth used for a given set of static traffic demands, the techniques to maximize sharability for the shared-protection schemes, and various considerations in dynamic restoration.

We then discussed some advanced topics involved in network survivability, such as service availability, service reliability, restoration time, and service restorability — their definitions, the factors that affect them, and the techniques that can improve them. In particular, a framework for cost-effective availability-aware connection provisioning was presented to provide differentiated services in WDM mesh networks. More studies are required in designing and operating a survivable WDM mesh network to provide differentiated services and to efficiently provide service-quality guarantees.

The research literature in survivable optical WDM networks is very large and also growing rapidly. Thus, only a subset of the major concepts could be captured in this chapter. For further reading on this topic, we refer the reader to two books on this subject [Grov03, OuMu05].

Besides protecting provisioned bandwidth (as has been discussed so far), there is emerging a parallel body of research on "Internet Protocol (IP) resilience". The objective of IP Resilience is to precompute or dynamically discover alternate routes for data packet traffic, to effectively deal with network failures. Note that capacity provisioning may not be necessary now, but routing is. A review of the literature on IP resilience can be found in [RaDM05].

Exercises

11.1. (Dedicated-Path Protection) Consider the network topology in Fig. 11.20.

(a) Find the shortest primary path from source node 3 to destination node 21.

(b) Find the shortest backup path from node 3 to node 21 so that this path is link disjoint to the path found in part (a). Calculate the total cost of this two-step approach (sum of primary and backup paths) for dedicated-path protection between nodes 3 and 21? Is the backup path found by you node disjoint to the primary path?

(c) We now describe a *one-step* algorithm for finding the primary and the backup paths concurrently for dedicated-path protection:

- Reverse the links of the primary path found in part (a). We will refer to this path as *reverse* path.
- Find another shortest path from the source node to the destination node on the updated network with the above *reverse* path.
- Remove the links on this newly found shortest route which use links on the *reverse* path.
- Reverse the links of the *reverse* path.

Use the above one-step algorithm to find two link-disjoint paths between nodes 3 and 21. Compare the cost of the one-step algorithm with the two-step approach. Is the new backup path found by you node disjoint to its primary?

11.2. (Shared-Path Protection) Consider the same network topology in the previous problem.

(a) We now wish to establish another connection from node 5 to node 15. Use the one-step algorithm to find the primary and backup paths between nodes 5 and 15. What is the total cost of establishing the two connections (3-21 and 5-15) with dedicated-path protection? Is this cost optimal for the two connections?

(b) On the network found in Problem 11.1, find the primary path for the connection from node 5 to node 15. Also find its backup while allowing it to share links with the backup path of the connection

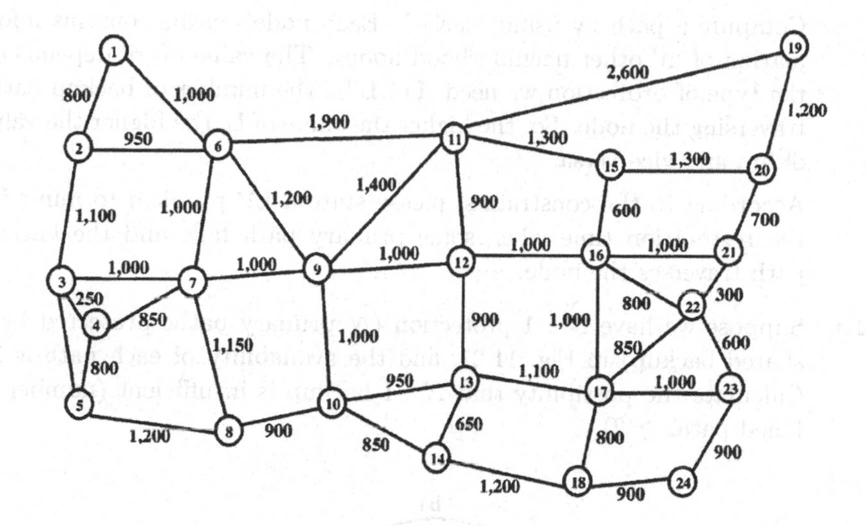

Figure 11.20 A sample 24-node network topology.

between nodes 3 and 21. What is the cost of establishing the two connections (3-21 and 5-15) with shared-path protection? In a WDM network, when can two connections not share their backup paths?

11.3. (Dedicated-Link Protection) Consider the same network topology in Problem 11.1. Suppose now we wish to separately protect every link along the primary path between nodes 3 and 21. Show the backup routes around each link for dedicated-link protection? What is the total cost of this dedicated-link-protection method?

11.4. (Shared-Link Protection) Consider the same network topology in Problems 11.1 and 11.2. If we allow sharing among the backups of the links, we call this method shared-link protection. Show the backup routes for protecting the two primary paths (between nodes 3 and 21 and between nodes 5 and 15) using shared-link protection. Compare the cost of shared-link protection with dedicated-link protection.

11.5. (Path restoration) Survivability means how quickly and efficiently one can restore a link failure. A faster online solution can be proposed.

Compute a path by using "cache": Each node's cache contains information of 'n' other neighborhood nodes. The value of 'n' depends on the type of protection we need. Let L be the number of backup paths traversing the node. So, the higher the value of L, the higher the value of 'n', and vice versa.

According to the constraints, please state a LP problem to minimize the restoration time when some primary path fails and the backup path traverses the node.

11.6. Suppose we have $N : 1$ protection (N primary paths protected by 1 shared backup) in Fig. 11.21, and the availability of each path is P. Calculate the possibility that $N : 1$ backup is insufficient (number of failed paths ≥ 2).

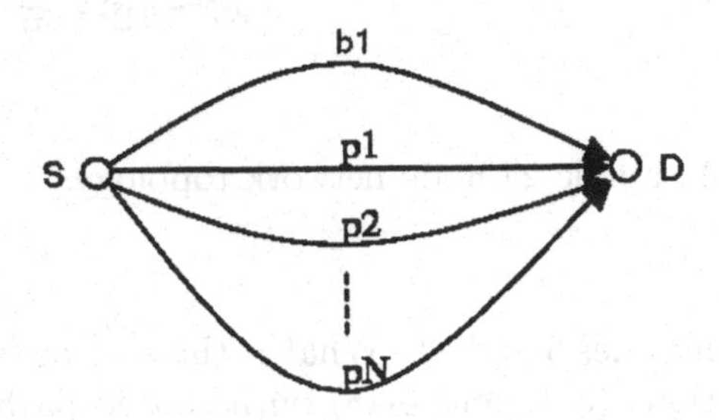

Figure 11.21 N:1 backup.

11.7. Consider the network topology in Fig. 11.22. A lightpath is set up between nodes 1 and 8 along the link $1 \rightarrow 2 \rightarrow 5 \rightarrow 8$. If link $2 \rightarrow 5$ goes down, which of the following restoration strategies will lead to a path with minimum number of hops? State the number of hopes in each case.

a) Path restoration
b) Sub-path restoration
c) Link restoration

11.8. Given the NSF network in Fig. 8.2,
a) find two lightpaths that can provide protection for the lightpath

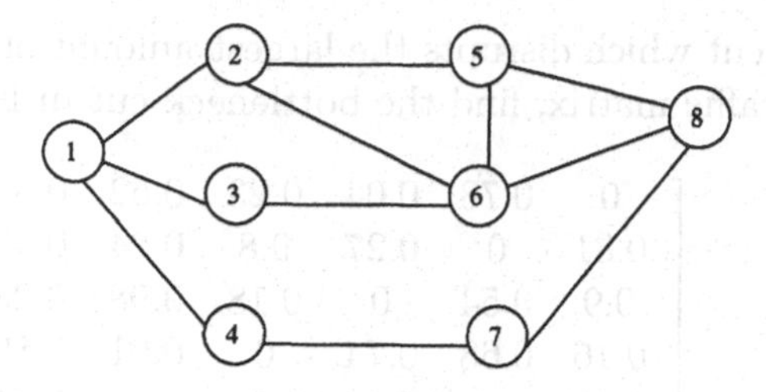

Figure 11.22 Restoration.

from CA1 to NY.

b) calculate recovery time for a failure on link $UT \rightarrow MI$ (link distances are shown in the diagram in km).

11.9. In Fig. 11.23, a primary path between s and d is protected by N backup paths, which is known as $1:N$ protection. Assume that the availability of path $P1$ is A_{P1}, and availabilities of paths $b1, b2, ..., bN$ are $A_{b1}, A_{b2}, ..., A_{bN}$, respectively. Derive an expression for a connection's availability from s to d.

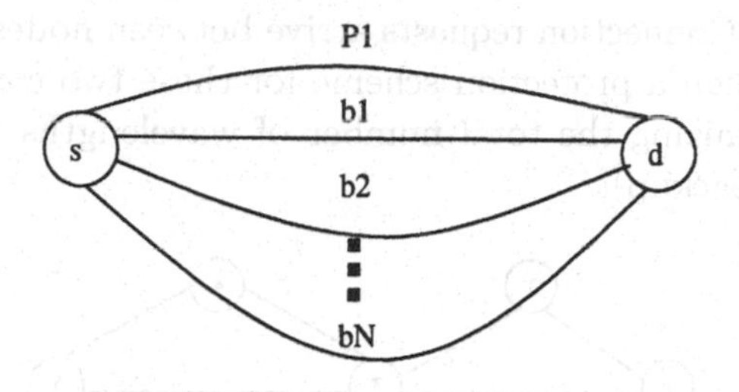

Figure 11.23 1:N backup.

11.10. Consider the bottleneck-cut identification problem for protection. A "cut" is defined as a set of links such that, if all of the links in this set fail, then the network will become partitioned into two disconnected fragments, e.g., links $1 \rightarrow 2$ and $3 \rightarrow 4$ in Fig. 11.24. A "bottleneck cut" is the cut which disrupts the largest amount of traffic. Given the following traffic matrix, find the bottleneck cut in Fig. 11.24.

$$\begin{bmatrix} 0 & 0.78 & 0.04 & 0.23 & 0.52 & 0.88 \\ 0.81 & 0 & 0.27 & 9.8 & 0.24 & 0.43 \\ 0.9 & 0.54 & 0 & 0.18 & 0.98 & 0.28 \\ 0.16 & 0.68 & 0.74 & 0 & 0.01 & 0.89 \\ 7.16 & 0.1 & 0.89 & 0.71 & 0 & 0.61 \\ 3.56 & 0.41 & 0.34 & 0.57 & 1.8 & 0 \end{bmatrix}$$

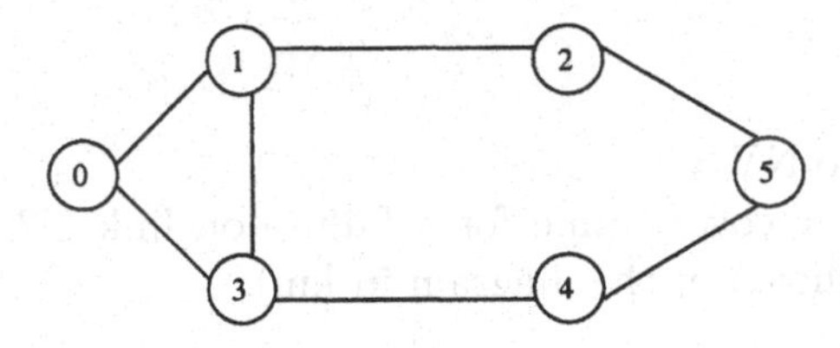

Figure 11.24 Bottleneck cut.

11.11. Consider the network topology in Fig. 11.25, where all links have the same cost. Connection requests arrive between nodes 1 → 5 and nodes 5 → 4. Design a protection scheme for these two connection requests, while minimizing the total number of wavelengths used (for primary as well as backup).

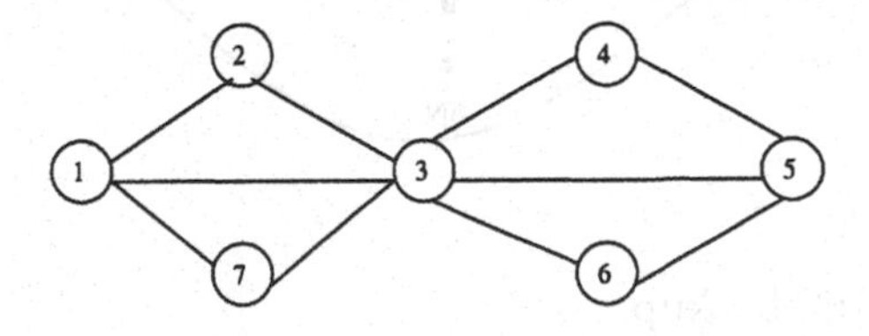

Figure 11.25 A sample network topology.

Chapter 12

Light-Tree: Optical Multicasting

12.1 Introduction

This chapter studies architectures and approaches for establishing multicast connections in a WDM mesh network using "light-trees". It also discusses schemes for protecting multicast trees in a mesh network.

Advances in optical WDM networking have made bandwidth-intensive multicast applications such as HDTV, interactive distance learning, live auctions, distributed games, movie broadcasts from studios, etc., widely popular [Paul98, Mill99, MaZQ98, SuGT01]. These applications require *point-to-multipoint* (PtMP) connections from a source node to the destination nodes in a network. Multicasting provides an easy means to deliver messages to multiple destinations without requiring too much message replication.

An optical signal passing through an optical wavelength-routing switch (WRS) may be routed from an input fiber to an output fiber without undergoing opto-electronic conversion. As we already know, a *lightpath* is a end-to-end wavelength-routed channel connecting a transmitter at a source node to a receiver at a destination node, which may be used to carry circuit-switched traffic, and it may span multiple fiber links.

Using lightpath communication, a significantly large number of lightpaths may be set up on the network in order to embed a logical (or virtual) topology. Now, a lightpath carries not only the direct traffic between the nodes it interconnects but also traffic from nodes upstream of the source (including the source) to nodes downstream of the destination (including the destination). One major advantage of lightpath communication is to reduce the number of hops (or lightpaths) a packet has to traverse because this reduction can, in turn, significantly improve the network's throughput.

Under lightpath communication, the network employs an equal number of transmitters and receivers because each lightpath operates on a point-to-point basis. However, this approach may not be able to fully utilize all of the wavelengths on all of the fiber links in the network; also, it may not be able to fully exploit all of the switching capability of each WRS (see Chapter 9).

Thus, we extend the lightpath concept by incorporating an optical multicasting capability. That is, if there is a connection from transmitter A to receiver B on a certain wavelength and there exist free resources (fiber link on the same wavelength) to direct A's transmission to some other receivers C and D as well, then why not do it? Now, A will have three logical downstream neighbors (B, C, and D) instead of one earlier (B). Thus, the logical (or virtual) connectivity of the network is increased and the hop distance is decreased. We refer to such a point-to-multipoint extension of a lightpath as a *light-tree*. A multicast request could also be served by using lightpaths or a hybrid of lightpaths and light-trees [SaSu03]. The chapter mainly focuses on light-tree multicasting.

Since a light-tree is a generalization of a lightpath, the set of light-tree-based virtual topologies is a *superset* of the set of lightpath-based virtual topologies. Hence, an "optimum" light-tree-based virtual topology is *guaranteed* to perform better than an "optimum" lightpath-based virtual topology. It should also be noted that optical multicasting (which is used to implement a light-tree) has some improved characteristics over electronic multicasting since "splitting light" is *conceptually* easier than copying a packet in electronic buffer.

In summary, a light-tree is a point-to-multipoint optical channel which may span multiple fiber links. Hence, a light-tree enables "single-hop" communication between a "source" node and a *set* of "destination" nodes; thus, a light-tree-based virtual topology can significantly *reduce the hop distance*, thereby *increasing the network throughput.* Figure 12.1 shows a light-tree (thick lines) which connects source node UT to destination nodes TX, NE,

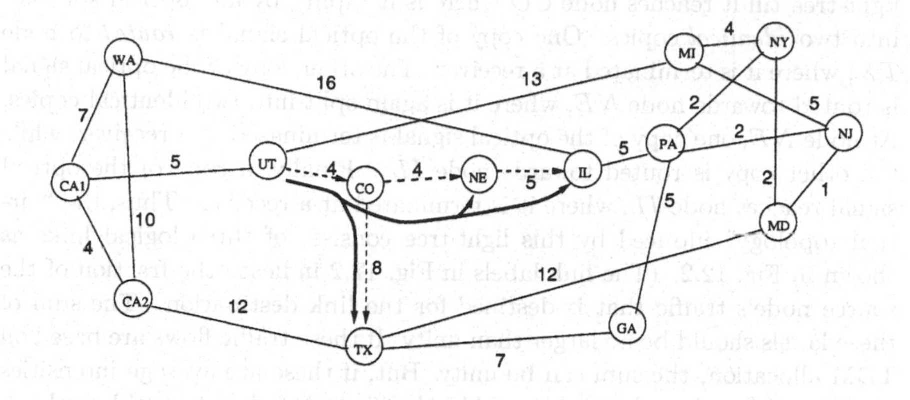

Figure 12.1 NSFNET backbone topology. (Link labels correspond to propagation delays.)

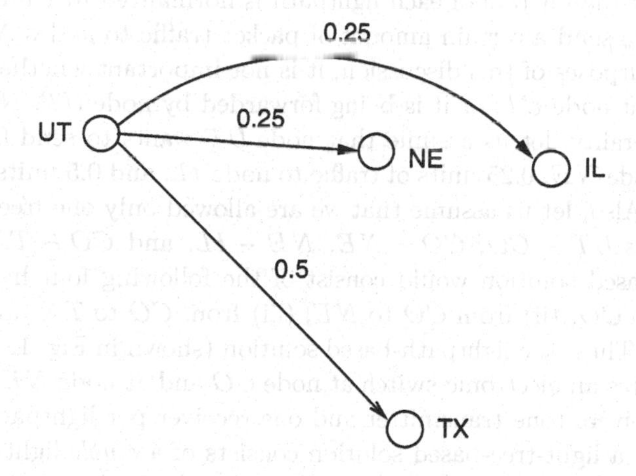

Figure 12.2 Virtual links induced by the light-tree consisting of source node UT, and destination nodes NE, TX, and IL.

and IL. Thus, an optical signal transmitted by node UT travels down the light-tree till it reaches node CO where is it "split" by an "optical splitter" into two *identical* copies. One copy of the optical signal is *routed* to node TX, where it is terminated at a receiver. The other copy of the optical signal is routed towards node NE, where it is again split into two identical copies. At node NE, one copy of the optical signal is terminated at a receiver, while the other copy is routed towards node IL[1]. Finally, a copy of the optical signal reaches node IL, where it is terminated at a receiver. Thus, the "virtual topology" induced by this light-tree consists of three logical links as shown in Fig. 12.2. (The link labels in Fig. 12.2 indicate the fraction of the source node's traffic that is destined for the link destination. The sum of these labels should be no larger than unity. If these traffic flows are based on TDM allocation, the sum can be unity. But, if these are average intensities of packet flows, then the sum should be significantly lower, to avoid overload, packet loss, unstable operation of the link, etc.)

12.1.1 Light-Tree for Unicast Traffic: An Illustrative Example

Let us consider the NSFNET backbone topology shown in Fig. 12.1. Let us assume that the bit rate of each lightpath is normalized to 1 unit, and node UT wants to send a certain amount of packet traffic to nodes NE, IL, and TX. For purposes of this discussion, it is not important whether this traffic originated at node UT, or it is being forwarded by node UT. Now, without loss of generality, let us assume that node UT wants to send 0.25 units of traffic to node NE, 0.25 units of traffic to node IL, and 0.5 units of traffic to node TX. Also, let us assume that we are allowed only one free wavelength on the links $UT - CO$, $CO - NE$, $NE - IL$, and $CO - TX$. Then, a lightpath-based solution would consist of the following four lightpaths: (i) from UT to CO, (ii) from CO to NE, (iii) from CO to TX, and (iv) from NE to IL. Thus, the lightpath-based solution (shown in Fig. 12.1 by dashed lines) requires an electronic switch at node CO and at node NE, and a total of 8 transceivers (one transmitter and one receiver per lightpath). One the other hand, a light-tree-based solution consists of a *single* light-tree (shown in Fig. 12.1 by thick lines), which requires a total of 4 transceivers (one transmitter at UT and one receiver per node at NE, IL, and TX) and it does not utilize the electronic switch at node CO or at node NE.

[1]This operation at node NE can also be performed by a "drop-and-continue" optical device.

Of course, the benefits of employing light-trees do not come for free. First, we need multicast-capable wavelength-routing switches (MWRS) at every node (or at least at some nodes) in the network (architectures of MWRSs are examined in Section 12.2). Second, we may need more optical amplifiers and wavelength converters in the network. The reason we need more amplifiers in the network is quite obvious: if we make n copies of an optical signal by using one or more optical splitters, then each copy will have $1/n$ times the original signal power (assuming equal power splitting); thus, more amplifiers may be required to maintain the optical signal power above a certain threshold so that the signal can be detected at their receivers. The reason we may need more wavelength converters is because a typical light-tree uses more fiber links than the number of fiber links used by a typical lightpath; hence, there may be more "color clashes" between different light-trees[2]. Another side-effect of using light-trees is that we may not be able to fully utilize the bandwidth of certain "branches" of a light-tree. For example, in Fig. 12.1, link $UT - CO$ is fully utilized since it is carrying 1 unit of traffic. On the other hand, link $CO - TX$ has only 0.5 unit of "useful" traffic; the remaining 0.5 units of traffic is actually meant for nodes NE and IL. However, by employing a light-tree-based virtual topology, the savings obtained by reducing the number of transceivers, terminating equipment, and electronic routers, may significantly offset the "extra" costs of implementing a light-tree-based virtual topology.

12.1.2 Layered-Graph model

Note that light-trees can not only provide improved performance for unicast traffic (as outlined above), but they can naturally better support broadcast traffic and multicast traffic because of their inherent point-to-multipoint nature. Efficient delivery of broadcast traffic may be required by a *WDM control network*. To understand why, let us first consider the wavelength-routed optical network shown in Fig. 12.3(a) which may be modeled as a *layered* graph, in which each layer represents a wavelength, and each physical fiber link has a corresponding link in each wavelength layer [LeLi93, ChBa95, ChBa96]. Figure 12.3(b) illustrates the layered-graph model of the optical network in Fig. 12.3(a). Now, the switching state of each wavelength-routing switch (WRS) is managed by a controller. Controllers communicate with one another using a control network, either *in-band*, *out-*

[2] Recall that, in the absence of wavelength converters, a light-tree must occupy the same wavelength on all fiber links.

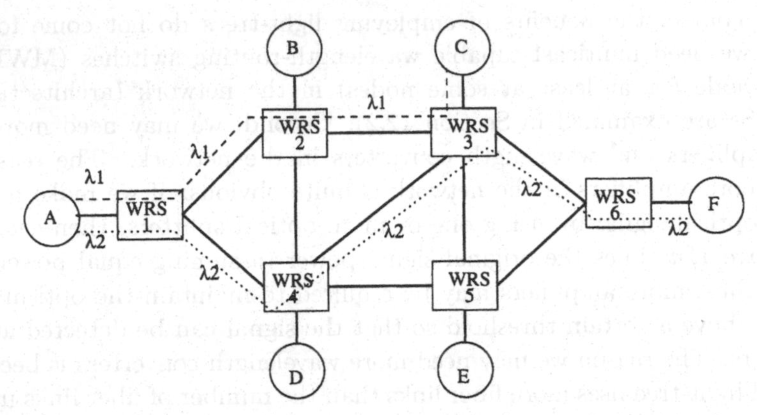

(a) Two lightpaths: one lightpath on wavelength λ_1 from node A to node C, and the other lightpath on wavelength λ_2 from node A to node F.

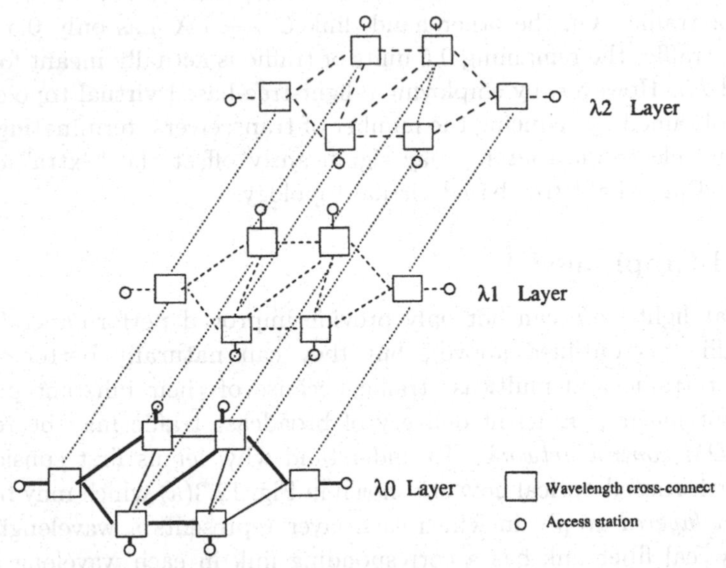

(b) Layered-graph model with three wavelength layers. (Wavelength λ_0 layer serves as the control network. For illustration, a broadcast tree is shown as the control network.)

Figure 12.3 Architecture of a wavelength-routed optical network and its layered-graph representation.

of-band, or *in-fiber, out-of-band.* In *in-fiber, out-of-band* signalling (which we advocate for a WDM mesh network), a wavelength layer is dedicated for the control network. For example, in Fig. 12.3(b), the wavelength λ_0 layer may be used for the control network, and controllers may employ multiple light-trees for fast information dissemination among themselves. Moreover, in the future, as multicast[3] applications become more and more popular and bandwidth intensive, there emerges a pressing need to provide multicasting support in such networks. Some multicast applications may have a large destination set which may be spread over a wide geographical area. For example, a live telecast of a popular music concert is one such application. A light-tree-based "broadcast layer" may provide an efficient transport mechanism for such multicast applications. Section 12.4.2 examines how to select an optimum light-tree-based virtual topology for broadcast traffic[4].

12.1.3 Steiner Trees

Consider the simple network topology in Fig. 12.4, on which we wish to establish a multicast session from the source node **F** to a set of destination nodes B, C, and D. This requires a "point-to-multipoint" connection from the source node **F** to every destination node. Such a session can be established using a *light-tree* approach, with the source node as root and the destination nodes as leaves, as shown by thick lines in the figure. In a network equipped with all-optical switches, an optical splitter is needed at node B to replicate the incoming bit stream into three copies: one for forwarding to node C, another to node D, and the third for local dropping at node B. In the absence of wavelength converters in a network, this light-tree based multicast session will exhibit the *wavelength-continuity constraint* [SaMu99].

A multicast session, as discussed in our above example, occupies a wavelength channel along the fiber links. The cost of carrying traffic between adjacent nodes which may be the number of hops, equipment cost, etc. is mapped as weights on the fiber links, as shown in Fig. 12.4. If we route a session on the links with higher number of hops or with higher equipment cost, communication cost increases, leaving fewer resources for other sessions. Thus, it is important to establish a session along minimum-cost links. We calculate the cost of a multicast session by summing the weights of the

[3]Multicasting is the ability of an application at a node to send a single message to the communication network and have it delivered to multiple recipients at different locations.

[4]A broadcast traffic "matrix" is a $N \times 1$ vector, where the entry in row i corresponds to the "broadcast" traffic generated by node i.

links occupied by the multicast session. It is desirable to minimize the cost of establishing a light-tree so that sufficient resources are available for other connections. For example, in Fig. 12.4, the same multicast session can be set up at a higher cost (28) by a light-tree occupying the links *F-E*, *E-D*, *D-C*, and *D-B*. (The tree shown in Fig. 12.4 has a cost of only 9 units.)

This problem of minimizing the cost of setting up a multicast tree in a network has been studied extensively, both with and without quality-of-service (QoS) guarantees such as bandwidth, reliability, delay, jitter, etc. [KoPP93, HeCT01, Pank99, TaMa80]. Such a minimum-cost multicast tree is called a minimum Steiner tree [HwRi92, Haki71] and the problem of finding a Steiner Minimum Tree (SMT) in a graph is NP-complete [GaJo79]. Heuristics are employed to solve SMT problems, e.g., two common heuristics are Pruned Prim's Heuristic (PPH) [CoLR01] and Minimum-cost Path Heuristic (MPH) [TaMa80]. In the first heuristic, a minimum spanning tree (MST) is constructed first using Prim's MST algorithm and then pruned by eliminating unwanted arcs. In the second heuristic, the closest destination nodes are picked one by one and added to a partially-built tree. A light-tree is a directed Steiner tree and the problem of finding a Directed Steiner Minimum Tree (DSMT) is NP-complete [Rama96], which follows from the fact that its special case SMT is NP-complete.

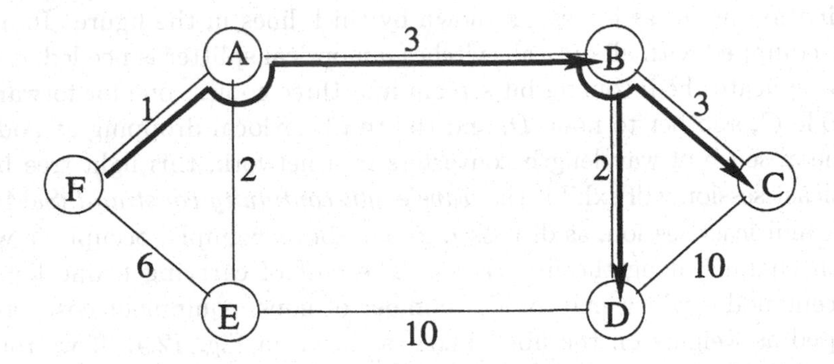

Figure 12.4 A six-node network topology with a light-tree shown in thick lines carrying traffic from the source node **F** to destination nodes *B*, *C*, and *D*. (Link labels, chosen arbitrarily for this example, may correspond to the number of hops, equipment cost, etc.)

12.2 Multicast-Capable Switch Architectures

There are two approaches to design switches capable of supporting multicasting. One approach is to use electronic crossconnects which perform switching in the electronic domain and the other is to use "all-optical" switches for switching in the optical domain. While switching in the latter is "transparent" to bit rate and bit-encoding schemes, switching in the former requires a knowledge of bit rate and bit-encoding strategies, and hence is "opaque".

12.2.1 Opaque Switch

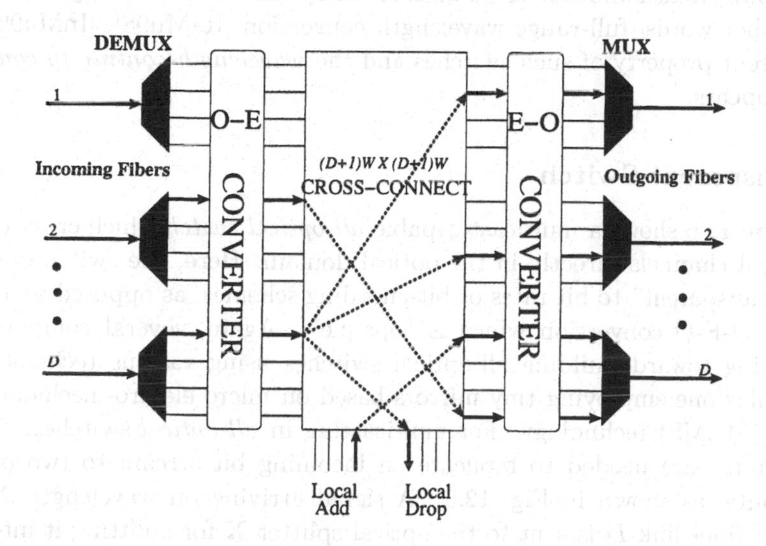

Figure 12.5 Opaque switch architecture for supporting multicast using electronic crossconnect and O-E-O conversion.

Figure 12.5 shows a *hybrid approach*, in which the incoming optical bit streams are converted to electronic data, the data is switched using an electronic crossconnect, and then the electronic bit streams are converted back to the optical domain. Observe that the signal in a channel arriving on the input fiber link D is replicated into three copies in the electronic domain. One copy is dropped locally at the node and the remaining two are switched to different channels on outgoing fiber links 1 and 2. (Along with the light-trees, the switch can also be used to establish lightpaths from a source to a destination as shown in the figure by a unicast connection from input fiber link 2 to output fiber link D.) This "opaque" switch architecture is currently very popular due to the existence of mature technology to design high-bandwidth multi-channel non-blocking electronic crossconnect fabrics at a low cost. The mature technology of optical crossconnects (OXCs) based on O-E-O conversion can be used for building a MWRS with O-E-O conversion.

Wavelength converters are not needed in a network where nodes are equipped with optical switches based on the *hybrid approach* because, once an incoming bit stream in optical domain is converted to electronic domain, it can be switched and converted back to the optical domain on any wavelength. In other words, full-range wavelength conversion [RaMu98], [InMu99] is an inherent property of such switches and the *wavelength-continuity constraint* disappears.

12.2.2 Transparent Switch

Figure 12.6 shows a multicast-capable *all-optical switch* which crossconnects optical channels directly in the optical domain. Here, the switch operation is "transparent" to bit rates or bit-encoding schemes, as opposed to a switch with O-E-O conversion which is "opaque". Again, several companies are working toward building all-optical switches using various technologies, a popular one employing tiny mirrors based on micro-electro-mechanical system (MEMS) technology. For multicasting in *all-optical* switches, "optical splitters" are needed to replicate an incoming bit stream to two or more outputs, as shown in Fig. 12.6. A signal arriving on wavelength λ_b from input fiber link D is sent to the optical splitter **X** for splitting it into three identical copies. One of the three replicas is dropped locally at the node while the other two are switched to output fiber links 1 and 2. Observe that the signal arriving on wavelength λ_a from input fiber link 2 bypasses the node.

This architecture employs two optical switches of different sizes. Given:

- Number of incoming/outgoing fibers = D,
- Number of wavelength channels carried by each fiber = W,

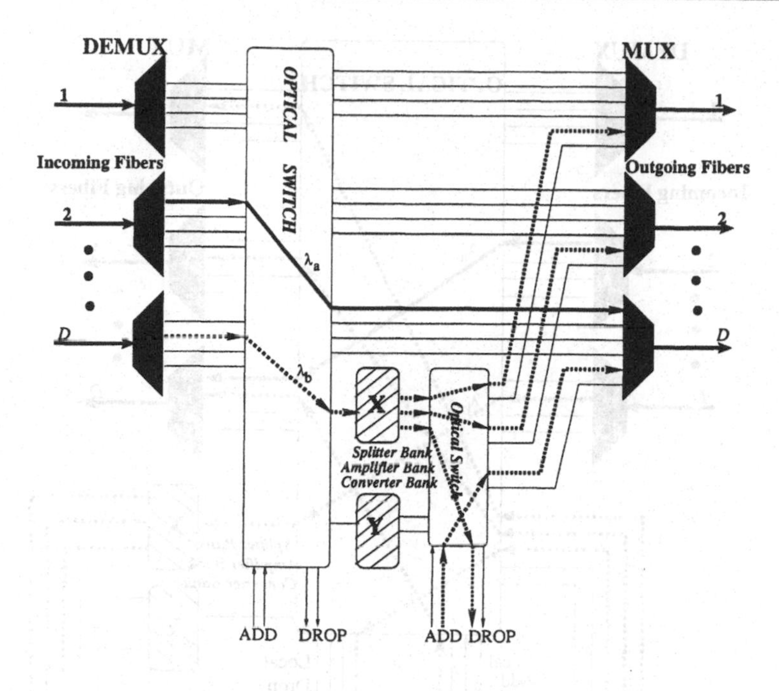

Figure 12.6 Switch architecture for supporting multicasting with all-optical crossconnects.

- Number of local add/drop ports to the switch $= L$; l on one switch and $L - l$ on the other,
- Splitting degree of each splitter $= \mathbf{d}$, and
- Number of splitters or the splitter-bank size $= \mathbf{B}$.

The size of the larger switch is $(D.W + l) \times (D.W + l + \mathbf{B})$ and the size of the other (smaller) switch is $(\mathbf{B}.\mathbf{d} + L - l) \times (\mathbf{B}.\mathbf{d} + L - l)$.

In this architecture, amplifiers are required because the output signal power weakens when the input signal is split, e.g., a 3-dB attenuation in power occurs for a two-way, equal-power splitting of an optical signal. Wavelength converters are useful in such switches to reduce the probability of blocking of multicast sessions. In the absence of wavelength converters, the light-tree-based multicast session would exhibit the *wavelength-continuity constraint.*

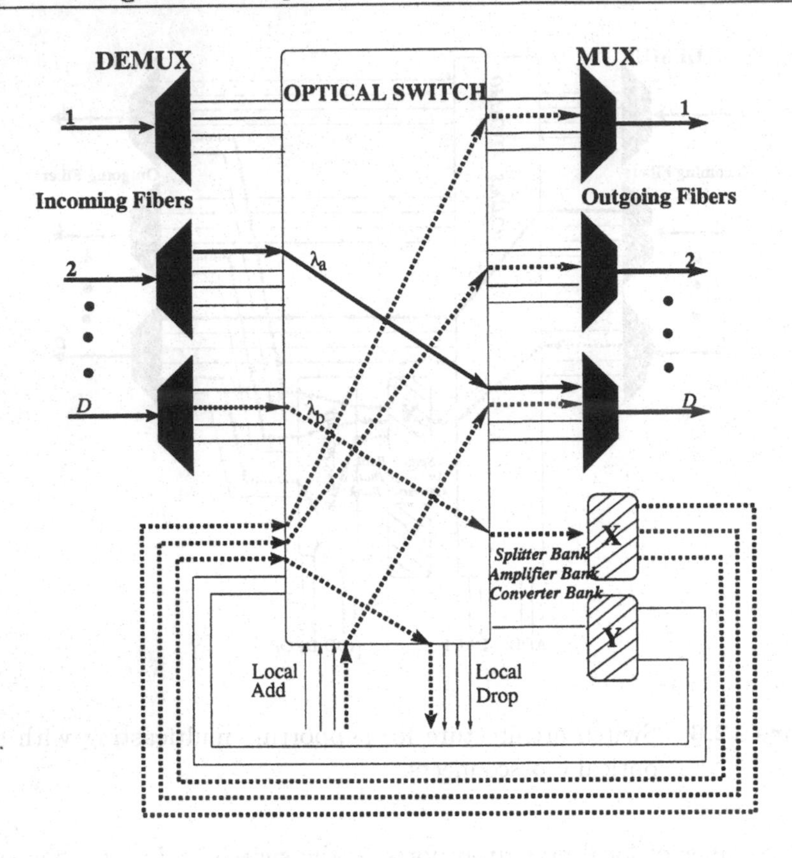

Figure 12.7 Modified switch architecture for supporting multicasting with all-optical crossconnects.

The switch shown in Fig. 12.6 can be built from off-the-shelf components. A modified architecture for a "transparent" switch is shown in Fig. 12.7. Here, instead of two optical switches, as was previously the case, we employ only one optical switch. But this optical switch would require a larger number of ports to switch the additional signals from the splitters, as shown in Fig. 12.7. In terms of the notations described earlier, the size of this switch is $(D.W + \mathbf{B}.\mathbf{d} + L) \times (D.W + \mathbf{B} + L)$. An advantage of this architecture is that the fanout of a signal is not limited by the splitting degree of a splitter. By proper switch configuration, output from one splitter can be fed as an input to another splitter, thus achieving a higher splitting degree of a signal.

In Fig. 12.7, an incoming signal on wavelength λ_b on input fiber D is split by a 3-way splitter **X**. One copy of the signal is dropped locally and the other two are switched to output fiber links 1 and 2. Again, the signal arriving from input fiber link 2 on wavelength λ_a bypasses the node without any local drop.

12.3 Light-Trees for Unicast Traffic

12.3.1 General Problem Statement

The problem of embedding a desired virtual topology on a given physical topology (fiber network) is formally stated below. Here, we state the problem for unicast traffic; the problems for broadcast and multicast traffic can be stated similarly, and they are tackled later in this chapter.

We are given the following inputs to the problem.

1. A physical topology $G_p = (V, E_p)$ consisting of a weighted undirected graph, where V is the set of network nodes, and E_p is the set of links connecting the nodes. Undirected means that each link in the physical topology is bidirectional. Nodes correspond to network nodes (packet switches) and links correspond to the fibers between nodes; since links are undirected, each link may consist of two fibers or two channels multiplexed (using any suitable mechanism) on the same fiber. Links are assigned weights, which may correspond to physical distances between nodes. A network node i is assumed to be equipped with a $D_p(i) \times D_p(i)$ wavelength-routing switch (WRS), where $D_p(i)$, called the physical degree of node i, equals the number of physical fiber links emanating out of (as well as terminating at) node i[5].

2. Number of wavelength channels carried by each fiber $= W$.

3. An $N \times N$ traffic matrix, where N is the number of network nodes, and the (i, j)-th element is the average rate of traffic flow from node i to node j. Note that the traffic flows may be asymmetric, i.e., flow from node i to node j may be different from the flow from node j to node i.

[5]Note that $D_p(i)$ includes the fiber(s) corresponding to local connections, i.e., for attaching an electronic router to the multicast-capable WRS.

4. The number of wavelength-tunable lasers (transmitters) (T_i) and wavelength-tunable filters (receivers) (R_i) at each node.

Our goal is to determine the following.

1. A virtual topology $G_v = (V, E_v)$ as another graph where the out-degree of a node is the number of transmitters at that node and the in-degree of a node is the number of receivers at that node. The nodes of the virtual topology correspond to the nodes in the physical topology. In the virtual topology, a link between nodes i and j corresponds to a light-tree rooted at node i with node j as one of the "leaves" on the light-tree. (Noting that each such link of the virtual topology may be routed over one of several possible paths on the physical topology, an important design issue is "optimal routing" of *all light-trees* so that the constraint on having a limited number of wavelengths per fiber is satisfied.)

12.3.2 Formulation of the Optimization Problem: Unicast Traffic

By extending the work in Chapter 8, we formulate the problem of finding an optimum light-tree-based virtual topology as an optimization problem, using principles of multicommodity flow for routing of light-trees on the physical topology and for routing of packets on the virtual topology, and using the following notation:

1. s and d used as subscript or superscript denote *source* and *destination* of a packet, respectively,

2. i and j denote the *originating node (or root)* and the *terminating node (or leaf)* in a light-tree, respectively, and

3. m and n denote *end-points of a physical link* that might occur in a light-tree.

- **Given:**
  - Number of nodes in the network = N.
  - Maximum number of wavelengths per fiber = W (a system-wide parameter).

- Physical topology P_{mn}, where $P_{mn} = P_{nm} = f$ (i.e., fiber links are assumed to be bidirectional) indicates that there are f direct physical fiber links between nodes m and n, where $f = 0, 1, 2, \ldots$ and $m, n = 1, 2, 3, \ldots, N$. If there is no fiber link between nodes m and n, then $P_{mn} = P_{nm} = 0$.
- Number of transmitters at node $i = T_i$.
- Number of receivers at node $j = R_j$.
- Traffic matrix Λ_{sd} which denotes the average rate of traffic flow from node s to node d, for s, d = 1, 2, ... , N.
- Maximum loading per channel $= \beta, 0 < \beta < 1$. β restricts the queueing delay on a lightpath from getting unbounded by avoiding excessive link congestion. We do not incorporate queueing delays explicitly in the problem formulation, under the assumption that they are negligible for suitably chosen values of β. Previous results in Chapter 8 indicate that queueing delays are negligible compared to propagation delays for a large network, such as the NSFNET, except under extremely heavy loading.
- Capacity of each channel = C (normally expressed in bits/second, but converted to units of packets/second by knowing the mean packet length).

- **Variables:**
  - Number of "busy" transmitters: The variable $V_i = 0, 1, 2, \ldots, T_i$ is the number of transmitters in use at node i.
  - Virtual topology: The variable $s_{ij} > 0$ if a transmitter at node i and a receiver at node j are members of the same light-tree in the virtual topology; $s_{ij} = 0$ otherwise. Note that s_{ij} is a real-valued variable which can range from 0 to V_i, where V_i is the number of "busy" transmitters at node i. For example, if $s_{ij} = 0.5$, then it implies that the virtual link i-j can use only 0.5 times the capacity of a light-tree channel. The integer-valued variable V_{ij} is used to take the "ceiling" (i.e., $V_{ij} \geq s_{ij}$) of the real-valued variable s_{ij}. Thus, if $V_{ij} > 0$, then link (i, j) exists in the virtual topology. Note that, if $s_{ij} > 0$ and $s_{ij} = V_{ij}$, then node i is connected to node j by a lightpath, a confirmation of the fact that a light-tree is a generalization of a lightpath.

- Traffic routing: The variable λ_{ij}^{sd} denotes the traffic flowing from node s to node d and employing s_{ij} as an intermediate virtual link. Note that traffic from node s to node d may be "bifurcated" with different components taking different sets of light-trees.
- Physical topology route: The variable $p_{mn}^{i} > 0$ if the fiber link P_{mn} is present in a light-tree rooted at node i; $p_{mn}^{i} = 0$ otherwise. Note that p_{mn}^{i} is a real-valued variable which can range from 0 to V_i, where V_i is the number of "busy" transmitters at node i. For example, in Fig. 12.1, for the light-tree rooted at node UT, $p_{UT,CO}^{UT} = 1$ and $p_{CO,TX}^{UT} = 0.5$. A fractional value of p_{mn}^{i}, say $p_{CO,TX}^{UT} = 0.5$, means the following: from the total amount of traffic transmitted on a light-tree rooted at node i, the physical fiber link m-n can carry an amount of traffic which is less than 0.5 times the capacity of a light-tree channel.
- Dummy integer variable: Variable $Q_{mn}^{i} \geq p_{mn}^{i}$ is employed to take the "ceiling" of variable p_{mn}^{i}.

- **Optimize:** Optimize one of two possible objective functions shown below.

1. Average packet hop distance minimization:

$$\textbf{Minimize}: \quad \frac{1}{\sum_{s,d} \Lambda_{sd}} \sum_{i,j} \sum_{s,d} \lambda_{ij}^{sd} \tag{12.1}$$

This is a linear objective function because $(\sum_{i,j} \sum_{s,d} \lambda_{ij}^{sd})$ is a linear sum of variables, while $\sum_{s,d} \Lambda_{sd}$ is a constant for a given traffic matrix.

2. Total number of transceivers in the network:

$$\textbf{Minimize}: \quad \sum_{i} V_i + \sum_{i,j} V_{ij} \tag{12.2}$$

$\sum_i V_i$ is the total number of transmitters in the network, while $\sum_{i,j} V_{ij}$ is the total number of receivers in the network. Thus, the objective function minimizes the total number of transceivers in the network.

- **Constraints:**

– On virtual-topology connection variables:

$$V_i \leq T_i \tag{12.3}$$
$$s_{ij} \leq V_{ij} \tag{12.4}$$
$$V_{ij} \leq s_{ij} + 1 \tag{12.5}$$
$$\sum_j V_{ij} \geq V_i \tag{12.6}$$
$$\sum_i V_{ij} \leq R_j \tag{12.7}$$
$$int \quad V_i, V_{ij} \tag{12.8}$$

– On physical route variables:

$$\sum_n p_{in}^i = V_i \tag{12.9}$$
$$\sum_j s_{ij} = V_i \tag{12.10}$$
$$\sum_m p_{mk}^i = \sum_n p_{kn}^i + s_{ik}, \quad k \neq i \tag{12.11}$$
$$p_{mn}^i \leq Q_{mn}^i \tag{12.12}$$
$$Q_{mn}^i \leq V_i \tag{12.13}$$
$$\sum_i Q_{mn}^i \leq W \times P_{mn} \tag{12.14}$$
$$int \quad Q_{mn}^i \tag{12.15}$$

– On virtual-topology traffic variables λ_{ij}^{sd}:

$$\lambda_{ij}^{sd} \geq 0 \tag{12.16}$$

$$\sum_j \lambda_{sj}^{sd} = \Lambda_{sd} \tag{12.17}$$

$$\sum_i \lambda_{id}^{sd} = \Lambda_{sd} \tag{12.18}$$

$$\sum_i \lambda_{ik}^{sd} = \sum_j \lambda_{kj}^{sd} \quad if \quad k \neq s, d \tag{12.19}$$

$$\sum_{s,d} \lambda_{ij}^{sd} \leq s_{ij} \times \beta \times C \tag{12.20}$$

– Optional constraints: physical topology as a *subset* of virtual topology:

$$P_{mn} = 1 \Rightarrow V_{mn} \geq 1, p_{mn}^{m} \geq 1, Q_{mn}^{m} \geq 1 \tag{12.21}$$

- **Explanation of Equations:** The above equations are based on principles of conservation of flows and resources (transceivers, wavelengths, etc.). Equation (12.3) ensures that the number of busy transmitters (i.e., transmitters from which a light-tree originates) at node i is less than the total number of transmitters at node i. Equations (12.4)-(12.5) ensure that the integer variable V_{ij} is always equal to the "ceiling" of s_{ij}. Although Equation (12.6) can be derived from Equations (12.10) and (12.4), it has been stated in order to improve the convergence of the Mixed Integer Linear Program (MILP) solver. Equation (12.7) ensures that the number of light-trees terminating at node j is less than the number of receivers at node j.

  Equations (12.9)-(12.11) are the flow-conservation equations that account for the routing of a light-tree from its origin to its termination. The flow-conservation equations have been formulated in two different ways [RaSi96]: (i) disaggregate formulation and (ii) aggregate formulation. In the disaggregate formulation, every i-j (or s-d) pair corresponds to a commodity, while, in the aggregate formulation, all the traffic that is "sourced" from node i (or node s) corresponds to the same commodity, regardless of the traffic's destination. The aggregate formulation is considered more tractable because it has fewer variables and constraints. We employ the aggregate formulation for the flow-conservation equations. Equation (12.9) ensures that the number of "light-streams" (a path within a light-tree) originating at node i is equal to the number of busy transmitters at node i. Equation (12.10) states that the sum of all "light-streams" at the terminating nodes (or leaf nodes) must equal the number of busy transmitters at the originating node (i). Note that s_{ij} values may not be integers. For example, consider the light-tree shown in Fig. 12.1. Recall that the traffic from UT to NE is equal to 0.25 unit, while the traffic from UT to TX is equal to 0.5 unit. Thus, the value of $s_{UT,NE}$ is greater than or equal to 0.25, while the value of $s_{UT,TX}$ is greater than or equal to 0.5 as shown in Fig. 12.2. Equation (12.11) ensures that the difference between the sum of incoming "light-streams" to node k and the sum of outgoing "light-streams" from node k equals the amount of "light-streams" "sinking" at node k. Equation (12.12) requires that one wavelength be allocated to each "light-stream" passing through link P_{mn}. Equation (12.13) is redundant, i.e., it can be derived from earlier equations. It has been added to improve the running time of a MILP solver.

Equation (12.14) ensures that the number of wavelengths used on any physical link is less than the maximum number of wavelengths per fiber; however, this equation does not ensure the wavelength-continuity constraint. Thus, the optimum virtual topology found by our formulation may require wavelength converters at some nodes in the network. One method for incorporating the wavelength-continuity constraint in the MILP formulation is to introduce a "wavelength" subscript (or superscript) to some of the variables. Although this method results in a MILP formulation which incorporates the wavelength-continuity constraint, it increases the number of variables and the number of equations in the formulation, which may result in a large increase in the running time of the MILP solver. Hence, in this study, we have not considered the wavelength-continuity constraint, which also corresponds to the case where the switches in the network are opaque.

Equations (12.16)-(12.20) are responsible for the routing of packet traffic on the virtual topology, and they take into account the fact that the combined traffic flowing through a channel cannot exceed the channel capacity. Equation (12.21) embeds the physical topology as a subset of the virtual topology, i.e., every link in the physical topology is also a lightpath in the virtual topology. This equation is optional, but we have incorporated it to reduce the solution space of the problem.

To make the formulation tractable, we "prune" the number of variables as follows. We track a limited number of alternate paths, denoted by K, between source-destination pairs, such that the selected routes are within a constant factor ($\alpha \geq 1$) of the shortest-path distance between the given source-destination pair, i.e., all variables of type p^i_{mn} and λ^s_{ij} are not enumerated. In our experiments, we chose $\alpha = 2$ and $K = 2$.

12.3.3 Illustrative Numerical Examples

This section presents numerical examples of the light-tree-based optimum virtual-topology design problem using the NSFNET backbone as our physical topology. The traffic matrix is randomly generated, such that a certain fraction F of the traffic is uniformly distributed over the range $[0, \frac{C}{a}]$ and the remaining traffic is uniformly distributed over the range $[0, \frac{C \times \Upsilon}{a}]$, where C is the light-tree channel capacity which is the same as the rate at which a transmitter can transmit, a is an arbitrary integer which may be one or greater, and Υ denotes the average ratio of traffic intensities between node-pairs with high traffic values and node-pairs with low traffic values. Table 12.1 compares the network-wide average hop distance for a lightpath-based virtual topology (obtained from the methods in Chapter 9) vs. a light-tree-based virtual topology using a MILP solver, *CPLEX*[6].

The values shown in the table were calculated by taking the average over 10 random traffic matrices, obtained with the parameters $C = 1250, a = 20, \Upsilon = 10, F = 0.7, \beta = 0.8, K = 2$, and $\alpha = 2$. At each node, the number of receivers was three times the number of transmitters. T_i was varied from 4 to 6, while W took values 4, 6, and 8. As expected, the average packet hop distance in a light-tree-based virtual topology is much *lower* than the average packet hop distance in a lightpath-based virtual topology, thereby demonstrating the advantage of using light-trees.

Table 12.2 compares the number of transceivers (opto-electronic components) required in a light-tree-based virtual topology vs. the number of transceivers required in a lightpath-based virtual topology. It is interesting to note that, if we use random traffic matrices as explained in the previous paragraph, then the number of transceivers required by a light-tree-based solution is only a little lower than the number of transceivers required by a lightpath-based solution. On the other hand, if we use an "ordered", i.e., non-random traffic matrix, then the number of transceivers required in a

[6]For each data point, the MILP solver was run on an unloaded HP-UX 9000/778 until it produced a solution within 5% of the LP solution or up to a maximum of 30 minutes.

light-tree-based solution may be significantly less. For example, consider the following traffic matrix: every node sends a fixed amount of data to its neighbors and does not send any data to any other node in the network. Note that the traffic matrix of a network control layer may have this pattern. Then, for such a "one-hop" traffic matrix, a light-tree-based solution requires significantly fewer transceivers. For example, let $C = 1250, \beta = 0.8$, and let each node in Fig. 12.1 send 250 units of data to each of its neighbors. Then, by solving the MILP, we find that the number of transceivers required by a lightpath-based solution is 70, while the number of transceivers required by a light-tree-based solution is 54, of which 21 are transmitters and 33 are receivers.

Table 12.1 Average packet hop distance for a lightpath-based virtual topology and a light-tree-based virtual topology.

Lightpath Solution			
Transceivers	Wavelengths		
	4	6	8
4	1.59	1.58	1.58
5	1.48	1.38	1.38
6	1.42	1.32	1.30

Light-Tree Solution			
Transceivers	Wavelengths		
	4	6	8
4	1.23	1.13	1.08
5	1.21	1.12	1.07
6	1.19	1.09	1.07

12.4 Light-Trees for Broadcast Traffic

12.4.1 General Problem Statement

In case of broadcast traffic, we are mainly interested in finding a set of light-trees which implements a "broadcast layer" over a *single* wavelength, e.g., for control and management signalling information. A general problem

Table 12.2 The number of transceivers required by a lightpath-based virtual topology and a light-tree-based virtual topology for different traffic matrices.

Traffic Matrix	Virtual Topology	
	Light-tree	Lightpath
Random	102	104
"One-Hop"	54	70

description for broadcast traffic is the same as that for unicast traffic, as was discussed in Section 12.3.1.

12.4.2 Formulation of the Optimization Problem: Broadcast Traffic

The formulation of the problem for broadcast traffic is similar to the formulation of the problem for unicast traffic, which was presented in Section 12.3.2. Note that the only difference between the formulation for broadcast traffic and the formulation for unicast traffic is in the routing of traffic on the virtual topology. The equations for routing broadcast traffic consists of two parts: (i) equations for constructing a spanning tree on the virtual topology for each node in the network; thus, the spanning tree rooted at node s will route the broadcast traffic originating at node s, and (ii) equations for the capacity constraint of the virtual links. The formulation is as follows.

- **Variables:**
  - Definitions of the variables V_i, T_i, s_{ij}, p^i_{mn}, V_{ij}, and Q^i_{mn} remain unchanged.
  - Spanning tree variable $0 \leq \lambda^s_{ij} \leq N$, which determines the spanning tree rooted at node s. $\lambda^s_{ij} > 0$ if virtual link V_{ij} belongs to the spanning tree rooted at node s. Note that variable λ^s_{ij} is used only to "construct" a spanning tree on the virtual topology; it does not represent the broadcast traffic flowing on the spanning tree. The traffic on each link of a spanning tree which is rooted at node s is equal to the amount of broadcast traffic that is sourced at node s (denoted by Λ^s). Variable $0 \leq \Lambda^s_{ij} \leq 1$ is a 0-1 inte-

ger variable, such that $\Lambda_{ij}^s = 1$ if virtual link V_{ij} belongs to the spanning tree rooted at node s; otherwise, $\Lambda_{ij}^s = 0$.
  - Broadcast traffic Λ^s originating from node s.

- **Given:**
  - Definitions of the constants N, W, P_{mn}, T_i, R_i, β, and C remain unchanged.
  - Broadcast traffic Λ^s denotes the average rate of broadcast traffic originating from source s, where $s = 1, 2, 3, \ldots, N$.

- **Constraints:**
  - On virtual-topology connection variables: same as in the previous section.
  - On physical route variables: same as in the previous section.
  - On "spanning tree" variables λ_{ij}^k and Λ_{ij}^k.

$$\sum_j \lambda_{sj}^s = N - 1 \quad (12.22)$$

$$\sum_m \lambda_{mk}^s = \sum_n \lambda_{kn}^s + 1, k \neq s \quad (12.23)$$

$$\Lambda_{ij}^s \leq V_{ij} \quad (12.24)$$

$$\lambda_{ij}^s \leq (N-1) \times \Lambda_{ij}^s \quad (12.25)$$

$$\sum_{i,j} \Lambda_{ij}^k = N - 1 \quad (12.26)$$

$$\sum_k \Lambda_{ij}^k \times \Lambda^k \leq s_{ij} \times \beta \times C \quad (12.27)$$

- **Optimize:**

$$\text{Minimize}: \quad \sum_i V_i + \sum_{i,j} V_{ij} \quad (12.28)$$

- **Explanation of Equations:** Equations (12.22)-(12.26) set up a spanning tree rooted at node s on the virtual topology, while Equation (12.27) ensures that the light-tree capacity constraints are satisfied. Equation (12.22) starts a flow of "commodity" s from node s. Equation (12.23) is the multi-commodity flow-conservation equation. This equation ensures that one unit of flow is "sunk" at every node in the network, except the node where the flow originated. Equation (12.24) ensures that, if link i-j belongs to the spanning tree, then it must necessarily exist in the virtual topology. Equation (12.25) ensures that variable $\Lambda_{ij}^{s} = 1$ for all virtual links s_{ij} which belong to the spanning tree rooted at node s. Equation (12.26) ensures that the number of virtual links in the spanning tree is equal to $N - 1$ (this constraint is required to avoid spanning-tree solutions which contain a cycle). Equation (12.27) ensures that the capacity constraint of virtual link s_{ij} is satisfied. To make the formulation tractable, we "prune" the number of variables in a fashion similar to the unicast case.

12.4.3 Illustrative Numerical Example

Figure 12.8 shows a light-tree-based "broadcast layer" on a single wavelength. In this example, $C = 1250, \beta = 0.8$, and each node "broadcasts" 100 units of traffic. Note that, in this example, a simple bidirectional spanning tree topology cannot support the broadcast traffic. A complete physical topology embedding, (i.e., for every fiber link, we use a transmitter-receiver pair) will require 84 transceivers. If we solve the MILP, then the resulting light-tree-based "broadcast layer" shown in Fig. 12.8 requires only 56 transceivers, of which 21 are transmitters and the remaining 35 are receivers.

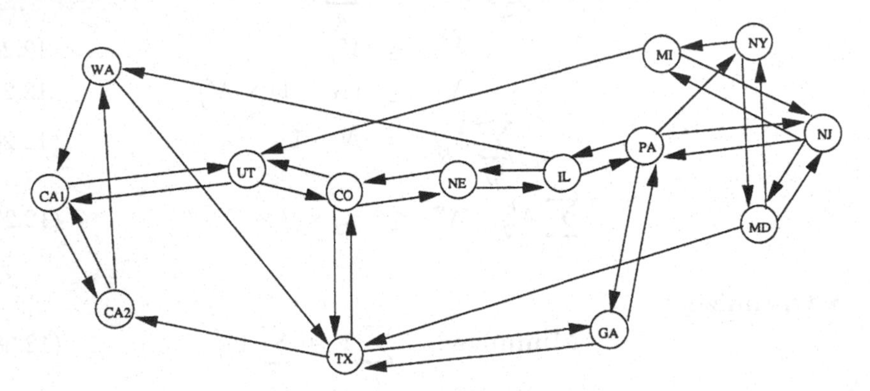

Figure 12.8 Virtual topology of the "broadcast layer", which consists of 21 transmitters and 35 receivers. Each node broadcasts its information over a spanning tree on the "broadcast layer".

12.5 Light-Trees for Multicast Traffic

12.5.1 General Problem Statement

The problem of setting up a group of multicast sessions at an overall minimum cost, using a *light-tree*-based approach on a given physical topology (fiber network) is formally stated below. Here, we provide a general problem statement. Different flavors of the problem are tackled in later subsections.

Note that, now, we are no longer considering packet traffic, as in the virtual-topology design problem; instead, we are now considering circuit traffic, in the sense that a set of multicast sessions are given and need to be set up such that each session requires the full capacity of a wavelength channel initially. Fractional-capacity sessions will also be considered later.

We are given the following inputs to the problem:

1. A physical topology $G_p = (V, E_p)$ consisting of a weighted undirected graph, where V is a set of network nodes, and E_p is the set of links connecting the nodes. Undirected means that each link in the physical topology is bidirectional. Nodes correspond to network nodes and the links correspond to the fibers between nodes. Each link is assigned a weight, which may correspond to the length of fiber between node pairs. A network node j is assumed to be equipped with a $D_p(j) \times D_p(j)$ MWRS, where $D_p(j)$, called the physical degree of node j, equals the number of physical fiber links emanating out of (as well as terminating at) node j.
2. Number of wavelength channels carried by each fiber = W.
3. A group of **k** multicast sessions.

Our goal is to set up (if possible) all the **k** multicast sessions on the given physical topology while minimizing the overall cost. Cost of a multicast session is the sum of the weights of the physical links occupied by it. Each multicast session forms a *light-tree* with the source as root and the destination nodes as leaves.

The problem of establishing several directed multicast trees at a minimum aggregate cost, which is a generalized version of the DSMT problem (Section 12.1.3), is an NP-complete problem. In our problem description, we assume that enough network resources are available to accommodate all connection requests. A dual version of this problem is to maximize the number of multicast connections established when network resources are insufficient and to minimize the blocking probability [RaRP02].

12.5.2 Problem Formulation for a Network with λ Converters

The switch architecture shown in Fig. 12.5 has implicit full-range wavelength conversion. Here, we formulate the problem of setting up a group of multicast sessions for a network equipped with such "opaque" switches. Following are a few notations which we shall be using:

1. s and d refer to source node and destination node, respectively, in a multicast session.
2. m and n denote *end-points of a physical link* that might occur in a light-tree.
3. i is used as an index for multicast session number, where $i = 1, 2, \ldots, \mathbf{k}$. Note that unicast sessions are easily accommodated by choosing a destination set size of unity.

- **Given:**
  - Number of nodes in the network = N.
  - Maximum number of wavelengths per fiber = W (a system-wide parameter).
  - Physical topology P_{mn}, where $P_{mn} = P_{nm} = f$ (i.e., fiber links are assumed to be bidirectional) indicates that there are f direct physical fiber links between nodes m and n, where $f = 0, 1, 2, \ldots$ and $m, n = 1, 2, 3, \ldots, N$. If there is no fiber link between nodes m and n, then $P_{mn} = P_{nm} = 0$.
  - Every physical link between nodes m and n is associated with a weight w_{mn} which may be the length of the fiber between them.
  - Capacity of each channel = C. For example, C may be OC-192 $\approx$ 10 Gbps.
  - A group of $\mathbf{k}$ multicast sessions S_i for $i = 1, 2, 3, \ldots, \mathbf{k}$. Each session S_i has a source node and a set of destination nodes denoted by $\{s_i, d_{i_1}, d_{i_2}, \ldots\}$. We will denote the cardinality of a multicast session i by L_i, which is equal to the number of source and destination nodes in that session.

- By default, every multicast session is assumed to operate at the full capacity of a channel, i.e., at C bps. This will be relaxed later for fractional-capacity sessions.
- Every node is equipped with wavelength converters capable of converting a wavelength to any other wavelength among W channels. These are not separate devices but is a "built-in" property of "opaque" crossconnects. Refer to Section 12.5.3 for the case where a node has no wavelength converters.

- **Variables:**
  - A boolean variable, M^i_{mn}, which is equal to one if the link between nodes m and n is occupied by multicast session i; otherwise, $M^i_{mn} = 0$.
  - A boolean variable, V^i_p, which is equal to one if node p belongs to multicast session i; otherwise, $V^i_p = 0$. A node belongs to a session if it is either the source or the destination or an intermediate node in the light-tree for the multicast session, e.g., $V^i_{s_i} = 1$ and $V^i_{d_{i_1}} = 1$.
  - An integer commodity-flow variable, F^i_{mn}. Each destination node for a session needs one unit of commodity. So, $L_i - 1$ units of commodity flow out of the source s_i for session i. F^i_{mn} is the number of units of commodity flowing on the link from node m to node n for session i. F^i_{mn} is also the number of destination nodes in session i downstream of the link between nodes m and n.

- **Optimize:**
  - Total cost of all multicast sessions:

$$Minimize: \sum_{i=1}^{i=k} \sum_{m,n} w_{mn} \cdot M^i_{mn} \qquad (12.29)$$

This is a linear objective function as it adds up the cost of individual multicast sessions.

- **Constraints:**

– Tree-creation constraints:

$$\forall i, \forall n \neq s_i : \sum_m M^i_{mn} = V^i_n \quad (12.30)$$

$$\forall i : \sum_m M^i_{ms_i} = 0 \quad (12.31)$$

$$\forall i, \forall j \in S_i : V^i_j = 1 \quad (12.32)$$

$$\forall i, \forall m \neq d_{i_j}, j = 1, \ldots, (L_i - 1) : \sum_n M^i_{mn} \geq V^i_m \quad (12.33)$$

$$\forall i, m : \sum_n M^i_{mn} \leq D_p(m).V^i_m \quad (12.34)$$

$$\forall m, n : \sum_i M^i_{mn} \leq P_{mn}.W \quad (12.35)$$

– Commodity-flow constraints:

$$\forall i, \forall m \notin S_i : \sum_n F^i_{nm} = \sum_n F^i_{mn} \quad (12.36)$$

$$\forall i, \forall m = s_i : \sum_n F^i_{s_i n} = L_i - 1 \quad (12.37)$$

$$\forall i, \forall m = s_i : \sum_n F^i_{ns_i} = 0 \quad (12.38)$$

$$\forall i, \forall m = d_{i_j}, j = 1, \ldots, (L_i - 1) : \sum_n F^i_{nm} = \sum_n F^i_{mn} + 1 \quad (12.39)$$

$$\forall i, m, n : M^i_{mn} \leq F^i_{mn} \quad (12.40)$$

$$\forall i, m, n : F^i_{mn} \leq N.M^i_{mn} \quad (12.41)$$

– Additional constraint:

$$\forall i, m, n : F^i_{mn} \leq (L_i - 1) \quad (12.42)$$

- **Explanation of Equations:** The equations are written for creating a connected tree for every multicast session. Equation (12.30) ensures that every node that belongs to a multicast session (except the source) has one incoming edge. Equation (12.31) states that the source node has no incoming edge since it is the root of the tree. Equation (12.32) ensures that every source node and the destination node of a multicast session belongs to the tree. Equation (12.33) ensures that every node (except destination nodes) belonging to the tree has at least one outgoing edge on the tree. Equation (12.34) ensures that every node with at least one outgoing edge belongs to the tree. Equation (12.35) restricts the number of light-tree segments between nodes m and n by $P_{mn}.W$ in either direction.

Equations (12.36) - (12.41) are flow-conservation equations to create a connected tree with the source having an end-to-end connection to every destination in the session. Equation (12.36) ensures that, at any intermediate node (which is neither a source nor a destination), the incoming flow is the same as the outgoing flow. However, the outgoing flow at the source node for a session is the number of destinations in the session and the incoming flow is zero. These are achieved by Equations (12.37) and (12.38), respectively. Equation (12.39) ensures that the total outgoing flow is one less than the incoming flow for destination nodes. Equations (12.40) and (12.41) ensure that every link occupied by a session has a positive flow and every link not occupied by the session has no flow. In Equation (12.41), N can be replaced by $L_i - 1$ without altering its meaning. A flow on any link for a multicast session is limited by the number of destinations in that session. This additional constraint is captured by Equation (12.42).

- **Number of Variables and Constraints:** Since the complexity of this MILP model is proportional to the number of variables in the system, we count them to provide an insight into the complexity of the formulation. It is a straightforward matter to verify that the number of variables in the formulation is $\mathbf{k}N+2\mathbf{k}N^2$, a count which grows linearly with the number of multicast sessions ($\mathbf{k}$) and quadratically with the number of nodes (N) in the network. The number of constraints in the above MILP formulation is bounded by $O(\mathbf{k}N^2)$, a number which again grows only linearly with the size of the set of multicast sessions and quadratically with the number of nodes in the network but is independent of the number of wavelength channels W supported in a fiber.

12.5.3 Variation of Problem Formulation with no λ Converters

In the absence of wavelength converters in the switch architectures of Figs. 12.6 and 12.7, the entire light-tree for a multicast session is on a common wavelength. The problem formulation for a network with the wavelength-continuity constraint (in an all-optical network) is almost the same as the problem with wavelength converters. Now, there are some additional variables and constraints. In the absence of wavelength converters, proper wavelength assignment of various multicast sessions becomes very important to minimize the overall cost [SuGT01].

1. Use of s, d, i, m, and n remain the same.

2. Use c as an index for the wavelength assigned to the multicast session.

- **Given:**
  - All the parameters in the previous problem remain the same except that, now, we do not have any wavelength converters.

- **Variables:**
  - A boolean variable, M_{mn}^{ic}, which is equal to one if the link between nodes m and n is occupied by multicast session i on wavelength c; otherwise, $M_{mn}^{ic} = 0$.
  - Definitions of the variables V_p^i and F_{mn}^i remain the same.
  - A boolean variable, C_c^i, which is equal to one if multicast session i is on wavelength c; otherwise, $C_c^i = 0$. The light-tree for a multicast session can occupy only one wavelength as there are no wavelength converters.

- **Optimize:**
  - Total cost of all multicast sessions:

$$Minimize: \sum_{i=1}^{i=k} \sum_{c=1}^{c=W} \sum_{m,n} w_{mn}.M_{mn}^{ic} \tag{12.43}$$

This is a linear objective function as it adds the cost of individual multicast sessions on all links and wavelengths.

- **Constraints:**

– Tree-creation constraints :

$$\forall i, \forall n \neq s_i : \sum_{m,c} M_{mn}^{ic} = V_n^i \quad (12.44)$$

$$\forall i : \sum_{m,c} M_{ms_i}^{ic} = 0 \quad (12.45)$$

$$\forall i, \forall j \in S_i : V_j^i = 1 \quad (12.46)$$

$$\forall i, \forall m \neq d_{i_j}, j = 1, \ldots, (L_i - 1) : \sum_{n,c} M_{mn}^{ic} \geq V_m^i \quad (12.47)$$

$$\forall i, m : \sum_{n,c} M_{mn}^{ic} \leq D_p(m).V_m^i (12.48)$$

$$\forall m, n : \sum_{i,c} M_{mn}^{ic} \leq P_{mn}.W \quad (12.49)$$

$$\forall m, n, c : \sum_{i} M_{mn}^{ic} \leq P_{mn} \quad (12.50)$$

– Commodity-flow constraints: The first four constraints are the same as Equations (12.36), (12.37), (12.38), and (12.39), respectively. Equations (12.40) and (12.41) are respectively modified as follows:

$$\forall i, m, n : \sum_{c} M_{mn}^{ic} \leq F_{mn}^i \quad (12.51)$$

$$\forall i, m, n : F_{mn}^i \leq N.\sum_{c} M_{mn}^{ic} \quad (12.52)$$

– Wavelength-related constraints:

$$\forall i : \sum_{c} C_c^i = 1 \quad (12.53)$$

$$\forall m, n (n > m) \forall i, c : M_{mn}^{ic} + M_{nm}^{ic} \leq C_c^i \quad (12.54)$$

- **Explanation of Equations:** Equation (12.50), (12.53), and (12.54) are new constraints. Other equations serve the same purpose as their corresponding ones in the previous problem formulation. Equation (12.50) restricts the number of sessions on the same wavelength between a node pair by P_{mn} (effectively ensuring that each fiber link can support no more than W wavelengths). Equation (12.53) ensures that a session chooses only one wavelength. Equation (12.54) ensures that no link is occupied by a session on the wavelength not chosen by it and all the links occupied by a session are on the same wavelength.

- **Number of Variables and Constraints:** The number of unknown variables in this system is $2\mathbf{k}N + \mathbf{k}N^2 + \mathbf{k}WN^2$, an additional $\mathbf{k}N + \mathbf{k}(W-1)N^2$ variables when compared with the formulation for a network with full wavelength-conversion capability. The number of additional constraints (when compared with the one in Section 12.5.2) in this MILP formulation is bounded by $\mathbf{k} + (\mathbf{k}+1)N^2$ and the total number of constraints is bounded by $O(\mathbf{k}N^2W)$.

  The constraint-set size grows linearly with the number of multicast sessions and number of channels per fiber, and it grows quadratically with the number of nodes in the network.

12.5.4 Variation of Problem Formulation with Fractional-Capacity Sessions

Now, we accommodate the more practical situation where some multicast sessions are operating at speeds lower than that of a wavelength channel. Hence, these sessions with different bandwidth granularities (say OC-1, OC-12, or OC-48) are *groomed* onto a single high-capacity wavelength (say OC-192) on a link. Grooming reduces the overall number of wavelength channels occupied by the sessions and hence the total cost of operating them. We denote the full channel capacity by C and the bit rate of low-speed session i by a fraction f_i (of the full capacity C). For example, if the full channel capacity is OC-192, then four multicast sessions, each at OC-48, can be "groomed" onto a common wavelength channel on a fiber link. Here, f_i will be 0.25 for all the four sessions.

In order to "groom" the signals carrying information on different bandwidth granularities, a digital cross-connect (DXC) is used, where signals are converted from the optical domain to the electronic domain (see [ZhMu02a] and Chapter 13). In a network equipped with DXCs, full wavelength conversion is implicit with no wavelength-continuity constraint. Hence, we present the formulation here only for a network equipped with full wavelength-conversion capability at every node.

- **Variables:**
  - All variables M^i_{mn}, F^i_{mn}, and V^i_n remain the same except the addition of a new variable f_i which is the fraction of the channel

capacity used by multicast session i.

- **Given:**
  - Along with the **k** multicast sessions, we are also given the capacity of each session i as a fraction of the capacity (C), denoted by f_i.
- **Optimize:**
  - Total cost of all multicast sessions:

$$Minimize: \quad \sum_{i=1}^{i=k} f_i . \sum_{m,n} w_{mn} . M^i_{mn} \quad (12.55)$$

- **Constraints:**
  - All constraint equations in Section 12.5.2 except Equation (12.35) remain the same. The altered link-capacity constraint is the following:

$$\forall m, n: \sum_i f_i . M^i_{mn} \quad \leq \quad P_{mn} . W \quad (12.56)$$

- **Number of Variables and Constraints:** The number of unknown variables in this formulation is $\mathbf{k} + \mathbf{k}N + 2\mathbf{k}N^2$, an additional **k** variables when compared with the formulation for a network with full wavelength-conversion capability. The number of constraints in this MILP formulation remain the same as in Section 12.5.2.

12.5.5 Variation of Problem Formulation with Splitters Constraints

Multicasting in the optical domain requires optical splitters to make multiple replicas of an incoming bit stream which are then transmitted to the destination nodes. Most nodes in a network have a limited number of splitters (because of cost) and these splitters have a limited fanout, i.e., splitting capacity because a higher fanout would result in an output signal with low power [Hada00]. By limiting the splitting degree of a signal at a node, the light-tree is more evenly distributed on the network which reduces the damage inflicted on the tree because of a node failure [BaVa95]. This problem of finding a Steiner tree in a network where the splitter fanout at a node is limited is known as the Degree-Constrained Steiner Problem (DCSP) and has been shown to be NP-complete [BaVa95, Voss90].

For the all-optical switch architectures (without electronic crossconnects) shown in Figs. 12.6 and 12.7, an array of optical splitters is necessary at a node to support several multicast sessions. Each splitter has a finite *splitting degree*, say **d**, which determines a signal fanout. Also, the size of the bank of splitters **B** at a node is limited, which is equal to the number of splitters at that node. For example, in Fig. 12.7, there are two splitters **X** and **Y**, one capable of splitting a signal 3-way and the other capable of 2-way splitting. Then, **d** = 3 for **X** and **d** = 2 for **Y**. The splitter-bank size, denoted by **B**, is 2. In the multicast-capable switch architectures presented in Section 12.2.2, the splitters available at a node are shared by all multicast sessions. The number of multicast sessions (requiring splitting) served by a node is limited by the splitter-bank size. Assuming *wavelength continuity*, the mathematical formulation for our problem requires some additional splitter-related constraints. Here, although we show a formulation using constant system-wide parameters **d** and **B**, the formulation can easily be adapted for a network where nodes have different splitter-bank size and optical splitters vary in their splitting degree. It is easy to see that this adaptation is achieved without increasing the number of constraint equations or the size of the problem. All sessions are operating at full channel capacity.

- **Variables:**
  - All variables $M_{mn}^{ic}, C_c^i, F_{mn}^i$, and V_n^i remain the same.
  - A boolean variable, A_j^i, which is equal to 1 if multicast session i requires a splitter at node j; otherwise, $A_j^i = 0$.

- **Given:**
  - Splitting degree of each splitter is **d**.
  - Splitter-bank size at each node is **B**.
  - A group of **k** multicast sessions.

- **Optimize:**
  - Total cost of all multicast sessions:

$$Minimize: \quad \sum_{i=1}^{i=k} \sum_{c=1}^{c=W} \sum_{m,n} w_{mn}.M_{mn}^{ic} \qquad (12.57)$$

- **Additional Constraints:**

$$\forall i, \forall n = d_{i_j}, j = 1, \ldots, (L_i - 1) : \sum_c \sum_m{}' M^{ic}_{nm} \ \leq \ \mathbf{d} - 1 \qquad (12.58)$$

$$\forall i, \forall n \neq d_{i_j}, j = 1, \ldots, (L_i - 1) : \sum_c \sum_m M^{ic}_{nm} \ \leq \ \mathbf{d} \qquad (12.59)$$

$$\forall i, \forall n = d_{i_j}, j = 1, \ldots, (L_i - 1) : \sum_c \sum_m M^{ic}_{nm} \ \geq \ A^i_n \qquad (12.60)$$

$$\forall i, \forall n = d_{i_j}, j = 1, \ldots, (L_i - 1) : \sum_c \sum_m M^{ic}_{nm} \ \leq \ D_p(n).A^i_n (12.61)$$

$$\forall i, \forall n \neq d_{i_j}, j = 1, \ldots, (L_i - 1) : \sum_c \sum_m M^{ic}_{nm} \ \geq \ 2.A^i_n \qquad (12.62)$$

$$\forall i, \forall n \neq d_{i_j}, j = 1, \ldots, (L_i - 1) : \sum_c \sum_m M^{ic}_{nm} \ \leq \ D_p(n).A^i_n (12.63)$$

$$\forall n : \sum_i A^i_n \ \leq \ \mathbf{B} \qquad (12.64)$$

  - Other equations from Section 12.5.3 remain unchanged.

- **Explanation of Equations:** Equation (12.58) restricts the number of outgoing paths at a destination node n for a session i by **d**-1, and Equation (12.59) achieves the same for other nodes in the network. Equations (12.60) through (12.63) accomplish the following: If there is at least one outgoing path of a session at a destination node, then a splitter is needed at this node for splitting the signal into two or more copies: one for local drop and the remaining for the downstream destination nodes. For all other intermediate nodes (including the source), a splitter is needed if they have at least two outgoing paths for a session. Equation (12.64) places a bound on the number of sessions using splitters available at node n.

- **Number of Variables and Constraints:** The number of unknown variables in this formulation is $3\mathbf{k}N + \mathbf{k}N^2 + \mathbf{k}WN^2$, an additional $\mathbf{k}N$ variables when compared with the formulation for a network with no wavelength-conversion capability (Section 12.5.3). The number of additional constraints (when compared with the constraints-set size in Section 12.5.3) in this MILP formulation is $3\mathbf{k}N$, and the bound on the total number of constraints is same as before which is $O(\mathbf{k}N^2W)$.

12.5.6 Illustrative Numerical Examples

Small Example

For illustration purpose, we first consider a small example, namely the network topology in Fig. 12.9, where the reader can verify the example visually with ease. Each link on the network carries $W = 2$ wavelengths. There are $P_{mn} = 1$ bidirectional fiber links between adjacent node pairs. We are given a group of five full-capacity multicast sessions $S_1 = \{$ **F**, A, B, C, D $\}$, $S_2 = \{$ **C**, A, E, F $\}$, $S_3 = \{$ **E**, A, B, C, F $\}$, $S_4 = \{$ **B**, D $\}$, and $S_5 = \{$ **A**, C $\}$ to establish on the network. The first element in the set is the source node and the remaining ones are the destination nodes. (Note that S_4 and S_5 are, essentially, unicast sessions.) The combined optimization problem formulations for routing and wavelength assignment of a group of multicast sessions is solved using a MILP solver, *CPLEX*.

Table 12.3 Optimal routing and wavelength assignment of multicast sessions in a network with and without wavelength converters.

Full-Capacity Multicast Sessions		Routing and Wavelength Assignment				
Source	Destination Nodes	Full λ Conversion		No λ Conversion		
		Route	Cost	Route	λ	Cost
F	{A, B, C, D }	F-E, E-A, E-D, D-C, C-B	18	F-E, E-A, E-D, D-C	λ_1	18
C	{A, E, F }	C-B, B-A, A-E, E-F	13	C-B, B-A, A-E, E-F	λ_0	13
E	{A, B, C, F }	E-F, E-A, A-B, B-C	13	E-A, A-F, A-B, B-C	λ_0	14
B	{ D }	B-D	7	B-D	λ_0	7
A	{ C }	A-B, B-C	8	A-B, B-C	λ_1	8
Total Cost			59			60

Table 12.3 shows that the optimal cost of establishing the above group of multicast sessions in a network with no wavelength converters is slightly more than the cost in a network where nodes are equipped with wavelength converters. Figures 12.9 and 12.10 show the light-trees for the five multicast sessions where nodes are equipped with and without wavelength converters, respectively. Observe that, in Fig. 12.9, there is a wavelength conversion from λ_1 (dashed line) to λ_0 (solid line) for the multicast session S_3 at node

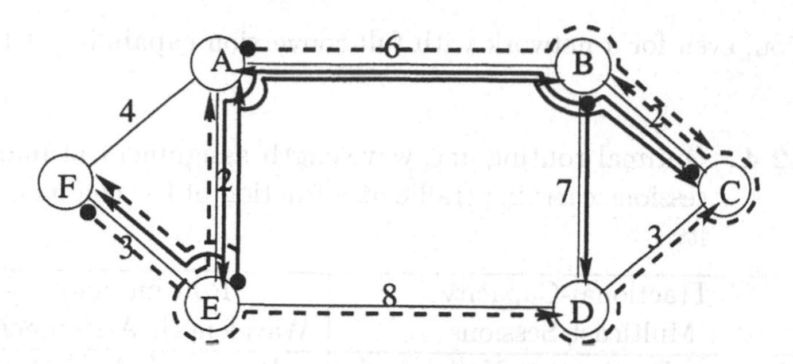

Figure 12.9 Optimal routing of the group of five multicast sessions with full-range wavelength converters. The dark circle represents the source node of a session and the nodes where arrows terminate are the destination nodes. Sessions on wavelength λ_0 (and λ_1) are shown with solid lines (and dashed lines) respectively. Observe that we need a wavelength converter at node **E** for the session with source **E**.

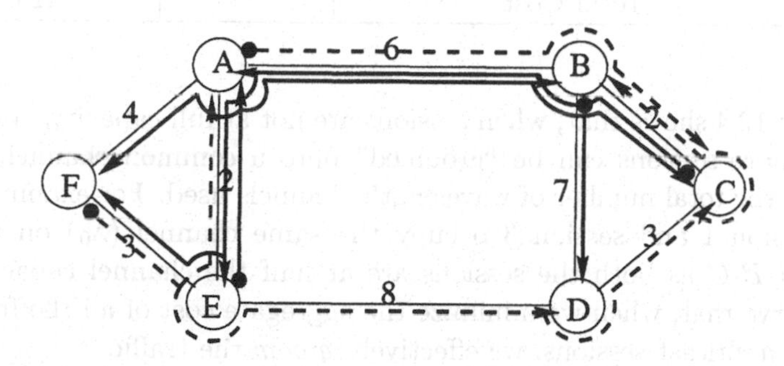

Figure 12.10 Optimal routing and wavelength assignment of the group of five multicast sessions with no wavelength conversion. The dark circles represent source nodes and the nodes where arrows terminate are the destination nodes. Sessions on solid lines occupy wavelength λ_0 while the ones on dashed lines occupy wavelength λ_1.

E in order to reduce the overall cost of setting up all the five sessions. The difference in the overall optimal cost for the two scenarios is not significant because, in our small example, the number of possible alternate routes for any session, even for a network with full-conversion capability, is limited.

Table 12.4 Optimal routing and wavelength assignment of multicast sessions carrying traffic at a fraction of the channel capacity.

Fractional-Capacity Multicast Sessions			Routing and Wavelength Assignment		
Source	Destination Nodes	f_i	Route	λ	Cost
F	{A, B, C, D}	0.5	F-A, A-B, B-C, C-D	λ_0	7.5
C	{A, E, F }	1.0	C-B, B-A, A-E, E-F	λ_1	13.0
E	{A, B, C, F }	0.5	E-F, E-A, A-B, B-C	λ_0	6.5
B	{ D }	1.0	B-D	λ_1	7.0
A	{ C }	1.0	A-B, B-C	λ_1	8.0
Total Cost					42.0

Table 12.4 shows that, when sessions are not at full capacity, traffic from two or more sessions can be "groomed" onto a common channel, thereby reducing the total number of wavelength-channels used. For example, traffic from session 1 and session 3 occupy the same channel (λ_0) on the links *A-B* and *B-C* as both the sessions are at half the channel capacity. One can observe that, when we minimize the aggregate cost of all the fractional-capacity multicast sessions, we effectively *groom* the traffic.

Table 12.5 compares the aggregate cost of optimally setting up the same group of five multicast sessions (at full capacity) by varying the splitter parameters (**d** and **B**) in the absence of wavelength converters. The optimal cost should now increase to compensate for limited resource availability. As expected, Table 12.5 shows that the total optimal cost of setting up the multicast sessions decreases with an increase in either the splitting degree (**d**) or the size of the splitter bank (**B**) at a node. The table also shows that it is not possible to establish all multicast sessions with limited network resources (**d** = 2 and **B** = 1) and some sessions will be blocked, which we indicate by using the term *Infeasible* in the table. Also, observe that the total optimal cost does not alter with an increase in splitting degree beyond

Table 12.5 Aggregate cost of optimally setting up the group of five multicast sessions for different values of **d** (splitting degree) and **B** (splitter-bank size) in a network with no wavelength converters.

Splitter Constraints	**B** = 1	**B** = 2	**B** = 3
d = 1	Infeasible	Infeasible	Infeasible
d = 2	Infeasible	61	61
d = 3	65	60	60
d = 4	65	60	60

the maximum nodal degree of the physical topology, namely 3.

A Larger Representative Example

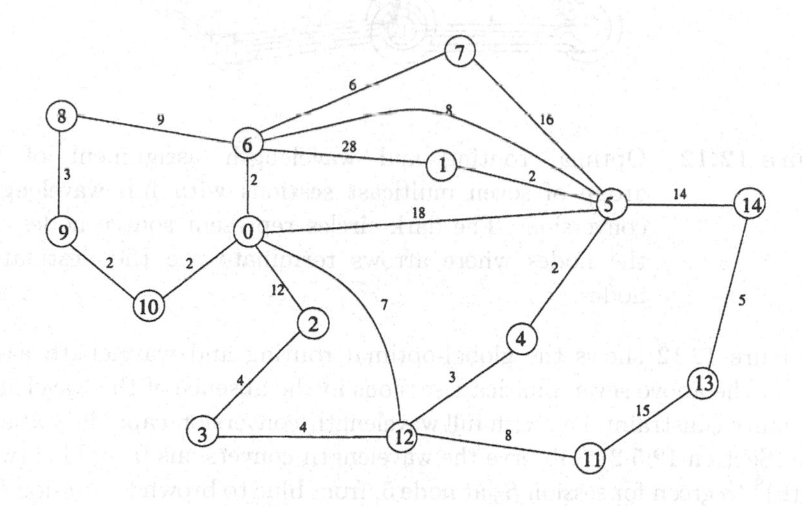

Figure 12.11 A typical 15-node telco network topology with bidirectional links.

We present solutions to the problem formulations for a larger, representative, 15-node wide-area network whose topology and link costs are shown in Fig. 12.11. Each link carries $W = 4$ wavelengths in both directions[7]. We are given a group of seven multicast sessions of varying clientele size $S_1 = \{$ **7**, 0, 6, 11 $\}$, $S_2 = \{$ **14**, 1, 2, 3, 4 $\}$, $S_3 = \{$ **12**, 0, 5, 8, 9, 10, 14 $\}$, $S_4 = \{$ **9**, 1, 2, 3, 4, 6, 10, 11 $\}$, $S_5 = \{$ **1**, 2, 4, 5, 6, 10, 12, 13, 14 $\}$, $S_6 = \{$ **0**, 1, 2, 4, 5, 6, 10, 11, 12, 13, 14 $\}$, and $S_7 = \{$ **11**, 0, 1, 2, 3, 4, 5, 6, 7, 8, 9, 10, 12, 13 $\}$. Note that S_7 is a broadcast session. (Recall that the first element in each set is the source and the remaining ones are destinations.)

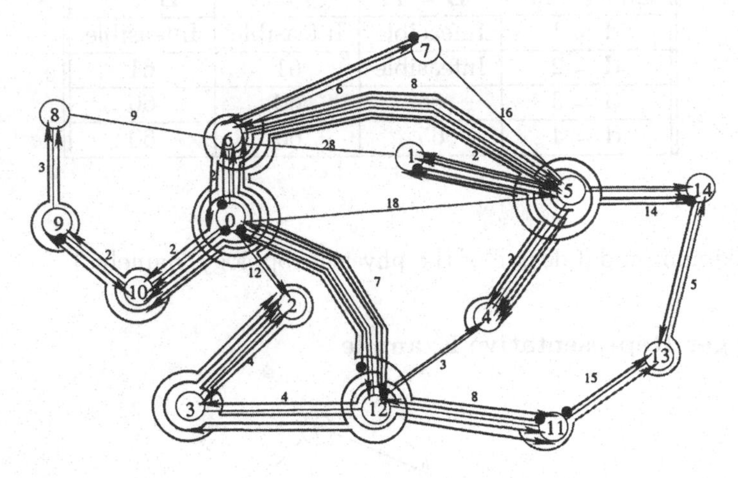

Figure 12.12 Optimal routing and wavelength assignment of the group of seven multicast sessions with full wavelength-conversion. The dark circles represent source nodes and the nodes where arrows terminate are the destination nodes.

Figure 12.12 shows the global-optimal routing and wavelength assignment of the above seven multicast sessions in the absence of the wavelength-continuity constraint, i.e., with full wavelength-conversion capability at every node (Section 12.5.2). Observe the wavelength conversions from blue (wavelength)[8] to green for session S_5 at node 5, from blue to brown for session S_4 at

[7]Some eight-wavelength results will be presented later in this subsection. The optimization problems that can utilize eight wavelengths require a very large amount of computation time, which will also be discussed later. They are also harder to visualize and verify manually; so we start with four-wavelength examples.

[8]Here, each color on a link represents a specific wavelength on the link, and they can be seen in the colored figures as well. But, in white and black figures, the colors on the

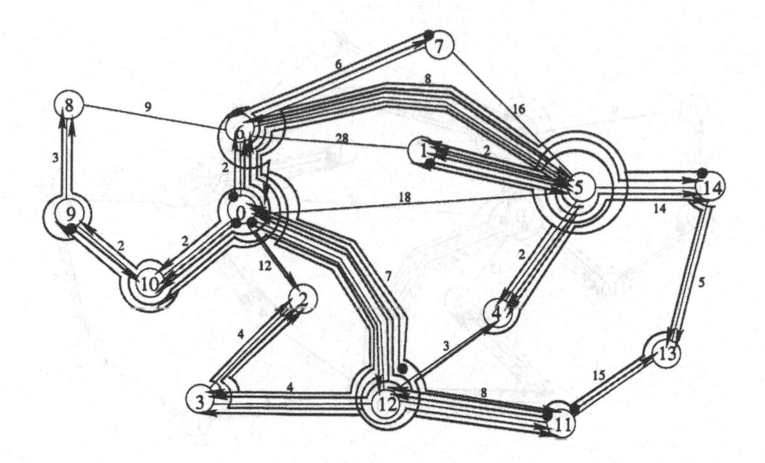

Figure 12.13 Optimal routing and wavelength assignment of the group of seven multicast sessions with no wavelength conversion. The dark circles represent source nodes and the nodes where arrows terminate are the destination nodes.

node 0, and from green to blue for session S_2 at node 5. The total minimized cost of setting up all seven sessions is 327 units. In case of the wavelength-continuity constraint, i.e., in the absence of wavelength-conversion capability at each node (Section 12.5.3), the optimal cost of setting up all seven sessions is 347 units as shown in Fig. 12.13. This shows a definite gain with global routing and wavelength assignment of several multicast sessions in the presence of wavelength converters.

Now, we consider the first six multicast sessions, where some sessions require a bandwidth lower than the full channel capacity. Specifically, three sessions operate at half the channel capacity and the remaining three at full channel capacity.

These six sessions are $S_1 = \{$ *1*, **7**, 0, 6, 11 $\}$, $S_2 = \{$ *1*, **14**, 1, 2, 3, 4 $\}$, $S_3 = \{$ *0.5*, **12**, 0, 5, 8, 9, 10, 14 $\}$, $S_4 = \{$ *0.5*, **9**, 1, 2, 3, 4, 6, 10, 11 $\}$, $S_5 = \{$ *1*, **1**, 2, 4, 5, 6, 10, 12, 13, 14 $\}$, and $S_6 = \{$*0.5*, **0**, 1, 2, 4, 5, 6, 10, 11, 12, 13, 14 $\}$. (Session S_7 has been dropped in this example to reduce cluttering.) In this modified notation, the first element in the set specifies the fraction of the channel capacity required by the multicast session, the second element is the source of the multicast session, and the remaining

links are shown in corresponding degrees of gray.

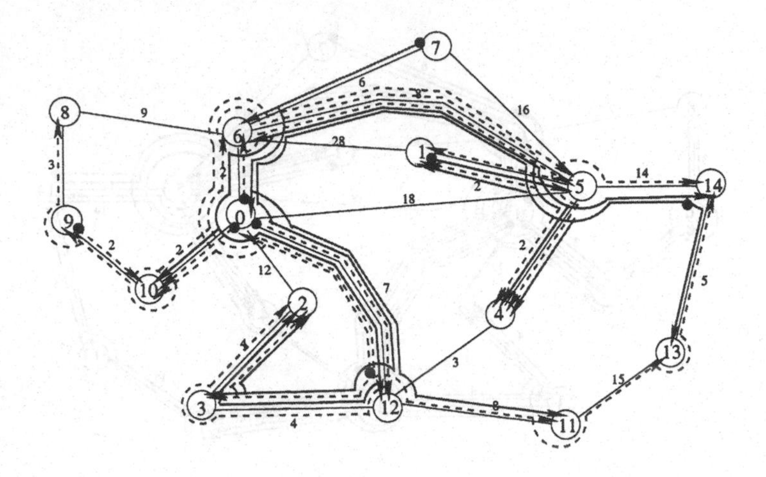

Figure 12.14 Optimal routing and wavelength assignment of the group of six fractional-capacity multicast sessions with full wavelength conversion. The solid lines represent sessions at full channel capacity and the dashed lines represent sessions at half the channel capacity.

elements are the destination nodes for that session. For example, the first element for S_3 is 0.5, which means it will occupy half the channel capacity $C/2$ (where C is the full channel capacity).

Figure 12.14 shows routing and wavelength assignment of the fractional-capacity sessions in the presence of wavelength converters (Section 12.5.4). The fractional-capacity sessions are shown with dashed lines, and full capacity sessions are shown with solid lines. Observe the "grooming" of the sessions on some links; for example, sessions on the red wavelength along the links 5 → 4, 0 → 6, 3 → 2, 12 → 3, 5 → 1, etc. in Fig. 12.14. The optimal cost of setting up the above six fractional-capacity sessions is 171.5 units.

Table 12.6 shows the optimal cost incurred to set up the seven multicast sessions, $\mathbf{S_1}$ through $\mathbf{S_7}$ (all occupying full channel capacity), on a network with varying sparse-splitting resources in a wavelength-continuous network. The table demonstrates the tradeoff between network-resource availability and the cost of setting up several sessions; the optimal cost decreases with either an increase in the splitting degree **d** or an increase in the splitter-bank size **B** at each node. The table also shows that all sessions cannot be established if sufficient number of splitters are not available or when splitters

Table 12.6 Cost incurred when optimally setting up a group of seven multicast sessions on a 15-node telco network for different values of **d** (splitting degree) and **B** (splitter-bank size) in a network with no wavelength converters. (Note: the mark "×" denotes the term of "infeasible".)

Splitter Constraints	B = 1	B = 2	B = 3	B = 4	B = 5	B = 6	B = 7
d = 1	×	×	×	×	×	×	×
d = 2	×	×	×	×	×	×	×
d = 3	×	×	691	452	362	353	347
d = 4	×	×	519	404	358	349	347
d = 5	×	×	516	404	351	349	347

have inadequate splitting degree, e.g., for $\mathbf{B} \leq 2$ or for $\mathbf{d} \leq 2$.

So far, we showed results for the 15-node network in Fig. 12.11 with $W = 4$ channels on each fiber link. Each of the solution points shown in Table 12.6 took an average of 28 hours on a Pentium IV, 1.7-GHz machine with 512 MB RAM running Linux operating system. We also ran the optimizations for $W = 8$ channels per fiber for a group of 15 multicast sessions (see Table 12.7) with average group size 11 where every node is a source node for a session.

The optimal cost of simultaneously setting up these 15 sessions on a network with wavelength converters is 1125 units and this solution was obtained in 4 hours on the same computing platform mentioned above. However, for a network with the wavelength-continuity constraint, the optimal cost of setting up the same group of 15 sessions (1319 units) was obtained after running the simulation for a significantly longer duration of 145 hours with the same computing platform.

12.6 Multicast Tree Protection

Fiber cuts in an optical network occur often enough to cause service disruption, and they may lead to significant information loss in the absence of adequate backup mechanisms. The loss could be heavy when the failed link in a "light-tree" carries traffic for multiple destinations.

Table 12.7 A group of 15 multicast sessions.

Session Id	{Source, Destination Nodes}
S_1	{**13**, 1, 2, 3, 4}
S_2	{**10**, 0, 5, 8, 9, 12, 14}
S_3	{**9**, 1, 2, 3, 4, 6, 10, 11}
S_4	{**2**, 3, 4, 5, 8, 9, 10, 12, 13}
S_5	{**4**, 2, 6, 7, 8, 9, 12, 13, 14}
S_6	{**1**, 2, 4, 5, 6, 10, 12, 13, 14}
S_7	{**12**, 3, 4, 5, 6, 7, 8, 10, 11, 13, 14}
S_8	{**6**, 1, 2, 4, 5, 9, 10, 11, 12, 13, 14}
S_9	{**5**, 1, 2, 3, 4, 6, 10, 11, 12, 13, 14}
S_{10}	{**0**, 1, 2, 4, 5, 6, 10, 11, 12, 13, 14}
S_{11}	{**14**, 0, 1, 2, 3, 4, 5, 6, 7, 8, 9, 10, 11, 12}
S_{12}	{**8**, 0, 1, 2, 3, 4, 5, 6, 7, 9, 10, 11, 12, 13}
S_{13}	{**7**, 0, 1, 2, 3, 4, 5, 6, 8, 9, 10, 11, 12, 13}
S_{14}	{**3**, 0, 1, 2, 4, 5, 6, 7, 8, 9, 10, 11, 12, 13}
S_{15}	{**11**, 0, 1, 2, 3, 4, 5, 6, 7, 8, 9, 10, 12, 13}

12.6.1 Protection Schemes

A straightforward approach to protecting a multicast tree is to compute a *link-disjoint* backup tree. Two trees are said to be *link disjoint* if they do not share any link along their edges. Such link-disjoint trees can be used to provide 1+1 dedicated protection (see Chapter 11) where both primary tree and backup tree carry identical bit streams to the destination nodes. When a link fails, the affected destination nodes reconfigure their switches to receive bit streams from the backup tree instead of the primary tree. Pitfalls of this approach include excessive use of resources and inability to discover link-disjoint trees in a mesh network, which may lead to the blocking of a significant number of multicast sessions while trying to establish them.

A more resource-efficient approach is to relax the protection constraint from link-disjointness to *directed-link-disjointness*[9], thus allowing the primary and backup tree to share links, but only in opposite directions. Such *directed-link-disjoint* trees provide 1:1 dedicated protection (see Chapter 11)

[9]In the literature, *directed-link-disjointness* has also been referred to as *arc-disjointness*.

where the resource for the backup tree is reserved and it is used to carry bit streams when failure occurs. We show below that *directed-link-disjointness* can be successfully exploited in an optical WDM network (where a failure may disrupt both primary and backup trees) to protect multicast sessions against single fiber cut.

Consider a primary light-tree carrying traffic to three destinations d_1, d_2, and d_3 along a link between nodes u and v as shown in Fig. 12.15. We examine the following two cases:

- **Case 1:** If the backup tree does not occupy the link $u \longrightarrow v$ and there is a cut on this link, the affected downstream nodes (d_1, d_2, and d_3 in our case) switch to a different incoming port (shown with dashed lines in Fig. 12.15), and continue to receive the bit stream from the backup tree.

- **Case 2:** If the backup tree occupies the link $v \longrightarrow u$ (in opposite direction to the primary), a cut in the link $u \longleftrightarrow v$ leads to failure of both the primary tree and the backup tree. Because the backup tree occupies the link $v \longrightarrow u$, node v should be reachable from the source shown with partial dashed lines in Fig. 12.15. All affected destination nodes downstream of node v are reachable along the primary tree even after the link failure. If the switch at node v is reconfigured to route the incoming bit stream from the backup tree (instead of from a primary-tree incoming port) to the original (primary) outgoing ports, the victimized destinations can continue to receive the traffic without having to perform any reconfiguration! Sometimes, a λ converter may be required at the downstream node (v) to convert a bit stream from a backup-tree wavelength to a primary-tree wavelength, if they are different. Because the two trees (primary and backup) do not share a link in the same direction, they can also occupy the same wavelength, if necessary, and then λ conversion will not be required. Also, in a network equipped with opaque crossconnects with O-E-O conversion, no explicit λ converter is needed.

12.6.2 General Problem Statement

The problem of setting up a group of multicast sessions (primary and their backup trees such that they do not share a link in the same direction) using a *light-tree*-based approach on a given physical topology (fiber network) is formally stated below. Here, we provide a general problem statement.

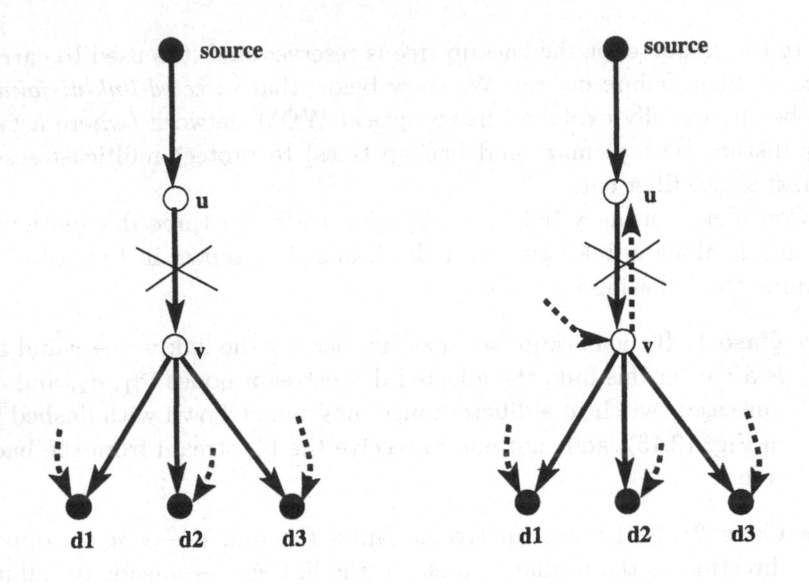

Figure 12.15 Fiber cuts along a light-tree from a source node to a set of destination nodes d_1, d_2, and d_3. Solid lines belong to primary tree and dashed lines form parts of the backup tree. A fiber cut may disrupt either the primary tree or both primary and backup trees.

Different flavors of the problem are tackled in later sections. We are given the following inputs:

1. A physical topology $G_p = (V, E_p)$ consisting of a weighted undirected graph, where V is a set of network nodes, and E_p is the set of links connecting the nodes. Undirected means that each link in the physical topology is bidirectional. Each link is assigned a weight to represent the cost (number of hops or equipment operating cost) of moving traffic from one end to the other. Node j is equipped with a $D_p(j) \times D_p(j)$ MWRS, where $D_p(j)$, called the physical degree of node j, equals the number of physical fiber links emanating out of (as well as terminating at) node j.

2. Number of wavelength channels on each fiber $= W$.

3. A group of k primary multicast sessions and a binary digit P_i (

$i = 1, 2, \ldots, k$) associated with each of them to indicate whether they require protection or not.

Our goal is to set up (if possible) all $2k$ primary and backup multicast sessions on the given physical topology while minimizing the total cost. Cost of a multicast session is the sum of the weights on the physical links occupied by it.

12.6.3 Problem Formulation for a Network without λ Continuity

The switch architecture shown in Fig. 12.5 has full-range wavelength conversion inherent in it. Here, we formulate the problem of setting up a group of multicast sessions for a network equipped with "opaque" switches. Following are a few notations which we shall use:

1. s and d refer to source node and destination nodes respectively, in a multicast session.
2. m and n denote *end-points of a physical link* that might occur in a light-tree.
3. i is used as an index for session number, where $i = 1, 2, \ldots, \mathbf{2k}$. Indices 1 through k are used for the primary trees and indices k+1 through $2k$ are used for the backup trees. If a primary session i requires protection, then its corresponding backup tree index is i+k. Otherwise, the index is left unused and the variables corresponding to that backup tree are ignored.

- **Given:**
  - Number of nodes in the network = N.
  - Maximum number of wavelengths per fiber = W.
  - Physical topology $P_{m,n}$, where $P_{m,n} = P_{n,m}$ (i.e., fiber links are bidirectional) indicates whether there is a direct physical fiber link between nodes m and n, where $m, n = 1, 2, 3, \ldots, N$. If there is no fiber link between nodes m and n, then $P_{m,n} = P_{n,m} = 0$.
  - Every physical link between nodes m and n is associated with a weight and cost $w_{m,n}$.
  - Capacity of each channel = C. For example, C may be OC-192 $\approx$ 10 Gbps.

- A group of k multicast sessions S_i for $i = 1, 2, 3, \ldots, \mathbf{k}$; and $P_i = 1$ or 0 for each session signifying whether session i requires protection or not, respectively. Each session S_i has a source node and a set of destination nodes denoted by $\{s_i, d_{i_1}, d_{i_2}, \ldots\}$. We denote the cardinality of a multicast session i by L_i, which is equal to the number of source and destination nodes in that session.
- Every multicast session is at full capacity of a channel, i.e., at C bps.
- Every node is equipped with wavelength converters capable of converting a wavelength to any other wavelength among W channels.

- **Variables:**
  - A boolean variable, $M^i_{m,n}$, which is equal to one if the link between nodes m and n is occupied by multicast session i; otherwise, $M^i_{m,n} = 0$.
  - A boolean variable, V^i_p, which is equal to one if node p belongs to multicast session i; otherwise, $V^i_p = 0$. A node belongs to a session if it is either the source or the destination or an intermediate node in the light-tree for the multicast session, e.g., $V^i_{s_i} = 1$ and $V^i_{d_{i_1}} = 1$.
  - An integer commodity-flow variable, $F^i_{m,n}$. Each destination node for a session needs one unit of commodity. So, there are L_i units of commodity flowing out of source s_i for session i. $F^i_{m,n}$ is the number of units of commodity flowing on the link from node m to node n for session i. $F^i_{m,n}$ is also the number of destination nodes in session i downstream of the link between nodes m and n.

- **Optimize:**
  - Minimize total cost of all multicast sessions:

$$Minimize: \quad \sum_{i=1}^{i=2k} \sum_{m,n} w_{m,n}.M^i_{m,n} \tag{12.65}$$

- **Constraints:**

– Tree-creation constraints:

$$\forall i, \forall n \neq s_i : \sum_m M^i_{m,n} = V^i_n \quad (12.66)$$

$$\forall i : \sum_m M^i_{m,s_i} = 0 \quad (12.67)$$

$$\forall i, \forall j \in S_i : V^i_j = 1 \quad (12.68)$$

$$\forall i, \forall m \neq d_{i_j}, j \geq 1 : \sum_n M^i_{m,n} \geq V^i_m \quad (12.69)$$

$$\forall i, m : \sum_n M^i_{m,n} \leq D_p(m).V^i_m \quad (12.70)$$

$$\forall m, n : \sum_i M^i_{m,n} \leq P_{m,n}.W \quad (12.71)$$

– Commodity-flow constraints:

$$\forall i, \forall m \notin S_i : \sum_n F^i_{m,n} = \sum_n F^i_{n,m} \quad (12.72)$$

$$\forall i, \forall m = s_i : \sum_n F^i_{s_i,n} = L_i \quad (12.73)$$

$$\forall i, \forall m = s_i : \sum_n F^i_{n,s_i} = 0 \quad (12.74)$$

$$\forall i, \forall m = d_{i_j}, j \geq 1 : \sum_n F^i_{n,m} = \sum_n F^i_{m,n} + 1 \quad (12.75)$$

$$\forall i, m, n : M^i_{m,n} \leq F^i_{m,n} \quad (12.76)$$

$$\forall i, m, n : F^i_{m,n} \leq N.M^i_{m,n} \quad (12.77)$$

– Directed-link-disjointness:

$$\forall i = 1 \ldots \mathbf{k} \forall m, n : M^i_{m,n} + M^{i+\mathbf{k}}_{m,n} \leq 1 \quad (12.78)$$

- **Explanation of Equations:** The equations are written for creating a tree for every multicast session. Equation (12.66) ensures that every node that belongs to a multicast session (except the source) has at least one incoming edge. Equation (12.67) states that the source node has no incoming edge as it is the root of the tree. Equation (12.68) ensures that every source node and the destination node of a multicast session belong to the tree. Equation (12.69) ensures that every node (except the destination nodes) belonging to the tree has at least one outgoing edge. Equation (12.70) ensures that every node with at least one outgoing edge belongs to the tree. Equation (12.71) restricts the number of lightpaths between nodes m and n by $P_{m,n}.W$ in each direction.

Equations (12.72) - (12.77) are flow-conservation equations to create a connected tree with the source having a lightpath to every destination in the session. Equation (12.72) ensures that, at any intermediate node (which is neither a source nor a destination), the incoming flow is same as the outgoing flow. However, outgoing flow at the source node for a session is the number of destinations in the session and the incoming flow is zero. These are achieved by Equations (12.73) and (12.74), respectively. Equation (12.75) ensures that the total outgoing flow is one less than the incoming flow for destination nodes. Equations (12.76) and (12.77) ensure that every link occupied by a session has a positive flow and links not occupied by the session have no flow. In Equation (12.77), N can be replaced by L_i without altering its meaning. A flow on any link for a multicast session is limited by the number of destinations in that session. Equation (12.78) ensures that the primary and backup tree share a link, if any, only in opposite directions.

12.6.4 Problem Formulation for a Network with λ Continuity

In the absence of wavelength converters in the switch architectures shown in Fig. 12.6 and Fig. 12.7, the entire light-tree for a multicast session is on a common wavelength. The problem formulation for a network with wavelength-continuity constraint (in an all-optical network) is almost the same as the problem with wavelength converters. Now, there are some additional variables and constraints. In the absence of wavelength converters, proper wavelength assignment of various multicast sessions [JDHM01] becomes very important to minimize the overall cost.

1. Use of s, d, i, m, and n remain the same.

2. Use c as an index for wavelength of a multicast session.

- **Given:**

- All previous parameters remain the same except for the wavelength-continuity constraint.

- **Variables:**
  - A boolean variable, $M_{m,n}^{i,c}$, which is equal to one if the link between nodes m and n is occupied by multicast session i (could be primary or backup) on wavelength c; otherwise, $M_{m,n}^{i,c} = 0$.
  - Definitions of variables V_m^i and $F_{m,n}^i$ remain same.
  - A boolean variable, C_c^i, which is equal to one if multicast session i is on wavelength c; otherwise, $C_c^i = 0$. The light-tree for a multicast session can occupy only one wavelength as there are no wavelength converters.

- **Optimize:**
  - Total cost of all multicast sessions:

$$Minimize : \sum_{i=1}^{i=2k} \sum_{c=1}^{c=W} \sum_{m,n} w_{m,n}.M_{m,n}^{i,c} \tag{12.79}$$

- **Constraints:**
  - Tree-creation constraints:

$$\forall i, \forall n \neq s_i : \sum_{m,c} M_{m,n}^{i,c} = V_n^i \tag{12.80}$$

$$\forall i : \sum_{m,c} M_{m,s_i}^{i,c} = 0 \tag{12.81}$$

$$\forall i, \forall j \in S_i : V_j^i = 1 \tag{12.82}$$

$$\forall i, \forall m \neq d_{i_j}, j \geq 1 : \sum_{n,c} M_{m,n}^{i,c} \geq V_m^i \tag{12.83}$$

$$\forall i, m : \sum_{n,c} M_{m,n}^{i,c} \leq D_p(m).V_m^i \tag{12.84}$$

$$\forall m, n : \sum_{i,c} M_{m,n}^{i,c} \leq P_{m,n}.W \tag{12.85}$$

$$\forall m, n, c : \sum_{i} M_{m,n}^{i,c} \leq P_{m,n} \tag{12.86}$$

– Commodity-flow constraints:
The first four constraints are the same as Equations (12.72), (12.73), (12.74), and (12.75) respectively. Additional equations are the following:

$$\forall i, m, n : \sum_c M_{m,n}^{i,c} \leq F_{m,n}^i \tag{12.87}$$

$$\forall i, m, n : F_{m,n}^i \leq N. \sum_c M_{m,n}^{i,c} \tag{12.88}$$

– Directed-link-disjointness:

$$\forall i = 1 \ldots \mathbf{k} \forall m, n : \sum_c M_{m,n}^{i,c} + \sum_c M_{m,n}^{i+\mathbf{k},c} \leq 1 \tag{12.89}$$

– Wavelength-related constraints:

$$\forall i : \sum_c C_c^i = 1 \tag{12.90}$$

$$\forall m, n (n > m) \forall i, c : M_{m,n}^{i,c} + M_{n,m}^{i,c} \leq C_c^i \tag{12.91}$$

- **Explanation of Equations:** Equation (12.86), (12.90), and (12.91) are new constraints. Other equations serve the same purpose as before. Equation (12.86) restricts the number of sessions on the same wavelength between a node pair by $P_{m,n}$ (effectively ensuring that each fiber link supports no more than W wavelengths). Equation (12.90) ensures that a session chooses only one wavelength. Equation (12.91) ensures that no link is occupied by a session on the wavelength not chosen by it and all links occupied by a session are on the same wavelength.

12.6.5 Illustrative Numerical Examples

We consider the 15-node network shown in Fig. 12.11. Each link carries $W = 4$ wavelengths in both directions. There are $P_{m,n} = 1$ bidirectional fiber links between adjacent node pairs. We are given a group of five multicast sessions: $S_1 = \{ \mathbf{0}, 1, 2, 4, 5, 6, 10, 11, 12, 13, 14 \}$, $S_2 = \{ \mathbf{9}, 1, 2, 3, 4, 5, 6, 10, 11\}$, $S_3 = \{ \mathbf{12}, 0, 5, 8, 9, 10, 14 \}$, $S_4 = \{ \mathbf{14}, 1, 2, 3, 4 \}$, and $S_5 = \{ \mathbf{7}, 0, 6 \}$ to be established on the network. The first element in each set is the source node and the remaining ones are destination nodes. Sessions S_1, S_4, and S_5 need single-link failure protection, i.e., each of them require a backup light-tree with session identifiers S_6, S_9, and S_{10}, respectively. The optimization problem formulations for routing and wavelength assignment for a group of primary and their backup sessions are solved using an ILP solver, **CPLEX**.

Figure 12.16 shows an optimal RWA of the above eight sessions in the absence of the wavelength-continuity constraint (Section 12.6.3). Primary trees are shown in solid lines and backup trees with dotted lines. Observe that the primary tree S_1 along links $0 \longrightarrow 10, 0 \longrightarrow 2, 0 \longrightarrow 12, 12 \longrightarrow 4, 4 \longrightarrow 5, 5 \longrightarrow 1, 5 \longrightarrow 6, 5 \longrightarrow 14, 14 \longrightarrow 13$, and $13 \longrightarrow 11$ and its backup S_6 along links $0 \longrightarrow 6, 6 \longrightarrow 8, 8 \longrightarrow 9, 9 \longrightarrow 10, 6 \longrightarrow 1, 1 \longrightarrow 5, 5 \longrightarrow 4, 4 \longrightarrow 12, 12 \longrightarrow 11, 11 \longrightarrow 13, 13 \longrightarrow 14, 12 \longrightarrow 3$, and $3 \longrightarrow 2$ share common links but only in opposite directions. The total cost of setting up all eight sessions is 386 units. In case of wavelength-continuity constraint (Section 12.6.4), the cost of setting up the same set of sessions is higher (391 units) whose RWA is shown in Fig. 12.17. Since any primary and its backup don't share a common link in the same direction, they can be on a common wavelength, e.g., S_5 and S_{10} are on the same wavelength in Fig. 12.17.

12.6.6 Other Protection Schemes

Although it may not always be possible to find a *directed-link-disjoint* backup tree once a primary tree has been discovered, it may be possible to protect each segment in the primary tree by finding a *segment-disjoint* path. A *segment* is defined as the sequence of edges from the source or any splitting point (on a tree) to a leaf node or to a downstream splitting point. A destination node is always considered as a *segment end-node* because it is either a leaf node in a tree or is a splitting point where a portion of a signal is dropped locally and the remainder continues (e.g., Drop-and-Continue [DaC] node) [ZhWQ00]. For example, in Fig. 12.18, a primary tree (shown in dotted lines) is found along edges $s \rightarrow u, s \rightarrow v, v \rightarrow d_1$, and $v \rightarrow d_2$. Here, node v is a *splitting point* and creates three segments, viz., $s \rightarrow u \rightarrow v, v \rightarrow d_1$, and $v \rightarrow d_2$. Observe that there is no other path, besides the paths along the primary tree, from source node s to either destination nodes d_1 or d_2; and, hence, it is not possible to discover a *directed-link-disjoint* backup tree. However, each of the three segments $s \rightarrow v, v \rightarrow d_1$, and $v \rightarrow d_2$ of the tree can be protected by *segment-disjoint* paths $< s \rightarrow w \rightarrow v >$, $< v \rightarrow d_2 \rightarrow d_1 >$, and $< v \rightarrow d_1 \rightarrow d_2 >$, respectively. Note that segment-disjoint paths can share their arcs with existing primary-tree arcs or other backup-segment arcs (see Fig. 12.18) which an arc-disjoint backup

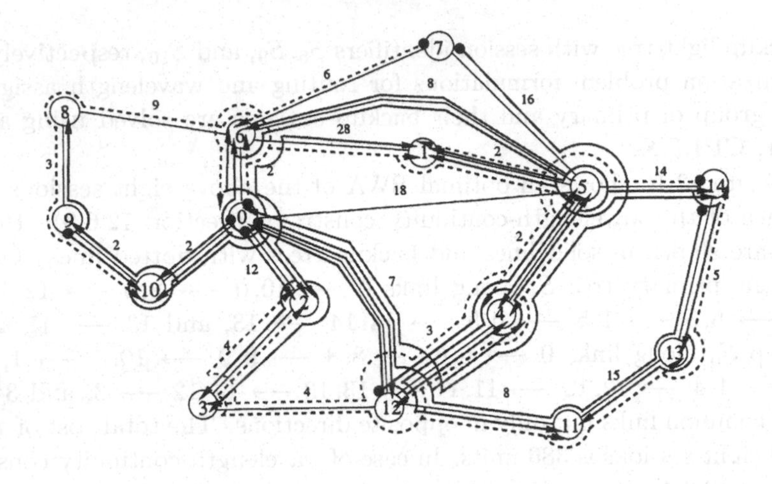

Figure 12.16 Optimal RWA of a group of eight multicast sessions (five primary sessions shown with solid arrows and three backup sessions shown with dotted arrows) in absence of λ-continuity constraint. Dark circles represent source nodes and terminating arrows indicate destination nodes.

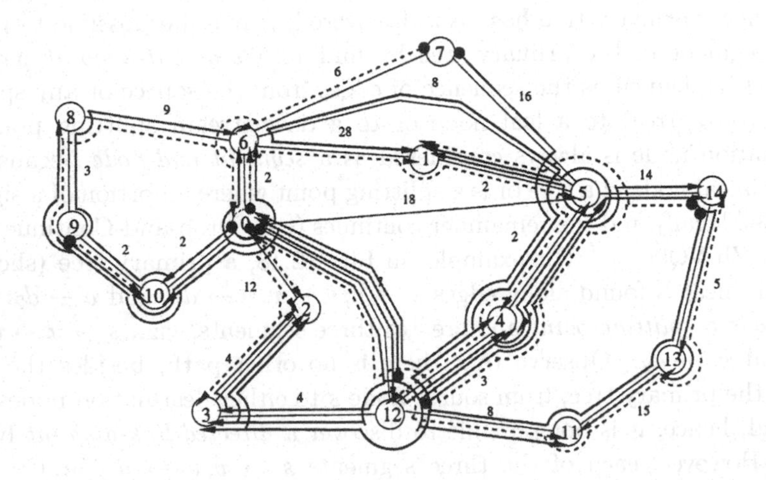

Figure 12.17 Optimal RWA of a group of eight multicast sessions (five primary sessions shown with solid arrows and three backup sessions shown with dotted arrows) in presence of λ-continuity constraint. Dark circles represent source node and terminating arrows indicate destination nodes.

tree cannot accomplish.

A basic requirement in disseminating information from a source node to all destination nodes is the existence of a path from the source to each destination node. In order to protect each such path (from source node to a destination node) against any cut on a (fiber) link along the primary path, a backup path from the source node to the destination node is essential which is link disjoint to its corresponding primary path. Thus, a link-disjoint path pair from the source node to every destination node is sufficient to handle any link failure in a directed graph. For example, Fig. 12.19 has link-disjoint paths from s to d_1 (primary path $< s \rightarrow u \rightarrow d_1 >$ and backup path $< s \rightarrow v \rightarrow w \rightarrow x \rightarrow d_1 >$ shown with dotted and dashed lines, respectively) and link-disjoint paths from s to d_2 (primary path $< s \rightarrow v \rightarrow d_2 >$ and backup path $< s \rightarrow u \rightarrow w \rightarrow x \rightarrow d_2 >$). Note that the backup path for d_1 shares a link $(w \rightarrow x)$ with the backup path for d_2. These link-disjoint paths from source node to a destination node are called a *path pair*. Readers can verify that a switch at a node can be appropriately configured to send bit streams from source s to both destination nodes when any link fails in the network.

For further reading on multicast-tree protection, we refer the interested reader to [SiOM05, SiMu03, SiSM03].

12.7 Other Related Issues on Light-Tree Multicasting

Multicast communication arises from a wide variety of applications such as video distribution and teleconferencing. Given a source node and a set of destination nodes, the multicast routing and wavelength assignment (MRWA) problem is to find a set of links and wavelengths on these links on which to establish the connection from the source to the destination nodes. Similarly, the wavelength assignment (WA) problem is to find a set of wavelengths to use on a predetermined multicast tree. It is desirable to find an RWA or a WA that is optimal with respect to some cost metric. In general, the RWA problem is to select a multicast tree, the wavelengths on the links in the tree, and the intermediate nodes that may perform wavelength conversion. In the WA problem, a multicast tree is given and the problem is to select the wavelengths on the links in the tree and the intermediate nodes for wavelength conversion. In general, it may not always be possible to find an RWA or a WA for a given multicast request because the needed wavelengths on particular links, as well as transmitters and receivers at intermediate nodes,

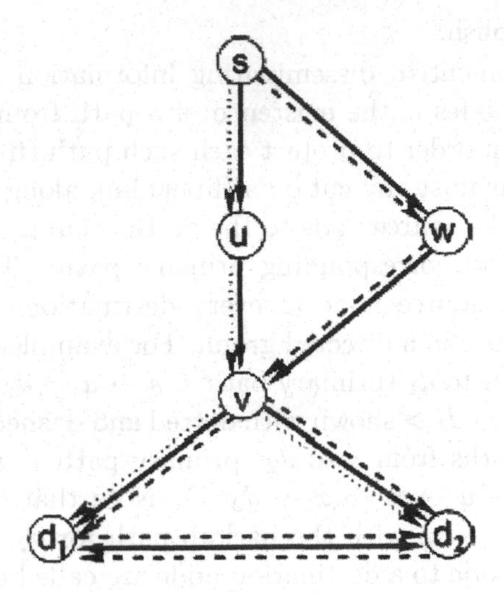

Figure 12.18 Using shared segment-disjoint approach to protect a multicast session where directed-link-disjoint tree approach fails.

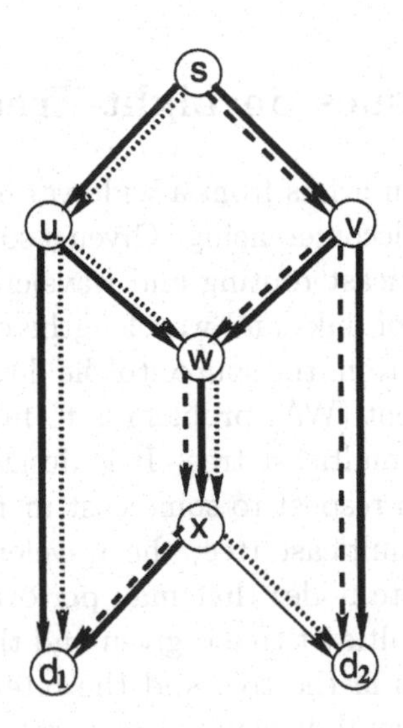

Figure 12.19 Using optimal path-pair approach to protect a multicast session where directed-link-disjoint tree approach fails.

are being used by other multicast connections. For more information, we refer the reader to [SaMu99, LiMe02].

Another issue is the light-tree routing problem under optical-layer power-budget constraints. Current light-tree routing generally requires the use of a number of optical power splitters, and it may have a high power budget. It would be reasonable to set up constraints on the end-to-end paths in order to guarantee an adequate signal quality and to ensure a measure of fairness among the destination nodes. These constraints require the light-tree to be balanced or distance-weighted balanced. Please see [XiRo04, AlDe00] for more information on this topic.

12.8 Summary

This chapter presented two types of switch architectures (opaque and transparent) for supporting multicasting in WDM networks. The concept of "light-tree" was introduced and we showed how a light-tree can reduce network-wide average packet hop distance and the total number of transceivers for unicast and broadcast traffic. Efficient algorithms to reduce network resources were presented for establishing multicast connections using light-trees.

The notions of *directed-link-disjointness*, *segment-protection*, and *path protection* were introduced for protecting multicast trees. These schemes were shown to protect a single tree. As an improvement, these schemes can be made more efficient by allowing *sharing* of backup edges across several trees; please see [SiOM05] for a detailed treatment of this problem.

Exercises

12.1. Consider the NSFNET backbone topology in Fig. 12.1. We need to establish a multicast session from source node $CA1$ to destination nodes MI, NY and NJ. Find the minimum Steiner Tree to set up the multicast session. What is the total cost of your light-tree?

Now, consider, the weight of the link $UT - MI$ is increased by 7 to 20 units. Find SMT on the new topology. What is the cost of the new light-tree?

12.2. For problem 12.1, please draw the virtual links and show the percentages of traffic from the source node to each of the destination nodes in both the cases.

12.3. For the network in Fig. 12.4, show the MWRS configuration at Node B using the architectures in Fig. 12.5 (opaque switch) and Figs. 12.6 and 12.7 (transparent switches).

12.4. In Fig. 12.4, suppose the cost of link A–B is increased to 30 units. How will the light-tree now be routed? Draw a new diagram. Show the MWRS configuration of Node D using switch architectures in Figs. 12.5 and 12.6.

12.5. A transparent switch, with two optical switches, has 12 incoming/outgoing fibers, having 8 wavelengths in each fiber. The sizes of the larger and the smaller optical switches are 98×102 and 20×20, respectively. Find the splitter-bank size. Assume that the number of local add/drop ports in the larger and smaller switches are 2 and 4, respectively. What is the splitting degree of each splitter?

Now, consider the modified switch having one optical cross-connect. Find the size of the switch. Use the data found before.

12.6. In a MILP formulation of multicast traffic with wavelength converters, there are 165 variables. Find the number of nodes in the network if there are 3 multicast sessions. Assume that each fiber link can support 8 wavelengths. What is the upper bound of the number of constraints in the above MILP? How does the number of variables change, assuming no wavelength converter? Also, calculate the number of constraints.

12.7. Consider the MILP formulation of multicast traffic with grooming. Find the number of variables and the number of constraints. How does your solution change if you have only 5 splitters at each node, with each having a splitting degree of 3? Use the data found in Problem 12.6.

12.8. For the multicast sessions in Table 12.3, draw the corresponding logical topology. Are the links in the logical topology directed or undirected? Why?

12.9. Consider the network topology in Fig. 12.9. There is only one direct physical bidirectional fiber link between any two nodes. Each link carries one wavelength. We need to establish 5 full-capacity multicast sessions $S_1 = \{\mathbf{E}, A, B, C\}$, $S_2 = \{\mathbf{D}, E, F, C\}$, $S_3 = \{\mathbf{C}, F, A, D\}$, $S_4 = \{\mathbf{A}, E, F\}$ and $S_5 = \{\mathbf{F}, E, D\}$. Solve the routing and wavelength assignment problem using an MILP. What is the total cost?

12.10. Consider the network topology in Fig. 12.9. There is one direct physical bidirectional fiber link between any two nodes. Each link carries 2 wavelengths. We need to establish 5 full-capacity multicast sessions $S_1 = \{\mathbf{E}, A, B, C\}$, $S_2 = \{\mathbf{D}, E, F, C\}$, $S_3 = \{\mathbf{C}, F, A, D\}$, $S_4 = \{\mathbf{A}, E, F\}$ and $S_5 = \{\mathbf{F}, E, D\}$. Solve the routing and wavelength assignment problem using an MILP. What is the total cost? Assume a no-λ-conversion network.
Solve this problem for a full λ-conversion network.

12.11. How would your solutions to Problems 12.9 and 12.10 change if connections S_1, S_2 and S_4 use 25%, and connections S_3 and S_5 use 50% of channel capacity?

12.12. Consider the network topology in Fig. 12.11. There is only one direct physical bidirectional fiber link between any two nodes. Each link carries 2 wavelengths. We need to establish 6 multicast sessions $S_1 = \{\mathit{1.0}, \mathbf{6}, 1, 5, 4\}$, $S_2 = \{\mathit{0.25}, \mathbf{0}, 6, 10, 2\}$, $S_3 = \{\mathit{1.0}, \mathbf{8}, 6, 7, 5, 14\}$, $S_4 = \{\mathit{1.0}, \mathbf{1}, 6, 8, 5\}$, $S_5 = \{\mathit{0.5}, \mathbf{12}, 4, 5, 8, 9, 10, 0, 2\}$, $S_6 = \{\mathit{0.5}, \mathbf{7}, 6, 5, 1, 14, 13\}$. Solve the routing and wavelength assignment problem using an MILP. What is the total cost? Assume a full λ-conversion network. How many wavelength converters are needed? At which nodes?

12.13. Formulate a dual problem for multicast tree protection without λ continuity.

Chapter 13

Traffic Grooming

13.1 Introduction

In an optical wavelength-division multiplexing (WDM) network, a lightpath provides a basic communication mechanism between two nodes. A lightpath is a wavelength circuit, which may span multiple fiber links and be routed by the intermediate optical switches between a given node pair. Although the bandwidth of a lightpath (i.e., a wavelength channel) in an optical WDM backbone network is quite high (10 Gbps (OC-192) today, and expected to grow to 40 Gbps (OC-768) soon), only a fraction of the customers are expected to have a need for such a high bandwidth. Many will be content with a lower bandwidth, e.g., STS-1[1](51.84 Mbps), OC-3, OC-12, OC-48, etc., for backbone applications. Since high-bandwidth wavelength channels are filled up by many low-speed traffic streams, efficiently provisioning customer connections with such diverse bandwidth needs is a very important problem, which is known as the traffic-grooming problem [MoLi01].

The traffic-grooming problem can be formulated as follows. Given a network configuration (including physical topology, where each edge is a physical link, number of transceivers at each node, number of wavelengths on each

[1]Synchronous Transport Signal (STS) is the electrical level transmission frame structure. STS-N frame is usually carried by an OC-N transmission link.

fiber, and the capacity of each wavelength) and a set of connection requests with different bandwidth granularities, such as OC-12, OC-48, etc., we need to determine how to set up lightpaths to satisfy the connection requests. Because of the sub-wavelength granularity of the connection requests, one or more connections can be multiplexed on the same lightpath.

The set of connection requests can all be given in advance (static traffic), or given one at a time (dynamic traffic). Traffic grooming with static traffic is a dual optimization problem. In a non-blocking scenario, where the network has enough resources to carry all of the connection requests, the objective is to minimize the network cost, e.g., total number of wavelength-links used in a WDM mesh network, while satisfying all the requests, where a wavelength-link is defined as a wavelength in a fiber-link. In a blocking scenario, where not all connections can be set up due to resource limitations, the objective is to maximize the network throughput. With dynamic traffic, where connections arrive one at a time, the objective is to minimize the network resources used for each request, which implicitly attempts to minimize the blocking probability for future requests.

Traffic grooming is usually divided into four sub-problems, which are not necessarily independent: (1) determining the virtual topology that consists of lightpaths, (2) routing the lightpaths over the physical topology, (3) performing wavelength assignment to the lightpaths, and (4) routing the traffic on the virtual topology. The virtual-topology design problem (Chapters 8 and 9) is conjectured to be NP-hard. In addition, routing and wavelength assignment (RWA) (Chapter 7) is also NP-hard. Therefore, traffic grooming in a mesh network is also a NP-hard problem [ZhMu02a].

To solve the traffic-grooming problem (static version), one approach is to deal with the four sub-problems separately. This approach first determines the virtual topology, then performs routing and wavelength assignment, and finally routes the traffic requests. There are considerable research results on each sub-problem already and they can be utilized to solve the traffic-grooming problem. Although this divide-and-conquer method makes traffic grooming easier to handle, it cannot achieve the optimal solution. Even if we can get the optimal solution for each sub-problem, these four sub-problems are not necessarily independent and the solution to one sub-problem might affect how optimally another sub-problem can be solved. Sometimes, using the optimal solution for one sub-problem might not lead to the optimal solution to the whole problem. Moreover, this approach requires all the traffic requests are known in advance, which is unachievable in dynamic grooming.

Another approach is to solve the four sub-problems as a whole. Since it can take into account all the constraints regarding the four sub-problems simultaneously, this approach has the potential to achieve better performance. With static traffic, the traffic-grooming problem can be formulated as an integer linear program (ILP) [ZhMu02a], and an optimal solution can be obtained for some relatively small networks. However, an ILP is not scalable and cannot be directly applied to large networks. One way to make the problem tractable is to develop heuristic algorithms and jointly solve the grooming problem for one connection request at a time.

13.1.1 Literature Review

Traffic grooming is an important and practical problem for designing WDM networks, and is receiving increased research attention both in academia and in industry. The work in [ZhMu02b] reviews most of the recent research work on traffic grooming in WDM ring and mesh networks.

Past research efforts on traffic grooming have mainly focused on SONET and WDM ring networks. The major cost of such a network is considered to be dominated by SONET add-drop multiplexers (ADMs). Therefore, minimizing the number of SONET ADMs has been the objective of static traffic grooming in ring networks.

As our fiber-optic backbone networks migrate from rings to mesh, traffic grooming on WDM mesh networks becomes an extremely important area of research. The work in [KoCh01] formulates the static traffic-grooming problem as an ILP and proposes a heuristic to minimize the number of transceivers. In [BrBa02], several lower bounds for regular topologies are presented, and greedy and iterative greedy schemes are developed. However, in both [KoCh01] and [BrBa02], the authors relax the physical-topology constraints, assuming all the virtual topologies are implementable on the given physical topology, i.e., they do not consider lightpath routing and wavelength assignment. The authors in [ZhMu02a] propose several node architectures for supporting traffic grooming in WDM mesh networks and formulate the static traffic-grooming problem as an ILP. They present two heuristics and compare the performance with that of the ILP.

The works in [ThSo01a, SrSo02, ZhMu02c, ThSo01b, CoSa01, LaGP01] consider a dynamic traffic pattern in WDM mesh networks. In [ThSo01a], the authors propose a connection admission control scheme to ensure fairness in terms of connection blocking. A theoretical capacity correlation model is presented in [SrSo02] to compute the blocking probability for WDM networks with constrained grooming capability. In [ZhMu02c], two route-computation algorithms are proposed and compared, and the results indicate that, in order to achieve good performance in a dynamic environment, different grooming policies and route-computation algorithms need to be used under different network states. The work in [ThSo01b] compares two schemes to dynamically establish reliable low-speed traffic in WDM mesh networks with traffic-grooming capability. In [CoSa01], the problem of planning and designing a WDM mesh network with certain forecast traffic demands, to satisfy all the connections as well as minimize the network cost, is studied. In [LaGP01], the authors investigate the design of multi-layer mesh networks to satisfy each connection's bandwidth and protection requirements while minimizing the overall network cost.

In the rest of the chapter, we will focus on traffic grooming in WDM mesh networks. We first investigate the static traffic-grooming problem, in which all the traffic demands are known in advance. Then, we explore the dynamic traffic-grooming problem, where connections arrive at a network, hold for a certain amount of time, and then depart from the network.

13.2 Static Traffic Grooming

Assigning network resources (e.g., wavelengths, transceivers) to successfully carry the connection requests (lightpaths) in an optical WDM mesh network is well known as the routing and wavelength assignment (RWA) problem (Chapter 7). It is also known as a lightpath-provisioning problem. A lot of RWA studies have been reported in the optical networking literature, either based on static traffic demands or based on dynamic traffic demands. Most previous studies have assumed that a connection requests bandwidth for an entire lightpath channel. In this study, it is assumed that the bandwidth of the connection requests can be some fraction of the lightpath capacity, which makes the problem more practical. For the static traffic-grooming problem, it is assumed that a set of connection demands are given, and they need to be established on the network.

Figure 13.1 shows an illustrative example of traffic grooming in a WDM mesh network. Figure 13.1(a) shows a small six-node network. Each fiber has two wavelength channels. The capacity of each wavelength channel in this example is OC-48, i.e., approximately 2.5 Gbps. Note that the bandwidth of an OC-n channel is approximately $n \times 51.84$ Mbps. Each node is equipped

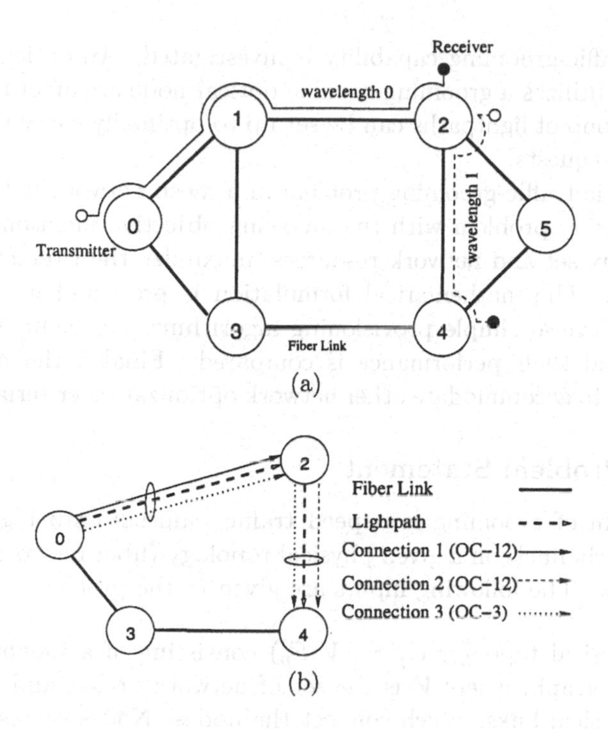

Figure 13.1 Illustrative example of traffic grooming.

with a tunable transmitter and a tunable receiver, both of which can be tuned to any wavelength. There are three connection requests: $(0, 2)$ with bandwidth requirement OC-12, $(2, 4)$ with bandwidth requirement OC-12, and $(0, 4)$ with bandwidth requirement OC-3. Two lightpaths have already been set up to carry these three connections, as shown in Fig. 13.1(a). Because of the resource limitations (transmitter in node 0 and receiver in node 4 are busy), we cannot set up a lightpath directly from node 0 to node 4; thus, connection 3 has to be carried by the spare capacity of the two existing lightpaths, as shown in Fig. 13.1(b). Different connection requests between the same node pair (s, d) can be either groomed on the same lightpath, which directly joins (s, d), using various multiplexing techniques, or routed separately through different virtual paths. A connection may traverse multiple lightpaths if no resources are available to set up a lightpath between the source and the destination directly.

In this chapter, the node architecture for the WDM optical network which supports traffic-grooming capability is investigated. An optical WDM network which utilizes a grooming-capable optical node architecture is studied, so that a group of lightpaths can be set up to optimally carry the low-speed connection requests.

The static traffic-grooming problem in a mesh network is formulated as an optimization problem with the following objective function: for a given traffic matrix set and network resources, maximize the (weighted) network throughput. The mathematical formulation is presented for static traffic demands. Several simple provisioning algorithms, i.e., heuristics, are also proposed and their performance is compared. Finally, the mathematical formulation to accommodate other network optimization criteria is extended.

13.2.1 General Problem Statement

The problem of grooming low-speed traffic requests onto high-bandwidth wavelength channels on a given physical topology (fiber network) is formally stated below. The following inputs are given to the problem:

1. A physical topology $G_p = (V, E_p)$ consisting of a weighted unidirectional graph, where V is the set of network nodes, and E_p is the set of physical links, which connect the nodes. Nodes correspond to network nodes and links correspond to the fibers between nodes. Though links are unidirectional, it is assumed that there are an equal number of fibers joining two nodes in different directions. Links are assigned weights, which may correspond to physical distance between nodes. In this study, it is assumed that all links have the same weight 1, which corresponds to the fiber hop distance. A network node i is assumed to be equipped with a $D_p(i) \times D_p(i)$ optical crossconnect ((OXC), also called wavelength-routing switch (WRS)), where $D_p(i)$ denotes the number of incoming fiber links to node i. For any node i, the number of incoming fiber links is equal to the number of outgoing fiber links.

2. Number of wavelength channels carried by each fiber is W. Capacity of a wavelength is C.

3. A set of $N \times N$ traffic matrices, where $N = |V|$. Each traffic matrix in the traffic-matrix set represents one particular group of low-speed connection requests between the nodes of the network. For example, if C is OC-48, there may exist four traffic matrices: an OC-1 traffic

matrix, an OC-3 traffic matrix, an OC-12 traffic matrix, and an OC-48 traffic matrix.

4. The number of lasers (transmitters) (TR_i) and filters (receivers) (RR_i) at each node i. Note that the transceiver can be either wavelength-tunable or part of a fixed-tuned array.

The goals of the formulation are to determine the following:

1. A virtual topology $G_v = (V, E_v)$. The nodes of the virtual topology correspond to the nodes in the physical topology. A link between nodes i and j corresponds to an unidirectional lightpath set up between node pair (i, j).

2. Routing connection requests on the virtual topology to either minimize the total network cost or maximize total throughput. In this study, maximizing total throughput is considered.

13.2.2 Node Architecture

To carry connection requests in a WDM network, lightpath connections may be established between pairs of nodes. A connection request may traverse through one or more lightpaths before it reaches the destination. Two important functionalities must be supported by the WDM network nodes: one is wavelength routing, and the other is multiplexing and demultiplexing. An OXC provides the wavelength-routing capability to the WDM network nodes. Optical multiplexer/demultiplexer can multiplex/demultiplex several wavelengths to the same fiber link. Low-speed connection requests will be multiplexed on the same lightpath channel by using an electronic-domain TDM-based multiplexing technique. Figures 13.2 and 13.3 show two sample node architectures in a WDM optical network.

The node architecture is composed of two components: WRS and access station. The WRS performs wavelength routing and wavelength multiplexing/demultiplexing. The access station performs local traffic adding/dropping and low-speed traffic-grooming functionalities. WRS is composed of an Optical Crossconnect (OXC), Network Control and Management Unit (NC&M), and Optical Multiplexer/Demultiplexer. In the NC&M unit, the network-to-network interface (NNI) will configure the OXC and exchange control messages with peer nodes on a dedicated wavelength channel

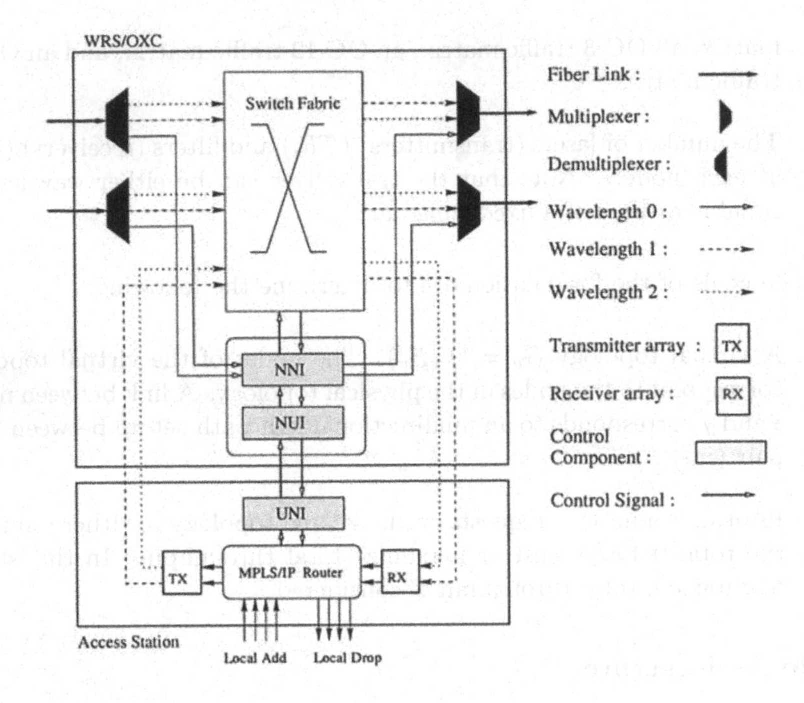

Figure 13.2 Node architecture 1: IP over WDM.

(shown as wavelength 0 in Figs. 13.2 and 13.3). The network-to-user interface (NUI) will communicate with the NNI and exchange control information with the user-to-network interface (UNI), the control component of the access station. The OXC provides wavelength-switching functionality. As shown in the simple example in Fig. 13.2, each fiber has three wavelengths. Wavelength 0 is used as a control channel for the NC&M to exchange control messages between network nodes. Other wavelengths are used to transmit data traffic.

In Fig. 13.2, each access station is equipped with some transmitters and receivers (transceivers). Traffic originating from an access station is sent out as an optical signal on one wavelength channel by a transmitter. Traffic destined to an access station is converted from an optical signal to electronic data by a receiver. Both tunable transceivers and fixed transceivers could be used in a WDM network. A tunable transceiver can be tuned between different wavelengths so that it can send out (or receive) an optical signal on

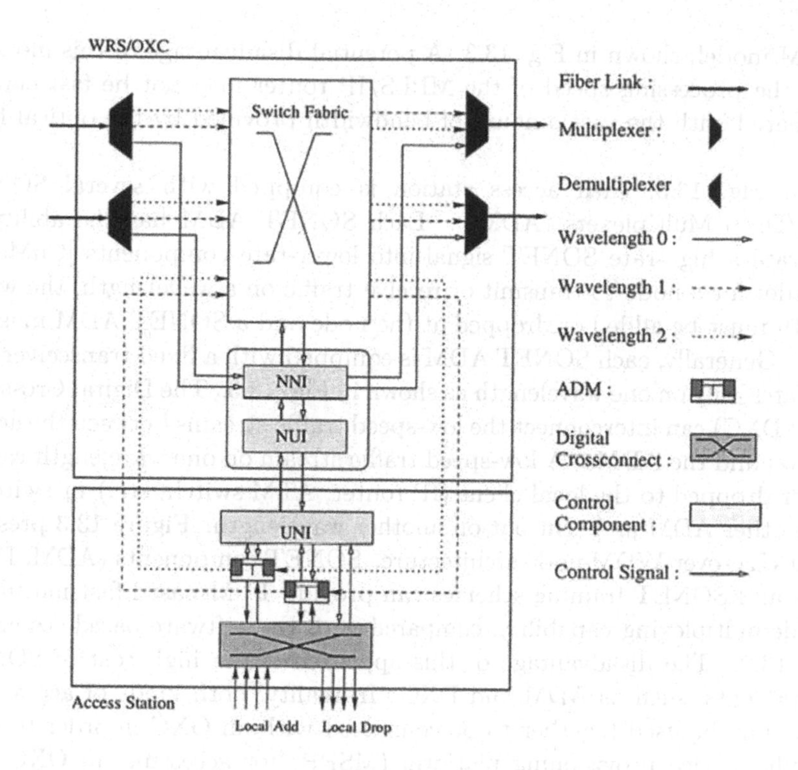

Figure 13.3 Node architecture 2: SONET over WDM.

any free wavelength in its tuning range. A fixed transceiver can only emit (or receive) an optical signal on one wavelength. To explore all of the wavelength channels on a fiber, a set of fixed transceivers, one per wavelength, can be grouped together to form a transceiver array. The size of a fixed transceiver array can be equal to or smaller than the number of wavelengths on a fiber, and the number of transceiver arrays can be equal to or smaller than the number of fibers joining a node.

The access station in Fig. 13.2 provides a flexible, software-based, bandwidth-provisioning capability to the network. Multiplexing low-speed connections to high-capacity lightpaths is done by the MPLS/IP router using a software-based queueing scheme. The advantages of this model are that 1) it provides flexible bandwidth granularity for the traffic requests and 2) this MPLS/IP-over-WDM model has much less overhead than the SONET-over-

WDM model, shown in Fig. 13.3. A potential disadvantage of this model is that the processing speed of the MPLS/IP router may not be fast enough compared with the vast amount of bandwidth provided by the optical fiber link.

In Fig. 13.3, each access station is equipped with several SONET Add/Drop Multiplexers (ADMs). Each SONET ADM has the ability to separate a high-rate SONET signal into lower-rate components [ChMo00]. In order for a node to transmit or receive traffic on a wavelength, the wavelength must be added or dropped at the node and a SONET ADM must be used. Generally, each SONET ADM is equipped with a fixed transceiver and operates only on one wavelength as shown in Fig. 13.3. The Digital Crossconnect (DXC) can interconnect the low-speed traffic streams between the access station and the ADMs. A low-speed traffic stream on one wavelength can be either dropped to the local client (IP router, ATM switch, etc.) or switched to another ADM and sent out on another wavelength. Figure 13.3 presents a SONET-over-WDM node architecture. SONET components (ADM, DXC, etc.) and SONET framing schemes can provide TDM-based fast multiplexing/demultiplexing capability, compared with the software-based scheme in Fig. 13.2. The disadvantage of this approach is the high cost of SONET components, such as ADM and DXC. In reality, both kinds of access stations may be used together to be connected with an OXC in order to have a multi-service provisioning platform (MSPP) for accessing an OXC in a carrier's network.

Optical crossconnects (OXCs) are known to be the most important network elements in a carrier's WDM backbone network. There are transparent and opaque approaches to build these OXCs. The transparent approach refers to all-optical (O-O-O) switching, and the opaque approach refers to switching with optical-electronic-optical (O-E-O) conversion. Depending on their architectures and technologies employed, different OXCs may have different multiplexing and switching capabilities, which result in different capabilities for grooming low-speed traffic streams onto high-capacity wavelength channels. Below, we give a brief introduction on several types of OXC.

- *Non-grooming OXC:* This type of OXC can be built with either transparent or opaque approach. It has wavelength-switching capability. If the transparent approach is used, it is possible for this type of OXC to switch at higher bandwidth granularity, such as waveband (a group of wavelengths) or fiber. There is no low-data-rate port on a non-grooming OXC. Thus, extra aggregation/de-aggregation network ele-

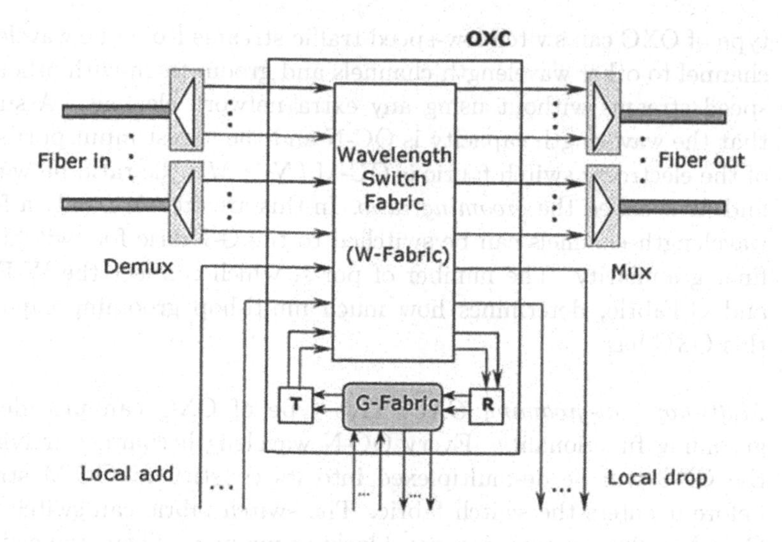

Figure 13.4 A multi-hop partial-grooming OXC.

ments are needed to work with this type of OXC if low-speed traffic streams need to be supported.

- *Single-hop grooming OXC:* Just like the non-grooming OXC, this type of OXC will only switch traffic at wavelength (or higher) granularity. On the other hand, it may have some lower-data-rate ports, which can directly support low-speed traffic streams. The traffic from these low-speed ports can be multiplexed onto a wavelength channel using a TDM scheme, before the traffic enters the switch fabric. Since this type of OXC does not have the capability to switch low-speed streams at intermediate nodes, all of the low-speed streams on one wavelength channel at the source node will be switched to the same destination node.

- *Multi-hop partial-grooming OXC:* As shown in Fig. 13.4, the switch fabric of this type of OXC is composed of two parts: a wavelength-switch fabric (W-Fabric), which can be either all-optical or electronic, and an electronic-switch fabric which can switch low-speed traffic streams. The electronic-switch fabric is also called grooming fabric (G-Fabric). With this hierarchical switching and multiplexing architecture, this type of OXC can switch low-speed traffic streams from one wavelength channel to other wavelength channels and groom them with other low-speed streams without using any extra network element. Assuming that the wavelength capacity is OC-N and the lowest input port speed of the electronic switch fabric is OC-M ($N \geq M$), the ratio between N and M is called the *grooming ratio*. In this architecture, only a few of wavelength channels can be switched to the G-Fabric for switching at finer granularity. The number of ports, which connect the W-Fabric and G-Fabric, determines how much multi-hop grooming capability this OXC has.

- *Multi-hop full-grooming OXC:* This type of OXC can provide full-grooming functionality. Every OC-N wavelength channel arriving at the OXC will be de-multiplexed into its constituent OC-M streams before it enters the switch fabric. The switch fabric can switch these OC-M traffic streams in a non-blocking manner. Then, the switched streams will be multiplexed back onto different wavelength channels. An OXC with full-grooming functionality has to be built using the opaque approach. Note that the switching fabric of this type of OXC can be viewed as a large grooming fabric.

- Light-tree-Based Source-Node Grooming OXC: Optical "light-tree" has been proposed to support multicast applications in optical WDM networks [SaMu99, SiMu01] (also see Chapter 12). A light-tree is a wavelength tree which connects one source node to multiple destination nodes. Through a light-tree, traffic from the source node will be delivered to all destination nodes of the tree. In a light-tree, the node which originates the traffic is called the "root" node, and the nodes which terminate the traffic are called the "leaf" nodes. Note that a leaf node can also serve as an intermediate multicast node since it can both receive traffic and forward the traffic to other nodes using its multicast capability, i.e., such a node may have the "Drop-and-Continue" property. In order to support multicast, an OXC needs to duplicate the traffic from one input port to multiple output ports. For an OXC using transparent switching technology, this duplication can be realized in the optical domain using an optical splitter by splitting the power of an optical signal from one input port to multiple output ports. For an OXC using opaque technology, the traffic duplication can be easily accomplished by copying the electronic bit stream from one input port

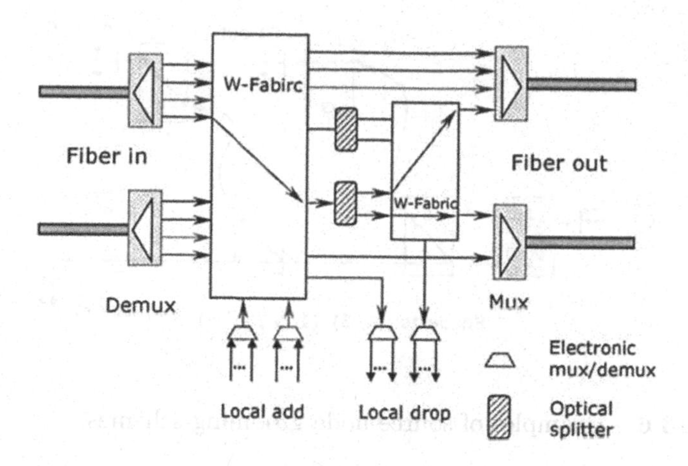

Figure 13.5 Sample source-node grooming OXC architecture.

to multiple output ports. Figure 13.5 shows a simplified architecture of a multicast-capable OXC using the transparent technology.

Figure 13.6 shows how the OXCs' multicast capability can be used to perform traffic grooming. There are three low-speed traffic steams from the same source node 1 to different destination nodes 3, 5, and 6, where the aggregated bandwidth requirement is lower than the capacity of a wavelength channel in this example. By setting up a light-tree, these three traffic streams can be packed onto the same wavelength channel, and delivered to all destination nodes (i.e., light-tree leaf nodes). At each destination node, only the appropriate traffic stream is picked up and relayed to the client equipment. In this way, the low-speed traffic from the same source node can be groomed to the same wavelength channel and be sent to different destination nodes. We call this grooming scheme light-tree-based source-node grooming scheme. Please note that, if a connection between a node pair requires full wavelength-channel capacity, a light-tree becomes a lightpath. From traffic-grooming perspective, the multicast-capable OXC can be called a light-tree-based source-node grooming OXC. Such an OXC can support the light-tree-based source-node grooming scheme as well as the single-hop grooming scheme.

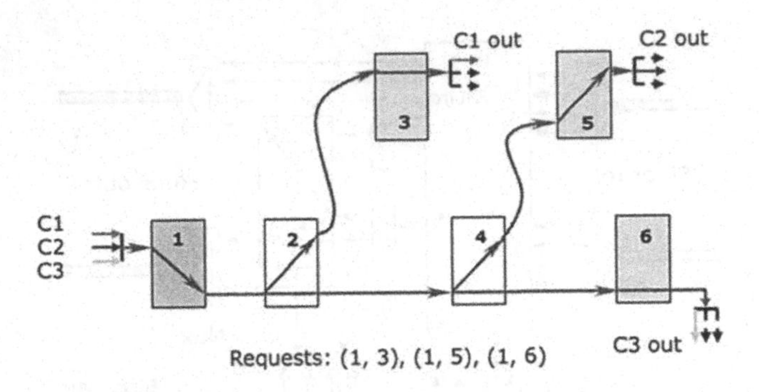

Figure 13.6 Examples of source-node grooming schemes.

13.2.3 Mathematical (ILP) Formulation of the Static Traffic-Grooming Problem

The traffic-grooming problem in a mesh network with static traffic pattern turns out to be an Integer Linear Program (ILP). The following assumptions are made in this study:

1. The network is a single-fiber, irregular mesh network, i.e., there is at most one pair of (bidirectional) fiber link between each node pair.

2. The wavelength-routing switches in the network nodes do not have wavelength-conversion capability. A lightpath connection must be set up on the same wavelength channel if it traverses through several fibers. Extension of this problem to include wavelength conversion is straightforward and it actually makes the problem simpler in terms of the number of variables and equations.

3. The transceivers in a network node are tunable to any wavelength on the fiber.

4. A connection request cannot be divided into several lower-speed connections and routed separately from the source to the destination. The data traffic on a connection request should always follow the same route.

5. Each node has unlimited multiplexing/demultiplexing capability and time-slot interchange capability. This means that the access station

of a network node can multiplex/demultiplex as many low-speed traffic streams to a lightpath as needed, as long as the aggregated traffic does not exceed the lightpath capacity. This may only be true for the software-based provisioning scheme in Fig. 13.2, which may support virtual-circuit connections. The grooming capability of the node architecture in Fig. 13.3 is limited by the number of output ports of SONET ADM, and the size and the functionality of the DXC.

Multi-Hop Traffic Grooming

In this section, we assume that a connection can traverse multiple lightpaths before it reaches the destination. So, a connection may be groomed with different connections on different lightpaths. By extending the work in Chapters 8 and 9, which define the collection of lightpaths in a WDM mesh network to form a virtual topology, we formulate the problem as an optimization problem. We will use the following notations in our mathematical formulation:

1. m and n denote end points of a physical fiber link that might occur in a lightpath.

2. i and j denote originating and terminating nodes for a lightpath. A lightpath may traverse single or multiple physical fiber links.

3. s and d denote source and destination of an end-to-end traffic request. The end-to-end traffic may traverse through a single lightpath or multiple lightpaths. Figure 13.7 shows how an end-to-end connection request may be carried.

4. y denotes the granularity of low-speed traffic requests. We assume $y \in \{1, 3, 12, 48\}$, which means that traffic demands between node pairs can be any of $OC-1$, $OC-3$, $OC-12$, and $OC-48$.

5. t denotes the index of $OC-y$ traffic request for any given node pair (s, d). For example, if there are ten $OC-1$ requests between node pair (s, d), then $t \in [1, 10]$.

- Given:
  - N: Number of nodes in the network.

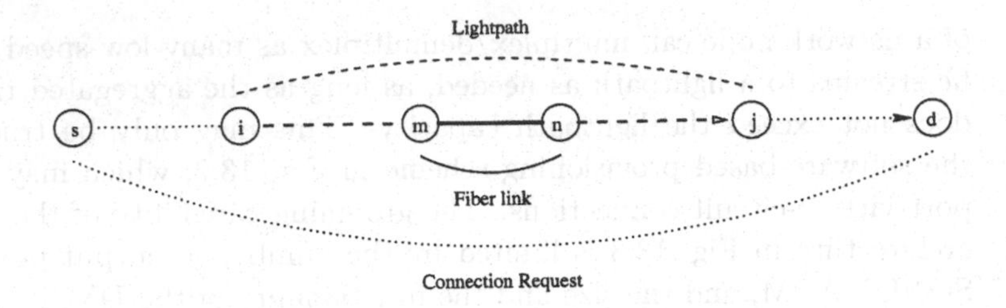

Figure 13.7 Illustrative example of a fiber link, a lightpath, and a connection request.

- W: Number of wavelengths per fiber. We assume all of the fibers in the network carry the same number of wavelengths.
- P_{mn}: Number of fibers interconnecting node m and node n. $P_{mn} = 0$ for the node pair which is not physically adjacent to each other. $P_{mn} = P_{nm} = 1$ if and only if there exists a direct physical fiber link between nodes m and n.
- P^w_{mn}: Wavelength w on fiber P_{mn}. $P^w_{mn} = P_{mn}$.
- TR_i: Number of transmitters at node i.
- RR_i: Number of receivers at node i. Note that, in this set of ILP formulation, we assume all the nodes are equipped with tunable transceivers, which can be tuned to any of W wavelengths.
- C: Capacity of each channel (wavelength).
- Λ: Traffic matrix set. $\Lambda = \{\Lambda_y\}$, where y can be any allowed low-speed streams, 1, 3, 12, etc. In our study, $y \in \{1, 3, 12, 48\}$. $\Lambda_{y,sd}$ is the number of $OC - y$ connection requests between node pair (s, d).

- Variables:

  - Virtual topology:
    - V_{ij}: Number of lightpaths from node i to node j in virtual topology. $V_{ij} = 0$ does not imply that $V_{ji} = 0$, i.e., lightpaths may be unidirectional.
    - V^w_{ij}: Number of lightpaths from node i to node j on wavelength w. Note that, if $V^w_{ij} > 1$, the lightpaths between nodes i and j on wavelength w may take different paths.

- Physical topology route:
  - * $P_{mn}^{ij,w}$: Number of lightpaths between node pair (i,j) routed through fiber link (m,n) on wavelength w.
- Traffic route:
  - * $\lambda_{ij,y}^{sd,t}$: The tth $OC-y$ low-speed traffic request from node s to node d employing lightpath (i,j) as an intermediate virtual link.
  - * $S_{sd}^{y,t}$: $S_{sd}^{y,t}=1$ if the tth $OC-y$ low-speed connection request from node s to node d has been successfully routed; otherwise, $S_{sd}^{y,t}=0$.

- Optimize: Maximize the total successfully-routed low-speed traffic, i.e.,

$$Maximize : \sum_{y,s,d,t} y \cdot S_{sd}^{y,t} \tag{13.1}$$

- Constraints:
  - On virtual-topology connection variables:

$$\sum_j V_{ij} \leq TR_i \quad \forall i \tag{13.2}$$

$$\sum_i V_{ij} \leq RR_j \quad \forall j \tag{13.3}$$

$$\sum_w V_{ij}^w = V_{ij} \quad \forall i,j \tag{13.4}$$

$$int \quad V_{ij}, V_{ij}^w \tag{13.5}$$

– On physical route variables:

$$\sum_m P_{mk}^{ij,w} = \sum_n P_{kn}^{ij,w} \quad \textit{if } k \neq i,j \quad \forall i,j,w,k \tag{13.6}$$

$$\sum_m P_{mi}^{ij,w} = 0 \quad \forall i,j,w \tag{13.7}$$

$$\sum_n P_{jn}^{ij,w} = 0 \quad \forall i,j,w \tag{13.8}$$

$$\sum_n P_{in}^{ij,w} = V_{ij}^{w} \quad \forall i,j,w \tag{13.9}$$

$$\sum_m P_{mj}^{ij,w} = V_{ij}^{w} \quad \forall i,j,w \tag{13.10}$$

$$\sum_{i,j} P_{mn}^{ij,w} \leq P_{mn}^{w} \quad \forall m,n,w \tag{13.11}$$

$$P_{mn}^{ij,w} \in \{0,1\} \tag{13.12}$$

– On virtual-topology traffic variables:

$$\sum_i \lambda_{id,y}^{sd,t} = S_{sd}^{y,t} \quad \forall s,d \;\; y \in \{1,3,12,48\} \;\; t \in [1,\Lambda_{y,sd}] \tag{13.13}$$

$$\sum_j \lambda_{sj,y}^{sd,t} = S_{sd}^{y,t} \quad \forall s,d,t \;\; y \in \{1,3,12,48\} \;\; t \in [1,\Lambda_{y,sd}] \tag{13.14}$$

$$\sum_i \lambda_{ik,y}^{sd,t} = \sum_j \lambda_{kj,n}^{sd,t} \quad \textit{if } \; k \neq s,d \;\; \forall s,d,k,t \tag{13.15}$$

$$\sum_i \lambda_{is,y}^{sd,t} = 0 \quad \forall s,d \;\; y \in \{1,3,12,48\} \;\; t \in [1,\Lambda_{y,sd}] \tag{13.16}$$

$$\sum_j \lambda_{dj,y}^{sd,t} = 0 \quad \forall s,d \;\; y \in \{1,3,12,48\} \;\; t \in [1,\Lambda_{y,sd}] \tag{13.17}$$

$$\sum_{y,t}\sum_{s,d} y \times \lambda_{ij,y}^{sd,t} \le V_{ij} \times C \quad \forall i,j \tag{13.18}$$

$$S_{sd}^{y,t} \in \{0,1\} \tag{13.19}$$

- Explanation of equations: The above equations are based on principles of conservation of flows and resources (transceivers, wavelengths, etc.).
  - Equation (13.1) shows the optimization objective function.
  - Equations (13.2)–(13.3) ensure that the number of lightpaths between node pair (i,j) is less than or equal to the number of transmitters at node i and the number of receivers at node j.
  - Equation (13.4) shows that the lightpaths between (i,j) are composed of lightpaths on different wavelengths between node pair (i,j). Note that the value of V_{ij}^{w} can be greater than 1. For example, in Fig. 13.1, two lightpaths on the same wavelength w can be set up between node 0 and node 5. One follows route $(0,1,2,5)$, while the other follows route $(0,3,4,5)$.
  - Equations (13.6)–(13.10) are the multicommodity equations (for flow conservation) that account for the routing of a lightpath from its origin to its termination. The flow-conservation equations have been formulated in two different ways [RaSi96]: (i) disaggregate formulation and (ii) aggregate formulation. In the disaggregate formulation, every i-j (or s-d) pair corresponds to a commodity, while in the aggregate formulation, all the traffic that is "sourced" from node i (or node s) corresponds to the same commodity, regardless of the traffic's destination. We employ the disaggregate formulation for the flow-conservation equations since it properly describes the traffic requests between different node pairs. Note that Equations (13.6)–(13.10) employ the wavelength-continuity constraint on the lightpath route.
    * Equation (13.6) ensures that, for an intermediate node k of lightpath (i,j) on wavelength w, the number of incoming lightpath streams is equal to the number of outgoing lightpath streams.
    * Equation (13.7) ensures that, for the origin node i of lightpath (i,j) on wavelength w, the number of incoming lightpath streams is 0.

  * Equation (13.8) ensures that, for the termination node j of lightpath (i,j) on wavelength w, the number of outgoing lightpath streams is 0.
  * Equation (13.9) ensures that, for the origin node i of lightpath (i,j) on wavelength w, the number of outgoing lightpath streams is equal to the total number of lightpaths between node pair (i,j) on wavelength w.
  * Equation (13.10) ensures that, for the termination node j of lightpath (i,j) on wavelength w, the number of incoming lightpath streams is equal to the total number of lightpaths between node pair (i,j) on wavelength w.
- Equations (13.11)–(13.12) ensure that wavelength w on one fiber link (m,n) can only be present in at most one lightpath in the virtual topology.
- Equations (13.13)–(13.19) are responsible for the routing of low-speed traffic requests on the virtual topology, and they take into account the fact that the aggregate traffic flowing through lightpaths cannot exceed the overall wavelength (channel) capacity.

Single-Hop Traffic Grooming

In this section, we assume that a connection can only traverse a single lightpath, i.e., only end-to-end traffic grooming is allowed. The formulation of the single-hop traffic-grooming problem is similar to the formulation of the multi-hop traffic-grooming problem, which was presented in the previous section, except for routing of connection requests on the virtual topology. We only present the different part as follows:

- On virtual-topology traffic variables:

$$\sum_{y,t} y \times S_{sd}^{y,t} \leq V_{sd} \times C \quad \forall s,d \tag{13.20}$$

$$S_{sd}^{y,t} \in \{0,1\} \tag{13.21}$$

Formulation Extension for Fixed-Transceiver Array

The mathematical formulations in the previous two sections are based on the assumption that the transceivers in a network node are tunable to any wavelength. If fixed-transceiver arrays are used at every network node and if M denotes the number of fixed-transceiver arrays used at each node, we can easily extend our formulation as follows:

- On virtual-topology connection variables:

$$\sum_j V_{ij}^w \leq M \quad \forall i, w \tag{13.22}$$

$$\sum_i V_{ij}^w \leq M \quad \forall j, w \tag{13.23}$$

$$\sum_w V_{ij}^w = V_{ij} \quad \forall i, j \tag{13.24}$$

$$int \quad V_{ij}, V_{ij}^w \tag{13.25}$$

The other parts of the formulations in the previous two sections are still the same. Equations (13.22)–(13.23) ensure that the number of lightpaths between node pair (i, j) on wavelength w is less than or equal to the number of transmitters at node i and the number of receivers at node j on wavelength w (every fixed-transceiver array only has one transceiver on each wavelength).

Computational Complexity

It is well known that the RWA optimization problem is *NP*-complete. If we assume that each connection request requires the full capacity of a lightpath, then the traffic-grooming problem we are studying becomes the standard RWA optimization problem. It is easy to see that the traffic-grooming problem in a mesh network is also a *NP*-complete problem since the RWA optimization problem is *NP*-complete. As the number of variables and equations increases exponentially with the size of network and with the number of wavelengths on each fiber, we use a small network topology as an example for obtaining ILP result. For large networks, we will use heuristic approaches.

13.2.4 Illustrative Numerical Results From ILP Formulations

This section presents numerical examples for the static traffic-grooming problem using Fig. 13.8(a) as our physical topology. The traffic matrices are randomly generated. As an example, we allow the traffic demand to be any one of OC-1, OC-3, and OC-12. The traffic matrices are generated as follows: 1) the number of OC-1 connection requests between each node

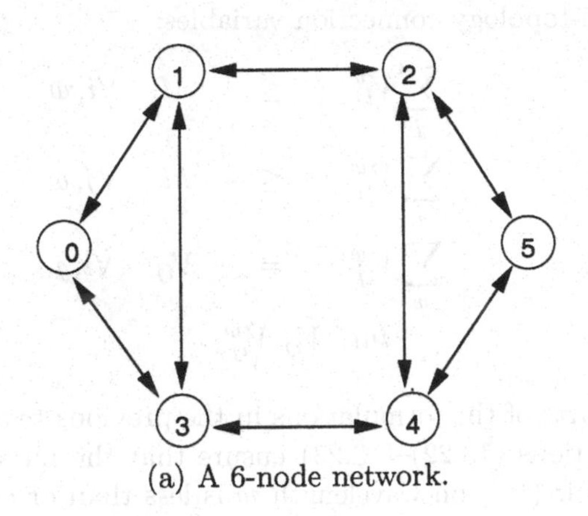

(a) A 6-node network.

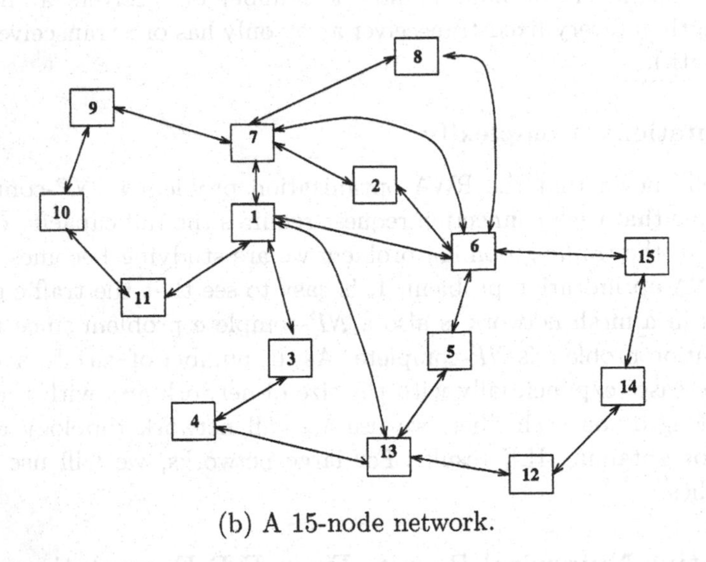

(b) A 15-node network.

Figure 13.8 Physical network topology.

pair is generated as an uniformly-distributed random number between 0 to 16; 2) the number of OC-3 connection requests between each node pair is generated as an uniformly-distributed random number between 0 to 8; and 3) the number of OC-12 connection requests between each node pair is generated as an uniformly-distributed random number between 0 to 2. These three traffic matrices are shown in Tables 13.1 through 13.3, and the total traffic demand turns out to be the equivalent of OC-988. The capacity of each wavelength (channel) is OC-48.

Table 13.4 shows the corresponding result for the network throughput obtained from a commercial ILP solver, "CPLEX", based on different network resource parameters. In Table 13.4, T denotes the number of transceivers and W denotes the number of wavelengths. In the single-hop case, we only allow a connection to transverse a single lightpath, which means that only end-to-end traffic-grooming (multiplexing) is allowed. In the multi-hop case, a connection is allowed to traverse multiple lightpaths, i.e., a connection can be dropped at intermediate nodes and groomed with other low-speed connections on different lightpaths before it reaches its destination. Figure 13.1(b) showed a multi-hop grooming case, where connection 3 traversed two lightpaths: it was groomed with connection 1 on lightpath $(0, 2)$ and with connection 2 on lightpath $(2, 4)$. As expected, the multi-hop case leads to higher throughput than the single-hop case.[2]

We can see from Table 13.4 that, when the number of tunable transceivers at each node is increased from 3 to 5, the network throughput increases significantly, both in multi-hop case and in single-hop case. But when the number of tunable transceivers at each node increases from 5 to 7, the network throughput does not improve. This is because there are not enough wavelengths to setup more lightpaths to carry the blocked connection requests. Some illustrative results of transceiver and wavelength utilization for the multi-hop case are shown in Tables 13.5 and 13.6.

In the multi-hop case, when the transceiver is not a limited resource compared with wavelength, more short lightpaths may be set up to carry connections through multiple lightpaths, but this scenario is less likely in the single-hop case. This is shown in Table 13.4 where $T = 5$, $W = 3$ and $W = 4$. When $T = 5$, $W = 4$, if multi-hop grooming is allowed, the network throughput is 100%; otherwise, some connections get blocked. In

[2]We should remark that, in obtaining these results, CPLEX was run for several hours for each ILP case. Not in all cases did CPLEX complete its optimization, so some further optimization may be possible if CPLEX were allowed to run longer.

Table 13.1 Traffic matrix of OC-1 connection requests.

	Node 0	Node 1	Node 2	Node 3	Node 4	Node 5
Node 0	0	5	4	11	12	9
Node 1	0	0	8	5	16	6
Node 2	14	12	0	9	6	16
Node 3	4	11	15	0	1	5
Node 4	10	2	3	3	0	9
Node 5	2	1	8	15	13	0

Table 13.2 Traffic matrix of OC-3 connection requests.

	Node 0	Node 1	Node 2	Node 3	Node 4	Node 5
Node 0	0	6	2	1	5	4
Node 1	8	0	8	6	7	8
Node 2	1	3	0	0	2	7
Node 3	5	7	3	0	2	6
Node 4	6	4	5	0	0	2
Node 5	5	4	4	2	0	0

Table 13.3 Traffic matrix of OC-12 connection requests.

	Node 0	Node 1	Node 2	Node 3	Node 4	Node 5
Node 0	0	1	1	1	0	0
Node 1	1	0	1	1	0	2
Node 2	0	1	0	2	1	0
Node 3	2	0	2	0	2	0
Node 4	1	2	0	2	0	1
Node 5	1	1	2	2	2	0

Table 13.4 Throughput and number of lightpaths established (total traffic demand is OC-988).

	Multi-hop		Single-hop	
	Throughput	Lightpath #	Throughput	Lightpath #
T=3, W=3	74.7% (OC-738)	18	68.0% (OC-672)	18
T=4, W=3	93.8% (OC-927)	24	84.1% (OC-831)	24
T=5, W=3	97.9% (OC-967)	28	85.7% (OC-847)	24
T=7, W=3	97.9% (OC-967)	28	85.7% (OC-847)	24
T=3, W=4	74.7% (OC-738)	18	68.0% (OC-672)	18
T=4, W=4	94.4% (OC-933)	24	84.7% (OC-837)	24
T=5, W=4	100% (OC-988)	29	95.5% (OC-944)	28

multi-hop case, 29 lightpaths are established compared with 28 lightpaths in the single-hop case.

Table 13.5 Results: transceiver utilization (multi-hop case).

		T=3, W=3	T=5, W=3	T=7, W=3
Node 0	Transmitter	100%	100%	71.4%
	Receiver	100%	100%	71.4%
Node 1	Transmitter	100%	100%	71.4%
	Receiver	100%	100%	71.4%
Node 2	Transmitter	100%	100%	71.4%
	Receiver	100%	100%	71.4%
Node 3	Transmitter	100%	100%	71.4%
	Receiver	100%	100%	71.4%
Node 4	Transmitter	100%	80%	57.4%
	Receiver	100%	80%	57.4%
Node 5	Transmitter	100%	80%	57.4%
	Receiver	100%	80%	57.4%

Tables 13.5–13.6 show some results for the node transceiver utilization and link wavelength utilization for the multi-hop case. When the number of transceivers is increased (from 3 to 5), the overall wavelength utilization is increased, as shown in Table 13.6. This is because more lightpaths have been established to carry the connection requests, shown in Table 13.4. As

Table 13.6 Results: wavelength utilization (multi-hop case).

	T=3, W=3	T=5, W=3	T=7, W=3
Link (0,1)	33.3%	100%	100%
Link (0,3)	100%	100%	100%
Link (1,0)	100%	100%	100%
Link (1,2)	100%	100%	100%
Link (1,3)	33.3%	66.7%	66.7%
Link (2,1)	100%	100%	100%
Link (2,4)	66.7%	100%	100%
Link (2,5)	66.7%	100%	100%
Link (3,0)	33.3%	100%	100%
Link (3,1)	100%	66.7%	66.7%
Link (3,4)	66.7%	100%	100%
Link (4,2)	66.7%	100%	100%
Link (4,3)	66.7%	100%	100%
Line (4,5)	66.7%	66.7%	66.7%
Link (5,2)	66.7%	100%	100%
Link (5,4)	66.7%	66.7%	66.7%

we mentioned before, when most of the links have fully utilized the available wavelengths, increasing the number of transceivers (from 5 to 7) will not help to improve the network throughput and will result in lower transceiver utilization, shown in Table 13.5 ($T = 7$ and $W = 3$).

Table 13.7 shows the virtual topology and the lightpath capacity utilization for the multi-hop case, when $T = 5$ and $W = 3$. As we can see, most of the lightpaths have high capacity utilization (above 90%). There are some node pairs ($(0, 1)$, $(1, 3)$, etc.), which have multiple lightpaths set up, though the aggregate traffic between them can be carried by a single lightpath. The extra lightpaths are used to carry multi-hop connection traffic.

In the ILP formulation, we treat the low-speed connection requests separately. The results from the ILP solutions show that, if there is a lightpath set up between (s, d), the low-speed connections between (s, d) tend to be packed on this lightpath channel directly. Based on this observation, we propose two simple heuristic algorithms for solving the traffic-grooming problem in large networks.

Table 13.7 Result: virtual topology and lightpath utilization (multi-hop case with T=5 and W=3).

	Node 0	Node 1	Node 2	Node 3	Node 4	Node 5
Node 0	0	2 (70%)	0 (100%)	1 (89%)	1 (100%)	1 (100%)
Node 1	1 (100%)	0	1 (100%)	2 (100%)	1 (100%)	0
Node 2	1 (100%)	1 (95%)	0	1 (100%)	1 (100%)	1 (70%)
Node 3	2 (100%)	1 (100%)	1 (100%)	0	0	1 (100%)
Node 4	1 (100%)	1 (100%)	0	0	0	1 (91%)
Node 5	0 (100%)	0	2 (98%)	1 (100%)	1 (100%)	0

13.2.5 Heuristic Approach

The optimization problem of traffic grooming is *NP*-complete. It can be partitioned into the following four subproblems, which are not necessarily independent:

1. Determine a virtual topology, i.e., determine the number of lightpaths between any node pair.
2. Route the lightpaths over the physical topology.
3. Assign wavelengths optimally to the lightpaths.
4. Route the low-speed connection requests on the virtual topology.

Routing

The routing and wavelength assignment problem (RWA) has received a lot of attention in the WDM networking literature. The current well-known routing approaches are fixed routing, fixed-alternate routing, and adaptive routing [ZaJM00a, RaMu02].

In fixed routing, the connections are always routed through a pre-defined fixed route for a given source-destination pair. One example of such an approach is fixed shortest-path routing. The shortest-path route for each source-destination pair is calculated off-line using standard shortest-path algorithms, such as Dijkstra's algorithm. If there are not enough resources to satisfy a connection request, the connection gets blocked.

In fixed-alternate routing, multiple fixed routes are considered when a connection request comes. In this approach, each node in the network is required to maintain a routing table that contains an ordered list of a number of fixed routes to each destination node. For example, these routes can be the first-shortest-path, the second-shortest-path, etc. When a connection request comes, the source node attempts to establish the connection on each of the routes from the routing table in sequence, until the connection is successfully established. Since fixed-alternate routing provides simplicity of control for setting up and tearing down connections, it is also widely used in the dynamic connection-provisioning case. It has been shown that, for certain networks, having as few as two alternate routes provides significantly lower blocking than having full wavelength conversion at each node with fixed routing [RaMu02].

In adaptive routing, the route from a source node to a destination node is chosen dynamically, depending on the current network state. The current network state is determined by the set of all connections that are currently in progress [ZaJM00a]. For example, when a connection request arrives, the current shortest path between the source and the destination is calculated based on the available resources in the network; then the connection is established through the route. If a connection gets blocked in the adaptive-routing approach, it will also be blocked in the fixed-alternate routing approach. Since each time a new connection request comes to a node, the route needs to be calculated based on the current network state, adaptive routing will require more computation and a longer setup time than fixed-alternate routing, but it is also more flexible than fixed-alternate routing.

In our heuristics, we will use adaptive routing.

Wavelength Assignment

Once the route has been chosen for each lightpath, the number of lightpaths traversing a physical fiber link defines the congestion on that particular link. With the wavelength-continuity constraint, we need to assign wavelengths to each lightpath such that any two lightpaths passing through the same physical link are assigned different wavelengths. We assume a single-fiber network system, i.e., there is only one fiber in each direction if two nodes are connected. Several wavelength-assignment approaches have been compared in Chapter 7, and all of them were found to perform similarly. We will use one simple approach, First-Fit (FF). In FF, all wavelengths are numbered. When searching for an available wavelength, a lower-numbered wavelength is considered before a higher-numbered wavelength. The first available wavelength is then selected. The idea behind this simple scheme is that it tries to pack all of the in-use wavelengths towards the lower end of the wavelength space.

Heuristics

We propose two heuristic algorithms for the traffic-grooming problem. Let $T(s,d)$ denote the aggregate traffic between node pair s and d, which has not been successfully carried. Let $t(s,d)$ denote one connection request between s and d, which has not been successfully carried yet. Let C denote the wavelength capacity.

- *Maximizing Single-Hop Traffic (MSHT).* The basic idea of this heuristic is introduced in Chapter 9 for the traditional virtual-topology design problem. This simple heuristic attempts to establish lightpaths between source-destination pairs with the highest $T(s,d)$ values, subject to constraints on the number of transceivers at the two end nodes, and the availability of a wavelength in the path connecting the two end nodes. The connection requests between s and d will be carried on the new established lightpath as much as possible. If there is enough capacity in the network, every connection will traverse a single lightpath hop. If there are not enough resources to establish enough lightpaths, the algorithm will try to carry the blocked connection requests using currently available spare capacity of the virtual topology. The pseudo-code for this heuristic is shown in Algorithm 13.1.
- *Maximizing Resource Utilization (MRU).* Let $H(s,d)$ denote the hop distance on physical topology between node pair (s,d). Define $T(s,d)/H(s,d)$ as the connection resource utilization value, which represents the average traffic per wavelength link. This quantity shows how efficiently the resources have been used to carry the traffic requests. This heuristic tries to establish the lightpaths between the node pairs with the maximum resource utilization values. When no lightpath can be set up, the remaining blocked traffic requests will be routed on the virtual topology based on their connection resource utilization value $t(s,d)/H^{'}(s,d)$, where $t(s,d)$ denotes a blocked connection request, and $H^{'}(s,d)$ denotes the hop distance between s and d on the virtual topology. The pseudo-code for this heuristic follows the same steps as the pseudo-code of MSHT (and, hence, it is not shown separately), except that the node pairs and blocked connections are sorted based on their resource utilization values.

Algorithm 13.1 MSHT

1. Construct virtual topology:
   (a) Sort all of the node pairs (s, d) according to the sum of uncarried traffic request $T(s, d)$ between (s, d) and put them into a list L in descending order.
   (b) Try to setup a lightpath between the first node pair (s', d') in L using first-fit wavelength assignment and shortest-path routing, subject to the wavelength and transceiver constraints. If it fails, delete (s', d') from L; otherwise, let $T(s, d) = Max[T(s, d) - C,\ 0]$ and go to Step 1a until L is empty.
2. Route the low-speed connections on the virtual topology constructed in Step 1.
   (a) Satisfy all of the connection requests which can be carried through a single lightpath hop, and update the virtual topology network state.
   (b) Route the remaining connection requests based on the current virtual topology network state, in the descending order of the connections' bandwidth requirement.

Both heuristic algorithms have two stages. Based on our observations from the ILP results, we find that packing different connections between the same node pair within the same existing lightpath, which directly joins the end points, is a very efficient grooming scheme. In the first stage, the algorithms try to establish lightpaths as much as possible to satisfy the aggregate end-to-end connection requests. If there are enough resources in the network, every connection request will be successfully routed through a single lightpath hop. This will minimize the traffic delay since the optical signals need not be converted into electronic domain. In the second stage, the spare capacities of the currently established lightpath channels are used to carry the connection requests blocked in the first stage, and the algorithms give single-hop groomable connections higher priority to be satisfied.

Heuristic Results and Comparison

Table 13.8 compares the results obtained from an ILP solver and the heuristic algorithms for the six-node network in Fig. 13.8(a). We observe that the MSHT and MRU heuristic algorithms show reasonable performance when compared with the results obtained from the ILP solver. The heuristic approaches have much less computation complexity than the ILP approach. The two proposed algorithms are relatively simple and straightforward; by using other RWA algorithms instead of adaptive routing and first-fit wavelength assignment, it may be possible to develop other complex heuristic algorithms to achieve better performance.

Table 13.8 Throughput results comparison between ILP and heuristic algorithms (total traffic demand is OC-988).

	Multi-hop (ILP)	Single-hop (ILP)	Heuristic (MSHT)	Heuristic (MRU)
T=3, W=3	74.7% (OC-738)	68.0% (OC-672)	71.0% (OC-701)	67.4% (OC-666)
T=4, W=3	93.8% (OC-927)	84.1% (OC-831)	89.4% (OC-883)	93.6% (OC-925)
T=5, W=3	97.9% (OC-967)	85.7% (OC-847)	94.4% (OC-933)	94.4% (OC-933)
T=7, W=3	97.9% (OC-967)	85.7% (OC-847)	94.4% (OC-933)	94.4% (OC-933)
T=3, W=4	74.7% (OC-738)	68.0% (OC-672)	71.0% (OC-701)	67.4% (OC-666)
T=4, W=4	94.4% (OC-933)	84.7% (OC-837)	93.1% (OC-920)	93.6% (OC-925)
T=5, W=4	100% (OC-988)	95.5% (OC-944)	100% (OC-988)	100% (OC-988)

Figures 13.9–13.11 show the results from the two heuristic algorithms, when applied to the larger network topology in Fig. 13.8(b). The traffic matrices follow the same distribution as we mentioned in Section 13.2.4.

In Fig. 13.9, we plot the network throughput versus the number of wavelengths on every fiber link when each node is equipped with 10 tunable transceivers. We observe that the MRU heuristic performs better than the MSHT algorithm with respect to network throughput. Since the number of tunable transceivers at each node is limited (10 in this case), when the number of wavelengths on each fiber link reaches a certain value (16 in this case), increasing the number of wavelengths does not help to increase the network throughput.

In Fig. 13.10, we plot the network throughput versus the number of transceivers at every node when each fiber link carries ten wavelengths. We compare the performance of the two heuristic algorithms. The results show that, when the number of transceivers is small (≤ 7 in this case), MSHT performs better than MRU. This is because MRU tries to utilize wavelengths efficiently. When the number of transceivers is small, wavelength is not the limiting resource in the network any more. So maximizing wavelength utilization will not help to improve the performance.

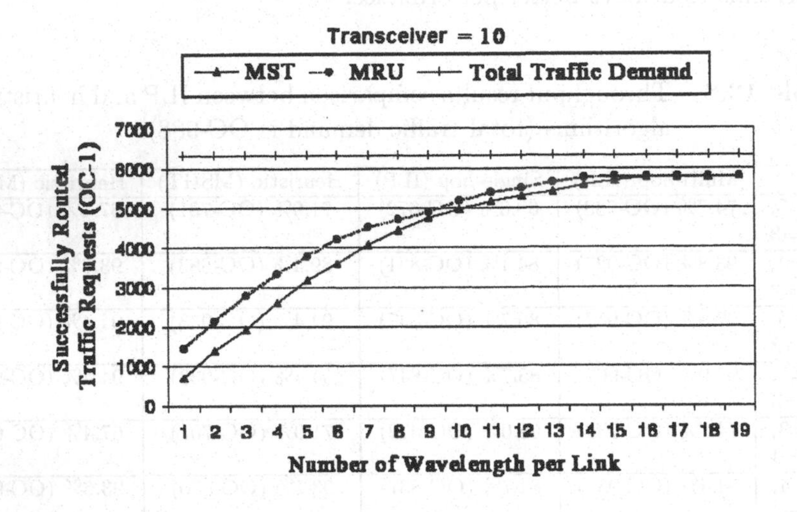

Figure 13.9 Network throughput vs. number of wavelengths for the network topology in Fig. 13.8(b) with 10 tunable transceivers at each node.

Figure 13.11 compares the performance using tunable transceiver and fixed transceiver in every network node. Each node is equipped with 12 transceivers if we use tunable transceiver. Each node is equipped with one transceiver array if we use fixed transceivers and the size of the transceiver array is equal to the number of wavelengths supported by every fiber link. The results in Fig. 13.11 indicate that, when nodes are equipped with the same number of transceivers, tunable-transceiver architecture has better performance. For the fixed-transceiver case, MSHT performs better than MRU.

13.2.6 Mathematical Formulation Extension for Other Optimization Criteria

In this section, we show how to extend our ILP formulations to handle different optimization criteria for the static traffic-grooming problem.

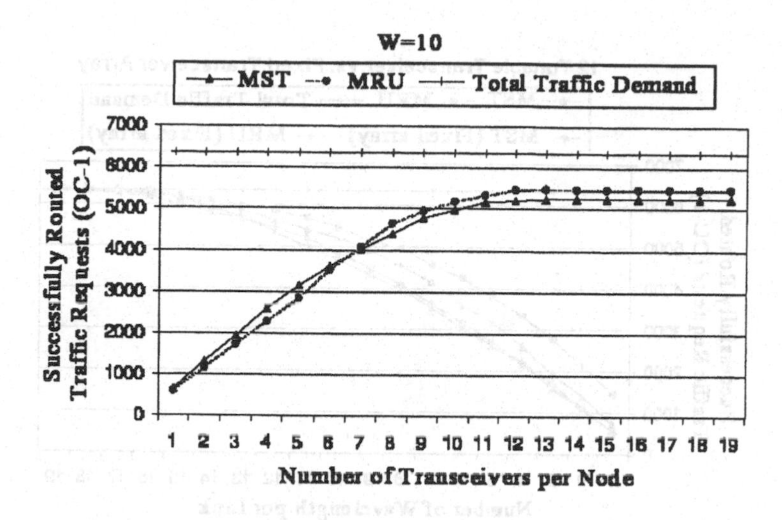

Figure 13.10 Network throughput vs. number of tunable transceivers for the network topology in Fig. 13.8(b) with 10 wavelengths on each fiber link.

Extension for Network Revenue Model

It has been shown earlier that the low-speed connection requests between the same node pair tend to be packed together on to the same lightpath channel. The connections which can be carried by a single lightpath channel are more likely to be satisfied than the connections which have to traverse multiple lightpaths, when they have the same bandwidth requirement and the optimization objective is to maximize network throughput. To make the problem more realistic, it is reasonable for us to assume that two connection requests may have different priority, even if they have the same bandwidth requirement. This is because different connection requests may have different end-node distance, quality-of-service requirement, etc. A connection's priority can be represented by a "weight" associated with it. In this section, we assume that the weight is determined by the bandwidth requirement and end-node distance of the connection request. For a given network topology and traffic demand, the objective is to maximize the weighted network throughput. Let W_i denote the weight of connection i, D_i denote the end-node distance of connection i, and C_i denote the bandwidth requirement of connection i. The connection's weight function is defined as:

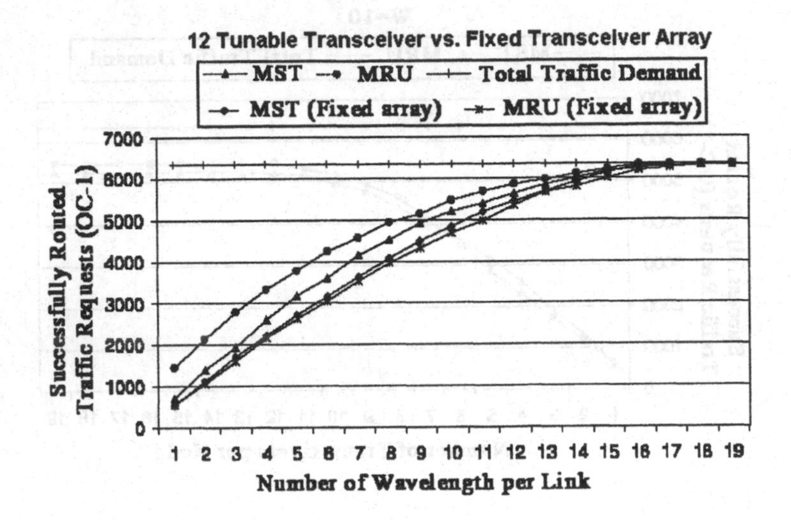

Figure 13.11 Network throughput vs. number of wavelengths (size of fixed transceiver array) for the network topology in Fig. 13.8(b) with 12 tunable transceivers at each node.

$$W_i = D_i \times C_i^{\alpha} \tag{13.26}$$

where $0 \leq \alpha \leq 1$ and D_i is measured by the shortest-path distance of the connection's end nodes on the physical topology. Equation (13.26) is called the power-law cost function [Klei70]. It is used to study the actual tariffs demanded by communications services for high-speed telecommunication channels, and there is effectively a "quantity discount" (controlled by α) in that capacity cost (per unit of channel capacity) decreases as the capacity increases. Thus, the network weighted throughput becomes:

$$T = \sum_{i=1}^{K} D_i \cdot C_i^{\alpha} \cdot S_i \tag{13.27}$$

where $S_i = 1$ if connection request i has been satisfied; otherwise $S_i = 0$, and the total number of connection requests is K. T can also be called *"network revenue"*. We can easily modify our ILP formulation to optimize T. The only part of the equations which should be modified is shown as follows:

• Optimize: Maximize network revenue.

$$Maximize : \sum_{y,s,d,t} D_{sd} \cdot y^{\alpha} \cdot S_{sd}^{y,t} \tag{13.28}$$

where D_{sd} denotes the distance between node pair (s, d).

Illustrative Results

In this section, we show some illustrative results to optimize network revenue using our ILP formulation extension. We use the same network topology and traffic matrix set as in Section 13.2.4. In Equation (13.28), D_{sd} is measured by the shortest-path hop distance between node s and d on the physical topology, and α is equal to 0.8.

Table 13.9 Results of comparison between revenue model and network throughput model.

	Optimize Revenue		Optimize Throughput
	Revenue	Throughput	Throughput
T=3, W=3	83.7%	72.4%	74.7%
T=5, W=3	98.5%	97.2%	97.9%
T=7, W=3	98.5%	97.2%	97.9%
T=3, W=4	83.7%	72.4%	74.7%
T=4, W=4	94.3%	91.7%	94.4%
T=5, W=4	100%	100%	100%

Table 13.9 compares the results between the two optimization models. In Table 13.9, T denotes the number of tunable transceivers per node and W denotes the number of wavelengths per fiber link. Multi-hop grooming is allowed in both models. It is shown that, in the revenue model, when $T = 3$ and $W = 3$, the maximal achievable revenue is 83.7%, and 72.4% of traffic requests have been satisfied to achieve the revenue, while the maximal achievable traffic load the network can carry is 74.7%. In the revenue model, we find that, if there is a lightpath set up between (s, d), it may first be used to carry some long multi-hop connections (with higher weight) which will traverse this lightpath as an intermediate hop. Thus, some connections directly between (s, d) may be blocked. This means that packing different

connections between the same node pair within the same existing lightpath, which directly joins the end points, is not a good grooming scheme any more. We find that, because of the quantity-discount parameter α in Eqn. (13.26), lower-speed connections are more likely to be satisfied than higher-speed connection requests. It is obvious that different heuristics are needed based on the different optimization criteria.

13.3 Dynamic Traffic Grooming

This section is devoted to the dynamic traffic-grooming problem. We also assume a heterogeneous, multi-granularity, optical WDM network environment, where connections with different bandwidth-granularity requirements arrive one at a time and each such connection needs to be properly routed through the network based on the current network state.

A carrier's WDM backbone network is expected to be a multi-vendor, heterogeneous network since network equipment (NE) may need to be used from different vendors and new equipment has to co-exist with legacy equipment. For example: (1) network nodes may have optical crossconnects (OXCs) employing different architectures and technologies; (2) not all nodes may have wavelength conversion and traffic-grooming capabilities (i.e., sparse wavelength conversion and sparse grooming networks); (3) wavelength conversion and traffic grooming may only be available on certain wavelength channels (i.e., nodes may have partial wavelength conversion and partial grooming capability); and (4) different fiber links may support different numbers of wavelength channels, which may also operate at different speeds. This range of heterogeneity increases the challenge for the network operator to efficiently engineer the customers' traffic of different bandwidth granularities.

In this section, we investigate the problem of dynamically provisioning connections of different bandwidth granularities in mesh-based heterogeneous WDM networks. The network is considered to have two levels of hierarchical switching capability: (1) wavelength switching and (2) SONET/SDH based low-speed circuit (time-slot) switching. The ideas of this study can also be applied to networks with other hierarchical switching capabilities, such as packet-based switching, waveband switching, fiber switching, etc.

13.3.1 Provisioning Connections in Heterogeneous WDM Network

There are three important components in WDM network control, which determine how connections of different bandwidth granularities are provisioned. They are (1) resource-discovery protocol, (2) signaling protocol, and (3) route-computation algorithm.

Resource-discovery protocols determine how the network resources are discovered, represented, and maintained in the OXCs' link-state databases (for distributed control) or by the network control and management system (for centralized control).

Route-computation algorithms determine how the route of a low-speed connection request is computed and selected according to the carrier's grooming policy [ZhMu02c, ZhZM03]. Grooming policy reflects the intention of a network operator on how to allocate network resources to a given request if multiple routes are available. It is a traffic-engineering decision.

Signaling protocols determine how the connection is configured and how a network node allocates its local network resources to the connection, e.g., port mapping or label assignment for a connection.

A unified control plane for intelligent WDM mesh-based networks is being standardized by two standards bodies — the International Telecommunications Union (ITU), which is working on the architecture for Automatically Switched Optical Networks (ASON) [ITU01a], and the Internet Engineering Task Force (IETF), which is developing Generalized Multi-Protocol Label Switching (GMPLS) [Mann02]. The purpose of this network control plane is to provide an intelligent, automatic, end-to-end circuit (or virtual circuit) provisioning/signaling scheme throughout different network domains.

As we mentioned above, the network may also be heterogeneous within one service domain. Thus, further extensions on the GMPLS control plane are needed to adapt it for the network heterogeneity caused by the interoperation of OXCs from different vendors and the co-existence of new network equipment with legacy network equipment. These extensions will be elaborated in following sections. Note that, without loss of generality and for purposes of illustration, we will consider the WDM network to have full wavelength-conversion capability at every network node in our remaining sections.

Resource Discovery

As an abstract view, the network we are considering may have two types of network links: physical link (optical fiber) and virtual link (lightpath). A lightpath is also known as a traffic-engineering link (TE-link) in [Mann02]. The switching granularity of the lightpath at any intermediate node should be full wavelength-channel granularity. A low-speed connection request may traverse one or multiple lightpaths. Because of the possibility of the interconnection of OXCs with different grooming capabilities, there may exist four types of lightpaths in a carrier's optical WDM network, as follows.

- *Multi-hop un-groomable lightpath*: A lightpath (i, j) is a multi-hop un-groomable lightpath if it is not connected with a finer-granularity switching element at its end nodes. This lightpath can only be used to carry the traffic directly between node pair (i, j). A lightpath, which is established between two non-grooming OXCs or single-hop grooming OXCs, is an example of this type of lightpath.

- *Source-groomable lightpath*: A lightpath (i, j) is a source-groomable lightpath if it is only connected with a finer-granularity switching element at its source node. All traffic on this lightpath has to terminate at node j, but the traffic may originate from any other network node as well. As an example, if a lightpath is established from a multi-hop full-grooming OXC to a non-grooming OXC (or single-hop grooming OXC), it is a source-groomable lightpath.

- *Destination-groomable lightpath*: A lightpath (i, j) is a destination-groomable lightpath if it is only connected with a finer-granularity switching element at its destination node. All traffic on this lightpath has to originate from node i. At the lightpath destination node j, the traffic on lightpath (i, j) can either terminate at j or be groomed to other lightpaths and routed towards other nodes. As an example, if a lightpath is established from a non-grooming OXC (or single-hop grooming OXC) to a multi-hop full-grooming OXC, it is a source-groomable lightpath.

- *Full-groomable lightpath*: A lightpath (i, j) is a full-groomable lightpath if it connects to finer-granularity switching elements at both end nodes. This type of lightpath can be used to carry traffic between any node pair in the network. Note that source-groomable, destination-groomable, and full-groomable lightpaths can also be called multi-hop groomable lightpaths.

The states of both physical link (fiber link) and virtual link (lightpath) should be advertised by link-state advertisements (LSAs). Besides the neighbors connected by physical links, a node may also have other neighbors, which are connected to the node by virtual links and which are geographically far away in the network. As a simplified abstraction, we present the link state of each network link type as follows.

- *Fiber Link*: The representation of a fiber link (in a full wavelength-convertible network) can be denoted as $f(m, n, t, w, c)$, where m and n denote the end nodes of the fiber link, t denotes fiber index (for numbering multiple fibers between the same node pair), w denotes the available (free) wavelength channels on that fiber, and c denotes the administrative link cost. In a WDM network with wavelength-continuity constraint, more information is needed to identify the availability property of each individual wavelength channel. Note that this representation can be viewed as bundling multiple wavelength channels between the same pair of nodes as proposed in [Mann02, Komp02].

  If there are multiple fibers between the same node pair, they may be further bundled. The purpose of link bundling is to improve routing scalability by reducing the amount of information that has to be handled by the network control plane. This reduction is accomplished by performing information aggregation/abstraction [Mann02]. After fiber links between the same node pair are bundled, they can be advertised as a logical link for route-computation algorithms.

- *Virtual Link*: The representation of a lightpath can be denoted as $l(i, j, v, t, m_1, m_2, c)$, where i and j denote the end nodes of the lightpath, v denotes the lightpath type, t denotes the lightpath id, m_1 denotes the minimal reservable bandwidth on this lightpath, which is determined by the grooming ratio of the end nodes, m_2 denotes the maximal reservable bandwidth on this lightpath, which is bounded by the total available (free) capacity on the lightpath, and c denotes the administrative link cost. Multiple lightpaths (of the same type) between the same node pair can also be bundled and advertised as a logical link. Note that link bundling should also follow the network fault-management constraints, i.e., the links in the same bundle should belong to the same shared-risk link group (SRLG) [Komp02].

Besides advertising the states of the network links, the state of a network node may also need to be advertised. This is because a multi-hop partial-grooming OXC can only perform multi-hop grooming on a limited number of lightpaths. So, if a network node employs this type of OXC, the available multi-hop grooming capability of the OXC should also be advertised when the node's state changes.

Route Computation

In an intelligent WDM network which we are considering, the route of a connection request will be computed either by the source node or by the network control and management system. Let $Req(s, d, r)$ denote a connection request, where s denotes the source node, d denotes the destination node, and r denotes the capacity requirement of the connection. There exist the following possibilities to route the connection request [ZhZM03]:

- *Operation 1*: Carry *Req* using an existing lightpath $l(s, d, v, t, m_1, m_2, c)$ between nodes s and d, if $m_1 \leq r \leq m_2$.
- *Operation 2*: Carry *Req* using multiple existing groomable lightpaths.
- *Operation 3*: Carry *Req* by establishing a new lightpath (either groomable or non-groomable) between node pair (s, d) if enough resources exist.
- *Operation 4*: Carry *Req* using a combination of both existing groomable lightpaths and setting up new groomable lightpaths using available wavelength channels in fiber links and grooming resources in network nodes.

Since there are multiple ways to carry the connection request, multiple routes may be simultaneously available. The decision on how to choose a proper route from multiple candidate routes is a traffic-engineering decision and is known as the network operator's *grooming policy*.

In a dynamic traffic environment, connections with various bandwidth requirements arrive in a network, stay for a certain period of time, and then leave the network. A grooming policy may have different performance under dynamic traffic environment than the static one. Moreover, dynamic traffic grooming might need to dynamically adjust the grooming policy according to the traffic pattern and current network state. Hence, grooming policies for dynamic traffic need to be investigated.

Below, we present four different grooming policies for dynamic traffic.

- *Minimize the Number of Traffic Hops on the Virtual Topology (MinTHV).*
  We first use Operation 1. If Operation 1 fails, we always try to set up a lightpath from s to d and route the traffic onto this lightpath (Operation 3). Only when such a direct lightpath cannot be set up, we use multi-hop grooming by either Operation 2 or Operation 4, and choose the one with fewer hops on the virtual topology (number of lightpaths). This policy chooses the route with the fewest lightpaths for a connection.

- *Minimize the Number of Traffic Hops on the Physical Topology (MinTHP).*
  We compare the number of wavelength-links used by all the four operations and choose the one with the fewest wavelength-links.

- *Minimize the Number of Lightpaths (MinLP).*
  This policy is similar to MinTHV but it tries to set up the minimal number of *new* lightpaths to carry the traffic. Operation 1 is attempted first. If it fails, we try to route the traffic using multiple existing lightpaths (Operation 2). If Operation 2 also fails, we try to set up one lightpath with the minimal number of wavelength-links either by Operation 3 or Operation 4. If such a lightpath is not feasible, we go with Operation 4 and set up two or more lightpaths.

- *Minimize the Number of Wavelength-Links (MinWL).*
  This policy is similar to MinTHP but it tries to consume the minimum number of *extra* wavelength-links, i.e., wavelength-links not being used by any lightpaths for now, to carry the traffic. The difference between MinLP and MinWL is that, if both Operations 1 and 2 fail, MinWL compares the number of wavelength-links used by Operations 3 and 4, and chooses the one requiring fewer wavelength-links; MinLP, on the other hand, compares the number of lightpaths used by Operations 3 and 4, chooses the one requiring fewer lightpaths, and uses the number of wavelength-links for tie-breaking.

In dynamic grooming, the network state changes as connection requests come and go. To achieve good performance, the grooming policy should be adjusted according to the current network state. For instance, if transceivers are becoming the more scarce resource, we should make full use of existing lightpaths to accommodate the new traffic and avoid setting up new lightpaths.

Signaling

After a route is successfully computed, every intermediate node along the route needs to be informed through appropriate signaling protocols. Two protocols — resource reservation protocol (RSVP) with traffic-engineering extensions [BZBH97, Awdu01] and constraint-based routing label distribution protocol (CR-LDP) [ADFF01, Jamo02] — have been proposed as the standard signaling protocols in the GMPLS control plane. In a heterogeneous WDM network, the route computed for a connection request will be composed of a sequence of intermediate node id as well as link bundle id. Since multiple candidate links may exist in a link bundle, an intermediate node needs to select one for the connection request when it needs to configure the OXC and establish the connection. If the link is a bundled lightpath link, then, based on the available capacity of each lightpath in the bundle and the bandwidth requirement of the request, different link-selection schemes can be used, e.g., random selection, first-fit selection, best-fit selection, etc.

13.3.2 A Generic Provisioning Model

Graph Model

The network heterogeneity increases the complexity of efficiently provisioning customers' connection requests. Hence, a generic bandwidth-provisioning model, which can incorporate various network elements (NEs) and accommodate different grooming policies, will enable network operators to manage their transport networks easily and efficiently, as well as reduce the overall cost (network cost and operation cost) significantly. This model should also be easy to implement and scale using a distributed control plane. Extending the work in [ZhZM03], we propose such a simple, generic provisioning model for heterogeneous optical WDM networks.

Figure 13.12 shows an illustrative example of the provisioning model, which we use to explain how the model works. Figures 13.12(a) and 13.12(b) show the network state for a simple three-node network. The shaded node (node 0) is the node which employs a multi-hop partial-grooming OXC and the un-shaded nodes (nodes 1 and 2) are equipped with single-hop grooming OXCs. Each link in Fig. 13.12(a) represents a free wavelength channel between a node pair and each link in Fig. 13.12(b) represents an established lightpath. The lightpath $(0, 2)$ is a source-groomable lightpath, the lightpath $(1, 0)$ is a destination-groomable lightpath, and the lightpath $(2, 1)$ is a multi-hop un-groomable lightpath. A low-speed connection request from node 1 to node 2 can be carried by lightpaths $(1, 0)$ and $(0, 2)$. On the other hand, a request from node 2 to node 0 cannot traverse lightpaths $(2, 1)$ and $(1, 0)$ since node 1 does not have multi-hop grooming capability.

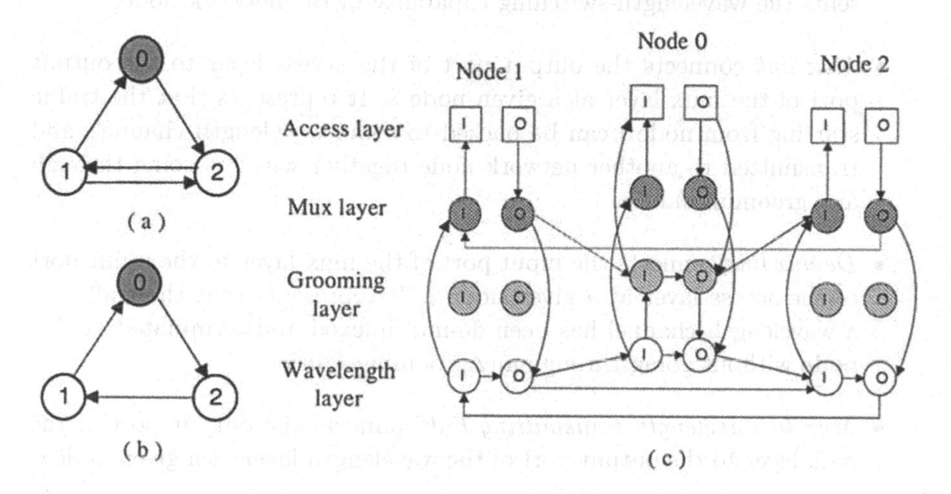

Figure 13.12 Network state for a simple three-node network and the corresponding auxiliary graph.

Figure 13.12(c) shows a graph representation of the network state in Figs. 13.12(a) and 13.12(b). The graph is divided into four layers, namely access layer, mux layer, grooming layer, and wavelength layer. The access layer represents the access point of a connection request, i.e., the point where a customer's connection starts and terminates. It can be an IP router, an ATM switch, or any other client equipment. The mux layer represents the OXC ports from which low-speed traffic streams are directly multiplexed (de-multiplexed) onto (from) wavelength channels without going through the grooming fabric. The grooming layer represents the grooming component of the network node. The wavelength layer represents the wavelength-switching capability and the link state of wavelength channels. A network node is divided into two vertices at each layer. These two vertices represent the input and output ports of the network node at that layer. The links in this graph model are named and work as follows.

- *Grooming switching link* connects the input port of the grooming layer to the output port of the grooming layer at a given node i, when node i has multi-hop traffic-grooming capability.

- *Wavelength switching link* connects the input port of the wavelength layer to the output port of the same layer at a given node i. It represents the wavelength-switching capability of the network node.

- *Mux link* connects the output port of the access layer to the output port of the mux layer at a given node i. It represents that the traffic starting from node i can be packed to some wavelength channels and transmitted to another network node together without going through any grooming fabric.

- *Demux link* connects the input port of the mux layer to the input port of the access layer at a given node i. It represents that the traffic on a wavelength channel has been de-multiplexed and terminated at this node without going through any grooming fabric.

- *Mux to wavelength transmitting link* connects the output port of the mux layer to the output port of the wavelength layer at a given node i.

- *Wavelength to mux receiving link* is the link which connects the input port of the wavelength layer to the input port of the mux layer at a given node i.

- *Grooming link* connects the output port of the access layer to the output port of the grooming layer at a given node i, when node i has multi-hop grooming capability (i.e., free outgoing grooming ports). It represents that the traffic starting from node i can be groomed with other traffic streams (originating from node i or from other network nodes) to the same wavelength channel and transmitted to the next network node together.

- *De-grooming link* connects the input port of the grooming layer to the input port of the access layer at a given node i, when node i has multi-hop grooming capability (i.e., free incoming grooming ports). It represents that the traffic streams on a wavelength channel have been de-multiplexed, and then they may be either terminated at node i or switched to other lightpaths.

- *Grooming to wavelength transmitting link* connects the output port of the grooming layer to the output port of the wavelength layer at a given node i, when node i has multi-hop grooming capability (i.e., free incoming grooming ports). It denotes that a multi-hop groomable lightpath (i.e., either a source-groomable lightpath or a multi-hop full-groomable lightpath) can be originated at node i.

- *Wavelength to grooming receiving link* connects the input port of the wavelength layer to the input port of the grooming layer at a given node i, when node i has multi-hop grooming capability (i.e., free incoming grooming ports). It denotes that a multi-hop groomable lightpath (i.e., either a destination-groomable lightpath or a multi-hop full-groomable lightpath) can be terminated at node i.

- *Wavelength link* connects the output port of the wavelength layer at node i to the input port of the wavelength layer at node j. It denotes the availability of the wavelength channels between node pair (i, j).

- *Lightpath link* can start at the output port of the mux layer (grooming layer) at node i, and terminate at the input port of the mux layer (grooming layer) at node j. The four combinations of the end points represent the four possible lightpath types between node pair (i, j) discussed in Section 13.3.1.

Each link in Fig. 13.12(c) represents the availability of the corresponding network resource. A link is removed if the corresponding network resource is not available (e.g., deleting a wavelength link), and a link is added if the corresponding network resource becomes available from unavailable state (e.g., adding a lightpath link). In this graph representation, a customer's connection request will always originate from the output port of the access layer at the source node and terminate at the input port of the access layer at the destination node. After properly adjusting the administrative link costs, suitable routes can be found according to different grooming policies for a request by simply applying standard shortest-path route-computation algorithms. Through some straightforward extensions (by stretching the single wavelength layer to multiple layers, one for each wavelength), the network without full wavelength-conversion capability can also be properly modeled.

Figure 13.13 shows the corresponding graph representation of network nodes employing different traffic-groomable OXC architectures. They are extended from the original graph model proposed in [ZhZM03]. In the original graph model, there is a lightpath layer in the auxiliary graph, and all lightpaths will originate/terminate at that lightpath layer. Hence, the auxiliary graph in [ZhZM03] cannot model a network node to differentiate the four lightpath types and to simultaneously support all lightpath types we presented in Section 13.3.1 even if the node has this capability. By splitting the lightpath layer into the mux layer and the grooming layer, the proposed extended model is able to simultaneously support all types of lightpaths and incorporate all possible network states in a heterogeneous network environment.

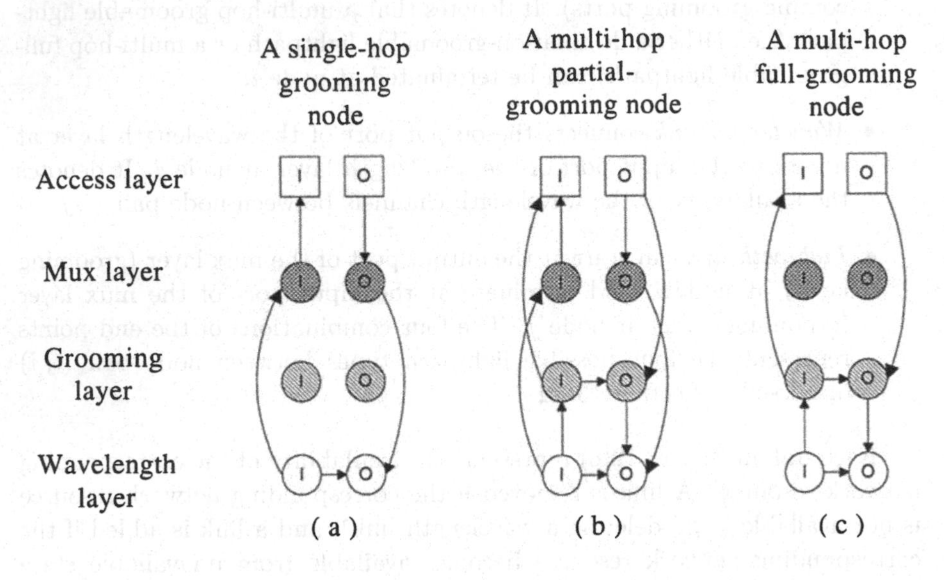

Figure 13.13 Different grooming OXCs and their representations in the auxiliary graph.

Engineering Network Traffic Using the Proposed Graph Model

We can observe that, by constructing this auxiliary graph and using it for connection provisioning, the network control software can easily incorporate different network elements (i.e., OXCs with different characteristics) and accurately represent different network states. This strategy provides a platform for network operators to realize different grooming policies, and to eventually improve the provisioning flexibility and network resource efficiency.

Figure 13.14 shows an example on how to achieve different traffic-engineering objectives through different grooming policies by using our generic graph model. The network state and the graph representation are shown in Fig. 13.12. Assuming that there is a new traffic request from node 1 to node 2, Fig. 13.14 shows two possible routes (in thick links) for this connection request. The route shown in Fig. 13.14(a) traverses two existing lightpath links, while the route shown in Fig. 13.14(b) will employ two new wavelength channels. If the connection requires full wavelength-channel capacity, or if the overall bandwidth requirement of the future traffic demands between the node pair is estimated to be close to full wavelength-channel capacity, the route in Fig. 13.14(b) is preferred since the wavelength channels are fully utilized and no grooming is needed at node 0; otherwise, the route in Fig. 13.14(a) may be preferred if enough free capacity is available on the existing lightpaths. By assigning proper weights to each link in the graph in Fig. 13.12(c), the corresponding routes will be computed through standard shortest-path computation algorithms according to different grooming policies.

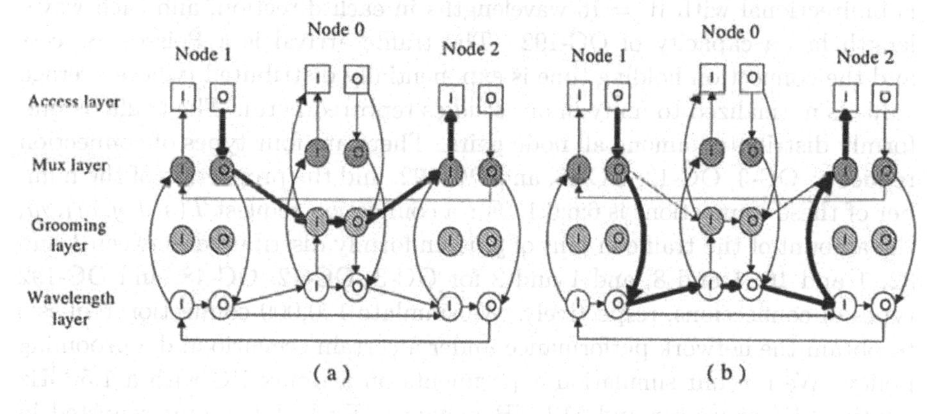

Figure 13.14 Two alternative routes for a new connection request (1, 2).

Computational Complexity

The computational complexity to construct such an auxiliary graph for a N-node WDM network is $O(N^2)$ for a full wavelength-convertible WDM network. This is because the auxiliary graph will consist of $N \times 4 \times 2$ nodes and at most $(N \times 4 \times 2)^2$ links. As we know, the computation complexity for a standard shortest-path algorithm in a N-node network is $O(N^2)$. Therefore, the computational complexity to provision a connection request using this model in a full wavelength-convertible WDM network is $O(N^2)$. If the WDM network does not have full wavelength-conversion capability, there will be $(2 \times (W+3) \times N)$ nodes and at most $(2 \times (W+3) \times N)^2$ links in the auxiliary graph, where W is the number of wavelength channels a fiber supports in the network. Hence, the computational complexity to provision a connection request using this model will be $O(N^2 \times W^2)$ in such a wavelength-continuous WDM network.

13.3.3 Illustrative Numerical Examples

Comparison of Grooming Policies

We compare the performance of different grooming policies on the network topology shown in Fig. 13.15, which has 19 nodes and 31 links. All the nodes have grooming capability but no wavelength-conversion capability. Each link is bidirectional with $W = 16$ wavelengths in each direction, and each wavelength has a capacity of OC-192. The traffic arrival is a Poisson process and the connection holding time is exponentially distributed (whose average value is normalized to unity in our studies reported here). The traffic is uniformly distributed among all node pairs. There are four types of connection requests: OC-3, OC-12, OC-48, and OC-192, and the proportion of the number of these connections is 6:6:6:1. For a connection request $T(s, d, g, m)$, m, the amount of the traffic in unit of g, is uniformly distributed between 1 and 32, 1 and 16, 1 and 8, and 1 and 2 for OC-3, OC-12, OC-48, and OC-192 types of connections, respectively. We simulate 100,000 connection requests to obtain the network performance under a certain scenario and a grooming policy. We ran our simulation experiments on a Linux PC with a 1.5-GHz Pentium IV processor and 512-MB memory. Each data point reported in the illustrations in this section took between 6-9 minutes of running time on this computer.

Table 13.10 shows the average utilization of wavelength-links (U_W) and the average utilization of transceivers (U_{Tx}) when the network load L is 300 Erlangs. When each node has only 16 transceivers, the utilization of transceivers is very high since they are the more constrained resources. When there are 32 transceivers at each node, the utilizations of both transceivers and wavelengths are quite balanced and high as well. If there are

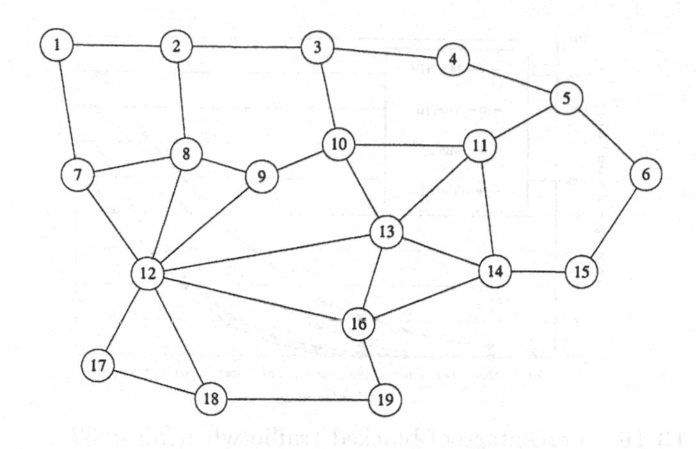

Figure 13.15 A 19-node network.

Table 13.10 Average utilization of wavelength-links and transceivers when W=16 and L=300 Erlangs.

	$Tx = 16$		$Tx = 32$		$Tx = 40$	
	U_W	U_{Tx}	U_W	U_{Tx}	U_W	U_{Tx}
MinTHV	0.7819	0.9858	0.8878	0.7264	0.8905	0.5884
MinTHP	0.5674	0.9901	0.7354	0.8165	0.7361	0.6910
MinLP	0.7403	0.9807	0.8890	0.8007	0.8918	0.6651
MinWL	0.6201	0.9859	0.8133	0.8683	0.8120	0.7825

40 transceivers at each node, we have relatively more transceivers; hence, wavelength-links become the more constrained resources, so the utilization of the transceivers is relatively lower.

Figure 13.16 shows the network performance under different grooming policies with Tx = 32 transceivers at each node. We observe that, as the network load increases, the percentage of blocked traffic also increases, but different grooming policies have different blocking probabilities. Grooming policy MinTHV performs best, followed by MinTHP, MinLP, and MinWL in sequence.

When we alter the network configuration by adding more transceivers at each node, the performance of each of the policies also changes, as shown in Fig. 13.17. In this scenario, each node has 40 transceivers instead of 32.

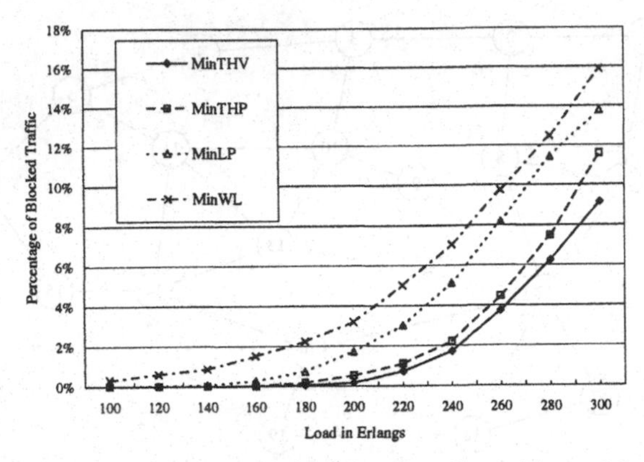

Figure 13.16 Percentage of blocked traffic when $Tx = 32$.

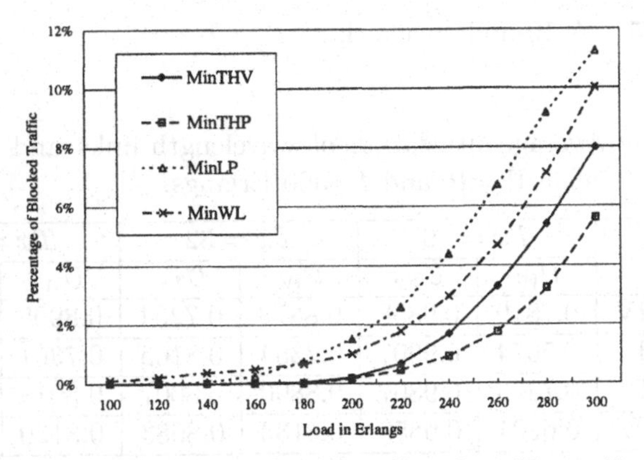

Figure 13.17 Percentage of blocked traffic when $Tx = 40$.

We observe that, now, MinTHP outperforms MinTHV and achieves the best results, and MinLP becomes the poorest-performing policy. This is because, in this network configuration, there are relatively more transceivers in the network so that wavelength-links become the more constrained resources. Recall that MinTHP utilizes wavelength-links more efficiently than other policies; hence, it performs the best in this case.

Another observation from both Figs. 13.16 and 13.17 is that MinTHV and MinTHP always perform better than MinLP and MinWL in terms of percentage of blocked traffic. This is because MinTHV and MinTHP examine the *overall* resource requirement of a given connection, while MinLP and MinWL only consider the new lightpaths to be set up or the extra wavelength-links used by these new lightpaths while setting up the connection. Therefore, MinTHV and MinTHP are more resource-efficient.

From the above results, we can observe that different grooming policies have different performance under various network configurations, which suggests that a grooming policy should be adjusted according to the current network state.

Adaptive Grooming Policy (AGP)

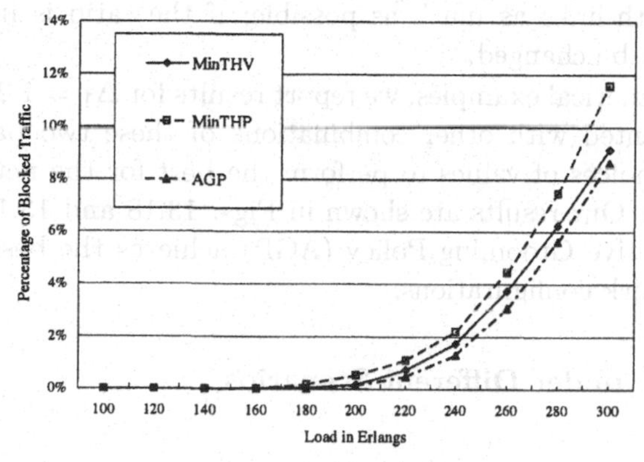

Figure 13.18 Performance of AGP when $Tx = 32$.

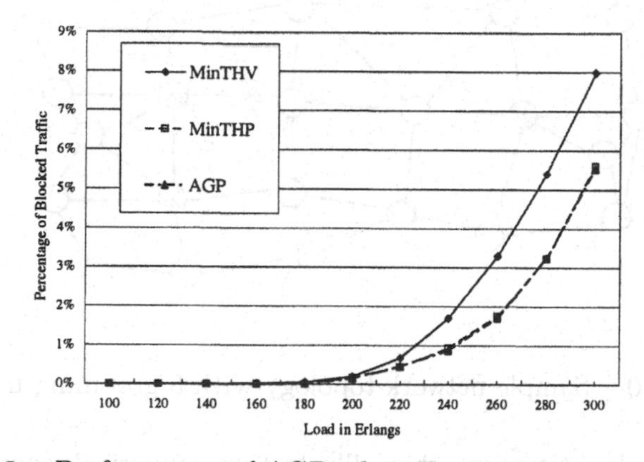

Figure 13.19 Performance of AGP when $Tx = 40$.

Since MinTHV performs best when transceivers are the more constrained resources and MinTHP gives the best results when wavelength-links become more scarce resources, we try to utilize the advantages of these two grooming policies by combining them together. Here, we present an Adaptive Grooming Policy (AGP) which, for each connection request, switches between MinTHV and MinTHP according to the current network state.

We use the ratio of the number of unused wavelength-links in the network to the total number of available transceivers at all nodes as an indicator of the network state. If the ratio is larger than Δ_1, MinTHV will be used to avoid setting up lightpaths since transceivers are more scarce resources at this time; if the ratio is less than Δ_2, MinTHP will be employed to try to save wavelength-links as much as possible; if the ratio is in between, the policy will not be changed.

For our numerical examples, we report results for $\Delta_1 = 1.2$ and $\Delta_2 = 1.0$. (We experimented with other combinations of these two parameters, and found these choices of values to perform the best for the network topology in Fig. 13.15.) Our results are shown in Figs. 13.18 and 13.19. We observe that the Adaptive Grooming Policy (AGP) achieves the best results under different network configurations.

Performance under Different Scenarios

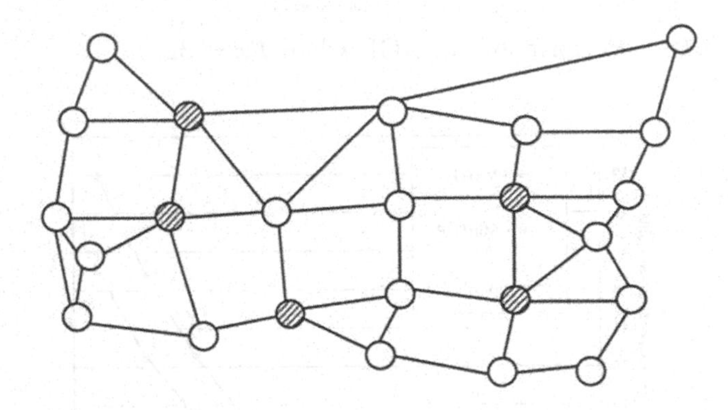

Figure 13.20 Sample network topology with 5 grooming nodes.

In this subsection, we show illustrative numerical examples on the network performance by using our proposed model to provision different bandwidth-granularity connection requests. The network topology used in our simulation experiments is shown in Fig. 13.20. It represents a typical operator's optical backbone network topology, which has 24 nodes and 43 bi-directional links. We simulated a dynamic traffic environment, where the traffic arrival process is Poisson and the connection-holding time follows a negative exponential distribution. The capacity of each wavelength channel is OC-192. For the illustrative results shown here, the bandwidth requirements of the connection requests follow an uniform distribution between OC-3, OC-12, OC-48, OC-192 (i.e., OC-3 : OC-12 : OC-48 : OC-192 = 1 : 1 : 1 : 1); connection requests are uniformly distributed among all node pairs; average connection-holding time is normalized to unity; the cost of a fiber link is modeled as unity; and load (in Erlang) is defined as the connection-arrival rate times the average connection-holding time, which denotes the average number of connections the network is carrying at any instant. Note that, from the connections' bandwidth distribution and Erlang load, it is straightforward to calculate the network offered load in the unit of OC-1.

We employ three metrics to evaluate the network performance, namely Traffic Blocking Ratio (TBR), Connection Blocking Probability (CBP), and Resource Efficiency Ratio (RER).

- *Traffic Blocking Ratio* represents the percentage of the amount of blocked traffic over the amount of bandwidth requirement of all traffic requests during the entire simulation period.
- *Connection Blocking Probability* represents the percentage of the total number of blocked connection requests over the number of all traffic requests during the entire simulation period. The connections with different bandwidth requirements may experience different connection blocking probabilities.
- *Resource Efficiency Ratio* represents how efficiently connections are routed and groomed. It can be computed as the average of network carried traffic (in terms of OC-1 unit) divided by the total allocated network capacity (i.e., total number of allocated wavelength links times 192) over the entire simulation period.

  Let β denote the RER. β can be computed as follows:

$$\beta = \frac{\sum_i \rho_i \times t_i}{\sum_i \gamma_i \times t_i},$$

  where t_i is the time period between the ith event (connection arrival or departure) and $(i+1)^{th}$ event, ρ_i is the network carried load during the time period t_i, and γ_i is the total number of wavelength links used during t_i. (Please note that ρ_i and γ_i do not change during time period t_i as there is no other event during that period.) If every connection required full wavelength-channel capacity (i.e., no capacity waste) and all connections were routed through the shortest path (in hop distance), the resource efficiency ratio will equal $\frac{1}{H}$, where H is the average hop distance of the network topology. This gives us an upper bound on the resource efficiency ratio. The average hop distance of the network topology in Fig. 13.20 is 3; hence, the upper bound for resource efficiency ratio for this network is approximately 33.3%.

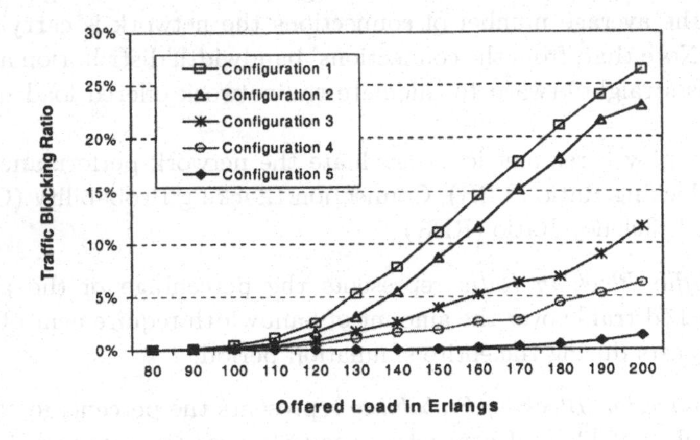

Figure 13.21 Traffic blocking ratio vs. offered load.

A connection is provisioned based on the following policies.

- The connection is provisioned through the least-cost route. The cost of a lightpath link is equal to the overall cost of the concatenated fiber links it traverses. Every fiber link has unit cost.
- If there are multiple least-cost routes and the connection does not require full wavelength-channel capacity, select the route employing the minimal number of free wavelength links.
- If there are multiple least-cost routes and the bandwidth requirement of the connection requires full wavelength-channel capacity, select the route traversing the minimal numbers of electronic grooming fabrics.

Note that it is also possible to apply other grooming policies by using the proposed provisioning model (see [ZhZM03]).

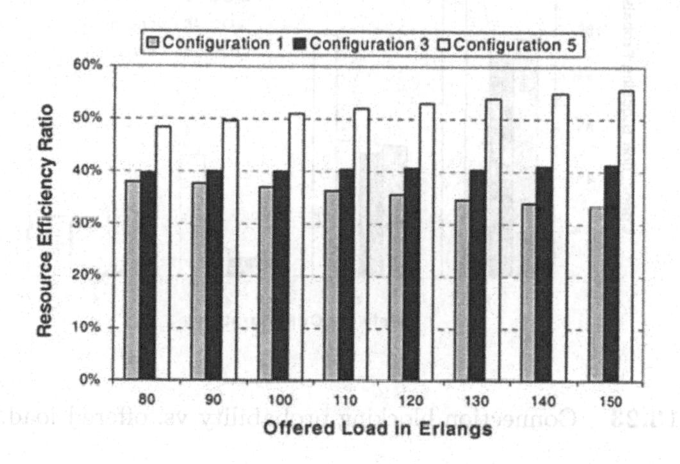

Figure 13.22 Normalized resource-efficiency ratio vs. offered load.

Figures 13.21–13.23 show illustrative numerical examples based on different network configurations. In Configuration 1, all network nodes are only equipped with single-hop grooming OXCs. In Configurations 2, 3, and 4, the shaded nodes in Fig. 13.20 are equipped with multi-hop partial-grooming OXCs. The numbers of grooming ports in multi-hop partial-grooming OXCs are 4, 8, and 16 in Configurations 2, 3, and 4, respectively. In Configuration 5, all shaded nodes in Fig. 13.20 are equipped with multi-hop full-grooming OXCs. Each bi-directional link in Fig. 13.20 contains two uni-directional fibers and each fiber supports eight wavelength channels.

In Fig. 13.21, we plot the traffic blocking ratio versus network offered load in Erlangs. We observe that, as the load increases, the traffic blocking ratio increases. When the network has more grooming capability, the traffic blocking ratio can be improved significantly. Figure 13.22 shows the normalized (to theoretical upper bound) resource efficiency ratio of different network configurations. We observe that multi-hop grooming capability can help to increase the resource efficiency ratio, which will, in turn, reduce the traffic blocking ratio (shown in Fig. 13.21). If there is no multi-hop groomable node in the network, at least one lightpath is needed between every node pair whenever there is any low-speed request. Hence, it may lead to a full-mesh logical connectivity. This tends to increase the prob-

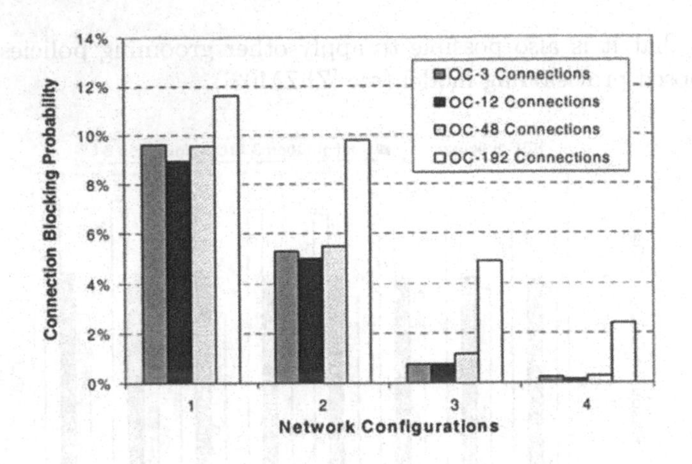

Figure 13.23 Connection blocking probability vs. offered load.

ability of connection blocking (due to lack of a free wavelength) and low wavelength-capacity utilization (due to under-utilization of the established lightpaths).

Figure 13.23 shows the blocking probabilities of connections with different bandwidth granularities. We can observe that connections requiring a full wavelength-channel capacity experience more chances to be blocked. This leads to a fairness concern. As Fig. 13.23 shows, the unfairness can become quite severe (relatively) when the network employs more grooming capability. This is because, when low-speed connections are carried by the network, the resources tend to be fragmented. Therefore, high-speed connections tend to be blocked more often since they cannot fit into the available trunk capacities. The authors in [ThSo01a] have addressed this unfairness issue and proposed a call-admission-control (CAC) algorithm to achieve the capacity fairness. This algorithm tries to keep the traffic blocking ratio of each connection bandwidth-granularity class at the same level by using statistical information to perform admission control. Another possible solution to resolve this fairness issue is trunk reservation or capacity reservation, i.e., pre-reserve a certain amount of resources for connections with high-granularity bandwidth requirement.

Different network topologies, different resource parameters (i.e., different wavelength number, different multi-hop grooming node locations, etc.) and different connection bandwidth-granularity distributions were also experimented with, and we observed similar results.

13.4 Summary

This chapter was devoted to the study of the traffic-grooming problem in a WDM mesh network. We examined the architecture of a node with grooming capability. We first presented the ILP formulation for the static traffic-grooming problem in such a WDM mesh network. We compared the performance of the single-hop grooming approach and multi-hop grooming approach on a small six-node network with randomly generated traffic pattern. Results from ILP showed that the end-to-end aggregate traffic between the same node pair tends to be groomed on to the same lightpath channel, which directly joins the end points, if the optimization objective is to maximize the network throughput. Two heuristic approaches were also proposed for solving the static traffic-grooming problem in large networks. We compared the performance of these two heuristic algorithms, MSHT and MRU, with different network resource parameters. The comparison results showed that MRU performs better if tunable transceivers are used and MSHT performs better if fixed transceivers are used. We extended the optimization problem to a network-revenue model and found a different grooming scheme, which can be used to design an efficient heuristic algorithm on the network-revenue model.

We also discussed the challenges to dynamically provision connections of different bandwidth granularities in heterogeneous WDM backbone networks. This heterogeneity may arise in a large, multi-vendor, and inter-operational nation-wide backbone network. We investigated how to properly extend the existing standards to incorporate this heterogeneity and proposed a simple, graph-based, provisioning model. Our proposed model is scalable and easy to implement in the existing control plane. Using this model, different traffic-engineering optimization objectives can be easily achieved through different grooming policies. Our results illustrated the effectiveness of the proposed model and the relationship between the nodal grooming capability and the network throughput, connection blocking probability, and resource efficiency. We observed that employing more grooming capacity can help a network operator to improve the network performance and the utilization of the link capacity. However, this may also increase the network cost since more electronic processing is needed. We verified the blocking performance unfairness between connections of different bandwidth-granularity classes due to resource-usage characteristics. We also observed that the unfairness problem can become more severe when a network has more grooming capability. This may be an interesting research topic in the future.

Exercises

13.1. Given the number of flow capacities as follows, how many OC-192 links are needed to carry all the traffic if we use packing/unpacking techniques? (Consider the three cases separately first, and then consider them together.)
4 OC-3
5 OC-12
6 OC-48

13.2. Construct an example of traffic grooming in a ring topology, showing that minimizing the number of ADMs may result in non-shortest-path routing.

13.3. Consider the topology shown in Fig. 13.24. Each link has two wavelengths of capacity OC-96 each. Determine the link utilization after satisfying the connection requests in the following traffic matrix. The unity traffic demand is considered to be OC-12.

$$A = \begin{bmatrix} 0 & 2 & 1 & 1 & 2 \\ 2 & 0 & 3 & 2 & 2 \\ 1 & 3 & 0 & 3 & 2 \\ 1 & 2 & 3 & 0 & 2 \\ 2 & 2 & 2 & 2 & 0 \end{bmatrix}$$

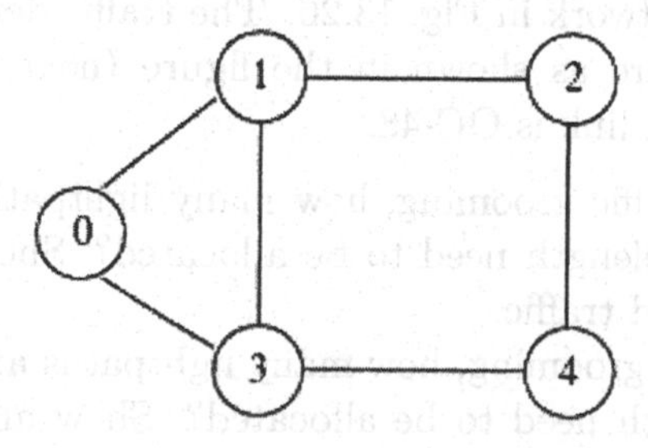

Figure 13.24 Graph representing network topology.

13.4. Construct a traffic-grooming example to show that, for a ring topology, minimizing the number of ADMs may need more wavelengths than the minimal.

13.5. Consider the network in Fig. 13.25. The corresponding link utilization matrix can be seen below (with 0.3 signifying 30% utilization of a wavelength):

$$\Lambda = \begin{bmatrix} 0.0 & 0.3 & 0.2 & 0.4 \\ 0.3 & 0.0 & 0.1 & 0.6 \\ 0.4 & 0.1 & 0.0 & 0.2 \\ 0.5 & 0.4 & 0.6 & 0.0 \end{bmatrix}$$

(1) How many wavelengths are saved by using traffic grooming? Draw the connections with and without grooming.
(2) What is the average utilization for each wavelength in the non-grooming mode? What is the utilization in traffic grooming mode? Which one is better?

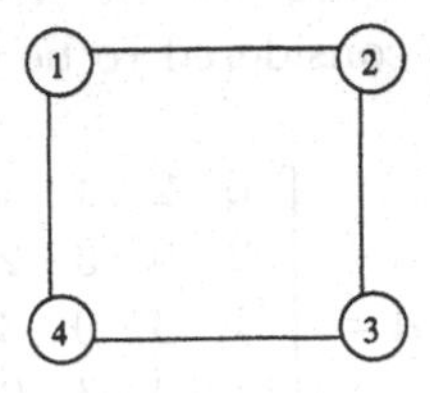

Figure 13.25 The optical network.

13.6. Consider the network in Fig. 13.26. The traffic demands between each pair of nodes are as shown in the figure (next to each link). The capacity of each link is OC-48.

(1) Without traffic grooming, how many lightpaths are required, and how many wavelength need to be allocated? Show all the lightpaths and their carried traffic.
(2) With traffic grooming, how many lightpaths are required, and how many wavelength need to be allocated? Show all the lightpaths and their carried traffic.

13.7. Consider the network topology in Fig. 13.27, with capacities shown on links (denoting the maximum "capacity" of the link).

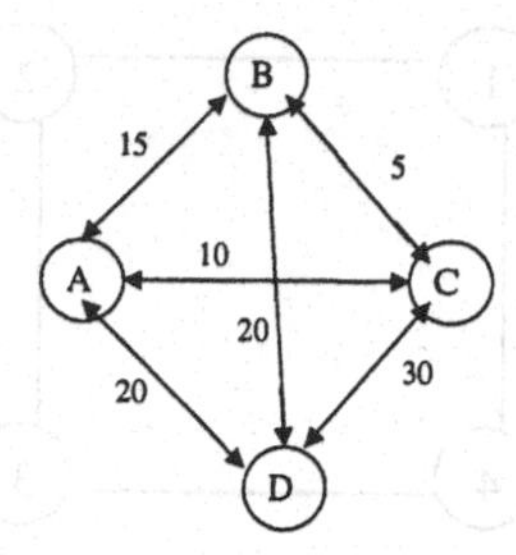

Figure 13.26 The network topology.

Given the following traffic matrix (the traffic is assumed to be symmetrical in both directions), try to accommodate the traffic optimally in terms of network throughput, by using traffic grooming (with multiple paths). Show the aggregation of traffic into the lightpaths.

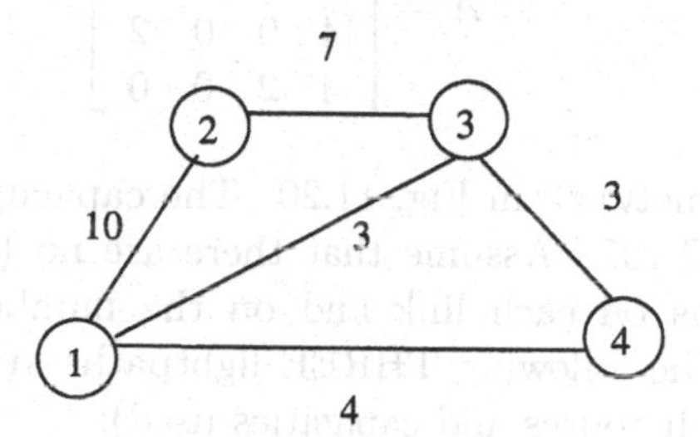

Figure 13.27 The network topology.

The traffic matrix can be seen below:

$$A = \begin{bmatrix} 0 & 1 & 3 & 6 \\ 1 & 0 & 4 & 0 \\ 3 & 4 & 0 & 1 \\ 6 & 0 & 1 & 0 \end{bmatrix}$$

13.8. Consider the network topology in Fig. 13.28. The capacity of each wavelength is 10 Gbps. Given the following traffic matrix in Gbps,

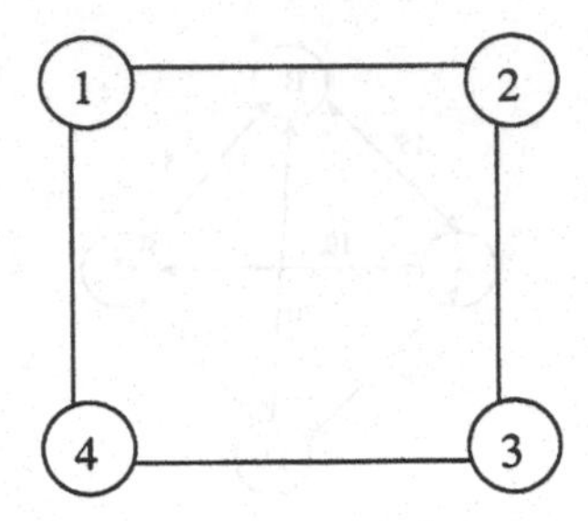

Figure 13.28 The network topology.

what is the number of wavelengths needed with traffic grooming and without traffic grooming?

$$A = \begin{bmatrix} 0 & 2 & 3 & 0 \\ 0 & 0 & 3 & 3 \\ 4 & 0 & 0 & 2 \\ 4 & 2 & 0 & 0 \end{bmatrix}$$

13.9. Consider the network in Fig. 11.20. The capacity of each wavelength channel is OC-192. Assume that there are no limits on the number of wavelengths on each link and on the number of transceivers at each node. The following THREE lightpaths are already established (including their routes and capacities used):
(1) From 8 to 12 at OC-96 (8-7-9-12)
(2) From 12 to 21 at OC-48 (12-16-22-21)
(3) From 10 to 16 at OC-144 (10-9-12-16)

When a new low-speed connection request arrives, one of the following four schemes can be used to establish it.
Scheme 1: Groom the new connection onto an existing lightpath with sufficient spare capacity.
Scheme 2: Create a new lightpath for the new connection.
Scheme 3: Groom the new connection onto existing lightpaths, thereby establishing a multihop connection.
Scheme 4: Use existing lightpath(s) for some part(s) of its route and set up new lightpaths for the remaining part(s).

We need to establish the following connections. Find the routes – if possible – for each of these connections, indicating the use of existing lightpaths, wherever applicable, and new lightpaths. If a connection cannot be established using a specified scheme, please say so, and justify.

a) 5 to 20 at OC-48 using Scheme 1
b) 5 to 20 at OC-48 using Scheme 2
c) 5 to 20 at OC-48 using Scheme 3
d) 5 to 20 at OC-48 using Scheme 4
e) 8 to 21 at OC-48 using Scheme 1
f) 8 to 21 at OC-48 using Scheme 3
g) 8 to 21 at OC-144 using Scheme 1
h) 8 to 21 at OC-144 using Scheme 3
i) 10 to 21 at OC-48 using Scheme 4

Chapter 14

Advanced Topics in Traffic Grooming

14.1 Introduction

As follow-on to the material in Chapter 13, we further investigate some advanced topics in traffic grooming in this chapter. First, a practical solution for an optical switch architecture is presented, which integrates all-optical waveband switching and OEO traffic grooming, and which exploits the advantages of mature technologies. Next, virtual concatenation (VC) and link-capacity adjustment scheme (LCAS) of SONET/SDH are investigated for traffic grooming. Noting that SONET/SDH is the dominant telecom network and that they need to efficiently carry packet data traffic, we study how to exploit the "inverse multiplexing" feature of next-generation SONET/SDH (NGS), and groom traffic flows over multiple paths. Finally, we investigate the survivable traffic-grooming problem, in which sub-wavelength-granularity connections need to be protected.

14.2 Hierarchical Switching and Waveband Grooming

14.2.1 Motivation

As optical networks evolve and the capacity of the network links continue to increase, there is a growing need for a new switching paradigm to route and provision connections in such networks. The current generation of optical switches, based on electrical switch fabrics, offer higher-speed ports (e.g., OC-192, approximately 10 Gbps) and are capable of switching lower-speed time-division multiplexed (TDM) circuits (e.g., STS-1, 51.84 Mbps) within the switch fabric. However, the existing electrical switch fabrics typically adopt an architecture that scales poorly with port count and data rate per port. Since such switch fabrics can only switch at most one wavelength per port, an extremely large port count (on the order of 1000×1000 or higher) will be needed as the number of wavelengths per fiber continues to grow. Besides the switch fabric itself, expensive optical-to-electrical and electrical-to-optical (OEO) conversion is needed for the switch to operate. All of these factors imply that the cost will be prohibitively high to switch all the bandwidth using only electrical switches in future optical networks.

All-optical switching, due to its transparency to wavelength and data rates, has emerged as the complementary technology to electrical switching. One of the viable all-optical switching technologies is based on micro-electro-mechanical systems (MEMS), which often utilize tilting micro-mirrors to redirect light beams. Since micro-mirrors are insensitive to the wavelength, they are capable of switching wavebands[1]. With waveband switching, the size of an optical switch can be significantly reduced. However, the limitation of all-optical switches is their lack of traffic grooming[2], arbitrary wavelength conversion[3], and multicast[4] capabilities, three functions necessary for a flexible network.

Recently, the concept of using both waveband-level and wavelength-level switching in a hierarchical manner has received growing attention. The au-

[1] A waveband, in general, is a group of wavelengths. We assume that all the wavelengths in a fiber are evenly grouped into a number of wavebands. In practice, the wavelengths within one waveband may or may not be physically neighboring wavelength channels, i.e., the term waveband may be used only in a logical sense.

[2] The process of multiplexing/demultiplexing/switching lower-speed connections onto/off from/between higher-speed channels.

[3] The technique to convert the data arriving on one wavelength into another arbitrary wavelength.

[4] Transmission of information from a single source node to multiple destination nodes.

thors in [HSKO99] presented a two-layer optical crossconnect (OXC) architecture and demonstrated a two-stage multiplexing/demultiplexing scheme. References [GeRa00a] and [GeRa02] provided qualitative discussions on similar ideas. The works in [LiVe02] showed how to reduce the OXC size by using waveband switching and fiber switching. The authors in [LYKK02] proposed a destination-based lightpath-grouping heuristic algorithm to take advantage of waveband switching. Reference [HuPS02] developed an MILP-based approach to design a generic two-layer network. The work in [SuNi02] presented heuristic approaches for designing two-layer (waveband and wavelength) networks.

Most of the previous quantitative work focused on a wavelength-to-waveband (or waveband-to-fiber) scenario. However, sub-wavelength traffic-grooming capability is required by most network operators, because some of their customers may not require the full bandwidth of a wavelength channel; hence, it is important to optimize the placement of OEO grooming switches during network planning. In this section, we present a hybrid (all-optical and OEO) hierarchical switch architecture which incorporates both all-optical waveband switching and OEO sub-wavelength grooming. This architecture combines the benefits of waveband switching and traffic grooming. We demonstrate this architecture's unprecedented scalability, excellent flexibility, and enormous savings over traditional OEO switches.

14.2.2 A Hybrid Node Architecture

The proposed architecture incorporates both all-optical waveband switching and OEO grooming switching. Below, we first discuss today's existing switch architectures, and then describe in detail the proposed architecture.

Figure 14.1 shows different types of node architectures. A backbone network node consists of WDM transmission equipment and switch equipment. Because WDM was first used as a point-to-point transmission technology, in most incumbent carriers' networks, the WDM transmission equipment and the switch equipment are usually not integrated. Typically, they are in separate bays co-located in the same facility. The inter-connection between the WDM transmission equipment and the switch equipment is carried by very-short-reach (VSR) optical interfaces [OIF05]. As shown in Fig. 14.1(a) (from [Gers00]), when the WDM signals arrive at an input fiber, they first encounter a WDM demultiplexer (DEMUX), and different wavelengths are separated. Each wavelength enters a receiver where the bits are converted to electrical signals, and then converted back to optical signals on a wavelength

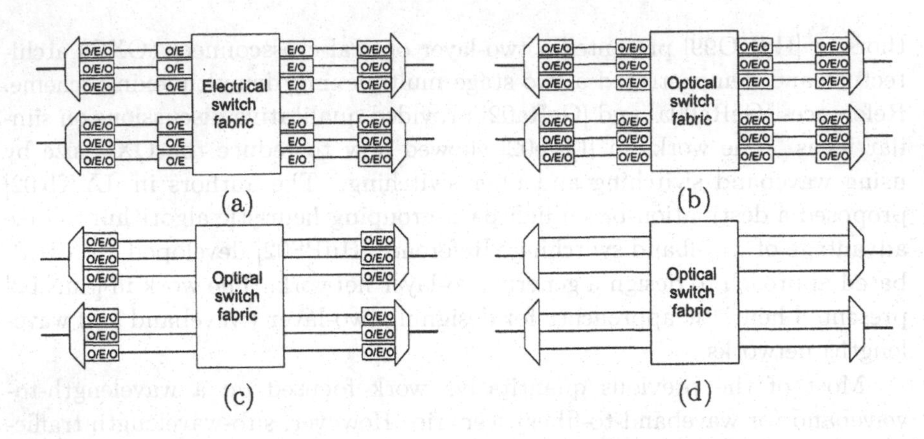

Figure 14.1 Different types of node architectures [Gers00].

specific for the VSR interface (e.g., 850 nm or 1310 nm). At the other end of the VSR interface, an optical-to-electrical (O-E) convertor converts the signals back to the electrical form and sends them into the switch fabric. On the output side of the switch fabric, the reciprocal E-O and O-E-O conversion processes are repeated. Note that the electrical switch fabric requires $4N$ O/E and E/O transponders, where N is the number of input/output ports. If an optical switch fabric is used instead, there are three different types of configurations. The configuration in Fig. 14.1(b) is similar to that of the electrical fabric, where $4N$ OEO transponders are required. The next configuration (Fig. 14.1(c)) removes the OEO transponders surrounding the switch fabric, and switches directly the optical signal from the VSR optical interface. The final configuration (Fig. 14.1(d)) removes all the OEO transponders in both the WDM transmission equipment and switching equipment, and switches directly on the wavelengths emerging from the fibers. This configuration is also referred to as a complete all-optical networking solution. All-optical switching/networking requires high-quality, long-reach transmission, since no OEO regeneration is available at each node.

An electrical switch fabric is usually constructed with a number of switch elements (single switch chips with a small number of ports). Today's commercially-available switch elements can provide on one single chip up to 72 OC-48 ports (i.e., 72 × 72) with STS-1 grooming granularity or up to 140 OC-48 ports without grooming [Veli05]. A common arrangement of the switch elements is the non-blocking three-stage Clos network (see Section 2.6.3). Besides the required number of switch elements in a Clos switch, another factor that dictates the size of a switch fabric is the maximum port data rate of a single switch chip. Currently, the highest data rate of commercial chips is approximately 3.25 Gbps per port, which means that, at data rates higher than OC-48, the input data streams need to be inverse-multiplexed into lower-speed streams (e.g., one OC-192 channel to four OC-48 channels) and multiple switch ports are used in parallel. As shown later in this section, the port data-rate limit and the Clos architecture make electrical switch fabrics very difficult to scale with growing number of wavelengths per fiber and increasing data rate per wavelength. Other limitations of electrical switch fabrics include space and power consumption. A line card that fits into a telecom equipment bay can only accommodate a limited number of switch chips, due to space and power constraints. Moreover, the signal traces on a single backplane can only travel a limited distance; thus, a large-scale switch with too many switching elements may lead to a large footprint in terms of power and size. Nevertheless, despite these limitations, the currently-deployed switches are mainly OEO switches because they can still meet today's capacity demand; they are deployable because mature switch chips are commercially available; and they can provide the valuable traffic-grooming, wavelength-conversion, and multicast functions.

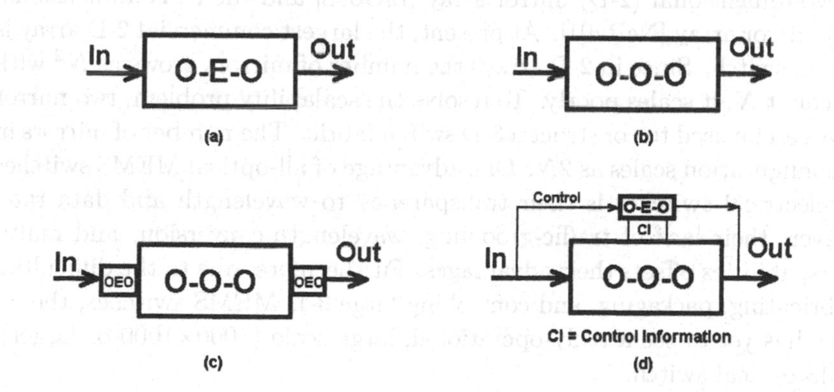

Figure 14.2 Comparison between switching types of OEO and OOO.

As a complementary technology to electrical switching, all-optical switching has gained increasing interest over the past several years. All-optical switching (OOO) (Fig. 14.2(b)) overcomes various limitations of OEO switching (Fig. 14.2(a)) in terms of optical scalability, bit rate, and protocol transparency. But, regarding signal quality at the receiver and maturity of available technologies, OEO has several advantages today. Thus, naturally, there is a very significant motivation to utilize the merits of the two technologies and avoid their disadvantages. Figures 14.2(c) and 14.2(d) show variants of OOO based on such a motivation. Optical packet switching (OPS) can be used with the switch architecture in Fig. 14.2(d) (see Chapter 17). Currently, Fig. 14.2(d) has attracted more attention from the research and industry communities. The strong points of this switch architecture are that OOO and OEO are combined in order to avoid the shortcomings of each; specifically, in this case, all-optical transportation of data traffic is directed by *control information* (CI)[5] which is derived from the OEO processing of the packet header (i.e., control packet) separately. Optical burst switching (OBS) and optical label switching (OLS) are both extensions of this case (see Chapter 18).

In particular, the last approach has become more popular after micro-electro-mechanical system (MEMS) devices were successfully demonstrated to re-direct light beams with actuated micro-mirrors [LiGo02]. Because MEMS utilizes fabrication processes that are also widely used in the integrated circuit (IC) industry, it can achieve relatively low fabrication cost. The two most dominant MEMS-based all-optical switch architectures are the two-dimensional (2-D) mirror array [LiGo98] and the three-dimensional (3-D) mirror array [NeRa01]. At present, the largest commercial 2-D array is a 32×32 switch. Since in 2-D arrays the number of mirrors grows as N^2 with port count N, it scales poorly. To resolve the scalability problem, two mirror arrays can be used to construct a 3-D switch fabric. The number of mirrors in this configuration scales as $2N$. One advantage of all-optical MEMS switches over electrical switches is their transparency to wavelength and data rate. However, their lack of traffic-grooming, wavelength-conversion, and multicast capabilities offsets their advantages. Furthermore, due to the difficulties in fabricating, packaging, and controlling large 3-D MEMS switches, the industry has yet to see a truly-operational, large-scale (1000×1000 or larger), MEMS optical switch.

An optical switch architecture is proposed, which incorporates both all-optical waveband switching and OEO traffic-grooming switching, as shown in Fig. 14.3. The entire node consists of a fiber patch panel, an all-optical waveband switch, an OEO grooming switch, traffic aggregation equipment, and waveband and wavelength multiplexers/demultiplexers. The fiber patch

[5] Control information (CI) consists of signaling and routing information, which is used to direct the data traffic through the corresponding switching node correctly.

panel is responsible for configuring the fiber topology. Since the fiber topology stays relatively static once configured, we only focus on the waveband switch and the grooming switch. Those fibers that are terminated at the node first pass through the waveband demultiplexers, where wavebands are separated and sent into the waveband switch. The waveband switch uses the entire waveband as its switching granularity. If a particular waveband needs to travel into a different output fiber, it can be switched all-optically at this level. In case a wavelength or a sub-wavelength TDM circuit needs to be switched to a different waveband or a different output fiber, the parent waveband is switched to the OEO grooming switch. The grooming switch has an electrical grooming switch fabric, and is capable of performing sub-wavelength TDM circuit switching (traffic grooming) and arbitrary wavelength conversion. The input and output ports of the grooming switch operate at a wavelength channel's data rate. Both the waveband switch and grooming switch have add and drop ports for locally-generated and terminated traffic. A waveband add/drop port can only add/drop a full waveband, while a grooming add/drop port can add/drop a single wavelength. The traffic aggregation/deaggregation equipment performs TDM multiplexing/demultiplexing for local lower-speed connections. If the local connections between a source-destination pair has enough load to fill a whole waveband, these connections will be aggregated and directly sent to the waveband add port. Otherwise, they will be aggregated into multiple wavelengths, sent to the grooming switch, and grouped with other wavelengths to fill a waveband. Without exception, all connections are transported across the fiber topology in one or more waveband paths. A waveband path is an end-to-end waveband connection that originates and terminates at a waveband add/drop port or a grooming switch. The waveband multiplexers/demultiplexers are responsible for separating/merging wavebands in a fiber, and the wavelength multiplexers/demultiplexers are responsible for separating/merging wavelengths in a waveband.

Since an all-optical waveband switch does not provide wavelength-conversion capability, it can be constructed in a waveband-layered approach, as shown in Fig. 14.3. Suppose each fiber contains B wavebands, then B individual all-optical switch fabrics can be used, each operating for a specific waveband. The number of input/output ports of a single switch fabric is determined by the number of input/output fibers, the number of waveband add/drop ports (which is decided by the network designer according to the traffic demand), and the number of ports that connect to the grooming

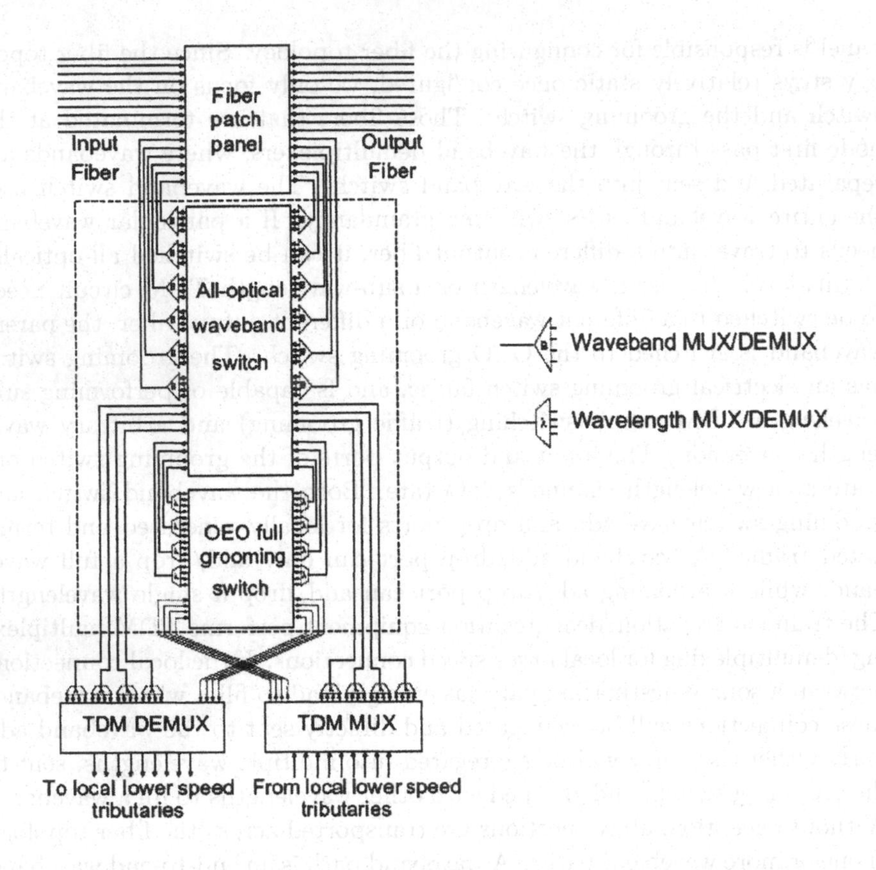

Figure 14.3 Architecture of proposed optical switch incorporating all-optical waveband switching and OEO grooming switching.

switch. This configuration significantly reduces the port-count requirement of individual all-optical switch device, since the number of input/output fiber pairs (which is equal to the nodal degree) is usually less than 10. Hence, it is possible to construct the entire all-optical waveband switch with inexpensive low-port-count MEMS devices (e.g., 2-D mirror arrays).

The OEO grooming switch can be constructed with commercial grooming switch chips. The size of the grooming switch is determined by how many waveband ports there are from/to the waveband switch, and how many local add/drop ports are required.

14.2.3 Comparison: Hybrid vs. OEO

Because all-optical waveband switching can group pass-through traffic into wavebands and avoid switching in the electrical domain, considerable savings can be obtained due to the fact that pass-through traffic dominates add/drop traffic in most networks. If a network's average hop distance is $\bar{H}$ and its traffic is uniformly distributed among all nodes, on average, at each node, only a fraction $1/\bar{H}$ of the total incoming and outgoing traffic is added or dropped. Since most networks have an average hop distance larger than 2, it is reasonable to assume that, in general, less than 50% is add/drop traffic at each node. By carefully grouping pass-through traffic, it is possible to significantly reduce the size of the electrical grooming switch and the number of associated OEO transponders. Moreover, the number of wavebands in one fiber does not need to be very large, as long as it provides a fine enough separation between the pass-through and the add/drop traffic. This means that the number of wavebands in a fiber can remain the same even when the total number of wavelengths increases. Consequently, when the network capacity is expanded with more wavelengths, the all-optical waveband switch can remain exactly the same in size and configuration, because the all-optical waveband switch is insensitive to how many wavelengths each waveband contains.

For purpose of illustration, we present the following comparison between a hybrid hierarchical switch and an OEO grooming switch, as shown in Fig. 14.4. The node in our example has degree of 6 ($F = 6$). We assume that 25% of the total traffic is added or dropped. In the hybrid hierarchical switch (Fig. 14.4(a)), we set the number of wavebands in a fiber B equal to 4, since approximately 1/4 of the traffic is add/drop traffic. BA and BD denote the number of waveband add and drop ports. BW denotes the number of ports on the waveband switch leading to the grooming switch. WB denotes the number of ports on the waveband switch receiving from the grooming switch. WA and WD denote the number of wavelength add/drop ports, respectively. Since there is 25% add/drop traffic, we ensure that at least 25% of all the capacity from input fibers can be dropped and added. Specifically, for 6 input fibers each containing 4 wavebands, all the drop ports (including both waveband and wavelength drop ports) should have a total of 6 wavebands of capacity. Similar rules apply for the waveband and wavelength add ports. Selection of BW and WB depends on how much traffic-grooming capability we desire the node to have. One extreme case is that BW and WB are both equal to the total number of input wavebands, which gives the node the capability to groom all the incoming traffic but will be very expensive. In general, only a portion of the incoming traffic needs to be groomed; therefore, BW and WB can be much smaller than the number of all incoming wavebands. For the all-optical waveband switch, we assume that the entire fabric is constructed with four 8×8 MEMS switches in a waveband-layered approach (Fig. 14.3). For the grooming switch, OE and EO transponders are necessary, as shown in Fig. 14.1(a). In the given example, we choose a specific set of these parameters as follows: $B = 4$, $BW = WB = 4$, and $BA = BD = 4$. The values of WA and WD depends on how many wavelengths there are in one fiber.

In the node architecture using only the OEO grooming switch (Fig. 14.4(b)), all incoming fibers are demultiplexed directly to wavelengths. The wavelengths are sent to the OEO grooming switch. The number of add/drop ports is determined by the amount of add/drop traffic, which in this case is set at 25%. Therefore, the number of add/drop ports should be equal to 25% of all the incoming wavelengths. Another factor to be considered is the port count and maximum port data rate of each individual switch chip. The switch chip is chosen to be 72×72 OC-48 switch chip with STS-1 grooming granularity, which is a high-capacity grooming chip commercially available today. Notice that, for wavelengths operating at data rates beyond OC-48, inverse multiplexing occurs, and extra ports are required. Thus, a 72×72 OC-48 switch chip can be used as a 18×18 OC-192 switch chip. These rules also apply to the OEO grooming switch in the hybrid hierarchical architecture. We assume that the switch requires $4N$ transponders where N is the number of input (output) ports connected to the demultiplexer (multiplexer).

14.2.4 Research Problems and Challenges

Challenging research problems for hierarchical switching include network design and provisioning. The goal of network design is to minimize the network cost by properly planning waveband paths and routing a set of known client connections over the waveband paths. This problem resembles the virtual-topology design and traffic-grooming problem in wavelength-routed networks (see Chapters 8 and 13). However, the hierarchical and hybrid nature of the switch, the variable number of wavebands per fiber and wavelengths per waveband, and the unique ways in which waveband paths are shared among connections all add new dimensions to the problem. Reference [YaOM03] studied the network-design problem in a hybrid network. The objective of

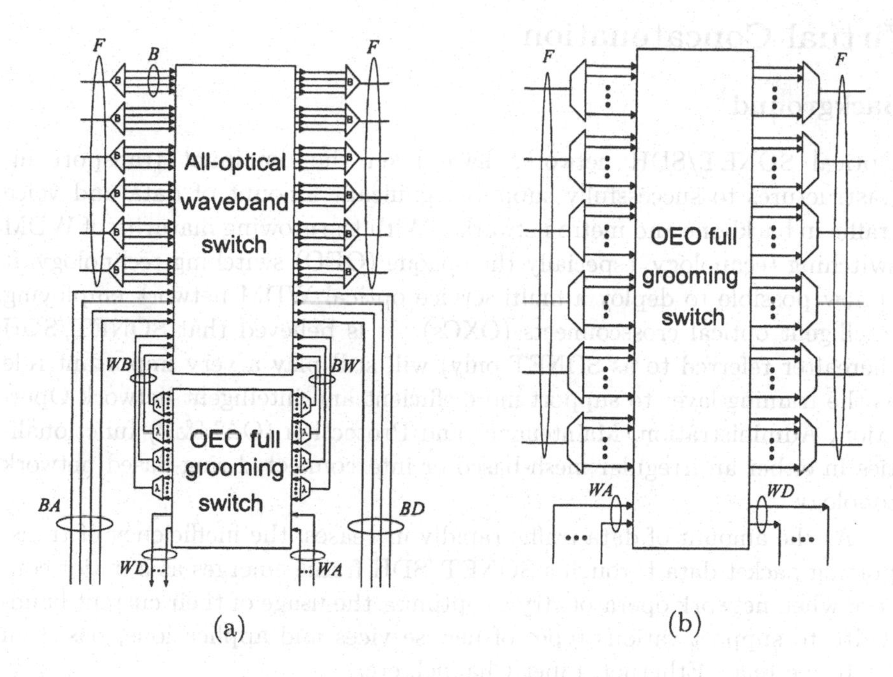

Figure 14.4 Comparison between hybrid hierarchical switch and OEO full-grooming switch.

network provisioning is to minimize the blocking probability by judiciously setting up waveband paths and routing dynamic connections over the waveband paths, given the network configuration which specifies the number of ports at each node and the number of wavelengths on each link.

So far, we have assumed an ideal network in which transmission impairment is not an issue. While the assumption holds in a full OEO network, it may not be the case in a hybrid network since a waveband is basically an all-optical channel. As a result, impairment-aware waveband routing in a hybrid network is another dimension of both network design and provisioning. In [YaMu03], we studied impairment-aware network design in a hybrid network by abstracting transmission impairment into hop-count constraint on a waveband.

14.3 Virtual Concatenation

14.3.1 Background

Optical SONET/SDH networks have been the dominant transport infrastructures to successfully support significant amount of data and voice traffic in backbone and metro networks. With the growing maturity of WDM switching technology, especially the opaque (OEO) switching technology, it is now possible to deploy a multi-service optical WDM network employing intelligent optical crossconnects (OXCs). It is believed that SONET/SDH (hereafter referred to as SONET only) will still play a very important role as the framing layer to support more efficient and intelligent network Operation, Administration, Maintenance and Protection (OAM&P) functionalities in either an irregular mesh-based or interconnected-ring-based network topology.

As the amount of data traffic rapidly increases, the inefficiency of transporting packet data through a SONET/SDH frame emerges as a major concern when network operators try to optimize the usage of their current bandwidth to support various types of new services and applications, based on IP, frame relay, Ethernet, Fiber Channel, etc.

In the traditional SONET/SDH multiplexing hierarchy, the frames of multiple low-speed traffic streams (say, STS-1 frame, approx. 51.84 Mbps) are combined to form the frame of a high-speed traffic stream. In order to support high-speed traffic from the same client source, e.g., a broadband ATM switch, N "contiguous" lower-order SONET/SDH payload containers are merged into one of greater capacity. This is called SONET/SDH's concatenation technique.

Usually, SONET/SDH's concatenation is implemented at certain speeds, such as STS-3c, STS-12c, STS-48c, etc., which leads to a tiered bandwidth-allocation mechanism for different client services. Unfortunately, although "contiguous" and "tiered" concatenation is simpler for implementation, it is not very flexible and not efficient, especially in a multi-service network environment. From a single network node perspective, traffic streams from different client network equipment are to be discretely mapped into different tiers of SONET/SDH bandwidth trunks (data containers), which may result in huge capacity waste. For example, carrying a Gigabit Ethernet connection using a concatenated OC-48 pipe (approx. 2.5 Gbps) will lead to 60% bandwidth wastage.

From a network perspective, the time-slot contiguous requirement of a

SONET/SDH concatenated channel imposes a constraint for traffic provisioning and may degrade network performance in a dynamic traffic environment where network resources are easy to be fragmented. The constraint also makes it more difficult for network operators to perform efficient traffic grooming, i.e., packing different low-speed traffic streams onto high-capacity wavelength channels.

14.3.2 Virtual Concatenation

Virtual concatenation (VC or VCAT) will help a SONET/SDH-based optical network to carry data traffic in a finer granularity and hence make the use of link capacity more efficiently. The basic principle of virtual concatenation [ANSI01, ITU02a] is quite simple. A number of smaller containers, which are not necessarily contiguous, are concatenated and assembled, to create a bigger container that carries more data per second. Depending on a network's switching granularity, virtual concatenation is possible for small container size from VT-1.5 upto STS-3c.

Figure 14.5 (from [Stan02]) shows an example of how to support multiple services using a single OC-48 channel through virtual concatenation. In Fig. 14.5, an OC-48 channel is used to carry two Gigabit Ethernet traffic streams and six STS-1 TDM voice traffic streams. Through a STS-1 switch, the traffic can be switched onto different OC-12 pipes, and these OC-12 pipes can be sent through the network over various routes. Figure 14.5 also illustrates the potential load-balancing benefit that VC can provide to a transport network.

One important thing to note is that, when multiple traffic streams from one client are sent over different routes, the VC mappers at the destination nodes need to compensate for the differential delay between the bifurcated streams when they are reconstructed as the destination node. Currently, such a typical, commercially-available device may support upto 50 ms (+/-25 ms) delay with external RAM, which is equivalent to a 10,000 km difference in route length [Hill02]. In general, a virtually-concatenated SONET/SDH channel made up of N × STS-1 is transported as individual STS-1s across the network; and, at the receiver, the individual STS-1s are re-aligned and sorted to recreate the original payload. Figure 14.6 shows an overview of a multi-service SONET/SDH-based optical network employing VC technology. VC can be supported at the edge OXCs (in port cards) or in separate traffic-aggregation network elements connecting client network equipment and OXCs.

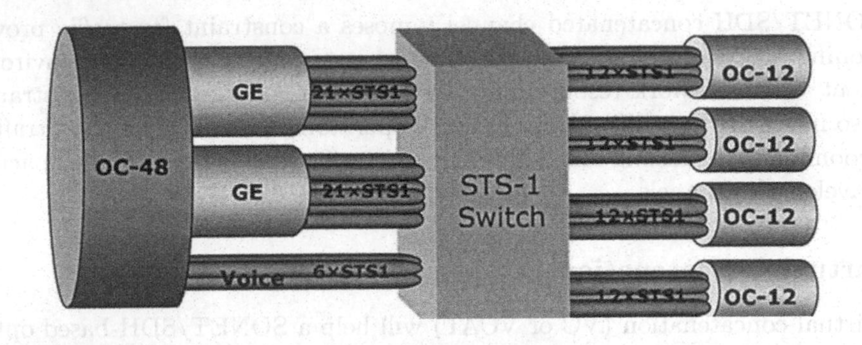

Figure 14.5 An example of using VC to support different network services.

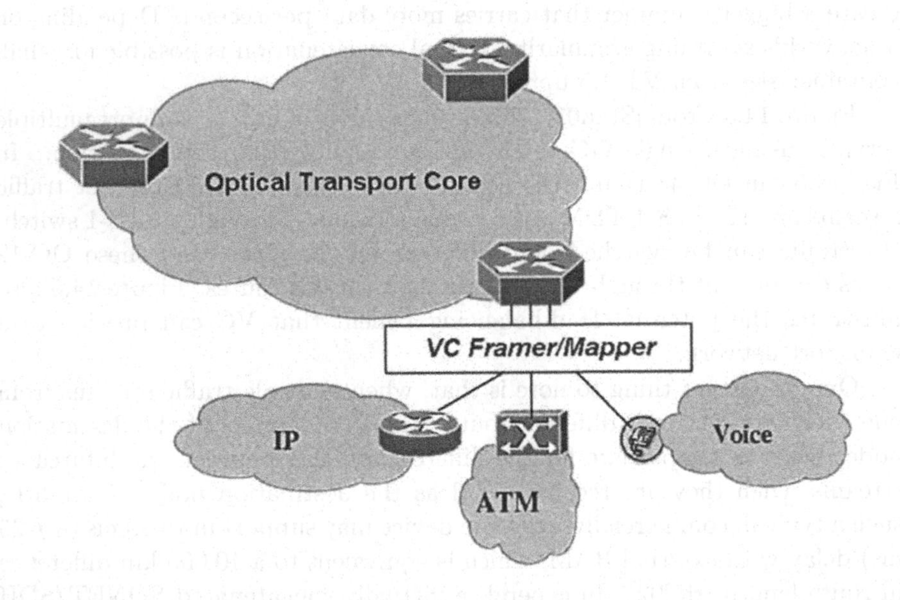

Figure 14.6 An overview of VC-enabled multi-service optical network.

14.3.3 Benefits of Virtual Concatenation

From a network perspective, the SONET/SDH VC technique can provide the following benefits to an optical network:

1. *Relax time-slot alignment and continuity constraints.* Instead of aligning to particular time-slots and consisting of N contiguous STS-1 time-slots within a wavelength, a high-speed STS-N channel could be constructed from any N STS-1 time-slots and carried by different wavelength channels.

2. *More efficiently utilize channel capacity to support multiple types of data and voice services.* Instead of mapping data traffic (packet/cell/frame) into SONET/SDH frames in a discrete tiered manner, optical networks can carry data traffic in a more resource-efficient way. Traffic granularity can be increased in the unit of 1.6 Mbps (VT-1.5) in a metro-area network, and 48 Mbps (STS-1) or 150 Mbps (STS-3c) in an optical backbone network.

3. *Bifurcate traffic streams to balance network load.* With VC, it is possible to split a high-speed traffic stream into multiple low-speed streams and route them separately through the network. This enables the traffic to be distributed across the network more evenly; and, hence, it can improve the network's blocking performance.

We investigate the benefits of SONET/SDH VC to an optical WDM mesh network (shown in Fig. 14.7) under dynamic traffic environment via simulations. Connections with different bandwidth granularities are assumed to come and leave the network, one at a time, following a Poisson arrival process and negative-exponential-distribution holding time.

To distinguish the benefit effects, two types of traffic pattern are studied (pattern I and II), one consisting of five service classes and the other having ten service classes. Capacity of each wavelength channel is assumed to be OC-192. In traffic pattern I, data rates for each service class are approximately 51 Mbps, 153 Mbps, 622 Mbps, 2.5 Gbps, and 10 Gbps, which can be perfectly mapped into the tiered SONET/SDH containers and their corresponding optical carriers, i.e., OC-1, OC-3c, OC-12c, OC-48c and OC-192c. The service characteristic of pattern II, i.e., service classes, service rate, and corresponding SONET/SDH containers with or without VC is shown in Table 14.1.

When traffic bifurcation is needed, a simple route-computation and traffic-bifurcation heuristic will be applied to a connection request. The problem of finding the set of minimal-cost routes with enough aggregated capacity for a request between a given node pair in a network is NP-complete. The heuristic works as follows:

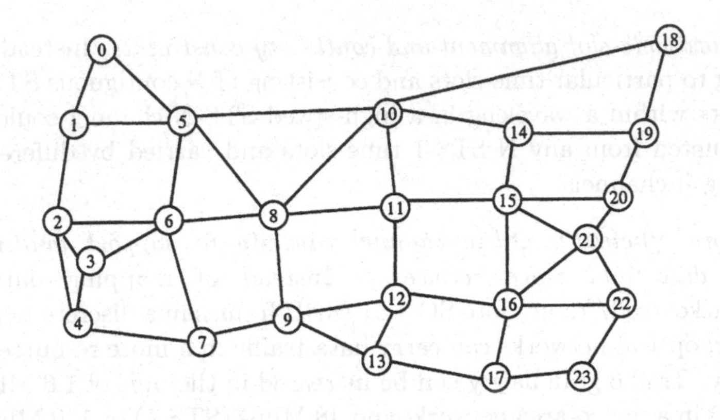

Figure 14.7 A 24-node example network topology.

Table 14.1 Traffic pattern II: ten service classes.

Class	Rate	Without VC	With VC	Class	Rate	Without VC	With VC
1	50 M	STS-1	STS-1	6	600 M	STS-12c	STS-12
2	100 M	STS-3c	STS-2	7	1 G	STS-48c	STS-21
3	150 M	STS-3c	STS-3	8	2.5 G	STS-48c	STS-48
4	200 M	STS-12c	STS-4	9	5 G	STS-192c	STS-96
5	400 M	STS-12c	STS-8	10	10 G	STS-192c	STS-192

Step 1 A shortest path is computed according to network administrative cost between the node pair.

Step 2 The bandwidth of the route is calculated. The bandwidth of the route is constrained by the link along the route, which has minimal free capacity. Then, update the available capacity of the links along the route (i.e., decrease the available capacity by the minimal capacity).

Step 3 Remove the link without free capacity and repeat Steps 1 and 2 until the connection can be carried by the set of routes computed or no more routes exist for the connection.

Note that, depending on the implementation, the VC mappers at the receiver end nodes may only be able to handle a limited number of routes (denoted by t) for a single connection, in which case, the connection will be blocked after t routes have been examined.

Figure 14.8 illustrates network performance in terms of bandwidth blocking probability (BBP) as a function of network offered load in Erlangs. BBP is considered as the measurement metric because connections from different service classes may have different bandwidth requirements. Network offered load is normalized to the unit of OC-192. A 24-node, 86-link network topology is used in our study (see Fig. 14.7).

Figure 14.8(a) shows the network performance for traffic pattern I with or without employing VC. Two types of network configuration are examined, i.e., all nodes are either equipped with STS-1 full-grooming switches or partial-grooming switches. Note that, in a partial-grooming switch, only a limited number of wavelength channels (6 in our simulation) can be switched to a separate grooming switch (or grooming fabric within an OXC) to perform traffic grooming.

As one can observe from Fig. 14.8(a), there is around 5-10 percent network performance gain through VC technique. In traffic pattern I, every service class can be perfectly mapped into one of the tiered SONET/SDH containers, and we assumed that no traffic bifurcation is allowed. Therefore, the performance improvement shown in Fig. 14.8(a) comes solely from the capability of eliminating the time-slot alignment and contiguity constraints provided by VC to the network. This observation is similar to the effect of the wavelength-conversion capability in a wavelength-routed WDM network, where each connection usually requires the capacity of a full wavelength channel.

Figure 14.8(b) shows how VC can significantly improve network performance when the network needs to support data-oriented services with different bandwidth requirements under the network configuration in which full-grooming OXCs are used at every node. It is straightforward to see that BBP is significantly reduced by employing VC. Meanwhile, more improvement can be achieved by allowing a simple traffic-bifurcation scheme. Note that, in our study, a connection will be bifurcated if and only if no single route with enough capacity exists. The results from different values of t have been examined and some of them are shown in Fig. 14.8(b), i.e., 4, 8, and unlimited. It can be expected that more advanced network load-balancing and traffic-bifurcation approaches can further optimize network throughput. Additional results based on different network configurations (e.g., different number of wavelengths, different network topologies, different OXC switching capabilities) are not included here to conserve space.

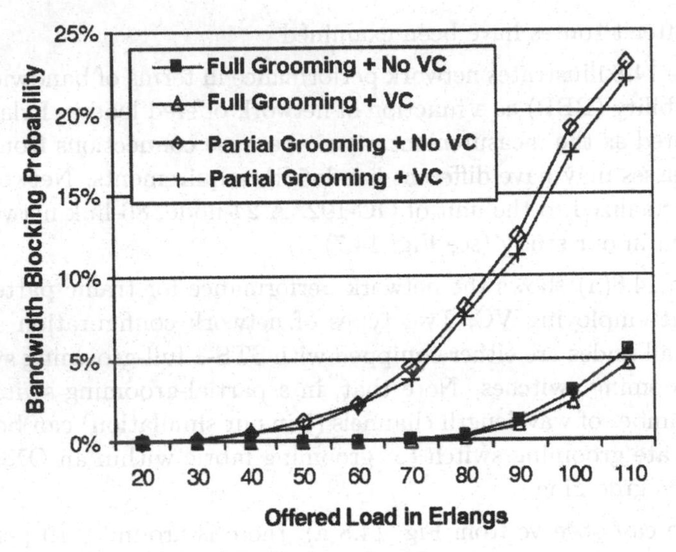

(a) Five-service class.

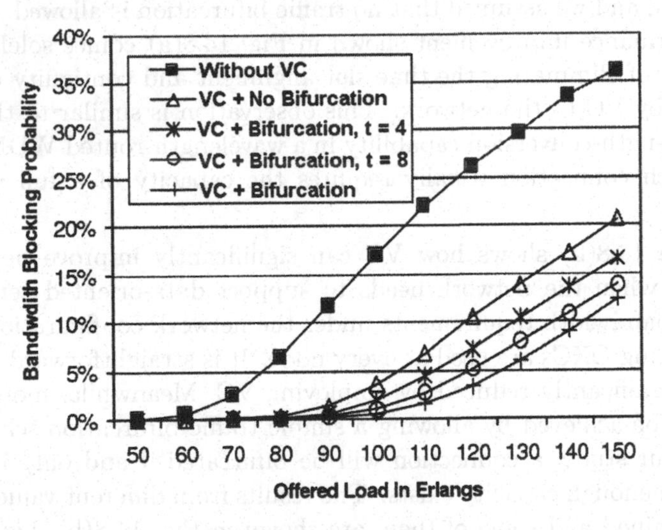

(b) Ten-service class.

Figure 14.8 Illustrative numerical results.

Besides the three benefits quantitatively demonstrated, SONET/SDH VC could also benefit an optical transport network on other aspects, such as network compatibility, network resiliency, network management and control, etc. VC works across legacy networks. Only the end nodes of the network are aware of the containers being virtually concatenated. In terms of resiliency, since individual members of a virtually-concatenated channel may be carried through different routes, a network failure may only affect partial bandwidth of a connection service. In this case, an unprotected, best-effort connection may still get the service under the reduced bandwidth before the network failure is fixed, and a high-priority connection can still be provided partial service before the network protection/restoration operation is active to restore the full service. VC may also help to simply some network control and management issues. For instance, link-state information can be maintained at a more aggregated level since time-slot alignment and continuity constraints will be handled by end nodes.

On the other hand, more intelligent algorithms and mechanisms are also needed in the network control plane in order to fully utilize the benefits provided by VC.

14.3.4 Related Technologies

Other related technologies include link-capacity adjustment scheme (LCAS) [ITU01b] and generic framing protocol (GFP) [ITU01c]. LCAS, built on virtual concatenation, is a two-way signaling protocol that runs continuously between the source and destination of a bandwidth pipe. LCAS allows network operators to adjust the pipe capacity while it is in use (on the fly). It increases the possibility for on-demand traffic provisioning and on-line traffic grooming/re-grooming and makes SONET/SDH-based optical WDM network more data friendly.

GFP is a traffic-adaptation protocol for broadband transport applications. It provides a standard mapping of either a physical layer or logical link layer signal to a byte-synchronous channel such as SONET/SDH links or wavelength channels in an optical transport network (OTN). There are two methods for mapping protocols onto GFP: frame-mapped GFP and transparent-mapped GFP. Frame-mapped GFP, optimized for a packet-switched environment, is the transport mode used for Point-to-Point Protocol (PPP), IP, and Ethernet traffic. Transparent-mapped GFP, intended for delay-sensitive storage-area network (SAN) applications, is the transport mode used for Fiber Channel (FC), Fiber Connection (FICON), and En-

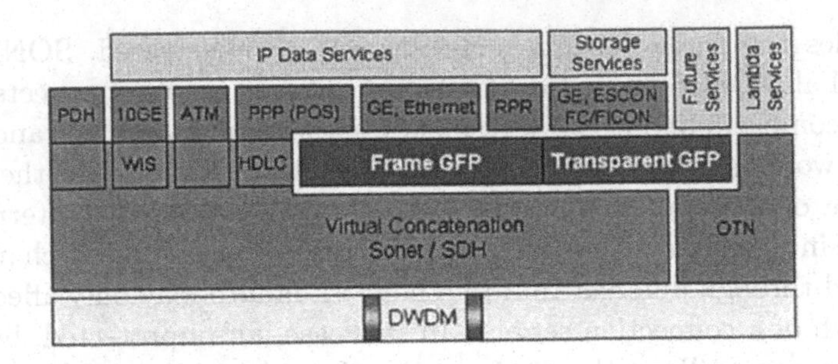

Figure 14.9 GFP mapping to SONET/SDH.

terprise Systems Connection (ESCON) traffic. Figure 14.9 (from [Stan03]) shows the mapping of GFP to SONET/SDH using VC.

14.4 Survivable Traffic Grooming

14.4.1 Introduction

While the transmission rate of a wavelength channel is high (typically STS-192 today and expected to grow to STS-768 in the near future), the bandwidth requirement of a typical connection request can vary from the full wavelength capacity down to STS-1 or lower. To efficiently utilize network resources, sub-wavelength-granularity connections can be groomed onto direct optical transmission channels, or lightpaths[6]. Meanwhile, the failure of a network element can cause the failure of several lightpaths, thereby leading to large data and revenue loss. Fault-management schemes such as protection are essential to survive such failures (see Chapter 11).

Different low-speed connections may request different bandwidth granularities as well as different protection schemes (dedicated or shared). How to

[6] We distinguish the terms "lightpath" and "connection" as follows. The bandwidth requirement of a lightpath is the full wavelength capacity (STS-192 in our present discussion). The bandwidth requirement of a connection can be any quantized value no more than the full wavelength capacity. Later in our examples and results, we use the quantized values STS-1, STS-3c, STS-12c, STS-48c, and STS-192c for illustration purposes since these values have been widely used in current systems (the "c" after the number implies this is a contiguous block of STS-1s that are part of the same connection). We use the term "STS-n" to refer to the payload carried within an OC-n optical interface ($n = 1, 3, 12$, etc.).

efficiently groom such low-speed connections while satisfying their protection requirements is the main focus of our present study. Since shared protection is more resource efficient than dedicated protection due to backup sharing, we focus on the problem of dynamic low-speed connection provisioning with shared protection against single-fiber failures. Single-fiber failures are the predominant type of failures in communication networks. Node failures are not considered here because most nodal equipments are 1+1 protected.

While both protection and grooming have been studied heavily (see Chapters 11 and 13), the survivable traffic-grooming problem, in which sub-wavelength-granularity connections need to be protected, is a relatively unexplored territory (see [OuZM04, OZZM03]).

Given a static traffic matrix and the protection requirement of each connection request, the work in [LaGP01] presents an integer linear program and a heuristic for satisfying the bandwidth and protection requirements of all the connection requests while minimizing the network cost in terms of transmission cost and switching cost.

For dynamically establishing low-speed connection requests with shared protection, the work in [ThSo01b] presents mixed working-backup grooming policy (MGP) and segregated working-backup grooming policy (SGP). With both schemes employing fixed-alternate routing [RaMu02], the work focuses on the effect of different wavelength-assignment algorithms and different topologies.

14.4.2 Proposed Schemes

We propose three approaches—protection-at-lightpath (PAL) level, mixed protection-at-connection (MPAC) level, and separate protection-at-connection (SPAC) level—for dynamically provisioning shared-protected subwavelength-granularity connection requests against single-fiber failures. We investigate their characteristics under a generic grooming-node architecture, and design efficient heuristics. Our work differs from previous work in that we focus on route computation, the impact of different backup-sharing approaches, and the tradeoff between wavelength and grooming capacity.

While different grooming-switch architectures are possible, as shown in Chapter 13, this chapter considers the partial-grooming switch architecture shown in Fig. 14.10. To provision a connection request, there are two types of resource constraints–wavelengths and grooming ports [ZaOM03]. Typically, the more the number of wavelengths the network has, the less the number of grooming ports a node needs, and vice versa.

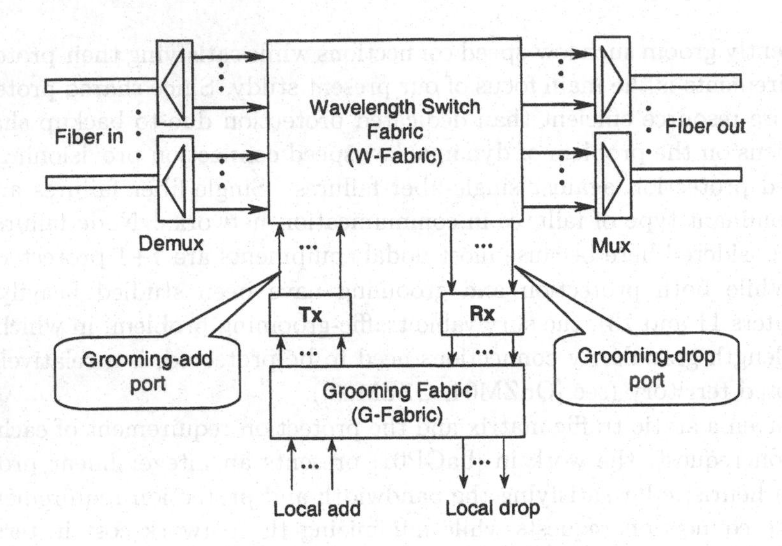

Figure 14.10 A simplified partial-grooming-node architecture.

Below, we shall illustrate the three schemes via an example. For the initial network configuration shown in Fig. 14.11, every edge corresponds to a bidirectional fiber; each fiber has two wavelengths; the wavelength capacity is STS-192; every node has three grooming ports (T and R represent the number of available grooming-add and grooming-drop ports, respectively).

14.4.3 Protection-at-Lightpath (PAL) Level

Basic Idea

PAL provides end-to-end protection with respect to lightpath. Under PAL, a connection is routed through a sequence of protected lightpaths, or *p*-lightpaths. A *p*-lightpath has a *lightpath* as working path and a link-disjoint *path* as backup path. For example, in Fig. 14.12(a), *p*-lightpath l_1 has lightpath $\langle 0, 1, 2 \rangle$ as working path and path $\langle 0, 5, 4, 2 \rangle$ as backup path. Please note the differences between the working path and the backup path of a *p*-lightpath. The working path of a *p*-lightpath is set up as a lightpath during normal operation. Therefore, as a lightpath does, the working path consumes a grooming-add port at the source node and a grooming-drop port at the destination node of a *p*-lightpath; and the working path of a *p*-lightpath

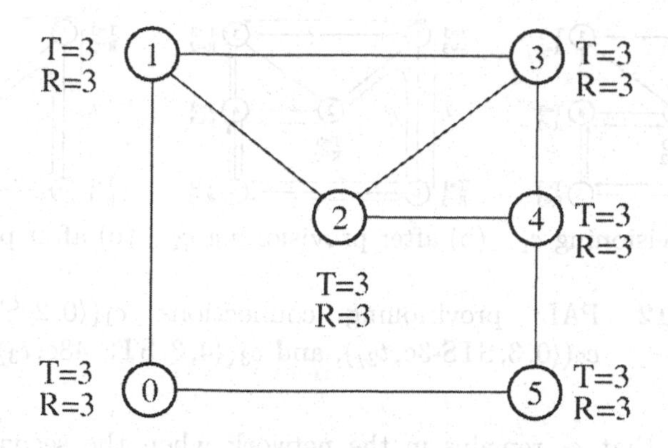

Figure 14.11 Example: initial network configuration.

bypasses any intermediate nodes along its path. However, the backup path of a p-lightpath is not set up as a lightpath during normal operation. Therefore, the backup path of a p-lightpath does not consume any grooming port; and wavelengths along a backup path are only reserved. In case the working path fails, protection switching occurs at lightpath level and the backup path is set up as a lightpath by utilizing the grooming ports previously used by the working path.

Two p-lightpaths can share wavelengths along common backup links if their working paths are link-disjoint. Clearly, a connection routed under PAL survives from single-link failures since each p-lightpath survives from single-link failures by definition. Since protection occurs at lightpath level, PAL has the advantages of low implementation complexity and low signaling overhead when a failure occurs. This will be further elaborated in Section 14.4.6. Below, we illustrate PAL in more detail by provisioning three connection requests.

Example

Upon the arrival of the first connection request $c_1, \langle 0, 2, \text{STS-12c}, t_1 \rangle$, one way of provisioning c_1 under PAL is shown in Fig. 14.12(a). Connection c_1 is routed via p-lightpath l_1, which has lightpath $\langle 0, 1, 2 \rangle$ as working path and path $\langle 0, 5, 4, 2 \rangle$ as backup path. p-lightpath l_1 consumes a grooming-add port at node 0 and a grooming-drop node at node 2. The free capacity of p-lightpath l_1 is STS-180.

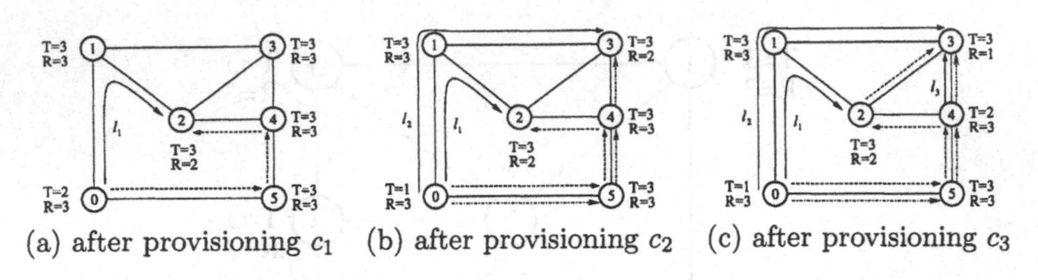

(a) after provisioning c_1 (b) after provisioning c_2 (c) after provisioning c_3

Figure 14.12 PAL: provisioning connections $c_1(\langle 0, 2, \text{STS-12c}, t_1\rangle)$, $c_2(\langle 0, 3, \text{STS-3c}, t_2\rangle)$, and $c_3(\langle 4, 3, \text{STS-48c}, t_3\rangle)$.

Suppose that c_1 remains in the network when the second connection request c_2, $\langle 0, 3, \text{STS-3c}, t_2\rangle$, arrives. One way of provisioning c_2 under the current network state is shown in Fig. 14.12(b). Connection c_2 is routed via p-lightpath l_2, which has lightpath $\langle 0, 1, 3\rangle$ as working path and path $\langle 0, 5, 4, 3\rangle$ as backup path. p-lightpath l_2 consumes a grooming-add port at node 0 and a grooming-drop port at node 3. The free capacity of p-lightpath l_2 is STS-189. Two wavelengths need to be reserved along links $\langle 0, 5\rangle$ and $\langle 5, 4\rangle$ because (1) the working paths of p-lightpaths l_1 and l_2 traverse common link $\langle 0, 1\rangle$, and (2) protection occurs at lightpath level, i.e., backup sharing only occurs at wavelength level.

Suppose that c_1 and c_2 remain in the network when the third connection request c_3, $\langle 4, 3, \text{STS-48c}, t_3\rangle$, arrives. One way of provisioning c_3 under the current network state is shown in Fig. 14.12(c). Connection c_3 is routed via p-lightpath l_3, which has lightpath $\langle 4, 3\rangle$ as working path and path $\langle 4, 2, 3\rangle$ as backup path. p-lightpath l_3 consumes a grooming-add port at node 4 and a grooming-drop port at node 3. The free capacity of p-lightpath l_3 is STS-144. Please note that the backup paths of p-lightpaths l_1 and l_3 share the wavelength reserved along link $\langle 4, 2\rangle$.

14.4.4 Mixed Protection-at-Connection (MPAC) Level

Basic Idea

MPAC and SPAC provide end-to-end protection with respect to connection. Under MPAC, a connection is routed via link-disjoint working and backup paths, each of which traverses a sequence of lightpaths. A lightpath traversed by a working path utilizes a portion of its capacity to carry traffic for that working path during normal operation. A lightpath traversed by a backup path reserves part of its capacity for that backup path. The backup capacity a lightpath reserves can be shared among multiple backup paths provided that their corresponding working paths are link-disjoint. In this context, "mixed" means that the capacity of one wavelength can be utilized by *both* working paths and backup paths; "separate", on the other hand, means that the capacity of a wavelength can be utilized by *either* working paths or backup paths, but not both. MPAC seems to be the most intuitive approach since it deals with individual connections and therefore can pack connections efficiently. Later in Section 14.4.6, we shall show that it may not achieve the best performance due to thc intricacy of backup sharing. Below, we illustrate MPAC in more detail using the same example as before.

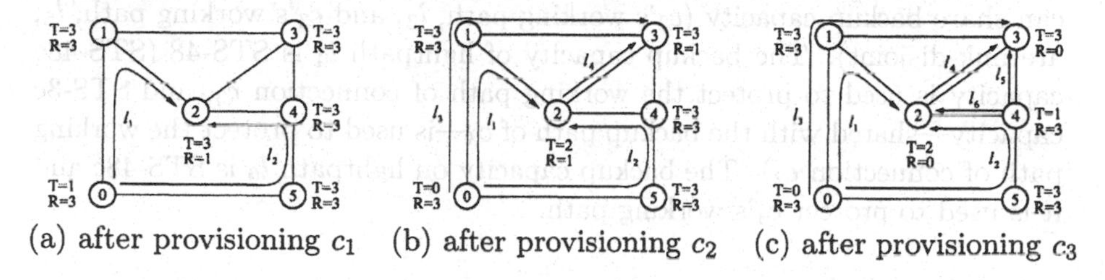

(a) after provisioning c_1 (b) after provisioning c_2 (c) after provisioning c_3

Figure 14.13 MPAC: provisioning connections $c_1(\langle 0, 2, \text{STS-12c}, t_1 \rangle)$, $c_2(\langle 0, 3, \text{STS-3c}, t_2 \rangle)$, and $c_3(\langle 4, 3, \text{STS-48c}, t_3 \rangle)$.

Example

When the first connection request $c_1, \langle 0, 2, \text{STS-12c}, t_1 \rangle$ arrives, one way of provisioning c_1 under MPAC is shown in Fig. 14.13(a). The working and backup paths of connection c_1 traverse lightpaths l_1 and l_2, respectively. The free capacity of both lightpaths l_1 and l_2 is STS-180. The backup capacity reserved on lightpath l_1 is zero. The backup capacity reserved on lightpath l_2 is STS-12c, and it is used to protect connection c_1's working path. Both lightpaths l_1 and l_2 consume a grooming-add port at node 0 and a grooming-drop port at node 2.

Suppose that connection c_1 remains in the network when connection request $c_2, \langle 0, 3, \text{STS-3c}, t_2 \rangle$, arrives. One possible solution of provisioning c_2 under MPAC is shown in Fig. 14.13(b). Connection c_2 is routed via two link-disjoint paths—lightpath l_3 and the two-lightpath sequence $\langle l_2, l_4 \rangle$. The working path can traverse either of the two paths, say lightpath l_3. The free capacity of both lightpaths l_3 and l_4 is STS-189. The free capacity of lightpath l_2 is STS-177 since lightpaths l_1 and l_3 traverse common link

$\langle 0,1\rangle$. The backup capacity on lightpath l_2 is STS-15 (STS-12c capacity is used to protect the working path of connection c_1, and STS-3c capacity is used to protect the working path of c_2). The backup capacity on lightpath l_4 is STS-3c and it is used to protect the working path of connection c_2.

Suppose that connections c_1 and c_2 remain in the network when connection request c_3, $\langle 4, 3, \text{STS-48c}, t_3\rangle$, arrives. One way of provisioning c_3 under MPAC is shown in Fig. 14.13(c). The working path of c_3 traverses lightpath l_5 and the backup path traverses the two-lightpath sequence $\langle l_6, l_4\rangle$. The free capacity of lightpaths l_5 and l_6 is STS-144. The free capacity of lightpath l_4 is STS-144 because the backup paths of connections c_2 and c_3 can share backup capacity (c_2's working path, l_3, and c_3's working path, l_5, are link-disjoint). The backup capacity of lightpath l_4 is STS-48 (STS-48c capacity is used to protect the working path of connection c_3, and STS-3c capacity—shared with the backup path of c_3—is used to protect the working path of connection c_2). The backup capacity on lightpath l_6 is STS-48c and it is used to protect c_3's working path.

14.4.5 Separate Protection-at-Connection (SPAC) Level

Basic Idea

SPAC provides end-to-end protection with respect to connection. Under SPAC, a connection is routed via link-disjoint working and backup paths. A working path traverses a sequence of lightpaths. A backup path traverses a sequence of links, each of which has judiciously reserved a number of wavelengths as backup resources. (This differs from MPAC, in which a backup path traverses a sequence of lightpaths.) In addition, a grooming-add port at the source end of the link and a grooming-drop port at the destination end of the link need to be reserved for each reserved wavelength because multiple backup paths groomed onto the same wavelength on a link may go to different next hops. In this context, "separate" means that the capacity of a wavelength can be utilized by *either* working paths or backup paths, but not both. SPAC is deliberately constructed in a way to trade grooming ports for increased backup sharing, as will be elaborated in Section 14.4.6. Below, we illustrate SPAC in more detail using the same example as before.

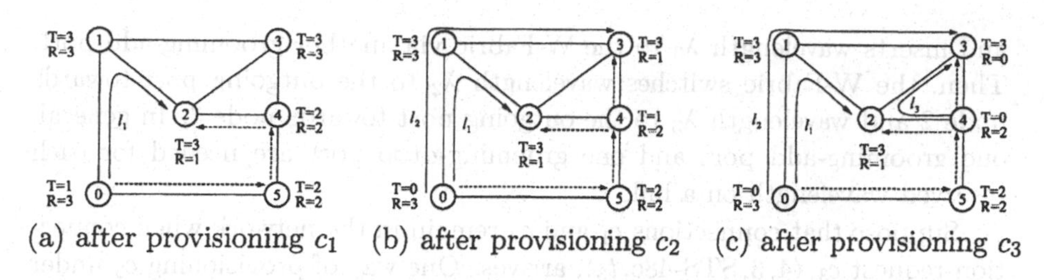

(a) after provisioning c_1 (b) after provisioning c_2 (c) after provisioning c_3

Figure 14.14 SPAC: provisioning connections $c_1(\langle 0, 2, \text{STS-12c}, t_1\rangle)$, $c_2(\langle 0, 3, \text{STS-3c}, t_2\rangle)$, and $c_3(\langle 4, 3, \text{STS-48c}, t_3\rangle)$.

Example

When the first connection request $c_1, \langle 0, 2, \text{STS-12c}, t_1\rangle$ arrives, one way of provisioning c_1 under SPAC is shown in Fig. 14.14(a). The working path of connection c_1 traverses lightpath l_1, and the backup path traverses path $\langle 0, 5, 4, 2\rangle$. The free capacity of lightpath l_1 is STS-180. Every link along the backup path needs to reserve one wavelength as backup capacity, while only STS-12c of the entire wavelength capacity is used to protect c_1's working path. For every link along the backup path, the upstream node needs to reserve one grooming-add port and the downstream node needs to reserve one grooming-drop port since one more wavelength has been reserved.

Suppose that connection c_1 remains in the network when connection request $c_2, \langle 0, 3, \text{STS-3c}, t_2\rangle$, arrives. One possible solution of provisioning c_2 under SPAC is shown in Fig. 14.14(b). The working path of connection c_2 traverses lightpath l_2, and the backup path traverses path $\langle 0, 5, 4, 3\rangle$. The free capacity of lightpath l_2 is STS-189. One wavelength along link $\langle 4, 3\rangle$ needs to be reserved as backup capacity, STS-3c capacity of which is used to protect c_2's working path. STS-15 capacity of the entire backup capacity along links $\langle 0, 5\rangle$ and $\langle 5, 4\rangle$ is used to protect the working paths of connections c_1 and c_2 (STS-12c for c_1 and STS-3c for c_2).

This step demonstrates why one grooming-drop port and two grooming-add ports need to be reserved at node 4. If link $\langle 0, 1\rangle$ fails, connection c_1 needs to be rerouted along $\langle 0, 5, 4, 2\rangle$, and connection c_2 needs to be rerouted along $\langle 0, 5, 4, 3\rangle$. As a result, node 4 needs to drop one wavelength, say λ_1, to the G-Fabric via one grooming-drop port. After unpacking wavelength λ_1, the G-Fabric grooms connection c_1 to an appropriate wavelength, say λ_2, and adds wavelength λ_2 to the W-Fabric via one grooming-add port; the G-Fabric also grooms connection c_2 to an appropriate wavelength, say λ_3, and inserts wavelength λ_3 to the W-Fabric via another grooming-add port. Then, the W-Fabric switches wavelength λ_2 to the outgoing port towards node 2 and wavelength λ_3 to the outgoing port towards node 3. In general, one grooming-add port and one grooming-drop port are needed for each reserved wavelength on a link.

Suppose that connections c_1 and c_2 remain in the network when connection request c_3, $\langle 4, 3, \text{STS-48c}, t_3 \rangle$, arrives. One way of provisioning c_3 under SPAC is shown in Fig. 14.14(c). The working path of connection c_3 traverses lightpath l_3 and the backup path traverses path $\langle 4, 3 \rangle$. The free capacity of lightpath l_3 is STS-144. STS-48c of the entire backup capacity along link $\langle 4, 3 \rangle$ is used to protect the working path of c_3; out of this STS-48c capacity, STS-3c is also used to protect the working path of c_2.

14.4.6 A Qualitative Comparison

The above illustrative examples indicate that the three schemes perform differently in terms of routing and the amount of resources required. Below, we qualitatively compare their characteristics with respect to routing, backup sharing, and operational complexity. For convenience, we will use the term "PAC" to refer to both MPAC and SPAC hereafter, whenever appropriate, because they have several similar properties.

Routing

The difference in routing between PAL and PAC is that PAL provides end-to-end protection with respect to lightpath while PAC provides end-to-end protection with respect to connection. Under PAL, when a failure occurs, the end nodes of the affected *p*-lightpaths first configure their backup paths and then switch over; the affected connections are oblivious to the protection-switching process. Under PAC, when a failure occurs, the end-nodes of the affected connections (which could be significantly more than the number of affected *p*-lightpaths) first configure their backup paths and then switch over.

Please note that a connection from node s to node d routed under PAL may not have two link-disjoint paths between node s and node d, while the connection still survives from single-link failures. For example, suppose that the connection is routed via two *p*-lightpaths l_1 and l_2 under PAL. The concatenation of the working paths of *p*-lightpaths l_1 and l_2 may not be link-disjoint from the concatenation of the backup paths of *p*-lightpaths l_1 and

l_2. This is because the working path of l_1 and the backup path of l_2 (or the working path of l_2 and the backup path of l_1) can traverse common links.

Routing-wise, PAL performs at an aggregate level (lightpath) and PAC performs on a per-flow (connection) basis. As a result, PAL trades the bandwidth efficiency in routing each specific sub-wavelength connection request for the savings in grooming-port usage. In PAL, the backup path of a *p*-lightpath does not require any grooming port. When a fiber along the working path of a *p*-lightpath fails, all of the traffic carried by the failed working path can be rerouted to the backup path of that *p*-lightpath, and the grooming ports (at the end nodes of the *p*-lightpath) previously used by the working path can be reused by the backup path. However, in SPAC, the end nodes of a link need to reserve a grooming-add/drop port for each reserved wavelength because multiple backup paths groomed onto the same wavelength on a link may go to different next hops; in MPAC, each lightpath reserves a portion of its bandwidth as backup capacity, thus backup capacity consumes a fraction of the grooming ports.

Backup Sharing

MPAC differs from PAL and SPAC in backup sharing. The backup path of a connection under MPAC is the concatenation of lightpaths. The backup path of a connection under SPAC (or the backup path of a *p*-lightpath under PAL) is the concatenation of links with reserved wavelengths. This difference in backup routing has two implications on backup sharing. First, since a lightpath may span multiple links, the backup capacity reserved on a lightpath (as in MPAC) is less likely to be shared among multiple connections than the backup capacity reserved on a link (as in PAL and SPAC).

The second implication applies to wavelength-convertible networks only. Under MPAC, the backup path of a connection traverses a sequence of lightpaths; thus, a backup path has both fixed route and fixed wavelength assignment [DoYa01]. Under SPAC, the backup path of a connection (or the backup path of a *p*-lightpath under PAL) traverses a sequence of links with a number of reserved wavelengths; thus, a backup path has only fixed route but not fixed wavelength assignment [DSBD01]. Basically, under SPAC and PAL, the reserved wavelengths on a link act like a "pool" for all the failure scenarios, and backup-capacity sharing among different wavelengths on a link is facilitated by the existence of wavelength converters. However, under MPAC, the backup-capacity sharing among different wavelengths on a link is not possible because backup capacity resides inside lightpaths, and multiple lightpaths cannot share their reserved backup capacity. We illustrate this difference in the following example.

Consider the changes in backup capacity on an arbitrary link $\langle u, v\rangle$ in a hypothetical network. Suppose that STS-156 capacity will be rerouted on link $\langle u, v\rangle$ when some other link $\langle x, y\rangle$ in this network fails; STS-108 capacity will be rerouted on link $\langle u, v\rangle$ when some other link $\langle i, j\rangle$ in this network fails; and no more than STS-108 capacity will be rerouted on link $\langle u, v\rangle$ when any other link fails. Clearly, STS-156 backup capacity needs to be reserved along link $\langle u, v\rangle$, assuming that any link is not in the same shared-risk-link group (SRLG)[7] as any other link. Under SPAC, link $\langle u, v\rangle$ needs to reserve one wavelength; under MPAC, a lightpath, l_1, from node u to node v needs to be set up. When a new connection request c_1, $\langle i, j, \text{STS-48c}, t_h\rangle$, arrives, suppose that its working path traverses link $\langle i, j\rangle$ and backup path traverses link $\langle u, v\rangle$ in this hypothetical network under both SPAC and MPAC. Since only STS-156 capacity needs to be rerouted when link $\langle i, j\rangle$ fails, no more backup capacity needs to be reserved under both SPAC and MPAC. Assume that connection c_1 remains in the network when another connection request c_2, $\langle i, j, \text{STS-48c}, t_h^{'}\rangle$, arrives. Suppose that connection c_2's working path traverses link $\langle i, j\rangle$ and backup path traverses link $\langle u, v\rangle$ in this hypothetical network under both SPAC and MPAC. Then, STS-204 capacity will be rerouted on link $\langle u, v\rangle$ if link $\langle i, j\rangle$ fails. As a result, under SPAC, link $\langle u, v\rangle$ needs to reserve two wavelengths (since wavelength capacity is STS-192), which combine to provide STS-204 backup capacity. Under MPAC, another lightpath, l_2, from node u to node v needs to be set up. Lightpath l_1 reserves STS-156 capacity and lightpath l_2 reserves STS-48 capacity as backup capacity.

The difference appears when connection c_1 leaves and connection c_2 remains in the network. Only STS-156 capacity will be rerouted on link $\langle u, v\rangle$ when either link $\langle x, y\rangle$ or link $\langle i, j\rangle$ fails after connection c_1 leaves. Consequently, under SPAC, only one wavelength needs to be reserved on link $\langle u, v\rangle$, and another previously reserved wavelength can be released. Under MPAC, however, lightpath l_1 still needs to reserve STS-156 backup capacity since STS-156 capacity will be rerouted on lightpath l_1 when link $\langle x, y\rangle$ fails. Without reconfiguring connection c_2's backup path, lightpath l_2 still needs to reserve STS-48 backup capacity. As a result, under MPAC, STS-204 capacity has been reserved while only STS-156 is really needed. (PAL will perform similarly to SPAC in this example.)

[7] A SRLG is a set of links which share the same risk [StCT01].

In short, PAL and SPAC trade the flexibility in grooming for the freedom in backup sharing. Under PAL and SPAC, working paths are groomed onto lightpaths while backup paths are groomed onto reserved wavelengths. However, MPAC has the flexibility in grooming working paths and backup paths (of different connections) onto the same lightpath.

Operational Complexity

From implementation point of view, PAL is simpler than PAC as PAL demands less information in route computation. While both PAL and PAC need the routing information of all the existing lightpaths to provision a shared-protected connection request, PAL does not require any information about the existing connections. PAC, however, does require the detailed routing information of all the existing connections. Under PAL, the routing information of the working paths of two p-lightpaths is sufficient to determine whether the backup paths of these two p-lightpaths can share wavelengths along common links. Under PAC, the routing information of the working paths of two connections, which includes lightpath-routing information, is needed to decide whether the backup paths of these two connections can share backup capacity along common lightpaths (in case of MPAC) or common links (in case of SPAC).

From network control point of view, PAL has lower signaling overhead. Assume that a lightpath can carry up to g connections. (In today's networks, g is typically 192 since wavelength capacity is STS-192 and the lowest bandwidth granularity is typically STS-1.) When a link fails, W lightpaths can be disrupted in the worst case. In PAL, at most W protection-switching processes are needed. However, in PAC, up to $W \times g$ protection-switching processes are required in the worst case. As protection-switching processes for shared protection typically require signaling, PAL requires lower control bandwidth and involves lower signaling complexity compared to PAC.

14.4.7 Illustrative Numerical Results

We simulate a dynamic network environment with the assumptions that the connection-arrival process is Poisson and the connection-holding time follows a negative exponential distribution[8]. For the illustrative results shown here, the capacity of each wavelength is STS-192; the number of the connection requests follows the distribution STS-1 : STS-3c : STS-12c : STS-48c :

[8]Please refer to [OZZM03] for the heuristic algorithms.

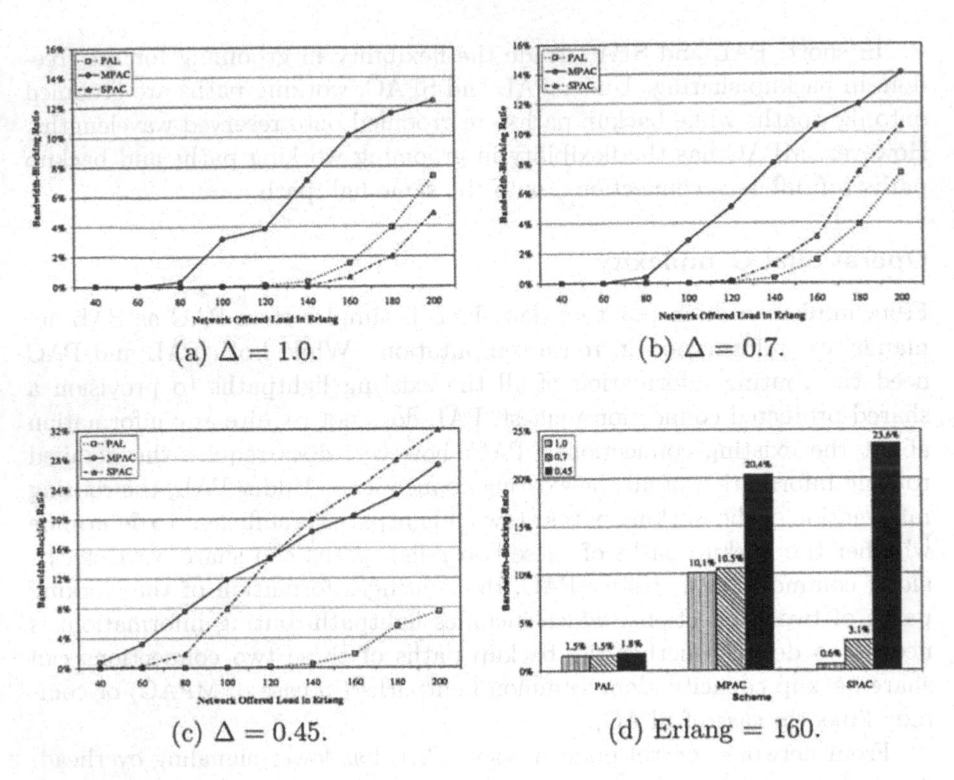

(a) $\Delta = 1.0$. (b) $\Delta = 0.7$.

(c) $\Delta = 0.45$. (d) Erlang = 160.

Figure 14.15 Bandwidth-blocking ratio versus network offered load.

STS-192c = 300 : 20 : 6 : 4 : 1 (which is close to the bandwidth distribution in a practical backbone network); connection requests are uniformly distributed among all node pairs; average connection-holding time is normalized to unity; the cost of any link is unity; load (in Erlang) is defined as connection-arrival rate times average holding time times a connection's average bandwidth normalized in the unit of STS-192; and our example network topology with 16 wavelengths per fiber is shown in Fig. 14.7. 100,000 connections were simulated in each experiment. The value of ϵ in the cost function is set to 10^{-6}; ϵ can trade off backup sharing and backup-path length, as shown in [BLRC02].

The number of grooming ports at a node is set as the number of wavelengths times its nodal degree times a scalar Δ ($0 \leq \Delta \leq 1$, $\Delta = 1$ implies that any incoming wavelength to the W-Fabric can be dropped to the G-Fabric). The number of alternate paths K for the three schemes is two.

We now quantitatively compare PAL to MPAC to SPAC using bandwidth-blocking ratio (BBR). BBR is defined as the amount of bandwidth blocked over the amount of bandwidth offered. Please note that pure blocking probability, defined as the percentage of the *number* of connections blocked, cannot reflect the effectiveness of the algorithm as connections have different bandwidth requirements. Figure 14.15 plots the BBR of the three schemes with $\Delta = 1.0, 0.7$, and 0.45. We make the following observations.

We find that PAL always has lower BBR than MPAC, and SPAC has lower BBR than MPAC when the number of grooming ports is large (e.g., $\Delta = 1.0$ and 0.7) or the number of grooming ports is small and the network offered load is moderate (e.g., $\Delta = 0.45$ and the network offered load is less than 120 Erlangs). This leads to our first observation: *It is beneficial to groom working paths and backup paths separately, as is the case in PAL and SPAC.*

Our second observation is that *SPAC has the lowest BBR when the number of grooming ports is sufficient (e.g.,* $\Delta = 1.0$*)*, as shown in Fig. 14.15(a). This is because SPAC has the maximum freedom in backup sharing when the number of grooming ports is sufficient (please see Section 14.4.6).

Our third observation is that *PAL achieves the lowest BBR when the number of grooming ports is moderate or small (e.g.,* $\Delta = 0.7$ *and* 0.45*)*, as shown in Figs. 14.15(b) and (c). The main reason for this is that backup paths do not consume grooming ports under PAL (Section 14.4.6).

Figure 14.15(d) shows the BBR of the three schemes with different values of Δ under the same network offered load, 160 Erlangs. Clearly, the decrease in grooming capacity impacts PAL the least and SPAC the most. Again, this is because PAL trades bandwidth efficiency in routing for grooming-port savings, and SPAC trades grooming ports for the flexibility in backup sharing (Section 14.4.6).

14.5 Summary

Following up on the material in Chapter 13, in this chapter, we further investigated three advanced topics in traffic grooming: (1) hierarchical switching and waveband grooming; (2) virtual concatenation (VC or VCAT); and (3) survivable traffic grooming.

We proposed a practical hierarchical optical switch architecture which integrates all-optical waveband switching and OEO traffic grooming. The proposed architecture was demonstrated to achieve large capacity, excellent scalability, high flexibility, and significantly lower cost. Different from previously-proposed hierarchical optical switch architecture, the proposed architecture can be built entirely with mature components, making it a very practical solution.

Virtual concatenation (VC) and link-capacity adjustment scheme (LCAS) can help a SONET/SDH-based optical network to evolve towards a data-centric intelligent automatically-switched optical network. They offer the following benefits: (a) relaxing the time-slot continuity and alignment constraints of traditional SONET/SDH concatenation; (b) improving bandwidth efficiency of a wavelength channel; (c) enabling inverse-multiplexing (i.e., traffic bifurcation) and load balancing; and (d) improving service resilience. Moreover, built on virtual concatenation, LCAS allows a network operator to adjust the pipe capacity while it is in use (on the fly). This increases the possibility for on-demand traffic provisioning and it makes SONET/SDH-based optical WDM networks more data friendly.

We also investigated the survivable traffic-grooming problem for optical WDM mesh networks in a dynamic context. Based on a generic grooming-node architecture, we explored three approaches—protection-at-lightpath (*PAL*) level, mixed protection-at-connection (*MPAC*) level, and separate protection-at-connection (*SPAC*) level—for grooming a connection request with *shared* protection against single-fiber failures. Our findings are as follows. Under today's typical connection-bandwidth distribution, 1) it is beneficial to groom working paths and backup paths separately, as in PAL and SPAC; 2) separately protecting each individual connection—i.e., SPAC—yields the best performance when the number of grooming ports is sufficient; and 3) protecting each specific lightpath—i.e., PAL—achieves the best performance when the number of grooming ports is moderate or small. For traffic grooming with *dedicated* protection [OuZM04], findings 2) and 3) hold, while finding 1) does not apply because we typically do not need to distinguish between working and backup paths in dedicated protection.

Exercises

14.1. Consider the network topology in Fig. 14.7, where every link is a bidirectional fiber. Each fiber has four wavelengths, each of capacity $STS-192$. Use protection-at-lightpath (PAL) scheme to setup the following connections, one at a time.

- $c0(<1, 10, STS-3c>)$
- $c1(<8, 14, STS-1>)$
- $c2(<7, 3, STS-3c>)$
- $c3(<7, 15, STS-12c>)$
- $c4(<14, 11, STS-12c>)$
- $c5(<3, 12, STS-3c>)$
- $c6(<18, 19, STS-48c>)$
- $c7(<22, 15, STS-1>)$
- $c8(<2, 20, STS-48c>)$
- $c9(<18, 17, STS-3c>)$

Assume that every node has four grooming ports. The connections, once set up, will remain in the network throughout. Specify free/used capacity along the links after all connections are set up. Also specify the number of grooming ports used/avaiable at the nodes after all connections are established.

14.2. Repeat Problem 14.1, assuming that every connection comes periodically, one after every 10 seconds (i.e., connection-interarrival time = 10 sec.), starting from connection $c0$ at $t=0$, and connection-holding time is 25 seconds.

14.3. Consider the network topology in Fig. 14.7, where every link is a bidirectional fiber. Each fiber has two wavelengths, each of capacity $STS-192$. Use mixed protection-at-connection (MPAC) scheme to setup the following connections, one at a time.

- $c0(<5, 23, STS-3c>)$
- $c1(<9, 13, STS-12c>)$
- $c2(<20, 8, STS-3c>)$

- $c3(< 23, 7, STS - 48c >)$
- $c4(< 3, 21, STS - 12c >)$

Assume that every node has four grooming ports; and connections, once set up, will remain in the network throughout. Specify free/used capacity along the links after all connections are set up. Also specify the number of grooming ports used/available at the nodes after all connections are established.

14.4. Now, consider that every connection in Problem 14.3 has two additional parameters: connection-arrival time and connection-holding time, i.e., $< t_s, t_h >$, in seconds. For the five connections in Problem 14.3, these parameter values are: $< 3, 10 >$, $< 7, 13 >$, $< 21, 7 >$, $< 23, 10 >$, and $< 24, 4 >$, respectively. How would your solution change? Show all the steps of your solution as time progresses.

14.5. Repeat Problem 14.4 using separate protection-at-connection (SPAC) scheme.

14.6. Consider the network topology in Fig. 14.7, where every link is a bidirectional fiber. Each fiber has eight wavelengths, each of capacity $STS - 192$. Every node has 16 grooming ports. New connections need to be setup using the following three schemes:
Scheme 1: PAL; **Scheme 2:** MPAC; and **Scheme 3:** SPAC.

The connections are:

- $c0(< 11, 2, STS - 3c >)$, use PAL
- $c1(< 8, 20, STS - 12c >)$, use PAL
- $c2(< 7, 4, STS - 1 >)$, use SPAC
- $c3(< 23, 7, STS - 48c >)$, use MPAC
- $c4(< 3, 21, STS - 96c >)$, use PAL
- $c5(< 5, 20, STS - 96c >)$, use MPAC
- $c6(< 10, 21, STS - 3c >)$, use SPAC
- $c7(< 13, 17, STS - 48c >)$, use PAL

Assume that the connections, once set up, will remain in the network throughout. Find the primary and backup routes for all connections, if possible. If no route exists, justify why? Specify free/used capacity along the links as before. Also specify the number of grooming ports used/avaiable at the nodes as before.

14.7. A waveband path can originate from a waveband add port (one of the BA ports), or from an OEO-switch to waveband-switch port (one of the WB ports) (see Fig. 14.4). Similarly, it can terminate at one of the BD ports or BW ports. How many types of waveband paths are there according to how such a path originates and terminates? What constraints does each type of waveband path impose on routing connection requests?

14.8. It is not desirable to route a connection to a large number of paths from network operation and management point of view. How can you modify the heuristic in Section 14.3.3 so that a connection will be routed on to no more than K path, where K is constant?

14.9. Based on Figs. 14.12-14.14, please show the network configuration after provisioning a fourth connection (5, 3, STS-12c, t4) using PAL, MPAC, and SPAC.

14.10. In Chapter 13, we introduced the generic graph model for traffic grooming. Can this model also handle the hierarchical optical switches shown in Fig. 14.3? Justify your answer. If not, how can we extend the model to cover this architecture? Justify your modification.

14.11. Another hierarchical optical switch architecture might contain a separate wavelength switch between the waveband switch and the OEO grooming switch. Does the architecture proposed in this chapter have wavelength switching capability? What are the advantages and disadvantages of each architecture?

14.12. When a traffic stream is broken into multiple components and routed through different paths towards a destination node using virtual concatenation, the destination node may need an external RAM to compensate for the differential delay between the bifurcated streams when they are reconstructed. What are the factors that determine the amount of RAM needed at the destination node?

14.13. Under a static network design case, in which all traffic requests are known and do not change, which virtual concatenation benefits listed in Section 14.3.3 may not apply?

Chapter 15

All-Optical Impairment-Aware Routing

15.1 Introduction

The problem of routing and wavelength assignment (RWA) in wavelength-routed optical networks has been presented in Chapter 7. Our discussions so far have assumed an ideal physical layer. However, in an all-optical network, transmission impairments due to non-ideal optical components in the physical layer may significantly affect the network performance. This chapter is devoted to a study on the impact of physical impairments on network performance (particularly with respect to blocking of connection requests due to poor signal quality) and to investigate intelligent RWA algorithms to combat impairment effects.

Recent technology advances in optical amplifiers, optical multiplexers and demultiplexers, optical switches, and other optical devices seem to make it possible to soon build an all-optical wavelength-division-multiplexing (WDM) network. In an all-optical network, optical-electrical-optical (OEO) conversions are not used at the intermediate nodes of a connection, resulting in a potentially lower cost to build the network. Such a transparent optical WDM network (see below) is a promising candidate for building the next-generation backbone network at low cost.

15.1.1 Transparent Optical Network

In a wavelength-routed WDM network, data traffic is carried via an optical channel, called a lightpath, traveling through network nodes interconnected by optical fibers. A node may provide optical-to-electronic (O/E) conversion and vice versa (E/O) to regenerate the data signal or to interface the optical network with electronic equipments [RFDM99].

In an *opaque* network, data transmission occurs over point-to-point links so that the signal is regenerated at every intermediate node along a lightpath via OEO conversion. The need for high transport capacity has been met by deploying the WDM technique. However, the operating expenses of such a point-to-point system may be quite high, mainly due to the large amount of regenerators required at the nodes of a national-scale network. The cost could be reduced in a *translucent* network [RFDM99] where the regeneration functionality is only employed at some nodes instead of at all nodes. But the eventual goal of electronics reduction leads to a trend in development of an all-optical *transparent* network [SSBN02, Tomk02, WCAS00].

In a transparent optical network, a data signal that is transmitted remains in the optical domain for the entire lightpath. Hence, the transparent network can eliminate the expensive OEO conversions. Moreover, it can offer transparency to bit rate, signal format, and protocols; thus, a transparent all-optical lightpath can carry an analog signal also, if so desired. Eventually, it may be conceivable that a user can dial for a "wavelength service path" which may carry any signal the user wants. The transparent network also can provide high-speed node-bypass functions and to reconfigure massive data pipes in response to failures.

However, the quality of an optical signal degrades as it travels through several optical components along the lightpath. The physical size of a transparent network is mainly determined by impairment effects such as attenuation, noise, crosstalk, chromatic/polarization-mode dispersion (CD/PMD), fiber non-linearities, filter concatenation, polarization-dependent loss/gain (PDL/PDG), signal transients, and so on [Tomk02]. Hence, impairment effects in the physical-layer transmission should not be ignored in all-optical lightpath routing.

This chapter focuses on how to take the effects of transmission impairments into consideration in RWA for an all-optical transparent WDM network [HuHM05a].

15.1.2 Network Models for RWA

The network models for RWA investigation fall into two categories, i.e., ideal network and realistic network. An ideal network is one in which transmission is error-free, i.e., network components are ideal and do not introduce any impairments. However, in a realistic network, impairments do exist, due to non-ideal components in the physical layer. There are two kinds of RWA algorithms, namely, RWA with and RWA without physical impairments consideration. They will be compared under two network models, as shown in Fig. 15.1.

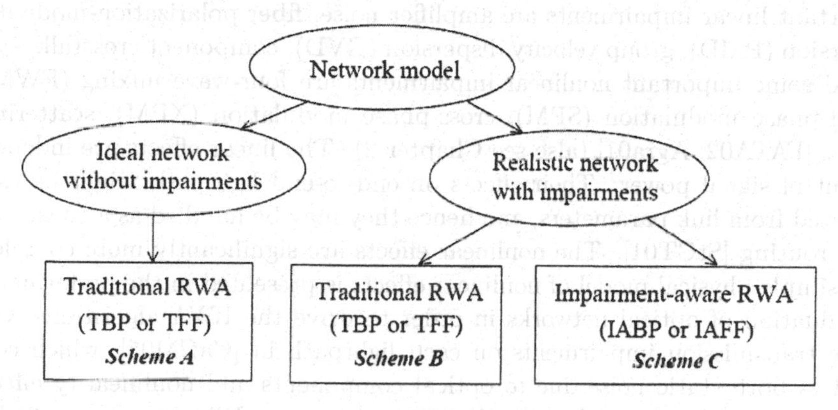

Figure 15.1 Three schemes for RWA comparisons.

15.2 Effects of Transmission Impairments on RWA

Many RWA problems have been investigated by assuming an ideal all-optical network where signal transmissions are considered to be error-free[1]. However, the transmission impairments, which are present in fibers and optical components in the physical layer, may significantly affect the quality of a lightpath [SILB98, RDFH99]. Hence, they must be taken into consideration during lightpath assignment.

[1]Actually, such studies apply directly in an OEO network also [as well as in a network with 3R (re-amplification, re-shaping, and re-timing) regeneration] because the signal is regenerated at each node, i.e., at the end of each fiber link. Thus, all impairments are arrested before the signal on the lightpath propagates on the next fiber link towards the destination.

In particular, a lightpath may traverse through a number of non-ideal transmission devices, such as optical fibers, optical amplifiers, and optical crossconnects (OXCs). Because of transparency of an optical network, noise and signal distortion due to linear and nonlinear effects accumulate along the lightpath and may cause significant signal degradation [SILB98, CeFP02]. At the destination node, the received signal quality may be so poor that the bit-error rate (BER) can reach an unacceptably high value, and thus the lightpath is not usable. We call this phenomenon *physical-layer blocking.*

The transmission impairments induced by non-ideal transmission components can be classified into two categories: linear and nonlinear. Some important linear impairments are amplifier noise, fiber polarization-mode dispersion (PMD), group velocity dispersion (GVD), component crosstalk, etc.; and some important nonlinear impairments are four-wave mixing (FWM), self-phase modulation (SPM), cross-phase modulation (XPM), scattering, etc. [FATA02, Agra01] (also see Chapter 2). The linear effects are independent of signal power. Their effects on end-to-end lightpath might be estimated from link parameters, and hence they may be handled as a constraint on routing [StCT01]. The nonlinear effects are significantly more complex. A simple physical model of nonlinear effects is presented in the performance evaluation of optical networks in order to drive the RWA algorithms with the transmission impairments on each lightpath in [CaCM05], which considers both static noise due to optical components and nonlinearity effects due to the current wavelength allocation and usage. When the transmission impairments come into play, an accurate selection of path and wavelength which is driven by the optical signal-to-noise ratio (OSNR) is mandatory. In particular, both static effects and nonlinearities can significantly affect the blocking probability: the first depends on the physical configuration and must be considered for any offered load to the network; the latter rapidly degrades the quality of the transmission layer when the number of lightpaths already established is large, i.e., when the offered load is high. Furthermore, the transmission impairments may become more severe at high bit rates, especially at 40 Gbps and beyond [CaCM05].

15.3 Incorporation of Network Layer and Physical Layer Effects

Without transmission-impairment awareness, a network-layer RWA algorithm might provide a lightpath which cannot meet the signal-quality requirement. Therefore, the control plane of a transparent optical network should incorporate the characteristics of the physical layer in setting up a lightpath for a new connection.

A lightpath is set up to carry data for a connection request by a RWA algorithm in the network layer. However, the data may encounter transmission impairments in the physical layer. How to incorporate the effects of physical impairments in network algorithms is still in the early stages of investigation. One approach to combine the effect of the two layers is to employ an impairment-aware algorithm consisting of two steps: *lightpath computation* and *lightpath verification.* We also refer to this as a cross-layer design principle.

A high-level structure of such impairment-aware RWA algorithms is presented in Fig. 15.2. The algorithm first uses a network-layer module to look for a candidate lightpath without considering the transmission impairments for the connection request. If no route or wavelength is available, the call is blocked due to lack of resources in the network layer. This type of blocking is referred to as *network-resource blocking.*

If one or more routes and wavelengths are available for the candidate lightpath, the physical-layer module is invoked to estimate the statistical signal quality of the candidate lightpath [CeFP02, HGHM02]. If the lightpath's quality is acceptable, i.e., if it satisfies a certain signal-quality requirement, the call will be admitted at the source node using this candidate lightpath. Otherwise, the physical-layer module notifies the network-layer module to reject the candidate lightpath, and the network-layer module will try to find another candidate lightpath from the available resources on another wavelength. In this way, if no available lightpath can have acceptable signal quality, the call is blocked. This kind of blocking is due to poor signal quality in the physical layer, and is called *physical-layer blocking.*

15.4 Constraints of Physical Impairments

There are several criteria which could be used to evaluate the signal quality of a lightpath [BeOS99]. BER is a very appropriate criterion because it is a comprehensive parameter which takes all impairment effects into consideration. BER is not readily available before the lightpath is actually set up. Instead, physical-layer impairments are modeled to statistically evaluate the BER in advance.

In this chapter, some important impairments, i.e., amplifier noise,

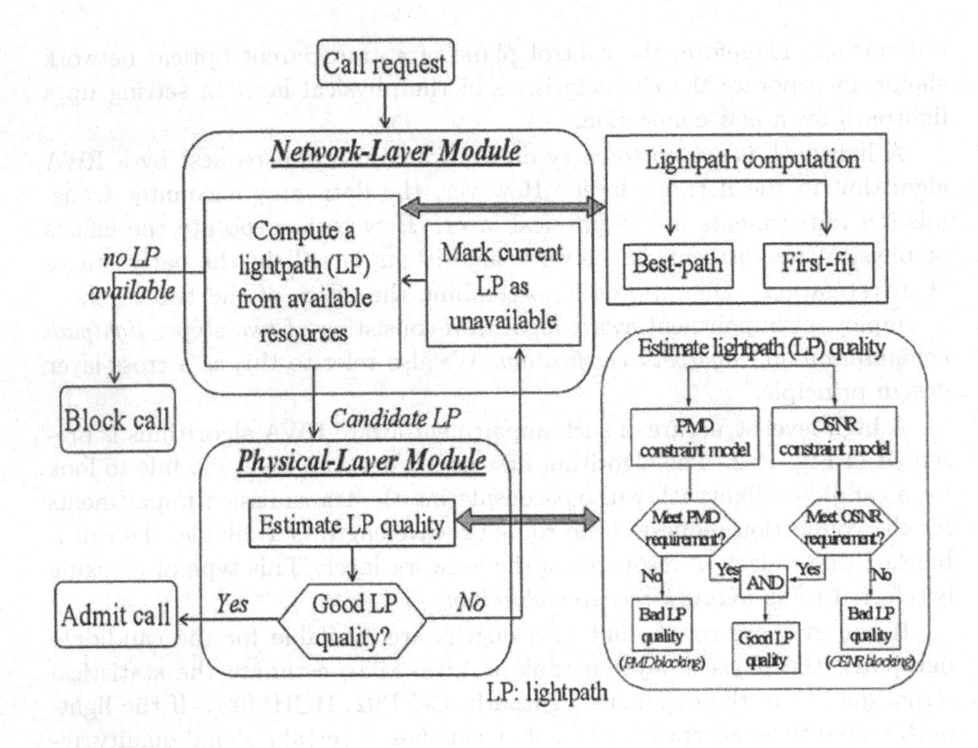

Figure 15.2 Integrated model of impairment-aware RWA algorithms.

crosstalk at OXC, and PMD, are considered as example impairments which affect the signal quality for the discussion of impairment-aware RWA. Constraints imposed by optical signal-to-noise ratio (OSNR) and PMD effects are used in our physical-layer module to evaluate the signal quality. The lightpath computed by taking only the OSNR and PMD into consideration might not guarantee that the BER requirement would be satisfied because of other types of physical impairments as well. However, a lightpath not satisfying the OSNR and PMD requirement will not be able to satisfy the BER requirement and should be blocked. Hence, at a minimum, OSNR and PMD must be taken into account. Also, additional impairments should be incorporated when necessary. A lightpath's signal-quality estimation in the physical-layer module is shown in Fig. 15.2. These approximate models are described below.

15.4.1 Polarization-Mode-Dispersion (PMD) Constraint Model

As the channel bit rate increases to 10 Gbps and beyond, PMD becomes one of the most critical limiting problems for data transmission in a high-speed network. PMD strongly affects the *transparent transmission length* as expressed below [StCT01]:

$$B \times \sqrt{\sum_{k=1}^{M} D_{PMD}^2(k) \times L(k)} \leq a \tag{15.1}$$

where B is the data rate, $D_{PMD}(k)$ is the fiber PMD parameter in the kth span of the transparent lightpath which consists of M fiber spans, and $L(k)$ is the fiber length of kth span. The parameter "a" which represents the fractional pulse broadening should typically be less than 10% of a bit's time slot for which the PMD can be tolerated [StCT01]. This transmission-length constraint is called the *PMD constraint* in this chapter. If a call needs to be routed farther than this PMD limit of transmission length, it should be blocked. This kind of blocking caused by the PMD impairment is referred to as *PMD blocking.*

15.4.2 Optical Signal-to-Noise Ratio (OSNR) Constraint Model

In this model, the impairments affecting the signal quality are various forms of noise, including amplifier noise and OXC crosstalk. In this regime, the Q factor [Agra02] can be used as a good intermediate parameter for BER and OSNR [BeOS99]. A BER of 10^{-9} corresponds to a Q factor equal to six with the Gaussian noise approximation. Q factor can be approximated as [BeOS99]:

$$Q = \sqrt{\frac{B_o}{B_e}} \times \frac{2OSNR}{\sqrt{4OSNR + 1} + 1} \tag{15.2}$$

where B_o is the optical bandwidth and B_e is the electrical bandwidth.

A lightpath's architecture, as shown in Fig. 15.3, is used to evaluate the OSNR of the lightpath. The in-line amplifier uses the backward-pumped distributed Raman amplifier (DRA) which is a promising technique for long-haul, high-speed ($\geq$ 10 Gbps) transmission systems [KRNM99, PeWi02]. For a fiber link between nodes k and $k+1$ on the lightpath, in-line optical amplification is employed, with an amplifier spacing of 82 km. Each amplification span consists of 70 km of standard single-mode fiber (SSMF)

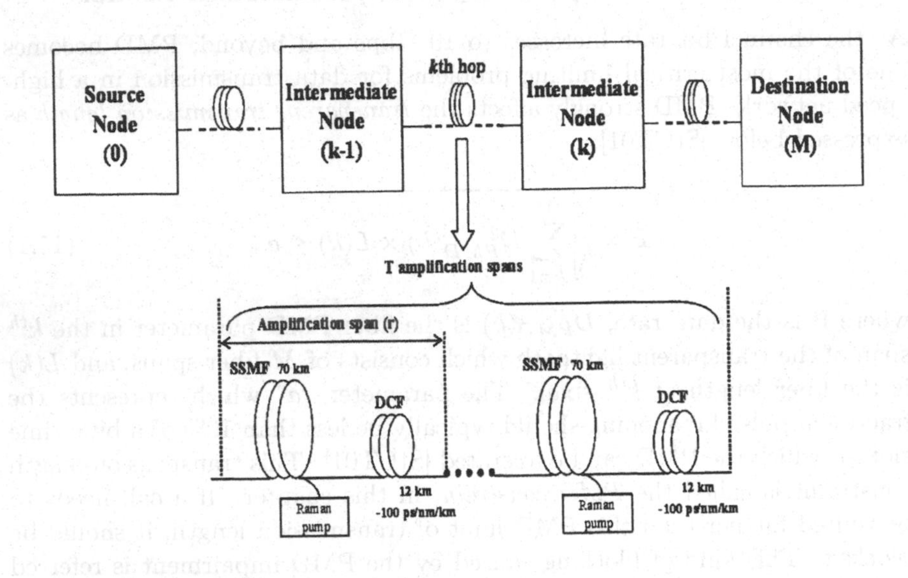

Figure 15.3 A sample of a lightpath architecture.

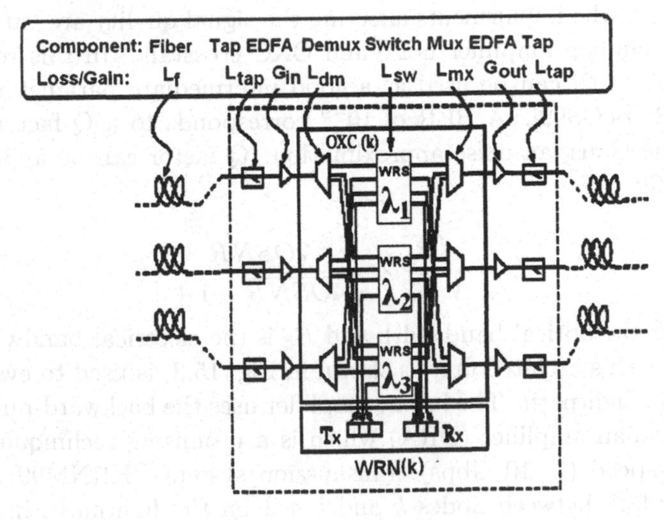

Figure 15.4 Architecture of a wavelength-routing node (WRN).

whose dispersion and dispersion slope are completely compensated by 12 km of dispersion-compensated fiber (DCF). The fiber attenuation of SSMF and DCF is 0.2 dB/km and 0.5 dB/km, respectively. The DRA exactly compensates for the fiber losses in an amplification span. Both the internal losses of the OXC and the loss of the fiber link between the last DRA and the OXC are compensated by the EDFAs at the node (see Fig. 15.3).

Since the OSNR on a lightpath varies with changes in network traffic, an iterative method, which is based on the *current network state*, is used to estimate the signal and noise power propagating through the lightpath. For a given lightpath from a source to a destination node, as shown in Fig. 15.3, we express the output power of signal (S), DRA's noise (N_{DRA}), and node noise (N_{node}), which includes crosstalk (N_{xt}) and ASE noise of EDFA (N_{EDFA}), at the $(k+1)^{th}$ intermediate node by the following recursive equations:

$$S(k+1) = S(k) \prod_{r=1}^{T} L_{tf}(r) G_a(r, \lambda) \tag{15.3}$$

$$N_{DRA}(k+1) = N_{DRA}(k) + P_a \sum_{r=2}^{T} \left(\prod_{j=r}^{T} L_{tf}(j) G_a(j, \lambda) \right) \tag{15.4}$$

$$N_{node}(k+1) = N_{xt}(k+1) + N_{EDFA}(k+1) \tag{15.5}$$

with: $N_{xt}(k+1) = N_{xt}(k) \prod_{r=1}^{T} L_{tf}(r) G_a(r, \lambda) + X_{sw} P_{xt}(k+1)$;

$$N_{EDFA}(k+1) = N_{EDFA}(k) \prod_{r=1}^{T} L_{tf}(r) G_a(r, \lambda) + P_{EDFA},$$

where $L_{tf}(r)$ is the total loss of the SSMF and DCF in the rth amplification span of the fiber segment between nodes k and $k+1$, which consists of T amplification spans (see Fig. 15.3); $G_a(r)$ is the amplifier gain of the rth span; and P_a is the noise generated by a DRA. Here, the DRA noise model is based on approximate models investigated in [ZLSC01, FlMe01]. P_{xt} is the total power of co-propagating signal shared with the desired signal on wavelength λ in the switch, and the X_{sw} is the switch crosstalk ratio. The details of the switch architecture (Fig. 15.4) and crosstalk generation are described in [RDFH99]. The P_{EDFA} is the total ASE noise[2] generated by

[2] $P_{edfa} = 2n_{sp}(G_{in} - 1)hf_{\lambda}B_o L_{dm} L_{sw} L_{mx} G_{out} L_{tap} + 2n_{sp}(G_{out} - 1)hf_{\lambda}B_o L_{tap}$, where h is Planck's constant; other parameters are defined in Table 15.1.

EDFAs at the output of the node. At the destination node, the OSNR is given as:

$$OSNR_{destination} = \frac{S(destination)}{N_{DRA}(destination) + N_{node}(destination)} \tag{15.6}$$

If the accumulated noise degrades the OSNR to fall below a required threshold, the lightpath should not be used. This form of physical-layer blocking is referred to as *OSNR blocking*.

Such an impairment-aware RWA algorithm not only requires each call to be routed within a certain transmission-length limitation for the sake of the PMD effect, but it also requires the OSNR at the destination node to be higher than a certain threshold before any lightpath is setup. Some typical parameters used by these constraint models are listed in Table 15.1.

15.4.3 Backward-Pumped Distributed Raman Amplifier (DRA) Model

A fiber-based Raman amplifier uses intrinsic properties of the silica fiber for amplification which is produced by stimulated Raman scattering (SRS) [Agra02]. When the same fiber used for signal transmission is also used for signal amplification, it is called distributed Raman amplification. Raman amplifier has been investigated experimentally and theoretically [Chra84, KRNM99, Hans99, ZLSC01, FlMe01, PeWi02, DaEi02, HeBe02]. Some key factors which need to be considered in designing a Raman amplifier include: amplifier spontaneous noise, Rayleigh scattering, interaction between pumps and signals (pump-pump, pump-signal, and signal-signal interactions), and pump depletion (saturation). A mathematical model to simulate the physical properties that affect these factors was developed in [KRNM99]. Some simplified models can approximate the amplifier behavior, such as fiber loss and ASE noise [ZLSC01, DaEi02]; multipath interference [FlMe01, HeBe02, DaEi02]; as well as SRS and its temperature dependence [HeBe02].

Our impairment-aware RWA algorithms require a reasonable amplifier model to simulate the most significant impairments from actual devices. The most important factors that limit the performance of distributed Raman amplifiers are Rayleigh scattering and Raman amplified spontaneous emission (ASE) [Hans99, Agra02, DaEi02]. To simplify our algorithm design and reduce its computation time, we capture the effects of ASE and

Table 15.1 System parameter used in the models.

Parameter	Value
Number of wavelengths	16
Wavelengths (in nm)	1542.6~1554.6 with 0.8 nm wavelength spacing
Channel bit rate	10 Gbps
Signal power per channel	1 mW
Switch crosstalk ratio (X_{sw})	-25 dB
Loss of multiplexer/demultiplexer (L_{mx}/L_{dm})	-4 dB
Loss of switch (Lsw)([RDFH99])	-8 dB
Loss of tap (L_{tap})	-1 dB
Gain of EDFA (G_{in}, G_{out})	12 dB, 6 dB
ASE factor of EDFA at node (n_{sp})	1.2
Pulse broadening factor (α)	0.1
Fiber PMD parameter ($D_{PMD}(k)$)	0.1 ps/$\sqrt{km}$
Desired DRA gain	20 dB
Three pump wavelengths (in nm)	1410, 1450, and 1500
Corresponding pump powers (in W)	0.607, 0.209, and 0.01
OSNR threshold	7.4 dB (BER= 10^{-9})
Polarization dependent factor (K_{eff})	2
fiber effective area (A_{eff})	50 μm^2
Fiber loss at pump wavelengths (α_p)	0.3 dB/km
Backscattering coefficient (S_c)	0.0022

double Rayleigh scattering (multipath interference) in our simulated DRA model in much the same way as in [ZLSC01, FlMe01]. They are summarized here for easier reference. The DRA model assumes that:

1. fiber losses for all pump lights are identical;
2. the energy losses, when a high-frequency photon is transformed into a low-frequency photon, are neglected;
3. small-signal case is considered for which gain and noise for each wavelength are independent of the number of signal wavelengths; and
4. the pump depletion caused by signal-pump coupling is negligible. Since

the pump supplies energy for signal amplification, it depletes as signal power increases.

As can be seen from equation:

$$g(w) = g_R(w)(\frac{P_{pump}}{a_{pump}}) \tag{15.7}$$

where g is the optical gain, g_R is Raman-gain coefficient, and a_{pump} is the cross-sectional area of the pump beam inside the fiber, the optical gain reduces when the pump power decreases. This reduction in gain is referred to as gain saturation [Agra02]. Since pump power is typically much larger than channel signal power, the Raman amplifier operates in the unsaturated regime [Agra02].

Gain and ASE Noise of Distributed Raman Amplifier

For a Y-wavelength backward-pumped DRA, we assume that the signal lightwaves and the pump lightwaves are launched into each amplification span (r) at the location $z = 0$ and at $z = L$, respectively. The power evolution of pump u, $P_u(z)$, is given by:

$$\frac{dP_u(Z)}{dz} = \alpha_p P_u(z) - P_u(z) \sum_{i=1}^{Y} (f_i - f_u) P_i \frac{g_{ui}}{K_{eff} A_{eff}(f_i - f_u})) \tag{15.8}$$

where α_p is the fiber loss at pump wavelength; A_{eff} is the fiber effective area; g_{ui} is the Raman gain coefficient between pump u and pump i at frequency f_u and f_i, respectively; K_{eff} is the polarization dependent factor; and Y is the number of pumps for amplification. g_{ui} is assumed to have a gain profile which closely approximates the measured Raman-gain profile in [Chra84]. The pump power along the fiber, $P_u(z)$, can be derived by integrating Eqn. (15.8) with respect to z.

In the small-signal case, according to [ZLSC01], at fiber length L, the small-signal optical gain, $G_w(L)$, and the ASE noise power, $P_{ASE}(L)$, for the wth WDM channel can be expressed as follows:

$$G_w(L) = exp\{\int_0^L B_w(z)dz\} \tag{15.9}$$

with

$$B_w(z) = -\alpha_s + \sum_{j=1}^{Y} \frac{g_{wj}}{K_{eff}A_{eff}} P_j(z)$$

$$P_{ASE}(L) = \int_0^L C_w(z) exp\{\int_z^L B_w(z_1)dz_1\}dz \quad (15.10)$$

with

$$C_w(z) = \sum_{i=1}^{Y} \frac{g_{wi}}{2A_{eff}} [hv_w \, \Delta_v \, (1 + \frac{1}{exp(\frac{h(f_i - f_w)}{KT}) - 1})] P_i(z)$$

where α_s is the fiber loss at signal frequency v_w; subscript w refers to the wth signal light; and h is Planck's constant.

Multipath Interference (MPI) in Distributed Raman Amplifier

An approximate analytical model [FlMe01] is used to calculate the double-Rayleigh scattering (DRS) power. The transmission fiber used for Raman amplification is considered as a sum of the effective length, L_g, which provides the gain, and the remainder length, L_l, which has fiber loss. Hence, the signal-to-DRS ratio, denoted as multipath interference (MPI) [FlMe01], is given by:

$$MPI(G_R) = MPI_{L_g}(G_R) + MPI_{L_l} \quad (15.11)$$

with

$$MPI_{L_g}(G_R) = (\frac{S_c \alpha_s}{2C_g})^2 [e^{2C_g L_g} - 1 - 2C_g L_g]$$

$$MPI_{L_l} = (\frac{S_c}{2})^2 [2L_l \alpha_s - 1 + e^{-2\alpha_s L_l}]$$

$$C_g = \frac{lnG_R - \alpha_s L_g}{L_g}$$

where MPI_{L_g} and MPI_{L_l} are the MPI in the effective length and fiber-loss length, respectively; S_c is the backscattering coefficient; C_g is the gain coefficient; and G_R is the Raman gain added to the fiber, i.e., the on-off gain of DRA. The on-off gain is defined as the output signal power when pumps are on divided by the output signal power with pumps off [FlMe01, DaEi02].

A key issue in the design of a broad-band amplifier is gain flattening. For DRAs, multiple pumps can be used to flatten the gain spectrum so that all signal channels could have the same MPI penalty. In our simulation model, the transmission fiber acts as a DRA and requires 70 km, and the DRA should provide a 20-dB on-off gain within the signal wavelength range from 1542.6 nm to 1554.6 nm. To achieve this design requirement, we use a semi-exhaustive search to select the frequencies and powers for the pump lights. We set a wide range of possible frequencies and powers for the pump signals, and coarse spacing of sample values; then, the gain is calculated according to Eqn. (15.9) for all samples. The pump signals with the minimum gain ripple, which is defined as in [ZLSC01], were determined as the optimal ones to meet our requirements.

Based on this optimization, three pump wavelengths are set to be 1410 nm, 1450 nm, and 1500 nm, and the corresponding pump powers are 0.607 W, 0.209 W, and 0.01 W, respectively. The gain ripple is 0.089, which is less than that in [ZLSC01]. Using these pump signals, the Raman gain, ASE noise, and MPI noise can be calculated according to Eqns. (15.9), (15.10), and (15.11), respectively. Since the pump power is larger than the signal power, the calculated gain and noise for each wavelength are assumed to be independent of the number of signal wavelengths existing in the DRA for simplicity, and they are put into a profile used in the physical-layer module. The signal power, ASE power, and MPI power are examined at the DRA output when the input signal power is 0 dBm. We find that the gain of the DRA is 6 dB, i.e., 20-dB on-off gain, within the interesting wavelength range, which meets the design goal for our DRA model. Also, these results show that ASE noise is the dominant noise and is about 10 dB larger than the MPI noise for the designed 70-km DRA.

15.5 Network-Layer Module

For a given physical network topology $PG(N, L)$, a set of auxiliary wavelength-layered topologies $WG_w(N, L)$ are created for each wavelength w, $w = 1, 2, ..., W$, where W is the maximum number of wavelengths supported by a fiber link, N is the set of nodes, and L is the set of bi-directional links. All wavelength-layered topology graphs (WG_s) are initialized to be the same as the physical network topology graph (PG) where the link weight corresponds to the fiber length. The routing decisions are made based on these auxiliary wavelength-layered graphs. Upon arrival of a connection request, the algorithms to compute a lightpath for the request are described as follows:

- **Given:** Current network state $WG_w(N, L)$, $w = 1, 2, ..., W$; a connection request R(source, destination); and signal-quality feedback (see below for detail).
- **Network-layer module of impairment-aware best-path algorithm (IABP)**
  - Step 1: Apply a shortest-path algorithm to find a path P_w in WG_w for $w = 1, 2, ..., W$. A vector of path distances is defined as $D = D_w | w = 1, 2, ..., W$. If no path is available in the wth wavelength-layered topology WG_w, the D_w is set to ∞. Otherwise, D_w is the total distance of path P_w.
  - Step 2: If not all elements of D are ∞, find the minimum distance $D_m \in D$, and mark the candidate wavelength $\lambda = m$; otherwise, the call is blocked; go to Step 5.
  - Step 3: Send the lightpath P_λ to the physical-layer module for signal-quality estimation, and wait for feedback from the physical-layer module.
  - Step 4: If the estimation of signal quality is "acceptable", (a) set up the call by using P_λ; (b) update WG_λ by specifying the links used by P_λ as occupied (can be done by changing the weights of all links along path P_λ to ∞); and (c) update the physical-layer information in the physical-layer module by recording the signal power as well as noise powers on each link along the lightpath P_λ. Otherwise, update D_m to ∞, and go to Step 2.
  - Step 5: Stop the procedure.
- **Network-layer module of impairment-aware first-fit algorithm (IAFF)**
  - Step 1: Initialize $w = 1$, i.e., consider the first wavelength first.
  - Step 2: Apply a shortest-path algorithm to find a path P_w in WG_w. If no path is available in WG_w, let $w = w + 1$ and repeat this step until we can find a path in one of the many wavelength-layered topology graphs. Mark the candidate wavelength as $\lambda = w$. If no path is available in all the wavelength-layered topology graphs, WG_s, the call is blocked; go to Step 5.

- Step 3: Output the lightpath P_λ to the physical-layer module for signal-quality estimation, and wait for feedback from the physical-layer module.
- Step 4: If the feedback of signal quality is "acceptable", (a) set up the lightpath by using P_λ; (b) update WG_λ by specifying the lightpath's links as occupied (can be done by changing the weight of all links along path P_λ to ∞); and (c) update the physical-layer information in the physical-layer module by recording the signal power as well as noise powers on each link along the lightpath P_λ. Otherwise, let $w = w + 1$, and go to Step 2.
- Step 5: Stop the procedure.

When a lightpath of a connection is torn down, the link states of all links along the lightpath will be changed and need to be updated. The wavelength resources are released in the corresponding WG. And the signal power and noise power are reset to zero on the wavelength channel used for this connection in the physical-layer module. Note that the routing technique in the lightpath-computation procedure could be optimized, such as using the multiple-shortest-paths technique. Using the multiple-shortest-paths technique, multiple routes are computed on each WG as candidates. In the worst case, the shortest paths on all wavelengths may have poor quality, so allowing multiple paths on each w will provide additional choices in the lightpath-computation procedure. Thus, it provides more chance to reduce the physical-layer blocking. Moreover, randomly choosing a lightpath among available candidate lightpaths might reduce some inter-channel impairment, e.g., channel crosstalk; however, such random-routing technique would not perform better than IABP. In IABP, for each connection request, the candidate lightpath is the shortest one among shortest paths on all wavelengths. Thus, IABP allocates wavelengths without order, but under a certain control that chooses the lightpath with minimum distance from all available shortest paths.

15.6 Characteristics of Impairment-Aware RWA

Compared with traditional impairment-unaware RWA, an impairment-aware RWA takes the physical-layer impairments into consideration while computing and setting up a lightpath. The criterion of resource acceptability is different for call admission. The call-admission criterion of the traditional algorithms depends on resource availability, whereas the criterion of the impairment-aware algorithms depends not only on resource availability but also on the lightpath's signal quality. In the impairment-aware algorithm, only the qualified available lightpaths can be used to set up a call request.

Note that impairment-aware RWA algorithms separate lightpath computation and lightpath verification in the network layer and physical layer, respectively. Thus, the network-layer module can compute routes with some performance optimization such as loading balancing and improving resource utilization.

In order to simplify our algorithm design and reduce its computation time, other potential impairments that are not treated here include: polarization-dependent loss/gain (PDL/PDG), residual-dispersion accumulation, fiber nonlinearities, filter concatenation, power divergence, signal transient, etc. With current transmission technologies, where up to 10-Gbps data traffic can be readily carried by each wavelength channel in today's backbone networks, the more accurate is the information from the physical-layer model, the more efficient will be the lightpath provisioning. Therefore, our algorithms, which combat some of the primary and dominant impairment effects, give a lower limit on network-performance improvement. Other algorithms could be more efficient when more physical impairments are incorporated in the lightpath-assignment procedure. It would be a challenge to determine how to combine all impairment effects and present them as an aggregate link parameter for lightpath routing since the effects from different kinds of impairments may not be additive. Moreover, some nonlinear effects on an end-to-end lightpath might not be estimatable from the link parameters along the lightpath [StCT01].

When the data rate increases to 40 Gbps and beyond, impairments become even more troublesome. The physical size of the transparent network is mainly limited by the impairments, requiring new transmission techniques for carrying data traffic in such higher-speed networks [Tomk02, Agra02]. While enabling the transfer of data traffic, with impairment consideration, the impairment-aware algorithms could make connection provisioning more efficient and hence lead to more efficient resource utilization.

Figure 15.6 shows connection blocking probability vs. network offered load (Erlangs) for algorithms using a sample network with 24 nodes and 16 wavelength on each fiber links, which is shown in Fig. 15.5. It takes into

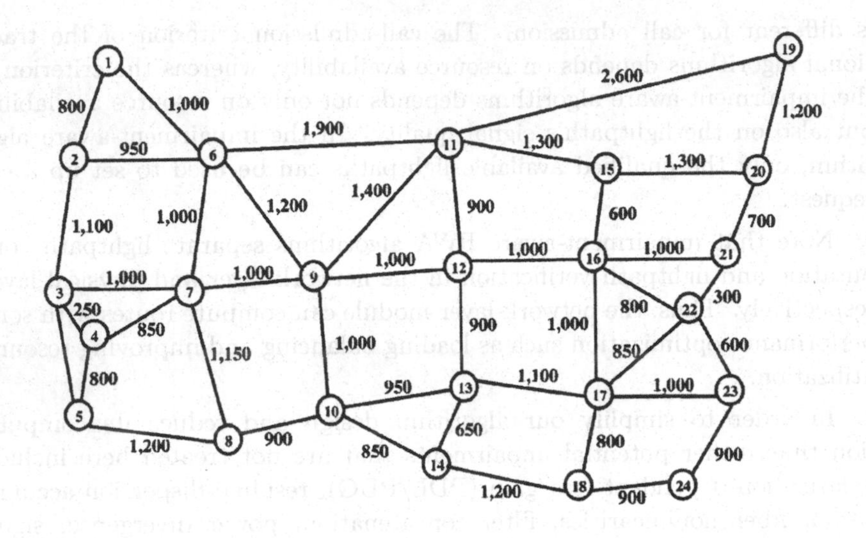

Figure 15.5 A sample mesh network with fiber lengths (in km) marked on each link.

account resource blocking, i.e., no free resource for setting up a connection, and physical-layer blocking, i.e., a connection cannot satisfy signal-quality requirement. The results show that: (1) impairment-unaware algorithms (TBP and TFF)[3] have higher blocking probability in a realistic network than in an ideal network due to the effect of transmission impairments; and (2) significant improvement in blocking can be achieved by our proposed algorithms (IABP and IAFF), as compared with impairment-unaware algorithms, in a realistic network.

As shown in Fig. 15.6, for example, at a load of 220 Erlangs, 14.9% and 34.4% improvement in blocking probability are achieved by using IABP and IAFF, respectively. This is because the proposed algorithms obtain information of a lightpath's quality from the physical layer, and they take impairment effects into consideration in the RWA stage. Hence, the algorithms automatically provision signal-quality-guaranteed connections and some unnecessary physical-layer blocking situations can be avoided. Furthermore, a lightpath with good signal quality is preferred through impairment constraints such that the network resources are under intelligent control and are

[3]TBP = Traditional Best Path; TFF = Traditional First Fit.

used more efficiently in the impairment-aware algorithms. Finally, we also observe that blocking probabilities of our proposed algorithms are very close to those of traditional RWA in an ideal network.

Note that the impairment-aware first-fit algorithm (IAFF) has lower blocking probability than the traditional first-fit algorithm (TFF) in an ideal network. This is because, for the traditional first-fit algorithm in an ideal network, lightpath assignment is impairment unaware, and only depends on resource availability; thus, the first lightpath on the free lowest-numbered wavelength will be chosen for a connection request. However, for impairment-aware first-fit algorithm (IAFF), if the current first-available lightpath does not satisfy the quality requirement, it will try to find the next available lightpath which might be shorter, and hence a better alternative for the call.

In current WDM networks, there are different network configurations, e.g., employing the current technology of EDFAs for in-line amplification. For example, instead of the DRA in Fig. 15.3, an EDFA which has 20-dB small-signal gain and 4-dB noise figure, is used for fiber-loss compensation in an amplification span, and each node has the crosstalk ratio of -30 dB. The corresponding simulation results are shown in Fig. 15.7. In this case, the proposed algorithms are not close to the result of the algorithm in an ideal network due to the limitations of the transmission technology. But the blocking probability can still be significantly reduced by using the impairment-aware algorithms as compared to the traditional algorithms in a realistic network.

Using impairment-aware RWA increases the computational complexity due to signal-quality estimation; however, it provides much better network performance (i.e., lower blocking as shown in Fig. 15.6). The tradeoffs between optimality of blocking probability and computational cost is considered as well.

Figure 15.8 shows that the average number of trials for processing the signal-quality estimation for each call request is very close to 1 for both impairment-aware algorithms (IABP and IAFF). This means that the computational cost of signal-quality estimation is reasonable for performance improvement, i.e., for reducing blocking probability (in Fig. 15.6). For a range of network offered loads, the number of trials for signal-quality estimation in our impairment-aware algorithms increases when the offered load is low, and decreases when offered load is high. This is because, in case of low traffic load, enough resources are available to ensure low call-blocking probability so that the algorithms can select alternate resources (route and/or

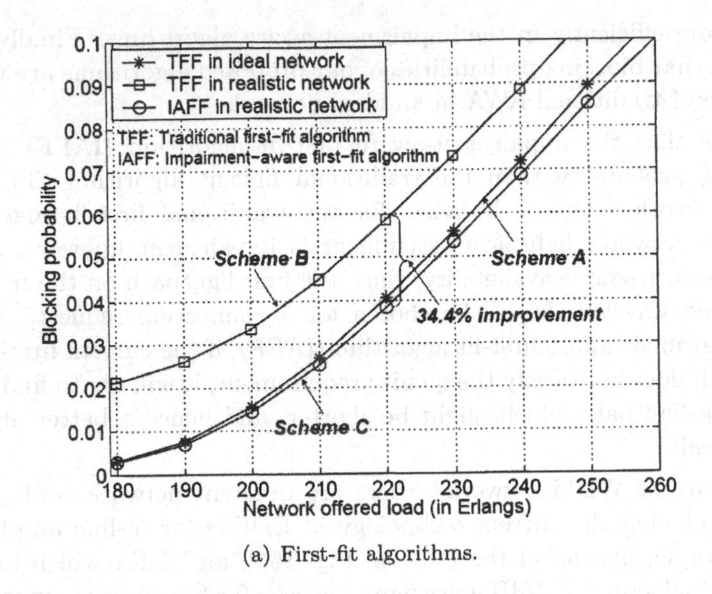

(a) First-fit algorithms.

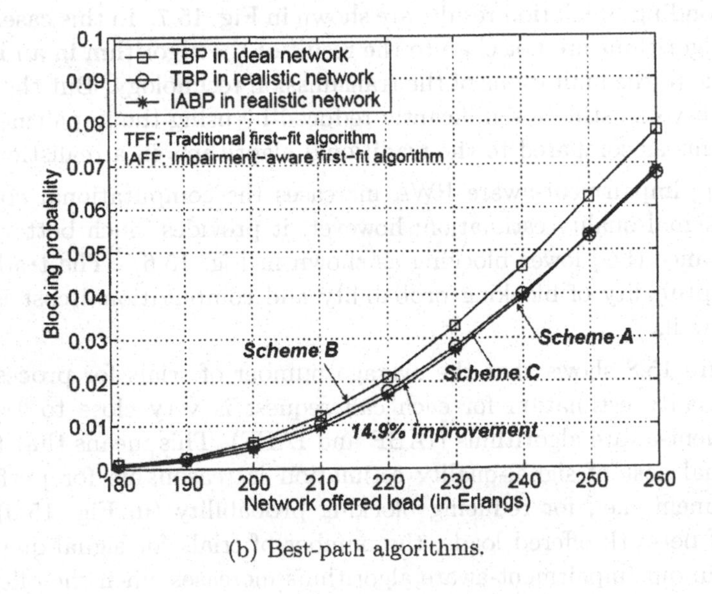

(b) Best-path algorithms.

Figure 15.6 Connection blocking probability vs. network offered load (in Erlangs).

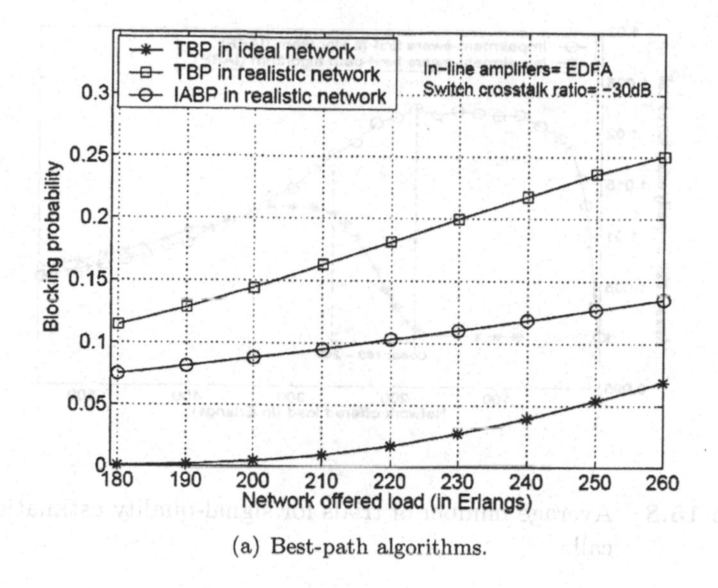

(a) Best-path algorithms.

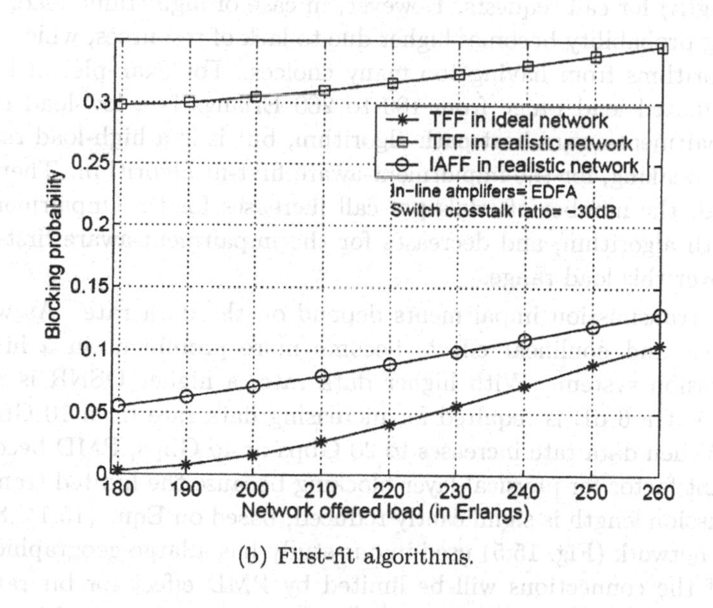

(b) First-fit algorithms.

Figure 15.7 Comparison of connection-blocking probability between algorithms employing in-line EDFA.

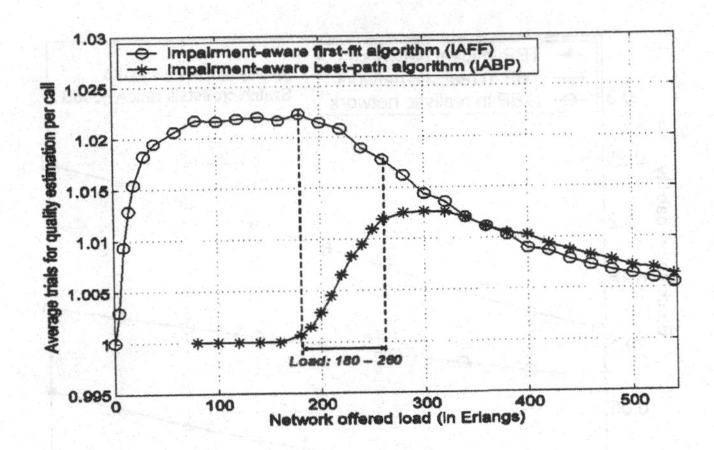

Figure 15.8 Average number of trials for signal-quality estimation per call.

wavelength) for call requests. However, in case of high traffic load, the call-blocking probability becomes higher due to lack of resources, which prevents the algorithms from having too many choices. For example, in Fig. 15.8, the simulated load range from 180 to 260 Erlangs is a low-load range for the impairment-aware best-path algorithm, but it is a high-load range, relatively speaking, for the impairment-aware first-fit algorithm. Therefore, as observed, the number of trials per call increases for the impairment-aware best-path algorithm, and decreases for the impairment-aware first-fit algorithm over this load range.

The transmission impairments depend on the data rate. As we know, the linear and nonlinear effects become more prominent in a high-speed transmission system. With higher data rate, a higher OSNR is required, e.g., an extra 6 dB is required for increasing data rate from 10 Gbps to 40 Gbps. When data rate increases to 20 Gbps or 40 Gbps, PMD becomes the dominant factor for physical-layer blocking because the limited transparent-transmission length is significantly reduced, based on Eqn. (15.1). Since the sample network (Fig. 15.5) used in our study has a large geographical scale, most of the connections will be limited by PMD effect for bit rates ≥ 20 Gbps. To illustrate the blocking performance of our proposed impairment-aware RWA algorithms for different channel bit rates, we study the proposed schemes for different network scales. We use a scalar β ($\beta \leq 1$), with which we multiply the original length of all fiber links inside the original network in Fig. 15.5, to change the span of the network for our study.

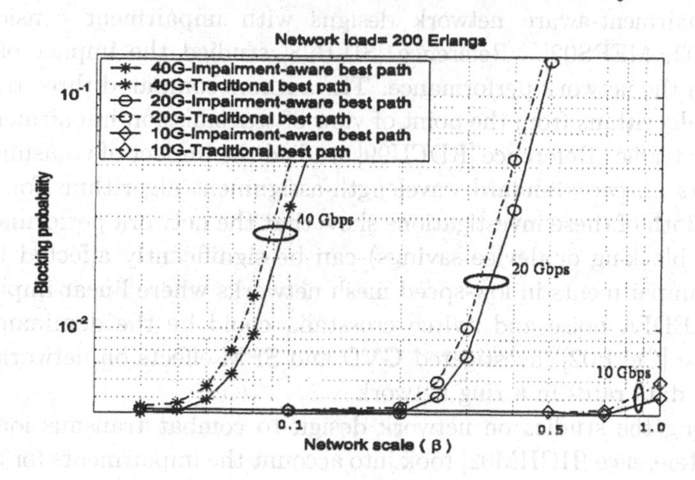

Figure 15.9 Effect of channel bit rate on algorithms in a realistic network.

In Fig. 15.9, we show the network scale vs. blocking probability for three different channel data rates, i.e., 10 Gbps, 20 Gbps, and 40 Gbps, using our proposed impairment-aware best-path RWA algorithm (IABP) and traditional best-path RWA algorithm (TBP). Results show that blocking probability for 40-Gbps data channel increases very fast when the network scale increases, whereas the blocking probability for 10 Gbps increases much slower when network size increases. This also indicates that the network may need new transmission technologies to enable it to carry high-speed data traffic. Overall, while enabling data traffic, the blocking probabilities for the proposed algorithm (IABP) are smaller than that of the traditional RWA algorithm (TBP) in a realistic network for all three different channel data rates.

15.7 Other Related Issues

15.7.1 Other Problems with Impairment Consideration

The study of the transmission impairments in an all-optical network is now attracting more attention from researchers for different network-design problems [SILB98, CeFP02, HGHM02, MEPS02]. There are two foci in these research works: effects of impairments on the performance [SILB98, CeFP02], and impairment-aware network designs with impairment considerations [HGHM02, MEPS02]. Reference [SILB98] studied the impact of impairments on the network performance. The authors compared three traditional routing algorithms from the point of view of transmission-impairment effects for static traffic. Reference [RDFH99] studied the impact of transmission impairments on two standard wavelength-assignment algorithms for dynamic traffic. Both of these investigations show that the network performance (i.e., network blocking or device savings) can be significantly affected by transmission impairments in low-speed mesh networks where linear impairments, such as EDFA noise and switch crosstalk, could be the dominant effects. Reference [CeFP02] investigated GVD and SPM effects on network cost for different data rates in a ring network.

Among the studies on network design to combat transmission impairments, Reference [HGHM02] took into account the impairments for lightpath assignment control for a low-speed optical network. With increase in channel bit rate, studies have been advanced to the high-speed transmission system (data rate $\geq$ 10 Gbps). In this regime, the linear (such as PMD, GVD, etc.) and nonlinear impairments become more prominent [StCT01, Will02]. In [AlET01b], the network cost, in terms of number of regenerators and fibers, was optimized by trying to carry traffic through "good" fibers which have lower PMD effects. Reference [MEPS02] showed that lightpath assignment strongly depends on transmission techniques for impairment compensation. In this work, a traffic-grooming problem was also investigated under signal-quality constraints.

Besides combating transmission impairments in all-optical transparent networks, some related studies focused on regenerator placement or savings for opaque or translucent networks, where some or all nodes perform signal regeneration to clean up the degraded signal [RFDM99, AlET01a, YaRa02].

15.7.2 Security Issues in All-Optical Networks

Network security is becoming a very sensitive and important topic for equipment manufacturers and network operators [MMBF97, ToVC02, MaTT03, RoKi03, DeAt04]. In a transparent optical network, security is even more complex since the optical signals are not regenerated as in an opaque network; therefore, the faults and attacks at the physical layer are more difficult to detect and isolate without significantly affecting the overall network performance [MMBF97, MaTT03]. As far as a secure solution is concerned, an optical network should support several critical security features, such as AAA (authentication, authorization, and accounting), the balance of power between lightpaths, as well as compliance with security protocols and standards [ToVC02, MaTT03, RoKi03, DeAt04]. To support security in all-optical networks, a security solution should be a process instead of a product, which should be able to continually evolve and change to accommodate new threats or requirements throughout the entire network infrastructure. In a security solution, all access points of the network would be potential targets, and must be protected accordingly. Thus, optical networks can deal with the prevention, detection, and reaction to potential attacks and faults of all access points and links. A key issue is to distinguish attacks from faults. One possible parameter can be the time scale in which faults/attacks occur, since faults due to natural fatigue and aging of the components may be slower than an attack which progresses much faster. Another way to distinguish faults from attacks is the frequency of their occurrence, i.e., a fault may occur once and remain as fault until it is repaired, whereas an attack may appear and disappear, thereby increasing the difficulty of its detection.

There is a possible solution, namely to employ optical label switching (see Chapter 18). An optical label could be applied to implement secure signaling, routing protocols, and control technologies in an optical network. In an all-optical network, security of access can be identified and empowered by intelligent control, depending on optical labels. Because an optical label is network-wide and globally unique, the concern of network security might be well solved by this characteristic. Furthermore, high-speed and convenient resource access should be guaranteed by using an optical label or an optical code. Moreover, it is helpful to balance power between lightpaths and wavelength assignment between links in quantity by using an optical label. In such a solution, security of optical networks would be modular in order to be scalable, flexible, and cost-effective.

15.8 Summary

In this chapter, we presented new RWA algorithms for efficient connection provisioning with signal-quality guarantees in an all-optical WDM mesh network operating with high-speed wavelength channels. Under high data rate, the impact of transmission impairments on a lightpath's quality can become very prominent, requiring appropriate techniques in both the physical layer and the network layer to mitigate the impairment effects on network performance. Therefore, focusing on network-layer techniques, a novel hierarchical RWA model is developed, where the OSNR and PMD effects were estimated in the physical layer, and regarded as metrics for high-speed connection provisioning in the network layer. Such impairment-aware-RWA algorithms automatically consider the effects of impairments when setting up a lightpath. The major network performance of connection-blocking probability was measured under different in-line amplification scenarios including all-DRA and all-EDFA amplifications. With signal-quality consideration, as compared to algorithms that are not impairment aware in a realistic optical network, the impairment-aware algorithms efficiently provide signal-quality-guaranteed connections while significantly reducing connection-blocking probability, better utilizing network resources, and having a reasonable computational requirement.

Exercises

15.1. From data-transparency perspective, what are the three network models that can be used for a WDM mesh network? What are their main characteristics?

15.2. What are the major issues in a transparent network?

15.3. What are physical-layer blocking and network-resource blocking?

15.4. What are PMD blocking and OSNR blocking?

15.5. What is transparent transmission length?

15.6. How could physical-layer devices affect the performance of a RWA algorithm?

15.7. Please describe the major characteristics of an impairment-aware RWA.

15.8. Besides IAFF and IABP, design another impairment-aware RWA. Specify the algorithm as shown for IAFF and IABP. Qualitatively compare your algorithm with IAFF and IABP.

15.9. What are the major impairment effects in an optical network with 40-Gbps data rate?

15.10. What is the transparent transmission length at 40 Gbps? Use appropriate data from Table 15.1. How does the transparent transmission length change at 20 Gbps? and at 10 Gbps?

15.11. Let us apply the results from the above problem to the network topology in Fig. 15.5. Let us consider *PMD only* and no other impairments. At 40 Gbps, which node pairs can communicate with each other?

15.12. Repeat the above problem for a data rate of 20 Gbps. Now, Node 1 can communicate with which other nodes (assuming no regeneration anywhere in the network)?

15.13. Repeat the above problem for Nodes 5, 19, and 24.

15.14. Based on your results in the last two problems, determine the nodes at which you will place the minimum number of regenerators so that all nodes can communicate with one another at a data rate of 20 Gbps.

Chapter

16

Network Control and Management

16.1 Introduction

One of the challenges involved in designing wavelength-routed networks with dynamic traffic demands is to develop efficient algorithms and protocols for establishing lightpaths. The algorithms must be able to select routes and assign wavelengths to connections in a manner which efficiently utilize network resources and which maximize the number of lightpaths that are established. Signaling protocols for setting up lightpaths must effectively manage the distribution of control messages and network-state information in order to establish a connection in a timely manner. Typically, a *network control and management protocol* is employed to perform the routing and wavelength assignment (RWA) and signaling tasks. Thus, the present chapter is complementary to Chapter 7, where *RWA algorithms* were studied.

Another issue in dynamic lightpath establishment is how to initiate requests for lightpath establishment and lightpath termination. There are a number of possible approaches for generating a connection request. For example, a customer can initiate a connection request by clicking on a web page or by calling up a service provider. A connection request can also be initiated by an Internet Protocol (IP) router or other network devices.

Recently, there has been much effort in developing the aforementioned protocols and standards for providing on-demand establishment of lightpaths. More specifically, a number of organizations are currently working on developing standards for facilitating dynamic lightpath establishment in optical networks. For example, the International Telecommunications Union (ITU) is developing the architecture for Automatically-Switched Optical Networks (ASON). The Internet Engineering Task Force (IETF) is developing the Generalized Multi-Protocol Label Switching (GMPLS) protocol. The Optical Internetworking Forum (OIF) blends together some of the work of IETF and ITU. For more information on these bodies, please refer to their websites – *http://www.ietf.org*, *http://www.itu.int*, and *http://www.oiforum.com*, respectively.

Optical Internetworking Forum (OIF) is working on the standardization of the user-to-network interface (UNI) and external network-to-network interface (E-NNI); and it has released UNI versions 1.0 and 2.0 and E-NNI 1.0. The Optical Domain Service Interconnect (ODSI), another standards forum, did some early work on standardizing the interfaces which would allow client networks and devices to interact with the optical network [Copl00], and lent some of its work to OIF. Note that UNI does not specify how the lightpaths are actually established within the optical network, but simply defines how a client would request a lightpath or release a lightpath from the optical network. E-NNI is also important to the telecommunications industry at large because it was derived directly from carrier requirements. For example, at a time when the industry needs direction and structure with rapidly changing technology, OIF's E-NNI 1.0 helps carriers establish a way for multiple networks to interconnect.

Automatically-Switched Optical Network (ASON) was developed by the ITU-T. The work was initiated in response to a demand from ITU members to create a complete definition of the operation of automatically-switched transport networks, management, control, data plane, etc. ASON is not a protocol or collection of protocols. But it draws on concepts from protocols used heavily in telecom transport networks, such as SONET/SDH, Signaling System 7 (SS7) and ATM. ASON cannot be directly implemented since it is a reference architecture which defines the components in an optical control plane and the interactions between those components. It also identifies which of those interactions will occur across a multi-vendor divide, and therefore require standardized protocols. Other areas are intentionally not standardized in order to allow vendors or operators to provide their own "value addition". As with most ITU projects, ASON was (and continues to be) developed in a top-down fashion, starting with a full and explicit list of requirements, moving on to high-level architecture, and then to individual component architecture. Only when the component architecture is defined in detail are protocols held up to the architecture to see if they fit. Any protocol that fits the requirements of the component architecture can potentially get the ASON "stamp of approval".

The Internet Engineering Task Force (IETF) is currently working on the Generalized Multi-Protocol Label Switching (GMPLS) mechanism, a generalized control framework for establishing various types of connections, including lightpaths, in IP-based networks. GMPLS is derived from *Multi-protocol label switching* (MPLS) which is a control framework currently being developed as a standard to enable fast switching in IP networks. MPLS control mechanisms can be used to establish a label-switched path (LSP) between two non-neighboring IP routers, enabling packets to bypass IP routers at intermediate nodes. The primary signaling mechanism for establishing an LSP in MPLS is the label-distribution protocol (LDP). The concept of MPLS can be extended to wavelength-routed optical networks as MPλS (multi-protocol lambda switching), which is further developed as GMPLS. Currently, there is much work being done by the IETF to modify existing routing and signaling protocols in order to support GMPLS. In particular, the IETF is focusing on enhancements to the Open Shortest Path First (OSPF) routing protocol, and the Constraint-Based Routing Label-Distribution Protocol (CR-LDP) and Resource Reservation Protocol (RSVP) signaling protocols. A summary of the proposed enhancements can be found in [BDLT01b]. OSPF is a link-state protocol in which the state of each link in the network is periodically broadcast to all nodes in the form of link-state advertisements (LSAs). The nodes may then make their routing decisions based on this information. RSVP is an IP-based protocol which is used for signaling the resource requirements of an application to intermediate routers. CR-LDP is a protocol which enables the distribution of control messages for establishing LSPs. CR-LDP utilizes constraint-based routing and runs on top of the Transmission Control Protocol (TCP) for reliability. While the current focus of the IETF is on a few specific protocols, GMPLS itself is not restricted to any single routing or signaling protocol. Furthermore, protocols such as OSPF, CR-LDP, and RSVP are flexible and can lend themselves to the implementation of various routing and signaling schemes for lightpath establishment. In this chapter, we will discuss some of these underlying routing and signaling schemes which can be implemented within a GMPLS framework, and we discuss these schemes from a performance perspective.

The rest of this chapter is organized as follows. We first present the basic functions of network control and management. Then, we discuss dynamic routing and wavelength assignment algorithms for WDM networks. Various signaling protocols for reserving resources and setting up lightpaths in a wavelength-routed network are presented next, and two control mechanisms are compared and evaluated. Fault management is discussed thereafter. Finally, we summarize this chapter.

16.2 Basic Functions of Network Control and Management

Basic functions of network control and management include resource management, route computation, signaling, and fault management, as outlined below.

1. **Resource management**:
   Resource-management protocols determine how the network resources are discovered, updated, and maintained in the link-state databases (for distributed control) of a network node or by the network control and management system (for centralized control). Resource discovery can be achieved via a Link-Management Protocol (LMP), which allows neighboring nodes to exchange identities and link information, and to also negotiate the functions to be supported between the nodes. Resource update is usually performed via the LSAs of a routing protocol such as OSPF.

2. **Route computation**:
   Route computation determines how the route of a connection request is selected according to a carrier's traffic-engineering objectives. Traffic-engineering policies reflects the intention of a network operator on how to allocate network resources to a given request if multiple routes are available.

3. **Signaling**:
   Signaling protocols determine how to set up, modify, and tear down a

connection. The receipt of a signaling message usually triggers some local action, e.g., allocation of time slots or wavelengths when receiving a connection-setup message. Two protocols – resource reservation protocol (RSVP) and constraint-based routing label distribution protocol (CR-LDP) with traffic-engineering (TE) extensions – have been proposed as the standard signaling protocols in the GMPLS control plane. In a heterogeneous WDM network, the route computed for a connection request will be composed of a sequence of intermediate node identifiers as well as link bundle id. Since multiple candidate links may exist in a link bundle, an intermediate node needs to select one for the connection request when it needs to configure the optical crossconnect (OXC) and establish the connection. If the link is a bundled lightpath link, then, based on the available capacity of each lightpath in the bundle and the bandwidth requirement of the request, different link-selection schemes can be used, e.g., random selection, first-fit selection, best-fit selection, etc.

4. **Fault management**:
   In an optical network, the high capacity of a link has the problem that a link failure can potentially lead to the loss of a large amount of data (and revenue). So, we need to develop appropriate protection and restoration schemes which minimize the data loss when a link failure occurs (see Chapter 11). Relative to the optical layer, upper layers of protocols (such as ATM, IP, and MPLS) have their own procedures to recover from link failures [ADDH94, Huit95, MSOH99]. However, the recovery time for upper layers is significantly larger (on the order of seconds), whereas we prefer that the fault-recovery times at the optical layer should be on the order of milliseconds in order to minimize data losses. Furthermore, it is beneficial to consider fault-recovery mechanisms in the optical layer for the following reasons [Gers98]: (a) the optical layer can efficiently multiplex protection resources (such as spare wavelengths and fibers) among several higher-layer network applications, and (b) survivability at the optical layer provides protection to higher-layer protocols which may not have built-in fault recovery.

   Essentially, there are two types of fault-recovery mechanisms [Gers98, Wu92, Wu98] (see also Chapter 11). If backup resources (routes and wavelengths) are pre-computed and reserved in advance, we call it a *protection scheme* [RaMu99a, FCMJ99]. Otherwise, when a failure occurs, if another route and a free wavelength have to be discovered dynamically for each interrupted connection, we call it a *restoration scheme* [RaMu99b, IrGr00]. A restoration scheme is usually more resource-efficient [IrGr00], while a protection scheme has a faster recovery time and provides guaranteed recovery ability. We consider protection schemes in this chapter.

In the following sections, we shall investigate route computation, signaling, and fault management.

16.3 Dynamic Routing and Wavelength Assignment

When lightpaths are established and taken down dynamically, routing and wavelength assignment (RWA) decisions must be made as connection requests arrive to the network. It is possible that, for a given connection request, there may be insufficient network resources to set up a lightpath, in which case the connection request will be blocked. The connection may also be blocked if there is no common wavelength available on all of the links along the chosen route. Thus, the objective in the dynamic situation is to choose a route and a wavelength which maximizes the probability of setting up a given connection, while at the same time attempting to minimize the blocking for future connections. Similar to the case of static lightpaths, the dynamic RWA problem can also be decomposed into a routing subproblem and a corresponding wavelength-assignment subproblem (see Chapter 7).

Approaches to solving the routing subproblem can be categorized as being either static or adaptive, and as utilizing either global or local network-state information.

16.3.1 Fixed Routing and Fixed-Alternate-Path Routing

Two examples of algorithms which utilize static routes are fixed routing and fixed-alternate-path routing.

In *fixed routing*, a single fixed route is predetermined for each source-destination pair.

In *fixed-alternate-path routing*, multiple fixed routes are pre-computed for each source-destination pair and stored in an ordered list at the source node's routing table. As a connection request arrives, one route is selected from the set of pre-computed routes.

Both of these approaches are much simpler to implement compared to *adaptive routing schemes*, but may suffer from higher connection blocking.

16.3.2 Adaptive Routing Based on Global Information

Adaptive routing approaches increase the likelihood of establishing a connection by taking into account the network-state information. For the case in which global information is available, routing decisions may be made with full information as to which wavelengths are available on each link. In order to find an optimal route, a cost may be assigned to each link based on wavelength availability, and a *least-cost routing algorithm* may be executed.

Adaptive routing based on global information may be implemented in either a centralized or distributed manner. In a centralized algorithm, a single entity, such as a network management system (NMS), maintains complete network-state information, and is responsible for finding routes and setting up lightpaths for connection requests. Since a centralized entity manages the entire network, there does not need to be a high degree of coordination among nodes; however, a centralized entity becomes a possible single point of failure.

A distributed, adaptive routing algorithm based on global information may be implemented in a number of ways. In a *link-state approach*, each node in the network must maintain complete network-state information [RaSe97]. Each node may then find a route for a connection request in a distributed manner. Whenever the state of the network changes, all of the nodes must be informed. Therefore, the establishment or removal of a lightpath in the network may result in the broadcast of update messages to all nodes in the network. The need to broadcast update messages may result in significant control overhead, especially if lightpaths are being established and removed at a high rate. Furthermore, it is possible for a node to have outdated information, and for the node to make an incorrect routing decision based on this information.

A distance-vector approach, namely, a *distributed-routing approach* to routing with global information is also possible [ZJSM01]. This approach does not require that each node maintain complete link-state information as in [RaSe97]; instead, it requires that each node maintain a routing table which indicates, for each destination and on each wavelength, the next hop to the destination and the distance to the destination. The approach relies on a distributed Bellman-Ford algorithm to maintain the routing tables. Similar to [RaSe97], this scheme also requires nodes to update their routing-table information whenever a connection is established or taken down. This update is accomplished by having each node send routing updates to their neighbors periodically or whenever the status of the node's outgoing links changes. We evaluate the above two approaches using several performance metrics in a later section.

Another form of adaptive routing is *least-congested-path (LCP) routing* [ChYu94]. The congestion on a link is measured by the number of wavelengths available on the link. Links which have fewer available wavelengths are considered to be more congested. The congestion on a path is indicated by the congestion on the most-congested link in the path. Similar to alternate routing, for each source-destination pair, a sequence of routes is preselected. Upon the arrival of a connection request, the least-congested path among the pre-determined routes is chosen. It has been shown in [ChYu94] that using shortest-path routing first and LCP second for tie-breaking works better than using LCP alone.

Although routing schemes based on global knowledge must deal with the task of maintaining a potentially large amount of state information which may be changing constantly, these schemes often make the most optimal routing decisions if the state information is up-to-date. Thus, global-knowledge-based schemes may be well suited for networks in which lightpaths are fairly static and do not change much with time.

16.3.3 Adaptive Routing Based on Neighborhood Information

In the global-information-based LCP approach, all links on all candidate paths have to be examined in choosing the least-congested path. Each node would be required to either maintain complete state information, or the information would need to be gathered in real time, as the lightpath is being established. A variant of LCP is proposed in [LiSo99], which only examines the first k links on each path (referred to as the source node's *neighborhood* information), where k is a parameter in the algorithm. It has been shown that, when $k = 2$, this algorithm can achieve similar performance as fixed-alternate routing.

16.3.4 Adaptive Routing Based on Local Information – Deflection Routing

Another approach to adaptive routing with limited information is *deflection routing*, or *alternate-link routing* [JuXi00]. This routing scheme chooses among alternate links on a hop-by-hop basis rather than choosing from alternate routes on an end-to-end basis. The routing is implemented by having each node maintain a routing table which indicates, for each destination, one or more alternate outgoing links to reach that destination. These alternate outgoing links are pre-computed and may be ordered such that a connection request will preferentially choose certain links over other links as long as wavelength resources are available on those links. If resources are unavailable on the preferred link, then an alternate link is chosen for the route. Other than a static routing table, each node will only maintain information regarding the status of wavelength usage on its own outgoing links. Hence, there are no update messages in the network and control bandwidth demand is greatly decreased.

16.3.5 Wavelength Assignment

In general, in a wavelength-continuous network, if there are multiple feasible wavelengths between a source node and a destination node, then a wavelength-assignment algorithm is required to select a wavelength for a given lightpath.

One example of a simple, but effective, wavelength-assignment heuristic is *first-fit* (see Chapter 7). In first-fit, the wavelengths are indexed, and a lightpath will attempt to select the wavelength with the lowest index before attempting to select a wavelength with a higher index. By selecting wavelengths in this manner, existing connections will be packed into a smaller number of total wavelengths, leaving a larger number of wavelengths available for longer lightpaths.

A more detailed review of wavelength-assignment algorithms can be found in [ZaJM00b] and Chapter 7.

16.4 Signaling and Resource Reservation

In order to set up a lightpath, a signaling protocol is required to exchange control information among nodes and to reserve resources along the path. In many cases, the signaling protocol is closely integrated with the routing and wavelength-assignment protocols. Signaling and reservation protocols may be categorized based on whether the resources are reserved on each link in parallel, reserved on a hop-by-hop basis along the forward path, or reserved on a hop-by-hop basis along the reverse path. Protocols will also differ depending on whether global information is available or not.

16.4.1 Parallel Reservation

In [RaSe97], the control scheme reserves wavelengths on multiple links in *parallel.* The scheme, which is based on link-state routing, assumes that each node maintains global information on the network topology and on the current state of the network, including information regarding which wavelengths are being used on each link. Based on this global information, the node can calculate an optimal route to a destination on a given wavelength. The source node then attempts to reserve the desired wavelength on each link in the route by sending a separate control message to each node in the route. Each node that receives a reservation-request message will attempt to reserve the specified wavelength, and will send either a positive or negative acknowledgment back to the source. If the source node receives positive acknowledgments from all of the nodes, it can establish the lightpath and begin communicating with the destination.

The advantage of a parallel reservation scheme is that it shortens the lightpath establishment time by having nodes process reservation requests in parallel. The disadvantage is that it requires global knowledge, since both the path and the wavelength must be known ahead of time.

16.4.2 Hop-by-Hop Reservation

An alternative to parallel reservation is hop-by-hop reservation in which a control message is sent along the selected route one hop at a time. At each intermediate node, the control message is processed before being forwarded to the next node. When the control message reaches the destination, it is processed and sent back towards the source node. The actual reservation of link resources may be performed either while the control message is traveling in the *forward* direction towards the destination, or while the control message is traveling in the *reverse (or backward)* direction back towards the source.

Forward Reservation

In a *forward-reservation scheme*, wavelength resources are reserved along the forward path to the destination on a hop-by-hop basis. The method of reserving wavelengths depends on whether or not global information is available at the source node. If the source node is maintaining complete state information, then it will be aware of which wavelengths are available on each link. Assuming that the state information is current, the source node may then send a connection set-up message along the forward path, reserving the same available wavelength on each link in the path. The distributed-routing approach [ZJSM01] is an example of this scheme.

For the case in which a node only knows the status of its immediate links, the source node may utilize a conservative reservation approach, choosing a single wavelength and sending out a control message to the next node in order to attempt to reserve this wavelength along the entire path; however, there is no guarantee that the selected wavelength will be available along every link in the path. If the wavelength is blocked, the source node may select a different wavelength and re-attempt the connection. The limitation of this approach is that it may result in a long set-up time, since it may take several attempts before a node can establish a lightpath.

An alternate approach to maximizing the likelihood of establishing a lightpath in a forward-reservation scheme is to use an aggressive reservation scheme which over-reserves resources [YMGM99]. When a reservation message arrives at a node, the node reserves all wavelengths which are available on all of the links traversed so far. When the reservation message reaches the destination node, the destination then chooses one wavelength from the wavelengths reserved along the entire path and releases the reservations on the remaining wavelengths.

The drawback of this forward-reservation scheme is that network resources are being over reserved for a short period of time, which may lead to the blocking of subsequent connection requests and lower network utilization.

Backward Reservation

To prevent the over-reservation of resources altogether, reservations may be made after the control message has reached the destination and is headed back to the source. Such reservation schemes are referred to as *backward-reservation* schemes [YMGM99]. In backward reservation, the source node sends control packets to the destination without reserving any resources. These control packets will collect information about wavelength usage along one or more paths, and the destination will then utilize this information to decide on a route and a wavelength. The destination then sends a reservation message to the source node along the chosen route, and this reservation message will reserve the appropriate network resources along the way [YMGM99, LiSo99].

One possible drawback of a backward-reservation scheme is that, if multiple connection are being set up nearly simultaneously, it is possible that a wavelength that was available on a link in the forward direction may be taken by another connection request and may no longer be available when the reservation message traverses the link in the reverse direction.

16.5 Two Connection-Management Approaches

16.5.1 Protocols

We compare two connection-management approaches using various metrics. The first approach, namely, the *link-state approach*, is proposed in [RaSe97]. The second approach, namely, the *distributed-routing approach*, is presented in [ZJSM01]. Both approaches have been described in the previous sections.

We assume that the signaling messages are delivered by a packet-switched control network. This control network is implemented on an out-of-band supervisory channel which can potentially operate on its own wavelength. (Alternately, the control network may employ sub-wavelength-granularity channels on the various links of the network using *traffic grooming* (see Chapter 13)). The control layer has the same topology as the physical network[1] and all packets are routed by shortest paths.

Since the control network is packet-switched, the signaling method described in [RaSe97] consumes much control bandwidth and involves longer connection set-up delay compared to a hop-by-hop signaling method [ZJSM01]. In order to examine the routing issues more closely, we modify the signaling method in [RaSe97] to be hop-by-hop signaling. Moreover, in the modified link-state approach, each node computes *only* its next-hop based on the topology information. This is different from [RaSe97] in which source routing is applied, i.e., the full route is determined at the source node. Therefore, the only difference between the compared approaches is how the routing information is updated and how RWA is performed.

The following summarizes the two approaches under consideration:

- **Routing**: In both approaches, routing is done with global information. However, in the link-state approach, each node maintains a database of network topology and the wavelength state on each link; LSAs are used for updating the topology and wavelength-usage information. In the distributed-routing approach, a distance-vector protocol is executed to keep the routing tables up-to-date.

[1] In the control layer, we can set up the lightpaths in a way that every lightpath occupies a single fiber link. Hence, the virtual topology in the control layer is exactly same as the physical topology.

- **Wavelength assignment**: A first-fit approach is used in both cases (see Chapter 7).

- **Signaling procedures**: A similar signaling procedure is used for both approaches. After the source node determines the route or the next hop, it sends out a RESERVE message to its next hop. Each intermediate node will examine the requested resources. If the resources are available, the node will reserve the resource and send a RESERVE message to the next hop; otherwise, the node sends a RESERVE-NACK back to the source. After the destination receives the RESERVE message, it checks whether or not it has a spare receiver on the requested wavelength. If the node has a spare receiver, it sends a RESERVE-ACK back to its previous hop. Otherwise, it sends a RESERVE-NACK. The switches (OXCs) are configured when a node receives a RESERVE-ACK. The node is also responsible for delivering the RESERVE-ACK to its previous hop on the route. If a node receives a RESERVE-NACK, it releases the reserved resources. When the source receives a RESERVE-NACK, it performs RWA again, and attempts to set up the connection on another route and wavelength. If such a route and a wavelength cannot be found, the connection is blocked. When the source receives a RESERVE-ACK, the connection set-up is successful, and the source may begin sending data over the connection.

  In order to prevent a connection request from being re-attempted too many times, a parameter M is used to control the maximum number of times that a connection may be attempted. A connection is blocked after the Mth attempt fails.

- **Update procedures**: Both approaches use incremental updates. In the link-state approach, each LSA contains the information about one channel on one link. In the distributed-routing approach, each node keeps a copy of every neighbor's routing table, and each update message only contains the recently-changed entries in the sending node's routing table.

16.5.2 Comparison

We compare the performance of the two approaches via simulation on the NSFNET, shown in Fig. 16.1. NSFNET has 16 nodes, 25 links, and the link lengths range from 750 *km* to 3000 *km*. Each link is a bidirectional fiber link, and the number on the each link in Fig. 16.1 represents the length of the link in units of 10 *km*.

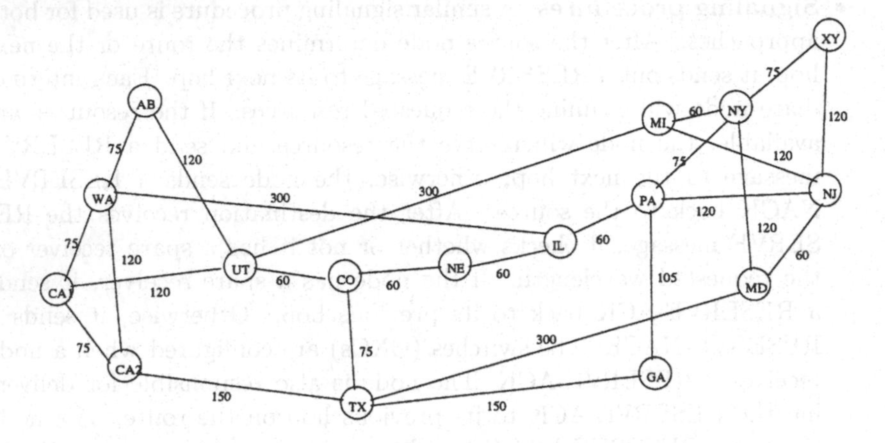

Figure 16.1 NSFNET: a nation-wide backbone network.

We assume the following:

- Number of wavelengths on each link, W, is 8.
- Traffic is uniformly distributed among all node pairs.
- Connection-holding time is exponentially distributed with mean 100 *ms*.
- Message processing time at a node, P, is 10 μs.
- Time to configure, test, and set up an OXC, C, is 500 μs.
- Time to transmit or switch a packet in the control network, R, is 0.
- The shortest path obtained in adaptive routing is defined as the path with the minimum number of hops. Under uniform traffic and low load, the average propagation delay between two nodes is $D = 14.7$ *ms* and the average hop distance is $H = 2.28$. Signaling messages are routed via the path with the shortest propagation delay in the control network.

- The number of transceivers on each wavelength at each node, TR, is a parameter to the simulation. $TR = 1, 2, 3$.
- No re-attempt is performed when a connection is blocked.

We obtained simulation results over the NSFNET topology by simulating a total of 10^4 connection requests which arrive and depart from the network over time. To study the network's behavior under different loads, the arrival rate of connection requests is varied as a parameter in the simulation. Load is measured in Erlangs, which can be calculated by multiplying the connection-arrival rate with the average connection-holding time. Hence, the load refers to the average number of connections measured at any instant of time in the network if there is no blocking.

Connection Set-Up Delay

Connection set-up delay is the time required to establish a connection once a connection request arrives. When the network is very lightly loaded (i.e., the shortest path is available for all connections), and there are no re-attempts, then the following elements contribute to the average set-up delays for both connection-management approaches:

- 2 propagation delays from source to destination node, $2D$,
- $2H + 1$ message processing delays, $(2H + 1)P$, and
- OXC configuration time, C.

Hence, the connection set-up time under light load is: $T_D = 2D + (2H+1)P + C = 30.02\ ms$. Figure 16.2 plots the set-up delays versus load in NSFNET, with different number of transceivers (TR). We observe that, when the load is very low, the set-up delays are fairly close to this lower bound. As the load increases, the shortest paths may become unavailable and longer paths must be selected. Therefore, the connection set-up delay may increase as load increases. However, the call-blocking rate also increases when the load increases. A connection which spans more hops is more likely to be blocked than a connection which spans fewer hops; thus, as the load continues to increase, the connection set-up delays will decrease.

It is interesting to observe that the distributed-routing approach gives lower connection set-up delays than the link-state approach. Both approaches attempt to find the path with the minimal hop-count. In the link-state approach, if there exists multiple paths to the destination, one is chosen randomly. However, in the distributed-routing approach, the routing tables are exchanged between neighboring nodes, thus the routing table from a path with a shorter propagation delay will reach a node first. Therefore, among the paths of the same hop-count, the path with the shortest propagation delay will be saved in the routing table.

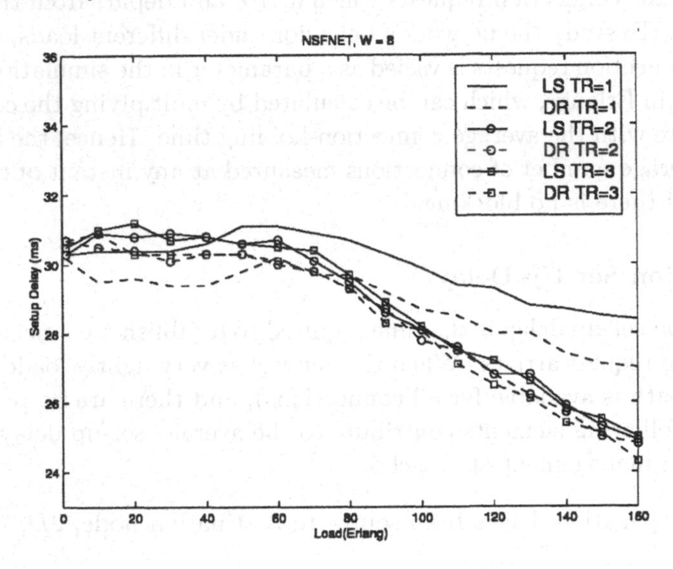

Figure 16.2 Connection set-up delay versus load for NSFNET and 8 wavelengths.

Blocking Probability

Blocking probability refers to the probability that a connection cannot be established due to resource contention along the desired route. Figure 16.3 plots the blocking probabilities versus load for the two approaches. It is observed that blocking in the link-state approach is slightly lower than that in the distributed-routing approach under low load but slightly higher under a certain high load. These differences are due to the facts that, under low load, the link-state approach has more accurate routing information which comes from shorter stabilizing delays (Fig. 16.4); under high load, both approaches may not have up-to-date information on routing but set-up delays in the link-state approach are higher. Hence, resources are reserved for a longer period of time for each connection in the link-state approach under high load.

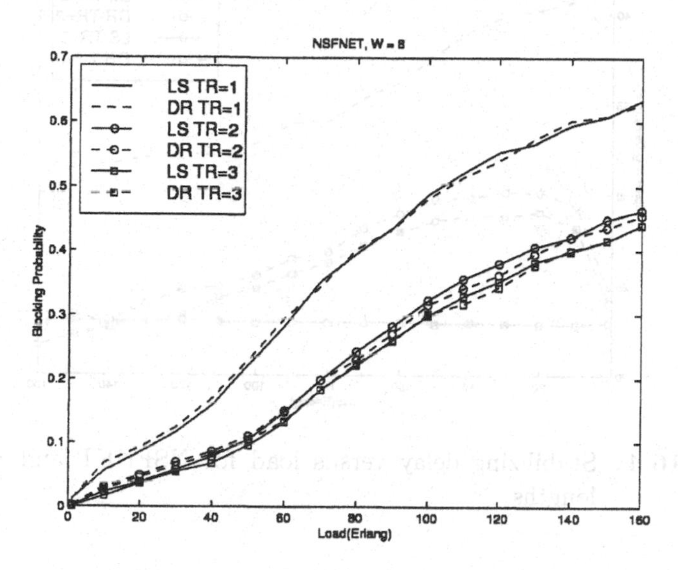

Figure 16.3 Blocking probability versus load for NSFNET and 8 wavelengths.

Figure 16.5 plots the network utilization versus load obtained through simulation. When each node has only one transceiver on each wavelength ($TR = 1$), we observe that the network saturates at around 50% for a load of 160 Erlangs (where around 60% of the connections are blocked). When more transceivers are available ($TR > 1$), the network utilization is close to 70% for a load of 160 Erlangs (where around 45% of the connections are blocked). This performance is not a limitation of the routing approaches, but rather a limitation of the number of transceivers at each node (when $TR = 1$), as well as the wavelength-continuity constraint.

Stabilizing Time

Stabilizing time is the time required for nodes to update topology information after a connection has been established or taken down. In the link-state approach, the stabilizing time is equal to the time it takes for a node's up-

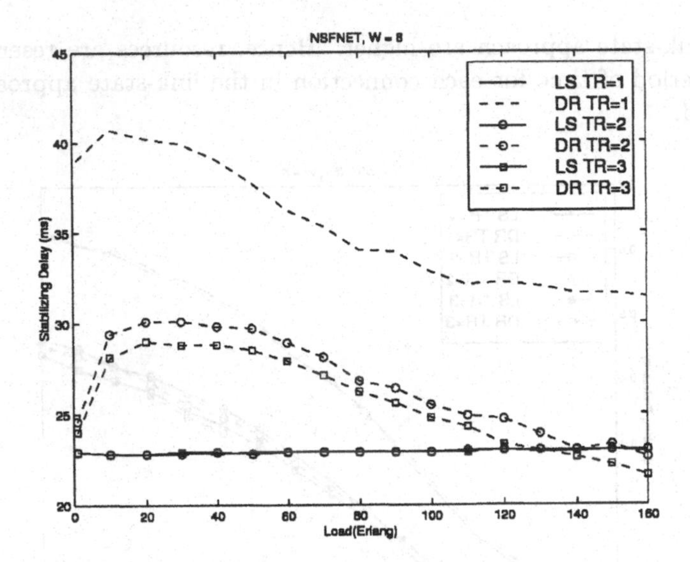

Figure 16.4 Stabilizing delay versus load for NSFNET and 8 wavelengths.

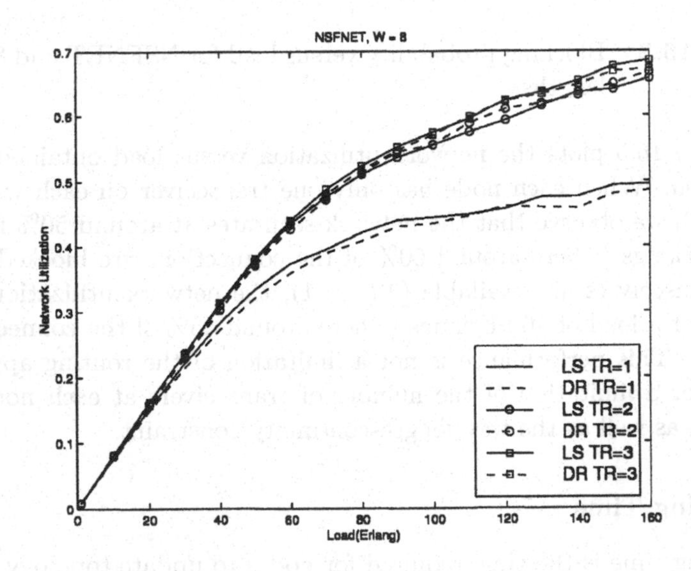

Figure 16.5 Network utilization versus load for NSFNET and 8 wavelengths.

date message (LSA) to be delivered to the farthest node, which we denote as T_i for node i. T_i can be computed as follows. For each node i in the network, and $j \neq i$, find the shortest route by the minimal propagation delay from i to j. Denote the number of hops on this route by H'_{ij} and the propagation delay by d_{ij}. If the time to transmit/switch an LSA is R, then the time it takes for node i's LSA to reach j is $H'_{ij}R + d_{ij} = d_{ij}$, since we assume $R = 0$.

We then find j for each node i such that d_{ij} is maximum. Thus, we have:

$$T_i = \max_j d_{ij}.$$

The average stabilizing time in the link-state approach will then be $\frac{1}{N}\Sigma_i T_i = 23.2\ ms$ for NSFNET, where N is the number of nodes in the network. From simulation, the average stabilizing delay is observed to be in the range $[22.8\ ms, 23\ ms]$ for NSFNET, as plotted in Fig. 16.4. The stabilizing delay for the distributed-routing approach is studied in simulation only, since the delay for this case is difficult to model. We notice from Fig. 16.4 that the distributed-routing approach has a larger stabilizing delay than the link-state approach in most cases. This is because the distributed-routing approach usually takes several rounds of exchange of information for the network to stabilize, especially in response to "bad" news. The stabilizing time for the distributed-routing scheme decreases as the load increases, and may be lower than that of the link-state approach under very high load. This is because the network-resource utilization is fairly high at these loads, and most routes are unavailable in the network. Hence, a change in wavelength usage does not affect the rest of the network as much as for lower loads.

We also notice that, in the distributed-routing approach, the stabilizing delay increases as the number of transceivers decreases. This is because, when each node has fewer transceivers, a change of state (connection being set up or taken down) will have a larger impact on the rest of the nodes. For example, when $TR = 1$, any connection being set up from or to a node means that other nodes cannot connect to this node on this wavelength; hence, this information has to be propagated to every node in the network. When $TR > 1$, this information may not be necessary for some nodes.

16.6 Fault Management

In an optical network, the high capacity of a link has the problem that a link failure can potentially lead to the loss of a large amount of data. So, we need to develop appropriate protection and restoration schemes which minimize the data loss when a link failure occurs. Upper layers of protocols (such as ATM, IP, and MPLS) have their own procedures to recover from link failures [ADDH94, Huit95, MSOH99]. However, the recovery time for upper layers may be significantly larger (on the order of seconds), whereas the fault-recovery time in the optical layer should be on the order of milliseconds in order to minimize data losses. Furthermore, it is beneficial to consider fault-recovery mechanisms in the optical layer for the following reasons [Gers98]: (a) the optical layer can efficiently multiplex protection resources (such as spare wavelengths and fibers) among several higher-layer network applications, and (b) survivability at the optical layer provides protection to higher-layer protocols that may not have built-in fault recovery.

Essentially, there are two types of fault-recovery mechanisms [Gers98, Wu92, Wu98] (see also Chapter 11; also some basic concepts from Chapter 11 are repeated below for completeness of our current discussion). If backup resources (routes and wavelengths) are pre-computed and reserved in advance, we call it a protection scheme [RaMu99a, FCMJ99]. Otherwise, when a failure occurs, if another route and a free wavelength have to be discovered dynamically for each interrupted connection, we call it a restoration scheme [RaMu99b, IrGr00]. A restoration scheme is usually more resource-efficient [IrGr00], while a protection scheme has a faster recovery time and provides guaranteed recovery ability. We consider protection schemes in this section. From the network-topology perspective, protection schemes can be classified as ring protection and mesh protection. Ring-protection schemes include Automatic Protection Switching (APS) and Self-Healing Rings (SHR) [RaSi01]. Both ring protection and mesh protection can be further divided into two groups: path protection and link protection. In path protection, the traffic is rerouted through a link-disjoint backup route once a link failure occurs on its working path. In link protection, the traffic is rerouted only around the failed link. Path protection usually has less resource requirement [RaMu99a] and lower end-to-end propagation delay for the recovered route. In this section, we consider path protection in networks with mesh topology.

In path protection, for each lightpath that is set up, there are two link-disjoint (and possibly node-disjoint) paths: a primary path and a backup path. The lightpath is set up on the primary path. In case of a link failure on the primary path, the lightpath is switched to the pre-reserved or pre-set-up backup path. The primary and the backup paths are link-disjoint, while the backup paths of different connections may or may not share common wavelengths on common links. If we do not allow sharing among backup paths, then we have a dedicated-path protection scheme. The OXCs on backup paths can be configured at the beginning, i.e., when the lightpath is set up on the primary path. Then, no OXC configuration is necessary when a failure occurs. This type of recovery can be very fast but the resources are not utilized very efficiently. There are two types of dedicated protection: 1+1 protection and 1:1 protection.

- In 1+1 protection, traffic is transmitted on both paths from the source to the destination. The destination receives data from the primary path first. If there is a failure on the primary path, the destination switches over to the backup path and continues to receive data. In order to avoid data loss, the source node should delay transmitting data on the backup path for some amount of time ϵ, depending on the propagation delay difference between the primary path and the backup path, as well as the failure-detection time, i.e., if the kth bit of data reaches the destination at time t_1 on the primary path, the same k^{th} bit should reach the destination at time $t_2 \geq t_1 + \epsilon$ on the backup path. If the destination receives the $(k-1)^{th}$ bit, detects there is a failure, and switches to the backup path, it should not miss the kth bit.

- In 1:1 protection, data is normally not transmitted on the backup path. Thus, we can use the backup path to carry some low-priority preemptable traffic. If there is a failure on the primary path, the source node is notified (by some protocol), and it switches over to transmit on the backup path. So, some data may be lost in the network, and the source must be able to retransmit those data, if necessary.

If sharing among backup paths is allowed as long as they satisfy certain constraints, the OXCs on the backup paths cannot be configured until the failure occurs. The recovery time in this scheme is longer, but the overall resource utilization is much better than in dedicated-path protection. Of course, more signaling is required to recover from the failure. We call this scheme the shared-path protection (or M:N) scheme. In this section, we consider both dedicated-path protection and shared-path protection. There have been some studies applying ring-protection schemes into a mesh-topology network. One such approach is to map a planar meshed graph into directed cycles, and each directed link is protected by a directed cycle [ERHS00]. So, basically, this approach is a ring-based, link-protection scheme. Another approach is also ring-based link protection but works differently [RaMu99b]. In [ERHS00], each fiber is unidirectional and is part of only one ring. In [FCMJ99], rings can share fibers. A ring cover is first decided for the meshed network, and shared-link protection is used within rings whereas protection wavelengths are not shared on different rings. This way, the OXCs can be pre-configured, and the approach in [FCMJ99] encourages a certain degree of sharing among protection wavelengths.

Figure 16.6 summarizes the classification of protection and restoration schemes.

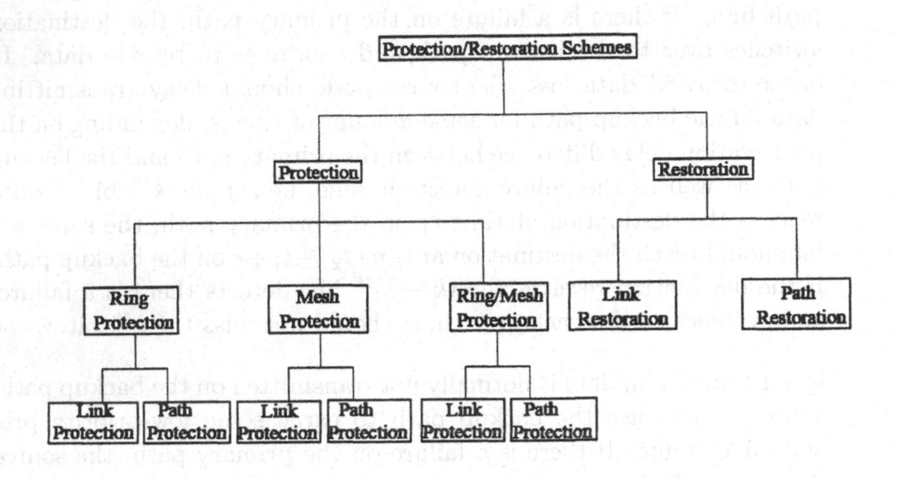

Figure 16.6 Different protection and restoration schemes.

In a wavelength-routed network, the traffic can be either static or dynamic. Under a static traffic pattern, the connection requests are available all at once. We call this the static lightpath establishment (SLE) problem which was studied in [RaMu99a]. This problem can be solved by Integer Linear Programming (ILP). The constraints of this problem are the number of wavelengths on each link, the number of transmitters and receivers at each node, and the wavelength-continuity constraint (if no wavelength conversion is used). The objective is to minimize the total number of wavelengths used on all the links in the network, which is denoted by a quantity called wavelength-mileage in [FCMJ99]. An alternative objective is to maximize the carried load, i.e., block the fewest number of connection requests.

Under dynamic traffic, connection requests come in one at a time and each connection exists for only a finite duration, called the connection-holding time. Given a fixed number of wavelengths on each fiber link, and a fixed number of transmitters and receivers at each node, our objective is to minimize the overall connection-blocking probability. Also, we would like to achieve small end-to-end propagation delays for the connections which are set up. A control and management protocol is required to set up and take down lightpaths. This protocol must have the following capabilities:

1. **Routing and Wavelength-Assignment (RWA)**: upon the arrival of a connection request, the protocol must select two link-disjoint routes from the source to the destination, and assign a wavelength to each route; if this process is not successful, the connection request is blocked;

2. **Signaling**: after the routing and wavelength assignment process is completed, the protocol signals the appropriate nodes to reserve the wavelength on requested links and/or configure their OXCs;

3. **Fault detection**: if a link failure occurs, the end nodes of the failed link (which is two unidirectional fibers, going on opposite directions) must be able to detect the failure; and those which detect the failure must notify the end nodes of the connections which are going through the failed link that a failure has occurred; (node failures can also be handled similarly;)

4. **Fault recovery**: the end nodes of connections involved in a failure must be able to signal the nodes in the backup paths (if shared protection is used), and switch their transmission or reception to the backup paths;

5. **Reverting/Non-reverting**: once the fiber link is fixed (signaled by some higher-layer messages), the end nodes will/will not switch the connections back to their primary paths; reverting may be preferred in shared-path protection scheme because new failures can be better handled;

6. **Network-state update**: the mechanism must also be able to provide updates to reflect which wavelengths are currently being used on each link so that nodes may make routing decisions based on up-to-date information.

In this section, we develop a control protocol with the above capabilities.

The rest of this section is organized as follows. In Section 16.6.1, we describe the architecture of our network model. In Section 16.6.2, the proposed control and management protocol, and an algorithm are discussed to solve the transmitter/receiver sharing problem in shared-path protection. We present numerical examples in Section 16.6.3.

16.6.1 Network Architecture

Figure 16.7 shows the architecture of a wavelength-routed WDM network consisting of six wavelength-routing switches (WRS). Associated with each WRS, there is an OXC and an access station. The OXC is an intelligent switch and is reconfigured upon new connection requests. Our control and management protocol is executed in each OXC. An access station has one or more fixed transmitter arrays (a transmitter array is a set of transmitters, one on each wavelength) and one or more fixed receiver arrays. The link connecting an access station and a OXC is one or more fibers. In this section, we assume that there are M transmitter arrays and M receiver arrays at each access station. Also there are M fibers connecting an access station and its OXC. We will vary the value of M and study its relationship with blocking probability. Note that the maximum number of M, max(M) = the maximum nodal degree of the topology graph G of the network. For simplicity of later discussion, we combine an OXC with its associated access station as an integrated unit, which we refer to as a node. In the network in Fig. 16.7, we show two lightpaths: one from Node A to Node C on wavelength λ_1, and the other from Node A to Node C on wavelength λ_2. Since the two paths are link-disjoint, one can serve as a primary path and the other can serve as a backup path for a connection from Node A to Node C.

This wavelength-routed network (Fig. 16.7) can be modeled as a layered graph, as in Fig. 16.8. Each layer represents a wavelength, and a physical fiber has a corresponding link in each layer. Without wavelength conversion, each lightpath is routed inside one single layer. We use a single layer (e.g., λ_0 in this network) for control messages and connections are set up in other layers (e.g., λ_1 and λ_2 layers). The control layer is a packet-switched network and all packets are routed by shortest paths. Note that, whenever there is a link failure, the corresponding logical link in the control layer also fails. The control packets that are generated after a link failure must be routed on paths that do not go through the failed link.

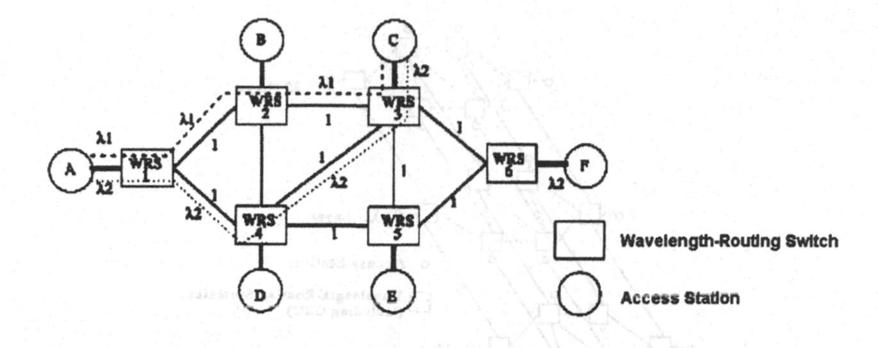

Figure 16.7 Architecture of a wavelength-routed WDM network.

16.6.2 Protocol Descriptions

Link-State Protocol

Two approaches for connection management were compared in Section 16.5: the link-state approach and the distributed-routing approach. The distributed-routing approach works well in terms of the amount of information stored at each node and the connection-setup delay. However, it is not very suitable for a path-protection network because the source node does not have enough knowledge to find two link-disjoint paths. Hence, we adopt the link-state approach for updating network-state information and signaling. Note that, because of the advantages of the distributed-routing approach [ZJSM01], we can combine link-state's updating network-state information protocol and distributed-routing's signaling protocol for connection management. In this chapter, we consider link-state protocol only.

The link-state approach is proposed in [RaSe97]. It is different from other approaches (such as the distributed-routing approach) in how it updates network-state information as well as its signaling protocol. We describe the basic link-state protocol first and then we propose modifications to make it work for path-protection networks.

In the original link-state approach, each node maintains the complete network topology, including information about the wavelengths that are in use on each link. Upon the arrival of a connection request, a node utilizes the topology information to select a route and a wavelength. Once the route and wavelength are selected, the node attempts to reserve the selected wavelength along each link on the route by sending simultaneous reservation requests

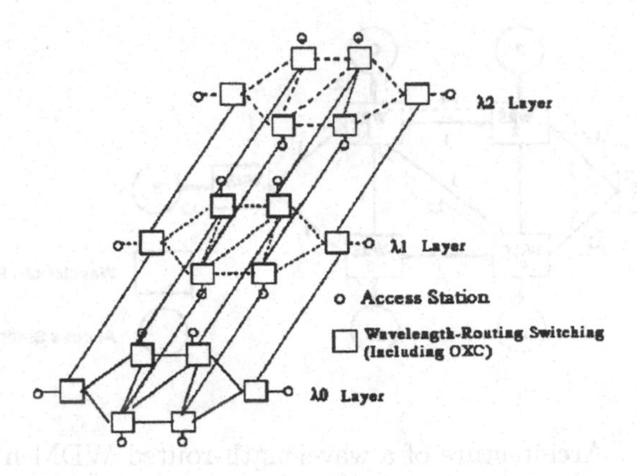

Figure 16.8 Layered-graph model of the wavelength-routed WDM network with three wavelengths.

to each node on the route. If an intermediate node is able to reserve the wavelength on the appropriate link, it sends an acknowledgement directly back to the source node. If all of the reservations are successful, the source sends a SETUP message to each of the nodes. The appropriate OXCs are then configured at each node, and the connection is established. If even one of the reservations is not successful, then the call is blocked and the source node sends a TAKEDOWN message to each node on the route in order to release the reserved resources. When a connection is established or taken down, each node involved in the connection broadcasts a topology-update message which indicates any changes in the status of wavelengths being used on the node's outgoing links.

To maintain a record of local connections' state information, each node has a Connection Switch Table (CST). In CST, each entry contains information about a connection path, including connection ID (a unique identifier in the network for a connection), incoming port, outgoing port, and state of the path (*reserved*, which means the connection on this path is reserved at this node but the switch is not configured yet, or *up*, which means the connection is set up on this path at this node, i.e., the OXC has been configured, or *idle*). In a path-protected network, each connection has two paths: a primary path and a backup path, routed link-disjointedly. They will have the same connection ID, but going through (not necessarily) different node(s)

Table 16.1 CST at Node 3.

Connection ID	Incoming Port	Outgoing Port	State	Type
1,3,0	2, λ_1	-1, λ_1	up	primary
1,3,0	4, λ_2	-1, λ_2	reserved	backup

in the network. We add one more bit to each entry to indicated whether it is a primary path or a backup path. We call this bit the type field. This information is required for routing and wavelength assignment (RWA) for future connection requests as well as for failure recovery if this connection is involved in a link failure.

Table 16.1 shows the CST at Node 3 (combination of Access Station C and OXC 3, which we refer to as Node 3) in the network illustrated in Fig. 16.7. Note that the incoming port is a previous-hop, wavelength pair, and the outgoing port is a next-hop, wavelength pair. For a connection source, the previous-hop in incoming port is -1. For a connection destination, the next-hop in outgoing port is -1.

The next subsection addresses the signaling algorithms in different path-protection schemes as well as other aspects of each scheme.

Path-Protection Schemes in Mesh Networks

Here, we describe three path-protection schemes in mesh networks: 1+1 dedicated-path protection, 1:1 dedicated-path protection, and shared-path protection. We will describe each scheme from the desirable capabilities listed in Section 16.2.

1+1 Dedicated-Path Protection and 1:1 Dedicated-Path Protection: The 1+1 dedicated-path protection and 1:1 dedicated-path protection schemes work very similarly; they only differ in the fault detection and recovery stages.

- **RWA**: The source node computes the shortest path on each wavelength, on which the source node has at least one free transmitter and the destination node has at least one free receiver. Then, it selects the wavelength leading to the shortest path for the primary path. If there is a tie, the wavelength with the lowest index is selected. This wavelength-assignment approach is called first-fit (see Chapter

7). Then, the source node removes every link that appears in the selected primary path (i.e., it removes the corresponding logical link in each of the data layers in Fig. 16.8) and repeats the computation again for the backup path. If the source node cannot find two link-disjoint paths, the connection is blocked. (This approach is called the *two-step approach*; see Chapter 11.)

- *Signaling*: After successfully finding a route and a wavelength for each of the two link-disjoint paths, the source node uses the link-state protocol to set up both lightpaths. Note that, each time the node sends out a request (RESERVE or SETUP), it sends the information to H1+H2 nodes, where H1 is the number of hops on the primary path and H2 is the number of hops on the backup path. When all H1+H2 RESERVE-ACK's are received, the source node sends out the SETUP requests. If there is a RESERVE-NACK, the source node takes down the connection by sending TAKEDOWN messages to the H1+ H2 nodes.

- *Fault detection and recovery*: Each link is bidirectional with a pair of uni-directional fibers going on opposite directions. Usually, these two fibers reside in the same cable and they may get cut at the same time. To accommodate scenarios in which only a single fiber is cut (in one direction), we develop the following detection scheme which works for both single-fiber-cut and fiber-pair-cut scenarios. Obviously, this approach also works for single bidirectional fiber cut. Before a link is cut, there are either some traffic going through that link or probing data in some special patterns just for keeping the line alive instead of idle. When there is a cut, the downstream node (both end nodes will be downstream as well as upstream if the link is bidirectional) of this link will detect the failure. If 1+1 protection is used, each connection destination simply switches to the other path to receive data once it detects that there is no signal coming from the primary path. No signaling is required for the recovery. However, in 1:1 protection, no data is transmitted on the backup path normally. When there is a failure on fiber $x \rightarrow y$, Node y looks up its CST, finds all the entries with incoming port (x, *) and type primary, and notifies the source of every such connection about this failure; the source nodes will subsequently switch to transmit on their backup paths.

  As an example, consider one connection. The source node that is notified about the failure may have transmitted some data which is lost in the network. So, it needs to retransmit the lost data. By the time the retransmitted data reaches the destination, the destination has already detected the failure and is waiting for data to come in from the backup path. So, the connection is recovered. If the link $x \rightarrow y$ is cut, both Node x and Node y will be responsible for sending notifications. Also, they must reflect this information in their state-update messages so that other nodes will not attempt to use this link in their routing algorithm until the failure is fixed. In previous work [RaMu99a], the upstream node of a failed link is usually assumed to take the role of notifying connection sources, and the downstream node of the failed link is assumed to notify connection destinations. Then, to combat single-direction link failures, we must have the downstream node of a failed link to notify the upstream node first, which can subsequently notify the connection sources. Compared to this approach, our approach is more efficient and gives better performance in recovery time. In Fig. 16.9, a lightpath from Node 0 to Node 3 goes through link 1 $\rightarrow$ 2. When link 1 $\rightarrow$ 2 breaks, Node 2 detects the failure, and notifies Node 0 via Node 4 and Node 5 (note that this message is delivered in a store-and-forward manner since the control layer is a packet-switched network). Node 0 can then switch to the backup path 0 $\rightarrow$ 5 $\rightarrow$ 4 $\rightarrow$ 3.

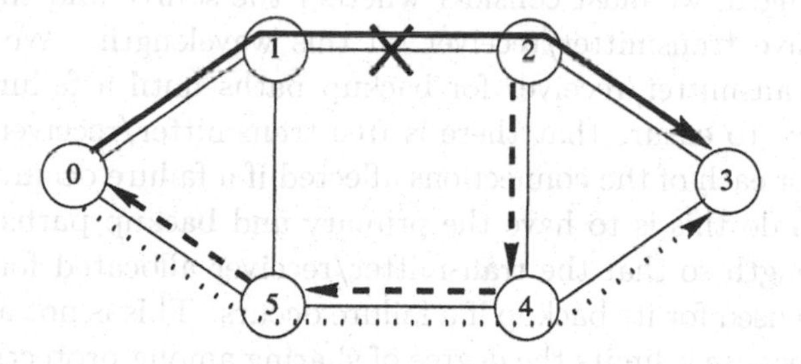

Figure 16.9 An example of the downstream node of a failed link notifying the connection source: the solid line shows the primary path, the dotted line shows the backup path, and the dashed line shows the failure notification to the source node of the connection.

- *Reverting/Non-Reverting*: Once the link is fixed, the source node and the destination node can switch back to their primary path because the primary path usually has fewer hops than the backup path. But reverting is not necessary in dedicated-path protection.

Shared-Path Protection

– **RWA**: Shared-path protection is more complex in the RWA algorithm because we must decide:

  which primary paths can share wavelength/link(s) in their backup paths, and

  how many transmitters/receivers are required for backup paths.

To solve these two problems, each node must not only have knowledge of the state of each link in the network, but also have knowledge of which connection is using which wavelength/link, and whether on its primary path or backup path. So, in each network-state update message, we let each node broadcast its CST instead of the wavelength states on each of its outgoing links.

The first problem is easier to solve than the second. If we know the routes of the two paths, we can easily determine whether or not they are link-disjoint. If the answer is yes, we allow them to share wavelength/link(s); otherwise, we do not. However, when we select a wavelength, we must consider whether the source and the destination still have transmitter/receiver on this wavelength. We do not allocate transmitter/receiver for backup paths until a failure occurs but we have to ensure that there is free transmitter/receiver that can be used for each of the connections affected if a failure occurs. One simple way to do this is to have the primary and backup paths on the same wavelength so that the transmitter/receiver allocated for the primary can be used for its backup if a failure occurs. This is not an ideal situation because it limits the degree of sharing among protection resources. Also, it makes the network vulnerable to transmitter/receiver failures. So, we allocate different transmitters/receivers to primary and backup paths.

If, on a wavelength, all the transmitters at the source or all the receivers at the destination are occupied, this wavelength is out of our consideration. We shall address how to decide the minimum number of transmitters/receivers on a certain wavelength at the source/destination node later in this section. Below, we introduce the RWA algorithms for primary paths and backup paths separately.

RWA for a primary path.

i. In each wavelength layer, remove the logical links that appear in either a primary or a backup path.

ii. Initialize the candidate wavelength set to be all the wavelengths: $\{\lambda_1, \lambda_2, ..., \lambda_W\}$, where W is the total number of wavelengths used for data connections in the network, i.e., we have $W + 1$ wavelengths in the network and λ_0 is used for control messages.

iii. For each wavelength λ_i, let the set of primary paths originating at the source node on wavelength λ_i be P_i, and the set of backup paths originating at the source node on wavelength λ_i be B_i. Apply the method described later in this section to B_i and decide the number of transmitters required T_i. If $|Pi| + Ti = M$ (recall that M is the number of transmitter arrays and receiver arrays at each access station), λ_i is removed from our candidate wavelength set. Otherwise, repeat this procedure for the set of primary paths terminating at the destination node on wavelength $\lambda_i(P_i')$, and the set of backup paths terminating at the destination node on wavelength $\lambda_i(B_i')$. Let R_i be the number of receivers required at the destination node for B_i'. If $|P_i'| + Ri = M$, λ_i is removed from our candidate wavelength set. Otherwise, λ_i stays in the candidate wavelength set.

iv. For each wavelength remaining in the candidate wavelength set, compute the shortest path. For the wavelengths giving the best shortest paths, we break the tie by First-Fit. Assume we get path p_1 (which is a set of links) and wavelength λ_1 for the primary path.

RWA for a backup path.

We denote the backup path by b^*. Note that we do not know b^* yet but we know its primary path p_1.

i. Eliminate all the logical links used in either a primary path, or a backup path if its primary path shares common physical link(s) with p_1. Also, eliminate all the corresponding logical links in all wavelength layers of the links appearing in p_1.

ii. Initialize our candidate wavelength set to be all the wavelengths: $\{\lambda_1, \lambda_2, ..., \lambda_W\}$.

iii. For each wavelength λ_i, let the set of primary paths originating at the source node on wavelength λ_i be P_i, and the set of backup paths originating at the source node on wavelength λ_i be B_i. Note that B_i does not contain b^* and P_i does not contain p_1 since they have not been reserved yet and are not contained in any CSTs. Now set $B_i = B_i + \{b^*\}$. Also, for $i = 1$ only, set $P_i = P_i + \{p_1\}$ because λ_1 is selected for p_1. Apply the method described in Section 16.3.3 to B_i and decide the number of transmitters T_i required. If $|P_i| + T_i > M$, λ_i is removed from our candidate wavelength set. Otherwise, repeat this procedure for the set of primary paths terminating at the destination node on wavelength $\lambda_i(P_i')$, and the set of backup paths terminating at the destination node on wavelength $\lambda_i(B_i')$. Set $B_i' = B_i' + \{b^*\}$ and $P_1' = P_1' + \{p_1\}$. Let R_i be the number of receivers required at the destination node for B_i'. If $|P_i'| + R_i > M$, λ_i is removed from our candidate wavelength set. Otherwise, λ_i stays in the candidate wavelength set. Note that, by setting $P_1 = P_1 + \{p_1\}$, $B_i = B_i + \{b^*\}$, $P_1' = P_1' + \{p_1\}$, and $B_i' = B_i' + \{b^*\}$, we do not allow the sharing of transmitter/receiver between the primary path and the backup path for a given connection. By avoiding the sharing of transmitter/receiver within the same connection, the network is immune to single-transmitter/receiver failures.

iv. For each wavelength remaining in the candidate wavelength set, compute the shortest path. For the wavelengths giving the best shortest paths, we break the tie by First-Fit, Last-Fit, or Max-Shared-First. We explained First-Fit earlier. The latter two work as follows.

Last-Fit (LF): the highest-indexed wavelength is selected.

Max-Shared-First (MSF): for each wavelength λ_i, let the links used in the backup path be $l_0, l_1, ..., l_{k-1}$, if the backup path is of k hops. Suppose wavelength λ_i is shared by s_j backup paths on link $l_j, 0 \leq j \leq k-1$. Then, compute the following function for λ_i: $MSF(i) = \sum_{j=0}^{j=k-1} s_j$. The λ_i of the largest $MSF(i)$ value is selected for the backup path. This is based on the idea of maximizing the sharing among protection resources. If there is a tie, either First-Fit or Last-Fit can be used to break the tie. Since wavelength assignment for primary paths is using First-Fit, for the wavelength assignment for backup paths, Last-Fit and Max-Shared-First with Last-Fit for tie-breaking should give better performances than First-Fit, because they pack the backup wavelengths at one end of the wavelength space. Thus, there can be more sharing among backup paths than the case where they are interleaved with wavelengths used by primary connections.

- **Signaling**: When the source node gets a connection request, it performs RWA for both paths, and then it sends RESERVE request to each node on the two paths. Upon reception of all the RESERVE-ACKs, it sends out SETUP request to nodes on the primary path only. When it gets all the SETUP-ACKs from the nodes to whom it sent SETUP request, the connection is set up and the source node starts transmitting its data.
- **Fault detection and recovery**: When there is a link cut, the downstream node of the fiber detects the cut immediately. As in 1:1 protection, the downstream node will notify each source node whose connections are going through this fiber about this failure. Each source node, as in the SETUP stage, sends SETUP-BACKUP request to each node along the backup path and tells it to configure its OXC. The nodes configure their OXCs, and send SETUP-BACKUP-ACK back to the source node. When the source node receives all SETUP-BACKUP-ACKs, it starts transmitting data on the backup path. Of course, it needs to estimate how much data has been lost due to this failure and retransmit the lost data as well.
- **Reverting**: Reverting may be preferable because, when a backup path b is being used, the other primary paths whose backup paths share wavelength/links with b are left un-protected. So, the connection may be switched back to the primary path once the link failure is fixed. However, there is a vulnerable period for the connections whose backup paths share wavelength/links with b. The length of this vulnerable period is equal to the time required to fix a link. If there is a second link failure during this vulnerable period, the affected connections may not be restored. (See [ZhZM05] for a study on reprovisioning of backup wavelengths.)

Determining the Number of Transmitters/Receivers Required for Backup Paths at a Source Node in Shared-Path Protection
Given a set of backup lightpaths originating at the same node n and on the same wavelength w, our objective here is to decide the minimum number of transmitters that should be reserved for them; or given a set of backup lightpaths destined for node n and on the same wavelength w, we need to decide the minimum number of receivers that should be reserved for them. This problem can be solved as follows:

- Find the set of primary paths for which these backup paths are reserved; let it be Pw.

- For each link l in the network, count the number of paths in Pw that traverse it. Note that each link consists of two fibers going on opposite directions. Hence, traversing from both directions should be counted. Let this number for link l be bwl. bwl denotes the number of broken paths we will have to restore on wavelength w if link l gets cut. In other words, it means the number of transmitters/receivers on wavelength w that will be working for backup paths once link l gets cut.

- The maximum number of bwl, among all links in the network, will be the number of transmitters (or receivers) required for the set of backup paths at source node n (or destination n) on wavelength w.

16.6.3 Illustrative Numerical Examples and Discussion

We study the performance of the three protection schemes using the sample network shown in Fig. 16.10, which represents a typical metropolitan-area telecom mesh network. The number on each link represents the length of the link in units of 10 km. Thus, the propagation delay of a link with length 10 units, i.e., 100 km, is 500 μs.

We assume the following parameters:

- Message processing time at a node (OXC), P, is 10 μs.

- Time to configure, test, and set up an OXC, C, is 500 μs.

- Link cuts occur at rate of 0.0015 cuts/ms and it takes 20 ms to fix a link[2].

[2]Usually fiber cuts occur at rate of 4.39 cuts/1000 sheath miles/year and it takes around 12 hours to fix a fiber failure [ToNe94]. In order to simulate enough cuts in our system, we have to increase the cut rate and shrink the cut-fixing delay.

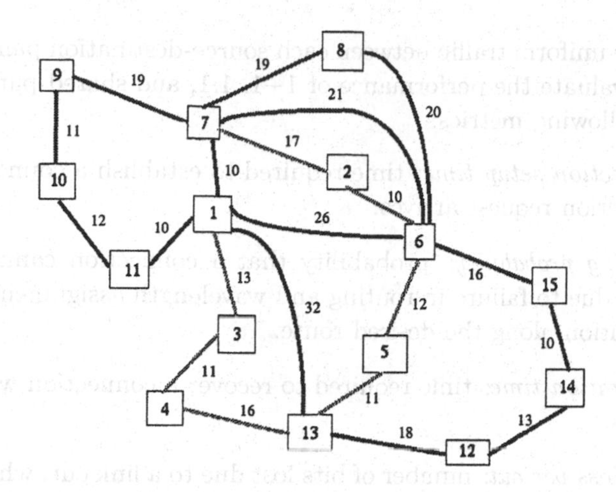

Figure 16.10 Sample network.

- Number of wavelengths on each link (in each direction) is 8.
- Bit rate per wavelength channel (per lightpath) is 2.5 Gbps, i.e., OC-48.
- Connection requests arrive as a Poisson process with mean arrival rate $\in$ [0.01, 0.3] arrivals/ms (given as a program parameter). Connection requests are uniformly distributed among all source-destination pairs.
- Connection-holding time is exponentially distributed with mean 100 ms. Note that, in the plots, load (Erlang) is calculated as: request arrival rate times mean connection-holding time.

The following are some characteristics about the sample network:

- Total number of nodes in the network, N= 15.
- Number of links (bidirectional) = 21.
- Average nodal degree = 2.8.
- Average link length = 153 km.
- Average propagation delay between two nodes D = 1.82381 ms.
- Average hop distance between two nodes H= 2.40952.

We simulate uniform traffic between each source-destination pair on this network and evaluate the performance of 1+1, 1:1, and shared-path protection using the following metrics:

- *Connection setup time*: time required to establish a connection once a connection request arrives.
- *Blocking probability*: probability that a connection cannot be established due to failure in routing and wavelength assignment or resource contention along the desired route.
- *Restoration time*: time required to recover a connection when a failure occurs.
- *Data loss per cut*: number of bits lost due to a link cut, which is related to both restoration time and number of connections carried by a link.

It is obvious that 1+1 and 1:1 protection have the same performance as to blocking probability as well as connection setup time. There is neither restoration time nor data loss per cut in 1+1, if we choose ϵ properly. (Recall that ϵ is the delay from the time the source transmits a bit on the primary path till it transmits the same bit on the backup path. Usually, the backup path is longer than the primary path and the switching at the destination is very fast upon detection of failure on the primary path. In such a case, ϵ can be 0.) So we present the results of 1:1 only for dedicated-path protection.

Figure 16.11 shows blocking probability versus load when dedicated-path protection is employed. The case without link failures is also plotted. We observe that the performance when M = 3 transmitter/receiver arrays are used at each node is very close to the performance of six transmitter/receiver arrays. Our results indicate that we do not need to equip each node with the maximum nodal-degree number of transmitter/receiver arrays, i.e., six in this network.

Figures 16.12 through 16.15 plot the blocking probability versus load for shared-path protection when different wavelength-assignment schemes for backup paths are applied. We notice that, in some cases, a system with three transmitter/receiver arrays gives better performance than a system with six. This may be due to the fact that, when only three transmitter/receiver arrays are existing at each node, intermediate nodes may have fewer chances to source/sink connections than in the six transmitter/receiver-array case, which implies that more connections can be routed through those nodes. Hence, the overall blocking is decreased.

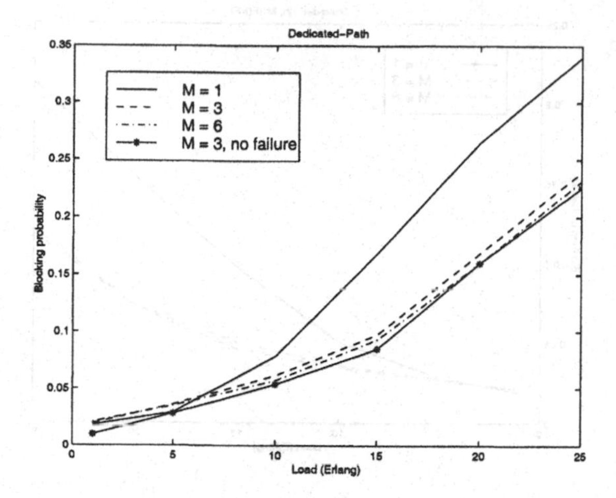

Figure 16.11 Blocking probability versus load for dedicated-path 1:1 and 1+1 protection, with $M = 1, 3, 6$.

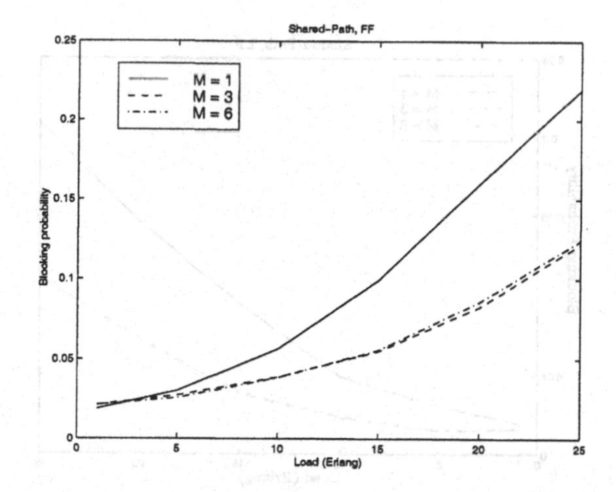

Figure 16.12 Blocking probability versus load for shared-path protection, FF for backup-path wavelength assignment, with $M = 1, 3, 6$.

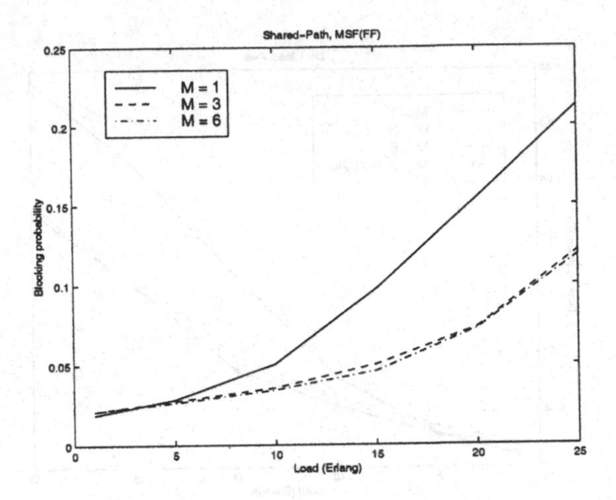

Figure 16.13 Blocking probability versus load for shared-path protection, MSF and FF for backup-path wavelength assignment, with $M = 1, 3, 6$.

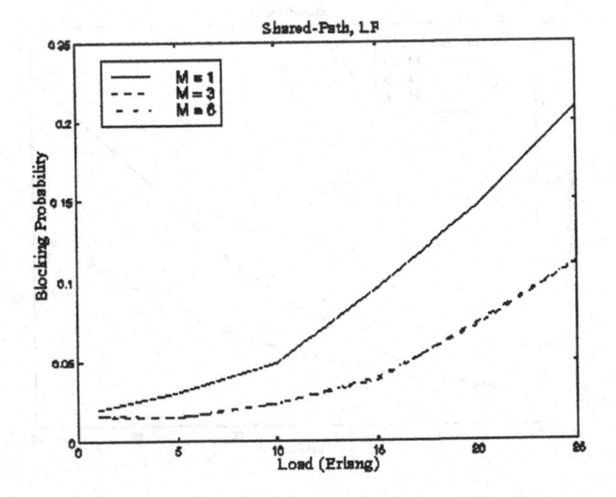

Figure 16.14 Blocking probability versus load for shared-path protection, LF for backup-path wavelength assignment, with $M = 1, 3, 6$.

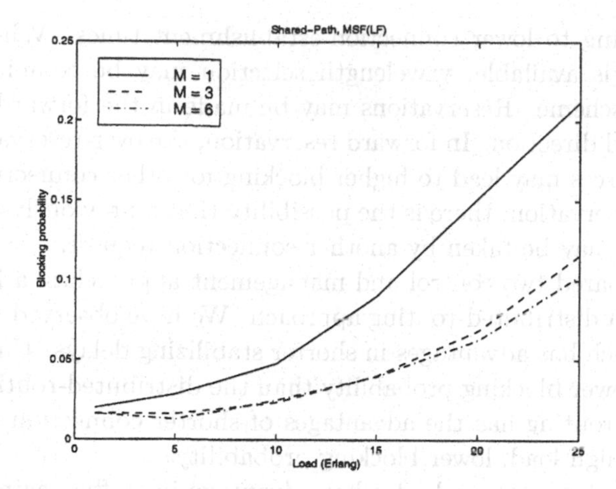

Figure 16.15 Blocking probability versus load for shared-path protection, MSF and FF for backup-path wavelength assignment, with $M = 1, 3, 6$.

16.7 Summary

The establishment of lightpaths in a wavelength-routed WDM network requires the implementation of fast and efficient control and management protocols to perform routing and wavelength assignment functions, as well as to exchange signaling information and to reserve resources. In this chapter, we have presented some of the routing, wavelength assignment, and signaling protocols for establishing lightpaths in a wavelength-routed network.

In routing and wavelength assignment (RWA) algorithms for dynamic lightpaths, the goal is to minimize the number of blocked connections. The performance of these algorithms depends on the amount of state information available to each node. If global information is known, then the routing and wavelength assignment decisions can be nearly optimal; however, it may be difficult to maintain complete up-to-date information in a very dynamic environment.

The performance of signaling protocols for reserving wavelengths along a lightpath will depend on whether or not global information is available, and whether or not multiple connection requests may be attempted simultaneously. For the case in which global information is available, reservations may be made either in parallel or on a hop-by-hop basis, with parallel reservations leading to lower connection-establishment times. When only local information is available, wavelength selection may be combined with the reservation scheme. Reservations may be made in the forward direction or the backward direction. In forward reservation, the over-reservation of wavelength resources may lead to higher blocking for other connections, while in backward reservation, there is the possibility that a previously-available link in the route may be taken by another connection request.

We compared two control and management approaches, a link-state approach and a distributed-routing approach. We have observed that the link-state approach has advantages in shorter stabilizing delays. Under low load, it also has lower blocking probability than the distributed-routing approach. Distributed routing has the advantages of shorter connection set-up delays and, under high load, lower blocking probability.

The link-state approach also has advantage in traffic engineering. Since each node maintains global information about the network, explicit routing can be implemented. This attribute can add more fault tolerance to the network. For example, it is simple and fast to compute two link-disjoint routes at the source node. It also makes shared protection possible with the knowledge of full network topology.

As fault management is another concern in network control and management, we proposed an on-line control and management protocol for setting up lightpaths with protection paths. The wavelength/links on the protection paths can be either dedicated to a certain connection, or shared among multiple connections. Dedicated-path protection has better performance in terms of connection-recovery time. However, it is not very resource efficient. It has higher blocking probability than the shared-path protection scheme. Under our current assumptions of message-processing time and OXC-configuration time, the connection-recovery time in the sample network when applying shared-path protection is under 10 ms. That is an acceptable recovery time.

Exercises

16.1. Explain the advantages and disadvantages of the three adaptive routing techniques that are based on global information: the completely centralized algorithm, the link-state approach, and the distance-vector approach.

16.2. Given the topology in Fig. 16.16 (where the numbers on the edges represent the "costs", say, distance in km), let us assume that we want to reserve a lightpath between the source (s) and the destination (d).

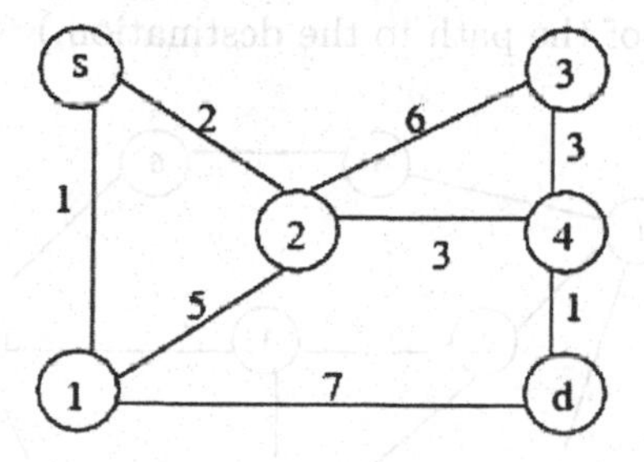

Figure 16.16 Network topology.

Let us assume that the lightpath of the lowest cost is tried first, and that it is available. Using the data below, compute the times needed to reserve the whole lightpath, using:

- parallel reservation
- hop-by-hop forward reservation.

Which time is better? What are the advantages and disadvantages of the two schemes?

Consider the following times:

- Time to process a message at a node: T=10 microseconds;
- Time to configure a node, and setup a crossconnect: C=100 microseconds;
- Propagation delay per km of fiber: P=5 microseconds;

16.3. (Path protection) Consider the following backbone network in Fig. 11.20 with a primary and backup path pair between node 3 and 21. Suppose that a failure occurs on link $9 \rightarrow 12$ on the primary path. Calculate the total failure-recovery time. Assume that the propagation delay in the network is 0.005 ms per kilometer.

16.4. Given the topology in Fig. 16.17, where we have the path 1-2-3-4 already setup, and assuming that there is a failure on link 2-3, show the recovery procedure for the three known restoration techniques – path restoration, sub-path restoration, and link restoration). (In sub-path restoration, the upstream end-node of the failed link discovers the latter part of the path to the destination.)

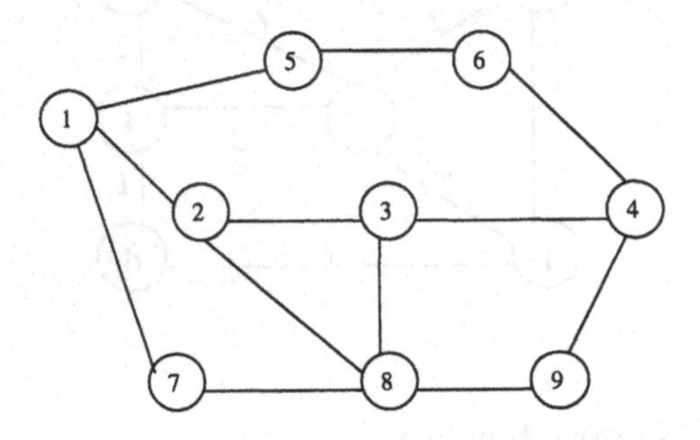

Figure 16.17 Another Network topology.

16.5. Given a network $G = (V, E)$, cost defined on each link $Cost : E \rightarrow Z^+$, congestion on each link $C : E \rightarrow \lambda$, where C gives the number of available wavelengths on a link. Given a source s and destination d, design an algorithm to find the least-cost path from s to d with number of available wavelengths on the path $\geq N$, where N is the desired number of available wavelengths. State the time complexity of your algorithm.

16.6. Consider a network where naive deflection routing is implemented. Nodes maintain a routing table indicating, for each destination, one or more alternate pre-computed outgoing links to reach that destination, which are preferentially ordered. If resources are unavailable on the preferred link, the next preferred link is chosen for the route. Nodes maintain only local information. Construct an example complete with

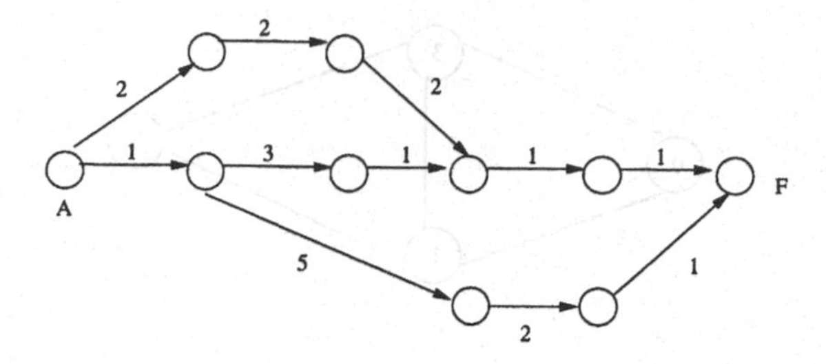

Figure 16.18 Disjoint primary and backup paths.

routing tables at each node and preferred links, where a link failure can lead to routing loops in a 4-node fully-connected network.

16.7. In Section 16.5.2, stabilizing time was defined. Find the average stabilizing time using the link-state approach for the network in Fig. 16.10. The number on each link represents its length in units of 10 km. The propagation delay of a link with length 10 units, i.e., 100 km, is 500 μs. Assume time to transmit/switch an LSA is 10 μs.

16.8. Consider the two-step approach to find link-disjoint primary and backup paths. For the network shown in Fig. 16.18, find link-disjoint primary and backup paths using this approach from node A to node F. Assume all links have equal number of available wavelengths, which is equal to the number of free transmitters and receivers at each node. The numbers on links denote the cost.

(a) Is the two-step approach successful in finding link-disjoint primary and backup paths?

(b) If not, find a pair of link-disjoint primary and backup paths from A to F without using the two-step approach.

16.9. Consider the network topology in Fig. 16.19. Assume connection request 0 $\rightarrow$ 2 arrives. Find a primary and a backup path for the request. Calculate the average protection-switching time in case of a failure (of either of the two links) on the primary path. Message processing time at each node D = 10 μs, propagation delay on each link P = 400 μs, time to set cross connection is C =100 μs, and time to detect a link failure F = 100 μs.

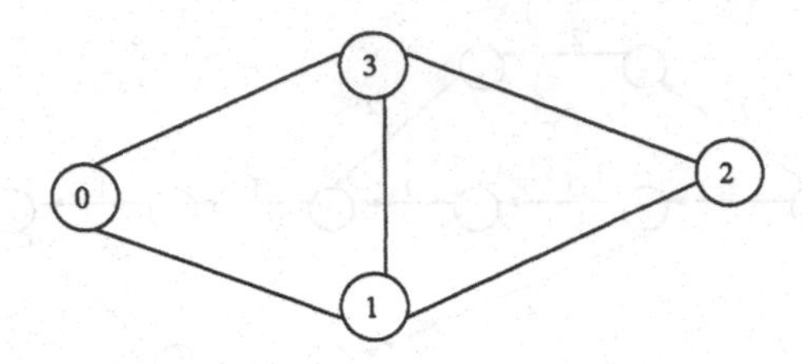

Figure 16.19 A sample network for protection switching.

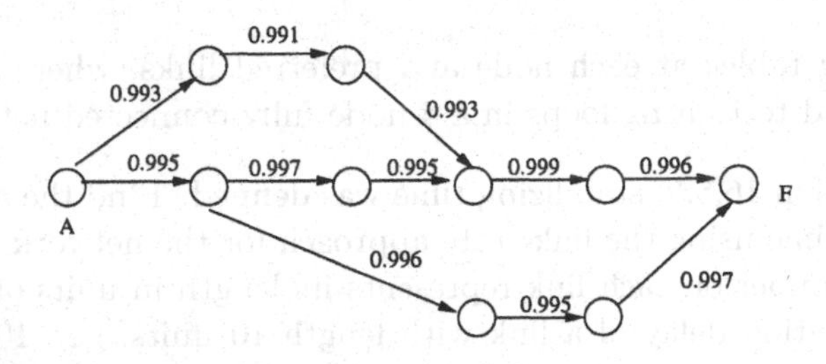

Figure 16.20 Availability of a connection carried over disjoint primary and backup paths.

16.10. Figure 16.20 shows the availability of various links in a network. Availability is the probability that a link is found in the operating state at any point of time. Assuming that the pair of link-disjoint primary and backup paths found in Problem 16.8 is used for a connection C, between nodes A and F, find the availability of connection C. Availability of a connection is the probability that it is found in the operating state at any time. Assume that the only failures are link failures and they are independent of one another.

16.11. Given a network with availability defined on links as in Problem 16.10, design an algorithm to find the most-reliable path (path with maximum availability) from source to destination. State the time complexity of your algorithm.

16.12. What are the different categories of signaling and reservation protocols? State the advantages and disadvantages of each.

16.13. Differentiate between the link-state and distributed-routing approaches for connection management.

16.14. Discuss the advantages and disadvantages of link vs. path protection.

16.15. What is the advantage of the Max-Shared-First heuristic over First-Fit and Last-Fit heuristics for wavelength assignment?

Chapter 17

Optical Packet Switching (OPS)

17.1 Introduction

Over the past few years, the Internet traffic has grown rapidly and the optical transport bandwidth has been continuously increasing. These changes are stimulating the evolution of data networks. In such a dynamic environment, a network architecture that accommodates multiple data formats, supports high data rates, and offers flexible bandwidth provisioning is the key feature of the next-generation Internet.

Among all the optical networking paradigms, wavelength-routed networks, in which lightpaths are setup on specific wavelengths, have been the focus of many studies (see Chapter 7). Over a short period of a few years, these networks have evolved from textbook subjects to real-life products. The current ongoing efforts to automate and expedite wavelength and bandwidth provisioning in the optical layer indicate the inevitable trends that lead to more intelligent optical networks [ARDC01, BDLT01b, BDLT01a]. Migration of certain switching functionalities from electronics to optics will remove the incumbent layers that impose unnecessary optical-electrical-optical conversions and unnecessary signal processing. A more flexible WDM optical network is desired by service providers to meet their versatile traffic demands.

From the demand side, the tremendous increase in the transport networks' bandwidth is stimulating the high volume of gigabit multimedia services. A robust network supporting various kinds of traffic is the cornerstone for the next-generation Internet. However, despite the high throughput the wavelength routers deliver, they still lack the flexibility in switching granularity. An optical crossconnect (OXC) can only switch whole wavelengths unless electrical grooming switch fabrics are used. On the other hand, the past evolution of the Internet has shown that a future-proof network not only needs high-capacity circuit switching, but also needs high-performance switching with finer granularities. The key to the successful accommodation of heterogeneous traffic lies in the deployment of cost-effective switching schemes that provide easy access to the large bandwidth WDM offers. These arguments serve as the main motivation for Optical Packet Switching (OPS).

17.2 Optical Packet Switching (OPS) Basics

Optical packet switching is optical switching with the finest granularity. Incoming packets are switched all-optically without being converted to electrical signal. It is the most flexible and also the most demanding switching scheme.

17.2.1 Slotted Networks

In general, there are two categories of optical packet-switched networks: slotted (synchronous) and unslotted (asynchronous) networks. When individual optical switches form a network, at the input ports of each node, packets can arrive at different times. Since the switch fabric can change its state incrementally (set up one input-output connection at an arbitrary time) or jointly (set up multiple input-output connections together at the same time), it is possible to switch multiple time-aligned packets together or to switch each packet individually 'on the fly'. In both cases, bit-level synchronization and fast clock recovery are necessary for packet-header recognition and packet delineation.

In general, all of the packets may have the same size in a slotted network (in a variation of the slotted network, it is possible that packets are of variable length, but each packet's length is an integral multiple of a slot, and all packet transmissions start (and end) at a slot boundary). A fixed-size time slot contains both the payload and the header. The time slot has a longer duration than the whole packet to provide some extra guard time. All the input packets arriving at the input ports need to be aligned in phase with one another before entering the switch fabric (Fig. 17.1). To successfully synchronize all the incoming packets, it is important to study what types of delay variation a packet could experience, which we will discuss below.

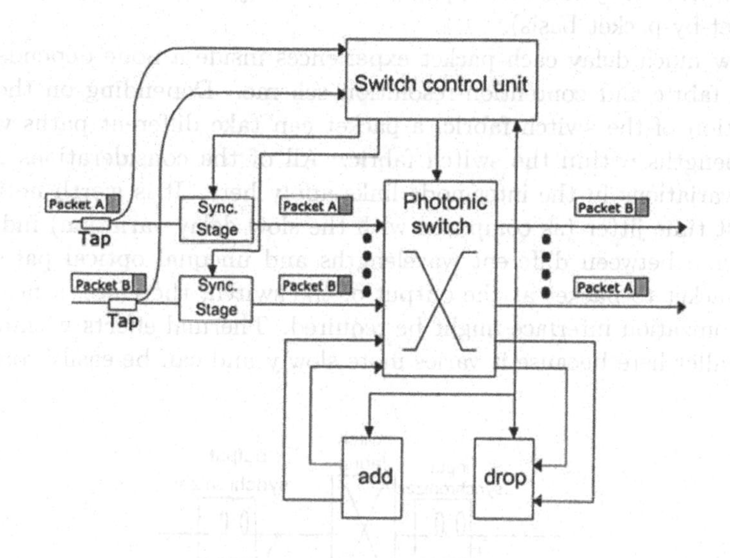

Figure 17.1 A generic node architecture of the slotted network.

The time for a packet to travel through a certain distance of the fiber depends on the fiber length, the chromatic dispersion, and the temperature variation. When WDM is used, the effect of chromatic dispersion has to be considered. Chromatic dispersion results in different propagation speed for packets transmitted on different wavelengths; therefore, different propagation delays occur. For example, with a typical fiber dispersion of 20 ps/nm/km (where ps is the time unit for delay variation, nm the unit for wavelength difference, and km the unit for propagation distance), a wavelength variation of 30 nm (consistent with the typical Erbium-Doped Fiber Amplifier's 1530 - 1560 nm window) and a propagation distance of 100 km, the propagation delay variation would be about 60 ns. If dispersion compensation fibers are used, the above delay variation can be reduced by one order of magnitude

The packet propagation speed also varies with temperature, with a typical figure of 40 ps/ ^{0}C/km. 100 km of fiber under temperature variation range of 0 - 25 ^{0}C introduces 100 ns of delay variation.

The delay variations mentioned above are relatively small with respect to time; thus, they can be compensated statically instead of dynamically (on a packet-by-packet basis).

How much delay each packet experiences inside a node depends on the switch fabric and contention-resolution scheme. Depending on the implementation of the switch fabric, a packet can take different paths with unequal lengths within the switch fabric. All of the considerations given to delay variations in the inter-node links apply here. It is worth noting that the fast time jitter (as compared with the slow delay variation) induced by dispersion between different wavelengths and unequal optical paths varies from packet to packet at the output of the switch; therefore, a fast output synchronization interface might be required. Thermal effects within a node are smaller here because it varies more slowly and can be easily controlled.

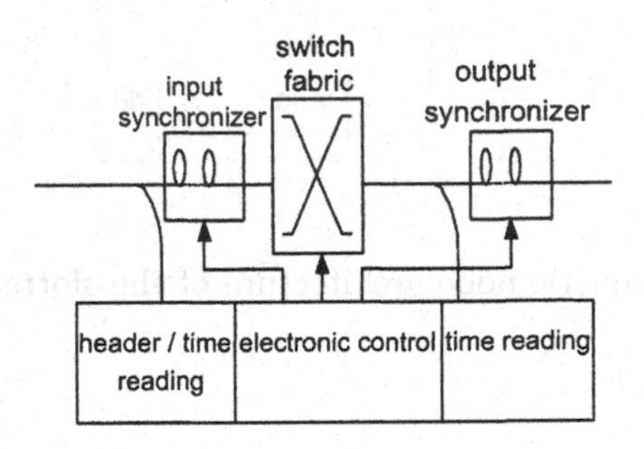

Figure 17.2 Functional block diagram for synchronization of packets.

Figure 17.2 shows a functional of block diagram for the node architecture of a slotted network. A passive tap splits a small amount of power from the incoming signal (or packet) for the header reading. The header-processing circuit recognizes a preamble at the beginning of the packet, and then it reads the header information. It also passes the timing information of the incoming packet to the control unit to configure the synchronization stages and the switch fabric. The input synchronization stage aligns packets before they enter the switch fabric. The output synchronization stage, which is not shown in Fig. 17.1, can further compensate for the fast time jitter that occurs inside the node. This may or may not be necessary, depending on the actual packet format and node architecture.

Packet delineation is essential for both header reading and switch configuration. During packet delineation, the incoming bits are locked in phase with the clock in order for the node to read the header information. Traditional phase-locked-loop approach is not applicable because it requires too many bits. Burst-mode receivers have been demonstrated to achieve bit-level synchronization within nanosecond time range [Taji99, YaMH99]. [SWWG00] describes another burst-mode receiver setup in which the transmitter frequency multiplexes its bit-clock with the baseband data and modulates the optical carrier with the composite signal. The data and clock travel along the fiber with negligible dispersion. At the receiver end, a photodiode detects the optical signal. Its RF output is amplified and split. The data and clock are first separated by a low-pass filter that cuts off the baseband and then by a narrow band-pass filter centered at the clock frequency. The retrieved clock is fed into the analog receiver. If the delay from the output of the photodiode to the input of the receiver is matched, the clock and data will be in bit-synchronization for all incoming packets.

Since packets are entering a node from different links, for all the previously-stated reasons, they could arrive totally out of phase with one another. Figure 17.3 [Pruc93] shows a typical synchronization stage consisting of a series of switches and delay lines, as they appear in the input synchronization stage of a node.

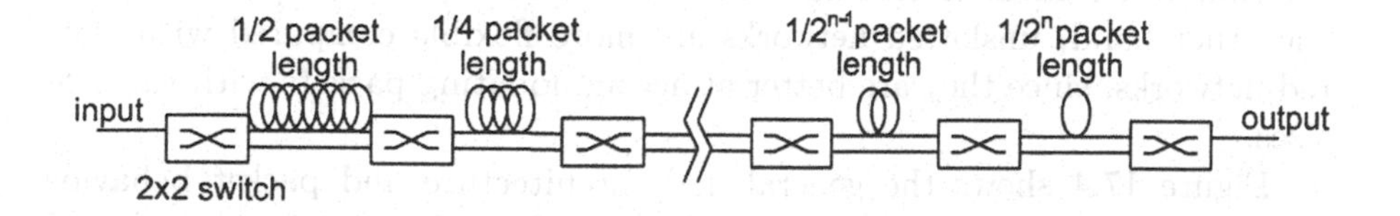

Figure 17.3 A scheme for input synchronization stage in a node.

Once the header processor recognizes the bit pattern and performs packet delineation, it identifies the packet start time, and the control unit will calculate the necessary delay and configure the correct path through these switched delay lines. The length of the delay lines are in an exponential sequence between the 2x2 switches, i.e., the first delay line is equal to 1/2 a time-slot duration, the second delay is equal to 1/4 a time-slot duration, etc. The resolution of this scheme is $1/2^n$ of the time-slot duration, where n is the number of delay lines. This type of synchronization scheme is suitable for both static (slow) and dynamic (fast) synchronization. At the system initialization, the synchronization is setup to compensate for delay variations between different inputs and to keep this configuration throughout the system operation time (static). For the packet-based (dynamic) synchronization, much faster switches are necessary to operate during the guard time.

From a physical point of view, such a packet-synchronization scheme introduces insertion loss and crosstalk due to the switches used. Cascading the switches will inevitably require optical amplification, which will result in degraded signal-to-noise ratio. Meanwhile, the crosstalk accumulated through the switches may also increase the bit-error rate. In a multi-node network, the power penalty brought by all the synchronization stages may significantly impair the system performance.

17.2.2 Unslotted Networks

In an unslotted network, the packets may or may not have the same size. Packets arrive and enter the switch without being aligned in time. Therefore, the packet-by-packet switch operation could take place at any point in time. Obviously, in an unslotted network, the chance for contention is larger because the behavior of the packets is more unpredictable (similar to the contention in the slotted and unslotted ALOHA networks [Tane96]). On the other hand, unslotted networks are more flexible compared with slotted networks, since they are better at accommodating packets with variable sizes.

Figure 17.4 shows the general node architecture and packet behavior for unslotted networks. (Note the absence of synchronization stages and packet alignment.) The fixed-length fiber delay lines hold the packet when header processing and switch reconfiguration are taking place. There is no packet-alignment stage, and all the packets experience the same amount of delay with the same relative position in which they arrived, provided there is no contention. The unslotted network circumvents the requirement of synchronization stages. However, given the same traffic load, the network throughput is lower than that of the slotted networks because contention is more likely to occur.

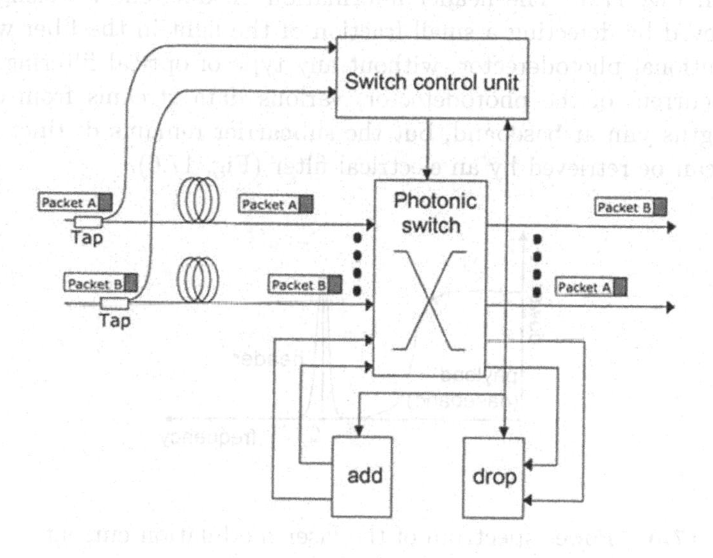

Figure 17.4 A generic node architecture of the unslotted network.

17.3 Header and Packet Format

In an electronic network, the packet header is transmitted serially with payload data at the same data rate, such as in IP packets and ATM cells. Electronic routers or switches will process the header information at the same data rate as the payload. In an optical network, the bandwidth may be much higher than their electronic counterparts. A typical wavelength channel has the line speed of 10 Gbps (OC-192) today (and increasing to higher speeds in the future). Although there are various techniques to detect and recognize packet headers at Gbps speed, either electronically or optically, it is costly to implement electronic header processors operating at such high speed.

Among several different proposed solutions, packet switching with subcarrier multiplexed (SCM) header is attracting increasing interest. In this approach, the header and payload data are multiplexed on the same wavelength (optical carrier). In the current that modulates the laser transmitter, payload data is encoded as the baseband signal, while header bits are encoded on a properly-chosen subcarrier frequency at a lower bit rate, as shown in Fig. 17.5. The header information on different wavelengths can be retrieved by detecting a small fraction of the light in the fiber with just a conventional photodetector, without any type of optical filtering. In the output current of the photodetector, various data streams from different wavelengths jam at baseband, but the subcarrier remains distinct and the header can be retrieved by an electrical filter (Fig. 17.6).

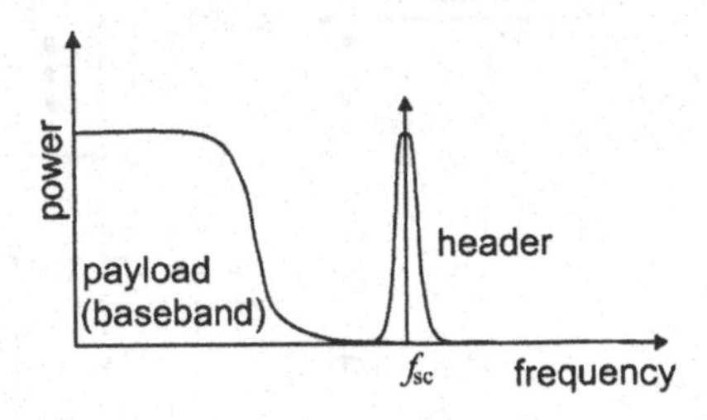

Figure 17.5 Power spectrum of the laser modulation current.

A nice feature of a subcarrier-multiplexed header is that the header can be transmitted in parallel with the payload data and it can take up the whole payload transmission time. Of course, the header can also be transmitted serially with the payload, if so desired. One potential pitfall of a subcarrier-multiplexed header is its possible limit on the payload data rate. If the payload data rate is increased, the baseband will expand and might eventually overlap with the subcarrier frequency, which is limited by the microwave electronics.

In many of the routing and switching protocols, packet headers have to be updated at each node. There have been several approaches proposed for optical header replacement for headers transmitted serially with the payload data stream. Optical header replacement could be done by blocking the old header with a fast optical switch and inserting the new header, generated locally by another laser, at the proper time. One important issue here is that,

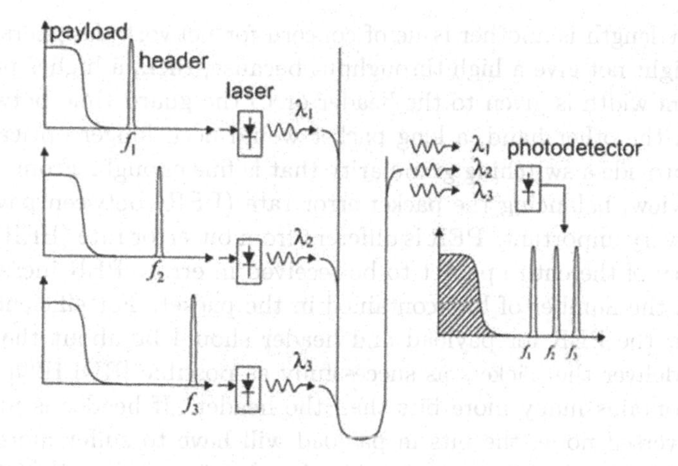

Figure 17.6 Header retrieval in SCM.

in a WDM network the new header should be precisely at the same wavelength as the payload data, otherwise serious problems could arise because of dispersion, nonlinearity, or wavelength-sensitive devices in the network.

It has been proposed that the header updating be done by transmitting the payload and the header on separate wavelengths, and demultiplexing the header for optoelectronic conversion, electronic processing and retransmission on the header wavelength. This approach suffers from fiber dispersion, which separates the header and payload as the packet propagates through the network. Subcarrier-multiplexed headers have far less dispersion problems since they are very close to the baseband frequency. SCM header could be removed by narrow-band optical filters, but it would be very sensitive to wavelength drift. Previous practical SCM header replacement schemes are limited to full optoelectronic conversion of the entire packet, followed by electronic filtering, re-modulation and re-transmission on a new laser. Reference [VaBl97] has proposed a technique to update the SCM header with simultaneous wavelength conversion of baseband payload using semiconductor optical amplifiers (SOAs). It involves a two-stage process: first, simultaneous SCM header suppression and wavelength conversion of the baseband payload is achieved due to the low-pass frequency response of cross-gain modulation in the SOAs, and then header replacement is achieved by optically re-modulating the wavelength-converted signal with a new header at the original subcarrier frequency.

Packet length is another issue of concern for network designers. A short packet might not give a high throughput because, then, a higher percentage of the bandwidth is given to the header or to the guard time between time slots. On the other hand, a long packet would need longer optical buffers, and not provide a switching granularity that is fine enough. From a physical point of view, balancing the packet-error rate (PER) between payload and header is very important. PER is different from bit-error rate (BER); it is the probability of the entire packet to be received in error. PER increases with BER and the number of bits contained in the packet. For efficient network operation, the PER for payload and header should be about the same, in order to deliver the packets as successfully as possible [RDFH99]. Payload usually contains many more bits than the header. If header is updated at every traversed node, the bits in payload will have to suffer more physical impairment than the bits in the header. Another fact is that, if SCM is used, the header is usually transmitted at a lower bit rate than the payload data. All of these factors lead to a big advantage of lower BER for header bits over the payload bits. Therefore, it is imperative to optimize the amount of power to be tapped from the packet at each node and the packet length in order to achieve a balanced PER for payload and header at the destination node.

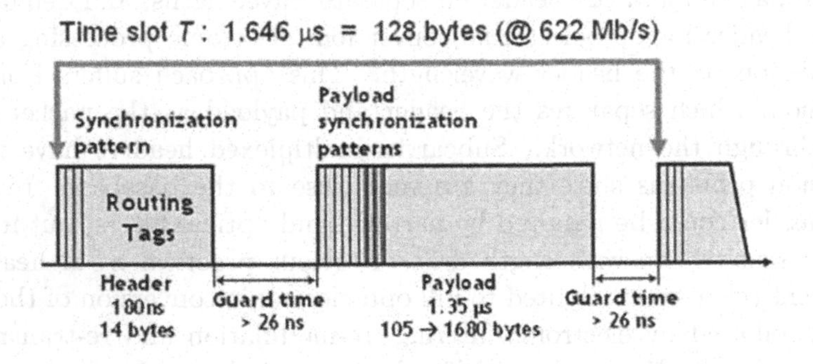

Figure 17.7 Packet format in KEOPS project.

The KEOPS packet format is presented in Fig. 17.7 [GRGJ98]. In the KEOPS project, the optical packets are placed into a fixed time slot, allowing for synchronous operation of the switching nodes and fiber-delay line packet buffering. These packets contain a 622 Mbps header (which is electronically processed at the nodes) and a payload with fixed duration and variable bit rate (e.g., up to 10 Gbps). In particular, guard times must be inserted to account for the optoelectronic device switching time, the jitter experienced by the payload in the nodes (mainly in the fiber delay lines), and the finite resolution of the synchronization units at the network/node interfaces. The key elements in defining a packet format are the time slot and payload duration. In addition, optical packet switching requires fast switching, compatible with packet guard times in the *ns* range.

17.4 Typical Contention Resolution in OPS Networks

In an OPS network, contention occurs at a switching node whenever two or more packets try to leave the switch fabric on the same output port, on the same wavelength, at the same time. In electrical packet-switched networks, contention is resolved with the store-and-forward technique, which requires the packets in contention to be stored in a memory bank, and to be sent out at a later time when the desired output port becomes available. This is possible because of the availability of electronic random-access memory (RAM). There is no equivalent optical RAM technology; therefore, the optical packet switches need to adopt different approaches for contention resolution. Meanwhile, WDM networks provide one additional dimension, namely wavelength, for contention resolution. This section explores all three dimensions of contention-resolution schemes: wavelength, time, and space, and provides a detailed discussion of how to combine these three dimensions into an effective contention-resolution scheme.

The contention-resolution mechanism of the three dimensions are outlined below:

- *Wavelength conversion* offers effective contention resolution without relying on buffer memory [DaHS98, ErLi00, HNCA99, HaDS98, CTYT00]. Wavelength converters can convert wavelengths of packets which are contending for the same wavelength of the same output port. It is a powerful and the most preferred contention-resolution scheme (as this study will demonstrate) that does not cause extra packet latency, jitter, and re-sequencing problems. The interested reader is referred to Chapter 10 for more details on wavelength conversion.

- *Optical delay line* (which provides sequential buffering) is a close imitation of the RAM in electrical routers, although it offers fixed and finite amount of delay. Many previously-proposed architectures employ optical delay lines to resolve contentions [HuCA98, TaGC01, TYCT00, GRGJ98, Haas93, HCGF98, CFKM96, CaCe01]. However, since optical delay lines rely on the propagation delay of the optical signal in silica to buffer the packet in time, i.e., due to their sequential access, they have more limitations than the electrical RAM. To implement large buffer capacity, the switch needs to include a large number of delay lines.

- *Space deflection* [FoBP95, CaTT99, BoFB93, AcSh92, FeSR94] is a multiple-path routing technique. Packets that lose the contention are routed to nodes other than their preferred next-hop nodes, with the expectation that they will eventually be routed to their destinations. The effectiveness of deflection routing depends heavily on the network topology and the offered traffic pattern.

Both wavelength conversion and optical buffering require extra hardware (wavelength converters and lasers for wavelength conversion; fibers and additional switch ports for optical buffering) and control software. Deflection routing can be implemented with extra control software only.

With an orthogonal classification, optical packet-switched networks can be divided into two categories: time-slotted networks with fixed-length packets [Prat97] and unslotted networks with fixed-size or variable-size packets [HaDS98, TYCT00]. In a slotted network, packets of fixed size are placed in time slots. When they arrive at a node, they are aligned before being switched jointly [GRGJ98] (see Section 17.2). In an unslotted network, the nodes do not align the packets and switch them one by one 'on the fly'; therefore, they do not need synchronization stages. Because of such unslotted operation, they can switch variable-length packets. However, unslotted networks have lower overall throughput than slotted networks, because of the increased packet contention probability [BoFB93]. Similar contrast exists between the unslotted and slotted version of the ALOHA network [Tane96]. Due to the lack of viable optical RAM technologies, all-optical networks find it difficult to provide packet-level synchronization, which is required in slotted networks. In addition, it is preferred that a network can accommodate natural IP packets with variable lengths. Therefore, this chapter primarily focuses on optical contention-resolution schemes in unslotted, asynchronous networks.

Figure 17.8 shows the generic node architectures with contention-resolution schemes in time, space, and wavelength domains. Every node has a number of add/drop ports, and this number will vary depending on the nodal degree (i.e., how many input/output fiber pairs the node has). Each add/drop fiber ports will correspond to multiple client interfaces reflecting multiple wavelengths on each fiber. Each client interface-input (-output) will be connected to a local receiver (transmitter). Different contention-resolution schemes give rise to different architectures.

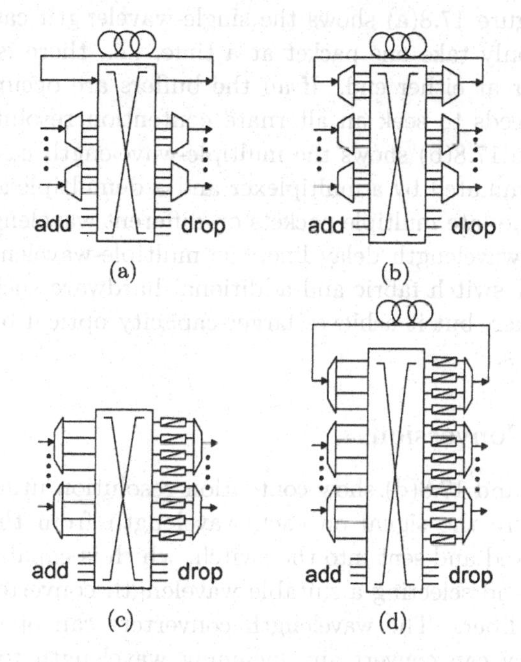

Figure 17.8 Node architectures for different contention-resolution schemes: (a) single-wavelength optical buffering; (b) multi-wavelength optical buffering; (c) wavelength conversion; (d) wavelength conversion and multi-wavelength optical buffering.

17.4.1 Optical Buffering

Optical buffering utilizes one or more optical fiber delay lines looping the signal from the output back to the input of the switch fabric. Figures 17.8(a) and 17.8(b) illustrate optical buffering. Time-domain contention resolution requires optical buffering wherein the packet causing the contention enters the optical delay line at the output of the switch fabric, traverses through the entire delay line, and re-enters the switch fabric. For unslotted operation, the delay line can consist of an arbitrary length of fiber. Simulation experiments find little improvement in network performance when delay lines are made longer than the maximum packet size.

We consider both cases of optical buffering: single-wavelength and multi-wavelength. Figure 17.8(a) shows the single-wavelength case, in which the delay line can only take one packet at a time, i.e., there is no multiplexer or demultiplexer at either end. If all the buffers are occupied, the packet in contention needs to seek an alternate contention resolution or must be dropped. Figure 17.8(b) shows the multiple-wavelength case in which each delay line is terminated by a multiplexer and a demultiplexer. Such a delay line can accommodate multiple packets on different wavelengths. Compared with the single-wavelength delay line, the multiple-wavelength counterpart requires a larger switch fabric and additional hardware such as multiplexer and demultiplexer, but it achieves larger-capacity optical buffering on multiple wavelengths.

17.4.2 Wavelength Conversion

Figures 17.8(c) and 17.8(d) show contention resolution utilizing wavelength conversion, where the signal on each wavelength from the input fiber is first demultiplexed and sent into the switch, which is capable of recognizing the contention and selecting a suitable wavelength converter leading to the desired output fiber. The wavelength converters can operate with a full degree (i.e., they can convert any incoming wavelength to a fixed desired wavelength) or with a limited range (i.e., they can convert one or several pre-determined incoming wavelengths to a fixed desired wavelength).

17.4.3 Space Deflection

Space deflection relies on another neighboring node to route the packet when contention occurs. As a result, the node itself can adopt any node architecture in Fig. 17.8. Space deflection resolves contention at the expense of the inefficient utilization of the network capacity and the switching capacity of another node. Obviously, this is not the best choice among contention-resolution schemes. As the later sections will discuss, the node will seek wavelength-domain contention resolution first, time-domain contention resolution second, and space-domain contention resolution third. In practice, the contention resolution will often employ a combination of wavelength-, time-, and space-domain contention resolution.

In most deflection networks, certain mechanisms need to be implemented to prevent looping (a packet being sent back to a node it has visited before), such as setting a maximum hop count and discarding all the packets that have passed more hops than this number. This is similar to the time-to-live (TTL) mechanism for routing IP packets.

17.4.4 Combination Schemes

By mixing the basic contention-resolution schemes discussed so far, one can create combination schemes. For example, Fig. 17.8(d) shows the node architecture for wavelength conversion combined with multi-wavelength buffering. Note that a packet can be dropped at any node under all of these schemes due to (1) unavailability of a free wavelength at the output port, (2) unavailability of a free buffer, and/or (3) the fact that the packet may have reached its maximum hop count.

We define the notations for these schemes as follows:

- *baseline*: No contention resolution. Packet in contention is dropped immediately.
- *Nbuf*, *Nbufwdm*: Buffering. The node has N delay lines, and each delay line can take one or multiple wavelengths at a time.
- *def*: Deflection.
- *wc*: Full-range wavelength conversion.
- *wclimC*: Limited wavelength conversion. C given wavelengths can be converted to one fixed wavelength.

In the following section, these notations indicate the different approaches and their priorities. For example, *4wav_wc+16buf+def* means a combination of full-range wavelength conversion, single-wavelength buffering with 16 delay lines, and deflection in a 4-wavelength system. The order of contention resolution is as follows: *A packet that loses the contention will first seek a vacant wavelength on the preferred output port. If no such wavelength exists, it will seek a vacant delay line. If no delay line is available, it will seek a vacant wavelength on the deflection output port. When all of the above options fail, the packet will be dropped.* This contention-resolution order provides the best performance in terms of the packet loss rate and the packet delay.

17.4.5 Simulation Experiments and Performance Comparison

Network Topology and Configuration

For purposes of illustration, the network topology under study is a part of a telco's optical mesh network, as shown in Fig. 17.9. Each link i is L_i km long and consists of two fibers to form a bi-directional link. Every fiber contains W wavelengths, each of which carries a data stream at data rate R. Each node is equipped with an array of W transmitters and an array of W receivers that operate on one of the W wavelengths independently at data rate R. By default, all packets are routed statically via the shortest path.

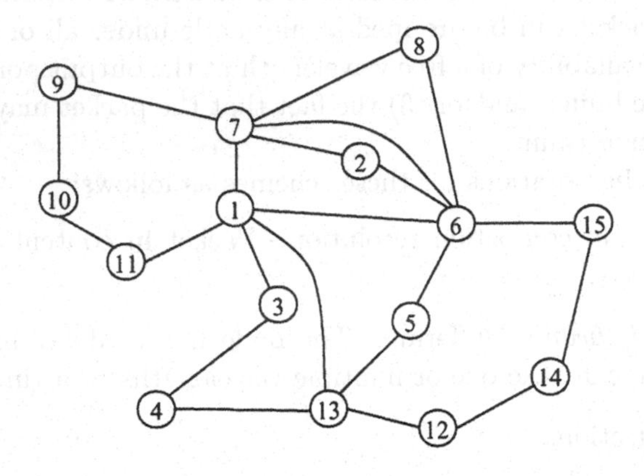

Figure 17.9 An example mesh network topology.

It is worth mentioning that, because some nodes have many neighbors, the deflection scheme should avoid 'blind deflections'. Deflection is carried out only in hub nodes, which are nodes that have much higher nodal degrees than other nodes, to serve as major routing nodes. Nodes 1, 6, 7, and 13 in Fig. 17.9 are examples of hub nodes. Most of the routing is done by the four hub nodes in this example network, and it is possible to have a deflection policy, with which a packet would only be deflected to a node that can eventually lead to the packet's original 'next-hop' node with no more than a few extra hops, say 2. If no such node exists, the packet will be dropped. Deflection routing requires the node to have a second entry in the routing table that contains preferred deflection ports for each destination.

For example, in Fig. 17.9, if a packet from node 3 is destined for node 9 via node 1, and there is contention at the output port in node 1 leading to node 7, the packet will be deflected to the port leading to node 11; if the port leading to 11 is also busy, the packet will be dropped immediately instead of being deflected to node 13 or node 6. Table 17.1 shows an example of the deflection entries in the routing table at node 7.

Table 17.1 Deflection table of node 7.

Next-hop node	Deflect to
1	6
2	6
6	2
8	6
9	Drop

With deflection routing, a loop-mitigation mechanism is necessary. A maximum hop count H is set to limit how many hops a packet can travel (each time the packet passes through a delay line or is transmitted from one node to another, it is counted as one hop). Nevertheless, the network topology of Fig. 17.9 and the aforementioned deflection policy can automatically eliminate looping (since the shortest path between any source-destination pair involves no more than two hub nodes, and we can ensure that the deflection table will not cause any looping). The purpose of setting the maximum hop count here is to limit physical impairments (signal-to-noise-ratio degradation, accumulated crosstalk, accumulated insertion loss, etc.) of packets, which is partly introduced by optical buffering.

Traffic Generation

One of the important requirements for the simulation experiment is to model the network traffic as close to reality as possible. A main characteristic of Internet traffic is its burstiness, or self-similarity. It has been shown in the literature that self-similar traffic can be generated by multiplexing multiple sources of Pareto-distributed ON/OFF periods. In the context of a packet-switched network, the ON periods correspond to a series of packets sent one after another, and OFF periods are the periods of silence [WTSW97].

The probability density function (pdf) and probability distribution function (PDF) of the Pareto distribution are:

$$p(x) = \frac{\alpha \cdot b^{\alpha}}{x^{\alpha+1}} \tag{17.1}$$

$$P(x) = \int_{b}^{x} \frac{\alpha \cdot b^{\alpha}}{x^{\alpha+1}} dx = 1 - \frac{b^{\alpha}}{x^{\alpha}} \tag{17.2}$$

where α is the shape parameter (tail index), and b is minimum value of x. When $\alpha \leq 2$, the variance of the distribution is infinite. When $\alpha \geq 2$, the mean value is infinite as well. For self-similar traffic, α should be between 1 and 2. The Hurst parameter H is given as $H = (3 - \alpha)/2$.

Since $0 < P(x) \leq 1$, the value of x can be generated from a random number $\mathcal{RND}$ with the range $(0, 1]$:

$$\frac{b^{\alpha}}{x^{\alpha}} = \mathcal{RND} \tag{17.3}$$

$$x = b \cdot (\frac{1}{\mathcal{RND}})^{\frac{1}{\alpha}} \tag{17.4}$$

The mean value of Pareto distribution is given by:

$$E(x) = \frac{\alpha \cdot b}{\alpha - 1} \tag{17.5}$$

Once $\alpha_{on}, \alpha_{off}, b_{on}, b_{off}$ are given, the distribution of the ON/OFF periods are determined. b_{on} is the minimum ON-period length, equal to the smallest packet size divided by the line rate. The average load of each ON/OFF source, L, is:

$$L = \frac{E_{on}}{E_{on} + E_{off}}$$

where E_{on} and E_{off} are the mean value of ON and OFF period. Therefore,

$$b_{off} = (\frac{1}{L} - 1) \frac{\alpha_{on} \cdot (\alpha_{off} - 1)}{(\alpha_{on} - 1) \cdot \alpha_{off}} \cdot b_{on} \tag{17.6}$$

During the ON period of the ON/OFF source, packets are sent back-to-back.

In the experiments, the network traffic is assumed to consist of IP packets. The nature of IP packets is known to be hard to capture [Caid05]. Statistical data indicates a predominance of small packets, with peaks at the common sizes of 44, 552, 576, and 1500 bytes. The small packets of 40-44 bytes in length include TCP acknowledgment segments, TCP control segments, and telnet packets carrying single characters. Many TCP implementations that do not implement path Maximum Transmission Unit (MTU) discovery use either 512 or 536 bytes as the default Maximum Segment Size (MSS) for non-local IP destinations, yielding a 552-byte or 576-byte packet size. An MTU size of 1500 bytes is the characteristic of Ethernet-attached hosts. The cumulative distribution of packet sizes in Fig. 17.10 [Caid05] shows that almost 75% of the packets are smaller than the typical TCP MSS of 552 bytes. Nearly 30% of the packets are 40 to 44 bytes in length. On the other hand, over half of the total traffic is carried in packets of size 1500 bytes or larger. This irregular packet-size distribution is difficult to express with a closed-form expression. We adopt a truncated 19-order polynomial, fitted from the statistical data, to faithfully reproduce the IP packet-size distribution (as shown in Fig. 17.10). The number of orders, 19, is the smallest number that can reproduce a visually close match with the steep turns in the statistical data. We set the maximum packet size to be 1500 bytes, since the percentage of packets larger than 1500 bytes is negligibly small.

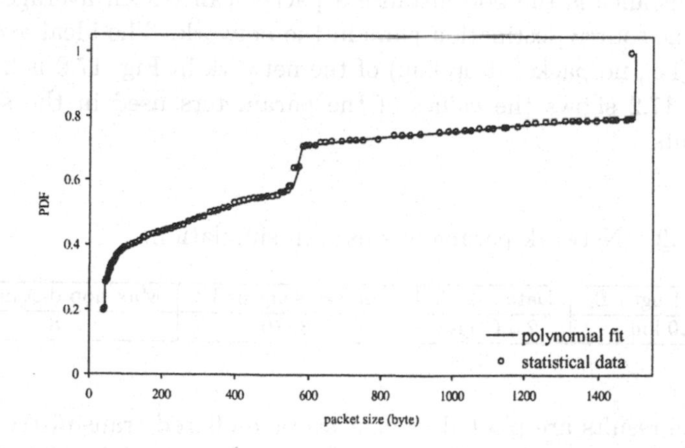

Figure 17.10 Probability distribution function (PDF) of IP packet sizes.

Simulation Metrics

We have chosen four metrics to evaluate the network performance with different contention-resolution schemes: network throughput, packet-loss rate, average end-to-end delay, and average hop distance. They indicate the network utilization, reliability, latency, and physical impairments to the signal. The packet-loss rate is the total number of lost packets divided by the total number of generated packets. The network throughput is defined as:

$$\text{Network throughput} = \frac{\text{total number of bits successfully delivered}}{\left(\frac{\text{network transmission capacity} \times \text{simulation time}}{\text{ideal average hop distance}}\right)}$$

$$\text{Network transmission capacity} = (\text{total \# of links}) \times (\text{\# of wavelengths}) \times (\text{data rate})$$

Network throughput is the fraction of the network resource that successfully delivers data. Because packets can be dropped, a part of the network capacity is wasted in transporting the bits that are dropped. In an ideal situation where no packets are dropped and there is no idle time in any links, the network will be fully utilized and the throughput will reach unity. Average hop distance is the hop distance a packet can travel, averaged over all the possible source-destination pairs in the network. The ideal average hop distance (i.e., no packet dropping) of the network in Fig. 17.9 is 2.42.

Table 17.2 shows the values of the parameters used in the simulation experiments.

Table 17.2 Network parameters used in simulation.

Link length L_i	Data rate R	# of wavelengths W	Max hop distance H
20 km	2.5 Gbps	4, 8, 16	8

All the results are plotted against average offered transmitter load, i.e., the total number of bits offered per unit time divided by the line speed. (For example, if the source is generating 0.5 Gbits of data per second and the transmitter/line capacity is 2.5 Gbps, the transmitter load would be 0.2.)

With a given average offered transmitter load and a uniform traffic matrix, the average offered link load per wavelength is:

$$\text{Ave. offered link load} = \frac{\text{ave. offered TX load} \times \text{total \# of TX in the network} \times \text{ave. hop distance}}{\text{\# of wavelengths} \times \text{total number of uni-directional links in the network}}$$

Comparison of the Basic Contention-Resolution Schemes

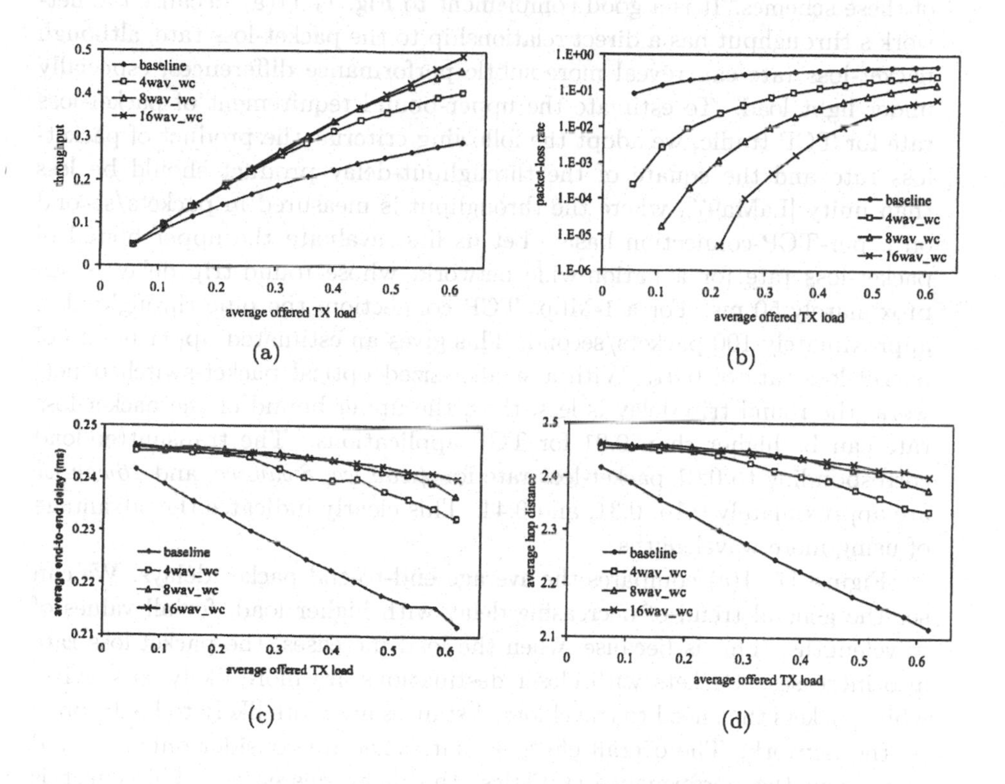

Figure 17.11 Performance comparison of the basic schemes – baseline and wavelength conversion.

Figure 17.11(a) compares the network throughput of the schemes incorporating wavelength conversion with different number of wavelengths. For reference, it also shows the throughput of *baseline*. We simulate four wavelengths in *baseline*, although the number of wavelengths does not affect the results of *baseline* since each wavelength plane operates independently. We find that more wavelengths provide better throughput performance for the wavelength-conversion scheme. Meanwhile, the margin of improvement in throughput decreases when the wavelength number increases; with 16 wavelengths, the network throughput is nearly linear to transmitter load.

Figure 17.11(b) compares the packet-loss rate (represented in fraction) of these schemes. It is a good complement to Fig. 17.11(a) because the network's throughput has a direct relationship to the packet-loss rate, although packet-loss rate can reveal more subtle performance differences, especially under light load. To estimate the upper-bound requirement of packet-loss rate for TCP traffic, we adopt the following criteria: the product of packet-loss rate and the square of the throughput-delay product should be less than unity [LaMa97], where the throughput is measured in packets/second on a per-TCP-connection basis. Let us first evaluate the upper bound of packet-loss rate for a nation-wide network, whose round-trip delay is approximately 50 ms. For a 1-Mbps TCP connection, the pipe throughput is approximately 100 packets/second. This gives an estimated upper bound of packet-loss rate of 0.01. With a smaller-sized optical packet-switched network, the round-trip delay is less; thus, the upper bound of the packet-loss rate can be higher than 0.01 for TCP applications. The transmitter load corresponding to 0.01 packet-loss rate for *4wav_wc*, *8wav_wc*, and *16wav_wc* are approximately 0.16, 0.31, and 0.44. This clearly indicates the advantage of using more wavelengths.

Figure 17.11(c) compares the average end-to-end packet delays. We can see the general trend of decreasing delay with higher load, for all values of wavelengths. This is because, when the load increases, the packet-loss rate also increases. Packets with closer destinations are more likely to survive, while packets that need to travel long distances are more likely to be dropped by the network. The overall effect is that, when we consider only survived packets in the performance statistics, the delay decreases. This effect is most prominent with *baseline*, because it has the highest packet-loss rate. The same reasoning can be applied to explain the lower delay of *4wav_wc*.

Figure 17.11(d) shows the average hop-distance comparison. Since neither *baseline* nor *wc* involves any buffering, the average hop distance is proportional to the average end-to-end delay.

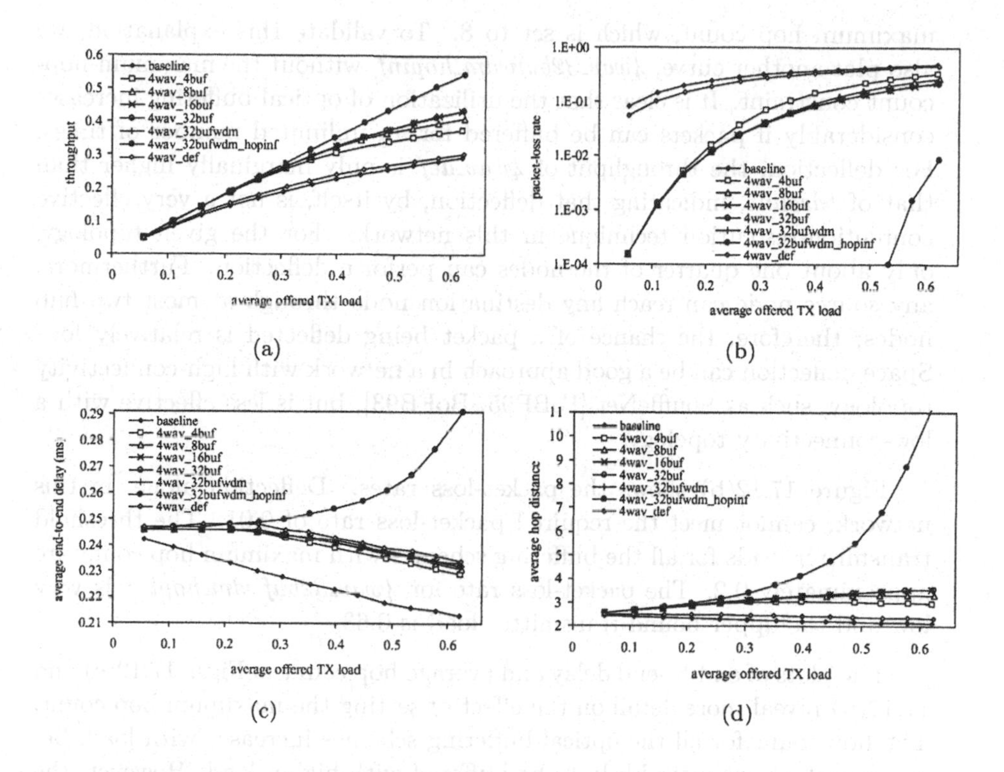

Figure 17.12 Performance comparison of the basic schemes – baseline, optical buffering, and deflection.

In the schemes incorporating optical buffering, all the optical delay lines are of length 1 km (with propagation delay of 5 μs), enough to hold a packet with maximum length (12000 bits). We simulate four wavelengths in the network, with different buffering and deflection settings; this is because other number of wavelengths renders similar results, and larger number of wavelengths requires much longer time to simulate. Figure 17.12(a) compares the throughput of different optical-buffer schemes and the deflection scheme. For the buffering schemes, we observe the difference in performance between different numbers of optical delay lines. Network throughput first increases with the number of delay lines, but it saturates after the number of delay lines reaches 16: the throughput curve for *4wav_16buf*, *4wav_32buf*, and *4wav_32bufwdm* are almost indistinguishable. This effect is due to the maximum hop count, which is set to 8. To validate this explanation, we also plot another curve, *4wav_32bufwdm_hopinf*, without the maximum-hop-count constraint. It is clear that the utilization of optical buffering increases considerably if packets can be buffered for an unlimited number of times. For deflection, the throughput of *4wav_def* is only marginally higher than that of *baseline*, indicating that deflection, by itself, is not a very effective contention-resolution technique in this network. For the given topology, only about one quarter of the nodes can perform deflection. Furthermore, any source node can reach any destination node through at most two hub nodes; therefore, the chance of a packet being deflected is relatively low. Space deflection can be a good approach in a network with high-connectivity topology, such as ShuffleNet [FoBP95, BoFB93], but is less effective with a low-connectivity topology.

Figure 17.12(b) shows the packet-loss rates. Deflection alone, in this network, cannot meet the required packet-loss rate of 0.01. The threshold transmitter loads for all the buffering schemes with maximum hop count are approximately 0.2. The packet-loss rate for *4wav_32bufwdm_hopinf* is very low and the upper-bound transmitter load is 0.63.

The plots of end-to-end delay and average hop count in Figs. 17.12(c) and 17.12(d) reveal more detail on the effect of setting the maximum hop count. The hop count for all the optical-buffering schemes increases with load, because packets are more likely to be buffered with higher load. However, the end-to-end delay decreases (except for *4wav_32bufwdm_hopinf*) with load. This is because (1) packets destined to closer nodes are more likely to survive; and (2) the buffer delay introduced by optical delay lines is small compared with link propagation delay. Without the maximum-hop-count constraint, both end-to-end delay and average hop count rise quickly with load, indicating that unrestricted buffering can effectively resolve the contention. The amount of extra delay caused by unrestricted buffering is less than 0.5 ms with transmitter load of 0.63, and this extra delay may be acceptable for most applications. The disadvantage of unrestricted buffering is the physical impairment of the signal (e.g., attenuation, cross-talk, etc.), since packets must traverse the switch fabric and delay lines many more times. In this case, optical regeneration may become necessary. We also notice that, when the load is light, the end-to-end delay increases for *4wav_def*; this is because, under light load, deflection is resolving the contention effectively by deflecting packets to nodes outside the shortest path, thus introducing extra propagation delay. For average end-to-end delay, the effect of the extra propagation delay due to deflections is more prominent than the effect of having more survived packets with closer destinations. This explains the initial increase of delay in the deflection scheme.

Table 17.3 lists the upper bound of average offered transmitter load with acceptable packet-loss rate set at 0.01. *4wav_32bufwdm_hopinf* offers the best packet-loss performance, but it may be more expensive to implement due to the large number of optical delay lines and switch ports, and it may also require optical amplification/regeneration. Wavelength conversion is also an effective approach to resolve contention, although its effectiveness depends on the number of wavelengths in the system. Deflection is the least-effective approach in the example network, but its benefit can be achieved when we combine it with other schemes.

Table 17.3 Comparison of upper-bound average offered transmitter load with packet-loss rate of 0.01.

Scheme name	Max. TX load
4wav_wc	0.16
16wav_wc	0.44
4wav_4buf	0.18
4wav_16buf	0.2
4wav_32buf_hopinf	0.63
4wav_def	<0.05

Comparison of Combination Schemes

This study chose four scenarios for different combinations of contention-resolution schemes: $16wav\_wc + 16buf$, $16wav\_wc + def$, $16wav\_16buf + def$, and $16wav\_wc + 16buf + def$.

Figure 17.13(a) shows the throughput comparison of these schemes. The scheme that incorporates all three dimensions (wavelength, time, and space) for contention resolution offers the best throughput. One can also observe the benefit of using wavelength converters: the schemes involving wavelength conversion perform better under heavy load.

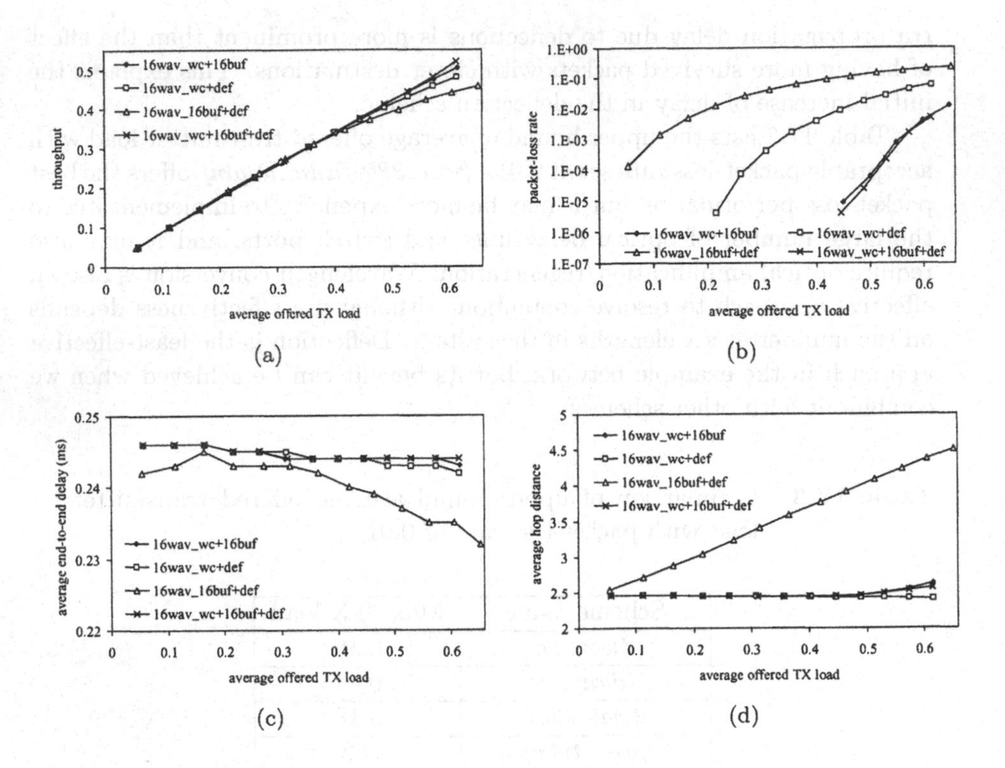

Figure 17.13 Performance comparison of the combination schemes.

Figure 17.13(b) compares the packet-loss rates. Although the throughput for all these schemes are quite close, their packet-loss rates have large differences. The best performer is $16wav\_wc + 16buf + def$, followed by $16wav\_wc + 16buf$, $16wav\_wc + def$, and $16wav\_wc + 16buf + def$. The benefit of wavelength conversion and buffering appears to be dominant. The upper-bound transmitter load for a packet-loss rate less than 0.01 is as follows: 0.2 for $16wav\_16buf + def$, 0.45 for $16wav\_wc + def$, 0.61 for $16wav\_wc + 16buf$, and 0.65 for $16wav\_wc + 16buf + def$.

Figures 17.13(c) and 17.13(d) show the average end-to-end delay and average hop distance of these schemes. The end-to-end delay presents a general trend of decrease for all four schemes, due to the dropping of packets with far-away destinations with increasing traffic load. $16wav\_16buf + def$ has a more prominent trend of decrease, indicating buffering and deflection alone can make the network prefer packets with closer destinations. $16wav\_16buf + def$ also introduces more extra hops due to the high utilization of optical delay lines.

Limited Wavelength Conversion

Both wavelength conversion and optical buffering can effectively resolve contention. However, they require extra hardware which may increase the system cost. Full-range wavelength conversion requires a fast tunable laser as the pump laser for every wavelength converter, and optical buffering requires extra ports on the switch fabric. Here, we consider the case of limited wavelength conversion, which can potentially save the amount of required hardware.

Figure 17.14 shows the throughput and packet-loss rate comparison of the three limited wavelength conversion cases–$16wav\_wclim1$, $16wav\_wclim2$, and $16wav\_wclim4$. For reference, it also shows $4wav\_wc$. The performance of these schemes ranks in the following order (from highest to lowest): $16wav\_wclim4$, $4wav\_wc$, $16wav\_wclim2$, $16wav\_wclim1$; their respective upper-bound transmitter load with packet-loss rate of 0.01 are 0.31, 0.17, 0.13 and 0.07. $16wav\_wclim4$ can accommodate nearly twice the load than $4wav\_wc$, indicating that limited wavelength conversion in a system with more wavelengths is better than full-range wavelength conversion in a system with fewer wavelengths.

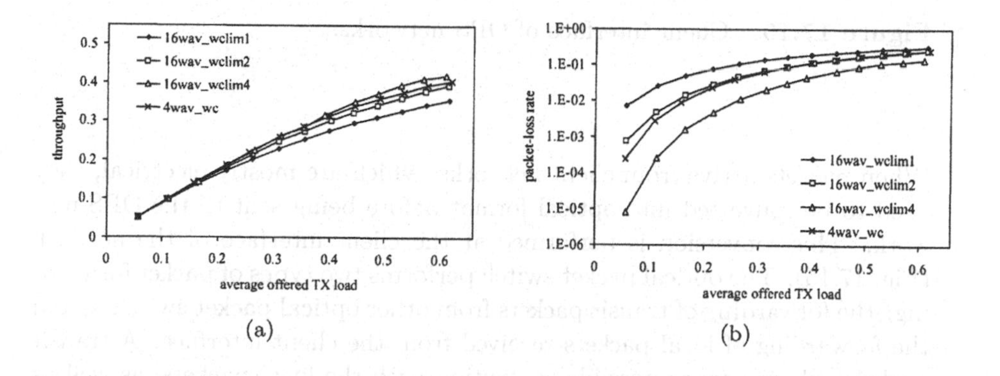

Figure 17.14 Performance comparison of limited wavelength conversion.

17.5 Hybrid Contention Resolution for OPS

Most existing studies on contention resolution have focused on the optical domain of one optical packet switch, or a network inter-connected by such a switch. An OPS network should perform packet-based switching optically. However, it still needs to interface with other types of networks to form an end-to-end connectivity. The other networks, especially the client networks, are often electrical. This means that there needs to be an interface at the edge of the OPS network. This section presents an architecture that takes advantage of the availability of electrical buffers at the edge, to resolve contentions and lower network cost.

17.5.1 Node Architecture

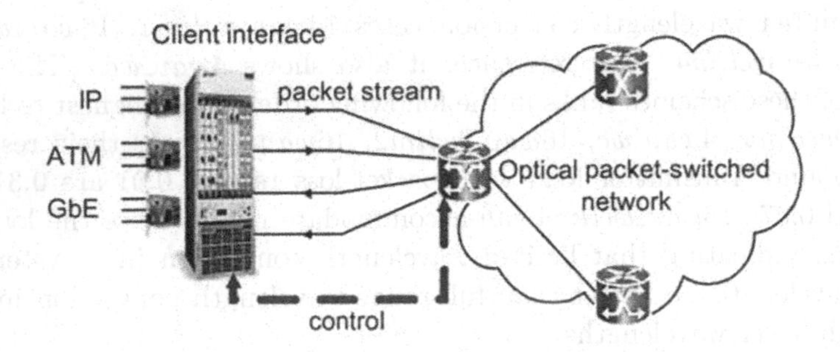

Figure 17.15 Client interface of OPS networks.

When packets arrive from client networks, which are mostly electrical, they need to be converted into optical format before being sent to the OPS network. This conversion is performed at the client interface of the network (Fig. 17.15). The optical packet switch performs two types of packet forwarding: the forwarding of transit packets from other optical packet switches, and the forwarding of local packets received from the client interface. A transit packet will experience possible contention with the local packets, as well as other transit packets. In most proposed architectures, the contention resolution usually requires a large amount of optical resources, such as wavelength converters and delay lines. In this architecture, the local packets are first queued in the electrical buffers, which can be easily implemented in the electrical part of the client interface. These packets enter the optical switch only when there is no transit packet occupying the preferred wavelength/output port. This buffering mechanism ensures that all the wavelength converters and delay lines are only used for transit packets. Since the switching is still carried out by the optical components and there is no O-E-O conversion in the network, the use of electrical buffers at ingress nodes for the client packets does not compromise the all-optical nature of the core network.

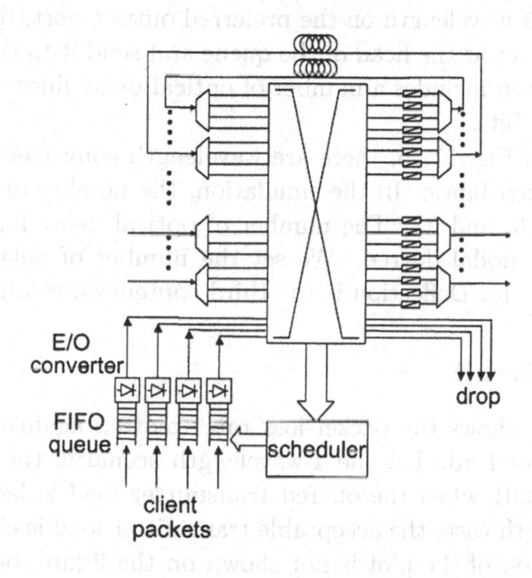

Figure 17.16 Architecture of an optical packet switch incorporating hybrid contention resolution.

Figure 17.16 shows the node architecture that implements hybrid contention resolution. In the optical portion of the switch, both optical delay lines and wavelength converters are used to resolve contention. In the electrical-optical interface portion of the switch, FIFO queues are included. All the packets from the client networks will be queued first. A scheduler observes the state of every wavelength at the output ports of the switch. When the packet's output port clears, the transmitter will convert the packet to optical format and send it to the optical switch fabric.

17.5.2 Simulation Configuration

The network topology is the same as in Fig. 17.9. Every node is equipped with W transmitters (where W is the number of wavelengths), with corresponding FIFO queues. A FIFO queue is fed with a self-similar traffic generator consisting of 12 ON/OFF sources generating IP packets. The traffic is uniformly distributed. Whenever the FIFO queue is not empty and there is a vacant wavelength on the preferred output port, the scheduler will retrieve the packet at the head of the queue and send it to the optical switch fabric. The switch includes a number of optical delay lines, which are of the length of 12500 bits.

As shown in Fig. 17.16, there are wavelength converters at each output port of the switch fabric. In the simulation, the number of wavelength, W, is set to 4, 8, 16, and 32. The number of optical delay lines at each node varies with the nodal degree. We set the number of delay lines equal to ($nodal\ degree - 1$). Deflection is the third contention-resolution dimension.

17.5.3 Illustrative Results

Figure 17.17(a) shows the packet-loss rates plotted against the average offered transmitter load. For the 4-wavelength scenario, the packet-loss rate is kept below 0.01 when the offered transmitter load is less than 0.5. For the 32-wavelength case, the acceptable transmitter load is close to 0.65. The light-load portion of the plot is not shown on the figure, because the simulation did not encounter any dropped packets during the simulated time.

Figure 17.17(b) shows the average end-to-end delay. With average transmitter load lower than 0.6, the end-to-end delay is dominated by propagation delay. The electrical queuing delay in the access FIFO queues are negligibly small. This is because the queued packet does not need to wait for more than a few packets' transmission delay to find a vacant wavelength. With 2.5 Gbps line rate and native IP packet sizes, the transmission delay of packets are on the order of microseconds. Since the network uses static routing, there will be a most-congested link in the network. In this topology, it is the link between nodes 1 and 11. There is traffic from totally 21 source-destination pairs passing this link. It is congested when the average transmitter load reaches 0.66. When the link is approaching congestion, electrical queuing delay and deflection delay start dominating. This explains the dramatic increase in delay in the figure.

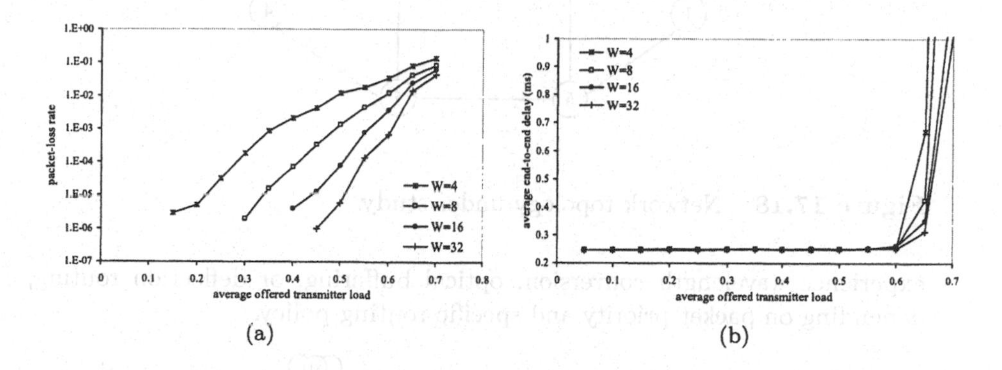

Figure 17.17 (a) Packet-loss rate, and (b) average end-to-end delay comparison of different simulation scenarios.

17.6 Priority-based Routing

A network built on top of an OPS network will need to not only provide IP-like connectionless services, but also higher-quality connection-oriented services. This will require the network to have differentiated service qualities at the packet level. Since most of the optical packet switches are still lab prototypes at this stage, it is early to delve into the realm of Quality of Service (QoS) studies of such networks; however it is worthwhile to study the network behavior with different packet priorities, which will lead us one step closer to the implementation of Class of Service (CoS).

17.6.1 Network Architecture and Routing Policies

Figure 17.18 shows the network topology under study. There are six nodes connected by 20-km long, bi-directional fiber links. Every fiber can accommodate four wavelengths, each of which is operating at 2.5 Gbps.

The node architectures are shown in Fig. 17.19. Each switch has four input/output ports used for local add/drop, since there are four available wavelengths. Between each source-destination pair, equal amount of traffic is generated, i.e., uniformly-distributed traffic. A packet in contention could

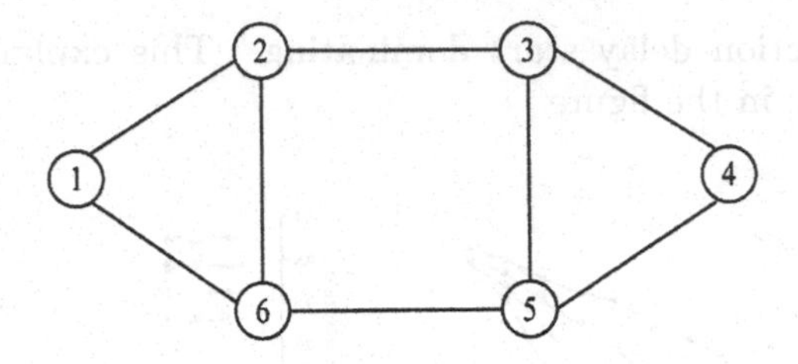

Figure 17.18 Network topology under study.

experience wavelength conversion, optical buffering, or deflection routing, depending on packet priority and specific routing policy.

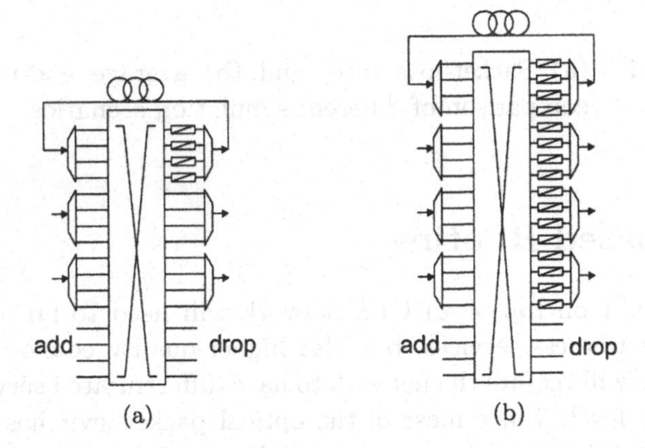

Figure 17.19 The architectures for (a) node 1, 4; (b) node 2, 3, 5, 6.

The wavelength converters in the study are based on degree-2 wavelength conversion. The four wavelengths are labeled as λ_1, λ_2, λ_3 and λ_4, with the wavelength conversion pattern: $\lambda_1 \leftrightarrow \lambda_2$, $\lambda_3 \leftrightarrow \lambda_4$.

Besides limited wavelength conversion, both optical buffering and deflection are used. Each node is equipped with one multi-wavelength delay line. Because of the specific topology of this network, when only shortest-path routing is considered, it can be easily observed that most of the packet forwarding happens at nodes 2, 3, 5 and 6; while nodes 1 and 4 only generate and receive packets. Therefore, for nodes 1 and 4, the contention mainly takes place at the local drop ports. For this reason, we place the wavelength converters only at the local drop ports, but not at the other output ports of the switch fabric, in order to more effectively resolve contentions. At the other nodes (2, 3, 5 and 6), the wavelength converters are placed at the output ports leading to other nodes, but not the local drop ports. The characteristics of each contention-resolution scheme are described in much more detail in [YMYD03].

There are three packet priority classes: class 3, 2 and 1, with class 3 being the highest. In case of contention, a packet will first seek an alternative vacant wavelength on the desired output port; if none is available, it will seek a vacant wavelength in the buffer. If there are no suitable wavelengths in the buffer, it will compare its priority class with the packets currently occupying the preferred port. If the new packet has higher priority, it can preempt a lower-priority packet and be transmitted successfully (either on its original wavelength or on a converted wavelength). If the two packets in contention are of the same priority, the one arriving later is switched according to the deflection next hop (from the secondary entry in the routing table) and the above process will be repeated. If the secondary routing entry fails as well, the packet will be dropped.

17.6.2 Illustrative Results

The packet-class distribution used in the first set of simulation is: 10% class 3, 30% class 2, and 60% class 1.

Figure 17.20(a) compares the packet-loss rates. Not surprisingly, class-1 packets have the highest packet-loss rate, which also increases rapidly with increasing load. Class-2 packets have reasonably-good packet-loss rate, which maintains below 0.01 until the transmitter load reaches 0.2. The increase of packet-loss rate for class-3 packets is barely noticeable, and, for the entire simulated range (transmitter load < 0.4), the packet-loss rate stays well below 0.01. The simulation results showed that, for load lower than 0.2, both class-3 and class-2 packets can maintain the packet-loss rate lower than 0.01. The low packet-loss rate achieved by higher-priority packets is at the cost of dropping more lower-priority packets.

Figure 17.20(b) shows the end-to-end delay comparison of the three classes. Because the propagation delay between two nodes is much larger than the delay introduced in the fiber delay line, a prominent increase in the end-to-end delay indicates that deflection routing is being applied to more packets. Class-1 packets experience the most increase in end-to-end delay as the load increases. This is due to the fact that they are not able to preempt any other class of packets, and need to resort to deflection more often. Class-2 and class-3 packets have similar end-to-end delays, which is because both these classes can preempt class-1 packets and can maintain a lower probability of using deflection routing to resolve contentions.

Figure 17.20(c) compares the average hop distance. From the figure it can be observed that all three classes have nearly the same average hop distances, which increase steadily with load. This is due to the effect of optical buffering. During contentions, optical buffering is attempted before preemption or deflection takes place. All three classes are utilizing buffering with similar probability.

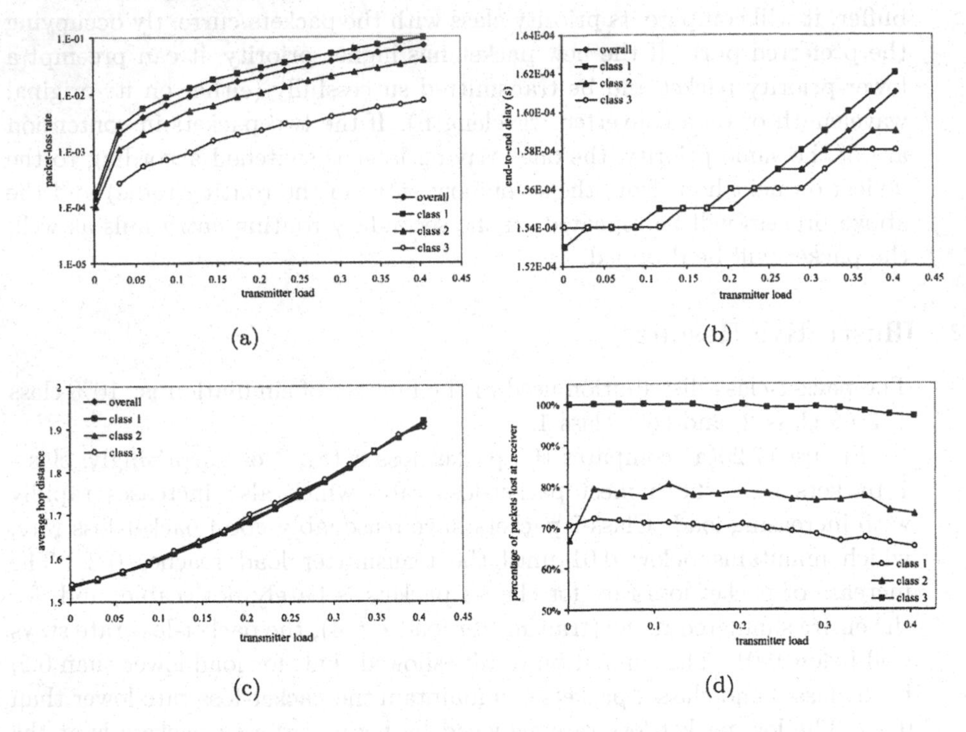

Figure 17.20 Performance comparison of priority-based routing (10%-30%-60% priority distribution).

The above simulation results indicate that it is possible to achieve desirable network performance in OPS networks. However, since there is no sophisticated queuing system as in the electrical domain, the efficiency of bandwidth utilization of such networks will be sub-optimal.

To better understand the source of contentions, we also measured the fraction of packets dropped due to contention at the receiver, as shown in Fig 17.20(d). All the lost class-3 packets are due to receiver contentions, implying that they can well survive the middle hops but eventually need to compete among themselves for the receivers. This is especially true under the case of uniform traffic, where each node is receiving packets from the remaining N-1 nodes. For both class-2 and class-1 packets, the fraction of packets lost at the receivers decrease gradually with the load, implying that in-transit contentions intensify faster than the contentions at the receivers.

In order to further examine the effect of priority distribution on the network performance, we have simulated a different priority distribution: 30% class 3, 30% class 2, and 40% class 1.

In the packet-loss rate comparison shown in Fig. 17.21(a), class-2 packets experience two times more packet loss than in the previous simulation. The cause is that there are twice as many class-3 packets, which will preempt both class-2 and class-1 packets, and that there are fewer class-1 packets for class-2 packets to preempt.

The end-to-end delay comparison in Fig 17.21(b) and average hop distance comparison in Fig. 17.21(c) look very similar to Fig. 17.20(b) and Fig. 17.20(c), indicating that, although packet-priority distribution has changed, the probability of packets being buffered or deflected did not change much.

Figure 17.21(d) shows the fraction of lost packets at the receivers of all the lost packets. Most of the dropped packets in class 3 are still at the receiver side, implying that class-3 packets only compete among themselves and the contentions mostly happen at the receiver. For class-2 and class-1, it appears that more fraction of the lost packets are due to contention at the transit switches, because of the increased number of class-3 packets. For class-3 packet-loss rate to maintain below 0.01, the transmitter load must be less than 0.25; and for class 2 and class 1, the transmitter load thresholds for less than 0.01 packet-loss rate are 0.15 and 0.1, respectively.

For the three different classes of priorities simulated, class 3 has the best performance in terms of latency, signal quality, and packet-loss rate. It will be suitable to carry various mission-critical, real-time traffic, and could be further explored to establish virtual-circuit-like connections to accommodate stringent user demands. Class 2, which has higher delay and packet-loss rate, appears to be a good candidate to carry medium-quality connection-oriented traffic (such as TCP or voice) or connection-less data traffic (such as UDP). Class 1 has the highest latency and loss rate. It can be used for applications and data services that do not require real-time connection

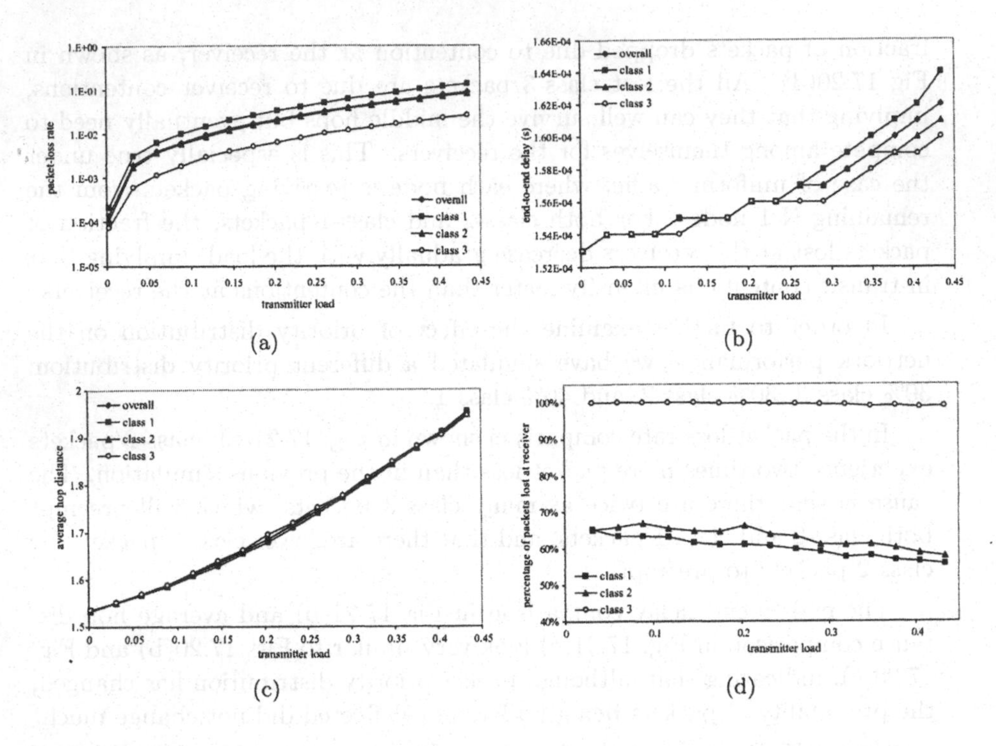

Figure 17.21 Performance comparison of priority-based routing (30%-30%-40% priority distribution).

and can recover from frequent packet losses. It is important to note that the performance of each class is not only related with the specific routing policies and the network topology, but also how the bandwidth is divided between the classes. The priority distribution is crucial to achieve the desired quality of packet delivery for each class. In general, the high quality of the higher-priority service is obtained at the cost of lower-priority services; therefore, the portion of bandwidth assigned to high-priority classes needs to be kept small. In other words, in such OPS networks, one needs to tradeoff bandwidth for better flexibility and service quality.

17.7 TCP Performance with OPS

The previous section proposed to use inexpensive electrical buffers at the ingress interfaces to reduce packet loss. Since our goal is not only to lower packet-loss rate, but also to provide a better transport for the TCP/IP traffic, this section extend the investigation to the TCP performance of OPS networks. To improve the TCP performance, the work in [HeSi01] proposed *optical flow-routing*, a type of aggregation that is supposed to reduce packet reordering. This section presents a packet-aggregation mechanism that allows many packets to be grouped together into a larger entity that can be more efficiently transported through the network. Such a mechanism can help reduce the traffic burstiness and improve the system performance.

17.7.1 Node Architecture

Figure 17.22 shows the node architecture. The packet aggregator assembles client packets into larger entities (referred to as aggregation packets) in a first-in-first-out (FIFO) manner. It directly interfaces with the client network elements (typically IP routers), and it consists of a number of FIFO sub-queues. Each sub-queue buffers packets going to the same destination. A sub-queue transmits all the buffered packets in an aggregation packet after a certain period of time, t_a. To avoid unnecessary delay, we set a packet-count threshold C such that, when the number of buffered packets reaches C, the sub-queue will transmit the aggregation packet even if the time from last transmission is less than t_a. This aggregation mechanism can be compared to a bus system (as in public transportation): At any time, there is one bus with one or more empty seats waiting for passengers for each destination. A bus has a maximum capacity of C passengers and it leaves every t_a seconds. If the bus is full before its scheduled departure time, it will leave early and the next empty bus will pull into the station. The aggregator not only preserves the order of packets, but also shapes the traffic by injecting more evenly-sized aggregation packets at more regular time intervals.

The ingress buffer controls when an aggregation packet can be injected into the optical switch in order to avoid contention. By using an ingress buffer, we can reduce its contention with transit packets by allowing local aggregation packets to be injected only when there is no transit packet occupying the preferred output port. The local aggregation packets are first stored in the ingress buffer electrically, and they are converted to optical format and then injected into the optical switch. A scheduler is used to constantly monitor the state of the switch fabric, and to control the transmission from the ingress buffer.

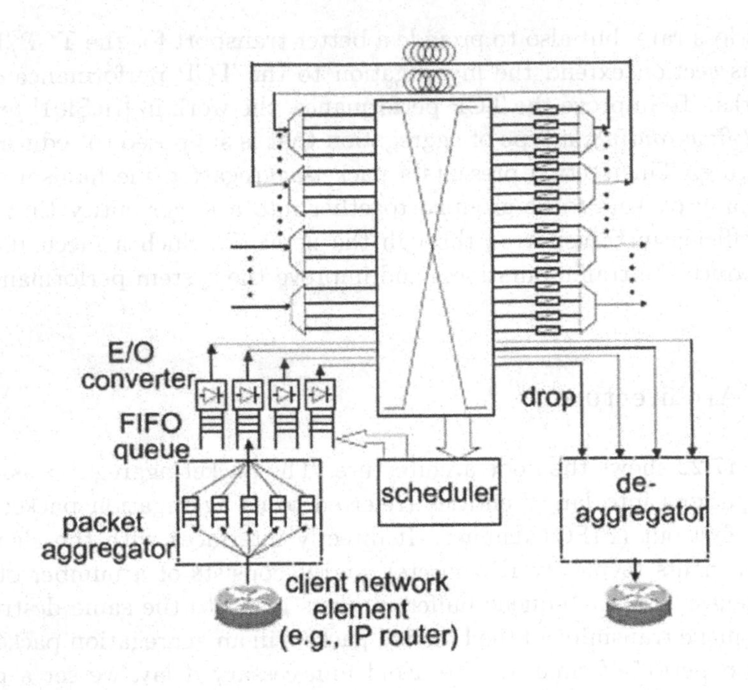

Figure 17.22 Proposed node architecture.

17.7.2 Simulation Configuration and Numerical Results

Figure 17.23 shows the network topology for the simulation. The simulation experiment uses a file-transfer protocol (FTP) session to measure the TCP performance. The main performance metric is the transfer time of a large file (assumed to be 1.6 Mbytes in this example). It is reasonable to assume that both hosts have Ethernet interfaces; therefore, the maximum transfer unit (MTU) is 1500 bytes. For the network scenario to be realistic, each link also carries some background IP traffic. Each node is equipped with four transmitters, fed by four traffic generators independently. The intensity of the background traffic is controlled by the average offered transmitter load.

One of the main factors that affect TCP performance is the receiver window size, whose typical values are 8, 32, or 64 Kbyte. The aggregation threshold C can also impact the TCP performance. With C values, the aggregation timer value t_a and the delay-line size should be adjusted ac-

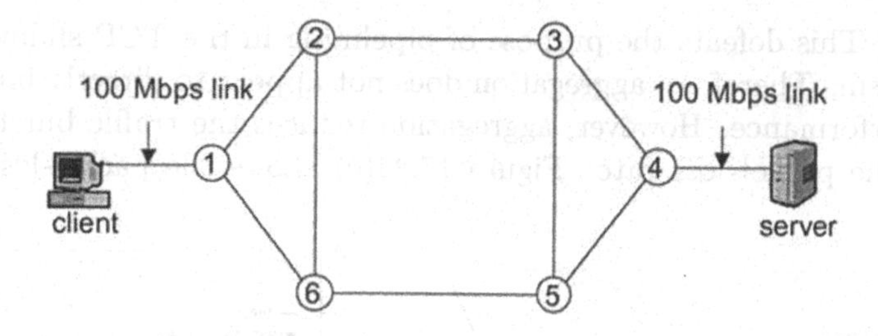

Figure 17.23 Network topology for TCP experiment.

cordingly. In the experiments, both t_a and the delay-line size are set to be equal to the transmission delay of C packets with maximum length (1500 bytes each). The running time for each data point varies between 4 and 75 hours on a 500-MHz Pentium III machine, depending on the TX load. The maximum TX load was 0.5 because larger values made the simulation time prohibitively long.

Figure 17.24(a) compares the file-transfer time T_{FTP} for different TCP window sizes and different values of C. For reference, it also shows the T_{FTP} without any background traffic for a client-server pair directly connected through a 100-Mbps link with the same propagation delay, i.e., a link length of 60 km. Without aggregation, a window size of 32 Kbyte provides the best result because the measured TCP round-trip time (RTT) is approximately 3 ms, and the TCP connection's data rate is 100 Mbps. (Note that the optimal window size should be the product of RTT and the data rate.)

Next, in Fig. 17.24(b), one can see the effect of the aggregation threshold with different C values of 10, 30, and 100 packets, while window size is equal to 8 Kbyte. For a TX load less than 0.2, the aggregation threshold does not have any effect on the system performance. As the TX load increases, the 10-packet aggregation scheme has the lowest T_{FTP}, followed by the 30-packet and the 100-packet schemes. The 10-packet scheme also performs better than the one without aggregation, indicating that aggregation improves TCP performance. However, with more packets aggregated, the performance deteriorates because more queuing delay is introduced in the packet aggregator and the ingress buffer. Intuitively, one would think the ideal aggregation packet should contain all the TCP segments sent within one window size. Unfortunately, the aggregator has to hold the first segment for at least the entire transmission delay of all the segments in that window. This defeats the purpose of pipelining in the TCP sliding-window mechanism. Therefore, aggregation does not appear to directly improve the TCP performance. However, aggregation reduces the traffic burstiness and lowers the packet-loss rate. Figure 17.24(c) shows the packet-loss rates of

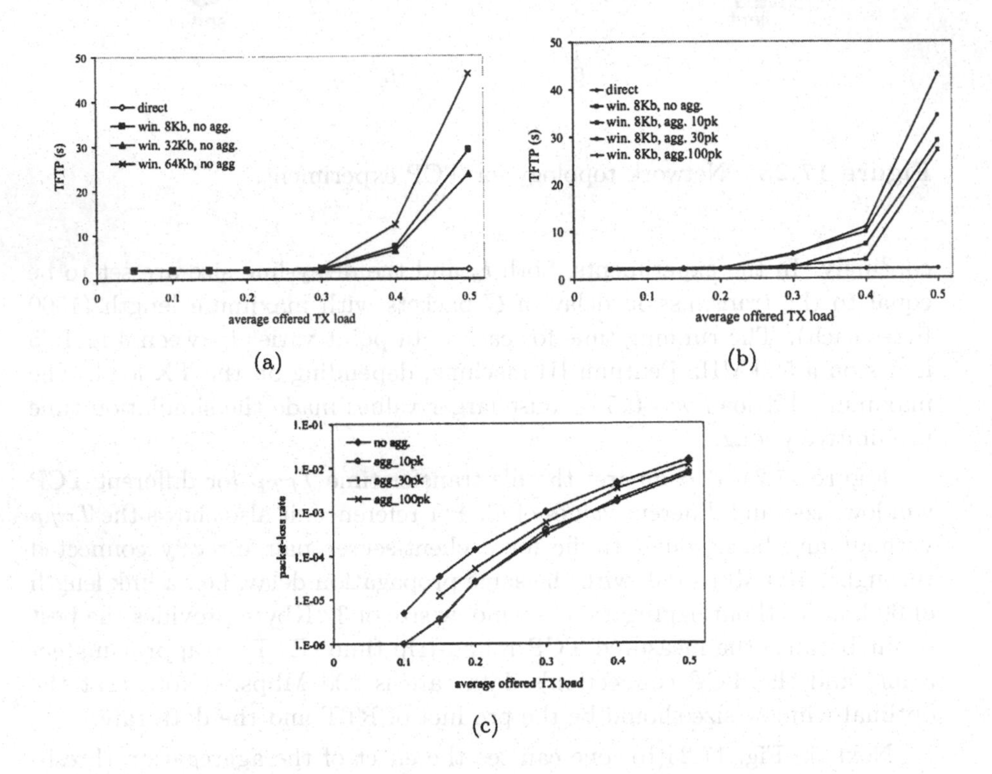

Figure 17.24 Comparison of T_{FTP} for (a) different TCP window sizes, and (b) different aggregation schemes. (c) Comparison of packet-loss rate with different aggregation thresholds.

aggregation schemes with different values of C. The 10-packet and 30-packet schemes offer the lowest packet-loss rate, followed by the 100-packet scheme and the no-aggregation scheme. Large aggregation appears to be less effective because the aggregation threshold C is based on the number of packets instead of bits. Since a large portion of IP traffic consists of very small packets, the size of the aggregation packet can vary dramatically when C is large, and the inter-departure time of the aggregation packets can still be quite bursty. Hence using bit count, instead of packet count, might be a better approach for aggregation.

17.8 Experimental OPS Networks

17.8.1 KEOPS Testbed

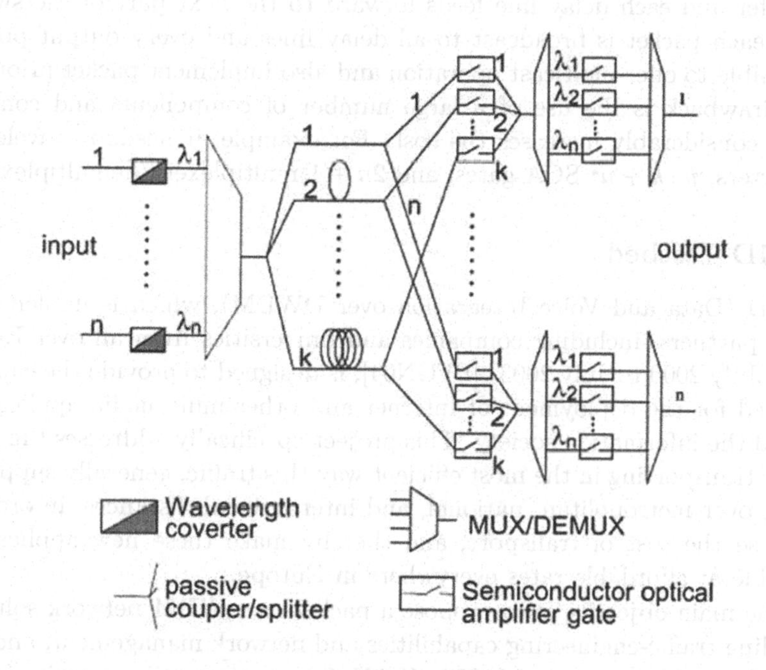

Figure 17.25 Broadcast-and-select switch proposed in the KEOPS project.

Keys to Optical Packet Switching of European Advanced Communications Technologies and Services (European ACTS KEOPS) (from September 1995 to September 1998) [CJRG98] is a typical OPS project, which is a broadcast-and-select space switch using single-stage forward-buffering scheme for contention resolution (Fig. 17.25). The wavelength converters encode the packet streams entering each input, therefore, the packets on each input are distinguished by a separate wavelength. The streams are then combined by a multiplexer and distributed to k groups of delay lines of different lengths, which give the packets necessary delays to resolve contention. By means of semiconductor optical amplifier (SOA) gates and passive couplers, each output port is able to select the packets with proper delays. At the final stage, the demultiplexer, SOA gates, and multiplexer can select one packet from a specific input port. In this architecture, there is only one stage of buffer and each delay line feeds forward to the next part of the switch. Since each packet is broadcast to all delay lines and every output port, it is possible to offer multicast operation and also implement packet priorities. The drawback is the use of a large number of components and controls, which considerably increases the cost. For example, it needs n wavelength converters, $n \cdot k + n^2$ SOA gates, and $2n + 1$ multiplexers/demultiplexers.

17.8.2 DAVID Testbed

DAVID (Data and Voice Integration over DWDM), which is implemented by 15 partners, including companies and universities from all over Europe (from July 2000 to July 2003) [BFGN04], is designed to provide the capacity required for the deployment of Internet and other multimedia applications behind the information society. This project specifically addresses the problem of transporting in the most efficient way this traffic, generally supported by IP, over metropolitan, national, and international distances, in order to decrease the cost of transport, and thereby make these new applications available at affordable rates everywhere in Europe.

The main objective is to propose a packet-over-WDM network solution, including traffic-engineering capabilities and network management, and covering the entire area from MAN to WAN. The project utilizes optics as well as electronics in order to find the optimum mix of technologies for future very-high-capacity networks.

The architecture of the switch is illustrated in Fig. 17.26. Each input fiber has several wavelengths. At the input ports, packets on these channels are demultiplexed and they are guided to the switching matrix. In the switching matrix, wavelength conversion is performed in order to handle the situation where there are several packets at the same output port or in the same delay line at the same time. Each packet travels through the switch with the same wavelength. At the output port, the packet's wavelength is converted to the desired output wavelength. The wavelength which the packet has inside the switch is independent of the external wavelength which the packet has outside the switch. In this switch, packet length can vary. In the switching matrix, packets are first converted to a wavelength that guarantees that there will be no contention. The number of wavelengths used inside the switch, W, can be less than or equal to the number of external wavelengths, M. The delay line buffer is implemented by cascading of k stages of m delay lines. At each state, packets from the delay lines are directed to a space switch through a wavelength splitter. The splitter consists of a combiner and a wavelength demultiplexer. This architecture guarantees that any output of stage i can be switched to any input of stage $i+1$. Each stage has different delays such that delay line i at state k produces delay im^{k-1}. The output stage chooses the correct packets to the appropriate outputs according to the wavelength and the delay line. When the correct output is reached, packets are converted to the correct wavelength and directed to the output fiber. The architecture is FIFO scheduled, i.e., packets arrive at output ports in FIFO order.

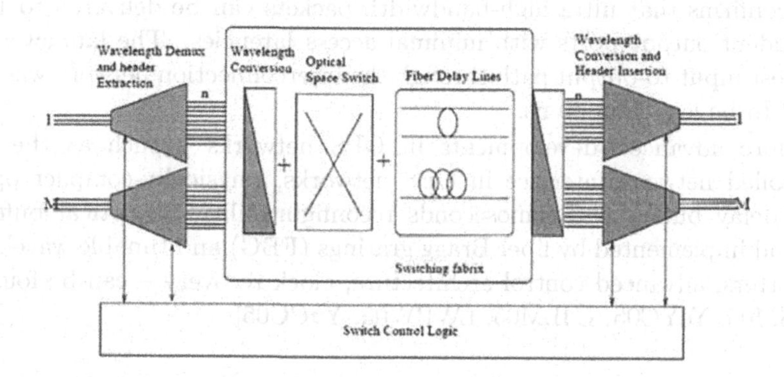

Figure 17.26 The architecture in the DAVID project.

Three techniques to support quality of service (QoS) levels were also proposed. The first technique to support the QoS classes is based on wavelength allocation. Some wavelength converters can be used by all the classes, some only by the higher classes, and the rest only by the highest class. The second alternative is a combination of wavelength allocation and threshold dropping. Here, only the packets of the highest class have their own wavelength converters. The two lower classes are separated by dropping the packets of the lowest class when the content of the buffer exceeds a certain threshold. The third way is wavelength allocation with scheduling. However, repeated wavelength search is used for the higher traffic classes, i.e., if a desired wavelength is not found, and the packet is to be lost, its delay is increased and the search is performed again. For the packets in the higher class, the search is done more times. For more detailed information, we refer the reader to [BFGN04, WLSB04].

17.8.3 Advanced Developments in OPS Testbeds

In [SLSM05], the implementation of a complete 12-port Data Vortex optical packet-switching fabric containing 36 fully-interconnected nodes is reported. The unique absence of internal buffers enables the utilization of practically the entire C-band capacity in each port since the WDM packet payload is transparently routed through the network. Importantly, the demonstration also confirms that ultra-high-bandwidth packets can be delivered to 12 independent output ports with minimal access latencies. The latency of the shortest input-to-output path through the interconnection network was measured to be less than 60 ns.

More advanced developments in OPS networks – such as the user-controlled network interface in OPS networks, physically-compact optical time delay buffer with nanoseconds reconfigurability, all-optical buffering method implemented by fiber Bragg gratings (FBG) and tunable wavelength converters, advanced control architecture, clock recovery – can be found in [NKSM05, YeYC05, CJLM05, LWPW05, YuCC05].

17.9 Summary

This chapter presented a unified study of optical packet switching. Slotted and unslotted networks were discussed. Effective contention resolution can be obtained with a combination of wavelength conversion, optical buffering, and space deflection. Priority-based routing with preemption is one possible solution to implement Classes of Service (CoS). Moreover, it was shown that, with the use of electrical buffer at the ingress nodes, an optical-electrical hybrid contention resolution may improve the network performance. The network performance can be further improved by using traffic aggregation at the ingress nodes.

Optical packet switching is promising to offer large capacity and data transparency. However, after many years of research, this technology has not yet been applied in actual products, because of 1) the lack of deep and fast optical memories and 2) the poor level of integration. It is our belief that these issues will need to be overcome, not only through technical breakthroughs but also through clever network design, making optimal use of optics and electronics, wherever they fit best.

In the near future, the developments in OPS seem to lead to integration of optical and electronic networks and the use of optical burst switching (OBS), which is the topic of the next chapter.

Exercises

17.1. What is meant by the electronic bottleneck in optical networks? What is the main cause of the electronic bottleneck? Is there a remedy for the electronic bottleneck? Justify.

17.2. Please compare the characteristics of the various contention-resolution schemes in OPS networks? How can you efficiently solve the contention problem in OPS networks?

17.3. In an OPS network, fiber delay lines (FDLs) have been proposed to resolve contentions. Please compare contention solutions achieved by recirculating FDLs and feed-forward FDLs.

17.4. What is the difference between asynchronous and synchronous networks? Why does contention resolution perform better in a synchronous network than in an asynchronous network? For what price is this improved performance obtained?

17.5. Clarify the differences between a slotted network and an unslotted network. It is important to realize synchronization of optical packet in slotted networks. Please design a synchronization scheme for an OPS network.

17.6. Why can one set up a guard time between optical packets? Clarify the meaning and role of jitter in an optical network?

17.7. Please design a format of an optical packet, which may reduce the overhead in an OPS network with a comparative performance.

17.8. Which would be practical for contention resolution in an OPS network: hybrid contention resolution or optical domain resolution?

17.9. What factors should be used to classify different priority services in an OPS network? Please design a reasonable scheme.

17.10. Please design a contention resolution scheme by using FDLs, which would be helpful to reduce the possibilities of a second contention of optical packets in an OPS network.

17.11. Please design a scheme for clock recovery in an OPS network.

17.12. Why is it difficult to realize a practical optical logic device? When an all-optical network is built, how can it be made compatible with existing electronic telecommunication networks, or should we totally discard the outdated telecommunication networks? Please imagine what a future optical network will be like if optical logic devices would become mature in the future.

17.13. Consider an optical buffer in a optical packet switch. The buffer consists of four delay lines. An incoming packet can be switched to any one of the delay lines, or it can be dropped. Packet lengths are fixed to one time slot. Consider two cases. In the degenerate case, the delay lines provide delays of 1, 2, 3, and 4 slots. In the non-degenerate case, the delay lines provide delays of 1, 2, 3, and 6 slots.

(a) Show for each case the buffer contents in each time slot if 3 packets, A, B, and C arrive in the first time slot, and 3 packets, D, E, and F arrive in the second time slot.

(b) What are the advantages and disadvantages of degenerate versus non-degenerate buffers?

(c) Compare the situation in which 4 packets arrive every other time slot to the situation in which 2 packets arrive every time slot. For each case, attempt to design a buffer (degenerate or non-degenerate) that can accommodate all packets over a period of 4 time slots while using the minimum number of delay lines. How many delay lines would you need in each case if void filling was not allowed?

Chapter 18

Optical Burst Switching (OBS)

18.1 Introduction

Optical burst switching (OBS) is attracting increasing attention in the research literature today. In this chapter, several key issues are discussed, such as burst assembly and schedule, protocols for signaling and routing, architecture of edge node and core node, contention resolution, and new challenges from data-intensive traffic.

There are mainly three switching schemes for WDM networks: wavelength routing (or switching), packet switching, and burst switching. Wavelength Routing (WR) is quite popular in today's WDM networks. Since the data traffic is rapidly growing, wavelength routing by itself is not perfect at achieving statistical multiplexing of data traffic. A long-term strategy for the network evolution is to explore Optical Packet Switching (OPS), which provides more flexibility, efficient resource utilization, potential functionality and finer switching granularity. One of the biggest challenges for OPS is that there is no optical equivalent of the random access memory (RAM) and logic devices for optical signal processing. Optical Burst Switching (OBS) has been proposed as a compromise between optical circuit switching and optical packet switching, while avoiding their shortcomings [HXLX03]. Optical Label Switching (OLS) is another switching method. OLS processes and updates labels in control overhead to enable switching and forwarding flexibility. OLS could be implemented together with some of the switching granularities above. For example, we get Generalized Multi-Protocol Label Switching (GMPLS) when OLS is applied to WR networks.

In this chapter, some functions in OLS will also be imparted into OBS networks so as to direct bursts through an optical network without the need to pass these bursts through electronics whenever a routing decision is necessary. But there are several key issues which need to be addressed first before OBS can be deployed in an optical network. In particular, the technological demands and restrictions of electronic and optical components have to be considered with regard to an application in an OBS network as well as assessment of the architectural and economic aspects of implementing OBS. According to the comparison in Table 18.1, OBS can potentially achieve better performance over other switching methods.

Table 18.1 Comparison between various switching methods.

Property	*Wavelength Routing*	*Optical Burst Switching*	*Optical Packet Switching*	*Optical Label Switching*
Granularity	Large	Middle	Small	Option
Limits of Hardware	Low	Lower	Higher	High
Optical Buffer	No	No	Yes	No
Wavelength Converter	Yes/No	No	Yes	No
Electronic Bottleneck	Yes/No	No	Yes	No
Statistical Multiplexing	Low	High	Higher	High
Control Overhead	Lower	Low	Higher	High
Scalability	Low	Higher	High	Higher
Flexibility	Low	High	Higher	Higher
Cost	Low	Lower	Higher	High
Self-similar Traffic	No Support	Support	No Support	Partial Support

In 1983, the concept of burst switching was first proposed, which was applied to transfer voice and data traffic over TDM links [Amst83]. At that time, it did not receive much attention because it failed to achieve better performance over ATM. In 1990, the concept of Optical Burst Switching (OBS) was proposed in project Highball at University of Delaware [MBES90]. OBS refers to a broad class of sub-wavelength switching architectures [ChQY04], which assemble optical bursts at the network edge and transparently forward the bursts through the core network.

In the OBS network in Fig. 18.1, the ingress edge routers aggregate client data (e.g., Internet Protocol (IP) packets) into large bursts. Each burst is associated with a control packet containing the related control information

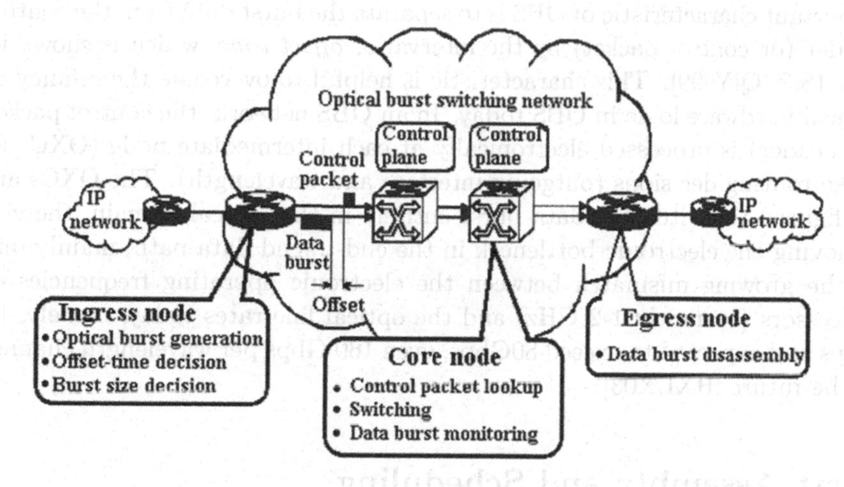

Figure 18.1 An OBS network.

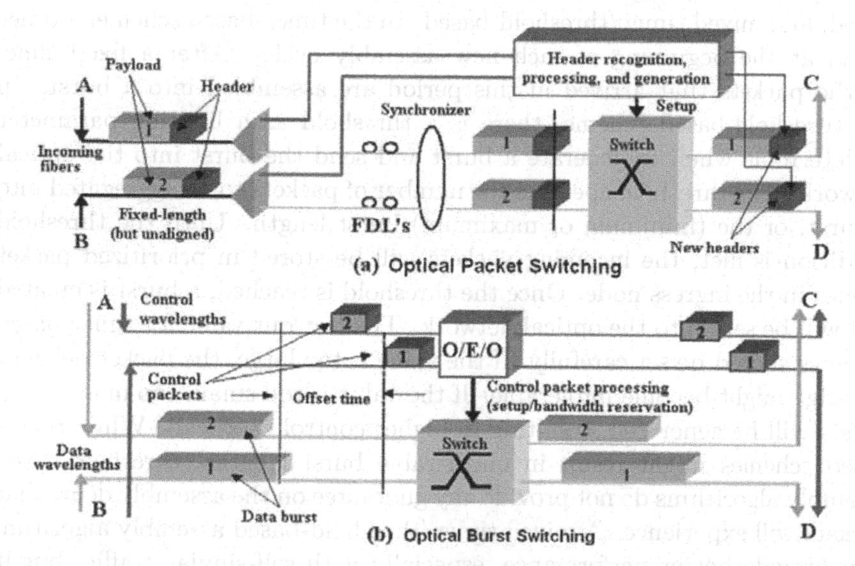

Figure 18.2 Comparison between OBS and OPS [QiCh03].

such as burst length and routing information. Compared to OPS, the most important characteristic of OBS is to separate the burst data from the control header (or control packet) by the interval of *offset time*, which is shown in Fig. 18.2 [QiYo99]. This characteristic is helpful to overcome the infancy of optical hardware logic in OBS today. In an OBS network, the control packet (or header) is processed electronically at each intermediate node (OXC) to make routing decisions (outgoing interface and wavelength). The OXCs are configured to switch the data burst entirely in the optical domain, thereby removing the electronic bottleneck in the end-to-end data path, mainly due to the growing mismatch between the electronic operating frequencies of processors (currently 1-2 GHz) and the optical line rates today, namely, 10 Gbps and expected to exceed 80Gbps, even 160 Gbps per wavelength channel in the future [HXLX03].

18.2 Burst Assembly and Scheduling

Burst assembly is the procedure for aggregating data packets from various sources, such as an IP router, into bursts at the edge of an OBS network. Usually, assembly algorithms can be classified as timer-based, threshold-based, and mixed timer/threshold-based. In the timer-based scheme, a timer starts at the beginning of each new assembly cycle. After a fixed time, all the packets that arrived in this period are assembled into a burst. In the threshold-based scheme, there is a threshold as a limiting parameter to determine when to generate a burst and send the burst into the optical network. The threshold specifies the number of packets to be aggregated into a burst, or the (minimum or maximum) burst length. Until the threshold condition is met, the incoming packets will be stored in prioritized packet queues in the ingress node. Once the threshold is reached, a burst is created and will be sent into the optical network. The timeout value for timer-based schemes should be set carefully. If the value is too large, the packet delay at the edge might become intolerable. If the value is too small, too many small bursts will be generated, resulting in higher control overhead. While timer-based schemes might result in undesirable burst lengths, threshold-based assembly algorithms do not provide any guarantee on the assembly delay that packets will experience. A mixed timer/threshold-based assembly algorithm may provide better performance, especially with self-similar traffic, but it may have higher operational complexity.

A burst-scheduling algorithm at an ingress node is in charge of adjusting the offset time for each burst, scheduling bursts on each output link, and forwarding the bursts and their control packets to the OBS network. First Fit (FF) and Latest Available Unused Channel (LAUC) are two reservation strategies appearing frequently in the literature [ChQY04]. FF will search for all the data channels in a fixed order, and assign the first eligible channel found to carry the data burst. The advantage of this technique is its simplicity: once a free wavelength is found, it is not necessary to scan the other wavelengths. In LAUC algorithm, for each wavelength, a single *scheduling horizon*[1] is maintained. Only the channels whose scheduling horizons precede the new burst's arrival time are considered "available" and the one with the latest scheduling horizon is chosen. The horizon is then updated after making the reservation for the next burst. The basic idea of this algorithm is to minimize the bandwidth gaps/voids created as a result of making a new reservation. A variant of LAUC is LAUC-VF (LAUC with void-filling). VF means that the voids between two earlier allocated bursts can be filled by a newly-arriving data burst. A scheduler without VF can only allocate a new burst on a specific channel after all the earlier reserved bursts on that channel. Thus, a strategy without VF is less efficient than one with VF. The disadvantage of VF is that a more complex scheduler is required for more processing time.

18.3 Signaling Protocols for OBS

A signaling protocol is the procedure through which services are provisioned. In a signaling protocol, service provisioning includes switching-path establishment, deletion and modification. By using a signaling protocol, a control packet can reserve resources for the corresponding data burst by guiding it through a routing path. In an optical network, there are one-way or two-way reservation signaling protocols.

In two-way reservation, which is also called Destination-Initiated Reservation Protocol [LJXC03], a control packet collects link and topology information instead of reserving resources for the data burst. The acknowledgement packet from the destination node to the source node reserves resources for the corresponding data burst while traveling along the reverse path. Then, the source node starts to launch the burst data along the reserved path after it receives the acknowledgement packet from the desti-

[1] The scheduling horizon is defined as the latest time at which the wavelength is currently scheduled to be in use.

nation node. Tell-and-Wait (TAW) is an example of a two-way reservation protocol [ChQY04].

In one-way reservation, which is also known as Source-Initiated Reservation Protocol [LJXC03], a control packet reserves resources for the corresponding data burst without an acknowledgment. Since one-way reservation is more flexible, has lower latency, and is more efficient compared with two-way reservation, one-way reservation is mainly adopted in OBS networks. There are three typical signaling protocols, namely, Just Enough Time (JET), Just In Time (JIT) and Time Label (TL).

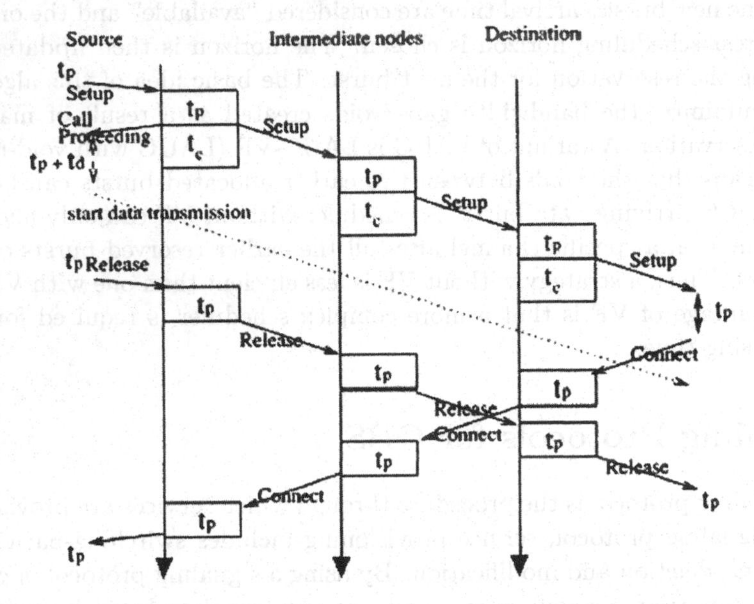

Figure 18.3 Just-In-Time (JIT) switching.

In 1990, JIT was proposed, which is shown in Fig. 18.3 [WeMc00]. Just-In-Time (JIT), which has been further developed after its initial proposal, can be considered as a variant of tell-and-wait as it requires each data burst transmission request to be sent to a central scheduler [MBES90]. The scheduler then informs each requesting node the exact time to transmit the data burst. Here, the term Just-In-Time means that, by the time a burst arrives at an intermediate node, the switch fabric has already been configured. This concept was later applied and extended to a wavelength-routed OBS network [DuBa02]. These schedulers are not only synchronized in time, but they also share the same global link-state information, which makes their implementation difficult.

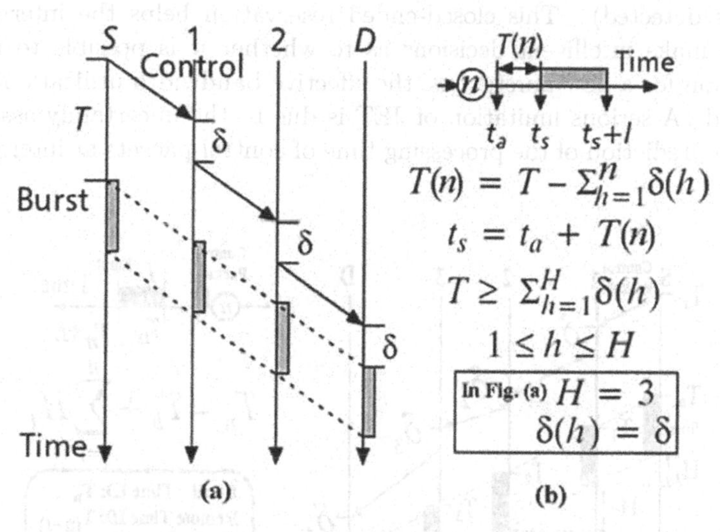

Figure 18.4 Just-Enough-Time (JET) switching.

The Just-Enough-Time (JET) protocol has received attention in the research community since it was proposed in 1997 [YoQi97, QiCh03]. JET is a distributed protocol for OBS networks which requires no optical buffer or delay at each intermediate node [QiYo99]. It accomplishes this by letting each control packet carry the offset-time information and makes the so-called delayed reservation for the corresponding burst, i.e., the reservation starts at the expected arrival time of the burst that is determined by a prediction mechanism. In the example shown in Fig. 18.4 [QiYo99], the bandwidth is reserved for the first burst starting from the burst-arrival time instead of the arrival time of control packet. At each intermediate node, the offset time is updated (reduced) to compensate for the actual control-packet processing time and switch-configuration time in Fig. 18.4. Note that the delay experienced by a control packet might vary for different reasons. In addition, when we consider deflection routing in an OBS network, the minimal offset time for the primary path might not be enough if the burst takes a longer alternate path. In such a case, an extra offset time can be added [QiYo99]. Another important feature of JET is that the burst-length information is also carried by the control packet, which enables it to make a closed-ended reservation (i.e., only for the burst duration with automatic release) instead of an open-ended reservation (which would not be terminated until a release signal is detected). This closed-ended reservation helps the intermediate node to make intelligent decisions as to whether it is possible to make a reservation for a new burst; thus, the effective bandwidth utilization can be increased. A serious limitation of JET is due to the uncertainly associated with the prediction of the processing time of control packets at intermediate nodes.

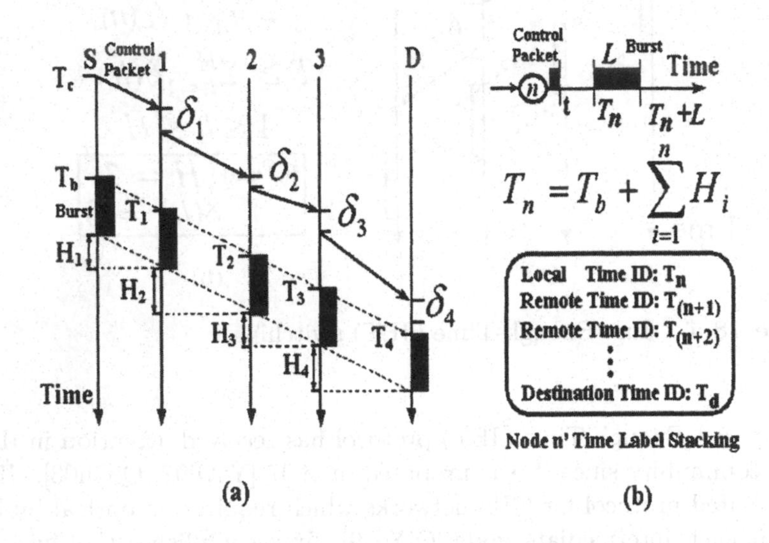

Figure 18.5 Time-Label (TL) switching.

Time Label (TL) in an OBS network was proposed in [HXLX03]. The instant that a data burst arrives at each node is defined as a *Time Label* (TL). Then, according to the Reserve-Fixed-Duration (RFD) (namely, burst length), the instant that the data burst departs from the node is also exactly determined by the corresponding TL. Because no optical buffer is introduced into an OBS network, data bursts are transparently propagated at light speed without extra delay. Thus, as long as the TL at the source node is determined, the TL at the next intermediate node is exactly determined according to the hop distance between the two nodes, as shown in Fig. 18.5. TL is free from influences caused by prediction in resource reservation. Label Stacking can be first operated in optical switching by employing TL. Precondition of TL is transparent switching of a data burst

Table 18.2 Comparison between the three signaling protocols in OBS.

Signaling Protocols	*JIT*	*JET*	*TL*
Optical Buffers	Yes	No	No
Wavelength Converters	Yes	No	No
Prediction Mechanism	No	Yes	No
Node Synchronization	Yes	Yes	Yes
Requirement of Link Status	Yes	Yes	Yes
Robustness	Low	Low	Higher
Guaranteed QoS	Yes	No	Yes
Flexibility	Low	High	Higher
Scalability	Low	High	Higher
Intelligent Functions	Lower	Low	High
Distributed Protocol	No	Yes	Yes

in an OBS network. TL must be network-wide and globally unique. TL needs node synchronization, which is easy to implement compared to bit or packet synchronization. Statistical multiplexing of resources is achieved by TL. Why do we say TL is more refined and an exact signaling protocol? JET is based on the prediction of control-processing time at each intermediate node, and TL is based on the propagation delay between nodes. However, in current optical networks, the deviation of optical propagation delay in each hop (or fiber link) is on the order of 10^{-8}s, which is much less than the deviation magnitude of 10^{-6}s which is caused by the control processing in JET. Another important advantage of TL is that the corresponding data burst will be triggered at the explicit TL at the source node, which avoids the prediction mechanism of the start and the end of the data burst at the source node in JET [QiYo99]. Thus, the TL approach is an exact mechanism to perform and refine reservation of network resources.

Performance comparison among the three signaling protocols is shown in Table 18.2.

18.4 Routing Protocols for OBS

In OBS networks, signaling protocols are more mature than routing protocols. Usually, routing algorithms for wavelength-routed networks are directly adopted in OBS networks. However, it may be necessary to develop OBS-specific routing algorithms. Recently, some researchers have put more efforts to develop routing protocols for the OBS environment. Dynamic routing with preplanned congestion avoidance for survivable OBS networks is proposed [HuHM05b].

From the view of resource provisioning, signaling and routing protocols correspond with time and space dimensions, respectively. However, it is worth noting that resource provisioning of conventional signaling and routing protocols is carried out in a single dimension, i.e., either time or space dimension. Resource provisioning in either of the two dimensions could lead to the contention of resource provisioning in the other dimension. Thus, some researchers have suggested that GMPLS could be applied into OBS networks [ChQY04]. In GMPLS, Link-Management Protocol (LMP) is used to coordinate relationship between routing and signaling [SAHA02] in optical networks. In fact, routing and signaling could be considered together instead of separated control components [Liu02]. Based on this motivation, the fundamental goal of Time-Space Label-Switching Protocol (TS-LSP) is to bind the signaling and routing functions closely together [HXLX03].

The TS-LSP is a new technology that can quickly and efficiently forward data with labels in an optical network. In order to illustrate the operation principles of TS-LSP, orthogonal time-space coordinates in which the vertical coordinate is the space label and the horizontal coordinate is the time label is shown in Fig. 18.6. Note that the TS-LSP role is for resource allocation and usage. In this figure, we consider a mesh network with 10 nodes, and four ports per node with two inputs and two outputs, and the nodes in the network are connected to each other by a single fiber (port ID: ij for node i, e.g., Port IDs 51, 52, 53, and 54 are all attached to node 5). In the orthogonal time-space coordinates, a crossing point between time label (or time ID) and space label (or port ID) is a concrete intermediate node. In the figure, each intermediate node has an input port and an output port at a TS-LSP. Because the propagation time of a burst through the intermediate node is neglected, at a TS-LSP, the input port ID and output port ID at an intermediate node correspond to the same time ID in the figure. In the intermediate node, cut-through switching is accomplished just before the corresponding data burst arrives. The corresponding data bursts are forwarded and transmitted along the TS-LSP. In the figure, a TS-LSP is determined by the corresponding time and space labels. There is a clear advantage of adopting TS-LSP: dramatically reducing routing and forwarding failure probability.

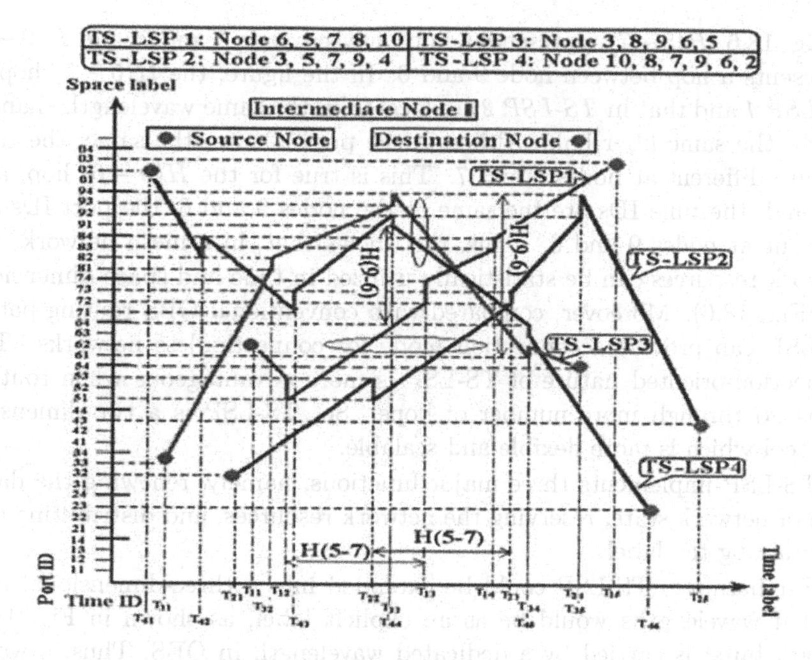

Figure 18.6 Time-Space Label-Switching Protocol.

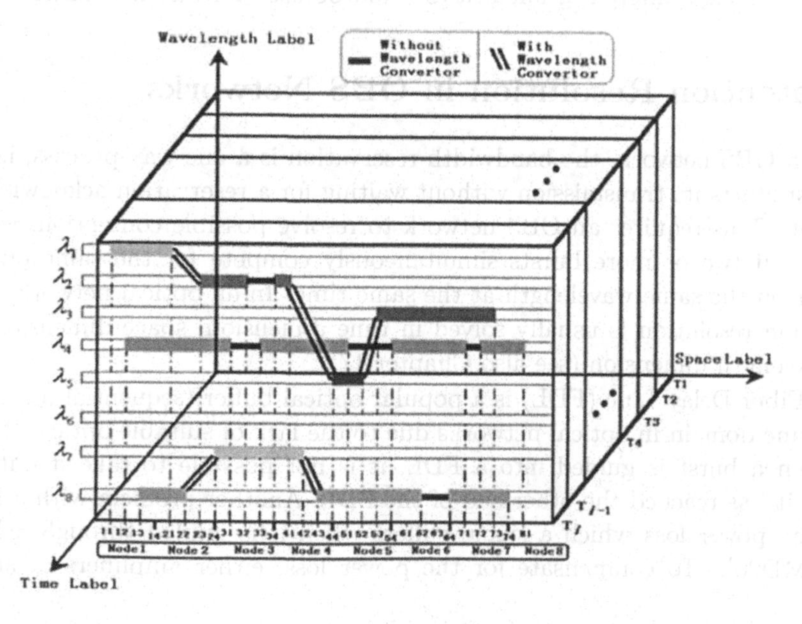

Figure 18.7 Three-dimensional Label-Switching Protocol.

In Fig. 18.6, $H(5-7)$ represents a hop between node 5 and 7, and $H(9-6)$ represents a hop between node 9 and 6. In the figure, the $H(5-7)$ hop in *TS-LSP 1* and that in *TS-LSP 2* are located in the same wavelength channel and in the same fiber link. Although the port IDs are the same, the time IDs are different at nodes 5 and 7. This is true for the $H(9-6)$ hop, and although the time IDs are the same at the nodes 9 and 6, the port IDs are different at nodes 9 and 6. Thus, this shows that, in a mesh network, the network resources can be statistically utilized in time and space dimensions (see Fig. 18.6). Moreover, compared with conventional OBS routing paths, TS-LSP can provide a connection mode for connectionless networks. The connection-oriented nature of TS-LSP is more advantageous when routing paths go through more number of hops. So, TS-LSP is a two-dimension protocol which is more flexible and scalable.

TS-LSP implements three major functions, namely, renewing the database of network state, reserving the network resources, and distributing and maintaining the labels.

Furthermore, TS-LSP could be extended into a three-dimensional protocol if wavelengths would be as an explicit label, as shown in Fig. 18.7. A data burst is carried by a dedicated wavelength in OBS. Thus, a wavelength could be viewed as an implicit label. The three-dimensional switching protocol could be further extended into a multi-dimensional label-switching protocol, e.g., different granularities could be used for another dimensions.

18.5 Contention Resolution in OBS Networks

In an OBS network, the bandwidth reservation is a one-way process, i.e., a burst starts its transmission without waiting for a reservation acknowledgement. This requires an OBS network to resolve possible contention, which arises if two or more bursts simultaneously compete for the same output fiber on the same wavelength at the same time. In an optical network, contention resolution is usually solved in time dimension, space dimension, or wavelength dimension (see also Chapter 17).

Fiber Delay Line (FDL) is a popular optical buffer (sequential memory) in time domain in optical networks due to the lack of suitable optical RAM. When a burst is guided into a FDL, it is not possible to take it out before it has reached the other end of the FDL. Another problem with a FDL is the power loss which a signal suffers when it is guided through a FDL [YaMD00]. To compensate for the power loss, either amplifiers or signal regenerators have to be used. The former alternative leads to noise, and the latter is expensive. There are two classes of FDL buffers: recirculating type and traveling type [YaMD00]. Additionally, traveling-type FDLs could be divided into two classes: normal delay lines and cascaded systems that contain several sets of delay lines. The recirculating type offers the possibility to use the same FDL several times, i.e., to recirculate a burst. The problem with the recirculating type is the power loss that leads to the need for amplification and therefore, also to noise. This limits the maximum buffering time or the need for regeneration between the buffering times. In the traveling-type FDL, the time a burst is buffered is determined beforehand; a delay line cannot be used several times. Of course, if there are several delay line groups serially connected, the delay can be determined in several pieces, but the length of the delay lines determines the maximum delay.

Deflection routing, also known as hot-potato routing, is an alternative where no buffers are implemented. If contention occurs and a burst cannot be switched to the correct output, it is routed to an alternate output instead. If the burst is lucky, it finds another path to the destination node. This works quite well in small networks with high connectivity, i.e., if the nodes have several neighbors. If the network has low connectivity, it is more probable that a deflected burst may not reach the destination node soon. Because this burst may use up a lot of resources (e.g., hops) by traveling in the network without finding the destination node, it is clear that other contention-resolution schemes are generally better in this situation. Additionally, deflection routing can only be used in networks with low load. If the traffic load is high, deflected bursts will create extra load and decrease the efficiency of the network. Deflection routing can be improved by only allowing certain alternate ports, i.e., if a burst cannot find a reasonable path to the destination, it is blocked although there were free outputs [TuZh99].

Optical networks have an additional dimension, namely, the wavelength. If contention occurs in a system using wavelength conversion, one packet is passed through the preferred (contending) wavelength and the other is guided to the same output port but on a different wavelength. This solution is optimal in the respect that neither burst is delayed. This approach is more suitable for circuit switching, because the need for fast tunable wavelength converters restricts their use in OBS networks.

If a contending burst cannot be deflected due to the unavailability of any wavelength, output port, or FDL, data loss becomes inevitable. More specifically, a common approach is to drop the incoming burst (which is a non-preemptive approach). In addition, it is possible for the incoming burst to preempt an existing burst based on priority or traffic profile.

Furthermore, burst segmentation was proposed in [VoJu02]. The approach is to segment an overlayed part between conflict bursts instead of dropping one [VoJu02, HuXi03]. Comparison among various contention-resolution schemes is shown in Table 18.3.

Table 18.3 Comparison of contention-resolution schemes in OBS.

Contention Resolutions	*FDL*	*Deflection*	*Wavelength Conversion*	*Segmentation*
Complicated Control	Lower	Lower	Low	High
Mature Technology	Yes	Yes	No	No
Stability	Yes	No	Yes	No
Delay	Longer	Longer	Shorter	Long
Loss Probabilities	Low	High	Lower	Low
Cost	High	Low	Higher	High
Efficiency	Low	Lower	High	Higher
Finer Granularity	No	No	No	Yes
Original Order of Bursts at Destination Node	Yes	No	Yes	No

The most effective combination of the above schemes is to use space deflection, buffering, and wavelength conversion together. However, there are some limits of signaling and routing protocols in OBS so that some methods are kept away from OBS networks, e.g., no wavelength converters in JET and TL. Recent research results have shown that it is helpful to reduce contentions by more reasonable signaling and routing protocols, e.g., TS-LSP [HXLX03].

18.6 Edge and Core Nodes in OBS and Testbeds

In general, transparent switching at an intermediate node is a basic requirement of an OBS network. It means that optical buffers and wavelength converters should be avoided, if possible. An architecture of a core node is shown in Fig. 18.8, which meets the above requirement. In Fig. 18.8, wavelength converter is avoided because no two same wavelengths are guided through an output link at the same time. Moveover, optical buffer is not required.

Obviously, an edge node is a key component of OBS. This is the reason that signaling and routing functions are usually initialized at an edge node in an OBS network. In general, an edge node consists of three stages, namely, (1) monitor or prediction, (2) classification or assembly, and (3) switch or schedule [Huan05].

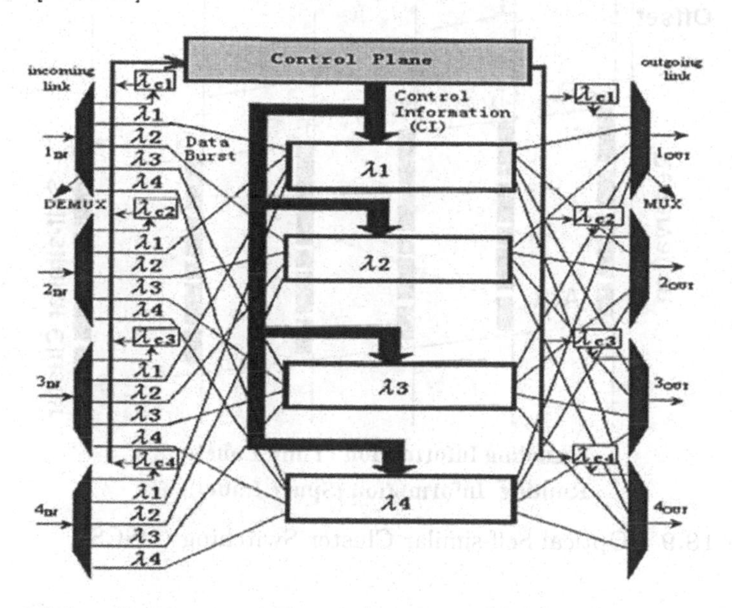

Figure 18.8 Architecture of a core node in OBS.

There are a few testbeds which have been implemented so far. Typically, the project of testbed for OBS network is funded by the High-Technology Research and Development Program of China [GuLa05]. Another testbed was developed in Japan in the period 2001-2005, which included network architecture, wavelength-reservation protocol, ultrafast processing of control packet and switching fabric [KKMS05].

18.7 New Challenges in OBS

As the traffic nature changes from being voice-dominant to data-dominant, traffic load moves to the self-similar property with traffic being bursty at all time scales, which is the reason that data-dominated traffic is also known as self-similar traffic [PaWi00]. Numerous experiments and measurements have shown that traffic of some applications, such as Ethernet LAN, WWW, variable bit rate (VBR) compressed video, WAN, TCP, and UDP, all exhibit self-similarity.

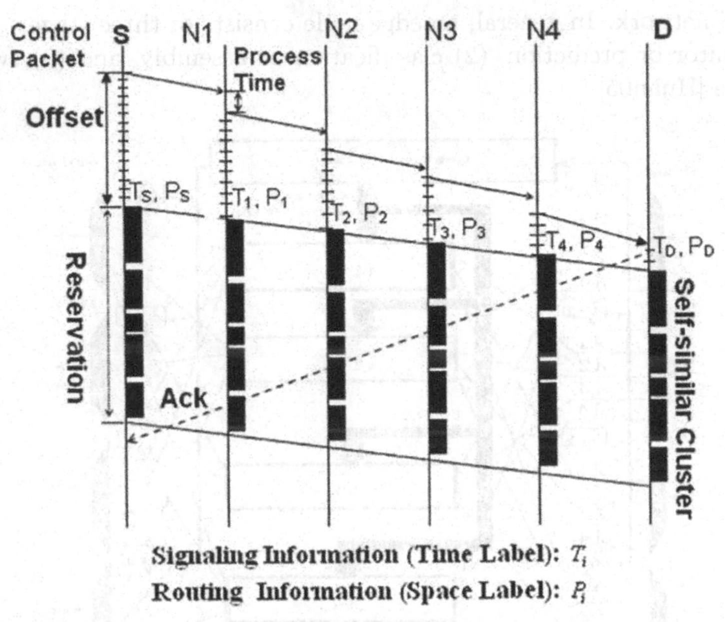

Figure 18.9 Optical Self-similar Cluster Switching (OSCS).

The self-similarity in traffic has been a challenge to the current network on various aspects of network design, such as network traffic engineering, performance evaluation, resource planning, and operational procedures established by the former studies over the past decades. Furthermore, the Noah Effect and Joseph Effect caused by self-similar traffic result in persistent congestion and frequent reconfiguration, which look like the flooding effect [Huan05]. In return, it leads to failure of efficient multiplexing of network resources [PaWi00]. Because OBS is a promising candidate for next-generation optical networks, it is worth noting whether OBS could overcome the challenges from self-similar traffic or not.

Several researchers have pointed out that burst assembly in OBS could overcome challenges from self-similar traffic by reducing its self-similar degree. In general, assembly algorithms can be classified as timer-based, burst-length-based, and mixed timer and burst-length-based [ChQY04]. However, there is no proof to support the claim whether self-similar degree could be changed or not by any one of the three burst-assembly mechanisms, neither is TCP that is employed to improve network performance in self-similar traffic by claiming to reduce its self-similar degree as well [PaWi00]. Thus, their effectiveness in overcoming the challenges from self-similar traffic is still questionable [Huan05]. As the volume of data traffic continues to grow, it is essential to consider the fundamental nature of traffic patterns in such circumstances when the next-generation optical switching technology is devised. Thus, Optical Self-similar Cluster Switching (OSCS) is proposed in [Huan05].

Due to the Long-Range Dependence (LRD) property of self-similar traffic, there is significant amount of data-burst trains with the same (or very similar) attributes in terms of QoS requirements (or Forward Equivalence Class (FEC)), ingress-egress nodal pairs, etc. In particular, these data-burst trains tend to access an ingress node with shorter inter-arrival time because of the heavy-tailed distribution of self-similar traffic. In self-similar traffic, data-burst trains like these above become bursty clusters between a pair of ingress-egress nodes. Such a bursty cluster under self-similar traffic, in which data bursts are of the same attribute, is defined as a self-similar cluster [Huan05]. By compensating non-predictability of traffic self-similarity, self-similar clusters are set up for a partial prediction of the burst arrival. The fundamental principle of OSCS is to utilize the partial predictable nature of a self-similar cluster to compensate for the unpredictability or high-variability nature of self-similar traffic, which is a key reason for network performance deterioration. It is helpful to avoid frequent reconfiguration and persistent congestion when self-similar clusters are separated from self-similar traffic and assigned some dedicated wavelengths. Three mechanisms are adopted to guarantee Quality of Service (QoS) of self-similar clusters, namely, qualification of offset time, Time-Space Label-Switching Protocol, and Acknowledgment mechanism [Huan05]. Thus, the motivation of OSCS is to improve network performance for self-similar traffic by taking full advantage of self-similarity instead of overcoming it, as shown in Fig. 18.9.

Furthermore, OBS will also encounter more challenges from new requirements and new applications in optical networks in the future.

18.8 Trends in OBS

The trend of optical switching and our research phase are shown in Fig. 18.10. The current status of our efforts is also shown in Fig. 18.10. By utilizing optics as well as electronics in order to find the optimum mix of technologies for future very-high-capacity networks, our main objective is to develop the commercial applications of OBS networks, including traffic-engineering capabilities and network management, and covering the entire area from WR to OPS.

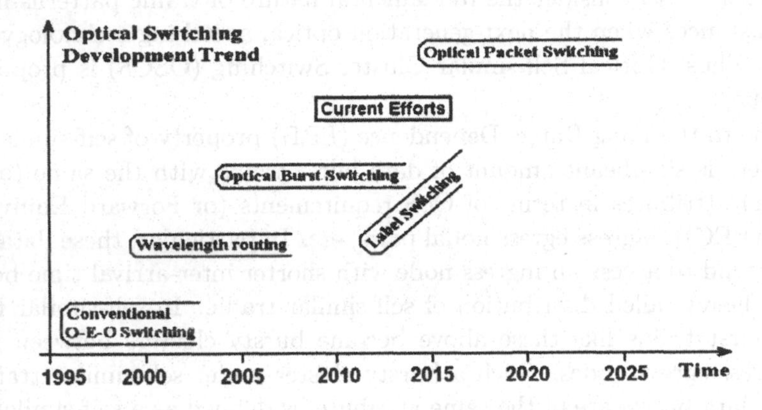

Figure 18.10 Trend of optical switching technologies.

Specifically, trends in OBS can be presented as follows. Firstly, the requirement of commercial applications is becoming stronger. According to a recent report released by Communications Industry Researchers Inc. (CIR) on June, 2004, the worldwide market for electronic and optical switching fabric will grow from US$ 951 million in sales in 2004 to US$ 1.5 billion in 2008 [CIR04]. A number of leading corporations in the world have started deployment of optical-switching technologies. These companies can enhance their R&D productivity gain by as much as 47% by using optical-switching technology [CIR04]. OBS may play a key role in the optical-switching market in the future because it can potentially meet the cost-effective requirements in the new equipments to be developed. Secondly, OBS will advance as the research in related fields is developing, such as burst grooming, OBS-specific routing protocols, burst classification, burst scheduling, QoS requirements, etc. Furthermore, it is beneficial for OBS networks that optical logical devices are becoming mature. Thirdly, new application requirements in optical networks are indispensable propellants toward the real network applications of OBS. An example is the development of GRID computing networks. GRID has received special attention from the research and industrial communities, such as NCSA National Technology Grid of USA, Euro Grid of EU, and CRL Grid of Japan. In 2003, the MIT Technology Review identified GRID as one of the "ten technologies that will change the world". There are various applications in GRID computing networks, such as E-science, E-business, E-health, E-government, etc. However, the GRID computing network design and corresponding types of applications are seriously affected in some aspects, such as throughput, priority, security, latency, cost, control and management, QoS, storage capacity, and so on. OBS seems be a potentially good choice to be an infrastructure for the GRID, namely, to offer huge capacity and relatively low latency, provide dynamic control and allocation of bandwidth, and reduce cost. To some extent, OBS is also a good choice for realizing seamless integration of optical switching and GRID computing networking [Simc03]. It is expected that new requirements from the scientific communities will push OBS forward into commercial applications.

18.9 Summary

Optical Burst Switching (OBS) is becoming a promising candidate for next-generation optical switching with contributions from many researchers worldwide. Even if optical buffer and bit-level optical processing may not be realizable soon, OBS can still be a prevailing switching technology because of its potential flexibility and scalability.

Exercises

18.1. Why is OBS considered to be a hybrid architecture? Please point out each component of OBS, how it can be implemented, and which domain the corresponding function can be realized in (electronic or optical)?

18.2. What are the differences between OPS and OBS? Considering current limits of optical hardware technologies, which approach (OPS or OBS) seem more promising for realization today?

18.3. If optical logical devices and optical random access memory would become mature in the future, will OBS become outdated? Please state and justify your views of OBS technology's future promise.

18.4. Compare JET and TL. In what aspects does TL surpass JET? Why?

18.5. As the volume of data traffic continues to grow, why is it essential to consider the nature of traffic in optical switching? If not, what will happen to optical switching?

18.6. What issues should be discussed in Optical Self-Similar Cluster Switching (OSCS)? Please justify your analysis.

18.7. Which stage is key in OSCS? Why? Can you determine any other solution for the key stage?

18.8. In an OBS network, nodes are classified into two types: edge nodes and intermediate (or core) nodes. What is required for intermediate nodes in order to guarantee that an OBS network can be implemented without any extra delay? Since edge nodes are important for OBS networks, what functions should be realized in edge nodes? Please design an edge node to implement its basic functions.

18.9. In an OBS network, what are the two kinds of signaling protocols used? Which kind does TL belong to? Please point out the differences between two kinds of signaling protocols.

18.10. What is the function of a routing protocol? What is the first consideration behind developing OBS-specific routing algorithms?

18.11. Is it possible to bind signaling and routing protocols together without any other extra protocols? Please try to compare TS-LSP with GMPLS.

18.12. How will the network operations, management, measurement, and control plane architectures exist across administrative domains in an OBS network? Is it possible to realize a seamless integration between OBS and ASON?

18.13. Is OBS an asynchronous or a synchronous network? Burst contention is a major challenge towards realization of OBS networks. How can one solve the possible contentions in an OBS network? Please discuss the issues from the views of contention and switching mechanisms.

18.14. Consider a GRID Network. What are the relationships between such a network and an OBS network?

18.15. About optical switching development, which is better: traffic-oriented switching or user-oriented switching?

18.16. How long is the required offset between a burst and its header in an OBS network using the JET (just enough time) protocol? How is the offset adjusted for bursts of different priorities?

18.17. In the last two chapters' references, various versions of optical switching are proposed. Please give your concerns about these versions, i.e., WR, OPS, OBS, Optical flow switching, Optical Self-similar Cluster Switching, Optical label switching, Optical code switching, and so on.

Appendix A
Further Reading

For further reading, one can refer to a large number of resources on the subject matter. For in-depth coverage on optical WDM device technology, please see [Gree92]. While the book's coverage of optical device technology is quite broad, its material on network architectures represents the state of the art in 1992. Recent books on optical WDM devices include [Agra04] and [DuDF02].

Other books dealing with optical networking in general include [StBa99] and [RaSi01]. Books dealing with some specific topics include: [JuVo05] for optical burst switching; [Grov03], [OuMu05] and [Zang02] for optical network survivability; [ZhZM05] for traffic grooming; [Kram05] for Ethernet Passive Optical Network (EPON); etc.

Early tutorial/review/survey articles on optical networking are the following: [Brac90], [Mukh92a], [Mukh92b], and [Rama93] although these articles are also quite outdated by now since they were written before wavelength routing had become mature. More recent review articles include [Mukh00] for general treatment of the topic; [ZaJM00a] for lightpath routing and wavelength assignment; [DuRo00] for virtual topology design; etc.

Most of the latest breakthroughs in optical network device technology are reported at the conferences [OFC] and [ECOC], and in the journals [JLT] and [PTL]. Papers reporting most of the latest developments on optical network architectures can be found in the conference [Info], [ICC], [Glob], and [ONDM]; the journals [ToN], [JSAC], [PNC], [OSN], and [JON]; and the magazines [Comm] and [Net]. (Papers on optical networking are published in other journals and conference proceedings also; please refer to the Bibliography for an extensive listing.)

Several specific journal special issues which deal with both device and network aspects and which are "collector's items" for anyone working in the field are [JSAC90], [JLT93], [JHSN95], [JSAC96], [JLT96], [JSAC98], [JSAC00], [JSAC02], [JSAC-OCN], [JLT02], [JLT03], and [JLT04].

A.1 General Resources and Publications

[Comm.] *IEEE Communications Magazine*, published monthly.

[ECOC.] *European Conference on Optical Communication*, held annually in September.

[Glob.] *IEEE Global Telecommunications Conference*, held annually in November/December.

[ICC.] *IEEE Internationl Conference on Communications*, held annually in May/June.

[Info.] *IEEE Infocom Conference*, held annually during Spring.

[JHSN95.] *Journal of High-Speed Networks*, Special issue on Optical Networks, vol. 3, nos. 1-2, January-April 1995.

[JLT.] *IEEE/OSA Journal of Lightwave Technology*, published monthly.

[JLT93.] *IEEE/OSA Journal of Lightwave Technology*, Special issue on Broadband Optical Networks, vol. 11, no. 5/6, May/June 1993.

[JLT96.] *IEEE/OSA Journal of Lightwave Technology*, Special issue on Multiwavelength Optical Networks, vol. 14, no. 6, June 1996.

[JLT02.] *IEEE/OSA Journal of Lightwave Technology*, Special issue on 40 Gb/s lightwave systems, vol. 20, no. 12, Dec 2002.

[JLT03.] *IEEE/OSA Journal of Lightwave Technology*, Special issue on Optical MEMS and Its Future Trends, vol. 21, no. 3, Mar 2003.

[JLT04.] *IEEE/OSA Journal of Lightwave Technology*, Special issue on Optical Interconnects, vol. 22, no. 9, Sep 2004.

[JON.] *OSA Journal of Optical Networking*, published monthly online at http://www.osa-jon.org/.

[JSAC90.] *IEEE Journal on Selected Areas in Communications*, Special issue on Dense Wavelength Division Multiplexing Techniques for High Capacity and Multiple Access Communication Systems, vol. 8, no. 6, Aug. 1990.

[JSAC96.] *IEEE Journal on Selected Areas in Communications*, Special issue on Optical Networks, vol. 14, no. 5, June 1996.

[JSAC98.] *IEEE Journal on Selected Areas in Communications*, Special issue on High-Capacity Optical Transport Networks, vol. 16, no. 7, Sep 1998

[JSAC00.] *IEEE Journal on Selected Areas in Communications*, Special issue on Protocols and Architectures for Next Generation Optical WDM Networks, vol. 18, no. 10, Oct 2000

[JSAC02.] *IEEE Journal on Selected Areas in Communications*, Special issue on WDM-based Network Architectures, vol. 20, no. 1, Oct 2002

[JSAC-OCN.] *IEEE Journal on Selected Areas in Communications*, Optical Communications and Networking Series.

[Net.] *IEEE Network Magazine*, published every other month.

[OFC.] *Optical Fiber Communications (OFC) Conference*, held annually in February/March.

[ONDM.] *Conference on Optical Network Design and Modelling*, held annually in February.

[OSN.] *Optical Switching and Networking Journal.*

[PNC.] *Photonic Network Communications.*

[PTL.] *IEEE Photonics Technology Letters*, published monthly.

[ToN.] *IEEE/ACM Transactions on Networking*, published every even-numbered month.

A.2 Enabling Technologies

[Agra04.] G. P. Agrawal, *Lightwave Technology: Components and Devices*, Wiley-Interscience, May 2004.

[Brac90.] C. A. Brackett, "Dense wavelength division multiplexing networks: Principles and applications," *IEEE Journal on Selected Areas in Communications*, vol. 8, pp. 948-964, Aug. 1990.

[DuDF02.] A. K. Dutta, N. K. Dutta, and M. Fujiwara, *WDM Technologies Active Optical Components: Active Optical Components*, Academic Press, Sept. 2002.

[Gree92.] P. E. Green, Jr., *Fiber Optic Networks*, Prentice-Hall, 1992.

A.3 Tutorials/Surveys/Reviews

[DuRo00.] R. Dutta and G. N. Rouskas, "A survey of virtual topology design algorithms for wavelength routed optical networks," *SPIE Optical Networks Magazine*, vol. 1, no. 1, pp. 73–89, Jan. 2000.

[Mukh92a.] B. Mukherjee, "WDM-Based Local Lightwave Networks – Part I: Single-Hop Systems," *IEEE Network Magazine*, vol. 6, no. 3, pp. 12-27, May 1992.

[Mukh92b.] B. Mukherjee, "WDM-Based Local Lightwave Networks – Part II: Multihop Systems," *IEEE Network Magazine*, vol. 6, no. 4, pp. 20-32, July 1992.

[Mukh00.] B. Mukherjee, "WDM Optical Communication Networks: Progress and Challenges" (Invited Paper), *IEEE Journal on Selected Areas in Communications*, vol. 18, no. 10, pp. 1810–1824, Oct. 2000.

[Rama93.] R. Ramaswami, "Multi-wavelength lightwave networks," *IEEE Communications Magazine*, vol. 31, no. 2, pp. 78-88, Feb. 1993.

[ZaJM00a.] H. Zang, J. P. Jue, and B. Mukherjee, "A review of routing and wavelength assignment approaches for wavelength-routed optical WDM networks," *SPIE Optical Networks Mag.*, vol. 1, no. 1, pp. 47–60, January 2000.

A.4 Other Books

[**Grov03.**] W. D. Grover, *Mesh-based Survivable Networks: Options and Strategies for Optical, MPLS, SONET and ATM Networking*, Prentice Hall PTR, New Jersey, 2003.

[**JuVo05.**] J. Jue and V. Vokkarane, *Optical Burst Switched Networks*, Optical Network Series, Springer, 2005.

[**Kram05.**] G. Kramer, *Ethernet Passive Optical Networks*, McGraw-Hill, 2005.

[**OuMu05.**] C. Ou and B. Mukherjee, *Survivable Optical WDM Networks*, Optical Network Series, Springer, 2005.

[**RaSi01.**] R. Ramaswami and K. Sivarajan, *Optical Networks: A Practical Perspective*, 2nd ed., Morgan Kaufmann Publishers, 2001.

[**StBa99.**] T. E. Stern and K. Bala, *Multiwavelength Optical Networks: A Layered Approach*, Prentice Hall PTR, 1999.

[**Zang02.**] H. Zang, *WDM Mesh Networks: Management and Survivability*, Kluwer Academic Publishers, 2002.

[**ZhZM05.**] K. Zhu, H. Zhu and B. Mukherjee, *Traffic Grooming in Optical WDM Mesh Networks*, Optical Network Series, Springer, 2005.

Appendix B

Glossary

all-optical cycle: in a wavelength-routed network, each input port of a wavelength-routing switch may always remain connected to some output port regardless of whether such connection is required or not. Due to such unspecified and uncontrolled connections, unintended optical paths may be set up in the network. Optical amplifiers in such paths may repeatedly amplify their own noise thereby saturating the gain of the amplifiers.

all-optical network: a wavelength-routed optical network wherein the information path between the source and the destination remains entirely optical. Such an all-optical path offers protocol transparency to the network.

arrayed waveguide guide (AWG) (a.k.a. passive router, waveguide grating router (WGR), Latin router, etc.): a passive network element which supports a *fixed routing matrix* to route an optical signal from a given input port to a given output port based on the wavelength of the signal. Signals of different wavelengths coming into an input port will each be routed to a different output port. Also, different signals using the same wavelength can be input simultaneously to different input ports, and they will not interfere with one another at the output ports. The AWG with N input and N output ports is capable of routing a maximum of N^2 connections, as opposed to a maximum of N connections in the passive-star coupler.

ATM PON (APON) [a.k.a. Broadband PON (BPON)]: a passive optical network (PON) based on Asynchronous Transfer Mode (ATM) as the MAC layer protocol.

broadcast-and-select network: a network in which information transmitted by any node passes by the interface of all other nodes, and each node independently selects which part of the information it wants to receive.

channel sharing: a technique used when the number of channels needed exceeds the number of wavelengths available so that some of the (WDM) channels must be shared to allow network operation (e.g., each WDM channel may be shared using TDM).

circuit switching: a type of transmission in which a dedicated path is established between two nodes (with zero or more intermediate switching nodes) in order to carry information. Most efficient for a connection that will remain up for a fairly long time duration.

code-division multiple access (CDMA): multiple channels on the same wavelength, but separated by the way they encode data. Usually has high overhead to encode data.

control channel: a channel used to conduct pretransmission coordination (e.g., on which data channel will data transmission occur, at what time will the transmission start, etc.)

crosstalk: the undesirable effect of a transmission on one channel interfering with the transmitted signal on another channel.

cross-layer design: an approach to combine the properties of two (or more) layers (e.g., physical layer and network layer in an optical network) so that the integrated approach is mindful of the strengths and weaknesses of each layer, thereby resulting in a more-efficient network design compared to the approach where each layer is considered independently. An example of cross-layer design in an optical WDM network is to employ an impairment-aware routing algorithm consisting of two steps, i.e., *lightpath computation* (at the network layer) and *lightpath verification* (at the optical layer to ensure that the lightpath has good signal quality, e.g., low bit-error rate (BER)).

dark fiber: fiber optic cable that is currently not being used.

deflection routing: a method of misrouting ("deflecting") packet(s) at a switching node at which two or more packets are competing for the same output channel. Packets that lose the contention are "deflected" to other parts of the network instead of being locally buffered at the switching node.

electronic speed (bandwidth): the maximum speed at which electronic components can operate.

electrooptic conversion: conversion technique employed when moving information from the electronic domain to the optical domain or vice versa. Typically employed at each hop in a multihop network.

embedding: a technique used to create a virtual (logical) topology on a fixed physical topology by properly configuring the channels in the network (viz., which node transmits to which other node(s) over which channel(s)).

erbium-doped fiber amplifier (EDFA): a amplifier that is able to amplify signals in the wavelength range of 1540 nm to 1570 nm which is within one of the low-loss operating regions of a fiber optic cable. However, gain is not uniform over the amplified region.

Ethernet PON (EPON): a passive optical network (PON) based on Ethernet technology, which carries data traffic encapsulated in Ethernet frames (defined in the IEEE 802.3 standard).

(wavelength-agile, tunable) filter (receiver): an optical filter that can be used to select a particular wavelength channel to receive information on, and, if tunable, it can be reconfigured to allow a multitude of possible network configurations.

fixed OADM (FOADM): an OADM which is made to operate permanently (statically or factory tuned) on a particular wavelength.

Generalized Framing Procedure PON (GFP-PON): a passive optical network (PON) which provides a generic mechanism to adapt traffic from higher layers (Ethernet/IP), over a transport layer such as Synchronous Optical Network/Synchronous Digital Hierarchy (SONET/SDH). It also offers dynamic bandwidth assignment, operation and maintenance, as in APON.

graph coloring: the algorithms that are used to ensure that, if two or more lightpaths share a common physical fiber link on their paths, they must necessarily be operated on different wavelengths on that link. Also referred to as a "color clash" constraint.

hop distance: the number of hops it takes for a packet to travel from its source to its final destination. A network's diameter is the maximum number of hops it takes for a packet to get from any node to any other node (assuming that shortest paths are always used).

inverse multiplexing: a mechanism to demultiplex high-rate input data streams into lower-speed streams (e.g., one OC-192 channel to four OC-48 channels) while using multiple switch ports in parallel. (Opposite of traffic grooming, possibly for multi-path routing.)

impairment-aware routing: the physical-layer impairments are taken into consideration while computing and setting up a lightpath.

(wavelength-agile, tunable) laser (transmitter): an electromagnetic transmission device that can be modulated to carry information over an optical fiber, and, if the laser is tunable, a multitude of possible virtual network configurations can be created.

laser array: a set of fixed-tuned lasers which are integrated into a single component, with each laser operating at a different wavelength so that, if each of the wavelengths in the array is modulated independently, then multiple transmissions may take place simultaneously.

lightpath: the optical path through which the information flows in a wavelength-routed optical network. A lightpath provides "single-hop" connectivity between its end points. A lightpath may be composed of a single wavelength (in a network without wavelength converters), or it may consist of multitude of wavelengths (in the presence of wavelength converters).

light-tree: a point-to-multipoint extension of a lightpath. Light-trees are suitable to carry multicast requests.

lightwave network: a network in which multiple channels, each capable of operating at peak electronic speed, are multiplexed onto a fiber optic cable. These networks have a theoretical capacity of tens of terabits per second (Tbps).

logical topology (a.k.a. virtual topology): the set of lightpaths that are embedded on a physical fiber network are said to form a multihop logical topology, which is the basis of network control and management.

medium-access protocol: a method to determine when a node has the right to use (transmit on and/or receive from) a given channel (e.g., random access, reservation, etc.)

micro-electro-mechanical systems (MEMS): the integration of mechanical elements, sensors, actuators, and electronics on a common silicon substrate through micro-fabrication technology. It can be employed to build transparent switches.

multicasting: the ability to transmit from a single source to multiple destinations.

multihop network: a network in which a packet may hop through zero or more intermediate nodes before it reaches its final destination.

network engineering (NE): "put the *bandwidth* where the *traffic* is".

network planning (NP): "put the *bandwidth* where the *traffic* is *forecasted to be*".

opaque switch (or network): data transmission occurs over point-to-point links so that the signal is regenerated at every intermediate node along a lightpath via OEO conversion.

optical add-drop multiplexer (OADM): a network element which can add and drop traffic (wavelengths) in the optical WDM network (similar to a SONET ADM which adds and drops time slots in a TDM network).

optical bandwidth: the large information-carrying capacity, measured in tens of Tbps, that occurs in the low-loss regions of an optical fiber.

optical burst switching (OBS): an optical switching technique which separates the burst data (possibly consisting of many packets and possibly quite long) from the control header (or control packet) by the interval of *offset time*, in which control packet is processed in the electronic domain to configure the switch fabric appropriately, and the burst data is switched all-optically.

optical crossconnect (OXC): a network element which switches optical signals from input ports to output ports. These elements are usually wavelength insensitive.

optical packet switching (OPS): a technique where incoming packets are switched all-optically without being converted to electrical signal. If all packets are of the same size and all inputs at a switch are synchronized, then the corresponding ability of the switch to route packets is called photonic slot routing.

optical power loss: losses that occur in a fiber optic network due to connections (splitting, coupling, and insertion loss) and propagation (loss due to signal attenuation). These losses must be taken into account while designing the network to ensure that it meets its power budget.

optic-electro-optic (OEO): the incoming optical bit streams are converted to electronic data, the data is switched using an electronic crossconnect, and then the electronic bit streams are converted back to the optical domain.

packet delay: the entire delay experienced by a packet, usually measured from the instant of its generation at the source node until it is completely delivered at its destination, including propagation delay(s) in the route to its destination, and queuing delays at the source and intermediate nodes.

packet switching: a technique in which each packet contains enough information in its header so that it can be independently switched at intermediate nodes in the network and routed to its final destination. At each intermediate node, the packet may be queued up while waiting for an output channel to clear up.

passive optical network (PON): a point-to-multipoint (PtMP) optical network suitable for broadband access with no active elements in the signal path from source to destination. The only interior elements in a PON are passive optical components, such as optical fiber, splices, splitters, and passive routers.

passive (wavelength) router: an N-input, N-output device that can separately route each of several wavelengths incident on an input fiber to the same wavelength on separate output fibers, based on a *fixed* routing matrix.

passive-star coupler: an N-input, N-output passive broadcast device that can divide the power from each of its N input fibers and direct them to each of its N output fibers simultaneously.

physical topology: the physical layout of the network (e.g., bus, star, multistar, mesh, tree, etc.)

pretransmission coordination: a technique in which some information is transmitted on a control channel before the data is transmitted on a data channel.

protection: a fault recovery and management mechanism, in which backup resources (routes and wavelengths) are precomputed and reserved in advance.

receiver collision: the loss of a transmission due to the fact that the receiver was occupied receiving another transmission on another channel.

reconfigurable OADM (ROADM) [a.k.a. dynamic OADM]: an OADM which can be dynamically tuned using some control mechanism.

regular structure: a structure where each node in the network has the same number of transmitters and the same number of receivers, and the entire network has well-defined (structured) nodal-interconnection pattern.

restoration: a fault recovery and management mechanism in which, after a failure occurs, an alternate route and a free wavelength have to be discovered dynamically for each interrupted connection.

routing and wavelength assignment (RWA): a process is to find a free path and a free wavelength to set up a lightpath in a WDM network.

routing mechanism: the method by which a packet travels from the source to the destination including what happens if there is contention for the output links (e.g., shortest path routing, adaptive routing, deflection routing, etc.)

scalability/modularity: the ability to grow (scale) a network to a larger size, but be able to do so in small increments (modules).

single-hop network: a network in which a packet travels from its source to its destination directly (in one hop). The packet does not encounter an electrooptic conversion before reaching its final destination.

soliton: soliton transmission is a technique to transmit ultrafast optical signals while maintaining the shape of the optical pulse even in the presence of fiber dispersion by utilizing an optical fiber nonlinearity.

splitter: a device that splits the power input from one optical fiber onto two or more output fibers.

time-division multiplexing (TDM): a medium-access technique in which a given wavelength is carved up into a number of different channels based on periodic time slices.

traffic engineering (TE): "put the *traffic* where the *bandwidth* is".

traffic grooming: a process for efficiently carrying low-speed traffic streams over high-capacity wavelength channels, including packing (multiplexing), unpacking (demultiplexing), and switching these low-speed connections at various nodes in a WDM network.

translucent network: in contrast with the opaque network, the regeneration functionality is only employed at some nodes instead of at all nodes.

transparent switch (or network): the data signal that is transmitted remains in the optical domain for the entire lightpath without OEO conversion. It can offer transparency to bit rate, signal format, and protocols.

tuning range: the range of wavelengths over which a tunable device (transmitter or receiver) can operate.

tuning time: the time it takes for a (tunable) transmitter or receiver to switch from one channel to another. Reciprocal of tuning speed.

wavelength assignment: a process to assign wavelengths (or colors) to one or more lightpaths at the same time.

wavelength channel (a.k.a. WDM channel): a channel which is centered at a specific wavelength within the low-loss region of an optical fiber.

wavelength-continuity property: in the absence of any wavelength conversion device, a lightpath is required to be on the same wavelength channel throughout its path in the network; this requirement is referred to as the *wavelength continuity* property of the lightpath.

wavelength converter (a.k.a. wavelength changer, wavelength shifter): a device that can *convert* the optical data arriving on one wavelength along a fiber link into another wavelength and forward it along the next fiber link.

wavelength-convertible switch [a.k.a. wavelength interchanging crossconnect (WIXC)]: a wavelength-routing switch that is also equipped with a wavelength-conversion facility.

wavelength-division multiplexing (WDM): a medium-access technique in which a single fiber can have multiple channels by having each channel operate at a different center wavelength (or frequency).

wavelength-routing switch (WRS) [a.k.a. wavelength-selective crossconnect (WSXC), wavelength crossconnect]: a space/wavelength division multiplexed optical switch used to route and add/drop multiwavelength optical signals at the nodes of all-optical networks.

WDM network control and management: the mechanisms to set up and tear down lightpaths, possibly in a distributed fashion, in a wavelength-routed network.

WDM-PON: a passive optical network (PON) in which multiple wavelengths may be supported in either or both upstream and downstream directions by employing (WDM).

Bibliography

[AaKo89] A. Aarts and J. Korst, *Simulated Annealing and Boltzmann Machines*, New York: John Wiley & Sons, 1989.

[Acam87] A. S. Acampora, "A multichannel multihop local lightwave network," *Proceedings, IEEE Globecom '87*, Tokyo, Japan, vol. 3, pp. 1459–1467, Nov. 1987.

[Acev92] J. J. Garcia-Luna-Aceves, "Distributed Routing with Labeled Distances," *Proceedings, IEEE INFOCOM '92*, Florence, Italy, pp. 633–643, May 1992.

[ACGK86] E. Arthurs, J. M. Cooper, M. S. Goodman, H. Kobrinski, M. Tur, and M. P. Vecchi, "Multiwavelength optical cross-connect for parallel-processing computers," *Electronic Letters*, vol. 24, pp. 119–120, 1986.

[AcKa89] A. S. Acampora and M. J. Karol, "An overview of lightwave packet networks," *IEEE Network Magazine*, vol. 3, pp. 29–41, Jan. 1989.

[AcSh92] A. S. Acampora and S. I. A. Shah, "Multihop lightwave networks: a comparison of store-and-forward and hot-potato routing," *IEEE Transactions on Communications*, vol. 40, pp. 1082–1090, June 1992.

[ADDH94] J. Anderson, B. T. Doshi, S. Dravida, and P. Harshavardhana, "Fast restoration of ATM netowrks," *IEEE Journal on Selected Areas in Communications*, vol. 12, pp. 128–138, Jan. 1994.

[ADFF01] L. Andersson, P. Doolan, N. Feldman, A. Fredette, and B. Thomas, "*LDP* specification," *RFC 3036*, Jan. 2001.

[AGKV88] E. Arthurs, M. S. Goodman, H. Kobrinski, and M. P. Vecchi, "Hypass: An optoelectronic hybrid packet-switching system," *IEEE Journal on Selected Areas in Communications*, vol. 6, pp. 1500–1510, Dec. 1988.

[Agra01] G. P. Agrawal, *Nonlinear Fiber Optics*, Third Edition, Academic Press, 2001.

[Agra02] G. P. Agrawal, *Fiber-Optic Communication Systems*, Third Edition, Wiley Interscience, 2002.

[Agra04] G. P. Agrawal, *Lightwave Technology: Components and Devices*, Wiley-Interscience, May 2004.

[Alba83] A. Albanese, "Star network with collision avoidance circuits," *Bell Systems Technical Journal*, vol. 62, pp. 631–638, March 1983.

[AlDe00] M. Ali and J. Deogun, "Power-efficient design of multicast wavelength routed networks," *IEEE Journal Selected Areas in Communications*, vol. 18, no. 10, pp. 1852–1862, Oct. 2000.

[AlET01a] M. Ali, D. Elie-Dit-Cosaque, and L. Tancevski, "Enhancements to multi-protocol lamda switch to accomodate transmission impairments," *Proceedings, IEEE GLOBECOM '01*, San Antonio, TX, pp. 70–75, Nov. 2001.

[AlET01b] M. Ali, D. Elie-Dit-Cosaque, and L. Tancevski, "Network optimization with transmission impairments-based routing," *Proceedings, European Conference on Optical Communications (ECOC) '01*, Amsterdam, Netherlands, pp. 42–43, Oct. 2001.

[Alfe88] R. C. Alferness, "Titanium-diffused lithium niobate waveguide devices," in *Guided-Wave Optoelectronics* (T. Tamir, ed.), Chapter 4, New York: Springer-Verlag, 1988.

[Alwa04] V. Alwayn, *Optical Network Design and Implementation*, Cisco Press, 2004.

[Amst83] S. Amstutz, "Burst Switching — An Update," *IEEE Communications Magazine*, vol. 27, no. 9, pp. 50–57, Sept. 1983.

[AnCQ01] V. Anand, S. Chauhan, and C. Qiao, "Sub-path protection: A new framework for optical layer survivability and its quantitative evaluation," *Technical Report*, Dept. of Computer Science and Engineering, State University of New York at Buffalo, Jan. 2002.

[Ande95] D. Anderson, "Low-cost mechanical fiber-optic switch," *OFC '95 Technical Digest*, San Diego, CA, vol. 8, pp. 185–186, Feb. 1995.

[ANSI01] ANSI T1X1.5, 2001-062, "*Synchronous Optical Networks (SONET)*," Jan. 2001.

[ANTS05] *Abilene Network Traffic Statistics*, http://www.abilene.iu.edu.

[ARDC01] D. O. Awduche, Y. Rekhter, J. Drake, and R. Coltun, "Multiprotocol lambda switching: combining MPLS traffic engineering control with optical crossconnects," *IETF Draft: draft-ietf-awduche-mpls-te-optical-03.txt*, Oct. 2001.

[ASLC02] A. A. Akyamac, S. Sengupta, J. Labourdette, S. Chaudhuri, and S. French, "Reliability in single domain vs. multi domain optical mesh networks," *Proceedings, National Fiber Optic Engineers Conference (NFOEC) '02*, pp. 240–249, 2002.

[ASWC96] V. Arya, D. W. Sherrer, A. Wang, R. O. Claus, and M. Jones, "Temperature compensation scheme for refractive index grating-based optical fiber devices," *Proceedings, SPIE*, vol. 2594, pp. 52–59, 1996.

[Awdu01] Awduche et al., "*RSVP-TE*: Extensions to RSVP for LSP tunnels," *RFC 3209*, Dec. 2001.

[AzBM96] M. Azizoglu, R. A. Barry, and A. Mokhtar, "Impact of tuning delay on the performance of bandwidth-limited optical broadcast networks with uniform traffic," *IEEE Journal on Selected Areas in Communications*, vol. 14, pp. 935–944, June 1996.

[Aziz91] M. Azizoglu, *Phase Noise in Coherent Optical Communications*, Ph.D. Dissertation, Department of Electrical Engineering and Computer Science, Massachusetts Institute of Technology, 1991.

[BaFG90] J. A. Bannister, L. Fratta, and M. Gerla, "Topological design of the wavelength-division optical network," *Proceedings, IEEE INFOCOM '90*, San Francisco, CA, pp. 1005–1013, June 1990.

[BaHu96] R. A. Barry and P. A. Humblet, "Models of blocking probability in all-optical networks with and without wavelength changers," *IEEE Journal on Selected Areas in Communications*, vol. 14, pp. 858–867, June 1996.

[BaKM05] A. Banerjee, G. Kramer, and B. Mukherjee, "Fair Queuing Using Service Level Agreements (SLAs) for Open Access in an Ethernet Passive Optical Network (EPON)," *Proceedings, IEEE International Conference on Communications (ICC) '05*, Seoul, Korea, May 2005.

[Bala96a] K. Bala et al., "Toward hitless reconfiguration in WDM optical networks for ATM transport," *Proceedings, IEEE GLOBECOM '96*, London, UK, pp. 316–320, Nov. 1996.

[Bala96b] K. Bala et al., "WDM network economics," *Proceedings, NFOEC '96*, Denver, CO, vol. 1, pp. 105–116, 1996.

[BaMS94] D. Banerjee, B. Mukherjee, and S. Ramamurthy, "The multidimensional torus: Analysis of average hop distance and application as a multihop lightwave network," *Proceedings, IEEE International Conference on Communications (ICC '94)*, New Orleans, pp. 1675–1680, May 1994.

[BaMu96] D. Banerjee and B. Mukherjee, "Practical approaches for routing and wavelength assignment in large all-optical wavelength-routed networks," *IEEE Journal on Selected Areas in Communications*, vol. 14, pp. 903–908, June 1996.

[BaMu00] D. Banerjee and B. Mukherjee, "Wavelength-routed optical networks: Linear formulation, resource budgeting tradeoffs, and a reconfiguration study," *IEEE/ACM Transactions on Networking*, vol. 8, no. 5, pp. 598–607, Oct. 2000.

[BaRo01] I. Baldine and G. N. Rouskas, "Traffic adaptive WDM networks: A study of reconfiguration issues," *IEEE/OSA Journal of Lightwave Technology*, vol. 19, no. 4, pp. 433–455, Apr. 2001.

[BaSB91] K. Bala, T. E. Stern, and K. Bala, "Algorithms for routing in a linear lightwave network," *Proceedings, IEEE INFOCOM '91*, Bal Harbour, FL, vol. 1, pp. 1–9, Apr. 1991.

[BaSu97] R. A. Barry and S. Subramaniam, "The MAX-SUM wavelength assignment algorithm for WDM ring networks," *Proceedings, OFC '97*, Dallas, TX, pp. 121-122, Feb. 1997.

[BaVa95] F. Bauer and A. Varma, "Degree-constrained multicasting in point-to-point networks," *Proceedings, IEEE INFOCOM '95*, Boston, MA, vol. 4, pp. 369–376, Apr. 1995.

[BBCG86] E.-J. Bachus, R.-P. Braun, C. Casper, and E. Grossman, "Ten-channel coherent optical fiber transmission," *Electronic Letters*, vol. 22, pp. 1002–1003, 1986.

[BDLT01a] A. Banerjee, J. Drake, J. P. Lang, B. Turner, D. Awduche, L. Berger, K. Kompella, and Y. Rekhter, "Generalized multiprotocol label switching: an overview of signaling enhancements and recovery techniques," *IEEE Communications Magazine*, vol. 39, pp. 144–151, July 2001.

[BDLT01b] A. Banerjee, J. Drake, J. Lang, B. Turner, K. Kompella, and Y. Rekhter, "Generalized multiprotocol label switching: An overview of routing and management enhancements," *IEEE Communications Magazine*, vol. 39, no. 1, pp. 144–150, Jan. 2001.

[BeMo00] R. Berry and E. Modiano, "Reducing electronic multiplexing costs in SONET/WDM rings with dynamically changing traffic," *IEEE Journal on Selected Areas in Communications*, vol. 18, no. 10, pp. 1961–1971, Oct. 2000.

[BeOS99] P. C. Becker, N. A. Olsson, and J. R. Simpson, *Erbium-Doped Fiber Amplifiers Fundamentals and Technology*, Academic Press: Optics and Photonics, 1999.

[Berk94] M. Berkelaar, "lpSolve: Readme file," *Documentation for the lpSolve program*, 1994.

[BESO92] A. Budman, E. Eichen, J. Schlafer, R. Olshansky, and F. McAleavey, "Multigigabit optical packet switch for self-routing networks with subcarrier addressing," *Technical Digest, Optical Fiber Communications Conference (OFC '92)*, San Jose, CA, pp. 90–91, Feb. 1992.

[BFGN04] A. Bianco, J. M. Finochietto, G. Galante, F. Neri, and V. Sarra, "Scheduling Variable-Size Packets in the DAVID Metropolitan Area Network," *Proceedings, IEEE International Conference on Communications (ICC) '04*, Paris, France, pp. 1750–1754, June 2004.

[Bhan99] R. Bhandari, *Survivable Networks: Algorithms for Diverse Routing*, Kluwer Academic Publishers, 1999.

[BiGu92] D. Bienstock and O. Gunluk, "A degree sequence problem related to network design," *Networks*, vol. 24, pp. 195–205, July 1994.

[BiKe95] A. Birman and A. Kershenbaum, "Routing and wavelength assignment methods in single-hop all- optical networks with blocking," *Proceedings, IEEE INFOCOM '95*, Boston, MA, pp. 431–438, April 1995.

[Birm96] A. Birman, "Computing approximate blocking probabilities for a class of all-optical networks," *IEEE Journal on Selected Areas in Communications*, vol. 14, pp. 852–857, June 1996.

[BLER02] E. Bouillet, J. F. Labourdette, G. Ellinas, R. Ramamurthy, and S. Chaudhuri, "Stochastic approaches to compute shared mesh restored lightpaths in optical network architectures," *Proceedings, IEEE INFOCOM '02*, New York, pp. 801–807, June 2002.

[BLMN01] A. Bianco, E. Leonardi, M. Mellia, and F. Neri, "Network Controller Design for SONATA: A Large-Scale All-Optical WDM Network," *Proceedings, IEEE International Conference on Communications (ICC) '01*, Helsinki, Finland, pp. 482–488, June 2001.

[BLRC02] E. Bouillet, J.-F. Labourdette, R. Ramamurthy, and S. Chaudhuri, "Enhanced algorithm cost model to control tradeoffs in provisioning shared mesh restored lightpaths," *Proceedings, OFC '02*, Anaheim, CA, pp. ThW2, March 2002.

[BoFB93] F. Brogonovo, L. Fratta, and J. Bannister, "Unslotted deflection routing in all-optical networks," *Proceedings, IEEE Globecom '93*, Houston, TX, pp. 119–125, Nov. 1993.

[BoMu96] M. S. Borella and B. Mukherjee, "Efficient scheduling of nonuniform packet traffic in a WDM/TDM local lightwave network with arbitrary transceiver tuning latencies," *IEEE Journal on Selected Areas in Communications*, vol. 14, pp. 923–934, June 1996.

[Bore95] M. S. Borella, *Scheduling and multicasting in wavelength-division multiplexed local optical networks*, Ph.D. Dissertation, University of California, Davis, Department of Computer Science, 1995.

[BoSD93] K. Bogineni, K. M. Sivalingham, and P. W. Dowd, "Low-complexity multiple access protocols for wavelength division multiplexed photonic networks," *IEEE Journal on Selected Areas in Communications*, vol. 11, pp. 590–604, May 1993.

[BPCM05] A. Banerjee, Y. Park, F. Clarke, H. Song, S. Yang, G. Kramer, K. Kim, and B. Mukherjee, "A review of wavelength-division multiplexed passive optical network (WDM-PON) technologies for broadband access," *preprint*, 2005.

[Brac90] C. A. Brackett, "Dense wavelength division multiplexing networks: Principles and applications," *IEEE Journal on Selected Areas in Communications*, vol. 8, pp. 948–964, Aug. 1990.

[BrBa02] M. Brunato and R. Battiti, "A multistart randomized greedy algorithm for traffic grooming on mesh logical topologies," *Proceedings, Workshop on Optical Network Design and Modeling (ONDM) '02*, Torino, Italy, pp. 417–430, Feb. 2002.

[BZBH97] R. Braden, L. Zhang, S. Berson, S. Herzog, and S. Jamin, "Resource reservation protocol (*RSVP*) - version 1, functional specification," *RFC 2205*, Sept. 1997.

[CaCe01] F. Callegati and W. Cerroni, "Wavelength allocation algorithms in optical buffers," *Proceedings, IEEE International Conference on Communications (ICC) '01*, Helsinki, Finland, pp. 499–503, June 2001.

[CaCM05] R. Cardillo, V. Curri, and M. Mellia, "Considering transmission impairments in wavelength routed networks," *Proceedings, 9th*

Conference on Optical Network Design and Modeling (ONDM), '05, Milan, Italy, pp. 421–429, Feb. 2005.

[Caid05] *Cooperative Association for Internet Data Analysis*, http://www.caida.org.

[CAMS96] T. K. Chiang, S. K. Agrawal, D. T. Mayweather, D. Sadot, et al., "Implementation of STARNET: a WDM computer communications network," *IEEE Journal on Selected Areas in Communications*, vol. 14, no. 5, pp. 824–39, June 1996.

[CaTT99] G. Castanon, L. Tancevski, and L. Tamil, "Routing in all-optical packet switched irregular mesh networks," *Proceedings, IEEE Globecom '99*, Rio de Janeiro, Brazil, pp. 1017–1022, Dec. 1999.

[CCEL04] E. Ciaramella, G. Contestabile, A. D. Errico, C. Loiacono, and M. Presi, "High-power widely tunable 40-GHz pulse source for 160-gb/s OTDM systems based on nonlinear fiber effects," *IEEE Photonics Technology Letters*, vol. 6, no. 3, pp. 753–755, March 2004.

[CeFP02] I. Cerutti, A. Fumagalli, and M. J. Potasek, "Effect of chromatic dispersion and self-phase modulation in multihop multirate WDM rings," *IEEE Photonics Technology Letters*, vol. 14, no. 3, pp. 411-413, March 2002.

[CFKM96] I. Chlamtac, A. Fumagalli, L. G. Kazovsky, P. Melman, et al., "CORD: contention resolution by delay lines," *IEEE Journal on Selected Areas in Communications*, vol. 14, no. 5, pp. 1014–1029, June 1996.

[ChBa95] C. Chen and S. Banerjee, "Optical switch configuration and lightpath assignment in wavelength routing multihop lightwave networks," *Proceedings, IEEE INFOCOM '95*, Boston, MA, pp. 1300–1307, June 1995.

[ChBa96] C. Chen and S. Banerjee, "A new model for optimal routing and wavelength assignment in wavelength division multiplexed optical networks," *Proceedings, IEEE INFOCOM '96*, San Francisco, CA, pp. 164–171, March 1996.

[ChDR90] M.-S. Chen, N. R. Dono, and R. Ramaswami, "A media access protocol for packet-switched wavelength division multiaccess metropolitan area networks," *IEEE Journal on Selected Areas in Communications*, vol. 8, pp. 1048–1057, Aug. 1990.

[ChFu91] I. Chlamtac and A. Fumagalli, "Quadro-stars: High performance optical WDM star networks," *Proceedings, IEEE Globecom '91*, Phoenix, AZ, pp. 1224–1229, Dec. 1991.

[ChFZ96] I. Chlamtac, A. Faragó, and T. Zhang, "Lightpath (wavelength) routing in large WDM networks," *IEEE Journal on Selected Areas in Communications*, vol. 14, pp. 909–913, June 1996.

[ChGa88] I. Chlamtac and A. Ganz, "Channel allocation protocols in frequency-time controlled high speed networks," *IEEE Transactions on Communications*, vol. 36, pp. 430–440, April 1988.

[ChGK89] I. Chlamtac, A. Ganz, and G. Karmi, "Purely optical networks for terabit communication," *Proceedings, IEEE INFOCOM '89*, Washington, DC, pp. 887–896, April 1989.

[ChGK92] I. Chlamtac, A. Ganz, and G. Karmi, "Lightpath communications: An approach to high bandwidth optical WANs," *IEEE Transactions on Communications*, vol. 40, pp. 1171–1182, July 1992.

[ChGK93] I. Chlamtac, A. Ganz, and G. Karmi, "Lightnets: Topologies for high speed optical networks," *IEEE/OSA Journal of Lightwave Technology*, vol. 11, pp. 951–961, May/June 1993.

[ChMo00] A. L. Chiu and E. H. Modiano, "Traffic grooming algorithms for reducing electronic multiplexing costs in WDM ring networks," *IEEE/OSA Journal of Lightwave Technology*, vol. 18, no. 1, pp. 2–12, Jan. 2000.

[ChMu03] W. Cho and B. Mukherjee, "Architecture and protocols for packet communication in optical WDM metropolitan-area ring networks using tunable wavelength add-drop multiplexors," *SPIE Optical Network Magazine*, vol. 4, no. 5, pp. 71–85, Sept./Oct. 2003.

[ChNT92] A. R. Chraplyvy, J. A. Nagel, and R. W. Tkach, "Equalization in amplified WDM lightwave transmission systems," *IEEE Photonics Technology Letters*, vol. 4, pp. 920–922, Aug. 1992.

[ChQY04] Y. Chen, C. Qiao, and X. Yu, "Optical burst switching: A new area in optical networking research," *IEEE Network*, vol. 18, no. 5, pp. 16–23, May 2004.

[Chra84] A. R. Chraplyvy, "Optical power limits in multi-channel wavelength-division-multiplexed systems due to stimulated Raman scattering," *Electronics Letters*, vol. 20, no. 2, pp. 58–59, 1984.

[Chra90] A. R. Chraplyvy, "Limits on lightwave communications imposed by optical-fiber nonlinearities," *IEEE/OSA Journal of Lightwave Technology*, vol. 8, pp. 1548–1557, Oct. 1990.

[ChWM01] W. Cho, J. Wang, and B. Mukherjee, "Improved approaches for cost-effective traffic grooming in WDM ring networks: Uniform-traffic case," *Photonic Network Communications*, vol. 3, no. 3, pp. 245–254, July 2001.

[ChYu91] M. Chen and T.-S. Yum, "A conflictfree protocol for optical WDM networks," *Proceedings, IEEE Globecom '91*, Phoenix, AZ, pp. 1276–1291, Dec. 1991.

[ChYu94] K. M. Chan and T. S. Yum, "Analysis of least congested path routing in WDM lightwave networks," *Proceedings, IEEE INFOCOM '94*, Toronto, Canada, pp. 962–969, June 1994.

[ChZA92] R. Chipalkatti, Z. Zhang, and A. S. Acampora, "High-speed communication protocols for optical star networks using WDM," *Proceedings, IEEE INFOCOM '92*, Florence, Italy, pp. 2124–2133, May 1992.

[CiOf93] I. Cidon and Y. Ofek, "MetaRing — A full-duplex ring with fairness and spatial reuse," *IEEE Transactions on Communications*, vol. 41, no. 1, pp. 110–120, January 1993.

[CIR04] Communications Industry Researchers, "*Electronic and optical switching fabric forecasts and market share analysis*," http://www.cir-inc.com, June 2004.

[CIR05] Communications Industry Researchers, "*The Future of WDM: Metro and Long Haul*," http://www.cir-inc.com, 2005.

[CJLM05] C. Chen, L. A. Johansson, V. Lal, M. L. Masanovic, et al., "Programmable optical buffering using fiber Bragg gratings combined with a widely-tunable wavelength converter," *Proceedings, OFC '05*, Anaheim, CA, pp. OWK4, March 2005.

[CJRG98] D. Chiaroni, C. Janz, M. Renaud, P. Gravey, et al., "*KEOPS Final Report*," ACTS Project Keys to Optical Switching AC043, Technical Report, Sept. 1998.

[ClGr02] M. Clouqueur and W. D. Grover, "Availability analysis of span-restorable mesh networks," *IEEE Journal on Selected Areas in Communications*, vol. 20, pp. 810–821, May 2002.

[Clos53] C. Clos, "A study of non-blocking switching networks," *The Bell System Technical Journal*, vol. 32, no. 2, pp. 406–424, March 1953.

[Coch95] P. Cochrane, *Optical Network Technology – Foreword*, New York: Chapman & Hall, 1995.

[CoLR01] T. H. Corman, C. E. Leiserson, and R. L. Rivest, *Introduction to Algorithms, Second Edition*, Cambridge, MA: MIT Press, 2001.

[Copl00] A. Copley, "Optical domain service interconnect (ODSI): Defining mechanisms for enabling on-demand high-speed capacity from the optical domain," *IEEE Communications Magazine*, vol. 38, no. 10, pp. 168–174, Oct. 2000.

[CoSa01] L. A. Cox and J. Sanchez, "Cost saving from optimized packing and grooming of optical circuits: mesh versus ring comparisons," *SPIE Optical Networks Magazine*, vol. 2, no. 3, pp. 72–90, May/June 2001.

[CPLE05] *CPLEX*, http://www.ilog.com

[CSBH89] K.-W. Cheung, D. A. Smith, J. E. Baron, and B. L. Heffner, "Multiple channel operation of an integrated acousto-optic tunable filter," *Electronics Letters*, vol. 25, pp. 375–376, 1989.

[CSLB05] K. K. Chow, C. Shu, C. Lin, and A. Bjarklev, "Polarization-insensitive widely tunable wavelength converter based on four-wave mixing in a dispersion-flattened nonlinear photonic crystal fiber," *IEEE Photonics Technology Letters*, vol. 17, no. 3, pp. 624–626, March 2005.

[CTYT00] G. Castanon, L. Tancevski, S. Yagnanarayanan, and L. Tamil, "Asymmetric WDM all-optical packet switched routers," *Proceedings, OFC '00*, Baltimore, MD, pp. 53–55, March 2000.

[DaEi02] D. Dahan and G. Eisenstein, "Numerical comparison between distributed and discrete amplification in a point-to-point 40Gb/s 40-WDM-based transmission system with three different modulation forms," *IEEE/OSA Journal Lightwave Technology*, vol. 20, no. 3, pp. 379–388, Marc 2002.

[DaHS98] S. L. Danielsen, P. B. Hansen, and K. E. Stubkjear, "Wavelength conversion in optical packet switching," *IEEE/OSA Journal of Lightwave Technology*, vol. 16, no. 12, pp. 2095–2108, Dec. 1998.

[DeAt04] C. DeCusatis and R. Atkins, "Security feature comparison for fibre channel storage area networks switches," *Proceedings, Fifth IEEE SMC Information Assurance Workshop*, pp. 203–209, June 2004.

[DeNa95] J.-M. P. Delavaux and J. A. Nagel, "Multi-stage erbium-doped fiber amplifier designs," *IEEE/OSA Journal of Lightwave Technology*, vol. 13, pp. 703–720, May 1995.

[Desa01] B. N. Desai, "An optical implementation of a packet-based (Ethernet) MAC in a WDM passive optical network overlay," *Proceedings, OFC '01*, Anaheim, CA, pp. WN5, March 2001.

[DGLR90] N. R. Dono, P. E. Green, K. Liu, R. Ramaswami, and F. F.-K. Tong, "A wavelength division multiple access network for computer communication," *IEEE Journal on Selected Areas in Communications*, vol. 8, pp. 983–993, Aug. 1990.

[DHKK89] C. Dragone, C. H. Henry, I. P. Kaminow, and R. C. Kistler, "Efficient multichannel integrated optics star coupler on silicon," *IEEE Photonics Technology Letters*, vol. 1, pp. 241–243, Aug. 1989.

[DJMP94] T. Durhuus et al., "All optical wavelength conversion by SOA's in a Mach-Zender configuration," *IEEE Photonics Technology Letters*, vol. 6, pp. 53–55, Jan 1994.

[DMJD96] T. Durhuus et al., "All-optical wavelength conversion by semiconductor optical amplifiers," *IEEE/OSA Journal of Lightwave Technology*, vol. 14, pp. 942–954, June 1996.

[DMNM04] L. Domash, W. Ming, N. Nemchuk, and E. Ma, "Tunable and switchable multiple-cavity thin film filters," *IEEE/OSA Journal of Lightwave Technology*, vol. 22, no. 1, pp. 126–135, Jan. 2004.

[Dowd91] P. W. Dowd, "Random access protocols for high speed interprocessor communication based on an optical passive star topology," *IEEE/OSA Journal of Lightwave Technology*, vol. 9, pp. 799–808, June 1991.

[Dowd92] P. W. Dowd, "Wavelength division multiple access channel hypercube processor interconnection," *IEEE Transactions on Computers*, vol. 41, no. 10, pp. 1223–1241, Oct. 1992.

[DoYa01] R. Doverspike and J. Yates, "Challenges for MPLS in optical network restoration," *IEEE Communications Magazine*, vol. 39, no. 2, pp. 89–96, Feb. 2001.

[DrEK91] C. Dragone, C. A. Edwards, and R. C. Kistler, "Integrated optics $n \times n$ multiplexer on silicon," *IEEE Photonics Technology Letters*, vol. 3, pp. 896–899, Oct. 1991.

[DSBD01] S. Datta, S. Sengupta, S. Biswas, and S. Datta, "Efficient channel reservation for backup paths in optical mesh networks," *Proceedings, IEEE Globecom '01*, San Antonio, TX, pp. 2104–2108, Nov. 2001.

[DuBa02] M. Duser and P. Bayvel, "Analysis of a dynamically wavelength-routed optical burst switched network architecture," *IEEE/OSA Journal of Lightwave Technology*, vol. 20, no. 4, pp. 574–585, April 2002.

[DuDF02] A. K. Dutta, N. K. Dutta, and M. Fujiwara, *WDM Technologies Active Optical Components: Active Optical Components*, Academic Press, Sept. 2002.

[DuRo00] R. Dutta and G. N. Rouskas, "A survey of virtual topology design algorithms for wavelength routed optical networks," *SPIE Optical Networks Magazine*, vol. 1, no. 1, pp. 73–89, Jan. 2000.

[EfIY01] F. J. Effenberger, H. Ichibangase, and H. Yamashita, "Advances in broadband passive optical networking technologies," *IEEE Communications Magazine*, vol. 39, no. 12, pp. 118–122, Dec. 2001.

[EGZJ93] A. F. Elrefaie, E. L. Goldstein, S. Zaidi, and N. Jackman, "Fiber-amplifier cascades with gain equalization in multiwavelength unidirectional inter-office ring network," *IEEE Photonics Technology Letters*, vol. 5, pp. 1026–1031, Sept. 1993.

[EiMe88] M. Eisenberg and N. Mehravari, "Performance of the multichannel multihop lightwave network under nonuniform traffic," *IEEE Journal on Selected Areas in Communications*, vol. 6, pp. 1063–1078, Aug. 1988.

[EiPW93] M. Eiselt, W. Pieper, and H. G. Weber, "Decision gate for all-optical retiming using a semiconductor laser amplifier in a loop mirror configuration," *IEE Electronic Letters*, vol. 29, pp. 107–109, Jan. 1993.

[ElAT02] D. Elie-Dit-Cosaque, M. Ali, and L. Tancevski, "Informed dynamic shared path protection," *Proceedings, OFC '02*, Anaheim, CA, pp. ThO4, March 2002.

[ElHS00] G. Ellinas, A. Hailemariam, and T. E. Stern, "Protection cycles in mesh WDM networks," *IEEE J. Selected Areas in Communications*, vol. 18, pp. 1924–1937, Oct. 2000.

[EPON02] *IEEE 802.3ah EFM EPON baseline technical proposals*, http://grouper.ieee.org/groups/802/3/efm/baseline/p2mpbaseline.html.

[ERHS00] G. Ellinas, S. Rong, A. Hailemariam, and T. E. Stern, "Protection cycle covers in optical networks with arbitrary mesh topologies," *Proceedings, OFC '00*, Baltimore, MD, pp. 213–215, March 2000.

[ErLi00] V. Eramo and M. Listanti, "Packet loss in a bufferless optical WDM switch employing shared tunable wavelength converters," *IEEE/OSA Journal of Lightwave Technology*, vol. 18, no. 12, pp. 1818–1833, Dec. 2000.

[EvIS76] S. Even, A. Itai, and A. Shamir, "On the complexity of timetable and multicommodity flow problems," *SIAM Journal of Computing*, vol. 5, pp. 691–703, 1976.

[FaAR05] C. Fan, S. Adams, and M. Reisslein, "The FTCFR AWG network: a practical single-hop Metro WDM network for efficient uni- and multicasting," *IEEE/OSA Journal of Lightwave Technology*, vol. 23, no. 3, pp. 937–954, March 2005.

[FATA02] M. Farahmand, D. Awduche, S. Tibuleac, and D. Atlas, "Characterization and representation of impairments for routing and path control in all-optical networks," *Proceedings, NFOEC '02*, Dallas, TX, pp. 279–289, July 2002.

[FCMJ99] A. Fumagalli, I. Cerutti, F. Masetti, R. Jagannathan, and S. Alagar, "Survivable networks based on optimal routing and WDM self-healing rings," *Proceedings, IEEE INFOCOM '99*, New York, pp. 726–733, March 1999.

[FeSR94] J. Fehrer, J. Sauer, and L. Ramfelt, "Design and implementation of a prototype optical deflection network," *Proceedings, ACM SIGCOMM*, vol. 24, pp. 191–200, 1994.

[FIMD94] N. J. Frigo, P. P. Iannone, P. D. Magill, T. E. Darcie, et al., "A wavelength-division multiplexed passive optical network with cost shared components," *IEEE Photonics Technology Letters*, vol. 6, no. 11, pp. 1365–1367, Nov. 1994.

[FlMe01] C. R. S. Fludger and R. J. Mears, "Electrical measurements of multipath interference in distributed Raman amplifiers," *IEEE/OSA Journal of Lightwave Technology*, vol. 19, no. 4, pp. 536–545, April 2001.

[FoBP95] F. Forghierri, A. Bononi, and P. R. Prucnal, "Analysis and comparison of hot-potato and single-buffer deflection routing in very high bit rate optical mesh networks," *IEEE Transactions on Communications*, vol. 43, no. 1, pp. 88–98, Jan. 1995.

[FPRT02] A. Fumagalli, A. Paradisi, S. M. Rossi, and M. Tacca, "Differentiated reliability (DiR) in mesh networks with shared path protection: theoretical and experimental results," *Proceedings, OFC '02*, Anaheim, CA, pp. 490–492, March 2002.

[FrGK73] L. Fratta, M. Gerla, and L. Kleinrock, "The flow deviation method: An approach to store-and-forward communication network design," *Networks*, vol. 3, pp. 97–133, 1973.

[FSAN05] *FSAN – Full Service Access Network*, Online at http://fsan.mblast.com/default.asp.

[FTCM94] F. Forghieri, R. W. Tkach, A. R. Chraplyvy, and D. Marcuse, "Reduction of four-wave mixing crosstalk in WDM systems using unequally spaced channels," *IEEE Photonics Technology Letters*, vol. 6, no. 6, pp. 754–756, 1994.

[FTUF02] A. Fumagalli, M. Tacca, F. Unghvary, and A. Farago, "Shared path protection with differentiated reliability," *Proceedings, IEEE ICC '02*, New York, pp. 2157–2161, April 2002.

[FuCT03] A. Fumagalli, I. Cerutti, and M. Tacca, "Optimal design of survivable mesh networks based on line switched WDM self-healing rings," *IEEE/ACM Transactions on Networking*, vol. 11, pp. 501–512, June 2003.

[Fuji88] M. Fujiwara et al., "A coherent photonic wavelength-division switching system for broadband networks," *Proceedings, European Conference on Communications (ECOC '88)*, pp. 139–142, 1988.

[Fuku98] M. Fukuda, *Optical Semiconductor Devices*, Wiley-Interscience, 1998.

[FuTa83] A. Fukuda and S. Tasaka, "The equilibrium point analysis – a unified analytic tool for packet broadcast networks," *Proceedings, IEEE GLOBECOM '83*, San Diego, CA, pp. 33.4.1–33.4.8, Nov. 1983.

[FuVa00] A. Fumagalli and L. Valcarenghi, "IP restoration vs. WDM protection: is there an optimal choice?" *IEEE Network*, vol. 14, pp. 34–41, Nov./Dec. 2000.

[GaGa92a] A. Ganz and Y. Gao, "A time-wavelength assignment algorithm for a WDM star network," *Proceedings, IEEE INFOCOM '92*, Florence, Italy, pp. 2144–2150, May 1992.

[GaGa92b] A. Ganz and Y. Gao, "Traffic scheduling in multiple WDM star systems," *Proceedings, IEEE International Conference on Communications (ICC) '92*, Chicago, IL, pp. 1468–1472, June 1992.

[GaJo79] M. R. Garey and D. S. Johnson, *Computers and Intractability: A Guide to the Theory of NP-Completeness*, New York: W. H. Freeman and Company, 1979.

[GaKo91] A. Ganz and Z. Koren, "WDM passive star protocols and performance analysis," *Proceedings, IEEE INFOCOM '91*, Bal Harbour, FL, pp. 991–1000, April 1991.

[GDCL02] W. D. Grover, J. Doucette, M. Clouqueur, D. Leung, and D. Stamatelakis, "New options and insights for survivable transport networks," *IEEE Communications Magazine*, vol. 40, pp. 34–41, Jan. 2002.

[GeLS98] O. Gerstel, P. Lin, and G. Sasaki, "Wavelength assignment in a WDM ring to minimize cost of embedded SONET rings," *Proceedings, IEEE INFOCOM '98*, San Francisco, CA, pp. 94–101, March 1998.

[GeMu03] A. Gencata and B. Mukherjee, "Virtual-topology adaptation for WDM mesh networks under dynamic traffic," *IEEE/ACM Transactions on Networking*, vol. 11, no. 2, pp. 236–247, April 2003.

[GeRa00a] O. Gerstel, R. Ramaswami, and W. K. Wang, "Making use of a two-stage multiplexing scheme in a WDM network," *Proceedings, OFC '00*, Baltimore, MD, pp. 44–46, March 2000.

[GeRa00b] O. Gerstel and R. Ramaswami, "Optical layer survivability–an implementation perspective," *IEEE Journal on Selected Areas in Communications*, vol. 18, pp. 1885–1899, Oct. 2000.

[GeRa00c] O. Gerstel and R. Ramaswami, "Optical layer survivability: A services perspective," *IEEE Communications Magazine*, vol. 38, pp. 104–113, March 2000.

[GeRa02] O. Gerstel, R. Ramaswami, and S. Foster, "Merits of hybrid optical networking," *Proceedings, OFC '02*, Anaheim, CA, pp. 33–34, March 2002.

[Gers98] O. Gerstel, "Opportunities for optical protection and restoration," *Proceedings, OFC '98*, San Jose, CA, pp. 269–270, Feb. 1998.

[Gers00] O. Gerstel, "Optical networking: a practical perspective (tutorial)," *IEEE Hot Interconnects*, Aug. 2000.

[GeRS00] O. Gerstel, R. Ramaswami, and G. H. Sasaki, "Cost-effective traffic grooming in WDM rings," *IEEE/ACM Transactions on Networking*, vol. 8, no. 5, pp. 618–630, October 2000.

[GGKV90] M. S. Goodman, J. L. Gimlett, H. Kobrinski, M. P. Vecchi, and R. M. Bulley, "The LAMBDANET multiwavelength network: Architecture, applications, and demonstrations," *IEEE Journal on Selected Areas in Communications*, vol. 8, pp. 995–1004, Aug. 1990.

[GhPC02] N. Ghani, J.-Y. Pan, and X. Cheng, "Metropolitan Optical Networks," *Optical Fiber Telecommunications*, Elsevier Press, vol. 4, pp. 329–403, March 2002.

[GiSp99] C. R. Giles and M. Spector, "The wavelength add/drop multiplexer for lightwave communication networks," *Bell Labs Technical Journal*, vol. 4, no. 1, pp. 207–229, January/March 1999.

[GiTe91] R. Gidron and A. Temple, "Teranet: A multihop multichannel ATM lightwave network," *Proceedings, IEEE International Conference on Communications (ICC) '91*, Denver, CO, pp. 602–608, June 1991.

[GlKW94] B. Glance, I. P. Kaminow, and R. W. Wilson, "Applications of the integrated waveguide grating router," *IEEE/OSA Journal of Lightwave Technology*, vol. 12, pp. 957–962, June 1994.

[GlSc90] B. S. Glance and O. Scaramucci, "High-performance dense FDM coherent optical network," *IEEE Journal on Selected Areas in Communications*, vol. 6, pp. 1043–1047, Aug. 1990.

[GoNT04] M. Gotoda, T. Nishimura, Y. Tokuda, "Widely tunable SOA-integrated DBR laser with combination of sampled-grating and superstructure grating," *Proceedings, 19th IEEE International Semiconductor Laser Conference*, Matsue, Japan, pp. 147–148, Sept. 2004.

[Gora02] W. Goralski, *SONET/SDH*, Third Edition, McGraw-Hill, Oct. 2002.

[Gree92] P. E. Green, "Misunderstood issues in lightwave networking," *OMAN Summer Topicals*, Santa Barbara, CA, pp. 47–48, Aug. 1992.

[Gree93] P. E. Green, *Fiber Optic Networks*, Englewood Cliffs, NJ: Prentice-Hall, 1993.

[GRGJ98] C. Guillemot, M. Renaud, P. Gambini, C. Janz, et al., "Transparent optical packet switching: the European ACTS KEOPS project approach," *IEEE/OSA Journal of Lightwave Technology*, vol. 16, pp. 2117–2133, Dec. 1998.

[Grov99] W. D. Grover, "High availability path design in ring-based optical networks," *IEEE/ACM Transactions on Networking*, vol. 7, no. 4, pp. 558–574, Aug. 1999.

[Grov03] W. D. Grover, *Mesh-Based Survivable Networks: Options and Strategies for Optical, MPLS, SONET and ATM Networking*, Prentice Hall, New Jersey, 2003.

[GuLa05] H. Guo and Z. Lan, "A testbed for optical burst switching network," *Proceedings, OFC '05*, Anaheim, CA, pp. OFA6, March 2005.

[GuPM03] K. Gummadi, M. Pradeep, and C. S. R. Murthy, "An efficient primary-segmented backup scheme for dependable real-time communication in multihop networks," *IEEE/ACM Transactions on Networking*, vol. 11, pp. 81–94, Feb. 2003.

[Haas93] Z. Haas, "The 'staggering switch': An electronically controlled optical packet switch," *IEEE/OSA Journal of Lightwave Technology*, vol. 11, pp. 925–936, May/June 1993.

[Hac94] A. Hac, "Improving reliability through architecture partitioning in telecommunication networks," *IEEE J. Selected Areas in Communications*, vol. 12, pp. 193–204, Jan. 1994.

[Hada00] R. Libeskind-Hadas, "Efficient Collective Communication in WDM Networks with a Power Budget," *Proceedings, Ninth IEEE International Conference on Computer Communications and Networks (ICCCN), Las Vegas, Nevada*, pp. 612–616, Oct. 2000.

[HaDL92] S. L. Hansen, K. Dybdal, and L. C. Larsen, "Gain limit in erbuim-doped fiber amplifiers due to internal Rayleigh backscattering," *IEEE Photonics Technology Letters*, vol. 4, pp. 559–561, June 1992.

[HaDS98] P. B. Hansen, S. L. Danielsen, and K. E. Studkjaer, "Optical packet switching without packet alignment," *Proceedings, ECOC '98*, Madrid, Spain, pp. WdD13, Sept. 1998.

[Haki71] S. L. Hakimi, "Steiner's problem in graphs and its implications," *Networks*, vol. 1, no. 2, pp. 113–133, 1971.

[HaKR96] E. Hall, J. Kravitz, R. Ramaswami, et al., "The Rainbow-II gigabit optical network," *IEEE Journal on Selected Areas in Communications*, vol. 14, pp. 814–823, June 1996.

[HaKS87] I. M. I. Habbab, M. Kavehrad, and C.-E. W. Sundberg, "Protocols for very high speed optical fiber local area networks using a passive star topology," *IEEE/OSA Journal of Lightwave Technology*, vol. LT-5, pp. 1782–1794, Dec. 1987.

[Hans99] P. B. Hansen, et al., "Rayleigh scattering limitation in distributed Raman pre-amplifiers," *IEEE Photonics Technology Letters*, vol. 10, no. 1, pp. 159–161, Jan. 1999.

[HCGF98] D. K. Hunter, W. D. Cornwell, T. H. Gilfedder, A. Franzen, and I. Andonovic, "SLOB: a switch with large optical buffers for packet switching," *IEEE/OSA Journal of Lightwave Technology*, vol. 16, no. 10, pp. 1725–1736, Oct. 1998.

[HeBe02] L. Helczynski and A. Berntson, "Comparison of EDFA and bidirectionally pumped Raman amplifier in a 40Gb/s RZ transmission system," *IEEE Photonics Technology Letters*, vol. 13, no. 7, pp. 669–761, July 2002.

[Hech04] J. Hecht, *Understanding Lasers: An Entry-Level Guide*, 2nd ed., Wiley US, 2004.

[Hech99] J. Hecht, *Understanding Fiber Optics*, 3rd ed., Prentice Hall, 1999.

[HeCT01] J. He, S. H. Chan, and D. H. K. Tsang, "Routing and Wavelength Assignment for WDM Multicast Networks," *Proceedings, IEEE Globecom '01*, San Antonio, TX, pp. 1536–1540, Nov. 2001.

[Henr85] P. S. Henry, "Lightwave primer," *IEEE Journal of Quantum Electronics*, vol. QE-21, pp. 1862–1879, Dec. 1985.

[HeSi01] J. He and D. Simeonidou, "A flow-routing approach for optical IP networks," *Proceedings, OFC '01*, Anaheim, CA, pp. MN2, March 2001.

[HGHM02] Y. Huang, A. Gencata, J. P. Heritage, and B. Mukherjee, "Routing and wavelength assignment with quality-of-signal constraints in WDM networks," *Proceedings, ECOC '02*, Copenhagen, Denmark, pp. P4-16, Sept. 2002.

[Hill02] T. Hills, "*Next-gen SONET*," Lightreading report, May 2002.

[Hill93] G. Hill et al., "A transport network layer based on optical network elements," *IEEE/OSA Journal of Lightwave Technology*, vol. 11, pp. 667–679, May/June 1993.

[Hint90] H. S. Hinton, "Photonic switching fabrics," *IEEE Communications Magazine*, vol. 28, pp. 71–89, Apr. 1990.

[Hira05] K. Hirabayashi, "PLZT electrooptic ceramic photonic devices for surface-normal operation in trenches cut across arrays of optical fiber," *IEEE/OSA Journal of Lightwave Technology*, vol. 23, no. 3, pp. 1393–1402, March 2005.

[HlKa91] M. G. Hluchyj and M. J. Karol, "ShuffleNet: An application of generalized perfect shuffles to multihop lightwave networks," *IEEE/OSA Journal of Lightwave Technology*, vol. 9, pp. 1386–1397, Oct. 1991.

[HNCA99] D. K. Hunter, M. H. M Nizam, M. C. Cia, I. Andonovic, et al., "WASPNET: a wavelength switched packet network," *IEEE Communications Magazine*, vol. 37, no. 3, pp. 120–129, March 1999.

[HSKO99] K. Harada, K. Shimizu, T. Kudou, and T. Ozeki, "Hierarchical optical path cross-connect systems for large scale WDM networks," *Proceedings, OFC '99*, San Diego, CA, pp. 356–358, Feb. 1999.

[Huan05] A. Huang et al., "Optical Self-similar Cluster Switching (OSCS) – A Novel Optical Switching Scheme by Detecting Self-similar Traffic," *Photonic Network Communications*, vol. 10, no. 3, pp. 297-C308, Nov. 2005.

[HuCA98] D. K. Hunter, M. C. Chia, and I. Andonovic, "Buffering in optical packet switches," *IEEE/OSA Journal of Lightwave Technology*, vol. 16, no. 12, pp. 2081–2094, Dec. 1998.

[HuHM05a] Y. Huang, J. P. Heritage and B. Mukherjee, "Connection Provisioning with Transmission Impairment Consideration in Optical WDM Networks with High-Speed Channels," *IEEE/OSA Journal of Lightwave Technology*, vol. 23, no. 3, pp. 982–993, March 2005.

[HuHM05b] Y. Huang, J. P. Heritage, and B. Mukherjee, "Dynamic Routing with Preplanned Congestion Avoidance for Survivable Optical Burst-Switched (OBS) Networks," *Proceedings, OFC '05*, Anaheim, CA, pp. OWC6, March 2005.

[Huit95] C. Huitema, *Routing in the Internet*, Englewood Cliffs, NJ: Prentice-Hall, 1995.

[HuPS02] G. Huiban, S. Perennes, and M. Syska, "Traffic grooming in WDM networks with multi-layer switches," *Proceedings, IEEE International Conference on Communications (ICC) '02*, New York, pp. 2896–2901, April 2002.

[HuXi03] A. Huang and L. Xie, "A Novel Segmentation and Feedback Model for Resolving Contention in Optical Burst Switching," *Photonic Network Communications*, vol. 6, no. 1, pp. 61–67, July 2003.

[HwRi92] F. K. Hwang and D. S. Richards, "Steiner tree problems," *Networks*, vol. 22, no. 1, pp. 55–89, 1992.

[HXLX03] A. Huang, L. Xie, Z. Li, and A. Xu, "Time-Space Label Switching Protocol (TSL-SP) - A New Paradigm of Network Resource Assignment," *Photonic Network Communications*, vol. 6, no. 2, pp. 169–178, Sept. 2003.

[IKSF04] K. Iwatsuki, J. Kani, H. Suzuki, and M. Fujiwara, "Access and Metro Networks Based on WDM Technologies," *IEEE/OSA Journal of Lightwave Technology*, vol. 22, no. 11, pp. 2623–2630, Nov. 2004.

[InBB95] J. Iness, S. Banerjee, and B. Mukherjee, "*GEMNET: A generalized, shuffle-exchange-based, regular, scalable, and modular multihop network based on WDM lightwave technology,*" *IEEE/ACM Transactions on Networking*, vol. 3, no. 4, pp. 470-476, Aug. 1995.

[Ines97] J. Iness, *Efficient Use of Optical Components in WDM-Based Optical Networks*, Ph.D. Dissertation, University of California, Davis, Department of Computer Science, 1997.

[InMu99] J. Iness and B. Mukherjee, "Sparse Wavelength Conversion in Wavelength-Routed WDM Networks," *Photonic Network Communications*, vol. 1, no. 3, pp. 183–205, Nov. 1999.

[IrGr00] R. R. Iraschko and W. D. Grover, "A highly efficient path-restoration protocol for management of optical network transport integrity," *IEEE Journal on Selected Areas in Communications*, vol. 18, pp. 779–794, May 2000.

[Ishi91] A. Ishimaru, *Electromagnetic Wave Propagation, Radiation, and Scattering*, Englewood Cliffs, NJ: Prentice Hall, 1991.

[ISII95] A. Inoue, M. Shigehara, M. Ito, M. Inai, Y. Hattori, and T. Mizunami, "Fabrication and application of fiber Bragg grating - A review," *Optoelectronics - Devices and Technologies*, vol. 10, pp. 119–130, Mar. 1995.

[ITU01a] ITU-T, Recommendation G.8080/Y.1304, "Architecture for the automatically switched optical network (ASON)," Nov. 2001.

[ITU01b] ITU-T, Recommendation G.7042/Y.1305, "Link capacity adjustment scheme (LCAS) for virtual concatenated signals," 2001.

[ITU01c] ITU-T, Recommendation G.704/Y.1303, "Generic Framing Procedure (GFP)," 2001.

[ITU02a] ITU-T, Recommendation G.707, "Network node interface for the synchronous digital hierarchy (SDH)," April 2002.

[ITU02b] ITU, Recommendation G.694.1, "Spectral grids for WDM applications," June 2002.

[Jake04] B. Jake, "Raman Amplification for Fiber Communications Systems," *IEEE/OSA Journal of Lightwave Technology*, vol. 22, no. 1, pp. 79–93, Jan. 2004.

[Jamo02] B. Jamoussi, et al., "Constrained-based *LSP* setup using *LDP*," *RFC 3212*, Jan. 2002.

[JaRS93] F. J. Janneillo, R. Ramaswami, and D. G. Steinberg, "A prototype circuit-switched multi-wavelength optical metropolitan-area network," *IEEE/OSA Journal of Lightwave Technology*, vol. 11, pp. 777–782, May/June 1993.

[JDHM01] X. Jia, D. Du, X. Hu, M. Lee, and J. Gu, "Optimization of Wavelength Assignment for QoS Multicast in WDM Networks," *IEEE Transactions on Communications*, vol. 49, no. 2, pp. 341–350, Feb. 2001.

[JeAy96] G. Jeong and E. Ayanoglu, "Comparison of wavelength-interchanging and wavelength-selective cross-connects in multiwavelength all-optical networks," *Proceedings, IEEE INFOCOM '96*, San Francisco, CA, pp. 156–163, March 1996.

[JHBC05] L. A. Johansson, Z. Hu, D. J. Blumenthal, L. A. Coldren, Y. A. Akulova, and G. A. Fish, "40-GHz Dual-Mode-Locked Widely Tunable Sampled-Grating DBR Laser," *IEEE Photonics Technology Letters*, vol. 17, no. 2, pp. 285–287, Feb. 2005.

[Jia93] F. Jia, *Architectures and Protocols for High-Speed Multichannel Networks*, Ph.D. Dissertation, University of California, Davis, Department of Computer Science, Sept. 1993.

[JiMI95] F. Jia, B. Mukherjee, and J. Iness, "Scheduling variable-length messages in a single-hop multichannel local lightwave network," *IEEE/ACM Transactions on Networking*, vol. 3, pp. 477–488, Aug. 1995.

[JiMu92a] F. Jia and B. Mukherjee, "Performance analysis of a generalized receiver collision avoidance (RCA) protocol for single-hop WDM lightwave networks," *Proceedings, SPIE '92*, Boston, MA, vol. 1784, pp. 229–240, Sept. 1992.

[JiMu92b] F. Jia and B. Mukherjee, "Bimodal throughput, nonmonotonic delay, optimal bandwidth dimensioning, and analysis of receiver collisions in a single-hop WDM local lightwave network," *Proceedings, IEEE Globecom '92*, Orlando, FL, pp. 1896–1900, Dec. 1992.

[JiMu93a] F. Jia and B. Mukherjee, "The receiver collision avoidance (RCA) protocol for a single-hop WDM lightwave network," *IEEE/OSA Journal of Lightwave Technology*, vol. 11, pp. 1053–1065, May/June 1993.

[JiMu93b] F. Jia and B. Mukherjee, "A high-capacity, packet switched, single-hop local lightwave network," *Proceedings, IEEE GLOBECOM '93*, Houston, TX, pp. 1110–1114, Dec. 1993.

[JLWF04] Q. Jiang, X. Liu, Q. Wang, and X. Feng, "Dynamically Gain Control in the Serial Structure C + L Wide-Band EDFA," *IEEE Photonics Technology Letters*, vol. 16, no. 1, pp. 87–89, Jan. 2004.

[JuBM96] J. P. Jue, M. S. Borella, and B. Mukherjee, "Performance analysis of the RAINBOW WDM optical network prototype,"

IEEE Journal on Selected Areas in Communications, vol. 14, pp. 945–951, June 1996.

[JuXi00] J. P. Jue and G. Xiao, "An adaptive routing algorithm with a distributed control scheme for wavelength-routed optical networks," *Proceedings, Ninth International Conference on Computer Communications and Networks*, Las Vegas, NV, pp. 192–197, Oct. 2000.

[KaAy98] E. Karasan and E. Ayanoglu, "Effects of wavelength routing and selection algorithms on wavelength conversion gain in WDM optical networks," *IEEE/ACM Transactions on Networking*, vol. 6, no. 2, pp. 186–196, April 1998.

[KaBW96] L. Kazovsky, S. Benedetto and A. Willner, *Optical Fiber Communication Systems*, Artech House, 1996.

[KaIw05] J. Kani and K. Iwatsuki, "A Wavelength-Tunable Optical Transmitter Using Semiconductor Optical Amplifiers and an Optical Tunable Filter for Metro/Access DWDM Applications," *IEEE/OSA Journal of Lightwave Technology*, vol. 23, no. 3, pp. 1164–1169, March 2005.

[Kami90] I. P. Kaminow, "FSK with direct detection in optical multiple-access FDM networks," *IEEE Journal on Selected Areas in Communications*, vol. 6, pp. 1005–1014, Aug. 1990.

[Kami05] N. Kamiyama, "Comparison of Single-Hop WDM Architectures for Local and Metropolitan Area Networks," *Proceedings, 9th Conference on Optical Network Design and Modeling*, Milan, Italy, pp. 260–271, Feb. 2005.

[Karp72] R. M. Karp, "Reducibility among combinatorial problems," *Complexity of Computer Computations*, New York: Plenum Press, pp. 85–104, 1972.

[KaSh91] M. J. Karol and S. Z. Shaikh, "A simple adaptive routing scheme for congestion control in ShuffleNet multihop lightwave networks," *IEEE Journal on Selected Areas in Communications*, vol. 9, pp. 1040–1051, Sept. 1991.

[KaTA03] J. Kani, M. Teshima, and K. Akimoto, "A WDM based optical access network for wide-area gigabit access services," *IEEE Optical Communications Magazine*, vol. 1, no. 1, pp. S43–S48, Feb. 2003.

[KBSM04] G. Kramer, A. Banerjee, N. Singhal, B. Mukherjee, S. Dixit, Y. Ye, "Fair Queuing with Service Envelopes (FQSE): A Cousin-Fair Hierarchical Scheduler for Subscriber Access Networks," *IEEE Journal on Selected Areas in Communications*, vol. 22, no. 8, pp. 1497–1513, Oct. 2004.

[Keis00] G. Keiser, *Optical Fiber Communications*, 3rd ed., McGraw-Hill, 2000.

[KKMS05] K. Kitayama, M. Koga, H. Morikawa, S. Hara, and M. Kawai, "Optical Burst Switching Network Testbed in Japan," *Proceedings, OFC '05*, Anaheim, CA, pp. OWC3, March 2005.

[Klei70] L. Kleinrock, "Analytic and simulation methods in computer network design," *AFIPS Conference Proceedings*, Montvale, NJ, vol. 42, pp. 569–579, 1970.

[KLKH05] Y. O. Kim, J. H. Lee, J. M. Kang, S. K. Han, "2R Limiter Circuit with Gain Clamped SOA for XGM Wavelength Converter," *IEEE Proceedings-Optoelectronics*, vol. 152, no. 1, pp. 11–15, Feb. 2005.

[KMDY02] G. Kramer, B. Mukherjee, S. Dixit, Y. Ye, and R. Hirth, "Supporting differentiated classes of service in Ethernet passive optical networks," *OSA Journal of Optical Networking*, vol. 1, no. 8/9, pp. 280–298, Aug. 2002.

[KoAc96] M. Kovačević and A. S. Acampora, "Benefits of wavelength translation in all-optical clear-channel networks," *IEEE Journal on Selected Areas in Communications*, vol. 14, pp. 868–880, June 1996.

[KoCh01] V. R. Konda and T. Y. Chow, "Algorithm for traffic grooming in optical networks to minimize the number of transceivers," *Proceedings, IEEE Workshop on High Performance Switching and Routing 2001*, Dallas, TX, pp. 218–221, May 2001.

[KoCh89] H. Kobrinski and K.-W. Cheung, "Wavelength-tunable optical filters: Applications and technologies," *IEEE Communications Magazine*, vol. 27, pp. 53–63, Oct. 1989.

[KoLa00] M. Kodialam and T. V. Lakshman, "Dynamic routing of bandwidth guaranteed tunnels with restoration," *Proceedings, IEEE INFOCOM '00*, Tel Aviv, Israel, pp. 902–911, March 2000.

[KoLa01] M. Kodialam and T. V. Lakshman, "Dynamic routing of locally restorable bandwidth guaranteed tunnels using aggregated link usage information," *Proceedings, IEEE INFOCOM '01*, Anchorage, AK, pp. 376–385, April 2001.

[Komp02] K. Kompella et al., "Link bundling in MPLS traffic engineering," *Internet Draft, Work in Progress, draft-ietf-mpls-bundle-03.txt*, May 2002.

[KoPP93] V. Kompella, J. Pasquale, and G. Polyzos, "Multicast Routing for Multimedia Communications," *IEEE/ACM Transactions on Networking*, vol. 1, no. 3, pp. 286–292, 1993.

[KOYM87] H. Kawaguchi et al., "Tunable optical wavelength conversion using a multielectrode distributed-feedback laser diode with a saturable absorber," *Electronic Letters*, vol. 23, no. 20, pp. 1088–1090, 1987.

[KrMP02] G. Kramer, B. Mukherjee, and G. Pesavento, "IPACT: A dynamic protocol for an Ethernet PON," *IEEE Communications Magazine*, vol. 40, no. 2, pp. 74–80, Feb. 2002.

[KRNM99] H. Kidorf, K. Rottwitt, M. Nissov, M. Ma, and E. Rabarijaona, "Pump interactions in a 100 nm bandwidth Raman amplifier," *IEEE Photonics Technology Letters*, vol. 11, no. 5, pp. 530–532, May 1999.

[KrSi01] R. M. Krishnaswamy and K. N. Sivarajan, "Design of logical topologies: A linear formulation for wavelength-routed optical networks with no wavelength changers," *IEEE/ACM Transactions on Networking*, vol. 9, no. 2, pp. 186–198, April 2001.

[LaAc90] J.-F. P. Labourdette and A. S. Acampora, "Partially reconfigurable multihop lightwave networks," *Proceedings, IEEE Globecom '90*, San Diego, CA, pp. 34–40, Dec. 1990.

[LaAc91] J.-F. P. Labourdette and A. S. Acampora, "Logically rearrangeable multihop lightwave networks," *IEEE Transactions on Communications*, vol. 39, pp. 1223–1230, Aug. 1991.

[Laar88] P. J. M. van Laarhoven, *Theoretical and Computational Aspects of Simulated Annealing*, Amsterdam, 1988.

[LaGP01] A. Lardies, R. Gupta, and R. A. Patterson, "Traffic grooming in a multi-layer network," *SPIE Optical Networks Magazine*, vol. 2, no. 3, pp. 91–99, May/June 2001.

[LaHA94] J.-F. P. Labourdette, G. W. Hart, and A. S. Acampora, "Branch-exchange sequences for reconfiguration of lightwave networks," *IEEE Transactions on Communications*, vol. 42, pp. 2822–2832, Oct. 1994.

[LaMa97] T. V. Lakshman and U. Madhow, "The performance of TCP/IP for networks with high bandwidth-dely products and random loss," *IEEE/ACM Transactions on Networking*, vol. 5, pp. 336–350, June 1997.

[LaPT96] J. P. R. Lacey, G. J. Pendock, and R. S. Tucker, "Gigabit-per-second all-optical 1300-nm to 1550-nm wavelength conversion using cross-phase modulation in a semiconductor optical amplifier.," *Proceedings, Optical Fiber Communication (OFC '96)*, San Jose, CA, vol. 2, pp. 125–126, 1996.

[LeLi93] K.-C. Lee and V. O. K. Li, "A wavelength-convertible optical network," *IEEE/OSA Journal of Lightwave Technology*, vol. 11, no. 5/6, pp. 962–970, May/June 1993.

[LeSi03] K. Lee and K.-Y. Siu, "On the reconfigurability of single-hub WDM ring networks," *IEEE/ACM Transactions on Networking*, vol. 11, no. 2, pp. 273–284, April 2003.

[LeSu94] C. Lee and T. Su, "2*2 single-mode zero-gap directional-coupler thermo-optic waveguide switch on glass," *Applied Optics*, vol. 33, pp. 7016–7022, Oct. 1994.

[LeZa89] T.-P. Lee and C.-E. Zah, "Wavelength-tunable and single-frequency lasers for photonic communications networks," *IEEE Communications Magazine*, vol. 27, pp. 42–52, Oct. 1989.

[LiGa92] B. Li and A. Ganz, "Virtual topologies for WDM star LANs: The regular structure approach," *Proceedings, IEEE INFOCOM '92*, Florence, Italy, pp. 2134–2143, May 1992.

[LiGo98] L. Y. Lin and E. L. Goldstein, "Free-space micromachined optical switches with submillisecond switching time for large-scale optical crossconnects," *IEEE Photonics Technology Letters*, vol. 10, pp. 525–527, April 1998.

[LiGo02] L. Y. Lin and E. L. Goldstein, "Opportunities and challenges for MEMS in lightwave communications," *IEEE Journal on Selected Topics in Quantum Electronics*, vol. 8, pp. 163–172, Jan./Feb. 2002.

[LiMe02] R. Libeskind-Hadas, and R. Melhem, "Multicast routing and wavelength assignment in multihop optical networks," *IEEE/ACM Transactions on Networking*, vol. 10, no. 5, pp. 621–629, Oct. 2002.

[LiSo99] L. Li and A. K. Somani, "Dynamic wavelength routing using congestion and neighborhood information," *IEEE/ACM Transactions on Networking*, vol. 7, no. 5, pp. 779–786, Oct. 1999.

[LiSp89] Y. K. Lin and D. R. Spears, "Passive optical subscriber loops with multi-access," *IEEE/OSA Journal of Lightwave Technology*, vol. 7, no. 11, pp. 1769–1777, Nov. 1989.

[LiTS01] Y. Liu, D. Tipper, and P. Siripongwutikorn, "Approximating optimal spare capacity allocation by successive survivable routing," *Proceedings, IEEE INFOCOM '01*, Anchorage, AK, pp. 699–708, April 2001.

[Liu02] K. H. Liu, *IP over WDM*, New York: John Wiley and Sons, pp. 147–151, 2002.

[LiVe02] R. Lingampalli and P. Vengalam, "Effect of wavelength and waveband grooming on all-optical networks with single layer photonic switching," *Proceedings, OFC '02*, Anaheim, CA, pp. 501–502, March 2002.

[LJXC03] K. Lu, J. P. Jue, G. Xiao, and I. Chlamtac, "Intermediate-Node Initiated Reservation (IIR): A New Signaling Scheme for Wavelength-Routed Networks," *IEEE Journal on Selected Areas in Communications*, vol. 21, no. 8, pp. 1285–1294, October 2003.

[LKSK05] V. V. Lysak, H. Kawaguchi, I. A. Sukhoivanov, T. Katayama, and A. V. Shulika, "Ultrafast Gain Dynamics in Asymmetrical Multiple Quantum-Well Semiconductor Optical Amplifiers," *IEEE Journal of Quantum Electronics*, vol. 41, no. 6, pp. 797–807, June 2005.

[LOLH05] H. H. Lee, J. M. Oh, D. Lee, J. Han, H. S. Chung, K. Kim, "A Variable-Gain Optical Amplifier for Metro WDM Networks With Mixed Span Losses: A Gain-Clamped Semiconductor Optical Amplifier Combined With a Raman Fiber Amplifier," *IEEE Photonics Technology Letters*, vol. 17, no. 6, pp. 1301–1303, June 2005.

[Lung99] B. Lung, "PON architecture futureproofs FTTH," *Lightwave, PennWell*, vol. 16, no. 10, pp. 104–107, Sept. 1999.

[LWKD02] G. Li, D. Wang, C. Kalmanek, and R. Doverspike, "Efficient distributed path selection for shared restoration connections," *Proceedings, IEEE INFOCOM '02*, New York, pp. 140–149, June 2002.

[LWPW05] T. Lin, K. A. Williams, R. V. Penty, and I. H. White, "Self-Configuring Intelligent Control for Short Reach 100Gb/s Optical Packet Routing," *Proceedings, OFC '05*, Anaheim, CA, pp. OWK5, March 2005.

[LYKK02] M. Lee, J. Yu, Y. Kim, C. Kang, and J. Park, "Design of hierarchical crossconnect WDM networks employing a two-stage multiplexing scheme of waveband and wavelength," *IEEE Journal on Selected Areas in Communications*, vol. 20, pp. 166–171, Jan. 2002.

[MaAT05] W. Mao, P. A. Andrekson, and J. Toulouse, "All-Optical Wavelength Conversion Based on Sinusoidal Cross-Phase Modulation in Optical Fibers," *IEEE Photonics Technology Letters*, vol. 17, no. 2, pp. 420–422, Feb. 2005.

[Maho93] M. J. O'Mahony, "Optical amplifiers," *Photonics in Switching* (J. E. Midwinter, ed.), San Diego, CA: Academic Press, vol. 1, pp. 147–167, 1993.

[Maho95] M. J. O'Mahony et al., "The design of a European optical network," *IEEE/OSA Journal of Lightwave Technology*, vol. 13, pp. 817–828, May 1995.

[MaKu03] X. Ma and G. Kuo, "Optical Switching Technology Comparison: Optical MEMS vs. Other Technologies," *IEEE Communications Magazine*, vol. 41, no. 11, pp. s16–s23, Nov. 2003.

[MaMI72] D. W. Matula, G. Marble, and J. D. Isaacson, "Graph coloring algorithms," *Graph Theory and Computing* (R. C. Read, ed.), New York and London: Academic Press, ch. 10, pp. 109–122, 1972.

[Mann02] E. Mannie et al., "Generalized multi-protocol label switching (GMPLS) architecture," *Internet Draft, Work in Progress, draft-ietf-ccamp-gmpls-architecture-02.txt*, March 2002.

[MaTT03] C. Mas, I. Tomkos, and O. K. Tonguz, "Optical Network Security: A Failure Management Framework," *Proceedings, SPIE International Symposium on Information Technologies and Communications (ITCOM '03)*, vol. 5047, pp. 230–241, Sept. 2003.

[Matu72] D. W. Matula, "k-components, clusters and slicings in graphs," *SIAM Journal of Applied Mathematics*, vol. 22, no. 3, pp. 459–480, 1972.

[Maxe85] N. F. Maxemchuk, "Regular mesh topologies in local and metropolitan area networks," *AT&T Technical Journal*, vol. 64, pp. 1659–1686, Sept. 1985.

[Maxe87] N. F. Maxemchuk, "Routing in the Manhattan street network," *IEEE Transactions on Communications*, vol. 35, pp. 503–512, May 1987.

[MaZC03] M. Ma, Y. Zhu, and T.-H. Cheng, "A Bandwidth Guaranteed Polling MAC Protocol for Ethernet Passive Optical Networks," *Proceedings, IEEE INFOCOM '03*, San Francisco, CA, pp. 22–31, April 2003.

[MaZQ98] R. Malli, X. Zhang, and C. Qiao, "Benefit of Multicasting in All-Optical Networks," *Proceedings, SPIE Conference on All-optical Networking*, vol. 2531, pp. 209–220, Nov. 1998.

[MBES90] D. L. Mills, C. G. Boncelet, J. G. Elias, P. A. Schragger, and A. W. Jackson, "Highball: A high speed, reserved access, wide-area Network," *Technical Report No. 90-9-3*, Electrical Engineering Department, University of Delaware, 1990.

[MBLM96] M. A. Marsan, A. Bianco, E. Leonardi, M. Meo, and F. Neri, "MAC protocols and fairness control in WDM multirings with tunable transmitters and fixed receivers," *IEEE/OSA Journal of Lightwave Technology*, vol. 14, no. 6, pp. 1230–1244, June 1996.

[MBRM96] B. Mukherjee, D. Banerjee, S. Ramamurthy, and A. Mukherjee, "Some principles for designing a wide-area optical network," *IEEE/ACM Transactions on Networking*, vol. 4, no. 5, pp. 684–696, Oct. 1996.

[McMR04] M. McGarry, M. Meier, and M. Reisslein, "Ethernet PONs: A Survey of Dynamic Bandwidth Allocation (DBA) algorithms," *IEEE Optical Communications*, vol. 42, no. 8, pp. S8 - S15, Aug. 2004.

[MDJD96] B. Mikkelsen et al., "Wavelength conversion devices," *Proceedings, Optical Fiber Communication (OFC '96)*, San Jose, CA, vol. 2, pp. 121–122, 1996.

[MDJP94] B. Mikkelsen et al., "Polarization insensitive wavelength conversion of 10 Gbit/s signals with SOAs in a Michelson interferometer," *Electronic Letters*, vol. 30, pp. 260–261, Feb 1994.

[MeBo96] P. M. Melliar-Smith and J. E. Bowers, "Thunder and lightning," *ARPA Networking PI Meeting*, Charleston, SC, Feb. 1996.

[Mehr90] N. Mehravari, "Performance and protocol improvements for very high speed optical fiber local area networks using a passive star topology," *IEEE/OSA Journal of Lightwave Technology*, vol. 8, pp. 520–530, April 1990.

[MePD95] S. Melle, C. P. Pfistner, and F. Diner, "Amplifier and multiplexing technologies expand network capacity," *Lightwave Magazine*, pp. 42–46, Dec. 1995.

[MEPS02] F. Matera, V. Eramo, A. Pizzinat, A. Schiffini, M. Guglielmcci, and M. Settembre, "Numerical investigation on wide geographical networks based on $n \times 40$Gb/s transmission," *Proceedings, OFC '02*, Anaheim, CA, pp. 162–163, March 2002.

[Mest95] D. J. G. Mestdagh, *Fundamentals of Multiaccess Optical Fiber Networks*, The Artech House Optoelectronics Library, Artech House, 1995.

[Mill99] C. K. Miller, *Multicast Networking and Applications*, Reading, MA: Addison-Wesley, 1999.

[MMBF97] M. Medard, D. Marquis, R. A. Barry, S. G. Finn, et al., "Security Issues in All-Optical Networks," *IEEE Network*, vol. 11, no. 3, pp. 42–48, May/June 1997.

[MMPE00] G. Mayer, M. Martinelli, A. Pattavina, and E. Salvadori, "Design and cost performance of the multistage WDM-PON access networks," *IEEE/OSA Journal of Lightwave Technology*, vol. 18, no. 2, pp. 121–142, Feb. 2000.

[MoAz95a] A. Mokhtar and M. Azizoglu, "Hybrid multiacess for all-optical LANs with nonzero tuning delays," *Proceedings, IEEE International Conference on Communications (ICC '95)*, Seattle, WA, pp. 1272–1276, June 1995.

[MoAz95b] A. Mokhtar and M. Azizoglu, "Multiacess in all-optical networks with wavelength and code concurrency," *Journal of Fiber and Integrated Optics*, vol. 14, no. 1, pp. 37–51, 1995.

[MoAz96] A. Mokhtar and M. Azizoglu, "Adaptive routing algorithms for wavelength-routed all-optical networks," *Proceedings, 1996 IEEE Midwest Symposium on Circuits and Systems*, Ames, IA, Aug. 1996.

[MoAz98] A. Mokhtar and M. Azizoglu, "Adaptive wavelength routing in all-optical networks," *IEEE/ACM Transactions on Networking*, vol. 6, no. 2, pp. 197–206, April 1998.

[Moha02] N. I. Mohammed, "Raman Amplifiers for Telecommunications," *IEEE Journal of Selected Topics in Quantum Electronics*, vol. 8, no. 3, pp. 548–559, May/June 2002.

[MoLi01] E. Modiano and P. Lin, "Traffic grooming in WDM networks," *IEEE Communications Magazine*, vol. 39, no. 7, pp. 124–129, July 2001.

[MoMS01] G. Mohan, C. S. R. Murthy, and A. K. Somani, "Efficient algorithms for routing dependable connections in WDM optical networks," *IEEE/ACM Transactions on Networking*, vol. 9, no. 5, pp. 553–566, Oct. 2001.

[MoTo93] J. B. Moore and D. E. Todd, "Recent developments in distributed feedback and distributed Bragg reflector lasers for wide-band long-haul fiber optic communication systems," *Proceedings, IEEE Southeastcon '93*, Charlotte, NC, pp. 9p, April 1993.

[Mouf02] H. Mouftah et al., "A framework for service-guaranteed shared protection in WDM mesh networks," *IEEE Communications Magazine*, vol. 40, pp. 97–103, Feb. 2002.

[MPBC04] G. Maier, A. Pattavina, L. Barbato, F. Cecini, and M. Martinelli, "Routing Algorithms in WDM Networks under Mixed Static and Dynamic Lambda-Traffic," *Photonic Network Communications*, vol. 8, no. 1, pp. 69–87, June 2004.

[MSOH99] S. Makam, V. Sharma, K. Owens, and C. Huang, "Protection/restoration of MPLS networks," *IETF draft*, Oct. 1999.

[MTTK05] K. Morito, S. Tanaka, S. Tomabechi, and A. Kuramata, "A Broad-Band MQW Semiconductor Optical Amplifier With High Saturation Output Power and Low Noise Figure," *IEEE Photonics Technology Letters*, vol. 17, no. 5, pp. 974–976, May 2005.

[Mukh92a] B. Mukherjee, "WDM-based local lightwave networks – Part I: Single-hop systems," *IEEE Network Magazine*, vol. 6, no. 3, pp. 12–27, May 1992.

[Mukh92b] B. Mukherjee, "WDM-based local lightwave networks – Part II: Multihop systems," *IEEE Network Magazine*, vol. 6, no. 4, pp. 20–32, July 1992.

[Mukh97] B. Mukherjee, *Optical Communication Networks*, New York: McGraw-Hill, 1997.

[Mukh00] B. Mukherjee, "WDM Optical Communication Networks: Progress and Challenges" (Invited Paper), *IEEE Journal on Selected Areas in Communications*, vol. 18, no. 10, pp. 1810–1824, Oct. 2000.

[MWWP93] M. W. Maeda, A. E. Willner, J. R. Wullert II, J. Patel, and M. Allersma, "Wavelength-division multiple-access network based on centralized common-wavelength control," *IEEE Photonics Technology Letters*, vol. 5, no. 1, pp. 83–85, Jan. 1993.

[NaMo00] A. Narula-Tam and E. Modiano, "Dynamic load balancing for WDM based packet networks," *Proceedings, IEEE INFOCOM '00*, Tel Aviv, Israel, pp. 1010–1019, March 2000.

[NaSi03] T. K. Nayak and K. N. Sivarajan, "Dimensioning optical networks under traffic growth models," *IEEE/ACM Transactions on Networking*, vol. 11, no. 6, pp. 935–947, Dec. 2003.

[NeRa01] A. Neukermans and R. Ramaswami, "MEMS technology for optical networking applications," *IEEE Communications Magazine*, vol. 39, pp. 62–69, Jan. 2001.

[NewF94] Personal communication with representatives from New Focus, Inc., April 1994.

[NKSM05] R. Nejabati, D. Klonidis, D. Simeonidou, and M. J. O'Mahony, "Demonstration of user-controlled network interface for subwavelength bandwidth-on-demand services," *Proceedings, OFC '05*, Anaheim, CA, pp. OWK2, March 2005.

[OHLJ85] N. A. Olsson, J. Hegarty, R. A. Logan, L. F. Johnson, K. L. Walker, L. G. Cohen, B. L. Kasper, and J. C. Campbell, "68.3 km transmission with 1.37 Tbit km/s capacity using wavelength division multiplexing of ten single-frequency lasers at 1.5 μm," *Electronic Letters*, vol. 21, pp. 105–106, 1985.

[OHTP02] M. Oksanen, O.-P. Hiironen, A. Tervonen, A. Pietilainen, et al., "Spectral slicing passive optical access network trial," *Proceedings, OFC '02*, Anaheim, CA, pp. 439–440, March 2002.

[OIF05] Optical Internetworking Forum Implementation Agreements, *Very Short Reach Interface*, http://www.oiforum.com/public/documents/VSRWhitePaper-.pdf.

[OkMS95] K. Okamoto, K. Moriwaki, and S. Suzuki, "Fabrication of 64×64 arrayed-waveguide grating multiplexer on silicon," *Electronic Letters*, vol. 31, pp. 184–186, Feb. 1995.

[OkSu96] K. Okamoto and A. Sugita, "Flat spectral response arrayed-waveguide grating multiplexer with parabolic waveguide horns," *Electronic Letters*, vol. 32, pp. 1661–1662, Aug. 1996.

[OkTa91] K. Okamoto, H. Takahashi, et al., "Design and fabrication of integrated-optic 8×8 star coupler," *Electronic Letters*, vol. 27, pp. 774–775, April 1991.

[Opti95] Optivision, Inc. home page, 1995, http://www.optivision.com/.

[OuMu05] C. Ou and B. Mukherjee, *Survivable Optical WDM Networks*, Springer, 2005.

[OuZM04] C. Ou, K. Zhu, B. Mukherjee, et al., " Traffic Grooming for Survivable WDM Networks - Dedicated Protection," *OSA Journal of Optical Networking*, vol. 3, pp. 50–74, Jan. 2004.

[OYTB05] W. Y. Oh, S. H. Yun, G. J. Tearney, B. E. Bouma, "Wide tuning range wavelength-swept laser with two semiconductor optical amplifiers," *IEEE Photonics Technology Letters*, vol. 17, no. 3, pp. 678–680, March 2005.

[OZSM04] C. Ou, H. Zang, N. Singhal, B. Mukherjee, et al., "Sub-Path Protection for Scalability and Fast Recovery in Optical WDM Mesh Networks," *IEEE Journal on Selected Areas in Communications*, vol. 22, no. 11, pp. 1859–1875, Nov. 2004

[OZZM03] C. S. Ou, K. Zhu, H. Zang, L. H. Sahasrabuddhe, and B. Mukherjee, "Traffic grooming for survivable WDM networks

– shared protection," *IEEE Journal on Selected Areas in Communications*, vol. 21, no. 11, pp. 1367–1383, Nov. 2003.

[OZZM04] C. Ou, J. Zhang, H. Zang, B. Mukherjee, et al., "New and improved approaches for shared-path protection in WDM mesh networks," *IEEE/OSA Journal of Lightwave Technology*, vol. 22, no. 5, pp. 1223–1232, May 2004.

[Pala04] J. C. Palais, *Fiber Optic Communications*, 5th ed., Prentice-Hall, Sept. 2004.

[Pank92] R. K. Pankaj, *Architectures for Linear Lightwave Networks*, Ph.D. Dissertation, Department of Electrical Engineering and Computer Science, MIT, Cambridge, MA, Sept. 1992.

[Pank99] R. K. Pankaj, "Wavelength requirements for multicasting in all-optical networks," *IEEE/ACM Transactions on Networking*, vol. 7, no. 3, pp. 414–424, June 1999.

[PaSe95] G. Panchapakesan and A. Sengupta, "On multihop optical network topology using Kautz digraph," *Proceedings, IEEE INFOCOM '95*, Boston, MA, pp. 675–682, April 1995.

[PaSt86] D. B. Payne and J. R. Stern, "Transparent single mode fiber optical networks," *IEEE/OSA Journal of Lightwave Technology*, vol. LT-4, pp. 864–869, 1986.

[Paul98] S. Paul, *Multicasting on the Internet and its Applications*, Boston, MA: Kluwer Academic Publishers, 1998.

[PaWi00] K. Park and W. Willinger, *Self-similar Network Traffic and Performance Evaluation*, New York: John Wiley and Sons Inc., 2000.

[PeKe99] G. Pesavento and M. Kelsey, "PONs for the broadband local loop," *Lightwave, PennWell*, vol. 16, no. 10, pp. 68–74, Sept. 1999.

[PeWi02] V. E. Perlin and H. G. Winful, "On distributed Raman amplification for ultrabroad-band long-haul WDM systems," *IEEE/OSA Journal of Lightwave Technology*, vol. 20, no. 3, pp. 409–416, March 2002.

[PiSa94] G. R. Pieris and G. H. Sasaki, "Scheduling transmissions in WDM broadcast-and-select networks," *IEEE/ACM Transactions on Networking*, vol. 2, pp. 105–110, April 1994.

[Powe93] J. P. Powers, *An Introduction to Fiber Optic Systems*, Homewood, IL: Irwin, 1993.

[Prat97] G. Prati, Ed., *Photonic Networks*, Springer Verlag, London, UK, 1997.

[Pruc93] P. R. Prucnal, "Optically processed self-routing synchronization, and contention resolution for 1D and 2D photonic switching architectures," *IEEE Journal of Quantum Electronics*, vol. 29, no. 2, pp. 600–612, Feb. 1993.

[QiCh03] C. Qiao and Y. Chen, "The potentials of optical burst switching (OBS)," *Proceedings, OFC '03*, Atlanta, GA, pp. TuJ5, March 2003.

[QiXu02] C. Qiao and D. Xu, "Distributed partial information management (DPIM) schemes for survivable networks - Part I," *Proceedings, IEEE INFOCOM '02*, New York, pp. 302–311, June 2002.

[QiXX02] C. Qiao, Y. Xiong, and D. Xu, "Novel models for efficient shared-path protection," *Proceedings, OFC '02*, Anaheim, CA, pp. 546–547, March 2002.

[QiYo99] C. Qiao and M. Yoo, "Optical Burst Switching (OBS)—A new paradigm for an optical Internet," *Journal of High Speed Networks*, vol. 8, no. 1, pp. 69–84, Jan. 1999.

[RAAE04] M. Raisi, S. Ahderom, K. Alameh and K. Eshraghian, "Dynamic Micro-photonic WDM equalizer," *Proceedings, 2nd IEEE International Workshop on Electronic Design, Test and Applications*, Perth, Australia, pp. 59–62, Jan. 2004.

[RaBS01] R. Ramamurthy, Z. Bogdanowicz, S. Samieian, et al., "Capacity performance of dynamic provisioning in optical networks," *IEEE/OSA Journal of Lightwave Technology*, vol. 19, pp. 40–48, Jan. 2001.

[RaDM05] S. Rai, O. Deshpande, and B. Mukherjee, "IP Resilience within an Autonomous System: Current Approach, Challenges, and Future Directions," *IEEE Communications Magazine*, Internet Technology Series, to appear, 2005.

[Rama96] S. Ramanathan, "Multicast Tree Generation in Networks with Asymmetric Links," *IEEE/ACM Transactions on Networking*, vol. 4, no. 4, pp. 558–568, Aug. 1996.

[Rama98] S. Ramamurthy, *Optical Design of WDM Network Architectures*, Ph.D. Dissertation, University of California, Davis, 1998.

[RaMu98] B. Ramamurthy and B. Mukherjee, "Wavelength Conversion in Optical Networks: Progress and Challenges," *IEEE Journal on Selected Areas in Communications*, vol. 16, no. 7, pp. 1040–1050, Sept. 1998.

[RaMu99a] S. Ramamurthy and B. Mukherjee, "Survivable WDM mesh networks, part I - protection," *Proceedings, IEEE INFOCOM '99*, New York, pp. 744–751, March 1999.

[RaMu99b] S. Ramamurthy and B. Mukherjee, "Survivable WDM mesh networks, part II - restoration," *Proceedings, IEEE International Conference on Communications (ICC '99)*, Vancouver, Canada, pp. 2023–2030, June 1999.

[RaMu02] R. Ramamurthy and B. Mukherjee, "Fixed-alternate routing and wavelength conversion in wavelength-routed optical networks," *IEEE/ACM Transactions on Networking*, vol. 10, no. 3, pp. 351–367, June 2002.

[RaRa00] B. Ramamurthy and A. Ramakrishnan, "Virtual topology reconfiguration of wavelength-routed optical WDM networks," *Proceedings, IEEE Globecom '00*, San Francisco, CA, pp. 1269–1275, Nov./Dec. 2000.

[RaRP02] S. Ramesh, G. N. Rouskas, and H. G. Perros, "Computing Blocking Probabilities in Multiclass Wavelength-Routing Networks With Multicast Calls," *IEEE Journal on Selected Areas in Communications*, vol. 20, no. 1, pp. 89–96, Jan. 2002.

[RaSa97] R. Ramaswami and G. H. Sasaki, "Multiwavelength optical networks with limited wavelength conversion," *Proceedings, IEEE INFOCOM '97*, Kobe, Japan, April 1997.

[RaSe97] R. Ramaswami and A. Segall, "Distributed network control for optical networks," *IEEE/ACM Transactions on Networking*, vol. 5, no. 6, pp. 936–943, Dec. 1997.

[RaSi95a] R. Ramaswami and K. Sivarajan, "Optimal routing and wavelength assignment in all-optical networks," *IEEE/ACM Transactions on Networking*, vol. 3, pp. 489–500, Oct. 1995.

[RaSi95b] R. Ramaswami and K. Sivarajan, "Design of logical topologies for wavelength-routed all-optical networks," *Proceedings, IEEE INFOCOM '95*, Boston, MA, pp. 1316–1325, April 1995.

[RaSi96] R. Ramaswami and K. N. Sivarajan, "Design of logical topologies for wavelength-routed optical networks," *IEEE Journal on Selected Areas in Communications*, vol. 14, no. 5, pp. 840–851, June 1996.

[RaSi01] R. Ramaswami, and K. Sivarajan, *Optical Networks: A Practical Perspective*, 2nd ed., Morgan Kaufmann Publishers, 2001.

[RaSM03] S. Ramamurthy, L. Sahasrabuddhe, and B. Mukherjee, "Survivable WDM mesh networks," *IEEE/OSA Journal of Lightwave Technology*, vol. 21, pp. 870–883, Apr. 2003.

[RaTh87] P. Raghavan and C. D. Thompson, "Randomized rounding: A technique for provably good algorithms and algorithmic proofs," *Combinatorica*, vol. 7, no. 4, pp. 365–374, 1987.

[RDFH99] B. Ramamurthy, D. Datta, H. Feng, J. P. Heritage, and B. Mukherjee, "Impact of transmission impairments on the teletraffic performance of wavelength-routed optical networks," *IEEE/OSA Journal of Lightwave Technology*, vol. 17, no. 10, pp. 1713–1723, 1999.

[RFDM99] B. Ramamurthy, H. Feng, D. Data, J. P. Heritage, and B. Mukherjee, "Transparent vs. opaque vs. translucent wavelength-routed optical networks," *Proceedings, OFC '99*, San Diego, CA, pp. 59–61, March 1999.

[RHK05] Ovum-RHK, "*Optical Networks: Global, and North America*," http://www.rhk.com, 2005.

[RoAm95] G. N. Rouskas and M. H. Ammar, "Dynamic reconfiguration in multihop WDM networks," *Journal of High Speed Networks*, vol. 4, no. 3, pp. 221–238, 1995.

[RoKi03] S.-S. Roh and S.-H. Kim, "Security model and authentication protocol in EPON-based optical access network," *Proceedings, 5th International Conference on Transparent Optical Networks '03*, vol. 1, pp. 99–102, June 2003.

[SAHA02] A. Shami, C. Assi, I. Habib, and M. A. Ali, "Performance Evaluation of Two GMPLS-Based Distributed Control and Management Protocols for Dynamic Lightpath Provisioning in Future IP Networks," *Proceedings, IEEE ICC '02*, New York, pp. 2289–2293, May 2002.

[SaIa96] R. Sabella and E. Iannone, "Wavelength conversion in optical transport networks," *Journal of Fiber and Integrated Optics*, vol. 15, no. 3, pp. 167–191, 1996.

[SaLu99] R. Sabella and P. Lugli, *High Speed Optical Communications*, Kluwer Academic Publishers, 1999.

[Sams97] "Single forward pumping EDFA," EDFA Datasheet, Samsung Electronics Co. Ltd., 1997.

[SaMu99] L. H. Sahasrabuddhe and B. Mukherjee, "Light-trees: optical multicasting for improved performance in wavelength-routed networks," *IEEE Communications Magazine*, vol. 37, no. 2, pp. 67–73, Feb. 1999.

[SaRM02] L. Sahasrabuddhe, S. Ramamurthy, and B. Mukherjee, "Fault management in IP-over-WDM networks: WDM protection versus IP restoration," *IEEE Journal on Selected Areas in Communications*, vol. 20, pp. 21–33, 2002.

[SaSu03] S. Sankaranarayanan and S. Subramaniam, "Comprehensive Performance Modeling and Analysis of Multicasting in Optical Networks," *IEEE Journal of Selected Areas in Communications*, vol. 21, no. 9, pp. 1399–1413, Nov. 2003.

[ScAl90] R. V. Schmidt and R. C. Alferness, "Directional coupler switches, modulators, and filters using alternating $\delta\beta$ techniques," *Photonic Switching* (H. S. Hinton and J. E. Midwinter, eds.), pp. 71–80, New York: IEEE Press, 1990.

[ShCS93] J. Sharony, K. Cheung, and T. E. Stern, "The wavelength dilation concept in lightwave networks: Implementation and system considerations," *IEEE/OSA Journal of Lightwave Technology*, vol. 11, pp. 900–907, May/June 1993.

[ShVe04] F. B. Shepherd and A. Vetta, "Lighting Fibers in a Dark Network," *IEEE Journal on Selected Areas in Communications*, vol. 22, no. 9, pp. 1583–1588, Nov. 2004.

[SiGS98] J. Simmons, E. Goldstein, and A. Saleh, "Quantifying the benefit of wavelength add-drop in WDM rings with independent and dependent traffic," *Proceedings, OFC '98*, San Jose, CA, pp. 361–362, Feb. 1998.

[SiGS99] J. Simmons, E. Goldstein, and A. Saleh, "Quantifying the benefit of wavelength add-drop in WDM rings with independent and dependent traffic," *IEEE/OSA Journal of Lightwave Technology*, vol. 17, no. 1, pp. 48–57, Jan. 1999.

[SILB98] R. Sabella, E. Iannone, M. Listanti, M. Berdusco, and S. Binetti, "Impact of transmission performance on path routing in all-optical transport networks," *IEEE/OSA Journal on Lightwave Technology*, vol. 16, no. 11, pp. 1965–1972, Nov. 1998.

[Sime03] D. Simeonidou (ed.), "*Optical network infrastructure for GRID*," Grid Forum Draft, http://www.projects.ghpn-rg.document.draft-ggf-ghpn-opticalnets-1.en.1, Sept. 2003.

[SiMu01] N. Singhal and B. Mukherjee, "Architectures and algorithms for multicasting in WDM optical mesh networks using opaque and transparent optical crossconnects," *Proceedings, OFC '01*, vol. 2, pp. TuG8-1 – TuG8-3, March 2001. (Expanded version in *IEEE/ACM Transactions on Networking*, to appear, 2006.)

[SiMu03] N. K. Singhal and B. Mukherjee, "Protecting Multicast Sessions in WDM Optical Mesh Networks," *IEEE/OSA Journal of Lightwave Technology*, vol. 21, no. 4, pp. 884–892, April 2003.

[SiOM05] N. K. Singhal, C. Ou, and B. Mukherjee, " Cross-Sharing vs. Self-Sharing Trees for Protecting Multicast Sessions in Mesh Networks," *Computer Networks Journal*, to appear, 2005.

[SiRa91] K. Sivarajan and R. Ramaswami, "Multihop networks based on de bruijn graphs," *Proceedings, IEEE INFOCOM '91*, Bal Harbour, FL, pp. 1001–1011, April 1991.

[SiRa94] K. Sivarajan and R. Ramaswami, "Lightwave networks based on de bruijn graphs," *IEEE/ACM Transactions on Networking*, vol. 2, no. 1, pp. 70–79, 1994.

[SiSM03] N. K. Singhal, L. H. Sahasrabuddhe and B. Mukherjee, "Provisioning of Survivable Multicast Sessions Against Single Link Failures in Optical WDM Mesh Networks," *IEEE/OSA Journal of Lightwave Technology*, vol. 21, no. 11, pp. 2587–2594, Nov. 2003.

[SlDL03] R. Slavik, S. Doucet, and S. LaRochelle, "High-Performance All-Fiber Fabry-Perot Filters With Superimposed Chirped Bragg Gratings," *IEEE/OSA Journal of Lightwave Technology*, vol. 21, no. 4, pp. 1059–1065, April 2003.

[SLSM05] B. A. Small, O. Liboiron-Ladouceur, A. Shacham, J. P. Mack, and K. Bergman, "Demonstration of a Complete 12-Port Terabit Capacity Optical Packet Switching Fabric," *Proceedings, OFC '05*, Anaheim, CA, pp. OWK1, March 2005.

[SMGM01] N. Sreenath, C. S. R. Murthy, B. H. Gurucharan, and G. Mohan, "A two-stage approach for virtual topology reconfiguration ofWDMoptical networks," *SPIE Optical Networks Magazine*, vol. 2, no. 5, pp. 58–71, May 2001.

[SMOI05] T. Segawa, S. Matsuo, Y. Ohiso, T. Ishii, Y. Shibata, and H. Suzuki, "Fast Tunable Optical Filter Using Cascaded Mach-Zehnder Interferometers with Apodized Sampled Gratings," *IEEE Photonics Technology Letters*, vol. 17, no. 1, pp. 139–141, Jan. 2005.

[SoAz97] A. Somani and M. Azizoglu, "All-optical LAN interconnection with a wavelength selective router," *Proceedings, IEEE INFOCOM '97*, Kobe, Japan, vol. 3, pp. 1278–1285, April 1997.

[SoLS04] H. Song, J. Lee, and J. Song, "Signal Up-Conversion by Using a Cross-Phase-Modulation in All-Optical SOA-MZI Wavelength Converter," *IEEE Photonics Technology Letters*, vol. 16, no. 2, pp. 593–595, Feb. 2004.

[SoZM05] L. Song, J. Zhang, and B. Mukherjee, "Dynamic Provisioning with Reliability Guarantee and Resource Optimization for Differentiated Services in WDM Mesh Networks," *Proceedings, OFC '05*, Anaheim, CA, pp. OWG5, March 2005.

[SrSo02] R. Srinivasan and A. K. Somani, "A generalized framework for analyzing time-space switched optical networks," *IEEE Journal on Selected Areas in Communications*, vol. 20, no. 1, pp. 202–215, Jan. 2002.

[SSBN02] M. Sharma, M. Soulliere, A. Boskovic, and L. Nederlof, "Value of agile transparent optical networks," *Proceedings, OFC '02*, Anaheim, CA, pp. 293–294, March 2002.

[Stan02] S. Stanley, "*Next-gen SONET silicon*," Lightreading report, June 2002.

[Stan03] S. Stanley, "*Making SONET Ethernet-friendly*," Lightreading report, March 2003.

[StCT01] J. Strand, A. L. Chiu, and R. Tkach, "Issues for routing in the optical layer," *IEEE Communications Magazine*, vol. 39, no. 2, pp. 81–87, Feb. 2001.

[Stev94] W. R. Stevens, *TCP/IP Illustrated*, Reading, MA: Addison-Wesley Publishing Co., 1994.

[StGr00] D. Stamatelakis and W. D. Grover, "IP layer restoration and network planning based on virtual protection cycles," *IEEE Journal on Selected Areas in Communications*, vol. 18, pp. 1938–1949, Oct. 2000.

[SuAS96] S. Subramaniam, M. Azizoglu, and A. K. Somani, "All-optical networks with sparse wavelength conversion," *IEEE/ACM Transactions on Networking*, vol. 4, pp. 544–557, Aug. 1996.

[SuBa97] S. Subramaniam and R. A. Barry, "Wavelength assignment in fixed routing WDM networks," *Proceedings, IEEE ICC '97*, Montreal, Canada, pp. 406–410, June 1997.

[SuGT01] Y. Sun, J. Gu, and D. H. K. Tsang, "Multicast routing in all-optical wavelength routed networks," *Optical Networks Magazine*, vol. 2, pp. 101–109, July/Aug. 2001.

[SuMo89] T. Suda and S. Morris, "Tree LANs with collision avoidance: Station and switch protocols," *Computer Networks and ISDN Systems*, vol. 17, pp. 101–110, 1989.

[SuNi02] Y. Suemura, I. Nishioka, Y. Maeno, S. Araki, R. Izmailov, and S. Ganguly, "Hierarchical routing in layered ring and mesh optical networks," *Proceedings, IEEE International Conference on Communications (ICC '02)*, New York, pp. 2727–2733, April 2002.

[SuSu01a] C. Su and X. Su, "Protection path routing on WDM networks," *Proceedings, OFC '01*, Anaheim, CA, pp. TuO2–T1–T3, March 2001.

[SuSu01b] X. Su and C. Su, "An online distributed protection algorithm in WDM networks," *Proceedings, IEEE ICC '01*, Helsinki, Finland, pp. 1571–1575, June 2001.

[SuTa84] J. W. Suurballe and R. E. Tarjan, "A quick method for finding shortest pairs of disjoint paths," *Networks*, vol. 14, pp. 325–336, 1984.

[SWWG00] K. V. Shrikhande, I. M. White, D. Wonglumsom, S. M. Gemelos, et al., "HORNET: A Packet-Over-WDM Multiple Access Metropolitan Area Ring Network," *IEEE Journal on Selected Areas in Communications*, vol. 18, no. 10, pp. 2004–2016, Oct. 2000.

[TaGC01] L. Tancevski, A. Ge, and G. Castanon, "Optical packet switch with partially shared buffers: design principles," *Proceedings, OFC '01*, Anaheim, CA, pp. TuK3-1–TuK3-3, March 2001.

[Taji99] A. Tajima, "A 10-Gbit/s optical asynchronous cell/packet receiver with a fast bit synchronization circuit," *Proceedings, OFC '99*, San Diego, CA, pp. 111–113, March 1999.

[Taka86] H. Takagi, *Analysis of Polling Systems*, Cambridge, MA: MIT Press, 1986.

[TaLa91] M. Tachibana, R. I. Laming, et al., "Erbium-doped fiber amplifier with flattened gain spectrum," *IEEE Photonics Technology Letters*, vol. 3, pp. 118–120, Feb. 1991.

[TaMa80] H. Takahashi and A. Matsuyama, "An Approximate Solution for the Steiner Problem in Graphs," *Math. Japonica*, vol. 24, pp. 573–577, 1980.

[Tane96] A. S. Tanenbaum, *Computer Networks*, Prentice Hall, 1996.

[Tang94] K. W. Tang, "CayletNet: A multihop WDM-based lightwave network," *Proceedings, IEEE INFOCOM '94*, Toronto, Canada, vol. 3, pp. 1260–1267, June 1994.

[TCFG95] R. W. Tkach et al., "Four-photon mixing and high-speed WDM systems," *IEEE/OSA Journal of Lightwave Technology*, vol. 13, pp. 841–849, May 1995.

[ThSo01a] S. Thiagarajan and A. K. Somani, "Capacity fairness of WDM networks with grooming capabilities," *SPIE Optical Networks Magazine*, vol. 2, no. 3, pp. 24–31, May/June 2001.

[ThSo01b] S. Thiagarajan and A. K. Somani, "Traffic grooming for survivable WDM mesh networks," *Proceedings, OptiComm '01*, Denver, CO, pp. 54–65, Aug. 2001.

[Tomk02] I. Tomkos, "Transport performance of WDM metropolitan area transparent optical networks," *Proceedings, OFC '02*, Anaheim, CA, pp. 350-352, March 2002.

[Toml03] W. J. Tomlinson, "Technologies for dynamic gain and channel power equalization," *Proceedings, OFC '03*, Atlanta, GA, pp. 244, March 2003.

[ToNe94] M. To and P. Neusy, "Unavailability analysis of long-haul networks," *IEEE Journal on Selected Areas in Communications*, vol. 12, pp. 100–109, Jan. 1994.

[TONT90] H. Toba, K. Oda, K. Nosu, and N. Takato, "Factors affecting the design of optical FDM information distribution systems," *IEEE Journal on Selected Areas in Communications*, vol. 6, pp. 965–972, Aug. 1990.

[ToVC02] P. Torres, L. C. G. Valente, M. C. R. Carvalho, "Security system for optical communication signals with fiber Bragg gratings," *IEEE Transactions on Microwave Theory and Techniques*, vol. 50, no. 1, pp. 13–16, Jan. 2002.

[TrBe97] D. Trouchet, A. Beguin, et al., "Passband flattening of PHASAR WDM using input and output star couplers designed with two focal points," *Proceedings, OFC '97*, Dallas, TX, pp. 302-303, Feb. 1997.

[TrBM96] S. B. Tridandapani, M. S. Borella, and B. Mukherjee, "A lower bound on the expected cost of minimum-delay multicast traffic," *Proceedings, International Conference on Computer Communications and Networks (IC^3N '96)*, Washington, DC, pp. 287-292, Oct. 1996.

[Triv82] K. S. Trivedi, *Probability and Statistics with Reliability, Queuing, and Computer Science Applications*, Englewood Cliffs, NJ: Prentice-Hall, 1982.

[TrMe95] S. B. Tridandapani and J. S. Meditch, "Supporting multipoint connections in multihop WDM optical networks," *Journal of High-Speed Networks*, vol. 4, no. 2, pp. 169–188, 1995.

[TrMu96] S. B. Tridandapani and B. Mukherjee, "Multicast traffic in multihop lightwave networks: Performance analysis and an argument for channel sharing," *Proceedings, IEEE INFOCOM '96*, San Francisco, CA, pp. 345–352, March 1996.

[TrMu97] S. B. Tridandapani and B. Mukherjee, "Channel sharing in multi-hop WDM lightwave networks: Realization, and performance of multicast traffic," *IEEE Journal on Selected Areas in Communications*, vol. 15, pp. 488–500, April 1997.

[TSKK05] T. Tanemura, J. Suzuki, K. Katoh, and K. Kikuchi, "Polarization-Insensitive All-Optical Wavelength Conversion Using Cross-Phase Modulation in Twisted Fiber and Optical

Filtering," *IEEE Phonotincs Technology Letters*, vol. 17, no. 5, pp. 1052–1054, May 2005

[TTHD04] K. Takabayashi, K. Takada, N. Hashimoto, M. Doi, et al., "Widely (132 nm) wavelength tunable laser using a semiconductor optical amplifier and an acousto-optic tunable filter," *IEE Electronics Letters*, vol. 40, no. 19, pp. 1187–1188, Sept. 2004.

[TuZh99] R. S. Tucker and W. D. Zhong, "Photonic Packet Switching," *IEICE Trans. Commun.*, vol. E82-B, no. 2, pp. 254–264, Feb. 1999.

[TYCT00] L. Tancevski, S. Yegnanarayanan, G. Castanon, L. Tamil, et al., "Optical routing of asynchronous, variable length packets," *IEEE Journal on Selected Areas in Communications*, vol. 18, pp. 2084–2093, Oct. 2000.

[TzZT04] A. Tzanakaki, I. Zacharopoulos, and I. Tomkos, "Broadband building blocks (optical networks)," *IEEE Circuits and Devices Magazine*, vol. 20, no. 2, pp. 32–37, March/April 2004.

[VaBl97] M. D. Vaughn and D. J. Blumenthal, "All-optical undating of subcarrier encoded packet header with simultaneous wavelength conversion of baseband payland in semiconductor optical amplifiers," *IEEE Photonics Technology Letters*, vol. 9, no. 6, pp. 827–829, June 1997.

[Varm01] E. L. Varma et al., "Architecting the Services in an Optical Network," *IEEE Communications Magazine*, vol. 39, no. 9, pp. 80–89, Sept. 2001.

[VaSZ02] E. Hernandez-Valencia, M. Scholten, and Z. Zhu, "The generic framing procedure (GFP): An overview," *IEEE Communications Magazine*, vol. 40, no. 5, pp. 63–71, May 2002.

[Veli05] *Product data sheet.* http://www.velio.com/product/product_index.html.

[VMVO00] I. van de Voorde, C. M. Martin, I. Vandewege, and X. Z. Oiu, "The superPON demonstrator: An exploration of possible evolution paths for optical networks," *IEEE Communications Magazine*, vol. 38, no. 2, pp. 74–82, Feb. 2000.

[VoJu02] V. M. Vokkarane and J. P. Jue, "Prioritized Routing and Burst Segmentation for QoS in Optical Burst-switched Networks," *Proceedings, OFC '02*, Atlanta, GA, pp. 221–222, March 2002.

[Voss90] S. Voss, *Steiner-Probleme in Graphen, Frankfurt/Main: Hain*, pp. 179–184, 1990.

[WaMu02] J. Wang and B. Mukherjee, "Interconnected WDM ring networks: Strategies for interconnection and traffic grooming," *SPIE Optical Networks Magazine*, vol. 3, no. 5, pp. 10–20, Sept./Oct. 2002.

[WaSM02] J. Wang, L. Sahasrabuddhe, and B. Mukherjee, "Path vs. subpath vs. link restoration for fault management in IP-over-WDM networks: Performance comparisons using GMPLS control signaling," *IEEE Communications Magazine*, vol. 40, pp. 2–9, Nov. 2002.

[WCAS00] A. Willner, M. C. Cardakli, O. H. Adamczyk, Y. Song, and D. Gurkan, "Key building blocks for all-optical networks," *IEICE Transactions in Communications*, vol. E83-B, no. 10, Oct. 2000.

[WCLF00] P. J. Wan, G. Calinescu, L. Liu, and O. Frieder, "Grooming of arbitrary traffic in SONET/WDM BLSRs," *IEEE Journal on Selected Areas in Communications*, vol. 18, no. 10, pp. 1995–2003, Oct. 2000.

[WCVM00] J. Wang, W. Cho, V. R. Vemuri, and B. Mukherjee, "Improved approaches for cost-effective traffic grooming in WDM ring networks: ILP formulations and single-hop and multihop connections," *IEEE/OSA Journal of Lightwave Technology*, vol. 19, no. 11, pp. 1645–1653, Nov. 2001.

[WeMc00] J. Y. Wei and R. I. McFarland, "Just-in-time signaling for WDM optical burst switching networks," *IEEE/OSA, Journal of Lightwave Technology*, vol. 18, no. 12, pp. 2019–2037, Dec. 2000.

[WeMY02] W. Wen, B. Mukherjee, and S. J. B. Yoo, "QoS-based protection in MPLS-controlled WDM mesh networks," *Photonic Network Communications*, vol. 4, pp. 297–320, July 2002.

[WeNL04] Y. Wen, A. Nirmalathas, and D. Lee, "RZ/CSRZ-DPSK and chirped NRZ signal generation using a single-stage dual-electrode Mach-Zehnder modulator," *IEEE Photonics Technology Letters*, vol. 16, no. 11, pp. 2466–2468, Nov. 2004.

[Whit95] T. J. Whitley, "A review of recent system demonstrations incorporating 1.3-μm praseodymium-doped fluoride fiber amplifiers," *IEEE/OSA Journal of Lightwave Technology*, vol. 13, pp. 744–760, May 1995.

[Wies96] J. M. Wiesenfeld, "Wavelength conversion techniques," *Proceedings, Optical Fiber Communication (OFC '96)*, San Jose, CA, vol. Tutorial TuP 1, pp. 71–72, 1996.

[WiHw93] A. E. Wilner and S. M. Hwang, "Passive equalization of nonuniform EDFA gain by optical filtering for megameter transmission of 20 WDMA channels through a cascade of EDFA's," *IEEE Photonics Technology Letters*, vol. 5, pp. 1023–1026, Sept. 1993.

[Will02] A. E. Willner, "Chromatic dispersion and polarization-mode dispersion," *OSA Optics and Photonics News*, pp. S16 - S21, 2002.

[WLSB04] H. Wessing, B. Lavigne, B. Sorensen, E. Balmefrezol, and O. Leclerc, "Combining control electronics with SOA to equalize packet-to-packet power variations for optical 3R regeneration in optical networks at 10 Gbit/s," *Proceedings, OFC '04*, Los Angeles, CA, pp. 23–27, Feb. 2004.

[WRBB05] A. J. Ward, D. J. Robbins, G. Busico, E. Barton, et al., "Widely Tunable DS-DBR Laser With Monolithically Integrated SOA: Design and Performance," *IEEE Journal of Selected Topics in Quantum Electronics*, vol. 11, no. 1, pp. 149–156, Jan. 2005.

[WrVa05] M. W. Wright and G. C. Valley, "Yb-Doped Fiber Amplifier for Deep-Space Optical Communications," *IEEE/OSA Journal of Lightwave Technology*, vol. 23, no. 3, pp. 1369–1374, March 2005.

[WTSW97] W. Willinger, M. S. Taqqu, R. Sherman, and D. V. Wilson, "Self-similarity through high-variability: statistical analysis of

Ethernet LAN traffic at the source level," *IEEE/ACM Transactions on Networking*, vol. 5, no. 1, pp. 71–86, Feb. 1997.

[Wu92] T. Wu, *Fiber Network Survivability*, Boston, MA: Artech House, 1992.

[Wu98] T. Wu, "Emerging technologies for fiber network survivability," *IEEE Communications Magazine*, vol. 36, no. 2, pp. 58–74, Feb. 1998.

[XiRo04] Y. Xin and G. N. Rouskas, "Light-tree routing under optical power budget constraints [Invited]," *Journal of Optical Networks*, vol. 3, no. 5, pp. 282–302, May 2004.

[XiXQ03] Y. Xiong, D. Xu, and C. Qiao, "Achieving fast and bandwidth-efficient shared-path protection," *IEEE/OSA Journal of Lightwave Technology*, vol. 21, no. 3, pp. 365–371, Feb. 2003.

[XYDQ01] C. Xin, Y. Ye, S. Dixit, and C. Qiao, "A joint working and protection path selection approach in WDM optical networks," *Proceedings, IEEE Globecom '01*, San Antonio, TX, pp. 2165–2168, Nov. 2001.

[YaMD00] S. Yao, B. Mukherjee, and S. Dixit, "Advances in Photonic Packet Switching: an Overview," *IEEE Communications Magazine*, vol. 38, no. 2, pp. 84–94, Feb. 2000.

[YaMH99] Y. Yamada, S. Mino, and K. Habara, "Ultrafast clock recovery for burst-mode optical packet communication," *Proceedings, OFC '99*, San Diego, CA, pp. 114–116, Feb. 1999.

[YaMu03] S. Yao and B. Mukherjee, "Design of hybrid waveband-switched networks with O-E-O traffic grooming," *Proceedings, OFC '03*, Atlanta, GA, pp. 357–358, March 2003.

[YaOM03] S. Yao, C. Ou, and B. Mukherjee, "Design of hybrid optical networks with waveband and electrical TDM switching," *Proceedings, IEEE Globecom '03*, San Francisco, CA, vol. 5, pp. 2803–2808, Dec. 2003.

[YaRa02] X. Yang and B. Ramamurthy, "Dynamic routing in translucent WDM optical networks," *Proceedings, IEEE ICC '02*, New York, pp. 2796–2802, April 2002.

[YaYT05] M. Yano, F. Yamagishi, T. Tsuda, "Optical MEMS for Photonic Switching – Compact and Stable Optical Crossconnect Switches for Simple, Fast, and Flexible Wavelength Applications in Recent Photonic Networks," *IEEE Journal of Selected Topics in Quantum Electronics*, vol. 11, no. 2, pp. 383–394, March/April 2005.

[YCBK95] S. J. B. Yoo, C. Caneau, R. Bhat, and M. A. Koza, "Wavelength conversion by quasi-phase-matched difference frequency generation in AlGaAs waveguides," *Proceedings, Optical Fiber Communication (OFC '95)*, San Diego, CA, vol. 8, pp. 377–380, Feb. 1995.

[YCBK96] S. J. B. Yoo, C. Caneau, R. Bhat, and M. A. Koza, "Transparent wavelength conversion by difference frequency generation in AlGaAs waveguides," *Proceedings, Optical Fiber Communication (OFC '96)*, San Jose, CA, vol. 2, pp. 129–131, 1996.

[YeYC05] Y. Yeo, J. Yu, and G. Chang, "Performance of DPSK and NRZ-OOK signals in a Novel Folded-Path Optical Packet Switch Buffer," *Proceedings, OFC '05*, Anaheim, CA, pp. OWK3, March 2005.

[YHMR04] H. Yang, M. Herzog, M. Maier, and M. Reisslein, "Metro WDM Networks: Performance Comparison of Slotted Ring and AWG Star Networks," *IEEE Journal on Selected Areas in Communications*, vol. 22, no. 8, pp. 1460–1473, Oct. 2004.

[YLES96] J. Yates, J. Lacey, D. Everitt, and M. Summerfield, "Limited-range wavelength translation in all-optical networks," *Proceedings, IEEE INFOCOM '96*, San Francisco, CA, pp. 954–961, March 1996.

[YMGM99] X. Yuan, R. Melhem, R. Gupta, Y. Mei, and C. Qiao, "Distributed control protocols for wavelength reservation and their performance evaluation," *Photonic Network Communications*, vol. 1, no. 3, pp. 207–218, 1999.

[YMYD03] S. Yao, B. Mukherjee, S. J. B. Yoo, and S. Dixit, "A unified study of contention-resolution schemes in optical packet-switched networks," *IEEE/OSA Journal of Lightwave Technology*, vol. 21, no. 3, pp. 672-683, March 2003.

[Yoo96] S. J. B. Yoo, "Wavelength conversion technologies for WDM network applications," *IEEE/OSA Journal of Lightwave Technology*, vol. 14, pp. 955–966, June 1996.

[YoQi97] M. Yoo and C. Qiao, "Just-Enough-Time (JET): A High Speed Protocol for Bursty Traffic in Optical Networks," *Proceedings, IEEE/LEOS Conference on Technologies for a Global Information Infrastructure*, pp. 26–27, Aug. 1997.

[YSIY96] H. Yasaka et al., "Finely tunable 10-Gb/s signal wavelength conversion from 1530 to 1560-nm region using a super structure grating distributed Bragg reflector laser," *IEEE Photonics Technology Letters*, vol. 8, pp. 764–766, June 1996.

[YuCC05] J. Yu, G. Chang, and A. Chowdhury, "Instantaneous clock recovery for burst-mode optical label and payload by using a conventional data receiver," *Proceedings, OFC '05*, Anaheim, CA, pp. OWK6, March 2005.

[YuGu04] S. Yu and W. Gu, "Wavelength Conversions in Quasi-Phase Matched $LiNbO_3$ Waveguide Based on Double-Pass Cascaded $\chi^{(2)}$ SFG + DFG Interactions," *IEEE Journal of Quantum Electronics*, vol. 40, no. 11, pp. 1548–1554, Nov. 2004.

[ZaJM00a] H. Zang, J. P. Jue, and B. Mukherjee, "A review of routing and wavelength assignment approaches for wavelength-routed optical WDM networks," *SPIE Optical Networks Magazine*, vol. 1, no. 1, pp. 47–60, January 2000.

[ZaJM00b] H. Zang, J. P. Jue, and B. Mukherjee, "Capacity allocation and contention resolution in a photonic slot routing all-optical WDM mesh network," *IEEE/OSA Journal of Lightwave Technology*, vol. 18, no. 12, pp. 1728–1741, Dec. 2000.

[ZaMu01] H. Zang and B. Mukherjee, "Connection management for survivable wavelength-routed WDM mesh networks," *SPIE Optical Networks Magazine*, vol. 2, no. 4, pp. 17–28, July 2001.

[ZaOM03] H. Zang, C. Ou, and B. Mukherjee, "Path-Protection Routing and Wavelength-Assignment (RWA) in WDM Mesh Networks under Duct-Layer Constraints," *IEEE/ACM Transactions on Networking*, vol. 11, no. 2, pp. 248–258, April 2003.

[ZhAc94] Z. Zhang and A. Acampora, "A heuristic wavelength assignment algorithm for multihop WDM networks with wavelength routing and wavelength reuse," *IEEE/ACM Transactions on Networking*, vol. 3, pp. 281–288, June 1995.

[Zhan03] L. Zhang et al., "Dual DEB-GPS Schedular for Delay-Constraint Applications in Ethernet Passive Optical Networks," *IEICE Transactions in Communications*, vol. E86-B, no. 5, pp. 1575–1584, May 2003.

[ZhMu02a] K. Zhu and B. Mukherjee, "Traffic grooming in an optical WDM mesh network," *IEEE Journal on Selected Areas in Communications*, vol. 20, no. 1, pp. 122–133, Jan. 2002.

[ZhMu02b] K. Zhu and B. Mukherjee, "A review of traffic grooming in WDM optical networks: Architectures and challenges," *SPIE Optical Networks Magazine*, vol. 4, no. 2, pp. 55–64, March/April 2003.

[ZhMu02c] K. Zhu and B. Mukherjee, "On-line approaches for provisioning connections of different bandwidth granularities in WDM mesh networks," *Proc., OFC '02*, Anaheim, CA, pp. 549–551, March 2002.

[ZhMu04] J. Zhang and B. Mukheriee, "A review of fault management in WDM mesh networks: basic concepts and research challenges," *IEEE Network*, vol. 18, no. 2, pp. 41–48, March/April 2004.

[ZhMu05] H. Zhu and B. Mukherjee, "Online Connection Provisioning in Metro Optical WDM Networks using Reconfigurable OADMs (ROADMs)," *IEEE/OSA Journal of Lightwave Technology*, to appear, Dec. 2005.

[ZhQi98] X. Zhang and C. Qiao, "Wavelength assignment for dynamic traffic in multi-fiber WDM networks," *Proceedings, 7th International Conference on Computer Communications and Networks*, Lafayette, LA, pp. 479–485, Oct. 1998.

[ZhQi00] X. Zhang and C. Qiao, "An effective and comprehensive approach for traffic grooming and wavelength assignment in SONET/WDM rings," *IEEE/ACM Transactions on Networking*, vol. 8, no. 5, pp. 608–617, Oct. 2000.

[ZhSu00] D. Zhou and S. Subramaniam, "Survivability in optical networks," *IEEE Network*, vol. 14, no. 6, pp. 16–23, Nov./Dec. 2000.

[ZhWQ00] X. Zhang, J. Wei, and C. Qiao, "Constrained Multicast Routing in WDM networks with Sparse Light Splitting," *Proceedings, IEEE INFOCOM '00*, Tel Aviv, Israel, pp. 1781–1790, March 2000.

[ZhYa02] H. Zhang and O. Yang, "Finding protection cycles in DWDM networks," *Proceedings, IEEE ICC '02*, New York, pp. 2756–2760, April 2002.

[ZhZM03] H. Zhu, H. Zang, K. Zhu, and B. Mukherjee, "A novel generic graph model for traffic grooming in heterogeneous WDM mesh networks," *IEEE/ACM Transactions on Networking*, vol. 11, no. 2, pp. 285–299, April 2003.

[ZhZM05] J. Zhang, K. Zhu, and B. Mukherjee, "Backup Reprovisioning to Remedy the Effect of Multiple Link Failures in WDM Mesh Networks," *IEEE Journal on Selected Areas in Communications*, to appear, 2005.

[ZiDJ92] M. Zirngibl, C. Dragone, and C. H. Joyner, "Demonstration of a 15 × 15 arrayed waveguide multiplexer on InP," *IEEE Photonics Technology Letters*, vol. 4, pp. 1250–1253, Nov. 1992.

[Zirn98] M. Zirngibl, "An evaluation of architectures incorporating wavelength division multiplexing for broad-band fiber access," *IEEE/OSA Journal of Lightwave Technology*, vol. 16, no. 9, pp. 1546–1558, Sept. 1998.

[ZJSD95] M. Zirngibl, C. H. Joyner, L. W. Stulz, C. Dragone, et al., "LARNet, a local access router network," *IEEE Photonics Technology Letters*, vol. 7, no. 2, pp. 215–217, Feb. 1995.

[ZJSM01] H. Zang, J. P. Jue, L. Sahasrabuddhe, S. Ramamurthy, and B. Mukherjee, "Dynamic lightpath establishment in wavelength routed WDM networks," *IEEE Communications Magazine*, vol. 39, no. 9, pp. 100-108, Sept. 2001.

[ZLSC01] X. Zhou, C. Lu, P. Shum, and T. H. Cheng, "A simplified model and optimal design of a multiwavelength backward-pumped fiber raman amplifier," *IEEE Photonics Technology Letters*, vol. 13, no. 9, pp. 945–947, Sep. 2001.

[ZZYM03] J. Zhang, K. Zhu, S. J. B. Yoo, L. Sahasrabuddhe, and B. Mukherjee, "On the study of routing and wavelength-assignment approaches for survivable wavelength-routed WDM mesh networks," *SPIE Optical Networks Magazine*, vol. 4, no. 6, pp. 16–28, Nov./Dec. 2003.

[ZZZM03] J. Zhang, K. Zhu, H. Zang, and B. Mukherjee, "Service provisioning to provide per-connection-based availability guarantee in WDM mesh networks," *Proceedings, OFC '03*, Atlanta, GA, pp. 622–624, March 2003.

Index

Printed in the United States
By Bookmasters